Climatology is, to a large degree, the study of the statistics of our climate. The powerful tools of mathematical statistics therefore find wide application in climatological research, ranging from simple methods for determining the uncertainty of a climatological mean to sophisticated techniques which reveal the dynamics of the climate system.

The purpose of this book is to help the climatologist understand the basic precepts of the statistician's art and to provide some of the background needed to apply statistical methodology correctly and usefully. The book is self contained: introductory material, standard advanced techniques, and the specialized techniques used specifically by climatologists are all contained within this one source. There is a wealth of real-world examples drawn from the climate literature to demonstrate the need, power and pitfalls of statistical analysis in climate research.

This book is suitable as a main text for graduate courses on statistics for climatic, atmospheric and oceanic science. It will also be valuable as a reference source for researchers in climatology, meteorology, atmospheric science, and oceanography.

Hans von Storch is Director of the Institute of Hydrophysics of the GKSS Research Centre in Geesthacht, Germany and a Professor at the Meteorological Institute of the University of Hamburg.

Francis W. Zwiers is Chief of the Canadian Centre for Climate Modelling and Analysis, Atmospheric Environment Service, Victoria, Canada, and an Adjunct Professor of the Department of Mathematics and Statistics of the University of Victoria.

Statistical Analysis in Climate Research

Hans von Storch
and Francis W. Zwiers

CAMBRIDGE
UNIVERSITY PRESS

PUBLISHED BY THE PRESS SYNDICATE OF THE UNIVERSITY OF CAMBRIDGE
The Pitt Building, Trumpington Street, Cambridge, United Kingdom

CAMBRIDGE UNIVERSITY PRESS
The Edinburgh Building, Cambridge CB2 2RU, UK
40 West 20th Street, New York, NY 10011–4211, USA
477 Williamstown Road, Port Melbourne, VIC 3207, Australia
Ruiz de Alarcón 13, 28014 Madrid, Spain
Dock House, The Waterfront, Cape Town 8001, South Africa

http://www.cambridge.org

First published 1999
First paperback edition (with corrections) 2001
Reprinted 2003

Typeset in Times 10/12pt [DBD]

A catalogue record for this book is available from the British Library

Library of Congress Cataloguing in Publication data

Storch, H. V. (Hans von), 1949–
Statistical analysis in climate research / Hans von Storch and
Francis W. Zwiers.
p. cm.
Includes index.
ISBN 0 521 45071 3
1. Climatology – Statistical methods. I. Title.
QC981.S735 1998
551.5′072–dc21 98-17416 CIP

ISBN 0 521 45071 3 hardback
ISBN 0 521 01230 9 paperback

Transferred to digital printing 2004

Contents

Preface

The tools of mathematical statistics find wide application in climatological research. Indeed, climatology is, to a large degree, the study of the statistics of our climate. Mathematical statistics provides powerful tools which are invaluable for this pursuit. Applications range from simple uses of sampling distributions to provide estimates of the uncertainty of a climatological mean to sophisticated statistical methodologies that form the basis of diagnostic calculations designed to reveal the dynamics of the climate system. However, even the simplest of statistical tools has limitations and pitfalls that may cause the climatologist to draw false conclusions from valid data if the tools are used inappropriately and without a proper understanding of their conceptual foundations. The purpose of this book is to help the climatologist understand the basic precepts of the statistician's art and to provide some of the background needed to apply statistical methodology correctly and usefully.

We do not claim that this volume is in any way an exhaustive or comprehensive guide to the use of statistics in climatology, nor do we claim that the methodology described here is a current reflection of the art of applied statistics as it is conducted by statisticians. Statistics as it is applied in climatology is far removed from the cutting edge of methodological development. This is partly because statistical research has not come yet to grips with many of the problems encountered by climatologists and partly because climatologists have not yet made very deep excursions into the world of mathematical statistics. Instead, this book presents a subjectively chosen discourse on the tools we have found useful in our own research on climate diagnostics.

We will discuss a variety of statistical concepts and tools which are useful for solving problems in climatological research, including the following.

- The concept of a sample.
- The notions of exploratory and confirmatory statistics.
- The concept of the statistical model. Such a model is implicit in every statistical analysis technique and has substantial implications for the conclusions drawn from the analysis.
- The differences between parametric and non-parametric approaches to statistical analysis.
- The estimation of 'parameters' that describe the properties of the geophysical process being studied. Examples of these 'parameters' include means and variances, temporal and spatial power spectra, correlation coefficients, empirical orthogonal functions and Principal Oscillation Patterns. The concept of parameter estimation includes not only point estimation (estimation of the specific value of a parameter) but also interval estimation which account for uncertainty.
- The concepts of hypothesis testing, significance, and power.

We do *not* deal with:

- *Bayesian statistics*, which is philosophically quite different from the more common *frequentist* approach to statistics we use in this book. Bayesians, as they are known, incorporate *a priori* beliefs into a statistical analysis of a sample in a rational manner (see Epstein [114], Casella [77], or Gelman et al. [139]).
- *Geostatistics*, which is widely used in geology and related fields. This approach deals with the analysis of spatial fields sampled at a relatively small number of locations. The most prominent technique is called *kriging* (see Journel and Huijbregts [207], Journel [206], or Wackernagel [406]), which is related to the *data assimilation* techniques used in atmospheric and oceanic science (see, e.g., Daley [98] and Lorenc [258]).

A collection of applications of many statistical techniques has been compiled by von Storch and Navarra [395]; we recommend this collection as complementary reading to this book and refer to its contributions throughout. This collection does not cover the field systematically; instead it offers examples of the exploitation of statistical methods in the analysis of climatic data and numerical experiments.

Cookbook recipes for a variety of standard statistical situations are not offered by this book because they are dangerous for anyone who does not understand the basic concepts of statistics. Therefore, we offer a course in the concepts and discuss cases we have encountered in our work. Some of these examples refer to standard situations, and others to more exotic cases. Only the understanding of the principles and concepts prevents the scientist from falling into the many pitfalls specific to our field, such as multiplicity in statistical tests, the serial dependence within samples, or the enormous size of the climate's phase space. If these dangers are not understood, then the use of simple recipes will often lead to erroneous conclusions. Literature describes many cases, both famous and infamous, in which this has occurred.

We have tried to use a consistent notation throughout the book, a summary of which is offered in Appendix A. Some elements of linear algebra are available in Appendix B, and some aspects of Fourier analysis and transform are listed in Appendix C. Proofs of statements, which we do not consider essential for the overall understanding, are in Appendix M.

Thanks

We are deeply indebted to a very large number of people for their generous assistance with this project. We have tried to acknowledge all who contributed, but we will inevitably have overlooked some. We apologize sincerely for these oversights.

- Thanks for her excellent editorial assistance: Robin Taylor.

- Thanks for discussion, review, advice and useful comments: Gerd Bürger, Bill Burrows, Ulrich Callies, Susan Chen, Christian Eckert, Claude Frankignoul, Marco Giorgetta, Silvio Gualdi, Stefan Güß, Klaus Hasselmann, Gabi Hegerl, Patrick Heimbach, Andreas Hense, Hauke Heyen, Martina Junge, Thomas Kaminski, Frank Kauker, Dennis Lettenmaier, Bob Livezey, Ute Luksch, Katrin Maak, Rol Madden, Ernst Maier-Reimer, Peter Müller, Dörthe Müller-Navarra, Matthias Münnich, Allan Murphy, Antonio Navarra, Peter Rayner, Mark Saunders, Reiner Schnur, Dennis Shea, Achim Stössel, Sylvia Venegas, Stefan Venzke, Koos Verbeeck, Jin-Song von Storch, Hans Wackernagel, Xiaolan Wang, Chris Wickle, Arne Winguth, Eduardo Zorita.

- Thanks for making diagrams available to us: Howard Barker, Anthony Barnston, Grant Branstator, Gerd Bürger, Bill Burrows, Klaus Fraedrich, Claude Frankignoul, Eugenia Kalnay, Viacheslaw Kharin, Kees Korevaar, Steve Lambert, Dennis Lettenmaier, Bob Livezey, Katrin Maak, Allan Murphy, Hisashi Nakamura, Reiner Schnur, Lucy Vincent, Jin-Song von Storch, Mike Wallace, Peter Wright, Eduardo Zorita.

- Thanks for preparing diagrams: Marion Grunert, Doris Lewandowski, Katrin Maak, Norbert Noreiks, and Hinrich Reichardt, who helped also to create some of the tables in the Appendices. For help with the LaTeX-system: Jörg Wegner. For help with the Hamburg computer network: Dierk Schriever. For help with the Canadian Centre for Climate Modelling and Analysis computer network in Victoria: Mike Berkley. For scanning diagrams: Mike Berkley, Jutta Bernlöhr, and Marion Grunert.

1 Introduction

1.1 The Statistical Description and Understanding of Climate

Climatology was originally a sub-discipline of geography, and was therefore mainly descriptive (see, e.g., Brückner [70], Hann [155], or Hann and Knoch [156]). Description of the climate consisted primarily of estimates of its mean state and estimates of its variability about that state, such as its standard deviations and other simple measures of variability. Much of climatology is still focused on these concerns today. The main purpose of this description is to define 'normals' and 'normal deviations,' which are eventually displayed as maps. These maps are then used for regionalization (in the sense of identifying homogeneous geographical units) and planning. The paradigm of climate research evolved from the purely descriptive approach towards an understanding of the dynamics of climate with the advent of computers and the ability to simulate the climatic state and its variability. Statistics plays an important role in this new paradigm.

The climate is a dynamical system influenced not only by immense external factors, such as solar radiation or the topography of the surface of the solid Earth, but also by seemingly insignificant phenomena, such as butterflies flapping their wings. Its evolution is controlled by more or less well-known physical principles, such as the conservation of angular momentum. If we knew all these factors, and the state of the full climate system (including the atmosphere, the ocean, the land surface, etc.), at a given time in full detail, then there would not be room for statistical uncertainty, nor a need for this book. Indeed, if we repeat a run of a General Circulation Model, which is supposedly a *model* of the real climate system, on the same computer with exactly the same code, operating system, and initial conditions, we obtain a second realization of the simulated climate that is identical to the first simulation.

Of course, there is a 'but.' We do not know all factors that control the trajectory of climate in its enormously large phase space.[1] Thus it is not possible to map the state of the atmosphere, the ocean, and the other components of the climate system in full detail. Also, the models are not deterministic in a practical sense: an insignificant change in a single digit in the model's initial conditions causes the model's trajectory through phase space to diverge quickly from the original trajectory (this is Lorenz's [260] famous discovery, which leads to the concept of chaotic systems).

Therefore, in a strict sense, we have a 'deterministic' system, but we do not have the ability to analyse and describe it with 'deterministic' tools, as in thermodynamics. Instead, we use probabilistic ideas and statistics to describe the 'climate' system.

Four factors ensure that the climate system is amenable to statistical thinking.

- The climate is controlled by innumerable factors. Only a small proportion of these factors can be considered, while the rest are necessarily interpreted as background noise. The details of the generation of this 'noise' are not important, but it is important to understand that this noise is an *internal* source of variation in the climate system (see also the discussion of 'stochastic climate models' in Section 10.4).

- The dynamics of climate are nonlinear. Nonlinear components of the *hydrodynamic* part include important advective terms, such as $u\frac{\partial u}{\partial x}$. The *thermodynamic* part contains various other nonlinear processes, including many that can be represented by step functions (such as condensation).

[1] We use the expression 'phase space' rather casually. It is the space spanned by the state variables x of a system $\frac{dx}{dt} = f(x)$. In the case of the climate system, the state variables consist of the collection of all climatic variables at all geographic locations (latitude, longitude, height/depth). At any given time, the state of the climate system is represented by one point in this space; its development in time is represented by a smooth curve ('trajectory').
This concept deviates from the classical mechanical definition where the phase space is the space of generalized coordinates. Perhaps it would be better to use the term 'state space.'

- The dynamics include linearly unstable processes, such as the baroclinic instability in the midlatitude troposphere.
- The dynamics of climate are dissipative. The hydrodynamic processes transport energy from large spatial scales to small spatial scales, while molecular diffusion takes place at the smallest spatial scales. Energy is dissipated through friction with the solid earth and by means of gravity wave drag at larger spatial scales.[2]

The nonlinearities and the instabilities make the climate system *unpredictable* beyond certain characteristic times. These characteristic time scales are different for different subsystems, such as the ocean, midlatitude troposphere, and tropical troposphere. The nonlinear processes in the system amplify minor disturbances, causing them to evolve irregularly in a way that allows their interpretation as finite-amplitude noise.

In general, the dissipative character of the system guarantees its 'stationarity.' That is, it does not 'run away' from the region of phase space that it currently occupies, an effect that can happen in general nonlinear systems or in linearly unstable systems. The two factors, noise and damping, are the elements required for the interpretation of climate as a stationary stochastic system (see also Section 10.4).

Under what circumstances should the output of climate models be considered stochastic? A major difference between the real climate and any climate model is the size of the phase space. The phase space of a model is much smaller than that of the real climate system because the model's phase space is truncated in both space and time. That is, the background noise, due to unknown factors, is missing. Therefore a model run can be repeated with identical results, provided that the computing environment is unchanged and the same initial conditions are used. To make the climate model output realistic we need to make the model unpredictable. Most Ocean General Circulation Models are strongly dissipative and behave almost linearly. Explicit noise must therefore be added to the system as an explicit forcing term to create statistical variations in the simulated system (see, for instance [276] or [418]). In dynamical atmospheric models (as opposed to energy-balance models) the nonlinearities are strong enough to create their own unpredictability. These models behave in such a way that a repeated run will diverge quickly from the original run even if only minimal changes are introduced into the initial conditions.

1.1.1 The Paradigms of the Chaotic and Stochastic Model of Climate. In the paradigm of the chaotic model of the climate, and particularly the atmosphere, a small difference introduced into the system at some *initial* time causes the system to diverge from the trajectory it would otherwise have travelled. This is the famous *Butterfly Effect*[3] in which infinitesimally small disturbances may provoke large reactions. In terms of climate, however, there is not just *one* small disturbance, but myriads of such disturbances at all times. In the metaphor of the butterfly: there are millions of butterflies that flap their wings all the time. The paradigm of the stochastic climate model is that this omnipresent noise causes the system to vary on all time and space scales, independently of the degree of nonlinearity of the climate's dynamics.

1.2 Some Typical Problems and Concepts

1.2.0 Introduction. The following examples, which we have subjectively chosen as being typical of problems encountered in climate research, illustrate the need for statistical analysis in atmospheric and climatic research. The order of the examples is somewhat random and it is certainly not a must to read all of them; the purpose of this 'potpourri' is to offer a flavour of typical questions, answers, and errors.

1.2.1 The Mean Climate State: Interpretation and Estimation. From the point of view of the climatologist, the most fundamental statistical parameter is the mean state. This seemingly trivial animal in the statistical zoo has considerable complexity in the climatological context.

First, the computed mean is not entirely reliable as an estimate of the climate system's true long-term mean state. The computed mean will contain errors caused by taking observations over a limited observing period, at discrete times and a finite number of locations. It may also be affected by the presence of instrumental, recording, and

[2]The gravity wave drag maintains an exchange of momentum between the solid earth and the atmosphere, which is transported by means of vertically propagating gravity waves. See McFarlane et al. [269] for details.

[3]Inaudil et al. [194] claimed to have identified a Lausanne butterfly that caused a rainfall in Paris.

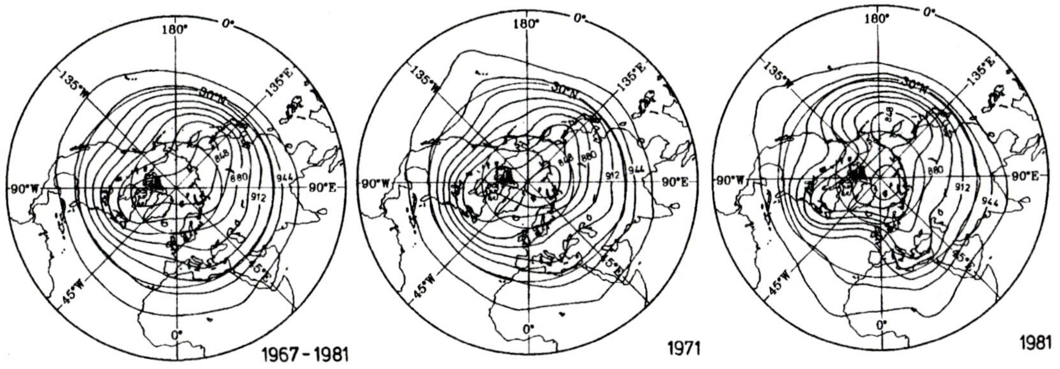

Figure 1.1: *The 300 hPa geopotential height fields in the Northern Hemisphere: the mean 1967–81 January field, the January 1971 field, which is closer to the mean field than most others, and the January 1981 field, which deviates significantly from the mean field. Units: 10 m [117].*

transmission errors. In addition, reliability is not likely to be uniform as a function of location.

Reliability may be compromised if the data has been 'analysed', that is, interpolated to a regular grid using techniques that make assumptions about atmospheric dynamics. The interpolation is performed either *subjectively* by someone who has experience and knowledge of the shape of dynamical structures typically observed in the atmosphere, or it is performed *objectively* using a combination of atmospheric and statistical models. Both kinds of analysis are apt to introduce biases not present in the 'raw' station data, and errors at one location in analysed data will likely be correlated with those at another. (See Daley [98] or Thiébaux and Pedder [362] for comprehensive treatments of objective analysis.)

Second, the mean state is *not* a typical state. To demonstrate this we consider the January Northern Hemisphere 300 hPa geopotential height field[4] (Figure 1.1). The mean January height field, obtained by averaging monthly mean analyses for each January between 1967 and 1981, has contours of equal height which are primarily circular with minor irregularities. Two troughs are situated over the eastern coasts of Siberia and North America. The Siberian trough extends slightly farther south than the North American trough. A secondary trough can be identified over eastern Europe and two minor ridges are located over the northeast Pacific and the east Atlantic.

Some individual January mean fields (e.g., 1971) are similar to the long-term mean field. There are differences in detail, but they share the zonal wavenumber 2 pattern[5] of the mean field. The secondary ridges and troughs have different intensities and longitudinal phases. Other Januaries (e.g., 1981) 300 hPa geopotential height fields are very different from the mean state. They are characterized by a zonal wavenumber 3 pattern rather than a zonal wavenumber 2 pattern.

The long-term mean masks a great deal of interannual variability. For example, the minimum of the long-term mean field is larger than the minima of all but one of the individual January states. Also, the spatial variability of each of the individual monthly means is larger than that of the long-term mean. Thus, the long-term mean field is not a 'typical' field, as it is very unlikely to be observed as an individual monthly mean. In that sense, the long-term mean field is a rare event.

Characterization of the 'typical' January requires more than the long-term mean. Specifically, it is necessary to describe the dominant patterns of spatial variability about the long-term mean and to say something about the range of patterns one is likely to see in a 'typical' January. This can be accomplished to a limited extent through the use of a technique called *Empirical Orthogonal Function analysis* (Chapter 13).

Third, a climatological mean should be understood to be a moving target. Today's climate is different from that which prevailed during the Holocene (6000 years before present) or even during the Little Ice Age a few hundred years ago.

[4] The *geopotential height field* is a parameter that is frequently used to describe the dynamical state of the atmosphere. It is the height of the surface of constant pressure at, e.g., 300 hPa and, being a length, is measured in metres. We will often simply refer to 'height' when we mean 'geopotential height'.

[5] A zonal wavenumber 2 pattern contains two ridges and two troughs in the zonal, or east–west, direction.

We therefore need a clear understanding of our interpretation of the 'true' mean state before interpreting an estimate computed from a set of observations.

To accomplish this, it is necessary to think of the 'January 300 hPa height field' as a *random field*, and we need to determine whether the observed height fields in our 15-year sample are representative of the 'true' mean state we have in mind (presumably that of the 'current' climate). From a statistical perspective, the answer is a conditional 'yes,' provided that:

1 the time series of January mean 300 hPa height fields is stationary (i.e., their statistical properties do not drift with time), and

2 the memory of this time series is short relative to the length of the 15-year sample.

Under these conditions, the mean state is representative of the random sample, in the sense that it lies in the 'centre' of the scatter of the individual points in the state space. As we noted above, however, it is not representative in many other ways.

The characteristics of the 15-year sample may not be representative of the properties of January mean 300 hPa height fields on longer time scales when assumption 1 is not satisfied. The uncertainty of the 15-year mean height field as an estimator of the long-term mean will be almost as great as the interannual variability of the individual January means when assumption 2 is not satisfied. We can have confidence in the 15-year mean as an estimator of the long-term mean January 300 hPa height field when assumptions 1 and 2 hold in the following sense: the *law of large numbers* dictates that a multi-year mean becomes an increasingly better estimator of the long-term mean as the number of years in the sample increases. However, there is still a considerable amount of uncertainty in an estimate based on a 15-year sample.

Statements to the effect that a certain estimate of the mean is 'wrong' or 'right' are often made in discussions of data sets and climatologies. Such an assessment indicates that the speakers do not really understand the art of estimation. An estimate is by definition an approximation, or guess, based on the available data. It is almost certain that the exact value will never be determined. Therefore estimates are never 'wrong' or 'right;' rather, some estimates will be closer to the truth than others on average.

To demonstrate the point, consider the following two procedures for estimating the long-term mean January air pressure in Hamburg (Germany). Two data sets, consisting of 104 observations each, are available. The first data set is taken at one minute intervals, the second is taken at weekly intervals, and a mean is computed from each. Both means are estimates of the long-term mean air pressure in Hamburg, and each tells us something about our parameter.

The reliability of the first estimate is questionable because air pressure varies on time scales considerably longer than the 104 minutes spanned by the data set. Nonetheless, the estimate does contain information useful to someone who has no prior information about the climate of locations near sea level: it indicates that the mean air pressure in Hamburg is neither 2000 mb nor 20 hPa but somewhere near 1000 mb.

The second data set provides us with a much more reliable estimate of long-term mean air pressure because it contains 104 almost independent observations of air pressure spanning two annual cycles. The first estimate is not 'wrong,' but it is not very informative; the second is not 'right,' but it is adequate for many purposes.

1.2.2 Correlation. In the statistical lexicon, the word *correlation* is used to describe a *linear statistical* relationship between two random variables. The phrase 'linear statistical' indicates that the mean of one of the random variables is linearly dependent upon the random component of the other (see Section 8.2). The stronger the linear relationship, the stronger the correlation. A correlation coefficient of $+1$ (-1) indicates a pair of variables that vary together precisely, one variable being related to the other by means of a positive (negative) scaling factor.

While this concept seems to be intuitively simple, it does warrant scrutiny. For example, consider a satellite instrument that makes radiance observations in two different frequency bands. Suppose that these radiometers have been designed in such a way that instrumental error in one channel is independent of that in the other. This means that knowledge of the noise in one channel provides no information about that in the other. However, suppose also that the radiometers drift (go out of calibration) together as they age because both share the same physical environment, share the same power supply and are exposed to the same physical abuse. Reasonable models for the total error as a function of time in the two radiometer

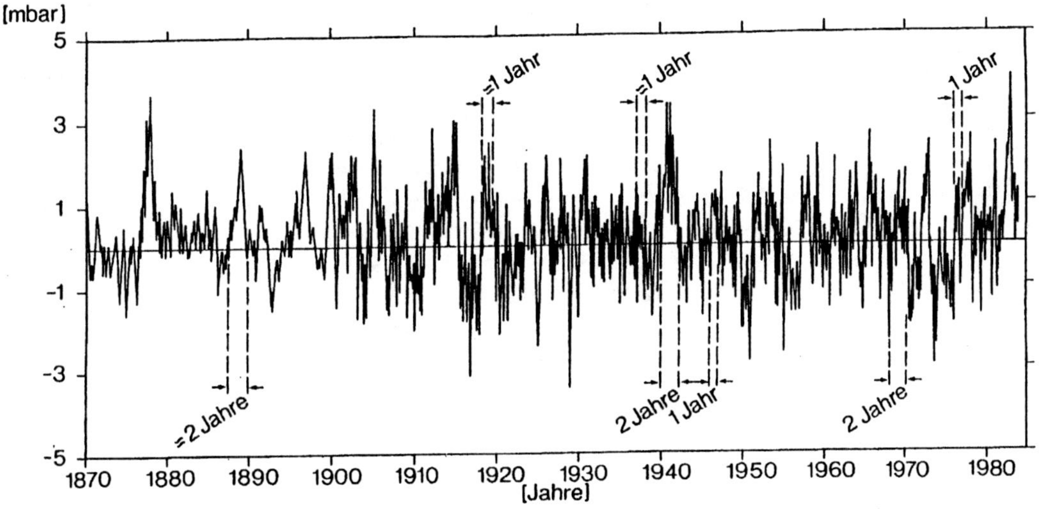

Figure 1.2: *The monthly mean Southern Oscillation Index, computed as the difference between Darwin (Australia) and Papeete (Tahiti) monthly mean sea-level pressure ('Jahr' is German for 'year').*

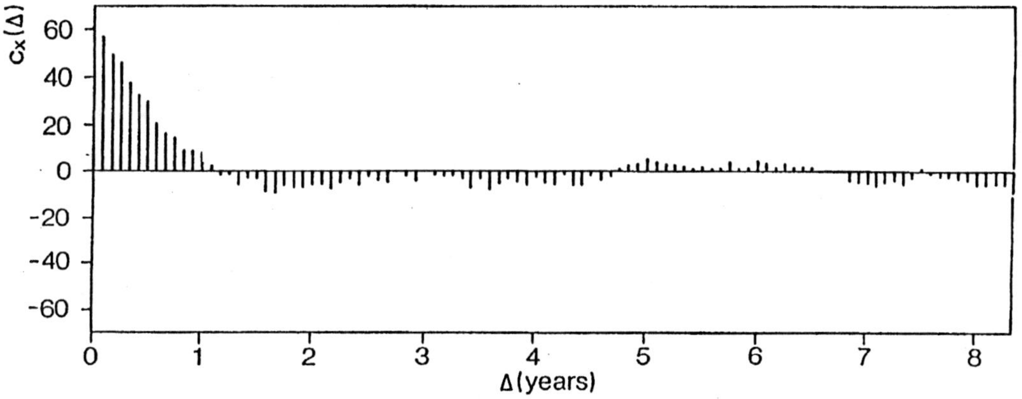

Figure 1.3: *Auto-correlation function of the index shown in Figure 1.2. Units: %.*

channels might be:

$$e_{1t} = \alpha_1(t - t_0) + \epsilon_{1t},$$
$$e_{2t} = \alpha_2(t - t_0) + \epsilon_{2t},$$

where t_0 is the launch time of the satellite and α_1 and α_2 are fixed constants describing the rates of drift of the two radiometers. The instrumental errors, ϵ_{1t} and ϵ_{2t}, are statistically independent of each other, implying that the correlation between the two, $\rho(\epsilon_{1t}, \epsilon_{2t})$, is zero. Consequently the total errors, e_{1t} and e_{2t}, are also statistically independent even though they share a common systematic component. However, simple estimates of correlation between e_{1t} and e_{2t} that do not account for the deterministic drift will suggest that these two quantities are correlated.

Correlations manifest themselves in several different ways in observed and simulated climates. Several adjectives are used to describe correlations depending upon whether they describe relationships in time (serial correlation, lagged correlation), space (spatial correlation, teleconnection), or between different climate variables (cross-correlation).

A good example of *serial correlation* is the monthly Southern Oscillation Index (SOI),[6] which

[6] The Southern Oscillation is the major mode of natural climate variability on the interannual time scale. It is frequently used as an example in this book.
It has been known since the end of the last century (Hildebrandson [177]; Walker, 1909–21) that sea-level pressure (SLP) in the Indonesian region is negatively correlated with that over the southeast tropical Pacific. A positive SLP anomaly

is defined as the anomalous monthly mean pressure difference between Darwin (Australia) and Papeete (Tahiti) (Figure 1.2).

The time series is basically stationary, although variability during the first 30 years seems to be somewhat weaker than that of late. Despite the noisy nature of the time series, there is a distinct tendency for the SOI to remain positive or negative for extended periods, some of which are indicated in Figure 1.2. This persistence in the sign of the index reflects the serial correlation of the SOI.

A quantitative measure of the serial correlation is the *auto-correlation function*, $\rho_{SOI}(t, t + \Delta)$, shown in Figure 1.3, which measures the similarity of the SOI at any time difference Δ. The auto-correlation is greater than 0.2 for lags up to about six months and varies smoothly around zero with typical magnitudes between 0.05 and 0.1 for lags greater than about a year. This tendency of *estimated* auto-correlation functions not to converge to zero at large lags, even though the real auto-correlation is zero at long lags, is a natural consequence of the uncertainty due to finite samples (see Section 11.1).

A good example of a *cross-correlation* is the relationship that exists between the SOI and various alternative indices of the Southern Oscillation [426]. The characteristic low-frequency variations in Figure 1.2 are also present in area-averaged Central Pacific sea-surface temperature (Figure 1.4).[7] The correlation between the two time series displayed in Figure 1.4 is 0.67.

Pattern analysis techniques, such as Empirical Orthogonal Function analysis (Chapter 13), Canonical Correlation Analysis (Chapter 14) and Principal Oscillation Patterns (Chapter 15), rely upon the assumption that the fields under study are *spatially correlated*. The Southern Oscillation Index (Figure 1.2) is a manifestation of the negative correlation between surface pressure at Papeete and that at Darwin. Variables such as pressure, height, wind, temperature, and specific humidity vary smoothly in the free atmosphere and consequently exhibit strong spatial interdependence. This correlation is present in each weather map (Figure 1.5, left). Indeed, without this feature, routine weather forecasts would be all but impossible given the sparseness of the global observing network as it exists even today. Variables derived from moisture, such as cloud cover, rainfall and snow amounts, and variables associated with land surface processes tend to have much smaller spatial scales (Figure 1.5, right), and also tend not to have normal distributions (Sections 3.1 and 3.2). While mean sea-level pressure (Figure 1.5, left) will be more or less constant on spatial scales of tens of kilometres, we may often travel in and out of localized rain showers in just a few kilometres. This dichotomy is illustrated in Figure 1.5, where we see a cold front over Ontario (Canada). The left panel, which displays mean sea-level pressure, shows the front as a smooth curve. The right panel displays a radar image of precipitation occurring in southern Ontario as the front passes through the region.

1.2.3 Stationarity, Cyclo-stationarity, and Non-stationarity. An important concept in statistical analysis is *stationarity*. A random variable, or a random process, is said to be stationary if all of its statistical parameters are independent of time. Most statistical techniques assume that the observed process is stationary.

However, most climate parameters that are sampled more frequently than one per year are not stationary but *cyclo-stationary*, simply because of the seasonal forcing of the climate system. Long-term averages of monthly mean sea-level pressure exhibit a marked annual cycle, which is almost sinusoidal (with one maximum and one minimum) in most locations. However, there are locations (Figure 1.6) where the annual cycle is dominated by a *semiannual* variation (with two maxima and minima). In most applications the mean annual cycle is simply subtracted from the data before the remaining *anomalies* are analysed. The process is *cyclo-stationary in the mean* if it is stationary after the annual cycle has been removed.

Other statistical parameters (e.g., the percentiles of rainfall) may also exhibit cyclo-stationary behaviour. Figure 1.7 shows the annual cycles

(i.e., a deviation from the long-term mean) over, say, Darwin (Northern Australia) tends to be associated with a negative SLP anomaly over Papeete (Tahiti). This seesaw is called the Southern Oscillation (SO). The SO is associated with large-scale and persistent anomalies of sea-surface temperature in the central and eastern tropical Pacific (El Niño and La Niña). Hence the phenomenon is often referred to as the 'El Niño/Southern Oscillation' (ENSO). Large zonal displacements of the centres of precipitation are also associated with ENSO. They reflect anomalies in the location and intensity of the meridionally (i.e., north–south) oriented Hadley cell and of the zonally oriented Walker cell.
The state of the Southern Oscillation may be monitored with the monthly SLP difference between observations taken at surface stations in Darwin, Australia and Papeete, Tahiti. It has become common practice to call this difference the Southern Oscillation Index (SOI) although there are also many other ways to define equivalent indices [426].

[7] Other definitions, such as West Pacific rainfall, sea-level pressure at Darwin alone or the surface zonal wind in the central Pacific, also yield indices that are highly correlated with the usual SOI. See Wright [427].

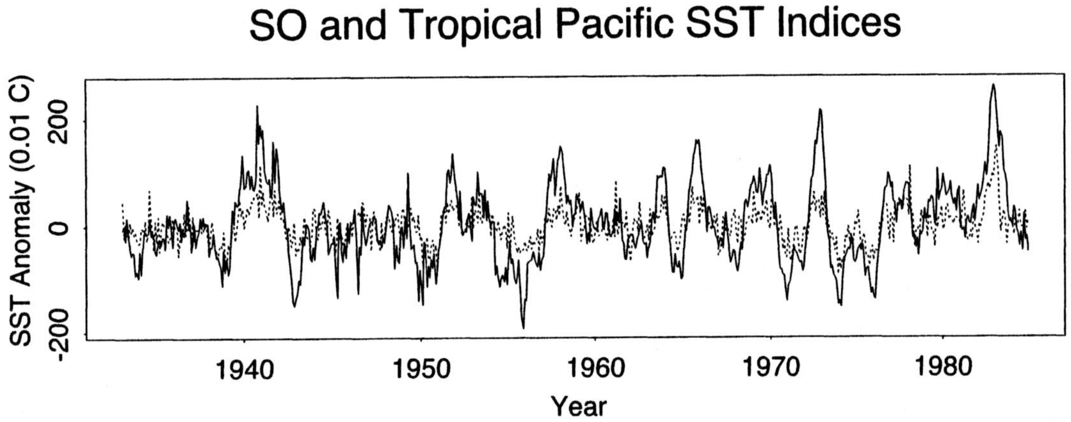

Figure 1.4: *The conventional Southern Oscillation Index (SOI = pressure difference between Darwin and Tahiti; dashed curve) and a sea-surface temperature (SST) index of the Southern Oscillation (solid curve) plotted as a function of time. The conventional SOI has been doubled in this figure.*

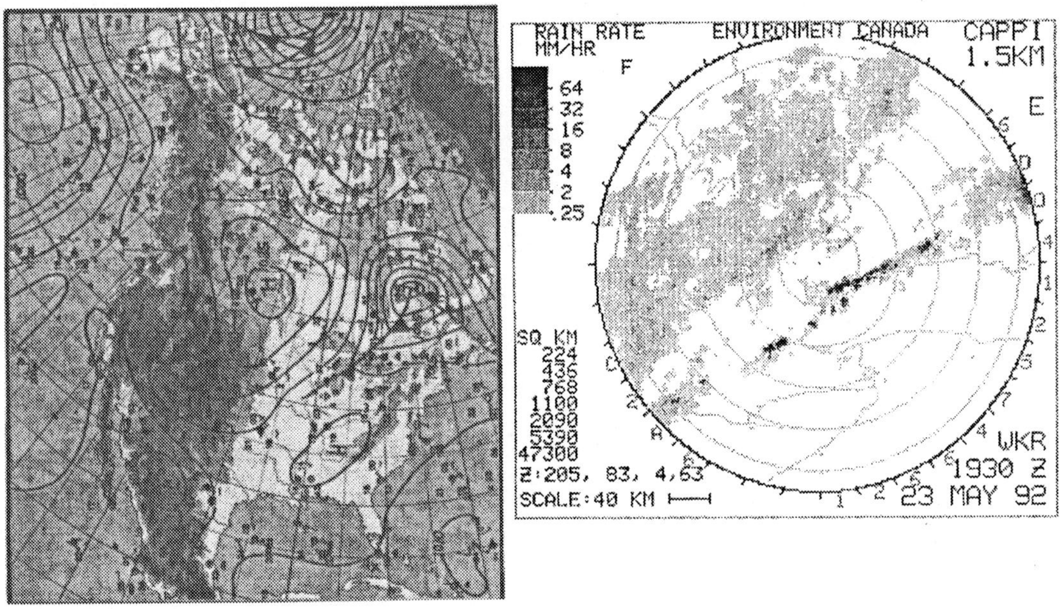

Figure 1.5: *State of the atmosphere over North America on 23 May 1992.*
Left: Analysis of the sea-level pressure field (12:00 UTC (Universal Time Coordinated); from Europäisher Wetterbericht 17, Band 144; with permission of the Deutsher Wetterdienst).
Right: Weather radar image, showing rainfall rates, for southern Ontario (19:30 local time; courtesy Paul Joe, AES Canada [94].)
Note that the radar image and the weather map refer to different times, namely 12:00 UTC on 23 May and 00:30 UTC on 24 May.

of the 70th, 80th, and 90th percentiles[8] of 24-hour rainfall amounts for each calendar month at Vancouver (British Columbia) and Sable Island (off the coast of Nova Scotia) [450].

The Southern Oscillation Index is not strictly stationary. Wright [427] showed that the linear serial correlation of the SOI depends upon the time

[8] Or 'quantiles,' that is, thresholds selected so that 70%, 80%, or 90% of all 24-hour rainfall amounts are less than the respective threshold [2.6.4].

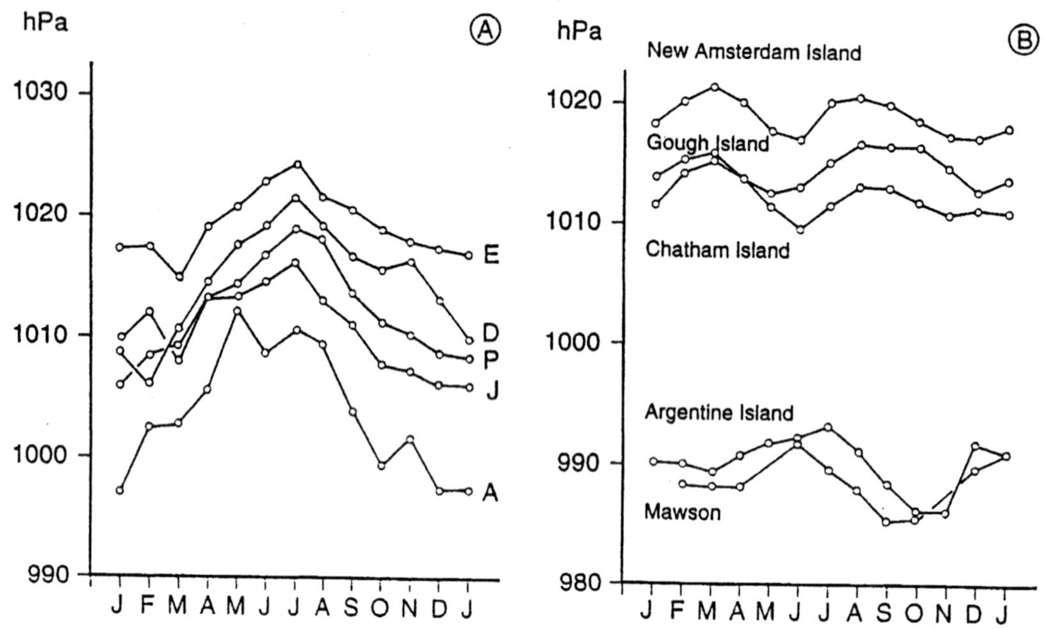

Figure 1.6: *Annual cycle of sea-level pressure at extratropical locations.*
a) Northern Hemisphere Ocean Weather Stations: A = 62° N, 33° W; D = 44° N, 41° W; E = 35° N, 48° W; J = 52° N, 25° W; P = 50° N, 145° W.
b) Southern Hemisphere.

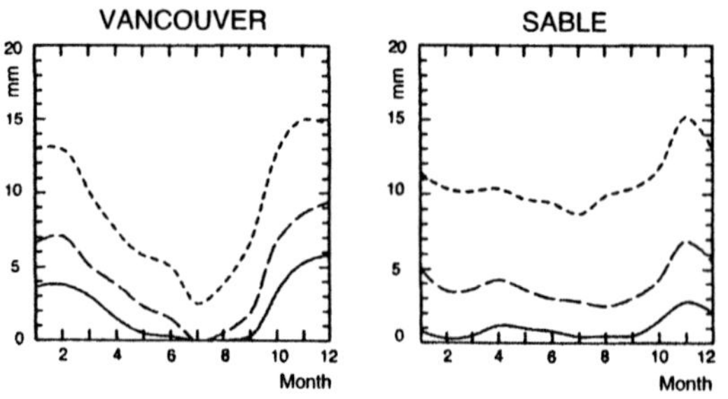

Figure 1.7: *Monthly 90th, 80th, and 70th percentiles (from top to bottom) of 24-hour rainfall amounts at Vancouver and Sable Island [450].*

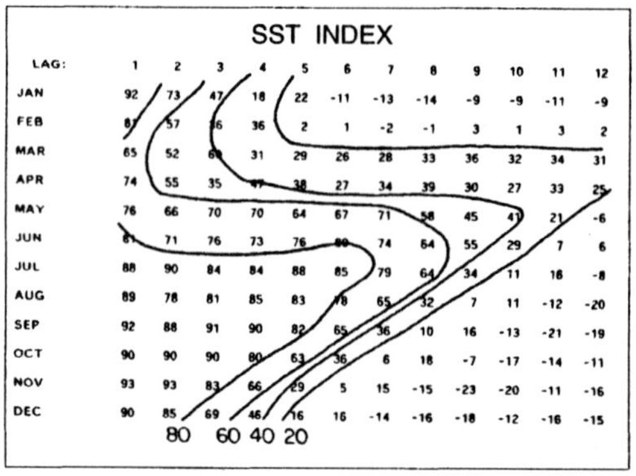

Figure 1.8: *Seasonal dependence of the lag correlations of the SST index of the Southern Oscillation. The correlations are given in hundreds so that isolines represent lag correlations of 0.8, 0.6, 0.4, and 0.2. The row labelled 'Jan' lists correlations between January values of the index and the index observed later 'lag' months [427].*

of the year. The serial correlation is plotted as a function of time of year and lag in Figure 1.8. Correlations between values of the SOI in May and values in subsequent months decay slowly with increasing lag, while similar correlations with values in April decay quickly. Because of this behaviour, Wright defined an ENSO year that begins in May and ends in April.

Regular observations taken over extended periods at a certain station sometimes exhibit changes in their statistical properties. These might be abrupt or gradual (such as changes that might occur when the exposure of a rain gauge changes slowly over time, as a consequence of the growth of vegetation or changes in local land use). Abrupt changes in the observational record may take place if the instrument (or the observer) changes, the site is moved,[9] or recording practices are changed. Such non-natural or artificial changes are

[9] Karl et al. [213] describe a case in which a precipitation gauge recorded significantly different values after being raised one metre from its original position.

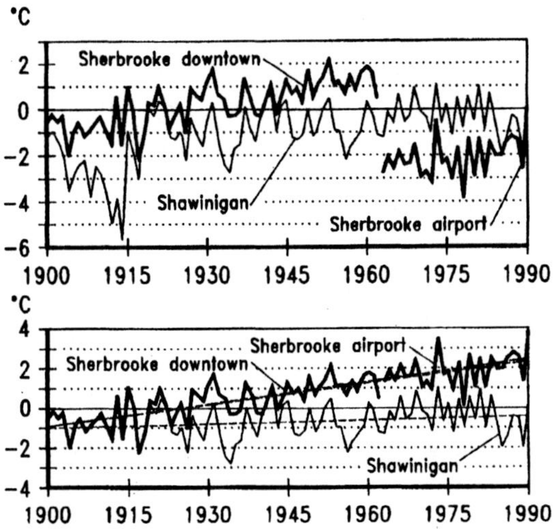

Figure 1.9: *Annual mean daily minimum temperature time series at two neighbouring sites in Quebec. Sherbrooke has experienced considerable urbanization since the beginning of the century whereas Shawinigan has maintained more of its rural character.*
Top: The raw records. The abrupt drop of several degrees in the Sherbrooke series in 1963 reflects the move of the instrument from downtown Sherbrooke to its suburban airport. The reason for the downward dip before 1915 in the Shawinigan record is unknown.
Bottom: Corrected time series for Sherbrooke and Shawinigan. The Sherbrooke data from 1963 onward are increased by 3.2 °C. *The straight lines are trend lines fitted to the corrected Sherbrooke data and the 1915–90 Shawinigan record.*
Courtesy L. Vincent, AES Canada.

called *inhomogeneities*. An example is contained in the temperature records of Sherbrooke and Shawinigan (Quebec) shown in the upper panel of Figure 1.9. The Sherbrooke observing site was moved from a downtown location to a suburban airport in 1963—and the recorded temperature abruptly dropped by more than 3 °C. The Shawinigan record may also be contaminated by observational errors made before 1915.

Geophysical time series often exhibit a trend. Such trends can originate from various sources. One source is urbanization, that is, the increasing density and height of buildings around an observation location and the corresponding changes in the properties of the land surface. The temperature at Sherbrooke, a location heavily affected by development, exhibits a marked upward trend after correction for the systematic change in 1963 (Figure 1.9, bottom). This temperature trend is much weaker for the neighbouring Shawinigan, perhaps due to a weaker urbanization effect at that site or natural variations of the climate system. Both temperature trends at Sherbrooke and Shawinigan are real, not observational artifacts. The strong trend at Sherbrooke must not be mistaken for an indication of *global warming*.

Trends in the large-scale state of the climate system may reflect systematic forcing changes of the climate system (such as variations in the Earth's orbit, or increased CO_2 concentration in the atmosphere) or low-frequency internally generated variability of the climate system. The latter may be deceptive because low-frequency variability, on short time series, may be mistakenly interpreted as trends. However, if the length of such time series is increased, a metamorphosis of the former 'trend' takes place and it becomes apparent that the trend is a part of the natural variation of the system.[10]

1.2.4 Quality of Forecasts. The *Old Farmer's Almanac* publishes regular outlooks for the climate for the coming year. The method used to prepare these outlooks is kept secret, and scientists question the existence of skill in the predictions. To determine whether these skeptics are right or wrong, measures of the skill of the forecasting scheme are needed. These *skill scores* can be used to compare forecasting schemes objectively.

The Almanac makes *categorical* forecasts of future temperature and precipitation amount in two categories, 'above' or 'below' normal. A suitable skill score in this case is the number of correct forecasts. Trivial forecasting schemes such as persistence (no change), climatology, or pure chance can be used as reference forecasts if no other forecasting scheme is available. Once we have counted the number of correct forecasts made with both the tested and the reference schemes, we can estimate the improvement (or degradation) of forecast skill by computing the difference in the counts. Relatively simple probabilistic methods can be used to make a judgement about the

[10]This is an example of the importance of time scales in climate research, an illustration that our interpretation of a given process depends on the time scales considered. A short-term trend may be just another swing in a slowly varying system. An example is the Madden-and-Julian Oscillation (MJO, [264]), which is the strongest intra-seasonal mode in the tropical troposphere. It consists of a wavenumber 1 pattern that travels eastward round the globe. The MJO has a mean period of 45 days and has significant memory on time scales of weeks; on time scales of months and years, however, the MJO has no temporal correlation.

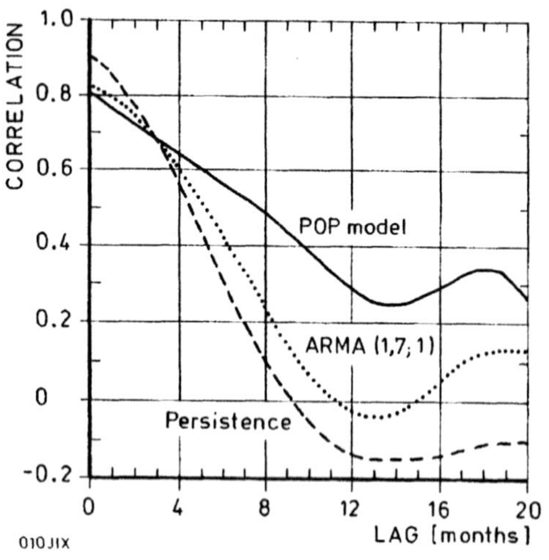

Figure 1.10: *Correlation skill scores for three forecasts of the low-frequency variations within the Southern Oscillation Index (Figure 1.2). A score of 1 indicates a perfect forecast, while a zero indicates a forecast unrelated to the predictand [432].*

significance of the change. We will return to the *Old Farmer's Almanac* in Section 18.1.

Now consider another forecasting scheme in which *quantitative* rather than categorical statements are made. For example, a forecast might consist of a statement such as: *'the SOI will be x standard deviations above normal next winter.'* One way to evaluate such forecasts is to use a measure called the *correlation skill score* ρ (Chapter 18). A score of $\rho = 1$ corresponds with a perfect forecasting scheme in the sense that forecast changes exactly mirror SOI changes even though the dynamic range of the forecast may be different from that of the SOI. In other words, the correlation skill score is one when there is an exact linear relationship between forecasts and reality. Forecasts that are (linearly) unrelated to the predictand yield zero correlation.

The correlation skill score for several methods of forecasting the SOI are displayed in Figure 1.10. Specifically, persistence forecasts (Chapter 18), POP forecasts (Chapter 15), and forecasts made with a univariate linear time series model (Chapters 11 and 12). Forecasts based on persistence and the univariate time series model are superior at one and two month lead times. The POP forecast becomes more skilful beyond that time scale.

Regretfully, forecasting schemes generally do not have the same skill under all circumstances. The skill often exhibits a marked annual cycle (e.g., skill may be high during the dry season, and low during the wet season). The skilfulness of a forecast also often depends on the low-frequency state of the atmospheric flow (e.g., blocking or westerly regime). Thus, in most forecasting problems there are physical considerations (state dependence and the memory of the system) that must be accounted for when using statistical tools to analyse forecast skill. This is done either by conducting a statistical analysis of skill that incorporates the effects of state dependence and serial correlation, or by using physical intuition to temper the precise interpretation of a simpler analysis that compromises the assumptions of stationarity and non-correlation.

There are various pitfalls in the art of forecast evaluation. An excellent overview is given by Livezey [255], who presents various examples in which forecast skill is overestimated. Chapter 18 is devoted to the art of forecast evaluation.

1.2.5 Characteristic Times and Characteristic Spatial Patterns. What are the temporal characteristics of the Southern Oscillation Index illustrated in Figure 1.2? Visual inspection suggests that the time series is dominated by at least two time scales: a high frequency mode that describes month-to-month variations, and a low-frequency mode associated with year-to-year variations. How can one objectively quantify these characteristic times and the amount of variance attributed to these time scales? The appropriate tool is referred to as time series analysis (Chapters 10 and 11).

Indices, such as the SOI, are commonly used in climate research to monitor the temporal development of a process. They can be thought of as filters that extract physical signals from a multivariate environment. In this environment the signal is masked by both spatial and temporal variability unrelated to the signal, that is, by spatial and temporal noise.

The conventional approach used to identify indices is largely subjective. The characteristic patterns of variation of the process are identified and associated with regions or points. Corresponding areal averages or point values are then used to indicate the state of the process.

Another approach is to extract characteristic patterns from the data by means of analytical techniques, and subsequently use the coefficients of these patterns as indices. The advantages of this approach are that it is based on an objective algorithm and that it yields the characteristic patterns explicitly. *Eigentechniques* such as Empirical Orthogonal Function (EOF)

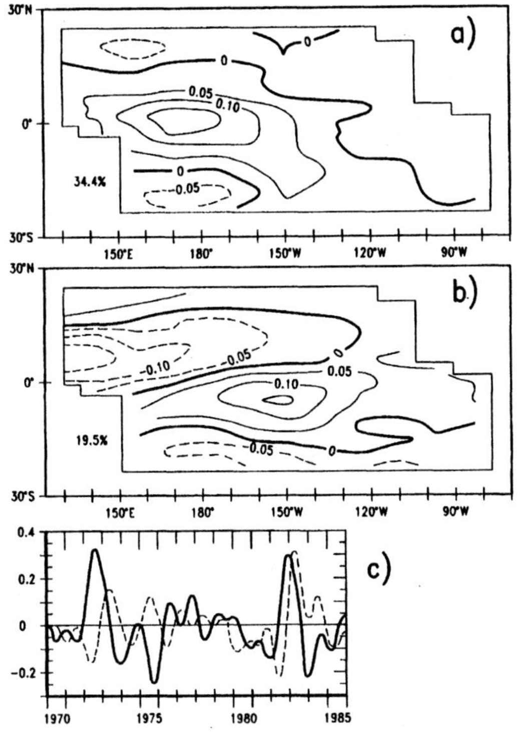

Figure 1.11: *Empirical Orthogonal Functions (EOFs; Chapter 13) of monthly mean wind stress over the tropical Pacific [394].*
a,b) The first two EOFs. The two patterns are spatially orthogonal.
c) Low-frequency filtered coefficient time series of the two EOFs shown in a,b). The solid curve corresponds to the first EOF, which is displayed in panel a). The two curves are orthogonal.

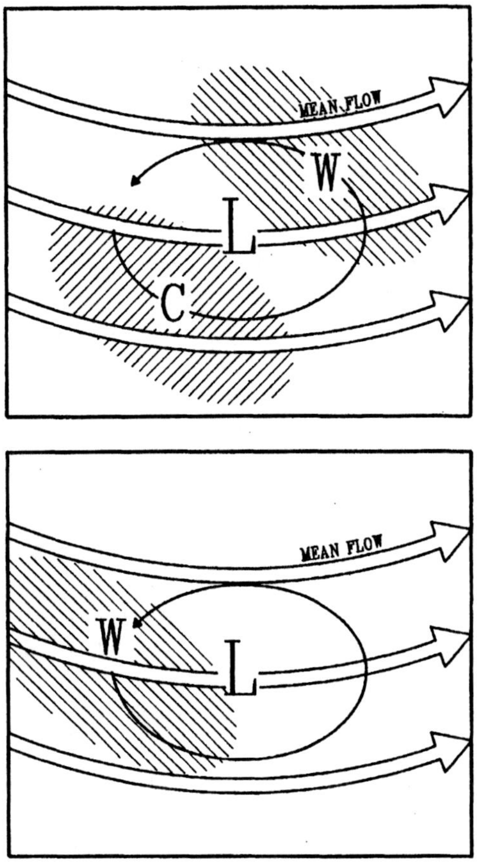

Figure 1.12: *A schematic representation of the spatial distributions of simultaneous SST and SLP anomalies at Northern Hemisphere midlatitudes in winter, when the SLP anomaly induces the SST anomaly (top), and when the SST anomaly excites the SLP anomaly (bottom).*
The large arrows represent the mean atmospheric flow. The 'L' is an atmospheric low-pressure system connected with geostrophic flow indicated by the circular arrow. The hatching represents warm (W) and cool (C) SST anomalies [438].

analysis and Principal Oscillation Pattern (POP) analysis are tools that can be used to define patterns and indices objectively (Chapters 13 and 15).

An example is the EOF analysis of monthly mean wind stress over the tropical Pacific [394]. The first two EOFs, shown in Figure 1.11a and Figure 1.11b, are primarily confined to the equator. The two fields are (by construction) orthogonal to each other. Figure 1.11c shows the time coefficients of the two fields. An analysis of the coefficient time series, using the techniques of cross-spectral analysis (Section 11.4), shows that they vary coherently on a time scale $T \approx$ 2 to 3 years. One curve leads the other by a time lag of approximately $T/4$ years. The temporal lag-relationship of the time coefficients together with the spatial quadrature leads to the interpretation that the two patterns and their time coefficients describe an eastward propagating signal that, in fact, may be associated with the Southern Oscillation.

1.2.6 Pairs of Characteristic Patterns. Almost all climate components are interrelated. When one component exhibits anomalous conditions, there will likely be characteristic anomalies in other components at the same time. The relative shapes of the patterns in related climate components are often indicative of the processes that dominate the coupling of the components.

To illustrate this idea we consider large-scale air–sea interactions on seasonal time scales at midlatitudes in winter [438] [312]. Figure 1.12

illustrates the two mechanisms that might be involved in air–sea interactions in the North Atlantic. The lower panel illustrates how a sea-surface temperature (SST) anomaly pattern might induce a simultaneous sea-level pressure (SLP) anomaly pattern. The argument is linear so we may assume that the SST anomaly is positive. This positive SST anomaly enhances the sensible and latent heat fluxes into the atmosphere above and downstream of the SST anomaly. Thus SLP is reduced in that area and anomalous cyclonic flow is induced.

The upper panel of Figure 1.12 illustrates how a SLP anomaly might induce an anomalous SST pattern. The anomalous SLP distribution alters the wind stress across the region by creating stronger zonal winds in the southwest part of the anomalous cyclonic circulation and weaker zonal winds in the northeast sector. This configuration induces anomalous mixing of the ocean's mixed layer and anomalous air–sea fluxes of sensible and latent heat (cf. [3.2.3]). Stronger winds intensify mixing and enhance the upward heat flux whereas weaker winds correspond to reduced mixing and weaker vertical fluxes. The result is anomalous cooling of the sea surface in the southwest sector and anomalous heating in the northeast sector of the cyclonic circulation.

One strategy for finding out which of the two proposed mechanisms dominates air–sea interaction is to identify the dominant patterns in SST and SLP that tend to occur simultaneously. This can be accomplished by performing a *Canonical Correlation Analysis* (CCA, Chapter 14). In the CCA two vector variables $\vec{\mathbf{X}}$ and $\vec{\mathbf{Y}}$ are considered, and sets of orthogonal patterns $\vec{p}_X^{\,i}$ and $\vec{p}_Y^{\,i}$ are constructed so that the expansion coefficients α_i^x and α_j^y in $\vec{\mathbf{X}} = \sum_i \alpha_i^x \vec{p}_X^{\,i}$ and $\vec{\mathbf{Y}} = \sum_j \alpha_j^y \vec{p}_Y^{\,j}$ are optimally correlated for $i = j$ or uncorrelated for $i \neq j$.

Zorita, Kharin, and von Storch [438] applied CCA to winter (DJF) mean anomalies of North Atlantic SST and SLP and found two pairs of CCA patterns $\vec{p}_{SST}^{\,i}$ and $\vec{p}_{SLP}^{\,j}$ that were associated with physically significant correlations. The pair of patterns with the largest correlation (0.56) is shown in Figure 1.13. The SLP pattern represents 21% of the total DJF SLP variance whereas the SST pattern explains 19% of the total SST variance.[11] Clearly the two patterns support the hypothesis that the anomalous atmospheric circulation is responsible for the generation of SST anomalies off the North American coast. Peng and Fyfe [312] refer to this as the 'atmosphere driving the ocean' mode. See also Luksch [261].

Canonical Correlation Analysis is explained in detail in Chapter 14 and we return to this example in [14.3.1–2].

1.2.7 Atmospheric General Circulation Model Experimentation: Evaluation of Paired Sensitivity Experiments and Verification of Control Simulation. Atmospheric General Circulation Models (AGCMs) are powerful tools used to simulate the dynamics of the atmospheric circulation. There are two main applications of these GCMs, one being the simulation of the present, past (e.g., paleoclimatic conditions), or future (e.g., climate change) statistics of the atmospheric circulation. The other involves the study of the simulated climate's sensitivity to the effect of different boundary conditions (e.g., sea-surface temperature) or parameterizations of sub-grid scale processes (e.g., planetary boundary layer).[12]

In both modes of operation two sets of statistics are compared. In the first, the statistics of the simulated climate are compared with those of the observed climate, or sometimes with those of another simulated climate. In the second mode of experimentation, the statistics obtained in the run with anomalous conditions are compared with those from the run with the *control* conditions. The simulated atmospheric circulation is turbulent as is that of the real atmosphere (see Section 1.1). Therefore the true signal (excited by the prescribed change in boundary conditions, parameterization, etc.) or the true model error is masked by random variations.

Even when the modifications in the experimental run have no effect on the simulated climate, the difference field will be nonzero and will show structure reflecting the random variations in the control and experimental runs. Similarly, the mean difference field between an observed distribution and its simulated counterpart will exhibit, possibly large scale, features, even if the model is perfect.

[11] The proportion of variance represented by the patterns is unrelated to the correlation.

[12] Sub-grid scale processes take place on spatial scales too small to be resolved by a climate model. Regardless of the resolution of the climate model, there are unresolved processes at smaller scales. Despite the small scale of these processes, they influence the large-scale evolution of the climate system because of the nonlinear character of the climate system. Climate modellers therefore attempt to specify the 'net effect' of such processes as a transfer function of the large-scale state itself. This effect is a forcing term for the resolved scales, and is usually expressed as an expected value which is conditional upon the large-scale state. The transfer function is called a 'parameterization.'

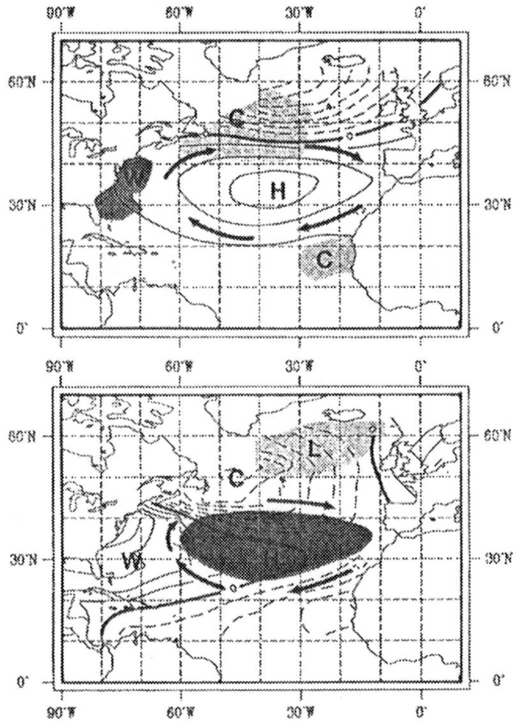

Figure 1.13: *The dominant pair of CCA patterns that describe the connection between simultaneous winter (DJF) mean anomalies of sea-level pressure (SLP, top) and sea-surface temperature (SST, bottom) in the North Atlantic. The largest features of the SLP field are indicated by shading in the SST map, and vice versa. See also [14.3.1]. From Zorita et al. [438].*

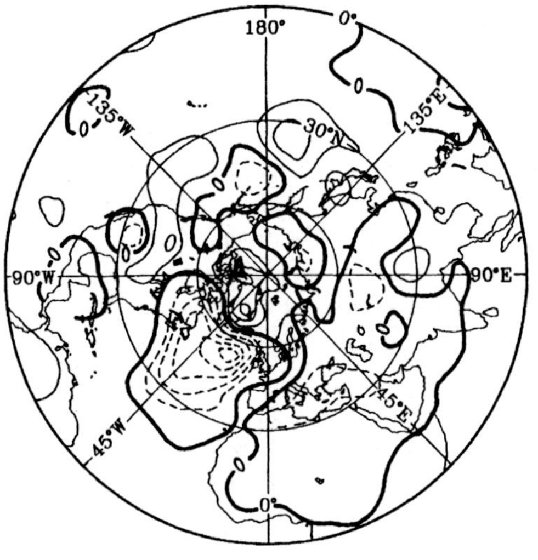

Figure 1.14: *The mean SLP difference field between control and experimental atmospheric GCM runs. Evaporation over the Iberian Peninsula was artificially suppressed in the experimental run. The signal is not statistically significant [402].*

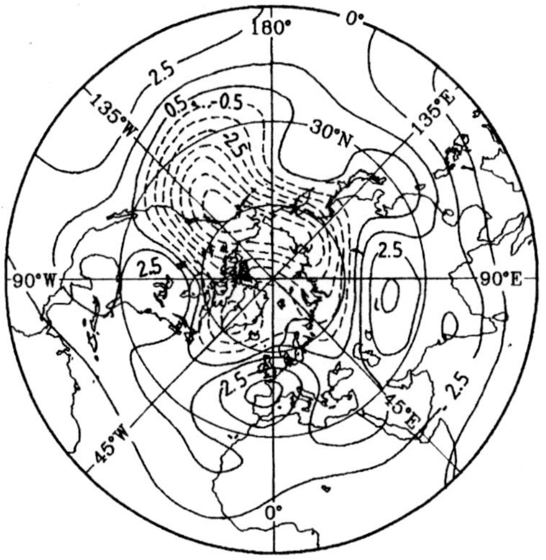

Figure 1.15: *The mean 500 hPa height difference field between a control run and an experimental run in which a positive (El Niño) SST anomaly was imposed in the equatorial Central and Eastern Pacific. The signal is statistically significant. See also Figures 9.1 and 9.2 [393].*

Therefore, it is necessary to apply statistical techniques to distinguish between the deterministic signal (or model error) and the internal noise.

Appropriate methodologies designed to diagnose the presence of a signal include the use of interval estimation methods (Section 5.4) or hypothesis testing methods (Chapter 6). Interval estimation methods use statistical models to produce a range of signal estimates consistent with the realizations of control and experimental mean fields obtained from the simulation. Hypothesis testing methods use statistical models to determine whether information in the realizations is consistent with the null hypothesis that the difference fields, such as in Figures 1.14 and 1.15, do not contain a deterministic signal and thus reflect only the effects of random variation.

We illustrate the problem with two examples: an experiment in which there is no significant signal, and another in which modifications to the model result in a strong change in the atmospheric flow.

In the first case, the surface properties of the Iberian peninsula were modified so as to turn it into a desert in the experimental climate. That is, evaporation at the grid points representing the Iberian peninsula was arbitrarily set to zero. The response, in terms of January Northern Hemisphere sea-level pressure, is shown in Figure 1.14 [402]. The statistical analysis revealed that the signal, which appears to be of very large scale, is mainly due to noise and is not statistically significant.

In the second case, anomalously warm sea-surface temperatures were prescribed in the tropical Pacific, in order to simulate the effect of the 1982/83 El Niño event on the atmosphere. The resulting anomalous mean January 500 hPa height field is shown in Figure 1.15. In this case the signal is statistically distinguishable from the background noise.

Before using statistical tests, we must account for several methodical considerations (see Chapter 6). Straightforward statistical assessments that compare the mean states of two simulated climates generally use simple statistical tests that are performed locally at grid points. More complex *field tests*, often called *field significance tests* in the climate literature, are used less frequently.

Grid point tests, while popular because of their simplicity, may have interpretation problems. The result of a set of statistical tests, one conducted at each grid point, is a field of decisions denoting where differences are, and are not, *statistically significant*. However, statistical tests cannot be conducted with absolute certainty. Rather, they are conducted in such a way that there is an *a priori* specified risk $1-\tilde{p}$ of rejecting the null hypothesis: 'no difference' when it is true.[13]

The specified risk $(1-\tilde{p}) \times 100\%$ is often referred to as the *significance level* of the test.[14]

A consequence of setting the risk of false rejection to $1-\tilde{p}$, $0 < \tilde{p} < 1$, is that we can expect approximately $(1-\tilde{p}) \times 100\%$ of the decisions to be *reject* decisions when the null hypothesis is valid. However, many fields of interest in climate experiments exhibit substantial spatial correlation (e.g., smooth fields such as the geopotential heights displayed in Figure 1.1).

The spatial coherence of these fields has two consequences for hypothesis testing at grid points. The first is that the proportion of the field covered by reject decisions becomes highly variable from one realization of the climate experiment to the next. In some problems a rejection rate of 20% may still be globally consistent with the null hypothesis at the 5% significance level. The second is that the spatial coherence of the studied fields also leads to fields of decisions that are spatially coherent: if the difference between two mean 500 hPa height fields is large at a particular point, it is also likely to be large at neighbouring points because of the spatial continuity of 500 hPa height. A decision made at one location is generally not statistically independent of decisions made at other locations. This makes regions of significant change difficult to identify. Methods that can be used to assess the field significance of a field of reject/retain decisions are discussed in Section 6.8. Local, or *univariate*, significance tests are discussed in Sections 6.6 and 6.7.

Another approach to the comparison of observed and simulated mean fields involves the use of classical *multivariate statistical tests* (Sections 6.6 and 6.7). The word *multivariate* is used somewhat differently in the statistical lexicon than it is in climatology: it describes tests and other inference procedures that operate on vector objects, such as the difference between two mean fields, rather than scalar objects, such as a difference of means at a grid point. Thus a multivariate test is a field significance test; it is used to make a single inference about a field of differences between the observed and simulated climate.

Classical multivariate inference methods can not generally be applied directly to difference of means or variance problems in climatology. These methods are usually unable to cope with fields under study, such as seasonal geopotential means, that are generally 'observed' at numbers of grid points one to three orders of magnitude greater than the number of realizations available.[15]

[13]The standard, rather mundane statistical nomenclature for this kind of error is *Type I* error; failure to reject the null hypothesis when it is false is termed a *Type II* error. Specifying a smaller risk reduces the chance of making a Type I error but also reduces the sensitivity of the test and hence increases the likelihood of a Type II error. More or less standard practice is to set the risk of a Type I error to $(1-\tilde{p}) \times 100\% = 5\%$ in tests of the mean and to $(1-\tilde{p}) \times 100\% = 10\%$ in tests of variability. A higher level of risk is usually felt to be acceptable in variance tests because they are generally less powerful than tests concerning the mean state. The reasons for specifying the risk in the form $1-\tilde{p}$, where $\tilde{p}$ is a large probability near 1, will become apparent later.

[14]There is some ambiguity in the climate literature about how to specify a 'significance level.' Many climatologists use the expression 'significant at the 95% level,' although standard statistical convention is to use the expression 'significant at the 5% level.' With the latter convention, which we use throughout this book, rejection at the 1% significance level indicates the presence of stronger evidence against the null hypothesis than rejection at the 10% significance level.

[15]A typical climate model validation problem involves the comparison of simulated monthly mean fields obtained from a 5–100 year simulation, with corresponding observed mean fields from a 20–50 year climatology. Such a problem therefore uses a combined total of $n = 25$ to 150 realizations of mean January 500 hPa height, for example. On the other hand, the horizontal resolution of typical present day climate models is such that these mean fields are represented on global grids with $m = 2000$ to 8000 points. Except on relatively small regional scales, the dimension of (or number of points in) the difference field is greater than the combined number of realizations from the simulated and observed climates.

One solution to this difficulty is to reduce the dimension of the observed and simulated fields to less than the number of realizations before using any inference procedure. This can be done using pattern analysis techniques, such as EOF analysis, that try to identify the climate's principal modes of variation empirically. Another solution is to abandon classical inference techniques and replace them with ad hoc methods, such as the 'PPP' test (Preisendorfer and Barnett [320]).

Both grid point and field significance tests are plagued with at least two other problems that result in interpretation difficulties. The first of these is that the word *significance* does not have a specific physical interpretation. The statistical significance of the difference between a simulated and observed climate depends upon both location and sample size. Location is a factor that affects interpretation because variability is not uniform in space. A 5 m difference between an observed and a simulated mean January 500 hPa height field may be statistically very significant in the tropics, but such a difference is not likely to be statistically, or physically, significant at midlatitudes where interannual variability is large. Sample size is a factor because the sensitivity of statistical tests is affected by the amount of information about the mean state contained in the observed and simulated realizations. Larger samples have greater information content and consequently result in more powerful tests. Thus, even though a 5 m difference at midlatitudes may not be physically important, it will be found to be significant given large enough simulated and observed climatologies. The statistical strength of the signal (or model error) may be quantified by a parameter called the *level of recurrence*, which is the probability that the signal's signature will not be masked by the noise in another identical but statistically independent run with the GCM (Sections 6.9–6.10).

The second problem is that objective statistical validation techniques are more honest than modellers would like them to be. GCMs and analysis systems have various biases that ensure that objective tests of their differences will reject the null hypothesis of no difference with certainty, given large enough samples. Modellers seem to have an intuitive grasp of the size and spatial structure of biases and seem to be able to discount their effects when making climate comparisons. If these biases can be quantified, statistical inference procedures can be adjusted to account for them (see Chapter 6).

Part I

Fundamentals

2 Probability Theory

2.1 Introduction

2.1.1 The General Idea. The basic ideas behind probability theory are as simple as those associated with making lists—the prospect of computing probabilities or thinking in a 'probabilistic' manner should not be intimidating.

Conceptually, the steps required to compute the chance of any particular event are as follows.

- Define an *experiment* and construct an exhaustive description of its possible outcomes.
- Determine the *relative likelihood* of each outcome.
- Determine the *probability* of each outcome by comparing its likelihood with that of every other possible outcome.

We demonstrate these steps with two simple examples. In the first we consider three tosses of an honest coin. The second example deals with the rainfall in winter at West Glacier in Washington State (USA).

2.1.2 Simple Events and the Sample Space. The *sample space*, denoted by $\mathcal{S}$, is a list of possible outcomes of an experiment, where each item in the list is a *simple event*, that is, an experimental outcome that cannot be decomposed into yet simpler outcomes.

For example, in the case of three consecutive tosses of a fair coin, the simple events are $\mathcal{S}$ = {HHH, HHT, HTH, THH, TTH, THT, HTT, TTT} with H = 'head' and T = 'tail.' Another description of the possible outcomes of the coin tossing experiment is {'three heads', 'two heads', 'one head', 'no heads'}. However, this is not a list of simple events since some of the outcomes, such as {'two heads'}, can occur in several ways.

It is not possible, though, to list the simple events that compose the West Glacier rainfall sample space. This is because a reasonable sample space for the atmosphere is the collection of all possible trajectories through its phase space, an uncountably large collection of 'events.' Here we are only able to describe *compound* events, such as the outcomes that the daily rainfall is more, or less, than a threshold of, say, 0.1 inch. While we are able to describe these compound events in terms of some of their characteristics, we do not know enough about the atmosphere's sample space or the processes that produce precipitation to describe precisely the proportion of the atmosphere's sample space that represents one of these two compound events.

2.1.3 Relative Likelihood and Probability. In the coin tossing experiment we use the physical characteristics of the coin to determine the relative likelihood of each outcome in $\mathcal{S}$. The chance of a head is the same as that of a tail on any toss, if we have no reason to doubt the fairness of the coin, so each of the eight outcomes is as likely to occur as any other.

The West Glacier rainfall outcomes are less obvious, as we do not have an explicit characterization of the atmosphere's sample space. Instead, we assume that our rainfall observations stem from a stationary process, that is, that the likelihood of observing more, or less, than 0.1 inch daily rainfall is the same for all days within a winter and the same for all winters. Observed records tell us that the daily rainfall is greater than the 0.1 inch threshold on about 38 out of every 100 days. We therefore *estimate* the relative likelihoods of the two compound events in $\mathcal{S}$.

As long as all outcomes are equally likely, assigning probabilities can be done by counting the number of outcomes in $\mathcal{S}$. The sum of all the probabilities must be unity because one of the events in $\mathcal{S}$ *must* occur every time the experiment is conducted. Therefore, if $\mathcal{S}$ contains M items, the probability of any simple event is just $1/M$. We see below that this process of assigning probabilities by counting the number of elements in $\mathcal{S}$ can often be extended to include simple events that do not have the same likelihood of occurrence.

Once the probability of each simple event has been determined, it is easy to determine the probability of a compound event. For example, the event {'Heads on exactly 2 out of 3 tosses'} is composed of the three simple events {HHT, HTH, THH} and thus occurs with probability 3/8 on any repetition of the experiment.

The word *repetition* is important because it underscores the basic idea of a probability. If an experiment is repeated *ad infinitum*, the proportion of the realizations resulting in a particular outcome is the probability of that outcome.

2.2 Probability

2.2.1 Discrete Sample Space. A *discrete* sample space consists of an enumerable collection of simple events. It can contain either a finite or a countably infinite number of elements.

An example of a *large finite* sample space occurs when a series of univariate statistical tests (see [6.8.1]) is used to validate a GCM. The test makes a decision about whether or not the simulated climate is similar to the observed climate in each model grid box (Chervin and Schneider [84]; Livezey and Chen [257]; Zwiers and Boer [446]). If there are m grid boxes (m is usually of order 10^3 or larger), then the number of possible outcomes of the decision making procedure is 2^m—a large but finite number. We could be exhaustive and list each of the 2^m possible fields of decisions, but it is easy and convenient to characterize more complex events by means of a numerical description and to count the number of ways each can occur.[1]

An example of an infinite discrete sample space occurs in the description of a precipitation climatology, where $\mathcal{S} = \{0, 1, 2, 3, \ldots\}$ lists the waiting times between rain days.[2]

2.2.2 Binomial Experiments. Experiments analogous to the coin tossing, rainfall threshold exceedance, and testing problems described above are particularly important. They are referred to as *binomial* experiments because each replication of the experiment consists of a number of *Bernoulli trials*; that is, trials with only two possible outcomes (which can be coded 'S' and 'F' for success and failure).

An experiment that consists of m Bernoulli trials has a corresponding sample space that contains 2^m entries. One way to describe $\mathcal{S}$ conveniently is to partition it into subsets of simple events according to the number of successes. These compound events are made up of varying numbers of sample space elements. The smallest events (0 successes and m successes) contain exactly one element each. The next smallest events (one success in m trials and $m-1$ successes in m trials) contain m elements each. In general, the event with n successes in m trials contains

$$\binom{m}{n} = \frac{m!}{n!(m-n)!}$$

simple events. These compound events do not contain any common elements, so it follows that $\sum_{n=1}^{m} \binom{m}{n} = 2^m$.

2.2.3 A Sample Space is More Than a Collection of Simple Events. A complete probabilistic description of an experiment must be more than just a list of simple events. We also need a rule, say $P(\cdot)$, that assigns probabilities to events. In simple situations, such as the coin tossing example of Section 2.1, $P(\cdot)$ can be based on the numbers of elements in an event.

Different experiments may generate the same set of possible outcomes but have different rules for assigning probabilities to events. For example, a fair and a biased coin, each tossed three times, generate the same list of possible outcomes but each outcome does not occur with the same likelihood. We can use the same threshold for daily rainfall at every station and will find different likelihoods for the exceedance of that threshold.

2.2.4 Probability of an Event. The probability of an event in a discrete sample space is computed by summing up the probabilities of the individual sample space elements that comprise the event. A list of the complete sample space is usually unnecessary. However, we do need to be able to enumerate events, that is, count elements in subsets of $\mathcal{S}$.

Some basic rules for probabilities are as follows.

- Probabilities are non-negative.
- When an experiment is conducted, one of the simple events in $\mathcal{S}$ *must* occur, so

  $$P(\mathcal{S}) = 1.$$

- It may be easier to compute the probability of the *complement of an event* than that of the event itself. If A denotes an event, then

[1] We have taken some liberties with the idea of a discrete sample space in this example. In reality, each of the 'simple events' in the sample space is a compound event in a very large (but discrete) space of GCM trajectories.

[2] We have taken additional liberties in this example. The events are really compound events in the uncountably large space of trajectories of the real atmosphere.

$\neg A$, its complement, is the collection of all elements in $\mathcal{S}$ that are not contained in A. That is, $\mathcal{S} = A \cup \neg A$. Also, $A \cap \neg A = \emptyset$. Therefore,

$$P(A) = 1 - P(\neg A).$$

- It is often useful to divide an event into smaller, mutually exclusive events. Two events A and B are *mutually exclusive* if they do not contain any common sample space elements, that is, if $A \cap B = \emptyset$. An experiment can not produce two mutually exclusive outcomes at the same time. Therefore, if A and B are mutually exclusive,

$$P(A \cup B) = P(A) + P(B). \quad (2.1)$$

- In general, the expression for the probability of observing one of two events A and B is

$$P(A \cup B) = P(A) + P(B) - P(A \cap B).$$

  The truth of this is easy to understand. The common part of the two events, $A \cap B$, is included in both A and B and thus $P(A \cap B)$ is included in the calculation of $P(A) + P(B)$ twice.

2.2.5 Conditional Probability. Consider a weather event A (such as the occurrence of severe convective activity) and suppose that the climatological probability of this event is $P(A)$. Now consider a 24-hour weather forecast that describes an event B within the daily weather sample space. If the forecast is skilful, our perception of the likelihood of A will change. That is, the probability of A *conditional* upon forecast B, which is written $P(A|B)$, will not be the same as the climatological probability $P(A)$.

The conditional probability of event A, given an event B for which $P(B) \neq 0$, is

$$P(A|B) = P(A \cap B)/P(B). \quad (2.2)$$

The interpretation is that only the part of A that is contained within B can take place, and thus the probability that this restricted version of A takes place must be scaled by $P(B)$ to account for the change of context. Note that all conditional probabilities range between 0 and 1, just as ordinary probabilities do. In particular, $P(\mathcal{S}|B) = P(B|B) = 1$.

2.2.6 Independence. Two events A and B are said to be *independent* of each other if

$$P(A \cap B) = P(A)P(B). \quad (2.3)$$

It follows from (2.2) that if A and B are independent, then $P(A|B) = P(A)$. That is, restriction of the sample space to B gives no additional information about whether or not A will occur.

Suppose A represents severe weather and B represents a 24-hour forecast of severe weather. If A and B are independent, then the forecasting system does not produce skilful severe weather forecasts: a severe weather forecast does not change our perception of the likelihood of severe weather tomorrow.

2.3 Discrete Random Variables

2.3.1 Random Variables. We are usually not really interested in the sample space $\mathcal{S}$ itself, but rather in the events in $\mathcal{S}$ that are characterized by functions defined on $\mathcal{S}$. For the three coin tosses in [2.1.2] the function could be the number of 'heads.' Such functions are referred to as *random variables*. We will usually use a bold face upper case character, such as $\mathbf{X}$, to denote the function and a bold face lower case variable $\mathbf{x}$ to denote a particular value taken by $\mathbf{X}$. This value is also often referred to as a *realization* of $\mathbf{X}$.

Random variables are *variable* because their values depend upon which event in $\mathcal{S}$ takes place when the experiment is conducted. They are *random* because the outcome in $\mathcal{S}$, and hence the value of the function, can not be predicted in advance.

Random variables are *discrete* if the collection of values they take is enumerable, and *continuous* otherwise. Discrete random variables will be discussed in this section and continuous random variables in Section 2.6.

The probability of observing any particular value $\mathbf{x}$ of a discrete random variable $\mathbf{X}$ is determined by characterizing the event $\{\mathbf{X} = \mathbf{x}\}$ and then calculating $P(\mathbf{X} = \mathbf{x})$. Thus, its *randomness* depends upon both $P(\cdot)$ and how $\mathbf{X}$ is defined on $\mathcal{S}$.

2.3.2 Probability and Distribution Functions. In general, it is cumbersome to use the sample space $\mathcal{S}$ and the probability rule $P(\cdot)$ to describe the random, or *stochastic* characteristics of a random variable $\mathbf{X}$. Instead, the stochastic properties of $\mathbf{X}$ are characterized by the *probability function* f_X and the *distribution function* F_X.

The *probability function* f_X of a discrete random variable $\mathbf{X}$ associates probabilities with values taken by $\mathbf{X}$. That is

$$f_X(\mathbf{x}) = P(\mathbf{X} = \mathbf{x}).$$

Two properties of the probability function are:

- $0 \le f_X(\mathbf{x}) \le 1$ for all $\mathbf{x}$, and
- $\sum_{\mathbf{x}} f_X(\mathbf{x}) = 1$, where the notation $\sum_{\mathbf{x}}$ indicates that the summation is taken over all possible values of $\mathbf{X}$.

The *distribution function* F_X of a discrete random variable $\mathbf{X}$ is defined as

$$F_X(\mathbf{x}) = \sum_{\mathbf{y} \le \mathbf{x}} f_X(\mathbf{y}).$$

Some properties of the distribution function are:

- $F_X(\mathbf{x}) \le F_X(\mathbf{y})$ if $\mathbf{x} \le \mathbf{y}$,
- $\lim_{\mathbf{x} \to -\infty} F_X(\mathbf{x}) = 0$, and
- $\lim_{\mathbf{x} \to +\infty} F_X(\mathbf{x}) = 1$.

The phrase *probability distribution* is often used to refer to either of these functions because the probability function can be derived from the distribution function and vice versa.

2.3.3 The Expectation Operator. A random variable $\mathbf{X}$ and its probability function f_X together constitute a model for the operation of an experiment: every time it is conducted we obtain a realization $\mathbf{x}$ of $\mathbf{X}$ with probability $f_X(\mathbf{x})$. A natural question is to ask what the average value of $\mathbf{X}$ will be in repeated operation of the experiment. For the coin tossing experiment, with $\mathbf{X}$ being the number of 'heads,' the answer is $0 \times \frac{1}{8} + 1 \times \frac{3}{8} + 2 \times \frac{3}{8} + 3 \times \frac{1}{8} = \frac{3}{2}$ because we expect to observe $\mathbf{X} = 0$ (no 'heads' in three tosses of the coin) 1/8 of the time, $\mathbf{X} = 1$ (one 'head' and two 'tails') 3/8 of the time, and so on. Thus, in this example, the *expected* value of $\mathbf{X}$ is 1.5.

In general, the *expected value* of $\mathbf{X}$ is given by

$$\mathcal{E}(\mathbf{X}) = \sum_{\mathbf{x}} \mathbf{x} f_X(\mathbf{x}).$$

The expected value of a random variable is also sometimes called its *first moment*, a term that has its roots in elementary physics. Think of a collection of particles distributed so that the mass of the particles at location $\mathbf{x}$ is $f_X(\mathbf{x})$. Then the expected value $\mathcal{E}(\mathbf{X})$ is the location of the centre of mass of the collection of particles.

The idea of expectation is easily extended to functions of random variables. Let $g(\cdot)$ be any function and let $\mathbf{X}$ be a random variable. The expected value of $g(\mathbf{X})$ is given by

$$\mathcal{E}\big(g(\mathbf{X})\big) = \sum_{\mathbf{x}} g(\mathbf{x}) f_X(\mathbf{x}).$$

The interpretation of the expected value as the average value of $g(\mathbf{X})$ remains the same.

We often use the phrase *expectation operator* to refer to the act of computing an expectation because we operate on a random variable (or a function of a random variable) with its probability function to derive one of its properties.

A very useful property of the expectation operator $\mathcal{E}$ is that the expectation of a sum is a sum of expectations. That is, if $g_1(\cdot)$ and $g_2(\cdot)$ are both functions defined on the random variable $\mathbf{X}$, then

$$\mathcal{E}\big(g_1(\mathbf{X}) + g_2(\mathbf{X})\big) = \mathcal{E}\big(g_1(\mathbf{X})\big) + \mathcal{E}\big(g_2(\mathbf{X})\big). \tag{2.4}$$

Another useful property is that if $g(\cdot)$ is a function of $\mathbf{X}$ and a and b are constants, then

$$\mathcal{E}\big(ag(\mathbf{X}) + b\big) = a\mathcal{E}\big(g(\mathbf{X})\big) + b. \tag{2.5}$$

As a special case, note that the expectation of a constant, say b, is that constant itself. This is, of course, quite reasonable. A constant can be viewed as an example of a degenerate random variable. It has the same value b after every repetition of an experiment. Thus, its average value in repeated sampling must also be b.

A special class of functions of a random variable is the collection of powers of the random variable. The expectation of the kth power of a random variable is known as the kth *moment* of $\mathbf{X}$. Probability distributions can often be identified by their moments. Therefore, the determination of the moments of a random variable sometimes proves useful when deriving the distribution of a random variable that is a function of other random variables.

2.3.4 The Mean and Variance. In the preceding subsection we defined the expected value $\mathcal{E}(\mathbf{X})$ of the random variable $\mathbf{X}$ as the *mean* of $\mathbf{X}$ itself. Frequently the symbol μ (μ_X when clarity is required) is used to represent the mean. The phrase *population mean* is often used to denote the expected value of a random variable; the *sample mean* is the mean of a sample of realizations of a random variable.

Another important part of the characterization of a random variable is *dispersion*. Random variables with little dispersion have realizations tightly clustered about the mean, and vice versa. There are many ways to describe dispersion, but it is usually characterized by *variance*.

The *population variance* (or simply the variance) of a discrete random variable $\mathbf{X}$ with probability distribution f_X is given by

$$\begin{aligned}\mathrm{Var}(\mathbf{X}) &= \mathcal{E}\big((\mathbf{X}-\mu_X)^2\big)\\ &= \sum_{\mathbf{x}}(\mathbf{x}-\mu_X)^2 f_X(\mathbf{x}).\end{aligned}$$

The variance is often denoted by σ^2 or σ_X^2.

The square root of the variance, denoted as σ_X, is known as the *standard deviation*.

In the coin tossing example above, in which $\mathbf{X}$ is the number of 'heads' in three tosses with an honest coin, the variance is given by

$$\begin{aligned}\sigma^2 &= \left(0-\frac{3}{2}\right)^2\times\frac{1}{8}+\left(3-\frac{3}{2}\right)^2\times\frac{1}{8}\\ &\quad+\left(1-\frac{3}{2}\right)^2\times\frac{3}{8}+\left(2-\frac{3}{2}\right)^2\times\frac{3}{8}=\frac{3}{4}.\end{aligned}$$

It will be useful to note a couple of the properties of the variance.

First,

$$\begin{aligned}\mathrm{Var}(\mathbf{X}) &= \mathcal{E}\big((\mathbf{X}-\mu_X)^2\big)\\ &= \mathcal{E}\big(\mathbf{X}^2-2\mathbf{X}\mu_X+\mu_X^2\big)\\ &= \mathcal{E}\big(\mathbf{X}^2\big)-2\mu_X\mathcal{E}(\mathbf{X})+\mathcal{E}\big(\mu_X^2\big)\\ &= \mathcal{E}\big(\mathbf{X}^2\big)-\mu_X^2.\end{aligned}$$

The third step in this derivation, distributing the expectation operator, is accomplished by applying properties (2.4) and (2.5). The last step is achieved by applying the expectation operator and simplifying the third line.

Second, if a random variable is shifted by a constant, its variance does not change. Adding a constant shifts the realizations of $\mathbf{X}$ to the left or right, but it does not change the dispersion of those realizations. On the other hand, multiplying a random variable by a constant does change the dispersion of its realizations. Thus, if a and b are constants, then

$$\mathrm{Var}(a\mathbf{X}+b)=a^2\mathrm{Var}(\mathbf{X}). \tag{2.6}$$

2.3.5 Random Vectors. Until now we have considered the case in which a single random variable is defined on a sample space. However, we are generally interested in situations in which more than one random variable is defined on a sample space. Such related random variables are conveniently organized into a random vector, defined as follows:

A random vector $\vec{\mathbf{X}}$ is a vector of scalar random variables that are the result of the same experiment.

All elements of a random vector are defined on the same sample space $\mathcal{S}$. They *do not* necessarily all have the same probability distribution, because their distributions depend not only on the generating experiment but also on the way in which the variables are defined on $\mathcal{S}$.

We will see in Section 2.8 that random vectors also have properties analogous to the probability function, mean, and variance.

The terms *univariate* and *multivariate* are often used in the statistical literature to distinguish between problems that involve a random variable and those that involve a random vector. In the context of climatology or meteorology, univariate means *a single variable at a single location.* Anything else, such as a single variable at multiple locations, or more than one variable at more than one location, is multivariate to the statistician.

2.4 Examples of Discrete Random Variables

2.4.1 Uniform Distribution. A discrete random variable $\mathbf{X}$ that takes the K different values in a set $\Omega = \{\mathbf{x}_1,\ldots,\mathbf{x}_K\}$ with equal likelihood is called a *uniform* random variable. Its probability function is given by

$$f_X(\mathbf{x})=\begin{cases}1/K & \text{if } \mathbf{x}\in\Omega\\ 0 & \text{otherwise.}\end{cases}$$

Note that the specification of this distribution depends upon K *parameters*, namely the K different values that can be taken. We use the shorthand notation

$$\mathbf{X}\sim\mathcal{U}(\Omega)$$

to indicate that $\mathbf{X}$ is uniformly distributed on Ω. If the K values are given by

$$\mathbf{x}_k = a+\frac{k-1}{K-1}(b-a),\quad \text{for } k=1,\ldots,K$$

for some $a<b$, then the parameters of the uniform distribution are the three numbers a, b, and K. It is readily shown that the mean and variance of a discrete uniform random variable are given by

$$\begin{aligned}\mathcal{E}\big(\mathcal{U}(a,b,K)\big) &= (a+b)/2\\ \mathrm{Var}\big(\mathcal{U}(a,b,K)\big) &= (b-a)^2/12.\end{aligned}$$

Note that the mean and variance do not depend on K.

2.4.2 Binomial Distribution. We have already discussed the *binomial distribution* in the coin tossing and model validation examples [2.2.2].

When an experiment consists of n independent tosses of a fair coin, the number of heads $\mathbf{H}$ that come up is a *binomial random variable*. Recall that the sample space for this experiment has 2^n equally likely elements and that there are $\binom{n}{h}$ ways to observe the event $\{\mathbf{H} = \mathbf{h}\}$. This random variable $\mathbf{H}$ has probability function

$$f_H(\mathbf{h}) = \binom{n}{h}\left(\frac{1}{2}\right)^n.$$

In general, the 'coin' is not fair. For example, consider sequences of n independent daily observations of West Glacier rainfall [2.1.2] and classify each observation into two categories depending upon whether the rainfall exceeds the 0.1 inch threshold. This natural experiment has the same number of possible outcomes as the coin tossing experiment (i.e., 2^n), but all outcomes are not equally likely.

The coin tossing and West Glacier experiments are both examples of *binomial* experiments. That is, they are experiments that:

- consist of n independent Bernoulli trials, and
- have the same probability of success on every trial.

A *binomial random variable* is defined as the number of successes obtained in a binomial experiment.

The probability distribution of a binomial random variable $\mathbf{H}$ is derived as follows. Let S denote a 'success' and assume that there are n trials and that $P(S) = p$ on any trial. What is the probability of observing $\mathbf{H} = \mathbf{h}$? One way to obtain $\{\mathbf{H} = \mathbf{h}\}$ is to observe

$$\overbrace{SSS\cdots S}^{h \text{ times}}\underbrace{FFF\cdots F}_{n-h \text{ times}}.$$

Since the trials are independent, we may apply (2.3) repeatedly to show that

$$P(SSS\cdots SFFF\cdots F) = p^h(1-p)^{n-h}.$$

Also, because of independence, we get the same result regardless of the order in which the successes and failures occur. Therefore all outcomes with exactly $\mathbf{h}$ successes have the same probability of occurrence. Since $\{\mathbf{H} = \mathbf{h}\}$ can occur in $\binom{n}{h}$ ways, the probability of observing this event is $\binom{n}{h}p^h(1-p)^{n-h}$.

Hence the *binomial distribution* is defined by

$$f_H(h) = \begin{cases} \binom{n}{h}p^h(1-p)^{n-h} & \text{for } 0 \le h \le n \\ 0 & \text{otherwise.} \end{cases} \tag{2.7}$$

We can readily verify that this is indeed a proper probability distribution. First, the condition that $f_H \ge 0$ is clearly satisfied. Second,

$$\begin{aligned} \sum_{h=0}^{n} f_H(h) &= \sum_{h=0}^{n} \binom{n}{h}p^h(1-p)^{n-h} \\ &= \big(p + (1-p)\big)^n = 1. \end{aligned}$$

Thus, the probabilities sum to 1 as required.

The shorthand $\mathbf{H} \sim \mathcal{B}(n, p)$ is used to indicate that $\mathbf{H}$ has a binomial distribution with two parameters: the number of trials n and the probability of success p. The mean and variance of $\mathbf{H}$ are given by

$$\mathcal{E}(\mathbf{H}) = np \tag{2.8}$$

$$\mathrm{Var}(\mathbf{H}) = np(1-p). \tag{2.9}$$

2.4.3 Example: Rainfall Forecast. Consider again the daily rainfall at West Glacier, Washington. Let R be the event that the daily rainfall exceeds the 0.1 inch threshold and let $\neg R$ be the complement (i.e., rain does not exceed the threshold).

Let us now suppose that a forecast scheme has been devised with two outcomes: R^f = *there will be more than* 0.1 *inch of precipitation* and $\neg R^f$. The binomial distribution can be used to assess the skill of categorical forecasts of this type.

The probability of threshold exceedance at West Glacier is 0.38 (i.e., $P(R) = 0.38$). Suppose that the forecasting procedure has been tuned so that $P(R^f) = P(R)$.

Assume first that the forecast has no skill, that is, that it is statistically independent of nature. Let C denote a correct forecast. Using (2.1) and (2.3) we see that the probability of a correct forecast when there is 'no skill' is

$$\begin{aligned} P(C) &= P(R^f) \times P(R) \\ &\quad + P(\neg R^f) \times P(\neg R) \\ &= 0.38^2 + 0.62^2 \approx 0.53. \end{aligned}$$

The forecasting scheme is allowed to operate for 30 days and a total of 19 correct forecasts are recorded. The forecasters claim that they have some useful skill. One way to substantiate this claim is to demonstrate that it is highly unlikely for unskilled forecasters to obtain 19 correct forecasts. We therefore assume that the forecasters are not skilful and compute the probability of obtaining 19 or more correct forecasts by accident.

The binomial distribution can be used if we make two assumptions. First, the probability of a 'success' (correct forecast) must be constant from day to day. This is likely to be a reasonable approximation during relatively short periods such as a month, although on longer time scales seasonal variations might affect the probability of a 'hit.' Second, the outcome on any one day must be independent of that on other days, an assumption that is approximately correct for precipitation in midlatitudes. Many other climate system variables change much more slowly than precipitation, however, and one would expect dependence amongst successive daily forecasts of such variables.

Once the assumptions have been made, the 30-day forecasting trial can be thought of as a sequence of $n = 30$ Bernoulli trials, and the number of successes $\mathbf{h}$ can be treated as a realization of a $\mathcal{B}(30, 0.53)$ random variable $\mathbf{H}$. The expected number of correct 'no skill' forecasts in a 30-day month is $\mathcal{E}(\mathbf{H}) = 15.9$. The observed 19 hits is greater than this, supporting the contention that the forecasts are skilful. However, $\mathbf{h}$ can vary substantially from one realization of the forecasting experiment to the next. It may be that 19 or more hits can occur randomly relatively frequently in a skill-less forecasting system. Therefore, assuming no skill, we compute the likelihood of an outcome at least as extreme as observed. This is given by

$$\begin{aligned} P(\mathbf{H} \geq 19) &= \sum_{h=19}^{30} f_H(\mathbf{h}) \\ &= \sum_{h=19}^{30} \binom{30}{h} 0.53^h 0.47^{(30-h)} \\ &\approx 0.22. \end{aligned}$$

The conclusion is that 19 or more hits are not that unlikely when there is no skill. Therefore the observed success rate is not strong evidence of forecast skill.

On the other hand, suppose 23 correct forecasts were observed. Then $P(\mathbf{H} \geq 23) \leq 0.007$ under the no-skill assumption. This is stronger evidence of forecast skill than the scenario with 19 hits, since 23 hits are unlikely under the no-skill assumption.

In summary, a probability model of a forecasting system was used to assess objectively a claim of forecasting skill. The model was built on two crucial assumptions: that daily verifications are independent, and that the likelihood of a correct forecast is constant. The quality of the assessment ultimately depends on the fidelity of those assumptions to nature.

2.4.4 Poisson Distribution. The Poisson distribution, an interesting relative of the binomial distribution, arises when we are interested in counting *rare events*. One application occurs in the 'peaks-over-threshold' approach to the extreme value analysis of, for example, wind speed data. The wind speed is observed for a fixed time interval t and the number of exceedances $\mathbf{X}$ of an established large wind speed threshold V_c is recorded. The problem is to derive the distribution of $\mathbf{X}$.

First, let λ be the rate per unit time at which exceedances occur. If t is measured in years, then λ will be expressed in units of exceedances per year. The latter is often referred to as the *intensity* of the exceedance process.

Next, we have to make some assumptions about the operation of the exceedance process so that we can develop a corresponding stochastic *model*.

For simplicity, we assume that λ is not a function of time.[3] We divide the base interval t into n equal length sub-intervals with n large enough so that the likelihood of two exceedances in any one sub-interval is negligible. Then the occurrence of an exceedance in any one sub-interval can be well approximated as a Bernoulli trial with probability $\lambda t/n$ of success. Furthermore, we assume that events in adjacent time sub-intervals are independent of each other.[4] That is, the likelihood of an exceedance in a given sub-interval is not affected by the occurrence or non-occurrence of an exceedance in the other sub-intervals. Thus, the number of exceedances $\mathbf{X}$ in the base interval is approximately binomially distributed. That is,

$$\mathbf{X} \sim \mathcal{B}\Big(n, \frac{\lambda t}{n}\Big).$$

[3] In reality, the intensity often depends on the annual cycle.

[4] In reality there is always dependence on short enough time scales. Fortunately, the model described here generalizes well to account for dependence (see Leadbetter, Lindgren, and Rootzen [246]).

By taking limits as the number of sub-intervals $n \to \infty$, we obtain the Poisson probability distribution:

$$f_X(x) = \frac{(\lambda t)^x}{x!} e^{-\lambda t} \quad \text{for } x = 0, 1, \ldots. \qquad (2.10)$$

We use the notation

$$\mathbf{X} \sim \mathcal{P}(\delta)$$

to indicate that $\mathbf{X}$ has a *Poisson distribution* with parameter $\delta = \lambda t$. The mean and the variance of the Poisson distribution are identical:

$$\mathcal{E}(\mathcal{P}(\delta)) = \text{Var}\big(\mathcal{P}(\delta)\big) = \delta.$$

We return to the Poisson distribution in [2.7.12] when we discuss the distribution of waiting times between events such as threshold exceedances.

2.4.5 Example: Rainfall Forecast Continued. Suppose that forecasts and observations are made in a number of categories (such as 'no rain', 'trace', 'up to 1 mm', ...) and that verification is made in three categories ('hit', 'near hit', and 'miss'), with 'near hit' indicating that the forecast and observations agree to within one category (see the example in [18.1.6]). Each day can still be considered analogous to a binomial trial, except that three outcomes are possible rather than two. At the end of a month, two verification quantities are available: the number of hits $\mathbf{H}$ and the number of near hits $\mathbf{N}$. These quantities can be thought of as a pair of random variables defined on the same sample space. (A third quantity, the number of misses, is a degenerate random variable because it is completely determined by $\mathbf{H}$ and $\mathbf{N}$.)

The *joint* probability function for $\mathbf{H}$ and $\mathbf{N}$ gives the likelihood of simultaneously observing a particular combination of hits and near-hits. The concepts introduced in Section 2.2 can be used to show that this function is given by

$$f_{HN}(h,n) = \begin{cases} C^{30}_{hn} p_H^h p_N^n p_M^{(30-h-n)} \\ \quad \text{for } h+n \le 30 \text{ and } h,n \ge 0 \\ 0 \quad \text{otherwise,} \end{cases}$$

where

$$C^{30}_{hn} = 30!/\big(h!n!(30-h-n)!\big),$$

p_H and p_N are the probabilities of a hit and a near hit respectively, and

$$p_M = (1 - p_H - p_N)$$

is the probability of a miss.

2.4.6 The Multinomial Distribution. The example above can be generalized to experiments having independent trials with k possible outcomes per trial if the probability of a particular outcome remains constant from trial to trial. Let $\mathbf{X}_1, \ldots, \mathbf{X}_{k-1}$ represent the number of each of the first $k-1$ outcomes that occur in n independent trials (we ignore the kth variate because it is again degenerate).

The $(k-1)$-dimensional random vector $\vec{\mathbf{X}} = (\mathbf{X}_1, \ldots, \mathbf{X}_{k-1})^{\mathrm{T}}$ is said to have a *multinomial* distribution with parameters n and $\vec{\theta} = (p_1, \ldots, p_{k-1})^{\mathrm{T}}$, and we write $\vec{\mathbf{X}} \sim \mathcal{M}_k(n, \vec{\theta})$. The general form of the multinomial probability function is given by

$$f_{\vec{X}}(\vec{x}) = \begin{cases} C^n_{x_1,\ldots,x_{k-1}} p_1^{x_1} \cdots p_k^{x_k} \\ \quad \text{if } x_i \ge 0 \text{ for } i = 1, \ldots, k \\ 0 \quad \text{otherwise} \end{cases}$$

where

$$C^n_{x_1,\ldots,x_{k-1}} = \frac{n!}{x_1! \cdots x_k!}$$

and

$$x_k = n - \sum_{i=1}^{k-1} x_i, \qquad p_k = 1 - \sum_{i=1}^{k-1} p_i.$$

With this notation, the distribution in [2.4.5] is $\mathcal{M}_3(30, (p_H, p_N)^{\mathrm{T}})$. The binomial distribution, $\mathcal{B}(n, p)$, is equivalent to $\mathcal{M}_2(n, p)$.

2.5 Discrete Multivariate Distributions

2.5.0 Introduction. The multinomial distribution is an example of a discrete multivariate distribution. The purpose of this section is to introduce concepts that can be used to understand the relationship between random variables in a multivariate setting. Marginal distributions [2.5.2] describe the properties of the individual random variables that make up a random vector when the influence of the other random variable in the random vector is ignored. Conditional distributions [2.5.4] describe the properties of some variable in a random vector when variation in other parts of the random variable is controlled.

For example, we might be interested in the distribution of rainfall when rainfall is forecast. If the forecast is skilful, this *conditional* distribution will be different from the marginal (i.e., climatological) distribution of rainfall. When the forecast is not skilful (i.e., when the forecast is independent of what actually happens) marginal

	$\mathbf{X}_1$			
$\mathbf{X}_2$	strong	normal	weak	all
weak	21	11	2	34
moderate	20	14	7	41
severe	4	4	6	14
very severe	0	3	8	11
all	45	32	23	100

Table 2.1: *Estimated probability distribution (in %) of $\vec{\mathbf{X}} = (\mathbf{X}_1, \mathbf{X}_2)$ = (strength of westerly flow, severity of Baltic Sea ice conditions), obtained from 104 years of data. Koslowski and Loewe [231]. See [2.5.1].*

and conditional distributions are identical. The effect of independence is described in [2.5.7].

2.5.1 Example. We will use the following example in this section. Let $\vec{\mathbf{X}} = (\mathbf{X}_1, \mathbf{X}_2)$ be a discrete bivariate random vector where $\mathbf{X}_1$ takes values *(strong, normal, weak)* describing the strength of the winter mean westerly flow in the Northeast Atlantic area, and $\mathbf{X}_2$ takes values *(weak, moderate, severe, very severe)* describing the sea ice conditions in the western Baltic Sea (from Koslowski and Loewe [231]). The probability distribution of the bivariate random variable is completely specified by Table 2.1. For example: $p(\mathbf{X}_1$ = *weak flow* and $\mathbf{X}_2$ = *very severe ice conditions*) = 0.08.

2.5.2 Marginal Probability Distributions. If $\vec{\mathbf{X}} = (\mathbf{X}_1, \ldots, \mathbf{X}_m)$ is an m-variate random vector, we might ask what the distribution of an individual random variable $\mathbf{X}_i$ is if we ignore the presence of the others. In the nomenclature of probability and statistics, this is the *marginal probability distribution*. It is given by

$$f_{X_i}(x_i) = \sum_{x_1,\ldots,x_{i-1},x_{i+1},\ldots,x_m} f(x_1 \ldots x_i \ldots x_m)$$

where the sum is taken over all possible realizations of $\vec{\mathbf{X}}$ for which $\mathbf{X}_i = \mathbf{x}_i$.

2.5.3 Examples. If $\vec{\mathbf{X}}$ has a multinomial distribution, the marginal probability distribution of $\mathbf{X}_i$ is the binomial distribution with n trials and probability p_i of success. Consequently, if $\vec{\mathbf{X}} \sim \mathcal{M}_m(n, \vec{\theta})$, with $\vec{\theta}$ defined as in [2.4.6], the mean and variance of $\mathbf{X}_i$ are given by

$$\mu_i = np_i \quad \text{and} \quad \sigma_i^2 = np_i(1 - p_i).$$

In example [2.5.1], the marginal distribution of $\mathbf{X}_1$ is given in the row at the lower *margin* of Table 2.1, and that of $\mathbf{X}_2$ is given in the column at the right hand *margin* (hence the nomenclature). The marginal distribution of $\mathbf{X}_2$ is

$$f_{X_2}(x_2) = \begin{cases} 0.34, & x_2 = \textit{weak} \\ 0.41, & x_2 = \textit{moderate} \\ 0.14, & x_2 = \textit{strong} \\ 0.11, & x_2 = \textit{very strong}. \end{cases}$$

Note that $f_{X_2}(\textit{weak})$, for example, is given by

$$\begin{aligned} f_{X_2}(\textit{weak}) &= f_{\vec{X}}(\textit{strong, weak}) \\ &\quad + f_{\vec{X}}(\textit{normal, weak}) \\ &\quad + f_{\vec{X}}(\textit{weak, weak}) \\ &= 0.21 + 0.11 + 0.02 \\ &= 0.34. \end{aligned}$$

2.5.4 Conditional Distributions. The concept of conditional probability [2.2.5] is extended to discrete random variables with the following definition.

Let $\mathbf{X}_1$ and $\mathbf{X}_2$ be a pair of discrete random variables. The conditional probability function of $\mathbf{X}_1$, given $\mathbf{X}_2 = \mathbf{x}_2$, is

$$f_{X_1|X_2=x_2}(x_1) = \frac{f_{X_1X_2}(x_1, x_2)}{f_{X_2}(x_2)} \tag{2.11}$$

provided that $f_{X_2}(x_2) \neq 0$.

Here $f_{X_2}(x_2)$ is the marginal distribution of $\mathbf{X}_2$ which is given by $f_{X_2}(x_2) = \sum f_{X_1X_2}(x_1, x_2)$. The sum is taken over all possible realizations of $(\mathbf{X}_1, \mathbf{X}_2)$ for which $\mathbf{X}_2 = \mathbf{x}_2$.

2.5.5 Examples. The conditional distributions for the example presented in Table 2.1 are derived by dividing row (or column) entries by the corresponding row (or column) sum. For example, the probability that the sea ice conditions are severe given that the westerly flow is strong is given by

$$\begin{aligned} f_{X_2|X_1=\textit{strong}}(\textit{severe}) &= \frac{f_{\vec{X}}(\textit{strong, severe})}{f_{X_1}(\textit{strong})} \\ &= \frac{0.04}{0.45} = 0.09\,. \end{aligned}$$

In the rainfall forecast verification example [2.4.5] the conditional distribution for the number of hits $\mathbf{H}$ given that there are $\mathbf{N} = m$ near hits is $\mathcal{B}(30 - m, p_H/(1 - p_N))$.

	$\mathbf{X}_1$			
$\mathbf{X}_2$	strong	normal	weak	all
weak	31	8	0	39
moderate	30	10	4	44
severe	6	3	3	12
very severe	0	2	4	6
all	67	23	11	101

Table 2.2: *Hypothetical future distribution of* $\vec{\mathbf{X}} = (\mathbf{X}_1, \mathbf{X}_2)$ = *(strength of westerly flow, severity of ice conditions), if the marginal distribution of the westerly flow is changed as indicated in the last row, assuming that no other factors control ice conditions. (The marginal distributions do not sum to exactly 100% because of rounding errors.) See [2.5.6].*

2.5.6 Example: Climate Change and Western Baltic Sea-ice Conditions. In [2.5.5] we supposed that sea-ice conditions depend on atmospheric flow. Here we assume that atmospheric flow controls the sea-ice conditions and that feedback from the sea-ice conditions in the Baltic Sea, which have small scales relative to that of the atmospheric flow, may be neglected. Then we can view the severity of the ice conditions, $\mathbf{X}_2$, as being dependent on the atmospheric flow, $\mathbf{X}_1$.

Table 2.1 seems to suggest that if stronger westerly flows were to occur in a future climate, we might expect relatively more frequent *moderate* and *weak* sea-ice conditions. The next few subsections examine this possibility.

We represent present day probabilities with the symbol f and those of a future climate, in say 2050, by $\tilde{f}$. We assume that conditional probabilities are unchanged in the future, that is,

$$f_{X_2|X_1=x_1}(x_2) = \tilde{f}_{X_2|X_1=x_1}(x_2).$$

Using (2.11) to express the joint present and future probabilities as products of the conditional and marginal distributions, we find

$$\tilde{f}_{\vec{X}}(x_1, x_2) = \frac{\tilde{f}_{X_1}(x_1)}{f_{X_1}(x_1)} f_{\vec{X}}(x_1, x_2).$$

Now suppose that the future marginal probabilities for the atmospheric flow are $\tilde{f}_{X_1}(\mathit{strong}) = 0.67$, $\tilde{f}_{X_1}(\mathit{normal}) = 0.22$ and $\tilde{f}_{X_1}(\mathit{weak}) = 0.11$. Then the future version of Table 2.1 is Table 2.2.[5] Note that the prescribed future marginal distribution for the strength of the atmospheric flow appears in the lowest row of Table 2.2. The changing climate is clearly reflected in the marginal distribution $\tilde{f}_{X_2}$, which is tabulated in the right hand column. This suggests that weak and moderate ice conditions will be more frequent in 2050 than at present, and that the frequency of severe or very severe ice conditions will be lowered from 25% to 18%.

[5] These numbers were derived from a 'doubled CO_2 experiment' [96]. Factors other than atmospheric circulation probably affect the sea ice significantly, so this example should not be taken seriously.

2.5.7 Independent Random Variables. The idea of independence is easily extended to random variables because they describe events in the sample space upon which they are defined. Two random variables are said to be independent if they always describe independent events in a sample space. More precisely:

Two random variables, $\mathbf{X}_1$ *and* $\mathbf{X}_2$*, are said to be 'independent' if*

$$f_{X_1,X_2}(x_1, x_2) = f_{X_1}(x_1) f_{X_2}(x_2) \qquad (2.12)$$

for all (x_1, x_2).

That is, two random variables are independent if their joint probability function can be written as the product of their marginal probability functions.

Using (2.11) and (2.12) we see that independence of $\mathbf{X}_1$ and $\mathbf{X}_2$ implies

$$f_{X_1|X_2=x_2}(x_1) = f_{X_1}(x_1).$$

Thus, knowledge of the value of $\mathbf{X}_2$ does not give us any information about the value of $\mathbf{X}_1$.[6]

A useful result of (2.12) is that, if $\mathbf{X}_1$ and $\mathbf{X}_2$ are independent random variables, then

$$\mathcal{E}(\mathbf{X}_1\mathbf{X}_2) = \mathcal{E}(\mathbf{X}_1)\mathcal{E}(\mathbf{X}_2). \qquad (2.13)$$

The reverse is not true: nothing can be said about the independence of $\mathbf{X}_1$ and $\mathbf{X}_2$ when (2.13) holds. However, if (2.13) does not hold, $\mathbf{X}_1$ and $\mathbf{X}_2$ are certainly *dependent*.

2.5.8 Examples. The two variables described in Table 2.1 are not independent of each other because the table entries are not equal to the product of the marginal entries. Thus, knowledge of the value of the westerly flow index, $\mathbf{X}_1$, tells you something useful about the relative likelihood that the different values of sea-ice intensity $\mathbf{X}_2$ will be observed.

What would Table 2.1 look like if the strength of the westerly flow, $\mathbf{X}_1$, and the severity of the Western Baltic sea-ice conditions, $\mathbf{X}_2$, were independent? The answer, assuming that there is

[6] Thus the present definition is consistent with [2.2.6].

	$\mathbf{X}_1$			
$\mathbf{X}_2$	strong	normal	weak	all
weak	15	11	8	34
moderate	18	13	9	40
severe	6	4	3	13
very severe	5	4	4	13
all	44	32	24	100

Table 2.3: *Distribution of* $\vec{\mathbf{X}} = (\mathbf{X}_1, \mathbf{X}_2) =$ *(strength of westerly flow, severity of ice conditions) assuming that the severity of the sea-ice conditions and the strength of the westerly flow are unrelated. See [2.5.8]. (Marginal distribution deviates from that of Table 2.1 because of rounding errors.)*

no change in the marginal distributions, is given in Table 2.3.

The two variables described by the bivariate multinomial distribution [2.4.5] are also dependent. One way to show this is to demonstrate that the product of the marginal distributions is not equal to the joint distribution. Another way to show this is to note that the set of values that can be taken by the random variable pair $(\mathbf{H}, \mathbf{N})$ is not equivalent to the cross-product of the sets of values that can be taken by $\mathbf{H}$ and $\mathbf{N}$ individually. For example, it is possible to observe $\mathbf{H} = n$ or $\mathbf{N} = n$ separately, but one cannot observe $(\mathbf{H}, \mathbf{N}) = (n, n)$ because this violates the condition that $0 \leq \mathbf{H} + \mathbf{N} \leq n$.

2.5.9 Sum of Identically Distributed Independent Random Variables. If $\mathbf{X}$ is a random variable from which n independent realizations $\mathbf{x}_i$ are drawn, then $\mathbf{y} = \sum_{i=1}^{n} \mathbf{x}_i$ is a realization of the random variable $\mathbf{Y} = \sum_{i=1}^{n} \mathbf{X}_i$, where the $\mathbf{X}_i$s are independent random variables, each distributed as $\mathbf{X}$. Using independence, it is easily shown that the mean and the variance of $\mathbf{Y}$ are given by

$$\begin{aligned}
\mathcal{E}(\mathbf{Y}) &= n\,\mathcal{E}(\mathbf{X}) \\
\mathrm{Var}(\mathbf{Y}) &= \mathcal{E}(\mathbf{Y}^2) - \mathcal{E}(\mathbf{Y})^2 \\
&= \sum_{i,j=1}^{n} \mathcal{E}(\mathbf{X}_i\mathbf{X}_j) - n^2\mathcal{E}(\mathbf{X})^2 \\
&= n\,\mathcal{E}(\mathbf{X}^2) + n(n-1)\mathcal{E}(\mathbf{X})^2 \\
&\quad - n^2\mathcal{E}(\mathbf{X})^2 \\
&= n\big(\mathcal{E}(\mathbf{X}^2) - \mathcal{E}(\mathbf{X})^2\big) \\
&= n\,\mathrm{Var}(\mathbf{X}).
\end{aligned}$$

Thus, the mean of the sum of n independent identically distributed random variables is n times the mean of the individual random variable. Likewise, the variance of the sum is n times the variance of $\mathbf{X}$.

2.6 Continuous Random Variables

2.6.0 Introduction. Up to this point we have discussed examples in which, at least conceptually, we can write down all the simple outcomes of an experiment, as in the coin tossing experiment or in Table 2.1. However, usually the sample space cannot be enumerated; temperature, for example, varies continuously.[7]

2.6.1 The Climate System's Phase Space. We have discussed temperature measurements in the context of a sample space to illustrate the idea of a continuous sample space—but the idea that these measurements define the sample space, no matter how fine the resolution, is fundamentally incorrect. Temperature (and all other physical parameters used to describe the state of the climate system) should really be thought of as functions defined on the climate's *phase space*.

The exact characteristics of phase space are not known. However, we assume that the points in the phase space that can be visited by the climate are not enumerable, and that all transitions from one part of phase space to another occur smoothly.

The path our climate is taking through phase space is conceptually one of innumerable paths. If we had the ability to reverse time, a small change, such as a slightly different concentration of tropospheric aerosols, would have sent us down a different path through phase space. Thus, it is perfectly valid to consider our climate a realization of a continuous stochastic process even though the time-evolution of any particular path is governed by physical laws. In order to apply this fact to our diagnostics of the observed and simulated climate we have to assume that the climate is *ergodic*. That is, we have to assume that every trajectory will eventually visit all parts of phase space and that sampling in time is equivalent to sampling different paths through phase space. Without this assumption about the operation of our physical system the study of the climate would be all but impossible.

[7] In reality, both the instrument used to take the measurement and the digital computing system used to store it operate at finite resolutions. However, it is mathematically convenient to approximate the observed discrete random variable with a continuous random variable.

The assumption of ergodicity is well founded, at least on shorter time scales, in the atmosphere and the ocean. In both media, the laws of physics describe turbulent fluids with limited predictability (i.e., small perturbations grow quickly, so two paths through phase space diverge quickly).

2.6.2 Continuous Random Variable. We have expanded the concept of the sample space S to the concept of a phase space $\mathcal{S}$. We must also expand the concept of the probability rule, $P(\cdot)$, used to compute the probability of events, by converting $P(\cdot)$ into a function that measures the relative size of an event.

The way events are measured is not uniform because measurements must reflect the likelihood of events. For example, let $\mathbf{T}$ represent temperature at a northern midlatitude location in January, and consider events A and B, where $A = \{\mathbf{T} \in (-5, 5)\,°\mathrm{C}\}$ and $B = \{\mathbf{T} \in (30, 40)\,°\mathrm{C}\}$. Both A and B describe $10\,°\mathrm{C}$ temperature ranges but $P(A) \neq P(B)$, that is, the *probability* measure of these events is not the same.

Now assume that we are able to observe temperature on a continuous scale (i.e., that the intervening instruments do not discretize the observed temperature) and consider the event $C = \{\mathbf{T} = 0.48\,°\mathrm{C}\}$. This event challenges our intuition because $P(C) = 0$. Why? Consider a sequence of events

$$C_k = \left\{\mathbf{T} \in \left(0.48 - \frac{1}{k}, 0.48 + \frac{1}{k}\right)°\mathrm{C}\right\}.$$

Note that $\lim_{k\to\infty} C_k = C$ and that the event C_{k+1} is a subset of C_k, or in mathematical terms $C_1 \supset C_2 \supset \cdots$. Therefore

$$P(C_1) > P(C_2) > \cdots.$$

Intuitively, we see that, for large k, the probability of event C_k is proportional to k^{-1}. It follows that $P(C) = 0$.

Let us consider another situation. Assume that the probability measure is continuous and that there is a point $\mathbf{x}$ and an $\epsilon > 0$ such that

$$P(\mathbf{X} = \mathbf{x}) = 2\epsilon.$$

Then, because of continuity, there must exist a $\delta > 0$ such that for all $\mathbf{y}$ with $|\mathbf{x} - \mathbf{y}| < \delta$

$$P(\mathbf{X} = \mathbf{y}) > \epsilon.$$

Now, if we choose $n > 1/\epsilon$ points $\mathbf{x}_1, \ldots, \mathbf{x}_n$ such that $|\mathbf{x} - \mathbf{x}_i| < \delta$, we obtain the contradiction that

$$P(\mathbf{X} \in \{\mathbf{x}_1, \ldots, \mathbf{x}_n\}) > 1.$$

The conclusion is that our initial assumption, that there is a point $\mathbf{x}$ for which $P(\mathbf{X} = \mathbf{x}) > 0$, is false. That is, *if* $\mathbf{X}$ *is a continuous random variable, then* $P(\mathbf{X} = \mathbf{x}) = 0$ *for all* $\mathbf{x}$.

While counter intuitive, the result is reasonable; the chance of observing a specific value is zero because innumerable different values can occur.

Finally, a continuous random variable is defined as follows:

Let $\mathcal{S}$ be a phase space and let $P(\cdot)$ be a continuous probability measure on $\mathcal{S}$. Then a continuous random variable $\mathbf{X}$ is a continuous function of $\mathcal{S}$ that takes values in an interval $\Omega \subseteq \mathbb{R}$, the real line, in such a way that

1 $P(\mathbf{X} \in \Theta) \geq 0$ *for all* $\Theta \subseteq \Omega$,

2 $P(\mathbf{X} \in \Omega) = 1$.

2.6.3 The Probability Density and Distribution Functions. Events described in terms of continuous random variables are expressed as open intervals on the real line, $\mathbb{R}$, and the probability of an event is expressed as the integral of a *probability density function* (pdf) taken over the interval that describes the event. In theory, the density function is derived from the definition of the random variable and the probability measure $P(\cdot)$. In practice, we will use intuition and simple mathematical arguments wherever possible.

Our working definition of the probability density function will be as follows:

Let $\mathbf{X}$ be a continuous random variable that takes values in the interval Ω. The probability density function for $\mathbf{X}$ is a continuous function $f_X(\cdot)$ defined on $\mathbb{R}$ with the following properties:

1 $f_X(x) \geq 0$ *for all* $x \in \Omega$,

2 $\int_\Omega f_X(x)\,dx = 1$,

3 $P(\mathbf{X} \in (a, b)) = \int_a^b f_X(x)\,dx$ *for all* $(a, b) \subseteq \Omega$.

An equivalent description of the stochastic characteristics of a continuous random variable is given by the distribution function, frequently referred to more descriptively as the *cumulative distribution function* (cdf).

The distribution function for $\mathbf{X}$ is a non-decreasing differentiable function $F_X(\cdot)$ defined on $\mathbb{R}$ with the following properties:

1 $\lim_{x\to-\infty} F_X(x) = 0$,

2 $\lim_{x\to+\infty} F_X(x) = 1$,

3 $\frac{d}{dx} F_X(x) = f_X(x)$.

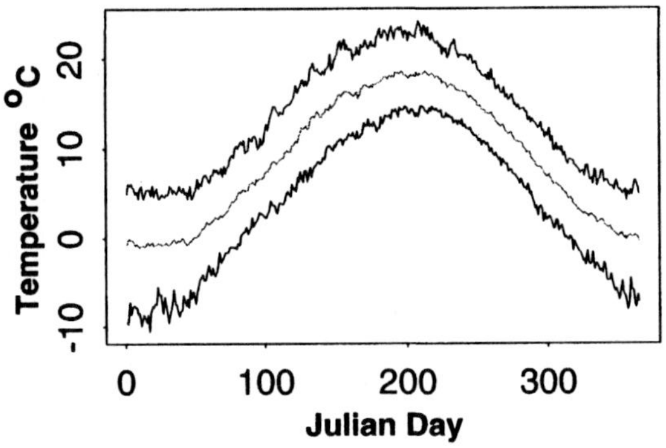

Figure 2.1: *The 10th, 50th, and 90th quantiles of daily mean temperature at Potsdam, Germany (1983–94).*

The last equation tells us that

$$F_X(x) = \int_{-\infty}^{x} f_X(r)\,dr. \tag{2.14}$$

The cumulative distribution function is often useful for computing probabilities because

$$P\big(\mathbf{X} \in (a,b)\big) = F_X(b) - F_X(a).$$

2.6.4 Median and Quantiles. The median, $x_{0.5}$, is the solution of

$$F_X(x_{0.5}) = 0.50.$$

It represents the *middle* of the distribution in the sense that

$$P(x \le x_{0.5}) = P(x \ge x_{0.5}) = 0.5.$$

Exactly 50% of all realizations will be less than the median, the other 50% will be greater.

The median is an example of a *p-quantile*, the point x_p on the real line such that

$$\begin{aligned} P\big(\mathbf{X} \in (-\infty, x_p)\big) &= p \\ P\big(\mathbf{X} \in [x_p, \infty)\big) &= 1 - p. \end{aligned}$$

That is, the p-quantile is the solution x_p of

$$F_X(x_p) = p.$$

An example of the annual cycle of the quantiles of daily mean temperature at Potsdam, Germany, is displayed in Figure 2.1. Note that the distribution is approximately symmetric during the transition seasons, but negatively skewed in winter, and slightly positively skewed in summer. The 'noise' evident in these curves is a consequence of estimating the quantiles from a finite sample of observations.

2.6.5 Expectation. The *expected value* of a continuous random variable $\mathbf{X}$ is given by

$$\mathcal{E}(\mathbf{X}) = \int_\Omega x f_X(x)\,dx.$$

If $g(\cdot)$ is a function then the definition of the expected value of $g(\mathbf{X})$ generalizes from the discrete case in the same way, and

$$\mathcal{E}\big(g(\mathbf{X})\big) = \int_\Omega g(x) f_X(x)\,dx.$$

Results (2.4) and (2.5), about the expectation of a sum of functions and about linear transformations of random variables, also apply in the continuous case:

$$\mathcal{E}\big(g_1(\mathbf{X}) + g_2(\mathbf{X})\big) = \mathcal{E}\big(g_1(\mathbf{X})\big) + \mathcal{E}\big(g_2(\mathbf{X})\big) \tag{2.15}$$

$$\mathcal{E}\big(a g(\mathbf{X}) + b\big) = a\mathcal{E}\big(g(\mathbf{X})\big) + b. \tag{2.16}$$

2.6.6 Interpreting Expectation as the Long-term Average. The expectation is often also named 'the mean' value, that is, this number is identified with the average of an infinite number of realizations of $\mathbf{X}$. We will show this here with an intuitive limit argument. Another heuristic argument is presented in [5.2.5].

First, we approximate the continuous random variable $\mathbf{X}$ with a discrete random variable $\mathbf{X}_\delta$ that takes values in the set $\{k\delta\colon k = 0, \pm1, \pm2, \dots\}$ for some small positive number δ and with probabilities

$$p_{k\delta} = \int_{(k-1/2)\delta}^{(k+1/2)\delta} f_X(\mathbf{x})\,dx \approx \delta f_X(k\delta).$$

The expected value of the discrete random variable $\mathbf{X}_\delta$ is given by

$$\mathcal{E}(\mathbf{X}_\delta) = \sum_{k=-\infty}^{\infty} k\delta p_{k\delta}.$$

By interpreting $p_{k\delta}$ as the frequency with which $\mathbf{X}$ takes a value in the neighbourhood of $\mathbf{x} = k\delta$, we see that the expectation of the approximating discrete random variable $\mathbf{X}_\delta$ is indeed a 'long-term' mean. Then, taking the limit as $\delta \to 0$, and noting that $p_{k\delta}$ tends to $\delta f_X(k\delta)$ as $\delta \to 0$, we obtain

$$\begin{aligned} \lim_{\delta\to 0} \mathcal{E}(\mathbf{X}_\delta) &= \lim_{\delta\to 0} \sum_{k=-\infty}^{\infty} k\delta \int_{(k-1/2)\delta}^{(k+1/2)\delta} f_X(\mathbf{x})\,dx \\ &= \int_{-\infty}^{\infty} x f_X(x)\,dx, \end{aligned}$$

thus concluding the argument. A rigorous proof is obtained by demonstrating that the sample mean is a consistent estimator of the expectation (see [5.2.5]).

2.6.7 The Central Moments: Location, Scale, and Shape Parameters. The kth moment $\mu^{(k)}$ of a continuous random variable $\mathbf{X}$ is also defined as in the discrete case. Specifically

$$\mu^{(k)} = \mathcal{E}(x^k) = \int_{-\infty}^{\infty} x^k f_X(x)\, dx.$$

The kth *central moment* $\mu'^{(k)}$ of a random variable $\mathbf{X}$ is the expectation of $(\mathbf{X} - \mu)^k$, given by

$$\mu'^{(k)} = \int_{-\infty}^{\infty} (x - \mu)^k f_X(x)\, dx.$$

Most characteristics of a distribution can be summarized through the use of simple functions of the first four moments. These slightly modified parameters are the mean, variance, *skewness*, and *kurtosis*:

- The *mean*, also known as the *location parameter*, is given by the first moment

  $$\mu = \mu^{(1)}.$$

- The *variance* is given by the second central moment

  $$\begin{aligned} \mathrm{Var}(\mathbf{X}) &= \mathcal{E}\big((\mathbf{X} - \mu)^2\big) \qquad (2.17)\\ &= \mathcal{E}(\mathbf{X}^2) - \big(\mathcal{E}(\mathbf{X})\big)^2\\ &= \mu^{(2)} - \big(\mu^{(1)}\big)^2. \end{aligned}$$

  The properties of the variance, discussed for the discrete case in [2.3.4], extend to the continuous case, in particular

  $$\mathrm{Var}(a\mathbf{X} + b) = a^2 \mathrm{Var}(\mathbf{X}). \qquad (2.18)$$

  The *standard deviation* $\sigma_X = \sqrt{\mathrm{Var}(\mathbf{X})}$ is also often described as a *scale parameter*.

- The *skewness* is a scaled version of the third central moment that is given by

  $$\gamma_1 = \int_{\mathbb{R}} \Big(\frac{x - \mu}{\sigma}\Big)^3 f_X(\mathbf{x})\, dx.$$

  Symmetric distributions (i.e., distributions for which $f_X(\mu - x) = f_X(\mu + x)$) have $\gamma_1 = 0$. Distributions for which $\gamma_1 < 0$ are said to be *negatively skewed* or *skewed to the left*, and distributions for which $\gamma_1 > 0$ are said to be *positively skewed* or *skewed to the right*.

  Daily rainfall distributions, bounded on the left by zero and unbounded on the right, are generally strongly skewed to the right—even though small amounts of rainfall occur considerably more often than large amounts. This occurs because rainfall distributions have a wide 'tail' that extends far to the right.

  On the other hand, geopotential height tends to be somewhat skewed to the left because lows tend to have greater amplitude than highs.[8]

- The *kurtosis*, a scaled and shifted version of the fourth central moment, is given by

  $$\gamma_2 = \int_{\mathbb{R}} \Big(\frac{x - \mu}{\sigma}\Big)^4 f_X(\mathbf{x})\, dx - 3. \qquad (2.19)$$

  Kurtosis is a measure of peakedness. *Platykurtic* distributions, such as the uniform distribution, have $\gamma_2 < 0$ and are less 'peaked' than the normal distribution (see [2.7.3]). Distributions with $\gamma_2 > 0$ are said to be *leptokurtic*, and are more 'peaked' than the normal distribution. The double exponential distribution, with density $f_X(\mathbf{x}) = \frac{1}{2}e^{-|x-\mu|}$, is leptokurtic.

The skewness and kurtosis are often referred to as *shape parameters*.[9]

Shape parameters can be useful aids in the identification of appropriate probability models. This seems to be especially true in *extreme value analysis* (Section 2.9) where debate over the merits of various distributions is often intense. However, skewness and kurtosis are often difficult to estimate well. In practice, it is advisable to use alternative shape parameters such as *L-moments* [2.6.9].

2.6.8 The Coefficient of Variation. When a random variable, such as precipitation, takes only positive values a scale parameter called the *coefficient of variation*,

$$c_X = \sigma_X / \mu_X,$$

is sometimes used. The standard deviation of such variables is often proportional to the mean and it is therefore useful to describe the scale parameter relative to the mean.

[8] Holzer [180] shows that this is due to the rectification of nonlinear interactions in the atmosphere's dynamics (see also [3.1.8]).

[9] The concept of skewness and kurtosis is not limited to continuous random variables. It carries over to discrete random variables in the obvious way: by replacing integration with summation in the definitions given above.

2.6.9 L-Moments. Hosking [183] introduced an alternative set of scale and shape statistics called *L-moments*, which are based on *order statistics*. The L-moments play a role similar to that of conventional moments; in particular, any distribution can be completely specified by either. The difference is that the higher ($j \geq 3$) L-moments can be *estimated* more reliably than conventional moments such as skewness and kurtosis. Robust estimators of higher moments are needed to identify and fit distributions such as the *Gumbel, Pareto,* or *Wakeby* distributions used in extreme value analysis (see Section 2.9).

To define the L-moments of a random variable $\mathbf{X}$ we must first define related random variables called order statistics. Let $\vec{\mathbf{X}} = (\mathbf{x}_1, \ldots, \mathbf{x}_n)^{\mathrm{T}}$ be a random vector that is made up of n independent, identically distributed random variables, each with the same distribution as $\mathbf{X}$. Suppose $\vec{\mathbf{x}} = (\mathbf{x}_1, \ldots, \mathbf{x}_n)^{\mathrm{T}}$ is a realization of $\vec{\mathbf{X}}$. Let $g(\cdot)$ be the function that sorts the elements of an n-dimensional vector in increasing order. That is

$$g(\vec{\mathbf{x}}) = (x_{(1|n)}, \ldots, x_{(n|n)})^{\mathrm{T}}$$

where $x_{(i|n)}$ is the ith smallest element of $\vec{\mathbf{x}}$. The random vector that corresponds to $g(\vec{\mathbf{x}})$ is

$$g(\vec{\mathbf{X}}) = (\mathbf{X}_{(1|1)}, \ldots, \mathbf{X}_{(n|n)})^{\mathrm{T}}.$$

Note that the elements of $g(\vec{\mathbf{X}})$ are no longer independent or identically distributed; their marginal distributions (see [2.8.3]) are complicated functions of the distribution of $\mathbf{X}$. The random variables $\mathbf{X}_{(j|n)}$ for $j = 1, \ldots, n$ are called order statistics. L-moments are defined as the expectations of linear combinations of these order statistics.

The first three L-moments are defined as

$$\begin{aligned}
\lambda^{(1)} &= \mathcal{E}(\mathbf{X}_{(1|1)}) \\
\lambda^{(2)} &= \frac{1}{2}\mathcal{E}(\mathbf{X}_{(2|2)} - \mathbf{X}_{(1|2)}) \\
\lambda^{(3)} &= \frac{1}{3}\mathcal{E}(\mathbf{X}_{(3|3)} - 2\mathbf{X}_{(2|3)} + \mathbf{X}_{(1|3)}).
\end{aligned} \tag{2.20}$$

The general kth L-moment is given by

$$\lambda^{(k)} = \frac{1}{k}\sum_{j=0}^{k-1}(-1)^j \binom{k-1}{j} \mathcal{E}(\mathbf{X}_{(k-j|k)}). \tag{2.21}$$

Thus, the first L-moment is the expected smallest value in a sample of one. Since there is only one value in such a sample, the first L-moment is equal to the conventional first moment. The second L-moment is the expected absolute difference between any two realizations (note that $\mathbf{X}_{2|2} \geq \mathbf{X}_{1|2}$ by definition). The third and fourth moments are shape parameters. Standardized L-moments are

- the *L-coefficient of variation*

$$c_X^L = \lambda^{(2)}/\lambda^{(1)}, \tag{2.22}$$

- the *L-skewness*

$$\gamma_1^L = \lambda^{(3)}/\lambda^{(2)}, \tag{2.23}$$

- the *L-kurtosis*

$$\gamma_2^L = \lambda^{(4)}/\lambda^{(2)}. \tag{2.24}$$

Examples of the application of L-moments in climate research include Guttmann [151] and Zwiers and Kharin [448].

2.7 Example of Continuous Random Variables

2.7.1 The Uniform Distribution. The simplest of all continuous distributions is the uniform distribution. A random variable that takes values in an interval (a, b) is said to be uniform if it has a probability density function that is constant inside the interval and zero outside. Such a density function is given by

$$f_X(x) = \begin{cases} 1/(b-a) & \text{for all } x \in (a, b) \\ 0 & \text{elsewhere,} \end{cases}$$

and the cumulative distribution function is given by

$$F_X(x) = \begin{cases} 0 & \text{for } x \leq a \\ (x-a)/(b-a) & \text{for } x \in (a, b) \\ 1 & \text{for } x \geq b. \end{cases}$$

We use the shorthand $\mathbf{X} \sim \mathcal{U}(a, b)$ to indicate that $\mathbf{X}$ has a uniform distribution.

It is readily shown that the mean, variance, skewness, and kurtosis of a $\mathcal{U}(a, b)$ random variable are given by

$$\begin{aligned}
\mathcal{E}(\mathcal{U}(a, b)) &= \frac{1}{2}(a+b) \\
\mathrm{Var}(\mathcal{U}(a, b)) &= \frac{1}{12}(b-a)^2 \\
\gamma_1(\mathcal{U}(a, b)) &= 0 \\
\gamma_2(\mathcal{U}(a, b)) &= -1.2\,.
\end{aligned}$$

Thus, the uniform distribution is symmetric (skewness = 0) and less peaked than a normal distribution (kurtosis < 0). The L-moments are [183]:

$$\begin{aligned}\lambda^{(1)} &= \frac{1}{2}(a+b)\\ \lambda^{(2)} &= \frac{1}{6}(b-a)\\ \gamma_1^L &= 0\\ \gamma_2^L &= 0.\end{aligned}$$

2.7.2 Probability and Likelihood. The uniform distribution illustrates very clearly that a probability *density* is not a probability. When the distribution is defined on an interval of length less than 1, the density is uniformly greater than 1 throughout the interval, even though probabilities are never greater than 1. Only integrated density functions provide probabilities.

Nevertheless, the density function describes the relative chances of observing specific events. In particular, when $f_X(\mathbf{x}_1) > f_X(\mathbf{x}_2)$ it is more likely that we will observe values of $\mathbf{X}$ near $\mathbf{x}_1$ than near $\mathbf{x}_2$. Therefore we call the values of the density function *likelihoods*. For the uniform distribution, all values of $\mathbf{X}$ in the range (a, b), including the mean, are equally likely. This is not true in the other distributions of continuous random variables.

2.7.3 The Normal Distribution. The distribution most frequently encountered in meteorology and climatology is the normal distribution. Many variables studied in climatology are averages or integrated quantities of some type. The law of large numbers, or *Central Limit Theorem* [2.7.5], states (under fairly broad regularity conditions) that random variables of this type are nearly normally distributed regardless of the distribution of the variables that are averaged or integrated.

The form of the normal distribution is entirely determined by the mean and the variance. Thus, we write $\mathbf{X} \sim \mathcal{N}(\mu, \sigma^2)$ to indicate that $\mathbf{X}$ has a normal distribution with parameters μ and σ^2.

In the climatological literature, the normal distribution is also often referred to as the *Gaussian distribution*, after C.F. Gauss who introduced the distribution some 200 years ago.

The normal density function is given by

$$f_{\mathcal{N}}(x) = \frac{1}{\sqrt{2\pi}\,\sigma} e^{-\frac{(x-\mu)^2}{2\sigma^2}} \quad \text{for all } x \in \mathbb{R}. \tag{2.25}$$

The density functions of normal random variables with different variances are illustrated in Figure 2.2. Note that the distribution is symmetric

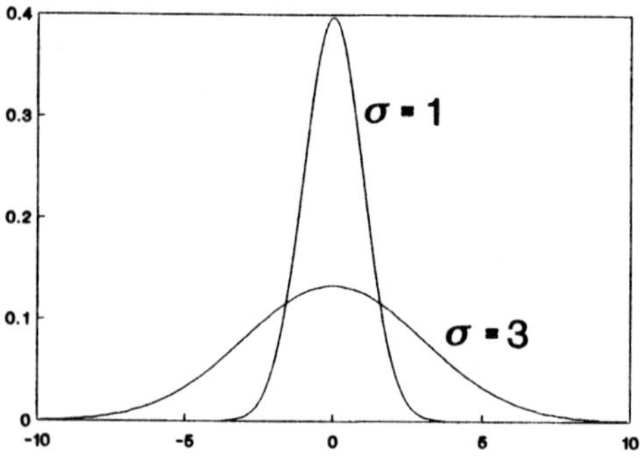

Figure 2.2: *Probability density functions for normal random variables with mean 0 and variances 1 and 9 (standard deviations $\sigma = 1$ and 3 respectively).*

about the mean, values near the mean are more likely than values elsewhere, and the spread of the distribution depends upon the variance. Larger variance is associated with greater spread. Changes in the mean shift the density to the left or right on the real line.

Also, note that the likelihood of obtaining a large realization of a normal random variable falls off quickly as the distance from the mean increases. Observations more than 1.96σ from the mean occur only 5% of the time, and observations more than 2.33σ from the mean occur only 1% of the time.

The mean, variance, skewness, and kurtosis of a normal random variable $\mathbf{X}$ are:

$$\begin{aligned}\mathcal{E}(\mathbf{X}) &= \mu\\ \mathrm{Var}(\mathbf{X}) &= \sigma^2\\ \gamma_1 &= 0\\ \gamma_2 &= 0,\end{aligned}$$

and the L-moments are:

$$\begin{aligned}\lambda^{(1)} &= \mu\\ \lambda^{(2)} &= \sigma/\pi\\ \gamma_1^L &= 0\\ \gamma_2^L &= 0.1226\,.\end{aligned}$$

The cumulative distribution function cannot be given explicitly because the analytical form of $\int_{-\infty}^{x} e^{-t^2/2}\,dt$ does not exist. But the cumulative distribution function is related in a simple manner to the *error function*, erf, which is available from subroutine libraries (for example in the *Numerical*

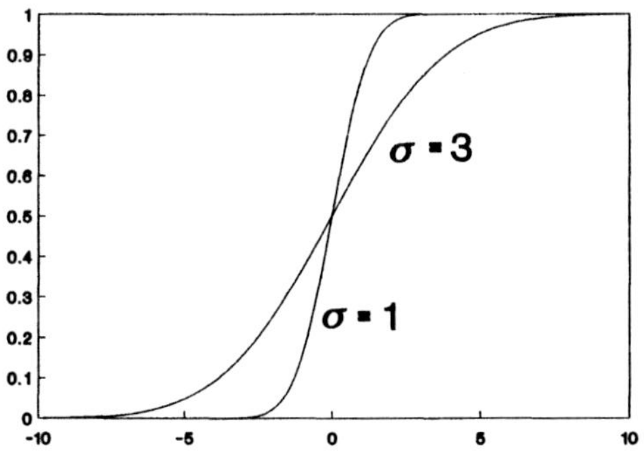

Figure 2.3: *Cumulative distribution functions of normal random variables for $\mu = 0$ and $\sigma = 1$ and 3.*

Recipes [322]). Specifically,

$$F_{\mathcal{N}}(x) = \frac{1}{\sqrt{2\pi}\,\sigma} \int_{-\infty}^{x} e^{-\frac{(t-\mu)^2}{2\sigma^2}}\,dt$$
$$= \frac{1}{\sqrt{\pi}} \int_{-\infty}^{\frac{x-\mu}{\sqrt{2}\sigma}} e^{-t^2}\,dt$$
$$= 0.5 + 0.5\,\mathrm{erf}\left(\frac{x-\mu}{\sqrt{2}\,\sigma}\right). \qquad (2.26)$$

The cumulative distribution functions for $\mu = 0$ and $\sigma = 1$ and 3 are plotted in Figure 2.3.

2.7.4 The Standard Normal Distribution. Any normal distribution can be transformed to the *standard* normal distribution, which has mean zero and variance one. In fact, if $\mathbf{X} \sim \mathcal{N}(\mu, \sigma^2)$, then $\mathbf{Z} = (\mathbf{X} - \mu)/\sigma \sim \mathcal{N}(0, 1)$.[10]

The proof, which is straight forward, illustrates the standard approach taken when deriving the distribution of a transformed random variable. First, suppose that $\mathbf{X} \sim \mathcal{N}(\mu, \sigma^2)$. Then, for any interval (a, b), we have

$$P\big(\mathbf{X} \in (a, b)\big) = \int_a^b \frac{1}{\sqrt{2\pi}\,\sigma} e^{-\frac{(y-\mu)^2}{2\sigma^2}}\,dy$$
$$= \int_{(a-\mu)/\sigma}^{(b-\mu)/\sigma} \frac{1}{\sqrt{2\pi}} e^{-z^2/2}\,dz$$

by a simple transformation of variable under the integral sign. However, the second expression is $P\big(\mathbf{Z} \in ((a-\mu)/\sigma, (b-\mu)/\sigma)\big)$ where $\mathbf{Z} \sim \mathcal{N}(0, 1)$.

The cumulative distribution function of the standard normal distribution is denoted by $F_{\mathcal{N}}$ in this book. This function, which is tabulated in Appendix D, can also be evaluated by numerical integration or by using simple approximations. For most purposes, the approximation

$$F_{\mathcal{N}}(x) \approx \left(1 + \mathrm{sgn}(x)\sqrt{1 - e^{-2x^2/\pi}}\,\right)/2 \qquad (2.27)$$

(where $\mathrm{sgn}(x) = 1$ if $x > 0$ and $\mathrm{sgn}(x) = -1$ if $x < 0$) is adequate and eliminates the use of tables.

2.7.5 The Central Limit Theorem. The *Central Limit Theorem* is of fundamental importance for statistics because it establishes the dominant role of the normal distribution.

If $\mathbf{X}_k$, $k = 1, 2, \ldots$, is an infinite series of independent and identically distributed random variables with $\mathcal{E}(\mathbf{X}_k) = \mu$ and $Var(\mathbf{X}_k) = \sigma^2$, then the average $\frac{1}{n}\sum_{k=1}^{n} \mathbf{X}_i$ is asymptotically normally distributed. That is,

$$\lim_{n\to\infty} \frac{\frac{1}{n}\sum_{k=1}^{n}(\mathbf{X}_k - \mu)}{\frac{1}{\sqrt{n}}\sigma} \sim \mathcal{N}(0, 1).$$

Note that the Central Limit Theorem holds regardless of the distribution of the $\mathbf{X}_k$.

According to the Central Limit Theorem, the distribution of a sum of independent and identically distributed random variables converges towards a normal distribution as the number, n, of random variables increases. Because the theorem makes an *asymptotic* statement nothing is known about when the convergence has made substantial progress. Sometimes n must be very large before near-normal conditions are reached [3.1.4]; other times the convergence is very fast and the distribution of a sum over a few random variables may be approximated by the normal distribution. Figure 3.2 in [3.1.2] demonstrates neatly the practical importance of the Central Limit Theorem.

2.7.6 The Log-Normal Distribution. A random variable $\mathbf{X}$ has a log-normal distribution with median θ if $\ln(\mathbf{X}) \sim \mathcal{N}(\ln(\theta), \sigma)$. The density function is given by

$$f_X(x) = \frac{1}{\sqrt{2\pi}\,\sigma}\frac{1}{x}\exp\left(-\frac{\big(\ln(x) - \ln(\theta)\big)^2}{2\sigma^2}\right).$$

Examples of this density function for various values of σ are displayed in Figure 2.4. The moments are given by

$$\mathcal{E}\big(\mathbf{X}^k\big) = \theta^k e^{(k\sigma)^2/2}.$$

[10] Germans can find a plot of the $\mathcal{N}(0, 1)$ probability density function in their wallets. It appears on the regular German 10 DM bank note together with a picture of its inventor.

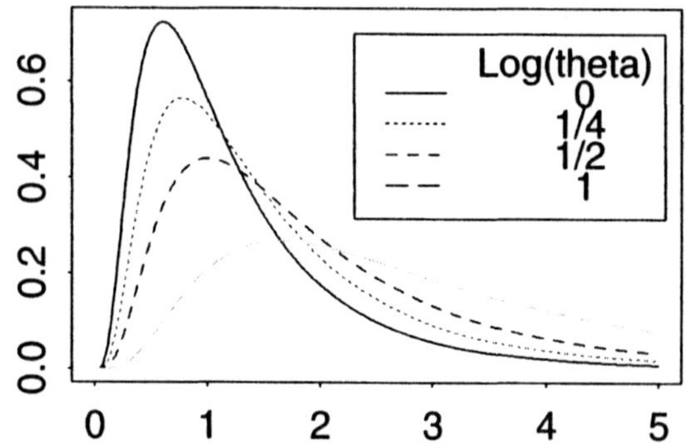

Figure 2.4: *Log-normal density functions for* $\sigma = \frac{1}{2}$ *and* $\ln(\theta) = 0, \frac{1}{4}, \frac{1}{2}$ *and* 1.

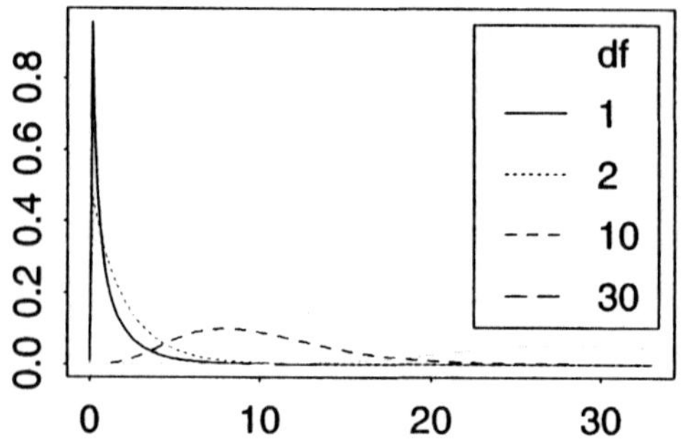

Figure 2.5: *Probability density functions for* $\chi^2(df)$ *random variables with 1, 2, 10, and 30 degrees of freedom.*

Therefore

$$\begin{aligned}\mathcal{E}(\mathbf{X}) &= \theta e^{\frac{1}{2}\sigma^2}\\ \mathrm{Var}(\mathbf{X}) &= \theta^2 e^{\sigma^2}\left(e^{\sigma^2}-1\right)\\ \gamma_1 &= \sqrt{e^{\sigma^2}-1}\left(e^{\sigma^2}+1\right).\end{aligned}$$

The distribution is skewed with a long tail to the right. The expectation is larger than the median.

The log-normal distribution is often useful when dealing with positive quantities such as precipitation.

2.7.7 Some Important Sampling Distributions. We now move on to the description of three important sampling distributions derived from the normal distribution: the χ^2 distribution, the t distribution, and the F distribution. We will see these distributions often in settings where we need to know about the uncertainty of an estimated mean or variance, or compare estimates of means or variances.

2.7.8 The χ^2 Distribution. The χ^2 distribution is defined as that of the sum of k independent squared $\mathcal{N}(0,1)$ random variables. It is therefore defined only on the positive half of the real line. The form of this distribution function depends upon a single parameter, k, referred to as the *degrees of freedom* (df).[11]

The probability density function of a $\chi^2(k)$ random variable $\mathbf{X}$ is given by

$$f_X(x)=\begin{cases}\dfrac{x^{(k-2)/2}e^{-x/2}}{\Gamma(k/2)2^{k/2}} & \text{if } x>0\\ 0 & \text{otherwise}\end{cases} \tag{2.28}$$

where Γ denotes the Gamma function. The derivation of (2.28) can be found in most standard mathematical statistics texts [335].

We write $\mathbf{X} \sim \chi^2(k)$ to indicate that a random variable $\mathbf{X}$ is χ^2 distributed with k degrees of freedom. Examples of the $\chi^2(k)$ distribution with $k = 1, 2, 10$, and 30 are shown in Figure 2.5. The distributions are partially tabulated in Appendix E.

The χ^2 distribution has a very important *additive* property: if $\mathbf{X}_1$ and $\mathbf{X}_2$ are independent χ^2 random variables with k_1 and k_2 df respectively, then $\mathbf{X}_1 + \mathbf{X}_2$ is a $\chi^2(k_1 + k_2)$ random variable. It follows then that a $\chi^2(k)$ random variable can be thought of as a sum of k independent $\chi^2(1)$ random variables.

Several characteristics of the χ^2 distribution can be noticed. First, all of the distributions are skewed to the left, but distributions with small numbers of degrees of freedom are more skewed than those with large numbers of degrees of freedom. In fact, the $\chi^2(30)$ distribution is very nearly normal, in accordance with the additive property and the Central Limit Theorem [2.7.5]. Second, only the distributions with one and two degrees of freedom have their *modes* (i.e., their most likely values) at the origin. Third, the spread of the distributions depends strongly upon the number of degrees of freedom.

[11] The expression *degrees of freedom* is used frequently in this book. Here it has two equivalent technical interpretations. Specifically, if $\mathbf{X}_1, \ldots, \mathbf{X}_n$ are independent, identically distributed $\mathcal{N}(\mu, \sigma^2)$ random variables, then $\chi^2 = \frac{1}{\sigma^2}\sum_{i=1}^n(\mathbf{X}_i = \overline{\mathbf{X}})^2$ is distributed $\chi^2(n-1)$. This sum of squared deviations can be re-expressed as a sum of $n-1$ squared $\mathcal{N}(0,1)$ random variables. This gives the first interpretation of *degrees of freedom*, which is frequently encountered in climate research: χ^2 contains information from $n-1$ independent, identically distributed random variables. The other interpretation is geometrical. The deviations $\mathbf{x}_i - \overline{\mathbf{x}}$ can be arranged in an n-dimensional random vector $(\mathbf{x}_1 - \overline{\mathbf{x}}, \ldots, \mathbf{x}_n - \overline{\mathbf{x}})^{\mathrm{T}}$. This vector takes values in an $(n-1)$-dimensional subspace since the deviations are constrained to sum to zero. See also [6.6.1] and Section 6.8.

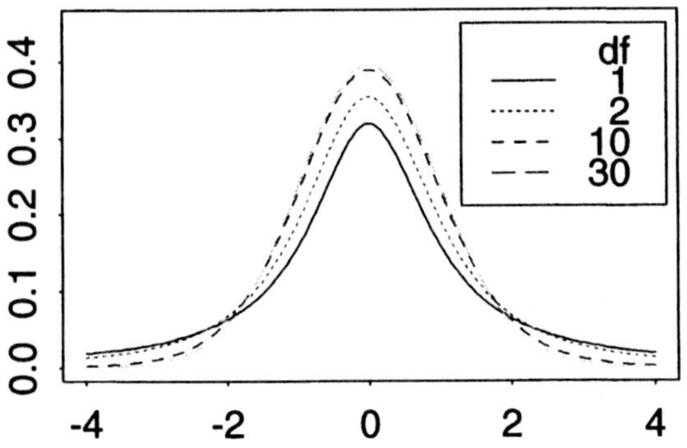

Figure 2.6: *Probability density functions for $t(k)$ random variables with 1, 2, 10, and 30 degrees of freedom.*

In general, if $\mathbf{X} \sim \chi^2(k)$, then

$$\mathcal{E}(\mathbf{X}) = k$$
$$\text{Var}(\mathbf{X}) = 2k.$$

2.7.9 The t distribution. A random variable $\mathbf{T}$ has the t distribution *with k degrees of freedom*, that is, $\mathbf{T} \sim t(k)$, if its probability density function is given by

$$f_T(t; k) = \frac{\Gamma((k+1)/2)(1+t^2/k)^{-(k+1)/2}}{\sqrt{k\pi}\,\Gamma(k/2)}.$$

$\mathbf{T}$ random variables are strongly related to normal and χ^2 random variables. In particular, if $\mathbf{A}$ and $\mathbf{B}$ are independent random variables such that

$$\mathbf{A} \sim \mathcal{N}(0, 1) \text{ and } \mathbf{B} \sim \chi^2(k),$$

then

$$\frac{\mathbf{A}}{\sqrt{\mathbf{B}/k}} \sim t(k).$$

The t distribution was introduced by W.L. Gosset under the pseudonym 'Student'—so is often called the *Student's t distribution.*

The t distribution is symmetric about zero. When $\mathbf{T}$ has more than one degree of freedom, the first central moment is zero (see e.g., Kalbfleisch [208]),

$$\mathcal{E}(\mathbf{T}) = 0 \text{ for } k \geq 2.$$

The first moment does not exist when $k = 1$.

Similarly, the second central moment exists for $k \geq 3$, where

$$\text{Var}(\mathbf{T}) = \frac{k}{k-2} \text{ for } k \geq 3.$$

It may be shown [208] that the jth moments of $\mathbf{T}$ for $j \geq k$ do *not* exist.

The $t(k)$ distribution is shown in Figure 2.6 for four values of the degrees of freedom parameter k. The density function $f_T(t; 1)$ for $\mathbf{T}$ with $k = 1$ degree of freedom does tend to zero as $t \to \pm\infty$, but too slowly for the integral $\int t f_T(t; 1)\, dt$ to exist. The convergence is faster when $k = 2$, so that the first moment exists but not the second moment. The convergence increases with the increasing numbers of degrees of freedom. Ultimately, the t distribution converges to the standard normal distribution. The difference between the distributions is small even when $k = 10$, and it becomes negligible for $k \geq 30$.

The $t(k)$ distribution is partially tabulated in Appendix F.

2.7.10 The F Distribution. Another of the sampling distributions closely related to the normal distribution is the F distribution. A random variable $\mathbf{F}$ is said to have an F distribution with k and l degrees of freedom, that is, $\mathbf{F} \sim F(k, l)$, if the density function of $\mathbf{F}$, $f_F(f; k, l)$, is given by

$$f_F(f; k, l) = \frac{(k/l)^{k/2}\Gamma((k+l)/2)}{\Gamma(k/2)\Gamma(l/2)} \times f^{(k-2)/2}\left(1 + \frac{k}{l}f\right)^{-(k+l)/2}.$$

This distribution arises in estimation and testing problems when statistics are developed that can be expressed as a constant times a ratio of independent χ^2 random variables (hence the connection to the normal distribution—see [2.7.8]).

In particular, if $\mathbf{X}$ and $\mathbf{Y}$ are independent random variables such that $\mathbf{X} \sim \chi^2(k)$ and $\mathbf{Y} \sim \chi^2(l)$, then

$$\frac{\mathbf{X}/k}{\mathbf{Y}/l} \sim F(k, l). \quad (2.29)$$

The first two central moments are

$$\mu = \mathcal{E}(\mathbf{F}) = \frac{l}{l-2}$$

for $l > 2$ and

$$\text{Var}(\mathbf{F}) = \frac{2l^2(k+l-2)}{k(l-2)^2(l-4)}$$

for $l > 4$. As for the t distribution, not all moments of the F distribution exist (see Kalbfleisch [208]).

The $F(k, l)$ density function is shown in Figure 2.7 for three combinations of (k, l). The distribution is skewed for all values of l. For fixed k, the skewness decreases slightly with

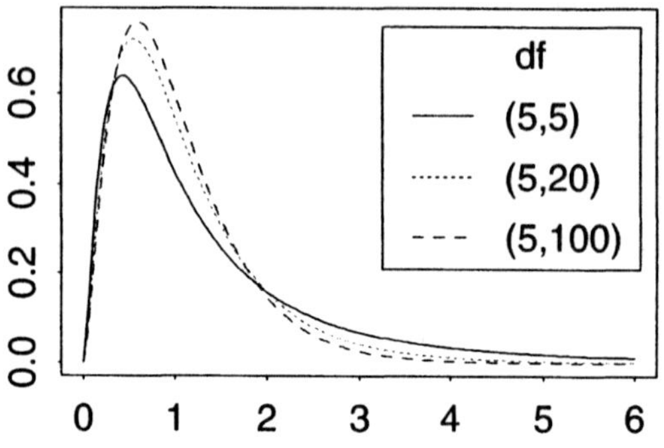

Figure 2.7: *Probability density functions for $F(k, l)$ random variables with $(k, l) = (5, 5)$, $(5, 20)$, and $(5, 100)$ degrees of freedom.*

increasing l. In fact, the distribution converges to a normalized χ^2 distribution as $l \to \infty$.

The F distribution is partially tabulated in Appendix G.

2.7.11 The Exponential Distribution. The distribution of wind energy, which is proportional to the square of wind speed, provides an interesting application of the χ^2 distribution. To a first order of approximation, the zonal and meridional components of the wind are normally distributed and independent (but see [2.6.6] and also Cook [89] and Holzer [180]). Thus the wind energy, when properly scaled, is approximately distributed $\chi^2(2)$. The latter distribution, illustrated in Figure 2.5, is also an example of an *exponential distribution*. The likelihood of observing a particular wind energy falls off exponentially with magnitude.

The density function of an exponential random variable $\mathbf{X}$ is given by

$$f_X(x) = \begin{cases} \theta^{-1} e^{-x/\theta} & \text{if } x > 0 \\ 0 & \text{otherwise,} \end{cases}$$

and the corresponding cumulative distribution function is given by

$$F_X(\mathbf{x}) = \begin{cases} 0 & \text{if } x \le 0 \\ 1 - e^{-x/\theta} & \text{if } x > 0. \end{cases}$$

The mean and variance are

$$\mu = \theta \text{ and } \sigma^2 = \theta^2.$$

The L-moments are

$$\begin{aligned} \lambda^{(1)} &= \theta \\ \lambda^{(2)} &= \theta/2 \\ \gamma_1^L &= 1/3 \\ \gamma_2^L &= 1/6. \end{aligned}$$

The χ^2 distribution with 2 df is an exponential distribution with $\theta = 2$.

2.7.12 Example: Waiting Times in a Poisson Process. The exponential distribution also arises when studying waiting times in a Poisson process. We used a Poisson process in [2.4.4] to model the occurrence of wind speed peaks over a threshold. If the threshold is large, the distribution of waiting times is useful for making inferences about the frequency with which we might expect damaging winds. Here, we will the derive the waiting time distribution for a Poisson process with intensity λ.

Let $\mathbf{T}$ be the waiting time for the first event in a Poisson process.[12] $\mathbf{T}$ is obviously a random variable because events in the Poisson process occur randomly. Let $F_T(\cdot)$ be the cumulative distribution function of $\mathbf{T}$. That is, $F_T(t) = P(\mathbf{T} < t) = 1 - P(\mathbf{T} \ge t)$. The event $\mathbf{T} \ge t$ occurs when no events take place in the time interval $(0, t)$. Equation (2.10) can be used to show that

$$P\big(\textit{no events in } (0, t)\big) = e^{-\lambda t}$$

and therefore that

$$F_T(t) = \begin{cases} 1 - e^{-\lambda t} & \text{if } t \ge 0 \\ 0 & \text{otherwise.} \end{cases}$$

Hence, the waiting time is exponentially distributed with $\theta = \lambda^{-1}$. Consequently, the mean waiting time is inversely proportional to the intensity of the Poisson process.

2.8 Random Vectors

2.8.1 Continuous Random Vectors. A *continuous random vector* $\vec{\mathbf{X}}$ is a vector of continuous random variables.

The climate system has a myriad of examples of continuous random vectors. One example is the monthly mean 300 hPa height field $\vec{\mathbf{Z}}$, either as simulated by a climate model, or as analysed from observations (Figure 1.1). In both cases, the random vector contains several hundred or thousand entries, each representing an observation at a different location. Another example is the surface temperature field $\vec{\mathbf{T}}$, which contains screen temperature[13] observations over the land and ocean surfaces. If we want to study relationships

[12] We can assume that we start observing the process just after the occurrence of an event, so the waiting time for the first event is equivalent to the waiting time between events.

[13] 'Screen temperature' is taken 2 m above the surface. The word 'screen' alludes to the enclosures—Stevenson screens—that are used to house land-based thermometers.

between geopotential and surface temperature, then we might form an even larger random vector by combining $\vec{\mathbf{Z}}$ and $\vec{\mathbf{T}}$.

2.8.2 Joint Probability Density Function. The *joint* probability density function of an m-dimensional random vector $\vec{\mathbf{X}}$ is a non-negative, continuous function defined on $\mathbb{R}^m$ for which $\int_{\mathbb{R}^m} f_{\vec{X}}(\vec{x})\, d\vec{x} = 1$.

The cumulative distribution function also extends to the multivariate case in a natural way. However, the concept is not as useful as in the univariate case, and therefore will not be discussed.

2.8.3 Marginal Distributions. In our discussion of discrete multivariate distributions [2.5.2], the marginal distribution of one variable was found by summing the joint probability function over all combinations of values taken by the remaining variables. Since integration is the continuous variable analogue to summation, the marginal probability density function for the kth variable in $\vec{\mathbf{X}}$, say $\mathbf{X}_k$, is defined by

$$\begin{aligned} f_{X_k}(x) &\\ &= \int_{\mathbb{R}^{m-1}} f_{\vec{X}}(x_1, \ldots, x_{k-1}, x, x_{k+1}, \ldots, x_m)\, d\vec{x}_{k'}, \end{aligned}$$

where $\vec{x}_{k'} = (x_1, \ldots, x_{k-1}, x_{k+1}, \ldots, x_m)$.

2.8.4 Expectation of a Weighted Sum of the Components of a Random Vector. The expected value of the kth component of $\vec{\mathbf{X}}$ is the mean of the marginal distribution

$$\begin{aligned} \mathcal{E}(\mathbf{X}_k) &= \int_{-\infty}^{\infty} x_k f_{X_k}(x_k) dx_k \\ &= \int_{\mathbb{R}^m} x_k f_{\vec{X}}(\vec{x})\, d\vec{x}, \end{aligned}$$

where $\vec{x} = (x_1, \ldots, x_m)^{\mathrm{T}}$.

The expected value of a linear combination of two components of $\vec{\mathbf{X}}$ is

$$\begin{aligned} &\mathcal{E}\big(a\mathbf{X}_k + b\mathbf{X}_j + c\big) \\ &\quad = \int_{\mathbb{R}^m} (ax_k + bx_j + c) f_{\vec{X}}(\vec{x})\, d\vec{x} \\ &\quad = a\,\mathcal{E}(\mathbf{X}_k) + b\,\mathcal{E}\big(\mathbf{X}_j\big) + c\,. \end{aligned} \tag{2.30}$$

The same result can be obtained directly from (2.15) and (2.16) as follows:

$$\begin{aligned} &\mathcal{E}\big(a\, g_1(\vec{\mathbf{X}}) + b\, g_2(\vec{\mathbf{X}}) + c\big) \\ &\quad = a\,\mathcal{E}\big(g_1(\vec{\mathbf{X}})\big) + b\,\mathcal{E}\big(g_2(\vec{\mathbf{X}})\big) + c, \end{aligned}$$

where the functions g_1 and g_2 select the kth and jth components of $\vec{\mathbf{X}}$ respectively.

Note that (2.30) holds regardless of the correlation between the components $\mathbf{X}_k$ and $\mathbf{X}_j$.

2.8.5 Independent Random Variables. The definition of independent random variables also extends smoothly from the discrete to the continuous case.

Let $\vec{\mathbf{X}}$ be a random vector and let $\mathbf{X}_i$ and $\mathbf{X}_j$ be any pair of elements in the vector. The components of $\vec{\mathbf{X}}$ are said to be pairwise independent *if for every (i, j) the joint density function of $\mathbf{X}_i$ and $\mathbf{X}_j$ can be written as the product of the marginal density functions of $\mathbf{X}_i$ and $\mathbf{X}_j$.*

The components of $\vec{\mathbf{X}}$ are said to be jointly independent *if $f_{\vec{X}}(\vec{x}) = \prod_{i=1}^{m} f_{X_i}(x_i)$.*

2.8.6 Conditional Density Functions. Finally, the concept of the conditional distribution is extended to the continuous case. However, here it is likelihoods, rather than probabilities, that are scaled. We saw that in the discrete case [2.5.4], the act of *conditioning* on the outcome of a variable reduced the number of outcomes that were possible by some finite proportion. Similarly, conditioning in the continuous case restricts possible realizations of the random vector to a hyper-space of the original m-dimensional vector space. The conditional probability density function is defined as follows.

Let $\vec{\mathbf{X}}$ be a random vector of the form $(\vec{\mathbf{X}}_1, \vec{\mathbf{X}}_2)$, where $\vec{\mathbf{X}}_1$ and $\vec{\mathbf{X}}_2$ are also both random vectors. The conditional probability density function *of $\vec{\mathbf{X}}_1$, given $\vec{\mathbf{X}}_2 = \vec{\mathbf{x}}_2$, is*

$$f_{\vec{X}_1|\vec{X}_2=\vec{x}_2}(\vec{x}_1) = \frac{f_{\vec{X}_1\vec{X}_2}(\vec{x}_1, \vec{x}_2)}{f_{\vec{X}_2}(\vec{x}_2)} \tag{2.31}$$

for all $\vec{x}_2$ such that $f_{\vec{X}_2}(\vec{x}_2)$ is nonzero.

2.8.7 The Multivariate Mean, the Covariance Matrix, and the Correlation Matrix. The long-term mean value of repeated realizations of an m-dimensional random vector $\vec{\mathbf{X}}$ is given by

$$\vec{\mu}_{\vec{X}} = \mathcal{E}(\vec{\mathbf{X}}) = \int_{\mathbb{R}^m} \vec{x} f_{\vec{X}}(\vec{x})\, d\vec{x}.$$

Note that the elements of $\vec{\mu}_{\vec{X}}$ are the means of the corresponding marginal distributions. We will usually refer only to $\vec{\mu}$ rather than $\vec{\mu}_{\vec{X}}$ unless clarity requires that specific reference be made to the random vector.

Jointly distributed random variables often have a tendency to vary jointly.[14] This 'co-variability' may be quantitatively described by the multivariate analogue of variance, namely the *covariance matrix*:

$$\Sigma_{\vec{X},\vec{X}} = \mathcal{E}\big((\vec{\mathbf{X}} - \vec{\mu})(\vec{\mathbf{X}} - \vec{\mu})^{\mathrm{T}}\big) \qquad (2.32)$$
$$= \int_{\mathbb{R}^m} (\vec{x} - \vec{\mu})(\vec{x} - \vec{\mu})^{\mathrm{T}} f_{\vec{X}}(\vec{x})\, d\vec{x}.$$

As above, we will drop the reference to the random vector in the notation for the covariance matrix unless a need for clarity dictates otherwise.

The (i, j)th element of Σ contains the *covariance*

$$\sigma_{ij} = \mathcal{E}\big((x_i - \mu_i)(x_j - \mu_j)\big)$$
$$= \iint_{\mathbb{R}^2} (x_i - \mu_i)(x_j - \mu_j) f_{x_i x_j}(x_i, x_j)\, dx_i dx_j$$

between the ith and jth elements of $\vec{\mathbf{X}}$. Note that the covariance matrix is symmetric: the covariance between $\mathbf{X}_i$ and $\mathbf{X}_j$ is the same as that between $\mathbf{X}_j$ and $\mathbf{X}_i$. The diagonal elements of Σ are the variances of the individual random variables that form the random vector $\vec{\mathbf{X}}$. That is, $\sigma_{ii}^2 = \mathrm{Var}(\mathbf{X}_i)$.

The covariance matrix is positive-definite.

Covariances describe the tendency of jointly distributed random variables to vary in concert. If the deviations of $\mathbf{X}_i$ and $\mathbf{X}_j$ from their respective means tend to be of the same sign, the covariance between $\mathbf{X}_i$ and $\mathbf{X}_j$ will be positive, and if the deviations tend to have opposite signs, the covariance will be negative.

As in the discrete variable case, the covariance is zero if $\mathbf{X}_i$ and $\mathbf{X}_j$ are independent. This occurs because the expectation of a product of independent random variables factors into a product of expectations. Note, however, that the reverse need not be true (see the example in [2.8.14]).

The effect of scaling on covariance is similar to that which occurs in the scalar case (see (2.6)). If $\mathcal{A}$ is a $k \times m$ matrix with $k \leq m$, then

$$\Sigma_{\mathcal{A}\vec{X},\mathcal{A}\vec{X}} = \mathcal{A}\Sigma\mathcal{A}^{\mathrm{T}}.$$

[14] Nearby values in all atmospheric and oceanic fields are related to one another. In fact, without this property initialization of numerical weather prediction models would require a much denser observing network than exists today. Objective analysis and data assimilation techniques, which are used to initialize forecast models, make extensive use of the *covariance* structure of the atmosphere. Climate forecast systems based on coupled ocean/atmosphere models also make extensive use of such techniques to initialize the oceanic components of these models.

A possible difficulty with covariance as a measure of the joint variability of a pair of random variables is that covariance is not *scale invariant.* As with the transports that covariances often represent in climate problems (see Section 8.2), a change in units has a profound effect on the size of the covariance. If all realizations of $\mathbf{X}_i$ and $\mathbf{X}_j$ are multiplied by constants c_i and c_j respectively, the covariance will increase by a factor of $c_i c_j$. However, the variances of $\mathbf{X}_i$ and $\mathbf{X}_j$ also increase by factors of c_i^2 and c_j^2.

Correlation (or cross-correlation) is a measure of covariability that is scale invariant.

The correlation *between two random variables* $\mathbf{X}_i$ *and* $\mathbf{X}_j$ *is given by*

$$\rho_{ij} = \frac{\mathrm{Cov}(\mathbf{X}_i, \mathbf{X}_j)}{\sqrt{\mathrm{Var}(\mathbf{X}_i)\mathrm{Var}(\mathbf{X}_j)}}. \qquad (2.33)$$

The correlation coefficient always takes values in the interval $[-1, 1]$. The absolute value of the coefficient is exactly 1 when $\mathbf{X}_i$ is linearly related to $\mathbf{X}_j$, that is, when constants a and b exist so that $\mathbf{X}_i = a + b\mathbf{X}_j$. Here the correlation is $+1$ if b is positive and -1 if b is negative. Values of the correlation coefficient between -1 and $+1$ are an indication of the extent to which there is a linear relationship between the two random variables. In fact, ρ_{ij}^2 can be interpreted usefully as the proportion of the variance of one of the variables that can be represented linearly by the other (see also [18.2.7] and Section 8.2).

As an example, consider the 1933–84 segment of the Southern Oscillation Index (SOI) (Figure 1.4). Superimposed on the graph is Wright's SST index of the SO [426]. The SST index carries roughly the same information about the Southern Oscillation on time scales of a year or more. The estimated correlation between the monthly mean values of these indices is 0.67. We will examine this example in more detail in Section 8.2.

A word of caution about the correlation coefficient: it is not always a measure of the extent to which there is a deterministic relationship between two random variables. In fact, two random variables may well be related through a deterministic, nonlinear function and yet have a correlation of zero.

2.8.8 Mapping the Correlation Matrix: Teleconnection Patterns.

The various combinations of correlations between the ith and jth components of a random vector $\vec{\mathbf{X}}$ form the *correlation matrix.* If $\vec{\mathbf{X}}$ represents a *field*, a (possibly gridded)

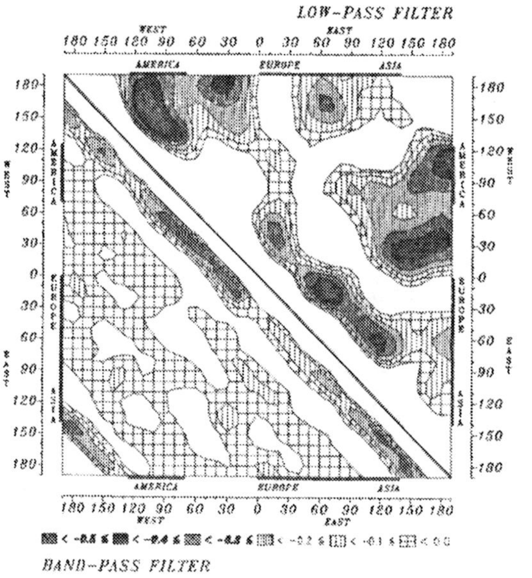

Figure 2.8: *Correlation matrices for the simultaneous variations of 500 hPa height along 50° N in the synoptic time scale (lower left) and the low-frequency transpose (upper right) band. Only negative correlations are shown. From Fraedrich et al. [127].*

set of observations in space, then the jth row (or column) of the correlation matrix contains the correlations between the field at the jth location and all other locations. When this row is mapped we obtain a spatial pattern of correlations that climatologists call a *teleconnection pattern* or *teleconnection map*. A map is considered 'interesting' if it exhibits large correlations at some distance from the 'base point' j, and if it suggests physically plausible mechanisms (such as wave propagation).

Such maps often unveil large-scale 'teleconnections' between a fixed base point and distant areas. We deal with these *teleconnection maps* in some detail in 17.4.

The entire correlation matrix may be plotted when the field is one-dimensional. For example, Fraedrich, Lutz, and Spekat [127] analysed daily 500 hPa geopotential height along 50° N. The annual cycle was removed from the data, and then two different time filters (see 17.5) were applied to separate the synoptic disturbances (2.5 to 6 days) from low-frequency atmospheric variability (time scales greater than 10 days).

The correlation matrices (only *negative* correlations are shown) of these two one-dimensional random vectors are shown in Figure 2.8. The upper right half shows spatial correlations of the low-frequency variations while the lower left half shows the longitudinal correlations at the synoptic time scale. The diagrams are read as follows. If we read across from 0° on the vertical scale and up from 40° E on the horizontal scale we see that the (simultaneous) correlation between low-frequency time scale variations at 0° and 40° E is about -0.3.

The banded structure in the lower left reflects the midlatitude stormtracks. The strongest (negative) correlations are found in a band that is about 30° off the diagonal. When there is a deep low at a given longitude, it is likely that there will be a high 30° to the east or west, and vice versa. The organization of the correlation minima in bands indicates that the disturbances propagate (the direction of this propagation cannot be read from this diagram).

The correlation structure is no longer banded at time scales of 10 or more days. On these time scales, height anomalies east of the dateline are strongly connected with anomalies of opposite sign over North America (this reflects the PNA-pattern, [3.1.7]); other links appear over Europe and Asia, and over the East Atlantic and Europe.

2.8.9 Multivariate Normal Distribution. The m-dimensional random vector $\vec{\mathbf{X}}$ has a *multivariate normal distribution* with mean $\vec{\mu}$ and covariance matrix Σ if its joint probability density function is given by

$$f_{\vec{X}}(\vec{\mathbf{x}}) = \frac{1}{(2\pi|\Sigma|)^{1/2}} e^{-(\vec{\mathbf{x}}-\vec{\mu})^{\mathrm{T}}\Sigma^{-1}(\vec{\mathbf{x}}-\vec{\mu})}. \quad (2.34)$$

A bivariate normal density function is shown in Figure 2.9. Like its univariate counterpart, the distribution is symmetric across all planes which pass through the mean. The spread, or dispersion, of the distribution is determined by the covariance matrix Σ.

An important property of the multivariate normal distribution is that linear combinations of normal random variables are again distributed as normal random variables. In particular, let $\mathcal{A}$ be a full rank $m' \times m$ matrix of constants with $m' \leq m$. Then $\vec{\mathbf{Y}} = \mathcal{A}\vec{\mathbf{X}}$ defines an m'-dimensional random vector which is distributed multivariate normal with mean vector $\mathcal{A}\vec{\mu}$ and covariance matrix $\mathcal{A}\Sigma\mathcal{A}^{\mathrm{T}}$ (Graybill [147]).

An immediate consequence of this result is that all marginal distributions of the multivariate normal distribution are also normal. That is, individual elements of a normal random vector $\vec{\mathbf{X}}$ are normally distributed and subsets of elements of

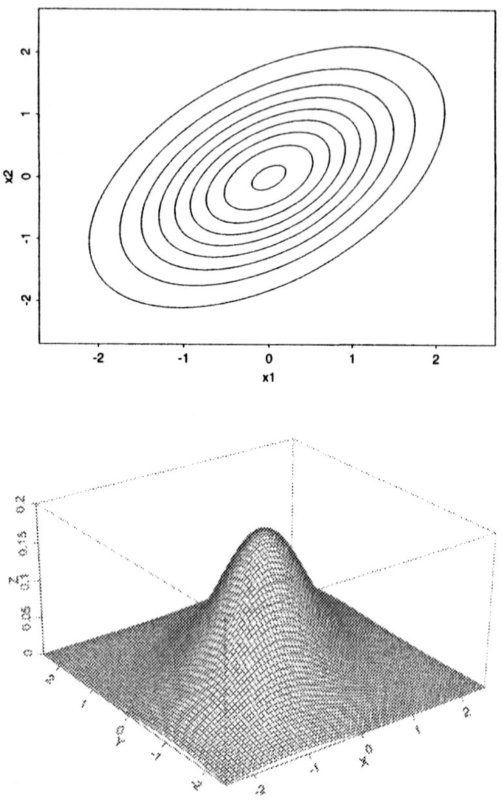

Figure 2.9: *A bivariate normal density function with variances* $\sigma_1^2 = \sigma_2^2 = 1$ *and covariances* $\sigma_{1,2} = \sigma_{2,1} = 0.5$.
Top: Contours of constant density;
Bottom: three-dimensional representation.

$\vec{\mathbf{X}}$ are multivariate normal. The reverse is not true in general.

2.8.10 Independence. The covariance matrix plays much the same role in the multivariate case as does the variance in the scalar case: it determines the spread of the distribution and the shape of the region occupied by the main body of the distribution.

Suppose that $\mathbf{\Sigma} = \sigma^2\mathcal{I}$, where $\mathcal{I}$ is the identity matrix. Then the contours of constant density are circular.

If $\mathbf{\Sigma} = \text{diag}(\sigma_1^2, \cdots, \sigma_m^2)$, the contours of the scaled random vector $\mathbf{\Sigma}^{-1/2}\vec{\mathbf{X}}$ are circular where $\mathbf{\Sigma}^{-1/2} = (\sigma_1^{-1}, \cdots, \sigma_m^{-1})$. The scaled random vector is identical to $\vec{\mathbf{X}}$ except that each element has been divided by its standard deviation. With such a diagonal covariance matrix, the density function of $\vec{\mathbf{X}}$ is given by

$$f_{\vec{X}}(\vec{x}) = \frac{1}{(2\pi)^{m/2}\prod_{i=1}^m \sigma_i} e^{-\sum_{i=1}^m \frac{(x_i-\mu_i)^2}{2\sigma_i^2}}$$

This can be rewritten as

$$f_{\vec{X}}(\vec{x}) = \prod_{i=1}^m \frac{1}{\sqrt{2\pi}\,\sigma_i} e^{-\frac{(x_i-\mu_i)^2}{2\sigma_i^2}} = \prod_{i=1}^m f_{X_i}(x_i).$$

Thus, if the non-diagonal elements $\mathbf{\Sigma}$ are zero, the joint density function factors as a product of marginal density functions and hence the components of $\vec{\mathbf{X}}$ are jointly independent. Furthermore, it can be shown that if some non-diagonal element, Σ_{ij}, of $\mathbf{\Sigma}$ is zero, the corresponding random variables $\mathbf{X}_i$ and $\mathbf{X}_j$ are pairwise independent.

2.8.11 Computing Probabilities. If Θ is any subset of $\mathbb{R}^m$, then the probability that a random outcome of the random vector $\vec{\mathbf{X}}$ occurs in Θ is given as the integral of the density $f_{\vec{X}}$ over the area Θ:

$$P(\vec{\mathbf{X}} \in \Theta) = \int_\Theta f_{\vec{X}}(\vec{x})\,d\vec{x}.$$

Of particular interest are those regions Θ_p chosen so that Θ_p is the smallest region for which

$$P(\vec{\mathbf{X}} \in \Theta_p) = p.$$

These areas turn out to be the interior regions bounded by contours of constant probability density. That is, for any given p, there is a constant κ_p such that Θ_p is given by

$$\Theta_p = \{\vec{x} : f_{\vec{X}}(\vec{x}) \geq \kappa_p\}.$$

For multivariate normal distributions, the region Θ_p is the interior of an ellipsoid (such as those shown in Figure 2.9).

The contours of constant density in a multivariate normal distribution are given by the contours of the *Mahalanobis distance*

$$\mathcal{D}^2(\vec{x}) = (\vec{x} - \vec{\mu})^{\mathrm{T}}\mathbf{\Sigma}^{-1}(\vec{x} - \vec{\mu}).$$

Assuming $\vec{\mathbf{X}} \sim \mathcal{N}(\vec{\mu}, \mathbf{\Sigma})$, it can be shown that $\mathcal{D}^2 \sim \chi^2(m)$, so that

$$P(\mathcal{D}^2(\vec{\mathbf{X}}) > \kappa_p) = \int_0^{\kappa_p} \chi_m^2(u)\,du = p.$$

Thus, the problem of calculating probabilities reduces to the inversion of the χ^2 distribution. We will return to this concept when introducing statistical tests in [6.2.2].

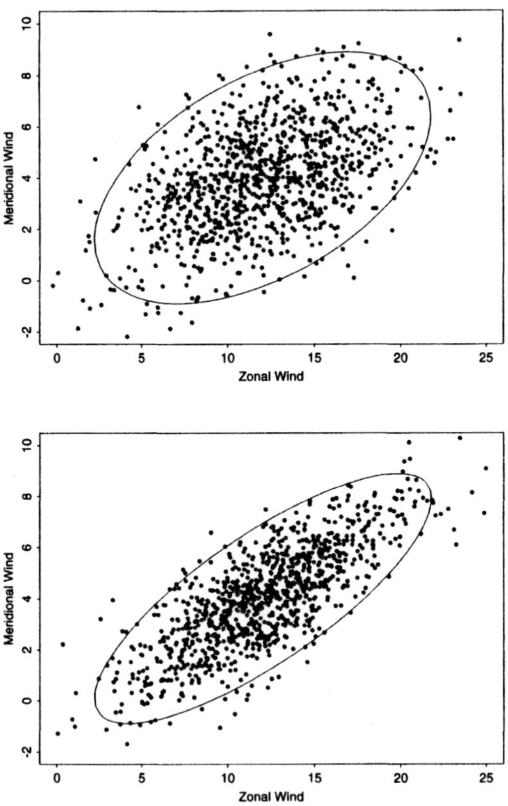

Figure 2.10: *Top: Example [2.8.12]: Ellipsoidal regions that are expected to contain the vector wind 95% of the time. Here the correlation is* $\rho_{UV} = 0.5$. *The dots represent 1000 realizations of the vector wind simulated from the corresponding normal distribution.*
Bottom: As above except $\rho_{UV} = 0.8$.

2.8.12 Example. Suppose that the vector wind $\vec{\mathbf{V}} = (\mathbf{U}, \mathbf{V})^{\mathrm{T}}$ at a particular location has a bivariate normal distribution with mean and covariance matrix

$$\vec{\mu} = (12, 4)^{\mathrm{T}} \quad \text{and} \quad \Sigma = \begin{pmatrix} 16 & 4 \\ 4 & 4 \end{pmatrix}.$$

The correlation between **U** and **V** is

$$\rho = \frac{\sigma_{UV}}{\sigma_U \sigma_U} = \frac{4}{\sqrt{16}\sqrt{4}} = \frac{1}{2}.$$

The quadratic form for the contours of constant density,

$$\mathcal{D}^2(\vec{v}) = (\vec{v} - \vec{\mu})^{\mathrm{T}} \Sigma^{-1} (\vec{v} - \vec{\mu}),$$

is distributed $\chi^2(2)$ and can be reduced to

$$\mathcal{D}^2(u, v) = \frac{u^2 - 2uv + 4v^2 - 16u - 8v + 108}{12}.$$

Suppose now that we wish characterize the 'normal' winds at our location by identifying the 95% of possible wind vectors that are closest to the mean. That is, we wish to exclude from our characterization the 5% of winds that are most extreme. The probability of obtaining a realization of a $\chi^2(2)$ random variable less than or equal to 5.99 is 0.95. Thus, the elliptical region $\mathcal{D}^2(u, v) \leq 5.99$, bounded by the solid curve in the top of Figure 2.10, is expected to contain 95% of all realizations of the vector wind. The points in the diagram represent 1000 realizations of a normal random vector with the mean and variance described above (see also [6.1.1]).

2.8.13 The Bivariate Normal Distribution. We describe the two-dimensional normal distribution in more detail because it comes up frequently. Suppose **U** and **V** are jointly normal with means μ_U and μ_V respectively. Suppose also that the covariance matrix is

$$\Sigma = \begin{pmatrix} \sigma_u^2 & \sigma_u \sigma_v \rho \\ \sigma_u \sigma_v \rho & \sigma_v^2 \end{pmatrix},$$

where ρ is the correlation between **U** and **V**. Using (2.34) we see that the joint density function is given by

$$f_{UV}(u, v) = \frac{1}{2\pi\sigma_U\sigma_V\sqrt{1-\rho^2}} \times \exp\left\{ -\frac{1}{2\sigma_U\sigma_V(1-\rho^2)} \left[\frac{(u-\mu_u)^2}{\sigma_u^2} - 2\rho\frac{(u-\mu_u)(v-\mu_v)}{\sigma_u\sigma_v} + \frac{(v-\mu_v)^2}{\sigma_v^2} \right] \right\}. \tag{2.35}$$

Ellipsoids of constant density, as shown in the top of Figure 2.10, are characterized by the equation

$$c(1 - \rho^2) = z_u^2 - 2\rho z_u z_v + z_v^2,$$

where the constant c is determined by the chosen density, and z_u and z_v are standardized deviations from the mean, given by

$$z_u = \frac{u - \mu_u}{\sigma_u}$$
$$z_v = \frac{v - \mu_v}{\sigma_v}.$$

This equation describes a circle if **U** and **V** are uncorrelated. When $\rho > 0$ ($\rho < 0$) the ellipsoid tips to the left (right), indicating that positive values of **U** tend to be associated with positive (negative) values of **V**. The principal axis of the ellipse is given by

$$u = \mu_u + \mathrm{sgn}(\rho)(\sigma_u/\sigma_v)(v - \mu_v)$$

and the minor axis is given by

$$u = \mu_u - \text{sgn}(\rho)(\sigma_u/\sigma_v)(v - \mu_v).$$

The ratio of the length of the principal axis to that of the minor axis in a given ellipsoid of constant density is

$$\left(\frac{1+|\rho|}{1-|\rho|}\right)^{1/2}.$$

Thus, the closer $|\rho|$ is to 1, the more concentrated the variation about the principal axis will be. The bottom of Figure 2.10 illustrates this with another hypothetical vector wind distribution, in which ρ_{UV} is increased to 0.8.

2.8.14 Example. Let us consider two univariate normal random variables,

$$\mathbf{U}, \mathbf{V} \sim \mathcal{N}(0, 1),$$

related through $\mathbf{U} = \mathbf{A}\mathbf{V}$, where $\mathbf{A}$ is a discrete random variable such that

$$P(\mathbf{A} = 1) = P(\mathbf{A} = -1) = 1/2.$$

Both random variables have a standard deviation of 1, so by (2.33) the correlation ρ_{UV} between $\mathbf{U}$ and $\mathbf{V}$ is equal to the covariance

$$\begin{aligned}\rho_{UV} &= \int_{-\infty}^{\infty} uv f_{UV}(u, v)\, du dv \\ &= \sum_a \left[\int_{-\infty}^{\infty} u^2 f_U(u)\, du\right] a \times P(\mathbf{A} = a) \\ &= \left[\sum_{a=\pm 1} a\, \frac{1}{2}\right] \int u^2 f_U(u)\, du \\ &= 0.\end{aligned}$$

This should not, however, lead us to the conclusion that they are independent, since $\mathbf{U}^2 = \mathbf{V}^2$. Examination of the probability density functions adds more satisfying evidence that these variables are dependent:

$$\begin{aligned}f_{U|V=v}(u) &= \frac{f_{UV}(u, v)}{f_V(v)} \\ &= \begin{cases} 0 & \text{if } u \neq \pm v \\ 1/2 & \text{if } u = \pm v. \end{cases}\end{aligned}$$

That is, $f_{U|V=v} \neq f_U$. The variables are dependent, since the joint (bivariate) density function f_{UV} cannot be represented as the product of the two marginal distributions f_U and f_V (see [2.8.5]).

Thus, we have found an example in which two dependent normal random variables have zero correlation. However, this does not contradict [2.8.10]; although both $\mathbf{U}$ and $\mathbf{V}$ are normal, they are not *jointly* normal. Independence and zero correlation are equivalent only when the random variables involved are jointly normal.

2.8.15 Conditional Distributions. Let $\vec{\mathbf{X}}$ be a normal random vector of the form $(\vec{\mathbf{X}}_1, \vec{\mathbf{X}}_2)$ where $\vec{\mathbf{X}}_1$ and $\vec{\mathbf{X}}_2$ are of dimension m_1 and m_2 respectively. The mean of $\vec{\mathbf{X}}$ is given by $\vec{\mu} = (\vec{\mu}_1, \vec{\mu}_2)$ where $\vec{\mu}_1$ and $\vec{\mu}_2$ are the means of $\vec{\mathbf{X}}_1$ and $\vec{\mathbf{X}}_2$ respectively. The covariance matrix of $\vec{\mathbf{X}}$ is given by

$$\Sigma = \begin{pmatrix} \Sigma_{11} & \Sigma_{12} \\ \Sigma_{21} & \Sigma_{22} \end{pmatrix}$$

where Σ_{11} is the covariance matrix of $\vec{\mathbf{X}}_1$, Σ_{22} is the covariance matrix of $\vec{\mathbf{X}}_2$, Σ_{12}, which is called the *cross-covariance matrix*, is the $m_1 \times m_2$ matrix of covariances of elements of $\vec{\mathbf{X}}_1$ with $\vec{\mathbf{X}}_2$, and $\Sigma_{21} = \Sigma_{12}^{\mathrm{T}}$. The marginal distribution of $\vec{\mathbf{X}}_1$ is $\mathcal{N}(\vec{\mu}_1, \Sigma_{11})$, and that of $\vec{\mathbf{X}}_2$ has a similar form. From (2.31), we obtain that the conditional distribution of $\vec{\mathbf{X}}_1$, given $\vec{\mathbf{X}}_2 = \vec{\mathbf{x}}_2$, is also multivariate normal with conditional mean

$$\vec{\mu_{1|2}} = \vec{\mu}_1 + \Sigma_{12}\Sigma_{22}^{-1}(\vec{\mathbf{x}}_2 - \vec{\mu}_2) \tag{2.36}$$

and conditional covariance matrix

$$\Sigma_{1|2} = \Sigma_{11} - \Sigma_{12}\Sigma_{22}^{-1}\Sigma_{12}^{\mathrm{T}}. \tag{2.37}$$

The proof may be found in [281] or [147].

It is interesting to note that the conditional mean of $\vec{\mathbf{X}}_1$ depends upon $\vec{\mathbf{X}}_2$ when $\Sigma_{12} \neq 0$ (i.e., when $\vec{\mathbf{X}}_1$ and $\vec{\mathbf{X}}_2$ are dependent upon each other).

2.8.16 More on Conditional Distributions—Optional.[15] The conditional mean (2.36) can be thought of as a *linear specification* of the value of $\vec{\mathbf{X}}_1$ that is based on $\vec{\mathbf{X}}_2$. The specification is linear because the conditional mean is a vector of linear combinations of the elements of $\vec{\mathbf{X}}_2$.

The specification skill can be determined by computing the cross-covariances between the vector of specification errors and random vectors $\vec{\mathbf{X}}_1$ and $\vec{\mathbf{X}}_2$. Useful specifications will have errors with near zero covariance with $\vec{\mathbf{X}}_1$ and exactly zero covariance with $\vec{\mathbf{X}}_2$. The interpretation in the first case is that the specification accounts for almost all of the variation in $\vec{\mathbf{X}}_1$ because the errors have little variation in common with $\vec{\mathbf{X}}_1$. In the second case, the interpretation is that all the information in

[15] Interested readers may want to return to this subsection after reading Chapter 8. This material is presented here because it flows naturally from the previous subsection.

$\vec{\mathbf{X}}_2$ about $\vec{\mathbf{X}}_1$ that is obtainable by linear methods is contained in the specification.

The specification errors are given by

$$\vec{\mathbf{X}}_{1|2} = \vec{\mathbf{X}}_1 - \left(\vec{\mu}_1 + \Sigma_{12}\Sigma_{22}^{-1}(\vec{\mathbf{X}}_2 - \vec{\mu}_2)\right).$$

The covariance between the specification errors and $\vec{\mathbf{X}}_2$ is zero as required:

$$\begin{aligned}&\mathrm{Cov}\left(\vec{\mathbf{X}}_2, \vec{\mathbf{X}}_{1|2}\right)\\ &= \mathcal{E}\left(\vec{\mathbf{X}}_2(\vec{\mathbf{X}}_1 - (\vec{\mu}_1 + \Sigma_{12}\Sigma_{22}^{-1}(\vec{\mathbf{X}}_2 - \vec{\mu}_2)))^{\mathrm{T}}\right)\\ &= 0.\end{aligned}$$

The covariance between the specification errors and $\vec{\mathbf{X}}_1$ is

$$\begin{aligned}&\mathrm{Cov}\left(\vec{\mathbf{X}}_1, \vec{\mathbf{X}}_{1|2}\right)\\ &= \mathcal{E}\left(\vec{\mathbf{X}}_1(\vec{\mathbf{X}}_1 - (\vec{\mu}_1 + \Sigma_{12}\Sigma_{22}^{-1}(\vec{\mathbf{X}}_2 - \vec{\mu}_2)))^{\mathrm{T}}\right)\\ &= \Sigma_{11|2}.\end{aligned}$$

To determine from this whether the specification is skilful, one could compute the proportion of the total variance of $\vec{\mathbf{X}}_1$ that is explained by $\vec{\mathbf{X}}_2$. The total variance of a random vector is simply the sum of the variances of the individual random variables that make up the vector. This is equal to the sum of the diagonal elements (or *trace*) of the covariance matrix. Thus, a measure of the skill, s, is

$$\begin{aligned}s &= 1 - \mathrm{tr}(\Sigma_{11|2})/\mathrm{tr}(\Sigma_{11})\\ &= \frac{\mathrm{tr}(\Sigma_{12}\Sigma_{22}^{-1}\Sigma_{12}^{\mathrm{T}})}{\mathrm{tr}(\Sigma_{11})}.\end{aligned}$$

Note that $s = 0$ if $\Sigma_{12} = 0$ and that $s = 1$ when $\Sigma_{12} = \Sigma_{11}^{1/2}\Sigma_{22}^{1/2}$. In fact, s cannot be greater than 1.

2.9 Extreme Value Distributions

2.9.0 Introduction. Many practical problems encountered in climatology and hydrology require us to make inferences about the *extremes* of a probability distribution. For example, the designs of emergency measures in river valleys, floodways, hydro-electric reservoirs, and bridges are all constrained in one way or another by the largest stream flow which is expected to occur over the life of the plan, floodway, reservoir, bridge, etc. The design of storm sewage systems, roads, and other structures in a city is constrained by the largest precipitation event anticipated during a fixed design period (typically 50 or 100 years). The design of electrical distribution systems, buildings and other free-standing structures must account for the extremes of wind pressure loading which are likely to occur during the life of the structures. The roofs of houses built in high latitudes must be able to withstand extreme snow loads. Insurers who underwrite the financial risk associated with these natural risks must have good estimates of the size and impact of extreme events in order to set their premiums at a profitable level.

Extreme value analysis is the branch of probability and statistics that is used to make inferences about the size and frequency of extreme events. The basic paradigm used varies with application but generally has the following components:

- data gathering;
- identification of a suitable family of probability distributions, one of which is to be used to represent the distribution of the observed extremes;
- estimation of the parameters of the selected model;
- estimation of *return values* for periods of fixed length. Return values are thresholds which are exceeded, on average, once per return period.

We will discuss each of these items briefly in the following subsections.

2.9.1 Data Gathering. Typically, the objects of study in extreme value analysis are collections of annual maxima of parameters that are observed daily, such as temperature, precipitation, wind speed and stream flow. Thus, observations are required on two time scales.

- Observations are taken on short time scales over a fixed time interval to obtain a single extreme value. For example, they might consist of daily precipitation accumulations for a year. The maximum of the 365 observations is retained as the extreme daily precipitation accumulation for the year, while the rest of the observations serve only to determine the extreme value.

  It is important to understand that the extreme is a realization of a random variable, namely the Nth order statistic (see[2.6.9]) of a sample of size N. The extreme value in a subsequent sample of equal size is another realization of the same random variable.

- This process is repeated over several time intervals in order to obtain the object of extreme value analysis: a sample consisting only of each interval's extreme value. In the previous example, if the daily precipitation accumulation is observed over a period of 50 years, then the sample of extreme values to be analysed is also of size 50, since each year yields one maximum.

Extreme value analysis requires some sort of assumption about the stationarity and ergodicity of the climate system, since only one realization of the past climate is available from climate archives. The implicit working assumption in most extreme value analyses is that the sample of n extremes are realizations of n independent and identically distributed random variables (we will discuss suitable distributions for extreme values shortly). Sometimes, though, it is clear that the climate system violates this assumption on certain time scales. For example, during an El Niño, the statistical characteristics of precipitation change on time scales of less than a season.

The following are examples of the context in which extreme value analyses are conducted.

Structural engineers designing a transmission tower may require knowledge about the extremes of the five-minute mean wind speed. They would extract daily, monthly or annual maxima of five-minute mean wind speed for a particular location from climatological archives for a nearby observing station.

Civil engineers designing a floodway around a city might require knowledge about the extremes of 24-hour precipitation, and will therefore extract daily, monthly and annual maxima of 24-hour precipitation from climatological archives.

A frequently encountered difficulty with archived precipitation data is that the archives generally contain the accumulation for a fixed 24-hour period (usually beginning at 00 UTC) as opposed to moving window 24-hour accumulations. This is of concern because often the critical quantity is not, for example, the maximum amount of rain that falls in a 24-hour time scale that consistently begins at 00 UTC (i.e., a fixed 24-hour window), but the maximum amount of rain that falls in a 24-hour period starting at any time of the day (i.e., a moving 24-hour window). Therefore a nuance of the analysis of extreme precipitation is that the fixed window accumulations must be multiplied by an empirically derived constant to ensure that the extremes of a sample of fixed window accumulations match those of a corresponding sample of moving window accumulations. Bruce [69] describes how the correction factor is estimated (see also Watt [416], p.76, and Hershfield and Wilson [176]).[16]

2.9.2 Model Identification. In extreme value analysis, the behaviour of the sample of extremes is almost always represented by a *parametric* model, a probability distribution selected for its ability to indicate the characteristics of the extreme values reasonably well.[17] Asymptotic arguments can be used to select the extreme value distribution if something is known about the distribution of the random variable observed on short time scales; an alternative approach is to use the extreme values themselves to identify a suitable model. Both methods will be briefly discussed here.

2.9.3 Model Identification: The Asymptotic Approach. Asymptotic arguments are often an important part of selecting an extreme value distribution. Under fairly general conditions it can be shown that, in samples of size n, the distribution of the extreme values converges, as $n \to \infty$, to one of three models: the *Gumbel* (or *Pearson type I*, or EV-I) distribution, the *Pearson type II* (or EV-II) distribution, and the *Pearson type III* (or EV-III) distribution.[18]

The rate of convergence is largely determined by the upper (sometimes lower) tail of the distribution of the short time scale variable (e.g., daily precipitation) that generates the extremes.[19] If the

[16] The correction factor used to convert fixed window 24-hour precipitation accumulations to moving window accumulations in Canada is 1.13 [69]. This factor will vary with location depending upon how and when precipitation is produced. The factor also depends upon the accumulation period.

[17] See Section 4.2 for a discussion of the difference between parametric and non-parametric statistics.

[18] In the classical treatment of extreme value analysis (see Gumbel [149]) it is necessary to assume that the extremes come from samples that can be represented by independent and identically distributed random variables. Leadbetter et al. [246] show that the independence assumption can be substantially relaxed. The same asymptotic results obtained in the classical setting are obtainable when the extremes are those of samples taken from a *weakly stationary*, ergodic time series (see [10.2.1]).

[19] When we speak of the 'convergence' of a sequence of random variables, say $\mathbf{Y}_i, i = 1, 2, \ldots$, to another random random variable $\mathbf{Z}$ we mean either *convergence in distribution* or *convergence in mean square*. We say $\mathbf{Y}_i$ converges to $\mathbf{Z}$ in distribution, and write $\mathbf{Y}_i \stackrel{d}{\to} \mathbf{Z}$ if $P(|\mathbf{Y}_i - \mathbf{Z}| > \epsilon) \to 0$ as $i \to \infty$ for every $\epsilon > 0$. We say $\mathbf{Y}_i$ converges to $\mathbf{Z}$ in mean square, and write $\mathbf{Y}_i \stackrel{ms}{\to} \mathbf{Z}$ if $\mathcal{E}\big((\mathbf{Y}_i - \mathbf{Z})^2\big) \to 0$ as $i \to \infty$. Convergence in mean square usually implies convergence in distribution.

distribution of the extreme values converges, to say the Gumbel distribution, then we say that the short time scale variable lies in *the domain of attraction* of the Gumbel distribution.

The EV-I distribution will be described briefly below. Descriptions of the EV-II and III distributions can be found in Gumbel [149] or Leadbetter et al. [246].

Both the exponential distribution and the normal distribution lie in the domain of attraction of the EV-I distribution. However, the distribution of the largest of a sample of n independent and identically distributed exponential random variables is closer to the EV-I distribution than the distribution of the largest of a sample of n independent and identically distributed normal random variables. Thus, Cook [89] argues that it is better to do extreme value analyses on wind pressure (which is proportional to wind speed squared) than on wind speed because the former has a distribution that is closer to exponential, and therefore closer to EV-I. Zwiers [439] makes use of this argument in his analysis of extreme wind speeds at several Canadian observing stations.

2.9.4 Model Identification: Using the Data. Unfortunately, the asymptotic EV distributions do not always fit the observed extremes well. This can occur for a variety of reasons, not the least of which is the cyclo-stationarity of the climate data under the best of conditions. Consider, for example, the daily precipitation accumulation. While the annual maximum daily precipitation accumulation is formally the maximum of 365 observations, the effect of the annual cycle may be such that only a small number of observations have any chance at all of attaining the status of annual maximum.

At Vancouver (British Columbia, Canada), for example (see Figure 1.7), the annual maximum is usually generated during winter when there is strong on-shore flow from the south-west. It is apparent from Figure 1.7 that only about 60 days of the year have the potential to generate the annual maximum at Vancouver. On the other hand, the annual maximum can occur with approximately equal likelihood on any day of the year on Sable Island (see Figure 1.7), located on the east coast of Canada.

Because the asymptotic distribution is not always obtained, other distributions such as the Generalized Extreme Value (GEV), Weibull, Pareto, and Wakeby distributions are also used in extreme value analysis.

Another frequently used method of model identification relies on estimates of the skewness and kurtosis of the extreme value distribution that are computed from the sample of extremes. The (skewness, kurtosis) pair is plotted on a chart of kurtosis as a function of skewness for various families of distributions, often called *Pearson curves* (see Elderton and Johnson [112]). A model is identified by the proximity of the plotted point to a distribution's curve (there is a unique Pearson curve for every distribution).

Model identification with Pearson curves is difficult and often not completely successful because the skewness and kurtosis estimates are subject to a great deal of sampling variability. Estimates often end up occupying a point in the (skewness, kurtosis) plane that can not be visited by adjusting parameters within known families of distributions.

A better alternative is to use L-moments in combination with L-moment versions of the Pearson curves [183] for model identification. L-moments are subject to less sampling variation, that is, they are more robust than conventional moments and discriminate better between competing models (see Hosking [183]).

2.9.5 Model Fitting. Once a model (i.e., extreme value distribution) has been selected the next step in the analysis is to 'fit' the chosen extreme value distribution to the sample of extremes. *Fitting* means estimating the unknown parameters of the chosen extreme value distribution.

Several methods may be used for parameter estimation. These methods may produce quite different results with the small sample of extremes that is usually available, even though their results become asymptotically identical as the number of observed extremes becomes large. The theoretical suitability of one method over another in repeated sampling has often been the subject of literal debate. However, these discussions are of little use when economic decisions strongly depend on the accuracy of the results, as is often true.[20]

The methods most often used for fitting (see Section 5.2) are

- the method of moments,
- the method of maximum likelihood,

[20]For example, estimates of the largest precipitation event expected to occur during a 25-year period will strongly influence the diameter, slope and other parameters of a city's storm sewer system. An estimate of the 25-year event that is too large will result in the building of a sewer system that has larger capacity, and therefore higher cost, than necessary.

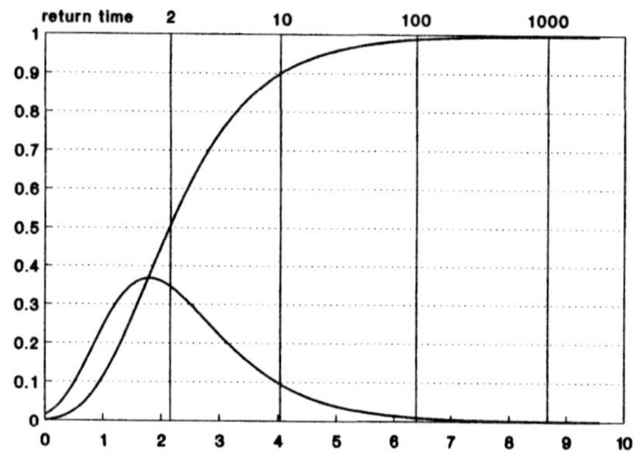

Figure 2.11: *An example of the probability density function* $f_Y(y)$ *and cumulative distribution function* $F_Y(y)$ *of an extreme value distribution (for annual maxima). This particular distribution is the Gumbel with parameters* $u = \ln 6$ *and* $\lambda = 1$ *(see [2.9.8]). The location of 2, 5, 10, 100, and 1000 year return values are indicated by the vertical bars. Note that the two-year return value corresponds to* $F_Y(y) = 1/2$*, the 10-year value corresponds to* $F_Y(y) = 9/10$*, etc. Also note that the distribution has a much wider right hand tail than that of distributions we have become familiar with.*

- the method of probability weighted moments, and
- the method of L-moments.

Optimality considerations in repeated sampling generally lead to the use of the method of maximum likelihood (see [5.3.8]). On the other hand, the method of L-moments is more *robust*. This method is less affected by occasional observational errors or data transcription errors (such as a misplaced decimal point) than other fitting methods. The method of probability weighted moments (see Hosking, Wallis, and Wood [184]) is closely related to the method of L-moments. The ordinary method of moments is also frequently used because of simplicity and convention considerations.[21]

2.9.6 Return Values. The last step in an extreme value analysis is usually to compute 'return values' for preset periods (e.g., 10, 50, 100 years). These values are thresholds that, according to the fitted model, will be exceeded on average once every return period.

Return values are simply the upper quantiles of the fitted extreme value distribution. For example, suppose that the random variable $\mathbf{Y}$ represents an annual extreme maximum and that $\mathbf{Y}$ has probability density function $f_Y(y)$. The 10-year return value for $\mathbf{Y}$ is the value $Y_{(10)}$ such that

$$P\left(\mathbf{Y} > Y_{(10)}\right) = \int_{\mathbf{y}^{(10)}}^{\infty} f_Y(y)\,dy = 1/10.$$

In general, the T-year return value for the annual maximum, say $Y_{(T)}$, is the solution of

$$\int_{Y_{(T)}}^{\infty} f_Y(y)\,dy = 1/T.$$

That is, the T-year return values are simply points on the abscissa such that the area under the right hand tail of the density function is $1/T$. The concept is illustrated in Figure 2.11.

Return values for extreme minima are similarly computed using the tail areas under the left hand tail of a suitable extreme value distribution.

2.9.7 Example: Daily Maximum Temperature. As an example, consider the change in the extremes of the daily maximum temperature at 2 m height that might occur as a result of a doubling of the atmospheric CO_2 concentration (see Zwiers and Kharin [448]). Zwiers and Kharin showed that the annual extremes of temperature can be well represented by the EV-I distributions in both the $1\times CO_2$ and $2\times CO_2$ climates of the 'CCC GCMII' General Circulation Model.[22] Estimates of the 10-year return values derived from the 'control run' $1\times CO_2$ are displayed in Figure 2.12 (top). These values verify reasonably well in general terms. However, values at specific locations should not be compared directly with return values estimated from station data because climate simulations can not be considered reliable at length scales shorter than a few grid lengths.

Figure 2.12 (bottom) illustrates the change induced in the 10-year return value by a doubling of CO_2. The globally averaged increase is about 3.1 °C. The corresponding value for the increase in the 10-year return value of the daily minimum temperature is 5.0 °C, indicating that the shape of the temperature distribution might change substantially with increasing CO_2 concentrations.

[21] The ordinary method of moments is similar to the method of L-moments [2.6.7, 2.6.9]. Instead of matching population L-moments to estimated L-moments, ordinary population moments (mean, variance, skewness, and kurtosis) are matched with corresponding estimates.

[22] The Canadian Climate Centre GCMII (McFarlane et al. [270]). The CCC $2\times CO_2$ experiment is described by Boer, McFarlane, and Lazare [52].

In addition to the overall warming caused by the change in the radiative balance of the model climate under CO_2 doubling, there are also a variety of interesting physical effects that contribute to the spatial structure of the changes in the return values. For example, daily maximum temperatures are no longer constrained by the effect of melting ice at the location of the $1 \times CO_2$ sea ice margin. Also, the soil dries and the albedo of the land surface increases over the Northern Hemisphere land masses, leading to a substantial increase in the extremes of modelled daily maximum temperature.

2.9.8 Gumbel Distribution. To conclude this section on extreme values, we provide a simple derivation of the Gumbel or EV-I distribution [149]. Let $\mathbf{X}_1, \ldots, \mathbf{X}_n$ represent n independent, identically distributed, exponential random variables observed on the short time scale. These random variables might, for example, represent a sample of n wind pressure measurements which, as we have noted previously, have a distribution that is close to exponential.[23] The distribution function for any one of these random variables is

$$F_X(x; \lambda) = P(\mathbf{X} < \mathbf{x}) = 1 - e^{-x/\lambda}.$$

Let $\mathbf{Y}$ be the maximum of $\{\mathbf{X}_1, \ldots, \mathbf{X}_n\}$. Then $\mathbf{Y} < \mathbf{y}$ if and only if $\mathbf{X}_i < \mathbf{y}$ for each $i = 1, \ldots, n$. Using independence, we obtain that

$$\begin{aligned} P(\mathbf{Y} < y) &= \prod_{i=1}^{n} P(\mathbf{X}_i < y) \\ &= F_X(y; \lambda)^n \\ &= (1 - e^{-y/\lambda})^n \\ &\approx \exp\{-ne^{-y/\lambda}\}. \end{aligned}$$

The quality of the approximation improves with increasing n, that is, if each extreme is obtained from a larger sample of observations collected on the short time scale. After a bit more manipulation, we see that, as n increases indefinitely, the distribution function of $\mathbf{Y}$ takes the form

$$F_Y(y; u, \lambda) = P(\mathbf{Y} < y) = \exp\{-e^{-(y-u)/\lambda}\}.$$

This is the distribution function of the *Gumbel* or *EV-I distribution*. Convergence to this distribution is achieved similarly for all distributions in the domain of attraction of the EV-I distribution. The essential element that controls convergence is simply the point at which the right hand tail of the distribution generating the individual observations begins to behave as the right hand tail of the exponential distribution. Distributions for which the maximum of a sample of size n converges to the Gumbel slowly exhibit exponential behaviour only for observations that are many standard deviations from the centre of the distribution.

The EV-I distribution is a two-parameter distribution with a location parameter u and a scale parameter λ. The density function of an EV-I random variable $\mathbf{Y}$ is given by

$$f_Y(y; u, \lambda) = \exp\{-[(y-u)/\lambda + e^{-(y-u)/\lambda}]\}.$$

The mean and the variance of $\mathbf{Y}$ are given by

$$\begin{aligned} \mu_Y &= u + \gamma\lambda \\ \mathrm{Var}(Y) &= \lambda^2\pi^2/6, \end{aligned}$$

where γ is Euler's constant. The L-moments are:

$$\begin{aligned} \lambda^{(1)} &= u + \gamma\lambda \\ \lambda^{(2)} &= \lambda \ln 2 \\ \gamma_1^L &= 0.1699 \\ \gamma_2^L &= 0.1504. \end{aligned}$$

As noted above, return values are obtained by inverting the distribution. For example, if the Gumbel distribution were fitted to annual maxima, then the T-year return value, say $Y_{(T)}$, is the solution of

$$\begin{aligned} 1/T &= P\big(\mathbf{Y} > Y_{(T)}\big) \\ &= 1 - F_Y(y_{(T)}; u, \lambda) \\ &= 1 - \exp\{-e^{-(Y_{(T)}-u)/\lambda}\}. \end{aligned} \tag{2.38}$$

Solving (2.38) yields

$$y_{(T)} = u - \lambda \ln\big(-\ln(1 - 1/T)\big).$$

2.9.9 Other Approaches. Another approach to extreme value analysis that we have not discussed is the so-called *peaks-over-threshold* approach. In contrast to analysing annual (or other period) maxima, the peaks-over-threshold approach sets a high threshold and then analyses all exceedances above that threshold. The appeal of this approach is that it may be possible to extract additional information about the extremes of a climate parameter by setting the threshold in such a way that more than one threshold crossing is observed per year. To apply the approach,

[23] We use exponential random variables in our derivation for mathematical convenience. We could use any collection of n independent and identically distributed random variables that have a distribution belonging to the 'domain of attraction' of the EV-I distribution and obtain identical results by using more sophisticated analytical techniques.

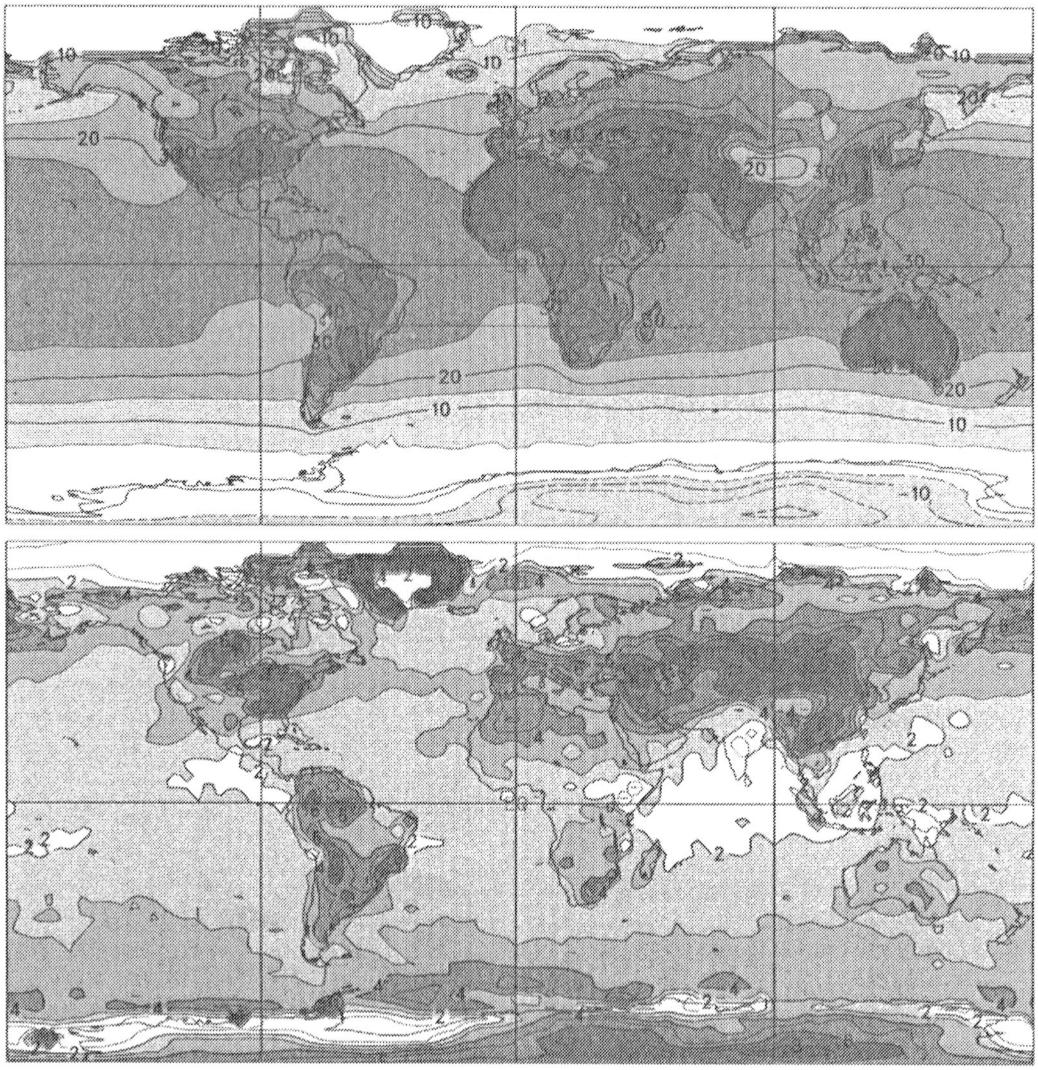

Figure 2.12: *Estimated 10-year return values of daily maximum temperature at 2 m height estimated from the output of model experiments with a General Circulation Model coupled to a mixed-layer ocean model and a sea-ice model. Units: °C. From [270].*
Top: Return values estimated from a 20-year control run.
Bottom: Change of the return values a derived from above and the output of a 10-year model experiment with doubled atmospheric CO_2 concentrations.

one must be careful about the placement of the threshold and also account for the effects of cyclo-stationarity. Ross [333] illustrates the peaks-over-threshold approach with an application to the analysis of wind speed data. Zwiers and Ross [450] describe an approach that provides more reliable estimates of return values and has been applied to precipitation data at a variety of Canadian locations. This method uses monthly extremes and standard extreme value distributions, while also accounting for cyclo-stationarity. References cited by [439], [333], [450], and [448] will provide the interested reader with entry points into the immense collection of extreme value literature.

3 Distributions of Climate Variables

3.0.1 The Components of the Climate System. The climate system is composed of all processes that directly or indirectly control the atmospheric environment of humans and ecosystems. The main components of the system are the hydro- and thermodynamic states of the atmosphere and the ocean. Sea ice affects the exchange of heat, momentum and fresh water between oceans and atmosphere. On longer time scales, the shelf ice and the land ice become relevant since these components are able to store and release large quantities of fresh water. The atmosphere, ocean, and land surface are interconnected by means of the hydrological cycle on a number of time scales. Precipitation falls on the land where it affects land surface properties such as albedo and heat capacity. Some of this precipitation evaporates into the atmosphere, and some flows to the ocean as runoff. Fresh water flux into the ocean by means of precipitation and runoff, and out of the ocean through evaporation, affects ocean variability, which in turn feeds back on atmospheric variability.

Changes in the chemical composition of the atmosphere also impact the climate system because the concentration of carbon dioxide, ozone, or other radiatively active gases affects the radiative balance of the atmosphere. These concentrations are controlled by the state of the atmosphere and the ocean, as well as the biospheric and anthropogenic sinks and sources of these chemicals. Clearly the components of the climate system cannot be defined exhaustively, since it is not a closed system in a strict sense.

In the following sections we describe several atmospheric, oceanic, cryospheric (ice and snow) and hydrologic variables.[1] The choice of the variables is subjective and biased towards those that are most easily observed. Shea et al. [348] list addresses of atmospheric and oceanographic data centres in the US, and give an overview of easily accessible atmospheric and oceanographic data sets at the National Center for Atmospheric Research (NCAR).

[1] Biospheric variables are beyond the scope of this text.

3.0.2 The Law of Large Numbers and Climate Time Scales. The instantaneous values or daily accumulations of many climate variables have skewed distributions. On the other hand, averages or accumulations taken over long periods tend to be 'near normal' because of the Central Limit Theorem [2.7.5].

3.0.3 Length and Time Scales. Two terms often used in climate research are *time scale* and *length scale*. Although these terms are vaguely defined, thinking about the temporal and spatial resolution needed to describe a phenomenon accurately will help us to select suitable variables for study and to find suitable approximations of the governing equations (see Pedlosky's book [310] on geophysical fluid dynamics).

A length scale is a characteristic length that is representative of the spatial variations relevant to the process under investigation. For instance, if this process is an extratropical storm, then its length scale may be taken as its diameter or as the distance between a pressure minimum and the closest pressure maximum. The length scale of a wind sea[2] may be the distance between a wave crest and a wave valley, or between two consecutive crests.

The term 'time scale' is defined similarly. Time scales are representative of the duration of the phenomenon of interest and the greater environment. For example, extratropical storms dissipate within a few days of cyclogenesis, so suitable time scales range from about an hour to perhaps two days. Convective storms (thunder storms), on the other hand, occur on much shorter spatial (up to tens of kilometres) and temporal scales (minutes to several hours). In both cases, the time scale gives an indication of the 'memory' of the process. A statistical measure of 'memory' is the *decorrelation time* described in Section 17.1. The decorrelation time for the mean sea-level pressure (SLP) is typically three to five days in

[2] The part of the ocean wave field that is in dynamical contact with the wind.

the extratropics, while that for convection is on the order of half a day.

We choose variables that describe the variation on the length and time scales of interest that are relevant to a problem. For example, to study extratropical cyclones, the divergence should be chosen rather than the velocity potential.[3] When we observe the process, we sample it at spatial and temporal increments that resolve the length and time scales.

An interesting feature of the climate system is that the length and times scales of climate variability are often related. Processes with long length scales have long time scales, and short time scales are associated with short length scales. This fact is illustrated for the atmosphere and the ocean in Figure 3.1. However, this rule is far from precise. Atmospheric tides are an example of a process with large spatial scales and short temporal scales.

3.1 Atmospheric Variables

3.1.1 Significant Variables. A myriad of variables can be used to monitor the physical state of the atmosphere; so understandably the following list of variables commonly used in climate research is not at all exhaustive.

Local climate is often monitored with *station data*: temperature (daily minimum and maximum, daily mean), precipitation (5 min, 10 min, hourly and daily amounts, monthly amount, number of wet days per month), pressure, humidity, cloudiness, sunshine, and wind (various time averaging intervals).

The large-scale climate is generally described with *gridded data*, such as: sea-level pressure, geopotential height, temperature, the vector wind, stream function and velocity potential, vorticity and divergence, relative humidity, and outgoing long-wave radiation. Some of these are based on observations (e.g., temperature) while others are derived quantities (e.g., vorticity).

The main problem with time series from station data is that the data are often not homogeneous; they exhibit trends or sudden jumps in the mean or variance that are caused by changes in the physical environment of the observing site, in the observing equipment, of the observing procedures and time, and of the responsible personnel (see Figure 1.9). More examples, for instance of the complicating effect of observation time on the daily temperature, the snow cover derived from satellite data, and the effect of lifting a precipitation gauge from the 1 m level to the 2 m level, are described in a review paper by Karl, Quayle, and Groisman [213].

Jones [201] discusses, in some detail, the problems in long time series of precipitation, temperature and other atmospheric data, and lists many relevant papers.

Gridded data have the advantage that they represent the full spatial distribution. However, in data sparse areas, the gridded value may be more representative of the forecast models and interpolation schemes that are used to do the *objective analysis*[4] than they are of the state of the climate system. Unfortunately, when used for diagnostic purposes, it is impossible to distinguish between observed and interpolated, or guessed, information. Difficulties also arise because most gridded data are a byproduct of numerical weather forecasting and therefore affected by changes in the forecast and analysis systems.[5] Such changes are made almost continually in an effort to improve forecast skill by incorporating the latest research and data sources and exploiting the latest computing hardware (see, e.g., Trenberth and Olsen [370], or Lambert [240] [241]).

Finally, we note in passing that climate model output is not generally affected by the kinds of problems described above, although it too can have its own idiosyncrasies (see, e.g., Zwiers [444]). However, simulations that are constrained by observations in some way can be affected.[6]

[3] The divergence is approximately the second derivative of the velocity potential and is sensitive to small scale features, as in extratropical storms. The velocity potential, on the other hand, displays planetary scale divergent, as in the large tropical overturning of the Hadley circulation.

[4] *Objective analysis* is used to initialize numerical weather forecasting models. Most weather forecasting centres re-initialize their forecasting models every six hours. Typically, the objective analysis system adjusts the latest six-hour forecast by comparing it with station, upper air, satellite, airline, and ship reports gathered during a six-hour window centred on the forecast time. The adjusted forecast becomes the initial condition for the next six-hour numerical forecast. Objective analysis systems are the source of most gridded data used in climate research. See Thiébaux and Pedder [362] or Daley [98] for comprehensive descriptions of objective analysis.

[5] *Re-analysis* projects (Kalnay et al. [210]) have done much to ameliorate this problem. These projects re-analysed archived observational data using a fixed analysis system. Note that re-analysis data are still affected by changes in, for example, the kind of observing systems used (e.g., many different satellite based sensors have been 'flown' for various lengths of time) or the distribution and number of surface stations.

[6] Examples include 'AMIP' (Atmospheric Model Intercomparison Project; see Gates [137]) and 'C20C' (Climate of the Twentieth Century; see Folland and Rowell [123]) simulations. Sea-surface temperature and sea-ice extent are prescribed from observations in both cases, and thus the models are forced with data that are affected by observing system changes.

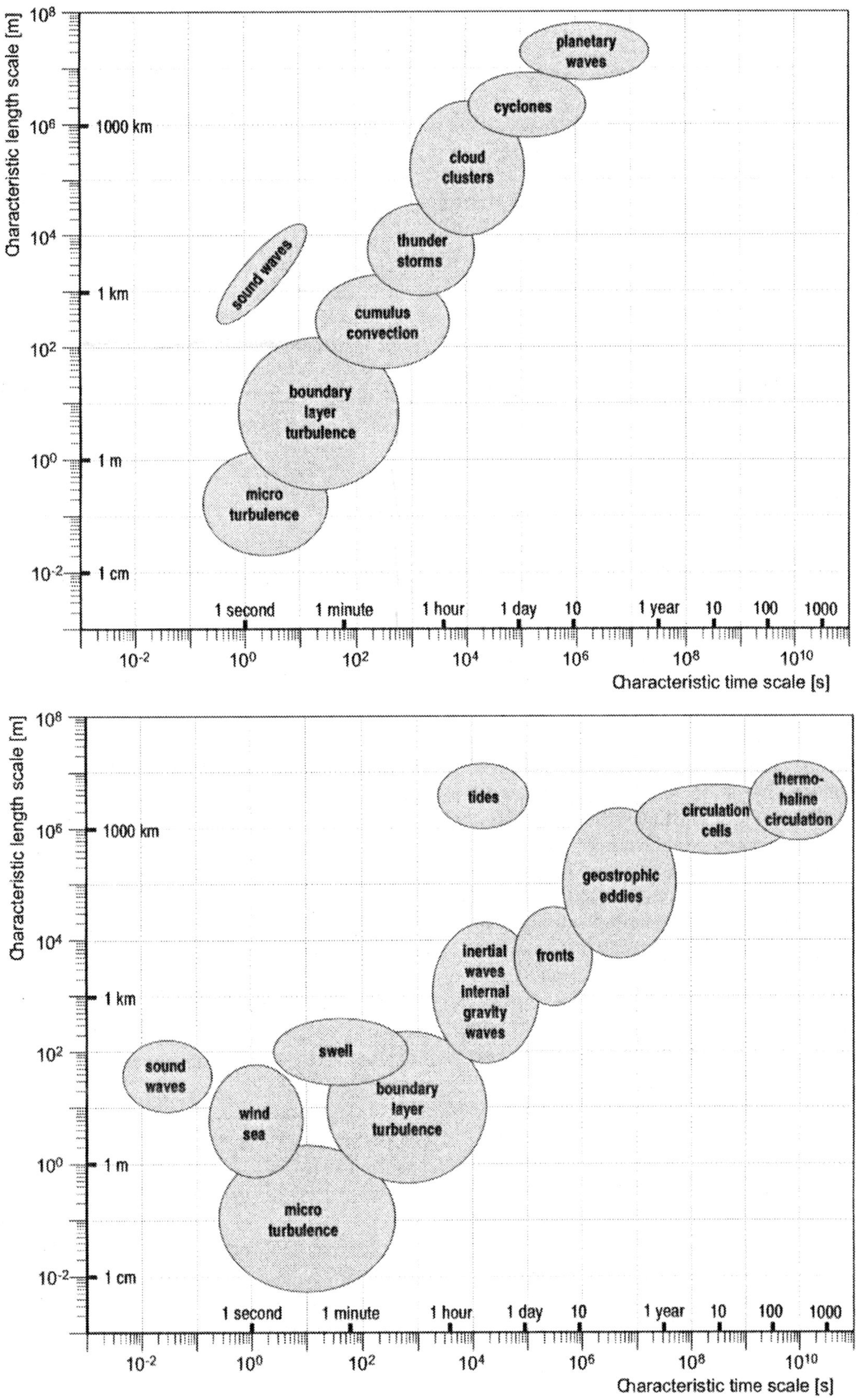

Figure 3.1: *Length and time scales in the atmosphere and ocean. After [390].*

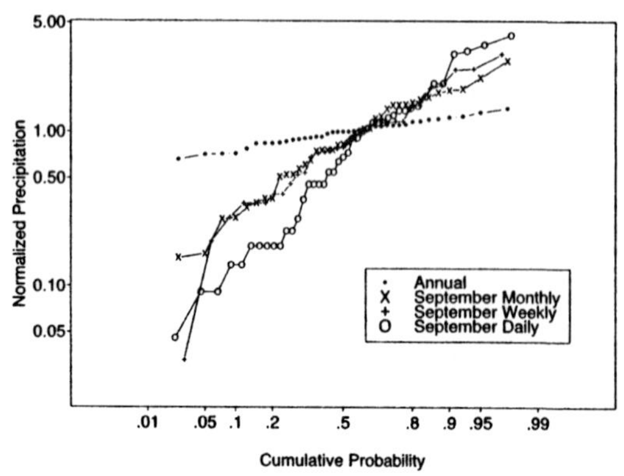

Figure 3.2: *Empirical distribution functions of the amount of precipitation, summed over a day, a week, a month, or a year, at West Glacier, Montana, USA. The amounts have been normalized by the respective means, and are plotted on a probability scale so that a normal distribution appears as a straight line. For further explanations see [3.1.3]. From Lettenmaier [252].*

3.1.2 Precipitation. Precipitation, in the form of rain or snow, is an extremely important climate variable: for the atmosphere, precipitation indicates the release of latent heat somewhere in the air column; for the ocean, precipitation represents a source of fresh water; on land, precipitation is the source of the hydrological cycle; for ecology, precipitation represents an important external controlling factor.

There are two different dynamical processes that yield precipitation. One is convection, which is the means by which the atmosphere deals with vertically unstable conditions. Thus, convection depends mostly on the local thermodynamic conditions. Convective rain is often connected with short durations and high rain rates. The other precipitation producing process is large-scale uplift of the air, which is associated with the large-scale circulation of the troposphere. Large-scale rain takes place over longer periods but is generally less intense than convective rain. Sansom and Thomson [338] and Bell and Suhasini [40] have proposed interesting approaches for the representation of rain-rate distributions, or the duration of rain-events, as a sum of two distributions: one representing the large-scale rain and the other the convective rain.

There are a number of relevant parameters that characterize the precipitation statistics at a location.

- The statistics of the *amount of precipitation* depend on the accumulation time, as demonstrated in Figure 3.2. The curves, which are empirical distribution functions of accumulated precipitation, are plotted so that a normal distribution appears as a straight line. For shorter accumulation times, such as days and weeks, the curves are markedly concave with medians (at probability 0.5) that are less than the mean (normalized precipitation = 1), indicating that these accumulations are not normally distributed. For the annual accumulation, the probability plot is a perfect straight line with coinciding mean and median. Thus, for long accumulation times the distribution is normal. Figure 3.2 is a practical demonstration of the Central Limit Theorem.

- The *number of rainy days per month* is often independent of the amount of precipitation.

- The time between any two rainfall events, or between two rainy days, is the *interarrival time*.

Lettenmaier [252] deals with the distribution aspects of precipitation and offers many references to relevant publications.

3.1.3 Probability Plots—a Diversion. Diagrams such as Figure 3.2 are called *probability plots*, a type of display we discuss in more detail here.

The diagram is a plot of the empirical distribution function, rotated so the possible outcomes y lie on the vertical axis, and the estimated cumulative probabilities $p(y) = \widehat{F}_Y(y)$ lie on the horizontal axis.[7] Alternatively, if we consider p the independent variable on the horizontal axis, then $y = \widehat{F}_Y^{-1}(p)$ is scaled by the vertical axis. For reasons outlined below, the variable p is re-scaled by $x = F_X^{-1}(p)$ with some chosen distribution function F_X. The horizontal axis is then plotted with a linear scale in x. The p-labels (which are given on a nonlinear scale) are retained. Thus, Figure 3.2 shows the function $x \to \widehat{F}_Y^{-1}[F_X(x)]$. If $\widehat{F}_Y = F_X$, the function is the identity and the graph is the straight line (x, x). The probability plot is therefore a handy visual tool that can be used to check whether the observed random variable $\mathbf{Y}$ has the postulated distribution F_X.

[7]The 'hat' notation, as in $\widehat{F}_Y(y)$, is used throughout this book to identify functions and parameters that are estimated.

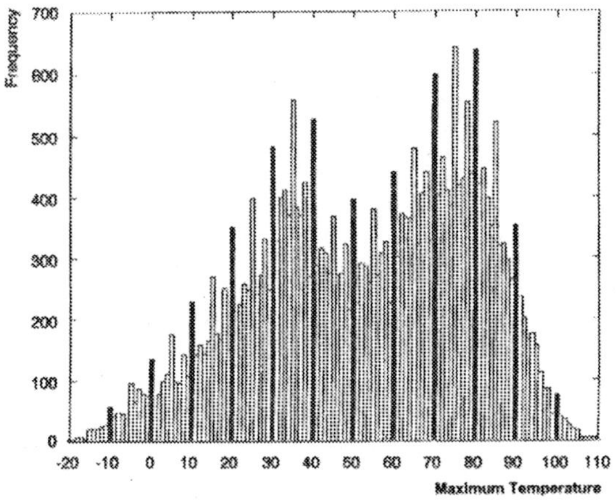

Figure 3.3: *Frequency distribution of daily maximum temperature in °F at Napoleon (North Dakota, USA) derived from daily observations from 1900 to 1986. From Nese [291].*

When the observed and postulated random variables both belong to a 'location-scale' family of distributions, such as the normal family, a straight line is also obtained when **Y** and **X** have different means and variances. In particular, if a random variable **X** has zero mean and unit variance such that $F_Y(y) = F_X(\frac{y-\mu}{\sigma})$, then

$$y = F_Y^{-1}(p) = \mu + \sigma F_X^{-1}(p) = \mu + \sigma x.$$

The line has the intercept μ at $x = 0$ and a slope of σ.

When we think **Y** has a normal distribution, the reference distribution F_X is the standard normal distribution. S-shaped probability plots indicate that the data come from a distribution with wider or narrower tails than the normal distribution. Probability plots with curvature all of one sign, as in Figure 3.2, indicate skewness. Other location scale families include the log-normal, exponential, Gumbel, and Weibull distributions.

3.1.4 Temperature. Generally, temperature is approximately normally distributed, particularly if averaged over a significant amount of time in the troposphere. However, daily values of near-surface temperature can have more complicated distributions.

The frequency distribution of daily maximum temperature at Napoleon (North Dakota) for 1900 to 1986 (Figure 3.3, Nese [291]) provides another interesting example.

- The distribution is skewed with a wide left hand tail. Cold temperature extremes apparently occur over a broad range (causing the long negative tail) whereas warm extremes are more tightly clustered.
- The distribution has two marked maxima at 35 °F and at 75 °F. This bimodality might be due to the interference of the annual cycle: summer and winter conditions are more stationary than the 'transient' spring and autumn seasons, so the two peaks may represent the summer and winter modes. The summer peak is taller than the winter peak because summer weather is less variable than winter weather. Also, the peak near the freezing point of 33 °F might reflect energy absorption by melting snow.
- There is a marked preference for temperatures ending with the digits 0 and 5. Nese [291] also found that the digits 2 and 8 were overrepresented. This is an example of psychology interfering with science.

Averages of daily mean air temperature[8] are also sometimes markedly non-normal, as is the case in Hamburg (Germany) in January and February. Weather in Hamburg is usually controlled by a westerly wind regime, which advects clouds and maritime air from the Atlantic Ocean. In this weather regime temperatures hover near the median. However, the westerly flow is blocked intermittently when a high pressure 'blocking' regime prevails. In this case, the temperature is primarily controlled by the local radiative balance. The absence of clouds and the frequent presence of snow cover cause the temperatures to drop significantly due to radiative cooling. Thus, daily temperatures are sometimes very low, but they usually vary moderately about the mean of −0.4 °C. Strong positive temperature deviations from the mean occur rarely. This behaviour is reflected in the empirical distribution function of the winter mean anomalies (Figure 3.4): the minimum two-month mean temperature in the 1901–80 record is −8.2 °C, the maximum is +3.2 °C, while the median is +0.2 °C.

The distribution function in Figure 3.4 is not well approximated by a normal distribution. It is markedly skewed (with an estimated skewness of −1.3 and an estimated third L-moment of −2.86). The degree of convergence towards the normal

[8] 'Daily means' are supposed to represent diurnal averages. In practice, they are obtained by averaging a small number of regularly spaced observations taken over the 24 hours of each day, or, more often, as the mean of the daily maximum and minimum temperatures.

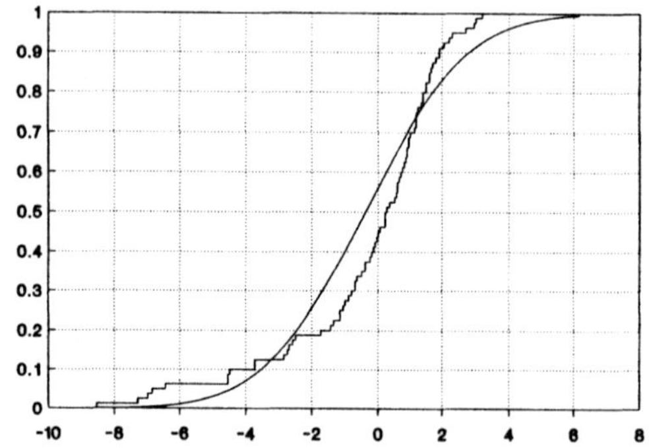

Figure 3.4: *Empirical distribution function of the January/February mean temperatures in Hamburg (Germany). The step function is the empirical distribution function derived from the years 1901–80, and the smooth line is the distribution function of the normal distribution fitted to the data.*

distribution that is predicted by the Central Limit Theorem [2.7.5], with $\mathbf{X}_k$ = *daily temperature at day k in one JF season* and $n = 60$), is still weak: more than 60 (non-independent) daily observations are required for the convergence to become significant.

3.1.5 Wind. Wind statistics are required for various climate-related purposes. For, example, wind statistics are needed to force a variety of ocean models that simulate regional or global circulation, the surface seas state, storm surges, or the flux of energy or momentum through the air–sea interface. Hasse [164] lists the 'surface wind' products that are frequently used in ocean-related studies: They include data sets derived from:

- the surface wind simulated by numerical weather forecasting models at their lowest computational levels;
- local wind measurement at 10 m height, (representative only for the immediate neighbourhood of the measurement);
- surface air-pressure maps (used to compute 'geostrophic winds');
- satellite microwave backscatter signals, which are transformed into wind estimates with empirically derived algorithms;
- a large fleet of *voluntary observing ships* (VOS), that provide either anemometer readings or visual wind speed estimates on the *Beaufort* scale.[9] These VOS are mostly regular merchant vessels, and the weather is reported by the crew. The COADS[10] data set is composed of all archived VOS reports.

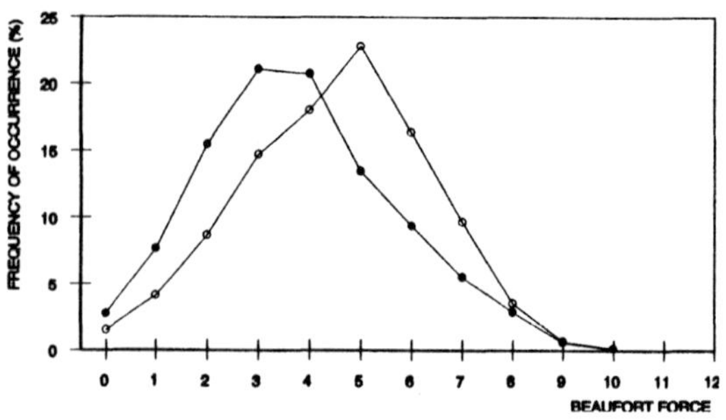

Figure 3.5: *Frequency distribution of wind estimates on the Beaufort scale, derived from voluntary ship reports in the English Channel after 1949. The solid dots are derived from 24 442 reports from ships* without *an anemometer, whereas the open dots stem from 981 observations made when an instrument was available. All 24 442 + 981 reports are visual assessments of the sea state. Peterson and Hasse [313].*

All these products have their problems. An example is shown in Figure 3.5, which discriminates between VOS reports based on visual estimates of observers who do or do not have an anemometer at their disposal (Peterson and Hasse [313]). Significantly higher Beaufort winds are reported when an instrument is available (Peterson and Hasse [313] rejected the null hypothesis of equal distributions at less than the 1% significance level).

It is generally believed that the Beaufort-estimates are more homogeneous than observations from shipborne instruments since the latter are affected by factors such as the height of the instrument, the motion of the ship and deformation of the flow as it passes over the ship's superstructure.

Peterson and Hasse [313] offer the following tentative explanation for the discrepancy in Figure 3.5:

> The reason for different Beaufort estimates of ships with and without an anemometer is not really known. A possible explanation is that anemometer

[9] The Beaufort scale is a visual measure of wind speed that is based on the state. It is reported as a force on a scale of 1–12.

[10] The *Comprehensive Ocean Atmosphere Data Set* (COADS) is an important collection of marine observations such as sea-surface temperature, air temperature, cloudiness and wind speed (see Woodruff et al. [425]). All available ship reports have been pooled together in 2° × 2° longitude × latitude bins. The data have not been corrected and they suffer from temporal inhomogeneities due to changes in instrumentation and observational practices.

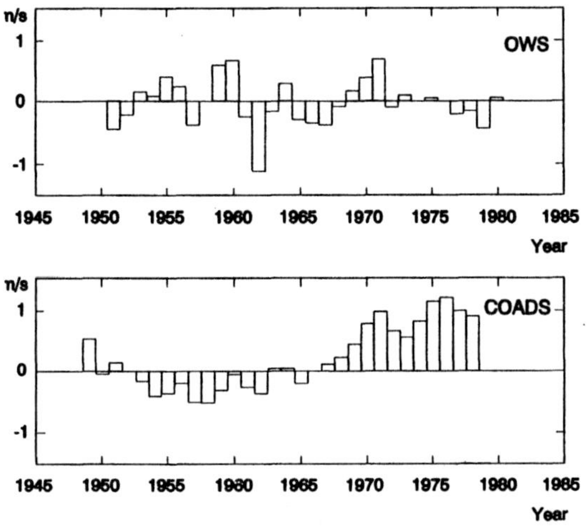

Figure 3.6: *Time series of anomalies of annual mean wind speeds in the North Pacific, as derived from two independent data sets. Upper panel: data from Ocean Weather Station P; lower panel: data from voluntary observing ships in the neighbourhood of the weather ship. From Isemer, unpublished.*

outputs ... often use simple dial displays showing instantaneous winds. On viewing the instruments one tends to be more impressed by the gusts instead of the lulls. The knowledge of gust speed may then inadvertently influence the estimation of the Beaufort force.

Another interesting case was investigated by Isemer (unpublished), who derived annual mean wind speeds for the area surrounding *Ocean Weather Station P* in the North Pacific using two different data sources. One data set contained the observations made at the weather ship, where the measurement was done in a fairly homogeneous manner (same position, same height, trained observers). The time series derived from this data, displayed in the upper panel of Figure 3.6, is stationary. The second data set consists of instrumental and visual reports from merchant vessels located in the neighbourhood of the Ocean Weather Station (OWS). The corresponding time series (lower panel of Figure 3.6) exhibits an upward trend of more than 1 m/s in 20 years. This trend is spurious, and is probably due to various factors such as the increasing height of ships.

A convenient way to present the distribution of the vector wind is by means of *wind roses.* The wind roses in Figure 3.7 describe the surface wind in the northern North Sea in January. At low wind speeds the vector wind distribution is almost *isotropic* (i.e., independent of direction).

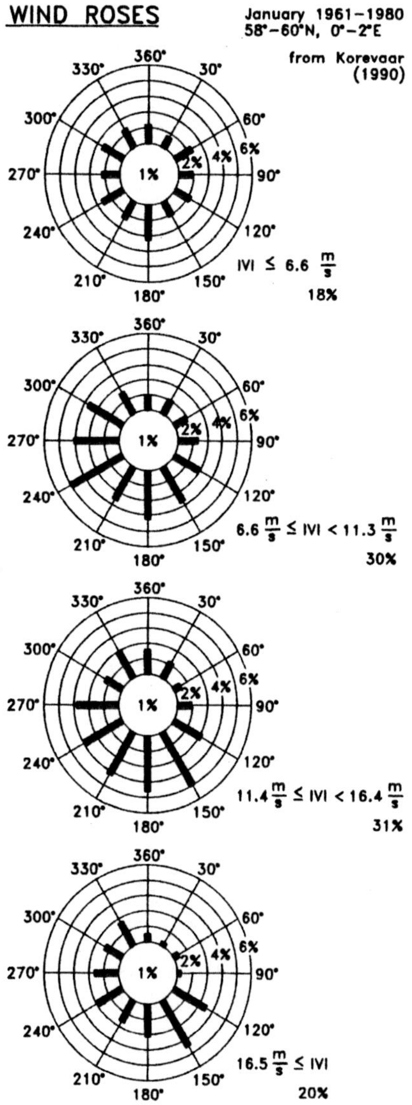

Figure 3.7: *Wind roses for the northern North Sea between 58–60° N latitude and 0–2° E longitude in January stratified by the wind speed. From Korevaar [230]. Reprinted by permission of Kluwer Academic Publishers.*

However, for stronger winds ($|\vec{v}| > 6.6$ m/s or $|\vec{v}| > 4$ on the Beaufort scale) the vector wind distribution is decidedly non-isotropic. In this case the most frequent (or *modal*) wind direction veers from southwest to southeast with increasing wind velocity.

3.1.6 Extratropical 500 hPa Height: Band-pass Filtered Variance. In a classical study, Blackmon [47] analysed the day-to-day winter (and summer) variability of 10 years of gridded

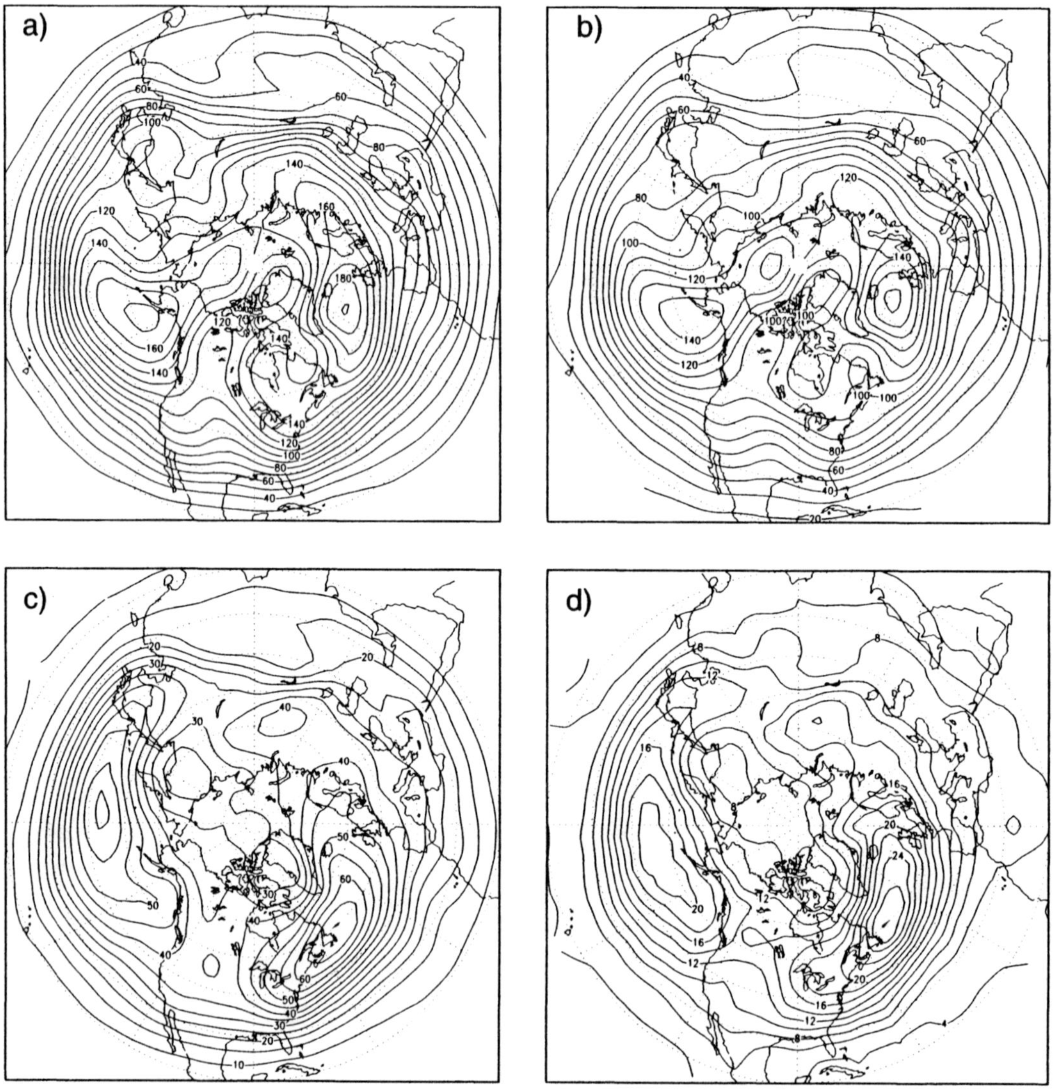

Figure 3.8: *Standard deviation of time-filtered 500 hPa geopotential height (in m) during winter. Courtesy V. Kharin.*
a) Variability of the original time series $\mathbf{X}_t$ *(contour interval: 10 m),*
b) 'Slow' variability of $\mathbf{X}_t^s$ *(longer than about 10 days; contour interval: 10 m),*
c) 'Baroclinic' variability of $\mathbf{X}_t^b$ *(between 2.5 and 6 days; contour interval: 5 m),*
d) 'Fast' variability of $\mathbf{X}_t^f$ *(between one and two days; contour interval: 2 m).*

daily Northern Hemisphere 500 hPa geopotential heights. After subtracting the annual cycle at each grid point (by calculating the first four harmonics of the annual cycle) he calculated first the overall standard deviation, and then separated the data into three components, each of which represents a different time scale. We repeated these calculations using 1979–87 analysis from the European Centre for Medium Range Weather Forecasts (ECMWF).

The overall standard deviation shown in Figure 3.8a is largest at about 50° N and smallest in the subtropics. Two centres of action, with standard deviations of about 175 m, are located over the Northeast Pacific, the Northeast Atlantic and North-Central Asia.

In order to determine how much of the variability depicted in Figure 3.8a comes from low-frequency[11] variability (10 days and longer)

[11] The term 'low-frequency' is not defined in absolute terms. Instead the meaning depends on the context. In the present variations on time scales of 10 and more days are 'slow' compared to the baroclinic and fast components. Slow variations are defined differently in [3.1.7].

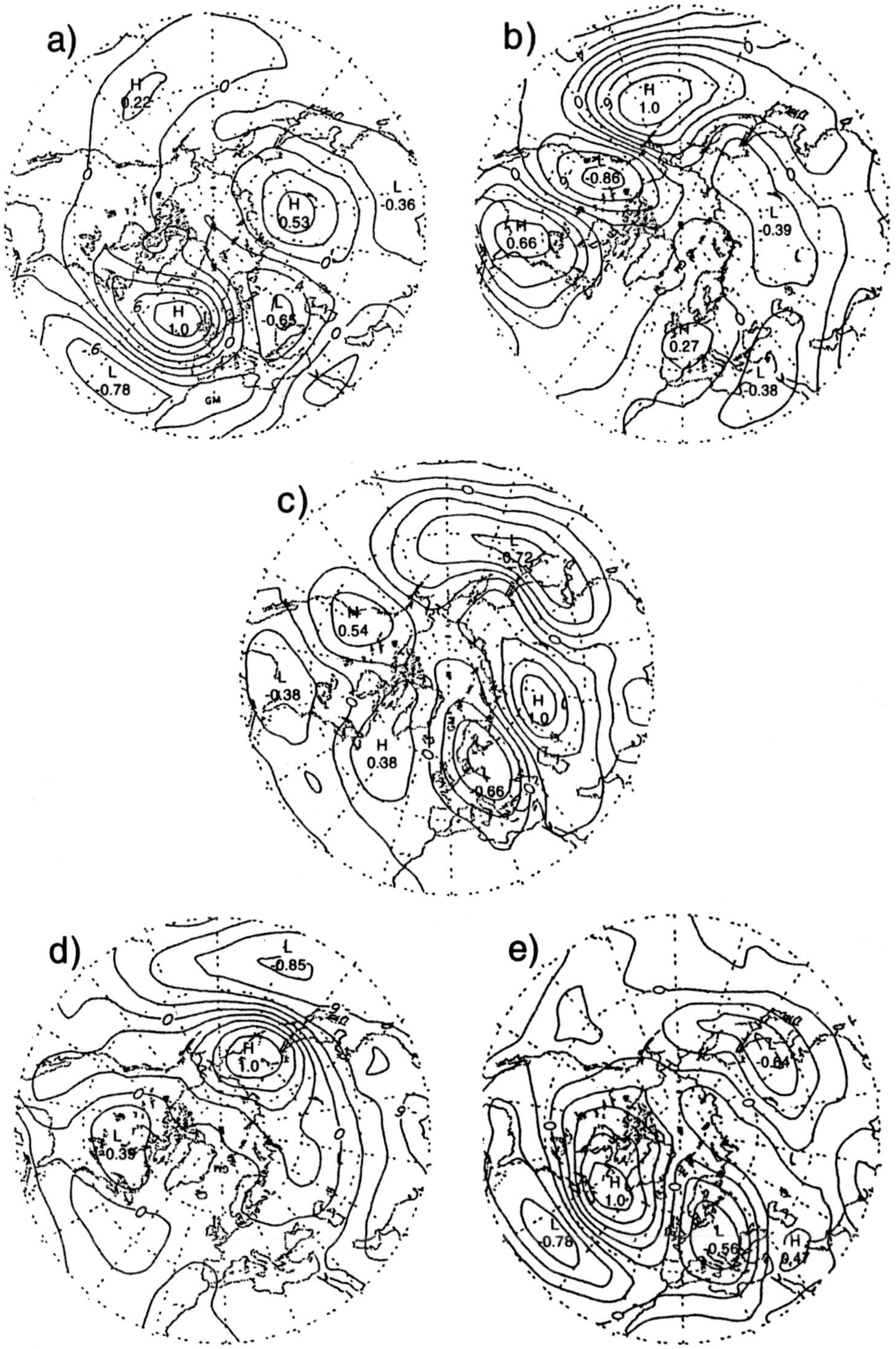

Figure 3.9: *The teleconnection patterns that represent a substantial part of the month-to-month variability of 500 hPa height during winter (Wallace and Gutzler [409]). Teleconnection patterns display correlations between a base point (given by a 1.0 in the maps) and all other points in the Northern Hemisphere extratropics. A maximum is marked by an 'H' and a minimum by an 'L'. The patterns are named (a)* Eastern Atlantic Pattern, *(b)* Pacific/North American Pattern, *(c)* Eurasian Pattern, *(d)* West Pacific Pattern, *and (e)* West Atlantic Pattern. *See also [13.5.5].*

or from baroclinic activity on a shorter time scale, the data are *time filtered*. That is, the original time series, say $\mathbf{X}_t$, is split up into $\mathbf{X}_t = \mathbf{X}_t^f + \mathbf{X}_t^b + \mathbf{X}_t^s$, with $\mathbf{X}^f, \mathbf{X}^b$, and $\mathbf{X}^s$ representing *fast*, *baroclinic*, and *slow* components. The 'fast' component varies on time scales between one and two days, the 'baroclinic' time scale covers the medium range between 2.5 and 6 days, and the 'slow' component contains all variability longer than about 10 days. The technical details of the separation are explained in Section 17.5.

The three components vary independently, as a result of the time scale separation. Thus the variance of the complete time series is distributed to the variances of the three components: $\mathrm{Var}(\mathbf{X}_t) \approx \mathrm{Var}(\mathbf{X}_t^f) + \mathrm{Var}(\mathbf{X}_t^b) + \mathrm{Var}(\mathbf{X}_t^s)$.

The spatial distributions of the standard deviations of the three components are shown in Figure 3.8. The largest contribution to the overall standard deviation in Figure 3.8a originates from the low-frequency variations (Figure 3.8b). In the North Pacific, the standard deviation due to low frequency variations is 145 m compared with 175 m in the unfiltered data, that is, about 70% of the total variance stems from the slow variations. An important contributor to this pool of variability is the process of 'blocking,' which often occurs on the west coast of continents and over eastern oceans. Another characterization of the low frequency variability 500 hPa height field is given in [3.1.7].

The baroclinic component (Figure 3.8c) is considerably less energetic than the slow processes with maximum standard deviations of about 70 m (representing about another 25% of the total variance). These variations may be traced back to the baroclinic waves, that is, extratropical storms. The regions of large variability in Figure 3.8c over the western and central part of the Pacific and Atlantic Ocean are called 'stormtracks.' (The same stormtracks are displayed by the shaded regions in Figure 3.10; there is a large circumpolar stormtrack in the Southern Hemisphere.)

The 'fast' component has small standard deviations, with maxima of the order of only 20 m (which is about 1–2% of the total variance; Figure 3.8d). Blackmon [47] argued that most of this variance is due to 'a spurious high-frequency component in the final analyses map.' However, the similarity of the structure of Figure 3.8d to Figure 3.8c, and the comparable results from the EOF analyses, suggest that at least some of the 'fast' variability is natural.

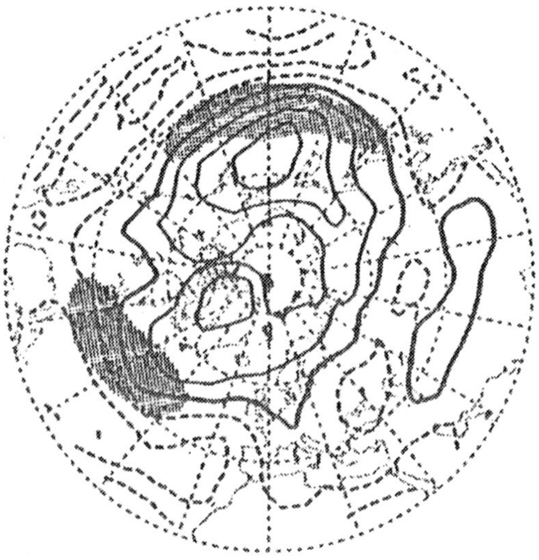

Figure 3.10: *Distribution of the skewness γ_1 of the low-pass filtered daily Northern Hemisphere 500 hPa geopotential height. All variability on time scales longer than six days was retained. Positive contours are dashed. The stormtracks are indicated by the stippling (compare with Figure 3.8b). From Nakamura and Wallace [287].*

3.1.7 Extratropical 500 hPa Height: Characteristic Low-Frequency Patterns. Wallace and Gutzler [409] examined the month-to-month variability of the 500 hPa height field during winter in the Northern Hemisphere extratropics. They calculated *teleconnection patterns*, that is, spatial distributions of the correlations at a base point with the height field everywhere else. The concept of teleconnection patterns and their identification is explained in some detail in Section 17.4. Wallace and Gutzler's study is further discussed in [17.4.2] and [17.4.3].

Five reproducible[12] patterns were identified (Figure 3.9). They were named after the regions they affect: *Eastern Atlantic (EA) Pattern*, *Pacific/North American (PNA) Pattern*, *Eurasian (EU) Pattern*, *West Pacific (WP) Pattern* and *West Atlantic (WA) Pattern*. Each pattern represents a fixed structure whose amplitude and sign are controlled by a time varying coefficient. The coefficient time series can be determined by projecting the monthly mean height fields onto the patterns. The coefficients for the five patterns are more or less statistically independent; that is, variations in one *mode* are not related to those in another. In space, the patterns have a wave-like appearance

[12] *Reproducible* means that essentially the same result is obtained when another independent data set is analysed with the same technique.

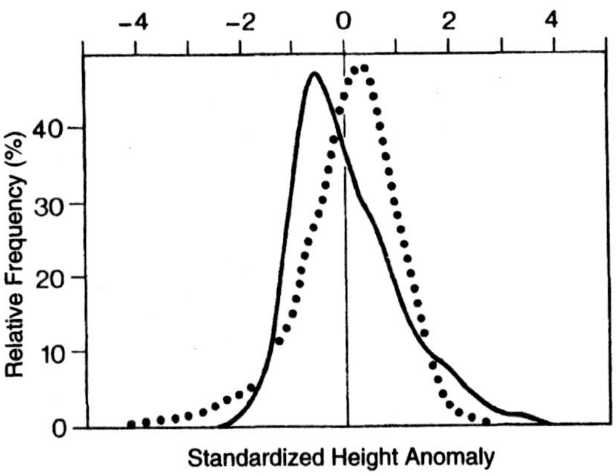

Figure 3.11: *Frequency distributions of the low-pass filtered daily 500 hPa geopotential height at a location poleward (solid line) and equatorward (dotted) of the Pacific stormtrack. From Nakamura and Wallace [287].*

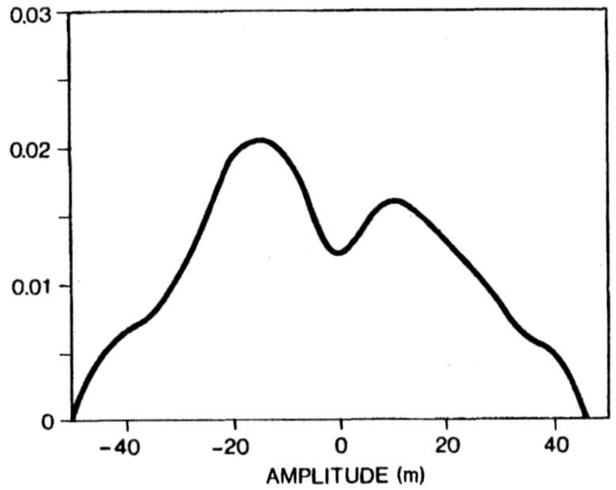

Figure 3.12: *Estimated probability density distribution f_Z of the 'wave-amplitude indicator'* **Z**. *Note the bimodality. From Hansen and Sutera [162].*

with sequences of two or more nodes of opposite signs indicating that a substantial part of the month-to-month variability in the extratropical midtropospheric height field could originate from standing oscillatory modes. The nodes in the patterns are sometimes named *centres of action.*

Barnston and Livezey [27] extended Wallace and Gutzler's study by analysing data from all seasons. By using *rotated EOFs* (Section 13.5) they were able to reproduce Wallace and Gutzler's results and increase the number of characteristic midtropospheric patterns.

3.1.8 Extratropical 500 hPa Height: Skewness. Nakamura and Wallace [287] analysed 30 years of daily anomalies (i.e., deviations from the mean annual cycle) of Northern Hemisphere 500 hPa height and derived frequency distributions for all grid points and for two different time scales. The 'high-frequency' variations, ranging from two to six days, are generally normally distributed; the 'low-frequency' variations, beyond six days, are not normal (Figure 3.10). North of the Pacific and North Atlantic 'stormtracks,' the skewness γ_1 (see [2.6.7]) is negative, but equatorward of the stormtracks the skewness is positive (Figure 3.11). Nakamura and Wallace suggest that the dynamical reason for this pattern is

> ...that quantities such as temperature and potential vorticity exhibit large meridional contrasts across the ...stormtracks, as if there were two different 'air masses' facing each other. It is conceivable that a piece of one air mass could become cut off to form an isolated vortex within the other air mass. ...Once cut off from the family of streamlines that trace out the westerly circumpolar flow, the anomaly is freed from the effect of advection and can remain stationary for a long time relative to the time scale of baroclinic waves. Such cut off flow configurations are identified with blocking anticyclones in high latitudes and cut off lows in lower latitudes. ...[We] suspect that the primary contributions to the observed skewness come from these anomalous circulations that occur relatively infrequently.

3.1.9 Bimodality of the Planetary-Scale Circulation. Even though the nonlinearity of the dynamics of the planetary-scale[13] atmospheric circulation was well known, atmospheric scientists only began to discuss the possibility of two or more stable states in the late 1970s. If such multiple stable states exist and are well separated, it should be possible to find bi- or multimodal distributions in the observed data.

Hansen and Sutera [162] identified a bimodal distribution in a variable characterizing the energy of the planetary-scale waves in the Northern Hemisphere winter (DJF). Daily amplitudes for the zonal wavenumbers $k = 2$ to 4 for 500 hPa height were averaged for midlatitudes. These were used to derive a 'wave-amplitude indicator' **Z** by subtracting the annual cycle and filtering out all variability on time scales shorter than five days. The probability density function f_Z was

[13] Often, the spatial scales of the atmospheric circulation are discussed in terms of *wavenumber* k in a zonal Fourier decomposition along latitudes. Long waves, for instance $k = 1, \ldots, 4$, represent *planetary* scales while shorter waves, $k \geq 5$, are called *baroclinic* scales.

estimated by applying the so-called *maximum penalty technique* to 16 winters of daily data. The resulting f_Z has two maxima separated by a minor minimum near zero (Figure 3.12).[14]

Hansen and Sutera conclude from the bimodality of their distribution that the nonlinear dynamics of the atmospheric general circulation yield two stable regimes. The 'zonal regime,' with $\mathbf{Z} < 0$, exhibits small amplitudes of the planetary waves. The 'wavy regime,' with $\mathbf{Z} > 0$, is characterized by enhanced planetary-scale zonal disturbances. The mean 500 hPa height field for the 62% of all days when the system is in the 'zonal' regime is indeed almost zonal (Figure 3.13a). The mean field for the 'wavy' regime, derived from the remaining 38% of all days, exhibits marked zonal asymmetries (Figure 3.13b).[15]

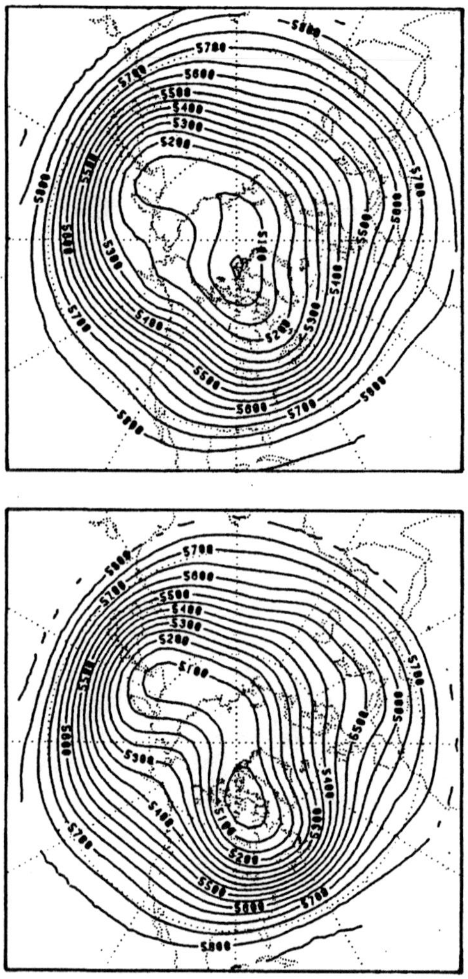

Figure 3.13: *Averages of 500 hPa Northern Hemisphere height fields in winter (DJF). Contour interval: 100 m. From [162].*
a) The 'zonal' regime: $\mathbf{Z} < 0$.
b) The 'wavy' regime: $\mathbf{Z} > 0$.

3.1.10 Biological Proxy Data. The effects of variation in, for example, temperature or precipitation, are often reflected in biological variables such as the width of tree rings (a detailed discussion of this type of data is offered by Briffa [65]), or the arrival of migrating birds. Records of plant flowering dates or similar events constitute *phenological data.*

An unusual example is the flowering date of wild snow drops in the rural town of Leck (northern Germany), which are plotted against the coefficient of the first EOF (Empirical Orthogonal Function; see Chapter 13) of Northwest European winter mean temperature in Figure 3.14. The flowering date varies between Julian day 16 (16 January) and 80 (21 March). The two variables, flowering date and the first EOF coefficient, are well correlated as indicated by the regression line in Figure 3.14. Thus, the flowering date of wild snow drops at Leck is a proxy of regional scale winter mean temperature.

There are other proxy data, some of them derived from historical archives, such as the yield of wine harvests or reports from courts and monasteries (e.g., Zhang and Crowley [436]), and others from tree rings (Briffa [65]), geological data such as sediments (e.g., van Andel [378]),

[14] There is an interesting story associated with Hansen and Sutera's bimodality:

Hansen and Sutera [162] conducted a 'Monte Carlo' experiment to evaluate the likelihood of fitting a bimodal *sample distribution* to the data when the *true distribution* is unimodal with the maximum penalty technique. It was erroneously concluded that the probability of such a misfit is small. The error in this conclusion was not at all obvious. Nitsche, Wallace, and Kooperberg [295] did a careful step-by-step re-analysis of the original data to find that the Monte Carlo experiments were inconsistent with the analysis of the observational data.

This is a very educational example, demonstrating a frequent pitfall of statistical analysis. Basic inconsistencies are sometimes hidden in a seemingly unimportant detail when sophisticated techniques, like the maximum penalty technique, are used. The error was found only because J. Wallace suspected that the finding could not be true.

Nitsche et al. reproduced the sample distribution shown in Figure 3.12, but showed that about 150 years of daily data would be required to exclude, with sufficient certainty, the possibility that the underlying distribution is unimodal. Essentially, then, reasonable *estimates* were made but the *test of the null hypothesis 'The sample distribution originates from a unimodal distribution'* was performed incorrectly. However, even without having rejected the null hypothesis, the possible implications incorporated in Figure 3.12 indicate that there *could* be two different stable atmospheric states.

[15] Compare with the monthly mean fields shown in Figure 1.1. January 1971 belongs to the zonal regime whereas January 1981 belongs to the wavy regime.

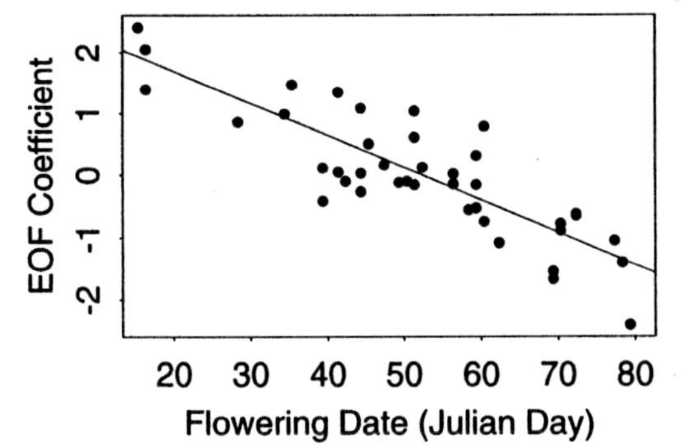

Figure 3.14: *The flowering date of wild snow drops at Leck (northern Germany) versus the coefficient of the first EOF of regional winter mean near-surface temperature. Courtesy K. Maak.*

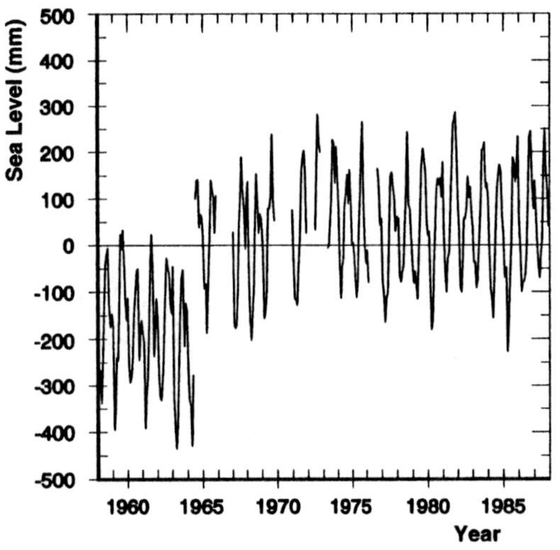

Figure 3.15: *Monthly mean of sea-level observations at Nezugaseki (Japan, 38.3° N, 139.4° E). The abrupt change in July 1964 is caused by an earthquake. From Cubasch et al. [97].*

or chemical composition data such as the oxygen isotope ratio of air trapped in the deep ice of Antarctica and Greenland.

Researchers 'translate' these proxy data into standard climate variables by means of (often nonlinear) *regression* relationships (see Chapter 8) that have been developed by relating contemporaneous proxy and instrumental observations as in Figure 3.14. The proxy data are often available for much longer periods than the instrumental record, so the proxy data together with the regression relationship may be used to estimate past climate states. Note that any such reconstruction is subject to uncertainty, as demonstrated by the scatter in Figure 3.14.

There are also many other limitations. The proxy data are typically regional in character. For example, trees that produce useful tree ring data tend to live in extreme climate zones where their growth is easily affected by relatively small changes in environmental conditions. Also, while proxy data typically yield one value per year, that value is often not representative of annual mean.

3.1.11 Missing Data. Observed data sets are often incomplete. Records of both station and analysed data contain numerous, and sometimes extended, gaps.[16] Information is often lost through data handling and management problems (e.g., paper records are lost, electronic transfers fail, tapes are inadvertently overwritten, computers or data assimilation systems crash). Most losses of this type are not related to the processes the data describe, but they are sometimes related to the calendar; 'procedural' losses seem to occur more frequently on weekends and holidays. Other types of losses, however, are the result of 'censoring' by the process that is being observed. For example, radiosonde data is often missing in the upper troposphere because strong winds have carried the balloon out of the tracking equipment's range, anemometer readings may be missing because strong winds have toppled the tower or generated large ocean waves that interfere with buoys, and so on.

[16]Trenberth and Olson's [371] description of missing National Meteorological Center analyses is typical.

3.2 Some Other Climate Variables

3.2.1 Ocean Temperatures. In oceanography, the *sea-surface temperature* (SST), and the *sub-surface temperature* are often regarded differently, even though they are closely related dynamically. One reason for this is that the sea-surface is the interface through which the atmosphere and ocean exchange energy and fresh water, whereas sub-surface temperature is internal to the ocean. The other reason is that SST is easily observed from ships as well as from satellites so that useful estimates of the mean SST, as well as its variability in the last 100 years, can be derived. Sub-surface temperature observation, using hydrographic sections or buoys, is difficult and expensive. Therefore, the data on sub-surface temperature, as well as all other sub-surface variables, is sparse, and little is known about the variability below the surface of the ocean.

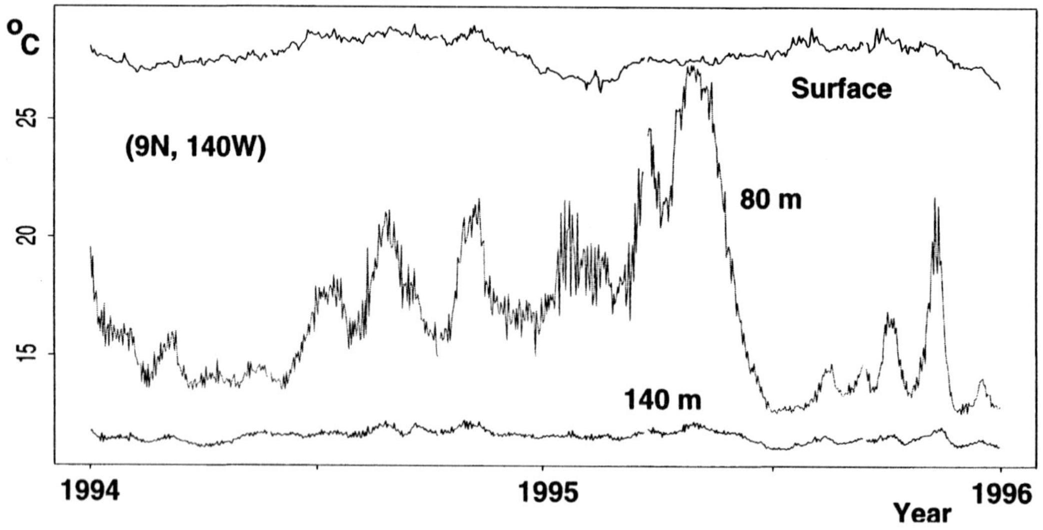

Figure 3.16: *Time series taken from daily 1994–95 observations of the ocean temperature at 9° N, 140° W in the Equatorial Pacific at the surface, at 80 m depth, and at 140 m depth.*

Historical SST data are compiled primarily from VOS reports (see [3.1.5] and footnote 10). The observations are scattered irregularly in both space and time. Coverage is heavy along the main shipping routes and non-existent in areas without shipping. There are systematic inhomogeneities in the observations that are caused by changes in instrumentation and operating procedures. For example, before 1945, SST was generally measured by hauling a bucket of water onto deck and taking its temperature with a mercury thermometer. These buckets were often designed differently for different countries, and some were insulated while others were not. After 1945, SSTs were generally obtained by measuring the temperature of the sea water used to cool the ship's engine ('engine intake temperature'). The temperature readings were also affected by the size and speed of the ships. The homogenization of SST is an art that requires not only detailed analysis of historical observational log books but also laboratory experiments in the wind tunnel and careful statistical correction schemes (see Folland and Parker [122] and Jones [201]).

3.2.2 Sea Level. The elevation of the ocean's surface relative to some benchmark is fairly easy to measure. However, the quantity that is measured reflects not only the real sea level but also the movement of the land-based observational platform relative to the geoid. Such movement can be caused by large-scale lifting and sinking of the Earth's crust associated with the process of equilibration after the retreat of the Ice Age glaciers. Earthquakes are another factor (Figure 3.15), causing abrupt changes of 10 cm or more in the reported sea level. More problems with the 'sea level' data are discussed by Emery and Aubrey [113], and Wyrtki [428]. An interesting case study, on the reports of sea level in the port of Shanghai (China), is given by Chen [81], who discusses the impact of various nuisance influences, such as changes in the discharge of rivers, ground subsidence due to ground water extraction, and the 'Cultural Revolution.'

3.2.3 Ocean Temperature: An Example. A buoy placed at 9° N, 140° W was used to monitor the near-surface atmospheric conditions as well as temperature at various levels in the ocean for several years. Time series of the temperature at the surface, at 80 m, and at 140 m are shown in Figure 3.16. The sea-surface temperature exhibits a marked annual cycle. Small variations with similar negative and positive anomalies occur on time scales shorter than one year. These *intra-seasonal* variations are almost normally distributed. The same holds for the temperature at 140 m and below (not shown), where small anomalies prevail.

At 80 m, however, the temperature variability exhibits features similar to the variability of rainfall. Minimum temperatures of 14 °C prevail most of the time but are overridden by large

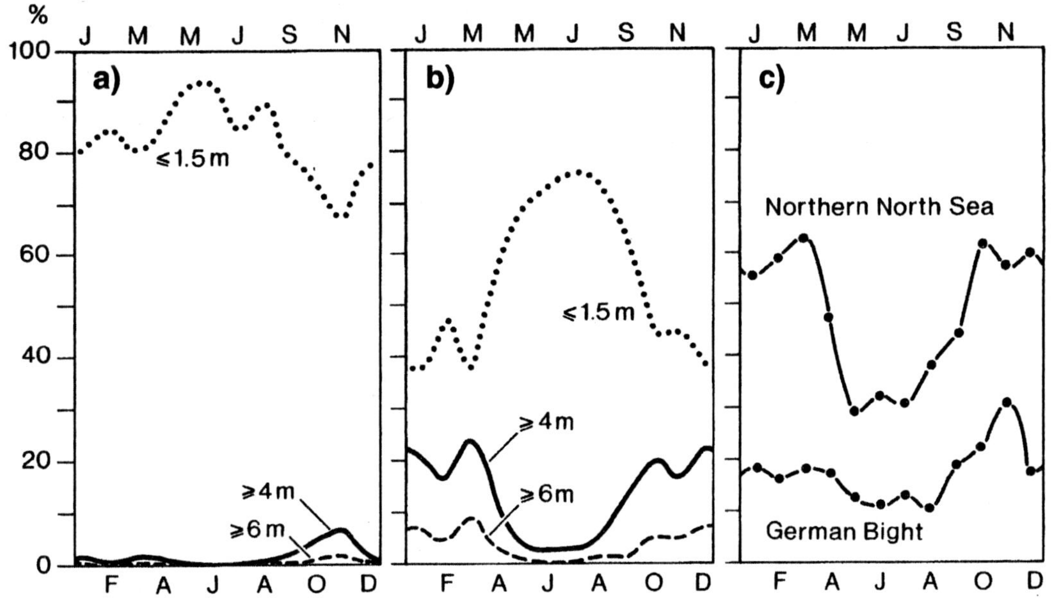

Figure 3.17: *a) Annual cycle of the frequency of exceedance for wave height in the German Bight. The curve labelled '≥ 4 m' displays the frequency of observing wave heights of 4 m or more.*
b) As above, except for the northern North Sea.
c) Annual cycle of the frequency of exceedance for wave periods of six seconds or longer in the German Bight and northern North Sea.
From Korevaar [230]. With permission of Kluwer Academic Press.

positive anomalies of up to 8 °C during shorter periods. Negative anomalies, on the other hand, are of similar magnitude to those at 140 m. Thus, the 80 m temperature, in contrast to both the surface temperature and the 140 m temperature, is strongly skewed.

The explanation lies in the vertical stratification of the ocean. The upper layer of the ocean is well mixed, because of continuous flow of mechanical energy from the atmosphere into the ocean, so that temperature and salinity are almost constant. Sometimes, when more mixing energy is available, the mixed layer is deepened, so water that is usually below the mixed layer has the same temperature as the surface. This deepening of the mixed layer is reflected by Figure 3.16.

3.2.4 Significant Wave Height and Mean Frequency. The waves on the sea surface modify the mechanical properties (roughness) of this surface and thus partly control the exchange of momentum and energy between the ocean and the atmosphere. The *wave height* and the *wave period* [230] are two variables that describe the state of the wave field, and are part of standard ship reports. Figure 3.17 displays the annual cycles of wave height and wave period for two areas in the North Sea. This is done by plotting the mean frequency of the waves that are lower than 1.5 m, or higher than 4 m or 6 m. The frequency of waves that have a period of more than 6 s is also given.

Wave heights in winter in the northern North Sea are less than 1.5 m 40% of the time and greater than 4 m 20% of the time. The waves are much lower in the German Bight where 80% of the waves are lower than 1.5 m. Most waves in the German Bight throughout the year and in the northern North Sea in summer have high frequency ($\leq (6\text{ s})^{-1}$), and 60% of the waves in the northern North Sea in winter have periods longer than 6 s.

3.2.5 Sea-ice Variables. Relevant variables describing the sea ice include the ice thickness, the thickness of the snow layer on top of the sea ice, the ice concentration (the percentage of area covered by sea ice), and the age of the ice. Further variables are the *freeboard* (the height by which the sea ice rises beyond the ocean surface[17]) and the *ice draft* (the downward extension of the ice-plus-snow column). Most of these variables are

[17] Freeboard might be negative if there is substantial snow cover.

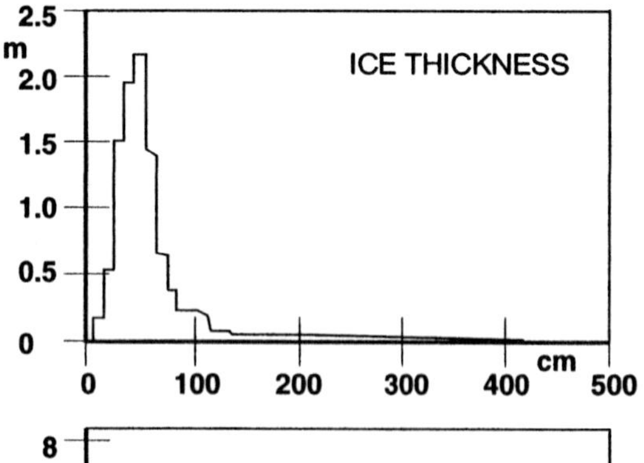

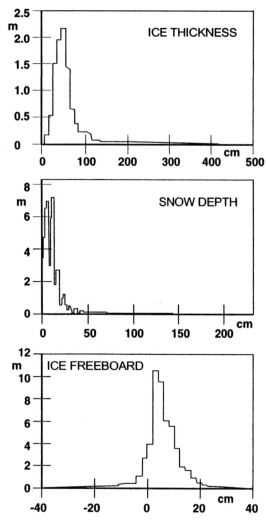

Figure 3.18: *Relative frequency distributions for three variables of Antarctic sea ice, after Wadhams, Lange, and Ackley [407].*
Top: ice thickness.
Middle: depth of the snow layer on the ice.
Bottom: freeboard.

difficult to monitor and many must be observed *in situ*,[18] although ice concentration may be inferred from satellites.[19] Here we present some empirically derived distribution functions of sea ice variables.

An example of an Antarctic distribution of ice thickness, the depth of the snow layer, and the freeboard is shown in Figure 3.18. The thickness

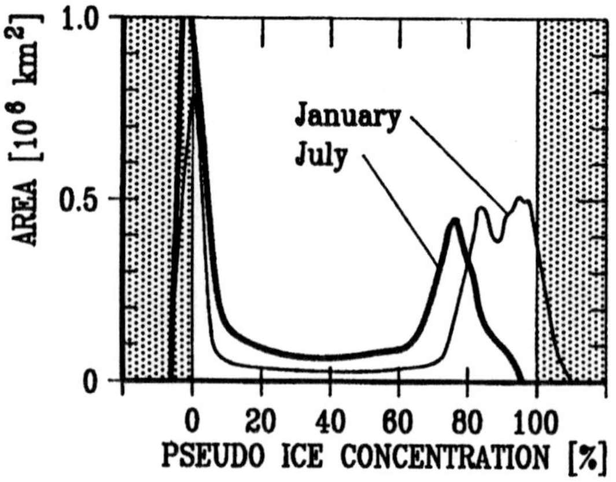

Figure 3.19: *Frequency distribution of pseudo sea-ice concentration for the Arctic Ocean in January and July. After Parkinson et al. [304].*

of the Antarctic sea ice, which varies primarily between zero and 1 m, is close to being normally distributed except for a very wide positive tail that contains extreme values of several metres (the latter indicating multi-year ice). The snow thickness is usually well below 50 cm and is strongly skewed. The freeboard is usually less than 20 cm and is also skewed with a long positive tail that contains maximum values up to several tenths of a metre.

The sea-ice concentration in the Arctic Ocean (Figure 3.19) is bimodal, with a pronounced maximum at very low ice concentrations representing the almost ice-free ocean and an other maximum at about 95%.[20] The distribution is almost uniform between these two extremes.

Figure 3.20 shows the distribution of the ice draft for two Arctic areas, namely the Beaufort Sea and the Fram Strait. Both distributions are strongly skewed, with a mode at 2–4 metres and a wide tail stretching out to 20 and more metres. The latter generally represents heavily ridged multi-year ice.

3.2.6 Hydrological Variables.[21] In this subsection we review the distributions of a number of hydrological variables. Hydrology is the science of the fresh water cycle, from precipitation to the eventual runoff into the oceans. Precipitation, a key hydrological variable, was discussed in [3.1.2]. We consider the *streamflow* of rivers in [3.2.7]. Other relevant variables are evaporation, the storage of

[18] As opposed to being *remotely* sensed from an aircraft or a satellite.

[19] The satellite measures radiation reflected from or generated by the surface. Ice concentration is indirectly derived from these readings. The result is referred to as *pseudo ice concentration*. Uncertainty about the transformation of radiation into ice concentration sometimes results in 'pseudo ice concentration' that is below 0% or above 100%.

[20] The sea is rarely fully ice covered. Instead *leads* open at least a small percentage of the surface.

[21] The material in [3.2.6] and [3.2.7] was supplied by Dennis Lettenmaier from the Department of Civil Engineering of the University of Washington in Seattle.

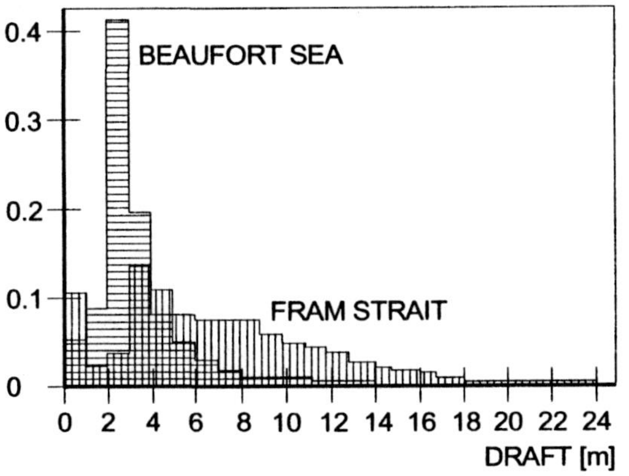

Figure 3.20: *Frequency distribution of sea-ice draft, in metres, for the Beaufort Sea (horizontal cross-hatching) and the Fram Strait (vertical cross-hatching). After Rothrock [334].*

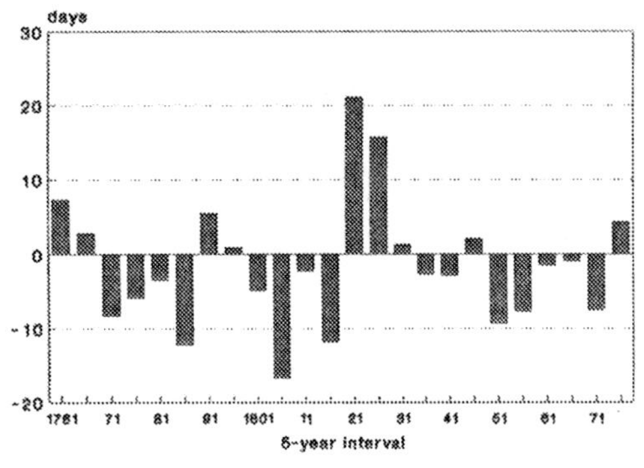

Figure 3.21: *The number of ice-free days on the River Newa in Saint Petersburg in the eighteenth and nineteenth centuries. The numbers are given as five-year mean deviations from the 1816–80 mean. The numbers on the abscissa are the numbers of the first of five years: 1761 represents the interval 1761–65, 71 represents 1771–80 and so on. Data taken from Brückner [70].*

water in the unsaturated soil (*soil moisture*) and the storage and transport in the saturated sub-surface (*ground water*) and the snow water equivalent.

Traditional measures of climate variability related to these hydrological variables include lake levels, the numbers of ice-free days on a river, and the dates of the break-up of ice on rivers in the spring. These measures are not as useful as indicators of climate variability and change as in the past because they are influenced by managerial activities, such as damming, re-routing and dredging of rivers, or the use of ice-breakers to keep water ways open. Long, homogeneous, historical records do exist and were used extensively in an earlier period of climate research (for instance Brückner [70]).

An example of such a record is displayed in Figure 3.21 which shows the number of ice-free days on the River Newa during the eighteenth and nineteenth centuries. The record contains substantial low-frequency fluctuations with amplitudes of 5 to 10 days and also exhibits changes in excess of 30 days between some five-year periods.

The level of lakes with many tributaries may be considered the result of a random process if the regulation of the various rivers is not coordinated. The Great Lakes in North America exhibit low-frequency variations that do not mirror planned human control but rather reflect the low-frequency climatic variations. Also, for long time scales, say tens of years, the effect of human control becomes weaker than the effect of the uncontrolled climate variations.

3.2.7 Streamflow. The streamflow of rivers that are regulated and manipulated by man can hardly be regarded as a random variable. Its variation on time scales of less than a year or so is not fully stochastic but rather is often influenced by deliberate human activities.

In the following we consider the 1948–87 records of streamflow of two unregulated rivers in the USA. There are essentially three processes that control the streamflow of unregulated rivers: precipitation in the drainage area, storage of water in the soil, and storage of water in the form of frozen water or snow. If the soil is wet, its storage capacity is low and most of the precipitated water will be routed to the river. In dry soil most of the water will be stored and streamflow will not be affected unless the amount of rain is substantial (see, e.g., [424]). Water that precipitates as snow or freezes at the ground is released to the streamflow at a later time, and at a more steady rate, when melting occurs (the details depend on temperature, net solar radiation and other variables).

The *Chehalis River* in Washington drains 113 square miles of low-lying coastal hills and flows to the Pacific. Moisture is steadily supplied by frontal storms primarily between November and May. *Cumulative probability functions* for monthly mean streamflows in February, June and October are given in Figure 3.22.[22] The distributions are displayed on 'normal probability paper' on

[22] The units are 'cubic feet per second': 1 cf/s corresponds to 0.028 m^3/s.

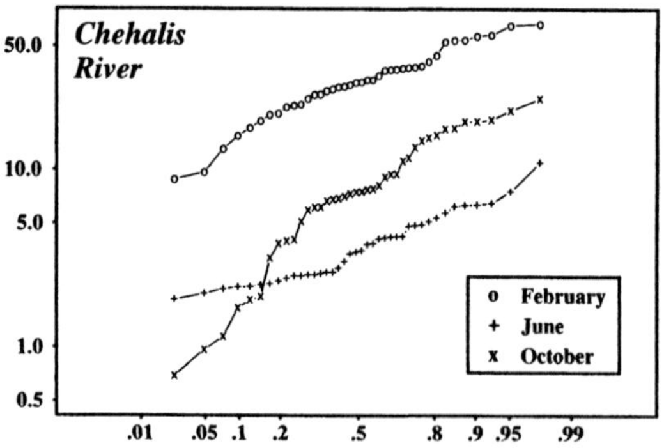

Figure 3.22: *The cumulative probability distribution of the monthly mean streamflow of the Chehalis River in Washington (USA) in cubic-feet per second (cf/s). The three distributions describe the February, June, and October 1948–87 averages. A straight line indicates a normal distribution. Courtesy D. Lettenmaier.*

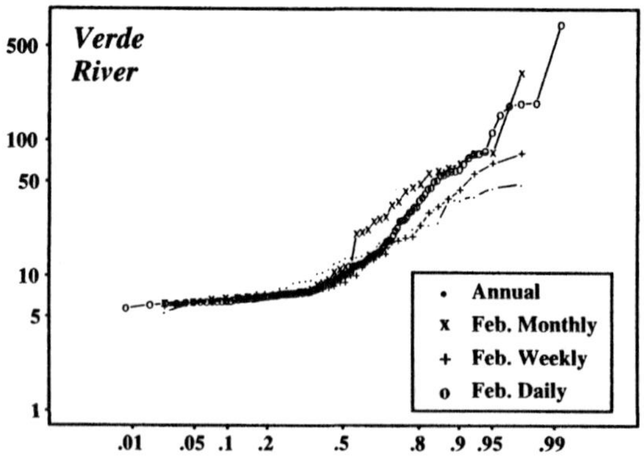

Figure 3.23: *The cumulative probability distribution of the streamflow of the Verde River in Arizona (USA) in cubic-feet per second. The four distributions describe the annual and February averages as well as the daily (sampled on the 1st, 11th, and the 21st) and weekly (first week of the month) streamflows in February. Courtesy D. Lettenmaier.*

which they would appear as straight lines if the distributions were normal.

The distribution of the February monthly averaged streamflow is almost normal. In winter, the wet soil has little capacity to store water so that almost all rainfall is directly transferred to streamflow. In June the occurrence of precipitation is much more variable and the dry soil is able to store a significant amount of water. Thus, minor rain events have little influence on the streamflow. As a consequence the distribution is substantially skewed. The mean is about 150 cf/s, the 10th percentile is about 80 cf/s while the 90th percentile is 250 cf/s. There are many Junes with weak streamflow and few with large streamflow. The October distribution is intermediate between the June and February distributions.

The other case is the *Verde River*, a tributary of the Salt River. It drains the White Mountains in central Arizona and flows westward. The moisture supply is obtained from extratropical storms in the winter season, and from convection and the 'Arizona Monsoon' in summer.

The latter are responsible for the most extreme streamflows, whereas snowmelt controls the winter and spring streamflows most of the time. This configuration leads to markedly non-normal distribution functions of annual and February mean streamflow (Figure 3.23) which indicate the presence of two different regimes. Weaker streamflows occur during the 'snowmelt' regime whereas the very large streamflows, connected with tropical storms, occur relatively infrequently.

The weekly and daily mean streamflows in Figure 3.23 mirror the non-normality of the rainfall rates: The river has almost no water 70% of the time. At other times the streamflow is quite variable, with a few very large extreme events (25 000 cf/s).

The shapes of the probability distributions obtained for different averaging intervals neatly demonstrate the *Central Limit Theorem* [2.7.5]. The distributions deviate from the normal distribution most strongly for the daily averages and least strongly for the annual averages. The annual averages of Verde River streamflow are clearly non-normal, a good illustration of the *asymptotic* nature of the Central Limit Theorem.

4 Concepts in Statistical Inference

4.0.1 Overview. Our purpose here is to introduce some basic ideas about how information is extracted from data. Section 4.1 deals with the fundamental concept of "inference.' The keywords here, "estimation' and 'hypothesis testing,' are introduced in a rather intuitive manner. The technicalities will be explained in detail in Chapters 5 and 6. However, special attention is given to the type of knowledge that can be gained under certain circumstances. This is done by presenting simple examples and discussing the logic that is applied. Two other fundamental concepts, *sampling* (i.e., gathering empirical evidence) and *statistics* (i.e., the condensation of the raw empirical evidence into a few useful quantities), are introduced in Sections 4.2 and 4.3.

4.1 General

4.1.1 Inference. The word *inference* is central in statistical analysis. A dictionary definition of inference [150] rephrases 'to infer' as 'to conclude by reasoning from something known or assumed.' A broad definition of statistical inference could be 'the procedure that involves extracting information from data about the process underlying the observations.'

There are two central steps in this process.

1 A statistical model is adopted that supposedly describes both the stochastic characteristics of the observed process and the properties of the method of observation. It is important to be aware of the models implicit in the chosen statistical method and the constraints those models necessarily impose on the extraction and interpretation of information.

2 The observations are analysed in the context of the adopted statistical model.

There are two major types of inference, namely *estimation* and *hypothesis testing*. The latter is a decision making process that tries to determine the truth of statements, called *hypotheses*, proposed before seeing the data.

There are also two major types of statistical data analysis, namely *exploratory analysis* and *confirmatory analysis* [375]. Exploratory data analysis is the art of extracting from a data set all possible information about the relationships between the variables represented in the data set. This information is used to develop hypotheses about the workings of the climate system. Then, in the best of all worlds, carefully designed experiments are conducted to produce data that can be used to *confirm* independently the hypotheses.

The opportunities for performing truly confirmatory analyses are very different when dealing with the *observational record* rather than a model simulation. We discuss this point in the next two subsections.

4.1.2 Confirmatory Analysis of the Observational Record. For obvious reasons, experiments cannot be done with the actual climate system (cf. Navarra [289]). Instead, special observing programs (such as the 'First GARP Global Experiment' [44]) are sometimes mounted to obtain the data required to address a particular scientific agenda.[1] However, even carefully designed observing programs are unable to eliminate the possibility that the effect the program is designed to observe is confounded (or contaminated) with other non-observed processes in the climate system.

Any confirmatory analysis of the observational record, that is, climate data observed in the past, is limited by two factors: the *lack of independent data* and the inability to separate completely the signal of interest from other sources of variation.

The presence of signals from various competing processes leads to an *open* observed record. That is, we cannot observe all state and forcing

[1] Such campaigns are often called 'experiments,' another case of bad scientific slang. They are not experiments because the investigators involved in these observing programs are unable to control the factors that affect climate variability. These programs are very useful, however, because the coverage and consistency of their observing networks (in space and time and also in terms of the observed variables) are greatly enhanced relative to the regular observing network.

variables. Even if we have enough data to establish a statistical link, we can not exclude the possibility that the repeated coincidence of two events is caused by another non-observed process. Our data coverage allows us to study only an open sub-system of the full system. In contrast, verifiable, 'confirmatory' statements require closed systems (for a discussion of this fundamental problem, see Oreskes, Schrader-Frechette, and Beltz [301]).

All observational data reflect the same *trajectory* of the climate system during the past tens or hundreds of years. Certainly, there are many different *data sets*, such as air pressure reported from land stations or sea-surface temperature reported from ships of opportunity (see Chapter 3). These data sets differ somewhat even if they purportedly represent the same variable—say near-surface wind (see [3.1.5])—but these differences are due to different observational, reporting and analysis practices. They *do not* represent the kind of independent information about the climate system that would be obtained by observing the same variables over a period of similar length at another point in time (e.g., beginning two centuries ago). In other words, such data sets do not offer the option for confirmatory analysis.

This limitation has a severe consequence: Many people, probably hundreds or thousands, have used different techniques to screen our 'one' observational record for rare events. Most of these 'unusual' results are eventually published in articles in scientific journals. Clearly, some of these 'unusual' facets are due to peculiar and rare circumstances that are, nevertheless, 'usual,'—they are 'Mexican Hats' (to use an analogy from Section 6.4) and can not be contested with a statistical test. We can identify an 'unusual' object by comparing it with all others in the observational record. Thus the statement, or null hypothesis, 'this object is not unusual' cannot be contested with a statistical test since independent data are unavailable. No statistical test, regardless of its power or elegance, can overcome this problem, although there are two possible solutions. The first is to extend the observational record backwards by creating new paleo data sets,[2] the second is to postpone testing the developed theories until nature generates enough independent data. Using suitably designed GCM experiments to test a hypothesis derived from the observational record is another approach to confirmatory analysis that is often satisfying.

[2] *Paleo data* are data derived from indirect evidence, such as sediments, that are believed to be representative of the state of climatic components before the current short instrumental period.

4.1.3 Confirmatory Analysis of Simulated Data. The situation is different when dealing with data generated in simulations with GCMs since new additional data can be created, and experiments can be designed to sort out different hypotheses. However, climate models can not be completely validated, which is a big limitation.[3] The answers given by GCMs could simply be an *artifact* of the model.

Experimentation with GCMs began in the 1960s, when pioneers such as Manabe and Bryan [265] examined the sensitivity of the climate to enhanced greenhouse gas concentrations. The standard methodology is to produce a pair of simulations that deviate from each other in only one aspect (such as different greenhouse gas concentrations or sea-surface temperature regimes). This type of experiment is well designed and can be used to confirm hypotheses derived from the observational record or other model experiments. (See Chapter 7 for examples.)

4.1.4 Estimation of Parameters. In estimation, a sample of realizations of a random variable is used to try to *infer* the value of a parameter that describes some property of the random variable. That is, a function of the observations is taken to be an educated guess of the true parameter value. This educated guess, the *estimator*, is either a number (*point estimator*) or an interval (*interval estimator*). Ideally, the point estimate is in the neighbourhood of the true value, and the neighbourhood becomes smaller with increasing sample size. Similarly, a good interval estimator uses the sample to select a range of parameter values that is likely to contain the true parameter. This interval is constructed to *cover* the true parameter with a fixed, high probability (typically

[3] General Circulation Models are *tuned* to reproduce, to the extent possible, the statistics of the observational record of the last few decades. Success in this regard is not a guarantee that the models can successfully simulate natural climate variability on longer time scales. It is also not a guarantee that the models will respond correctly to changes in, for example, the chemical composition or turbidity of the atmosphere. See Oreskes et al. [301].
However, GCMs are considered powerful tools for examining the sensitivity of the climate system since they are based, to a large extent, on physically robust concepts.

95%) in repeated sampling.[4] Thus, the length of the interval decreases with increasing sample size.

Various 'parameters' are subject to estimation, such as the conventional moments (see [2.6.7]) that characterize the probability distribution of the observed random variable. However, estimation is not limited to such elementary parameters; one may also want to estimate the entire probability distribution, or more exotic parameters such as the 'level of recurrence' of two random variables (Sections 6.9–6.10). The 'random variable' might really be a random field observed at m points and we might want to estimate the m^2 parameters that comprise the field's covariance matrix.

As discussed in [1.2.1], there are no 'right' or 'wrong' statements in the realm of estimation; rather, statements can only be considered in terms of precision and reliability. There are some well-defined concepts that can be used, in principle, to obtain estimators with desirable properties. For example, the *maximum likelihood method* can be used to construct estimators that are 'asymptotically optimal' under broad regularity conditions (see [5.3.8]). However, the complexity that is often encountered in climatology causes the design of estimators to be closer to art than sound craftsmanship.

4.1.5 Point Estimation: Examples. A simple example of a point estimation exercise can be found at the end of [2.8.7] where we report the estimated correlation between the standard Southern Oscillation Index (SOI) and an SST index developed by Wright [426]. Here the sample consists of the 624 realizations of the monthly mean SOI and the monthly mean SST index observed between 1933 and 1984. The correlation between corresponding random variables $\mathbf{I}_{SO}$ and $\mathbf{I}_{SST}$ is estimated to be

$$\widehat{\rho}_{SST,SOI} = 0.67.$$

A more involved example of an estimation exercise is found in [1.2.6], where optimally correlated patterns are identified in a sample of realizations of a paired random vector. The statistical model treats the SLP and SST fields as a paired random variable $(\vec{\mathbf{X}}, \vec{\mathbf{Y}})$ with covariance and cross-covariance matrices Σ_X, Σ_Y, and Σ_{XY}. These matrices are estimated in the conventional manner (see [5.2.7]) and the estimates are subsequently employed in a Canonical Correlation Analysis (see Chapter 14). The patterns shown in Figure 1.13 are a best guess rather than the true canonical correlation patterns. Note that these patterns represent simultaneous estimates of several hundred parameters.

[4] This statement must *not* be reduced, or changed, to the misleading statement 'the interval contains the true parameter with (the selected high) probability.' While the latter is technically equivalent, it encourages the mistake of regarding the parameter, rather than the endpoints of the interval, as being random.

4.1.6 Interval Estimators: An Example. We return to the example that deals with the correlation between the SOI and the SST based index of the Southern Oscillation [1.2.6]. In [8.2.3] we impose a model on the bivariate random variable, $\vec{\mathbf{X}} = (\mathbf{I}_{SO}, \mathbf{I}_{SST})$, and then use it to construct an interval estimator $(\widehat{\rho}_L, \widehat{\rho}_U)$ for $\rho_{SST,SOI}$. The estimator is designed so that the interval will cover the true value of $\rho_{SST,SOI}$ 19 out of 20 times if the 'experiment' that resulted in the 1933 to 1984 segments of the SO and SST indices is repeated infinitely often. Note that it is the endpoints of the interval that vary from one replication of the experiment to the next: the true value of $\rho_{SST,SOI}$ is fixed by the physical mechanism that connects SST variations in the Equatorial Pacific with the Southern Oscillation. Performing the computation with the observed indices yields $\widehat{\rho}_L = 0.621$ and $\widehat{\rho}_U = 0.708$. The confidence interval is therefore given by the inequality

$$0.621 < \rho_{SST,SOI} < 0.708\,. \tag{4.1}$$

Note that (4.1) does not include a probability statement about its correctness. In that sense, the 'confidence interval' (4.1) really provides no 'confidence.'

None the less, interval estimators are much more useful than point estimators because they give concrete expression to the idea that the estimator is but another random variable subject to sampling variation. Unfortunately, often in practice a confidence interval cannot be constructed. Then, an estimator is often considered useful if it performs well in some controlled laboratory setting, or returns 'physically reasonable' numbers or distributions. In this context 'physical significance' is the catch phrase that seems to be able to override most statistical scepticism.

4.1.7 The Test of a Null Hypothesis. We briefly touched on the subject of statistical hypothesis testing in [1.2.7]. Here, we continue to discuss the concept in an intuitive manner before using a more rigorous approach in Chapter 6.

A statistical test is a decision making procedure that attempts to determine whether a given set of observations contains information consistent with a concept that was formulated *a priori*. This 'concept' is known as the *null hypothesis* and is usually denoted with the symbol H_0.

In general, only two decisions are possible about H_0:

- *reject* H_0 (if sufficient evidence is found that it is false), or
- *do not reject* H_0 (if sufficient evidence can not be found that it is false).

The decision is a random variable because it is a function of the sample. Thus, there will be some sampling variability in the decision. The same decision about H_0 may not be made in every replication of the experiment that produced the sample.

The decision making rule used in hypothesis testing is constructed using a statistical model so that effects of the sampling variability on the average decision are known, and so that the rule extracts the strongest possible evidence against H_0 from the sample.

Since there is sampling variability, there is a chance of rejecting H_0 when H_0 is true. The probability, or risk $\tilde{p}$, of making this incorrect decision is called the *significance level*. The amount of risk can be controlled by the user of the test. The only way to avoid all risk is to set $\tilde{p} = 0$ so that H_0 is never rejected, which, of course, makes the test useless. However, the risk of false rejection can be set very near zero, at the expense of reducing the chances of rejecting H_0 when it is false.

It is important to remember that the concept of *significance* is an artifact of the conceptual model that we place around our data gathering. The significance level $\tilde{p}$ is realized only if the statistical model we are using is correct and only if the 'experiment' that generated the data is replicated *ad infinitum*. In the real world we need to base our decision about H_0 on a single sample.

The decision making mechanism often consists of a statistic $\mathbf{T}$ and an interval designed so that it contains $(1 - \tilde{p}) \times 100\%$ of the realizations of $\mathbf{T}$ when H_0 is true. Then H_0 is rejected at the $\tilde{p} \times 100\%$ significance level if the observed value of $\mathbf{T}$, say $\mathbf{T} = \mathbf{t}$, falls outside the interval.

The important aspects of a statistical test are as follows.

- The statistical model correctly reflects the stochastic properties of the observed random variables and the way in which they were observed. The actual significance level of the decision is different from the specified level $\tilde{p}$ if there is a problem with the statistical model.
- The decision rule should be constructed so that the chances of rejecting H_0 are optimized when H_0 is false. That is, the decision rule should maximize the *power* of the test.

Usually the model and the null hypothesis are separate but related entities.

The statistical model used to represent an experiment is expressed in terms of a random variable and the way in which it was observed. For example, if the null hypothesis is that the mean of a random variable is zero (i.e., H_0: $\mu = 0$), we might use a model that says that the sample was drawn at random from a normal distribution with known variance σ^2 and unknown mean μ. That is, the model describes, in statistical terms, the way in which observations were collected (they were drawn at random), and the probability distribution (normal, with known variance σ^2) of the random variable which is observed.

Note that it is often not necessary, or desirable, to prescribe a particular probability distribution. Our test of the mean can be conducted almost as efficiently if we assume only that the observations are drawn from a symmetric distribution with unknown mean μ.

The null hypothesis H_0 specifies a value of the unknown parameter in the statistical model of the experiment. Note that in general the model may have many parameters and H_0 might specify values for only a few of them. The parameters that are not specified are called *nuisance parameters* and must be estimated. The testing procedure must properly account for the uncertainty of any parameter estimates.

4.1.8 Example: Number of Hurricanes in a Pair of GCM Experiments. As an example, we consider Bengtsson, Botzet, and Esch's simulation [42, 43] of possible changes of the frequency of hurricanes due to increasing atmospheric concentrations of greenhouse gases. They dealt with hurricanes in both hemispheres, but we limit ourselves in the following to their results for the Northern Hemisphere.

Bengtsson et al. conducted a pair of 'time-slice experiments' with a high-resolution Atmospheric General Circulation Model. One experiment was performed with present-day sea ice and sea-surface temperature distributions, and atmospheric CO_2 concentration. In the other experiment, doubled

CO_2 concentrations were prescribed together with anomalous sea ice and SST conditions simulated in an earlier experiment in which the GCM was coupled to a low-resolution ocean.[5] The number of hurricanes in a model year is treated as a random variable.

The number of hurricanes in a year in the $1\times CO_2$ and the $2\times CO_2$ experiment is labelled $\mathbf{N}_1$ and $\mathbf{N}_2$, respectively. The question of whether the number of storms changes in the $2\times CO_2$ experiment can be expressed as the null hypothesis:

$$H_0: \mathcal{E}(\mathbf{N}_1) = \mathcal{E}(\mathbf{N}_2)$$

or, in words, 'the expected number of hurricanes in the $1\times CO_2$-model world equals the expected number of hurricanes in the $2\times CO_2$-model world.' We adopt a significance level of 5%, that is, we accept a 5% risk of incorrectly rejecting the null hypothesis.

To design a test strategy we consider the number of hurricanes in any model year as being statistically independent. We also assume that the shape of the distribution of the number of hurricanes is the same in both the $1\times CO_2$ and $2\times CO_2$ experiments. That is, we assume that the mean changes in response to CO_2 doubling but that the higher moments (see [2.6.7]) do not.

Given these assumptions we may then use the *Mann–Whitney test* [6.6.11]. This test operates with the sum of the *ranks* of the samples. Rank 1 is given to the smallest number of hurricanes found in all years from both time-slice experiments, rank 2 to the second smallest number and so on. Then the sum of the ranks of the yearly hurricane frequencies in the $2\times CO_2$-experiment $\mathbf{N}_2$ is formed. Very small or large rank sums give evidence that the null hypothesis is false because rank sums of this type occur when most of the yearly hurricane frequencies in one experiment are greater than those in the other experiment. Under the null hypothesis we would expect a roughly equal number of large frequencies in both experiments. Rank sum thresholds for making decisions about H_0 at various significance levels are listed in Appendix I.

In Bengtsson et al.'s case, the sample sizes are $n_1 = n_2 = 5$ since both simulations were run for five years. The yearly hurricane frequencies in the simulations are:

year	$1\times CO_2$	$2\times CO_2$
1	49	41
2	55	42
3	63	46
4	51	38
5	63	38

The rank sum for the $n_2 = 5$ realizations of $\mathbf{N}_2$ is 15. Note that all are smaller than any of the realizations of $\mathbf{N}_1$. When the null hypothesis is true, the 5% threshold value for the rank sum is 18; that is, if H_0 is true, the rank sum will be greater than or equal to 18 in 19 out of every 20 replications of this experiment, and it will be less than 18 only once. Since the actual rank sum of 15 is smaller than the 5% threshold of 18, we reject the null hypothesis at the 5% significance level.[6] We may conclude, at least in the framework of the GCM world, that an increase of the CO_2-concentration will reduce the frequency of Northern Hemisphere hurricanes.

4.1.9 Testing a Null Hypothesis: Interpretation of the Result. Given a particular sample, the decision to reject H_0 with a significance level of $\tilde{p}$ may occur for several reasons.

- We may have incorrectly rejected a true null hypothesis. *Occasional errors of this kind are unavoidable if we wish to make decisions.* We saw in the example above that unusual rank sums can occur even when there is no change in hurricane frequency.[7]
- The statistical model adopted for the observations may not be valid. The observations may not have been sampled in the way assumed by the model (e.g., they might not be independent) or they might not have the assumed distribution (e.g., it might not be symmetric about the mean). The resulting decision making procedure may reject H_0 much more frequently than specified by $\tilde{p}$ even when H_0 is true.

[5] Briefly, the rationale for this methodology is as follows: Hurricanes are not resolved in the low-resolution GCMs. It is, however, assumed that the low-resolution model simulates the large-scale SST and sea-ice distributions well. It is also assumed that the atmospheric circulation is, to a first order approximation, in equilibrium with its lower boundary conditions. These assumptions make it possible to assess the impact of the changed SST and sea-ice distributions and the enhanced CO_2 concentration on hurricanes in the high-resolution GCM.

[6] If the null hypothesis is true, the probability that the five years representative of $2\times CO_2$ conditions all have fewer storms than those representative of the $1\times CO_2$ conditions is 1/252 (0.49%).

[7] We reiterate that the significance level determines the frequency with which we will make this type of error (which statisticians call a 'type I' error). A testing procedure that operates at the 5% significance level will make a type I error 5% of the time when H_0 is true.

Protection against this type of error can be partly obtained by using *robust* statistical methods. Robust methods continue to perform reasonably well under moderate departures from the assumed model (see Section 6.6 and [8.3.17]). However, in general, there is no way to determine positively that the model underlying the test is valid. Instead additional physical arguments are required to support the model. Also, other statistical tests can sometimes be used to ensure that the data are not grossly inconsistent with the adopted model (e.g., one can test the null hypothesis that the observations come from a normal distribution).

- We may have correctly rejected a false H_0.

Similarly, the decision not to reject H_0 can happen for several reasons.

- H_0 may be false, but the test may not have sufficient evidence to reject H_0. The probability of this type of error depends upon the *power* of the test. The probability of not rejecting H_0 when it is false must also be nonzero to have a useful decision making mechanism.
- The model adopted for the observations may not be valid and the decision making procedure developed from this model rejects H_0 too infrequently even when H_0 is false. This error in the model results in a test with very low power.
- H_0 may be true and insufficient evidence was found to reject H_0. This is the desired outcome.

The relevant catch phrase in all of this is 'statistical significance,' which may be markedly different from 'physical significance.' The size of departure that is detectable by a statistical test is a function of the amount of information about the tested parameter available in the sample. Large samples contain more information than do small samples, and thus even physically trivial departures from H_0 will be found to be statistically significant given a large enough sample.

4.1.10 Source of Confusion: The Significance Level. The term *significance level* sometimes causes confusion. Some people, particularly climatologists, interpret the 'significance level' as 'one minus the probability of rejecting a correct null hypothesis.' With this convention large probabilities, for example, 99%, are associated with statistical significance. This usage is contrary to the convention used in the statistical literature. Here we follow the statistical convention and define the 'significance level' as the probability of incorrectly rejecting the null hypothesis. A *smaller* significance level implies *more* evidence that H_0 is false. If H_0 is rejected with a significance level of 1%, then there is 1 chance in 100 of obtaining the result by accident when the null hypothesis is true.

4.1.11 Source of Confusion: Confidence and Significance. One often reads statements that an author is '95% confident that the null hypothesis is false' or that 'the null hypothesis is rejected at the 95% confidence level.' These statements interpret rejection of the null hypothesis at the 5% significance level incorrectly. When we reject a null hypothesis we are simply stating that the value of the test statistic is unusual in the context of the null hypothesis (i.e., we have observed a value of the test statistic that occurs less than 5% of the time when H_0 is true). Because the value is unusual, we conclude that the null hypothesis is likely false. But we can not express this 'likelihood' as a probability.[8]

The precise logical statement in the argument is 'H_0 true $\Rightarrow$ 1 out of 20 decisions is 'reject H_0',' which is not at all related to the statement 'reject $H_0 \Rightarrow H_0$ false in 19 out of 20 cases.'

4.2 Random Samples

4.2.1 Sampling. The conceptual model for a simple random sample is that a simple, repeatable experiment is performed that has the effect of drawing elements from a sample space at random and with replacement.

The amount of imagination required to apply this paradigm depends upon the problem at hand. We will briefly consider three examples.

- Suppose one wanted to estimate the height of the average human living today. We can literally accomplish this by selecting humans at random from the global population (about five billion people) and recording their

[8] At least not in the 'frequentist' paradigm we use in this book. *Bayesian* statisticians extend the notion of probability to include subjective assessments of the likelihood that a parameter has one value as opposed to another. It then becomes possible to solve statistical decisions by comparing the odds in favour of one hypothesis with those in favour of another. See Gelman et al. [139] for an introduction to Bayesian analysis.

heights. With care, and a lot of preparation, it is at least conceptually possible to ensure that everyone has the same probability of being selected. Thus, we can be assured that, if we sample the population 1000 times, the resulting sample of 1000 heights will be *representative* of the entire global population.

Here, the concept of a simple random sample representative of the population is easy to comprehend because the population from which the sample is to be drawn is finite. The logistics required to obtain the sample (i.e., preparing a list of five billion names and selecting randomly from those names) are easily visualized.

- Suppose now that one wanted to estimate global mean temperature at 00 UTC on a given day: again, an easily imagined accomplishment. One approach would be to select randomly n locations on the globe and to measure the temperature at each location at precisely 00 UTC. Our thinking in this example is necessarily a bit more abstract than in the previous example. The number of points at which a temperature measurement can be taken is infinite, and the logistics of placing a thermometer are more difficult for some points than for others. None the less, given the desire and sufficient resources, this exercise could actually be performed.[9]

- Finally, suppose that one wanted to estimate the climatological mean temperature at a location such as Hamburg (Germany), or Victoria (Canada), without consulting historical temperature observations. The concept of the simple random sample does not serve us particularly well here. Our observations are necessarily confined to an interval of time near the present. Temperatures in the past and in the distant future cannot be sampled; only a finite number of observations will be taken so temperatures realized after the last observation will not be sampled. To treat the sample as a random sample, we must make some assumptions about the properties of the temperature process. In particular, we assume that the process is *stationary* or *cyclo-stationary* (meaning that its statistical properties are time invariant) and that the process is *ergodic* (meaning that sampling a given realization of the process in time yields information equivalent to randomly sampling independent realizations of the same process).

It is clear, then, that the concept of sampling a geophysical process is complex, and that very strong assumptions are implicit in the analysis of climate data.

4.2.2 Models for a Collection of Data. Usually, the *sampling* exercise can be represented by a collection of *independent and identically distributed* random variables, say $\{\mathbf{X}_1, \ldots, \mathbf{X}_n\}$. When the sample is taken, we end up with a set of *realizations* $\{\mathbf{x}_1, \ldots, \mathbf{x}_n\}$. Part of the conceptual baggage we carry is the idea that the sample could be taken again, resulting in another set of realizations, say $\{\mathbf{x}'_1, \ldots, \mathbf{x}'_n\}$ of $\{\mathbf{X}_1, \ldots, \mathbf{X}_n\}$. The statistical model describes the range of possible realizations of the sample and the relative likelihood of each realization.

The phrase independent and identically distributed represents two *sampling assumptions* that are almost always needed when using classical inference methods (see Chapters 5–9). The assumptions are as follows.

- The observations $\mathbf{x}_1, \ldots, \mathbf{x}_n$ are realizations of n independent random variables $\mathbf{X}_1, \ldots, \mathbf{X}_n$.

- The random variables $\mathbf{X}_1, \ldots, \mathbf{X}_n$ are identically distributed.

However, the independence assumption can not be made when making inferences about time series or stochastic processes (Chapter 12). Then models are required that account for the dependence between observations. One way to do this is to assume that the sample comes from a stationary and ergodic process. Some types of analysis (e.g., extreme value analysis, see Section 2.9) are able to cope with dependence quite well; others, such as hypothesis testing about the mean of a sample (see Section 6.6), cope with dependence very poorly.

In general, models are either *parametric* or *non-parametric*.

- Parametric models require a *distributional assumption*: that is, the assumption that the distribution of $\mathbf{X}_i, i = 1, \ldots, n$, belongs to a certain family of probability distributions (such as $\mathbf{X}_i$ is normal). The model is *parametric* because it specifies everything

[9] Shen et al., [349] have given careful thought to the problem of estimating the sampling error in the global mean temperature that arises from the density and distribution of the observing network (including the random network discussed above).

about the distribution function except for a few free parameters (for instance, the mean and variance in the case of a normal distribution). Provided that the distribution assumption is correct, the parametric model leads to very efficient statistical inference because it brings a substantial amount of information into the procedure in addition to that contained in the data.

- Non-parametric approaches to statistical inference are distinguished from parametric methods in that the distributional assumption is replaced by something more general. For example, instead of assuming that data come from a distribution having a specific form, such as the normal distribution, it might be assumed that the distribution is unimodal and symmetric. This includes not only the normal distribution, but many other families of distributions as well.

  Non-parametric methods are advantageous when it is not possible to make specific distributional assumptions. Frequently, non-parametric methods are only slightly less efficient than methods that use the correct parametric model, and generally more efficient compared with methods that use the incorrect parametric model. Non-parametric statistical inference is therefore relatively cheap insurance against moderate departures from the distributional assumptions. We will discuss a few non-parametric inference techniques in Chapter 6. A complete treatment of the subject can be found in Conover [88].

  While they allow us to relax the distributional assumption needed for parametric statistical inference, these procedures rely more heavily upon the sampling assumptions than do parametric procedures. Non-parametric models are heavily impacted by departures from the sampling assumptions (see Zwiers [442]), so their use is *not* advised when there may be dependence within a sample.

4.3 Statistics and Sampling Distributions

4.3.0 Introduction. In the rest of this chapter we will make the standard assumptions that a sample can be represented by a collection of independent and identically distributed (iid) random variables. The effects of dependence and methods used when there is dependence are addressed in Chapters 6, 8, 9, and 10–12.

We have seen that a random variable is a function defined on a sample space and that it inherits a probability distribution from the probabilities assigned to the sample space elements. In the same way, a statistic is a function defined on a sample, and it inherits its probability distribution from those of the random variables that represent the sample. Thus, a statistic is a random variable. Every time we replicate the 'experiment' that generates the sample, we get a different set of realizations of the random variables that constitute the sample, and thus a different realization of the statistic computed from the sample.

We describe here some basic statistics and their probability distributions under the *standard normal conditions*. That is, we assume that the random variables $\{\mathbf{X}_1, \ldots, \mathbf{X}_n\}$ that represent a sample are independent and identically distributed normal random variables with mean μ and variance σ^2.

4.3.1 The Sample Mean. An example of a simple statistic is the *sample mean*,

$$\bar{\mathbf{X}} = \frac{1}{n}\sum_{i=1}^{n}\mathbf{X}_i, \tag{4.2}$$

expressed here as a random variable. Once an experiment has been conducted and a particular sample $\{\mathbf{x}_1, \ldots, \mathbf{x}_n\}$ has been observed, we write

$$\bar{\mathbf{x}} = \frac{1}{n}\sum_{i=1}^{n}\mathbf{x}_i$$

to represent the corresponding realized value of $\bar{\mathbf{X}}$. By applying (2.16) we see that the random variable $\bar{\mathbf{X}}$ has mean and variance

$$\mathcal{E}(\bar{\mathbf{X}}) = \mu \tag{4.3}$$

$$\mathrm{Var}(\bar{\mathbf{X}}) = \sigma^2/n. \tag{4.4}$$

Thus, it is apparent that the sample mean can be regarded as an *estimator*[10] of the true mean and that the spread of the distribution of $\bar{\mathbf{X}}$, as well as the uncertainty of the estimator, decreases with increasing sample size.

The sample mean has a normal distribution when random variables $\mathbf{X}_i$ are normally distributed. When observations are not normally

[10] The concept of an estimator is discussed with more precision in Chapter 5.

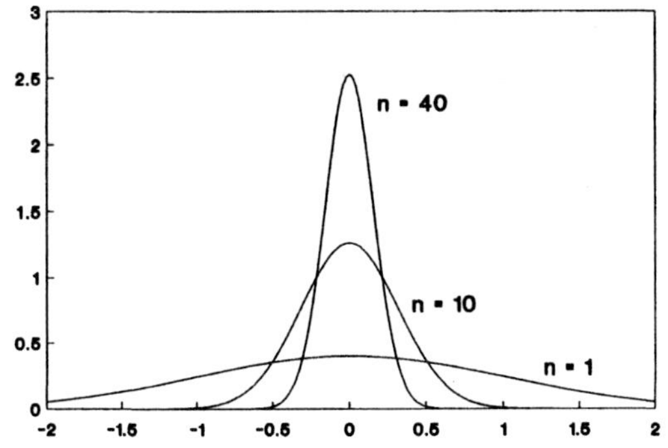

Figure 4.1: *The probability density function of the sample mean when the sample consists of $n = 10$ and $n = 40$ independent and identically distributed random variables. The distribution of the individual observations is labelled $n = 1$.*

distributed, the *Central Limit Theorem*[11] [2.7.5] assures us, under quite general conditions, that the sample mean will have a distribution that approaches a normal distribution as the sample size increases.

The effect of increasing sample size on the distribution of the sample mean is illustrated in Figure 4.1. The distribution becomes increasingly compact as the sample size increases. Consequently the true population mean, μ, becomes better known as sample size increases.

4.3.2 The Sample Variance. Another example of a relatively simple statistic is the *sample variance*, which is given by

$$\begin{aligned} \mathbf{S}^2 &= \frac{1}{n-1}\sum_{i=1}^{n}(\mathbf{X}_i - \bar{\mathbf{X}})^2 \qquad (4.5) \\ &= \frac{1}{n-1}\Big(\Big(\sum_{i=1}^{n}\mathbf{X}_i^2\Big) - n\bar{\mathbf{X}}^2\Big). \end{aligned}$$

By using (2.16) and (2.17) it can be shown that

$$\mathcal{E}(\mathbf{S}^2) = \sigma^2. \qquad (4.6)$$

When random variables $\mathbf{X}_i$ are normally distributed, it can be shown that $(n-1)\mathbf{S}^2/\sigma^2 \sim \chi^2(n-1)$.[12] Consequently,

$$\mathrm{Var}(\mathbf{S}^2) = \frac{2\sigma^4}{n-1}. \qquad (4.7)$$

Equation (4.7) shows that we can think of $\mathbf{S}^2$ as an estimator of σ^2 that has decreasing uncertainty with increasing sample size n. The uncertainty goes to zero in the limit as the sample becomes infinitely large.

It can also be shown that $\mathbf{S}^2$ is independent of $\bar{\mathbf{X}}$.

4.3.3 The t Statistic. It is natural to interpret the sample mean as a measure of the *location* of the sample. This measure is often expressed as a distance from some fixed point μ_0. This distance should be stated in dimensionless units so that the same inference can be made regardless of the scale of observation.[13]

Suppose, for now, that $\mu_0 = \mathcal{E}(\mathbf{X}_i)$. When the variance σ^2 is known, the distance between $\bar{\mathbf{X}}$ and μ_0, in dimensionless units, is

$$\mathbf{Z} = \sqrt{n}\,\frac{\bar{\mathbf{X}} - \mu_0}{\sigma}.$$

Random variable $\mathbf{Z}$ has mean zero and unit variance regardless of the scale on which the observations are made. It is normally distributed when random variables $\mathbf{X}_i$ are normal. When this is not true, the Central Limit Theorem [2.7.5] states that the distribution of $\mathbf{Z}$ will approach the standard normal distribution as the sample size grows large.

When that variance is not known, we can estimate it with S^2 and compute the *t statistic* or, as it is also often called, *Student's t statistic*

$$\mathbf{T} = \sqrt{n}\,\frac{\bar{\mathbf{X}} - \mu_0}{S}.$$

Again, we have a measure that is independent of the scale of measurement and it can be shown that the asymptotic distribution is normal with unit variance. When samples are finite and consist of independent, identically distributed normal random variables with mean μ_0, $\mathbf{T}$ has the t distribution

[11] Independence of the $\mathbf{X}_i$s is not a necessary condition for obtaining convergence results such as the Central Limit Theorem. Similar results can often be obtained when the $\mathbf{X}_i$s are dependent on one another, although in this case the asymptotic variance of $\bar{\mathbf{X}}$ will only be proportional rather than equal to σ^2/n. The constant of proportionality depends upon the nature of the dependence.

[12] The $\chi^2(k)$ distribution is discussed in [2.7.8]. Figure 2.5 shows the $\chi^2(k)$ distribution for four different degrees of freedom k.

[13] This is the *principle of invariance*. Statistical methods that are not invariant under transformations of scale should not be trusted because users can manipulate the inferences made with such methods by using a suitable transform.

with $n - 1$ degrees of freedom (see [2.7.9]). One way to show this is to factor $\mathbf{T}$ as

$$\begin{aligned}\mathbf{T} &= \sqrt{n}\,\frac{\bar{\mathbf{X}} - \mu_0}{S} \\ &= \frac{\sqrt{n}(\bar{\mathbf{X}} - \mu_0)/\sigma}{\left[\left((n-1)S^2/\sigma^2\right)/(n-1)\right]^{1/2}} \\ &= \frac{\mathbf{Z}}{\sqrt{\mathbf{Y}/(n-1)}}.\end{aligned}$$

Here

$$\begin{aligned}\mathbf{Z} &= \sqrt{n}\,\frac{\bar{\mathbf{X}} - \mu_0}{\sigma} \sim \mathcal{N}(0,1) \\ \mathbf{Y} &= \frac{(n-1)S^2}{\sigma^2} \sim \chi^2(n-1),\end{aligned}$$

and $\mathbf{Y}$ is independent of $\mathbf{Z}$. This exactly characterizes a t distributed random variable (see [2.7.9]). $\mathbf{T}$ has zero mean and variance $\frac{n-1}{n-3}$.

Figure 2.6 shows that the t distribution is slightly wider than the standard normal distribution and that it tends towards the normal distribution as sample size increases. Indeed, the two are essentially the same for samples of size $n > 30$. The extra width in the small sample case comes about because the distance between $\bar{\mathbf{X}}$ and μ_0 is measured in units of estimated rather than known standard deviations. The additional variability induced by this estimate is reflected in the slightly wider distribution.

4.3.4 The F-ratio. Suppose now that we have two collections of independent and identically distributed random variables $\mathbf{X}_1, \ldots, \mathbf{X}_{n_X}$ and $\mathbf{Y}_1, \ldots, \mathbf{Y}_{n_Y}$ representing two random samples of size n_X and n_Y respectively. A natural way to compare the *dispersion* of the two samples is to compute

$$F = \frac{S_X^2}{S_Y^2}, \tag{4.8}$$

where S_X^2 is the sample variance of the $\mathbf{X}$ sample and S_Y^2 is the sample variance of the $\mathbf{Y}$ sample. When both samples consist of independent and identically distributed normal random variables, and the random variables in one sample are independent of those in the other, the random variable $(\sigma_Y/\sigma_X)^2 F$ is independent of both scales of observation and $F \sim F(n_X - 1, n_Y - 1)$ (see [2.7.10]). This is shown by factoring F so that it can be expressed as a ratio of independent χ^2 random variables, each divided by its own degrees of freedom. In fact, by (2.29), we have

$$\begin{aligned}F &= \frac{\left[(n_X - 1)S_X^2/\sigma_X^2\right]/(n_X - 1)}{\left[(n_Y - 1)S_Y^2/\sigma_Y^2\right]/(n_Y - 1)} \\ &= \frac{\chi_X/(n_X - 1)}{\chi_Y/(n_Y - 1)},\end{aligned}$$

with $\chi_X \sim \chi^2(n_X - 1)$ and $\chi_Y \sim \chi^2(n_Y - 1)$. Several examples of the F distribution are displayed in Figure 2.7.

5 Estimation

5.1 General

In Chapter 4 we describe some of the general concepts of statistical inference, including the basic ideas underlying estimation and hypothesis testing. Our purpose here is to discuss estimation in more detail, while hypothesis testing is addressed further in Chapter 6.

5.1.1 The Art of Estimation. We stated in Chapter 4 that statistical inference is the process of extracting information from data about the processes underlying the observations. For example, suppose we have n realizations $\mathbf{x}_i$ of a random variable $\mathbf{X}$. How can we use these realizations to make inferences about the distribution of $\mathbf{X}$?

The first step is to adopt some kind of statistical model that describes how the sample $\{\mathbf{x}_1, \dots, \mathbf{x}_n\}$ was obtained. It is often possible to use the 'standard normal setup' introduced in Section 4.3. It represents the sample as a collection of n independent and identically distributed normal random variables $\{\mathbf{X}_1, \dots, \mathbf{X}_n\}$. Estimators of the mean (4.2) and the variance (4.5) are derived in this setting.

The standard notation used to differentiate a parameter p from its estimator is to indicate the estimator with a hat, as in $\widehat{p}$. Confusion can arise because the notation does not make it clear when $\widehat{p}$ represents a random variable and when it represents a realization of a random variable. Estimators should be viewed as random variables unless the context makes it clear that a particular value has been realized. The language we use also gives verbal cues that help to distinguish between the two; we generally think of an *estimator* as a function on a sample (and hence as a random variable) and an *estimate* as a particular value that is realized by an estimator. Just to exercise this notation, the estimators of the mean and variance that are introduced in Section 4.3 are $\widehat{\mu} = \overline{\mathbf{X}}$, and $\widehat{\sigma}^2 = \mathbf{S}^2$. Intuitively, these estimators behave as we would expect. They take values in the neighbourhoods of the true values of the parameters they are estimating, their scatter decreases with increasing sample size, and their scatter is related to the scatter within the sample.

For the moment, *estimators* are mere functions of the sample without any qualitative properties. The art is to find good estimators that yield estimates in a specified neighbourhood of the true value with some known likelihood. The objective of estimation theory is to offer concepts and measures useful for evaluating the performance of estimators.

Because estimators are random variables, they are subject to sampling variability. An estimator can not be right or wrong, but some estimators are better than others. Examples of admittedly silly estimators of the mean μ and the variance σ^2 are

$$\widehat{\mu}_s = \mathbf{X}_1$$
$$\widehat{\sigma}_s^2 = (\mathbf{X}_1 - \mathbf{X}_2)^2/2.$$

Note that $\widehat{\mu}_s$ has n times the variance of estimator (4.2), and $\widehat{\sigma}_s^2$ has $n-1$ times the variance of estimator (4.5).

5.1.2 Estimation and the 'iid' Assumptions. In Chapter 4 we stressed the importance of the 'iid' (or sampling) assumptions in the process of inference. However, these assumptions are often not satisfied in climate research. Even so, many estimators will still produce useful parameter estimates. But it is much more difficult (sometimes even impossible) to construct confidence intervals or other measures of the uncertainty of the point estimate.

5.1.3 Some ways in which to violate the 'iid' assumptions. The 'independence' assumption is violated when methods that require independence are applied to serially correlated data. A possible solution is to sub-sample the data, that is, remove data from the complete data set until the gaps between the remaining observations are long enough to ensure independence.

Information is generally lost by sub-sampling and the quality of the estimator is not improved (in terms of bias or mean squared error; see Section 5.3). The estimate computed from the sub-sampled data is generally less certain than that computed from all of the data.

However, sometimes the use of the entire data set leads to problems. For example, when serially correlated data are not evenly distributed in time, the use of all of the data can lead to severe biases (systematic errors).

For example, suppose that we want to estimate the expected (i.e., mean) daily summer rainfall at a location affected by the El Niño phenomenon using a 31-year data set of rainfall observations. A naive estimate could be constructed by averaging over all observations without accounting for the characteristics of the data set. Suppose that the data set contains 1 year of very good daily data (obtained during a special observing project) and 30 years of once weekly observations. Further, suppose that the special observing project took place during an El Niño year in which there was a marked lack of rain. If we average over all the available data, then the year of the special observing project has seven times more *influence* on the estimate than any of the other years. It is very likely, then, that the computed average *underestimates* the true expected (long-term mean) rainfall. Sub-sampling is an appropriate solution to this problem.

The 'identically distributed' assumption is violated when the sampled process is non-stationary. For example, if there are annual or diurnal cycles in the mean of the sampled process, the sampling method affects the way in which an estimated mean can be interpreted. A data set that contains observations taken at frequent, equally spaced intervals over an integral number of years or days will provide good estimates of the annual or daily mean respectively. On the other hand, if all the data come from winter, or from the early morning, then the estimate will not be representative of the true annual mean value.

5.2 Examples of Estimators

5.2.0 The Setting. We again assume that the result of the sampling process can be represented by a sample of n independent and identically distributed random variables $\{\mathbf{X}_1, \dots, \mathbf{X}_n\}$. In general, we use $\mathbf{X}$ to represent any of the iid random variables in the sample and assume that the (common) probability density function of $\mathbf{X}$ is $f_X(\cdot)$. The only difference between the current setup and the standard normal setup in Chapter 4 is that we do not yet assume a specific form for f_X.

Having now set the stage, we carry on to introduce a number of estimators. Whenever possible, we write the estimators in their random (rather than realized) form to emphasize that they are subject to sampling variability inherited from the sampling process.

5.2.1 Histograms. The *frequency histogram* is a crude estimator of the true probability density function, f_X, of $\mathbf{X}$. To obtain a frequency histogram or a *relative frequency distribution*, the real line, $\mathbb{R}$ (or the complex plane, or the multi-dimensional space), is partitioned into K subsets Θ_k such that

$$\begin{aligned} \bigcup_{k=1}^{K} \Theta_k &= \mathbb{R} \quad \text{and} \\ \Theta_k \cap \Theta_j &= \emptyset \quad \text{for } k \neq j. \end{aligned} \tag{5.1}$$

The number of observations that fall into each Θ_k is counted, and the total count is divided by the total number of observations so we obtain

$$\mathbf{H}(\Theta_k) = \frac{|\{\mathbf{X}_k : \mathbf{X}_k \in \Theta_k\}|}{n},$$

where $|\mathcal{S}|$ denotes the number of elements in set $\mathcal{S}$. $\mathbf{H}(\Theta_k)$ is an estimator of

$$P(\mathbf{X} \in \Theta_k) = \int_{\Theta_k} f_X(x)\, dx,$$

which in turn is a discretized approximation of the density function f_X. Consequently, the random step function

$$\widehat{f}_X(x) = \frac{H(\Theta_k)}{\int_{\Theta_k} dx} \quad \text{if } x \in \Theta_k, \tag{5.2}$$

is a crude estimator of the true density function.[1] The denominator in (5.2) is the area of subset Θ_k (or the length of the interval, if the partitions (5.1) are intervals, as is often true). The denominator in (5.2) has been introduced to ensure $\int_{\mathbb{R}} \widehat{f}_X(x)\, dx = 1$. It turns out, with suitable regularity conditions, that this estimator converges to the true density function as sample size $n \to \infty$ if the number of elements in each subset tends to infinity as the sample size $n \to \infty$, and if the number of subsets Θ_k also goes to infinity as $n \to \infty$.

[1] *Kernel* type density estimators produce much better density function estimates. See, for example, Silverman [350] or Jones, Marron, and Sheather [200].

The histogram is also an estimator of probabilities. The probability that $\mathbf{X} \in [a, b]$ is conveniently estimated by

$$\widehat{P}(\mathbf{X} \in [a, b]) = \int_a^b \widehat{f}_X(x)\, dx = \frac{|\{\mathbf{X}_k : \mathbf{X}_k \in [a, b]\}|}{n} = \mathbf{H}([a, b]).$$

That is, the probability of obtaining an observation in a given interval or region the next time the experiment (i.e., sampling exercise) is repeated is estimated by the frequency with which observations fall into that set in the available sample.

Several examples of histograms are shown in Chapter 3, for example, Figures 3.3, 3.5, 3.7, 3.18, or 3.20.

Note that the histogram depends on the details of the partitioning, and that the partitioning is chosen subjectively.

5.2.2 Empirical Distribution Function. Combining the definition of the cumulative distribution function in (2.14) with the definition of the estimated probability density function in (5.2) gives the following natural estimator of the distribution function

$$\widehat{F}_X(x) = \frac{|\{\mathbf{X}_k : \mathbf{X}_k \le x\}|}{n} = \widehat{P}(\mathbf{X} \le x) = \mathbf{H}([-\infty, x]). \tag{5.3}$$

$\widehat{F}_X$ is often called the *empirical distribution function*. It is a non-decreasing step function with $\widehat{F}_X(-\infty) = 0$ and $\widehat{F}_x(\infty) = 1$. The value of the function increases by a step of $1/n$ at each observation (or it increases by a multiple of $1/n$ if several observations have the same value). Note that $\widehat{F}_X(\mathbf{x}_{(n|n)}) = 1$, and that the estimated probability of observing a value larger than the largest value, $\mathbf{x}_{(n|n)}$, in the sample or a value smaller than the smallest value, $\mathbf{x}_{(1|n)}$, is zero.[2]

A slightly different estimator of the distribution function is described in [5.2.4].

The empirical distribution function of the monthly mean Southern Oscillation Index (see Figures 1.2 and 1.4, and subsections [1.2.2], [2.8.7], and [8.1.4]) is shown in Figure 5.1.

5.2.3 Goodness-of-fit Tests—a Diversion. The subject of goodness-of-fit tests arises naturally in the context of estimating the distribution function.

[2]A reminder: $\mathbf{x}_{(j|n)}$ is the jth order statistic of the sample $\{\mathbf{x}_1, \ldots, \mathbf{x}_n\}$, that is, the jth largest value in the sample.

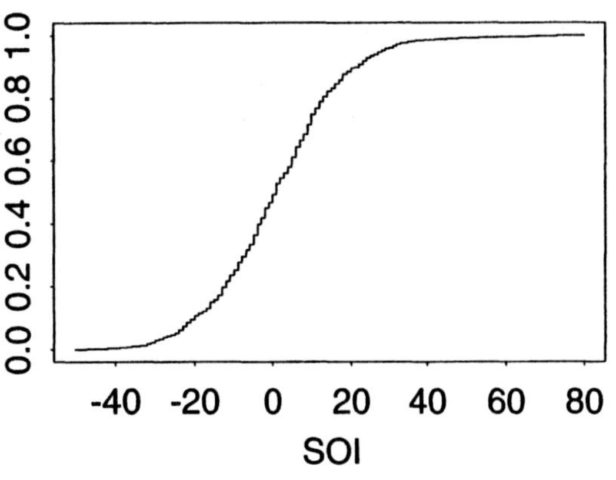

Figure 5.1: *The empirical distribution function of the monthly mean SO index as computed from 1933–84 observations.*

It is sometimes of interest to know whether a given sample $\{\mathbf{x}_1, \ldots, \mathbf{x}_n\}$ could be realizations of a random variable $\mathbf{Y}$, with a particular type of probability distribution, such as the normal distribution. One approach to this type of *goodness-of-fit* question compares the empirical distribution function $\widehat{F}_X$ with the proposed distribution function F_Y. The difference $\widehat{F}_X - F_Y$ is a random variable and it is therefore possible to construct goodness-of-fit tests that determine whether the difference is unlikely to be large under the null hypothesis H_0: $F_X = F_Y$. Conover [88] provides a good introduction to the subject. Stephens [356] [357] provides technical details of a variety of goodness-of-fit tests not discussed by Conover.

The *Kolmogorov–Smirnov test* is a popular goodness-of-fit statistic that compares an empirical distribution function with a specified distribution function F_Y. The Kolmogorov–Smirnov test statistic,

$$\mathbf{D}_{KS} = \max_x |\widehat{F}_X(x) - F_Y(x)|,$$

measures the distance between the empirical distribution function and the specified distribution. Obviously, a large difference indicates an inconsistency between the data and the statistical model F_Y.

There is a large family of related tests, some of which feature norms other than the *max-norm*.[3] The Kolmogorov–Smirnov test becomes 'conservative,' that is, rejects the null hypothesis

[3]Other tests, such as the Anderson–Darling test and the Cramer–von Mises test (see [356], [357], [307]) use statistics that are more difficult to compute, but they are also more powerful and more sensitive to departures from the hypothesized distribution in the tails of the distribution.

less frequently than indicated by the significance level, when F_Y has parameters that are estimated from the sample.

This problem often occurs when we want to test for normality in a set of data. The *Lilliefors test* [253], is a variation of the Kolmogorov–Smirnov test that accounts for the uncertainty of the estimate of the mean and variance. The Lilliefors test statistic is given by

$$\mathbf{D}_L = \max_x |\widehat{F}_X(x) - F^*_{\mathcal{N}}(x)|,$$

where $F^*_{\mathcal{N}} \sim \mathcal{N}(\widehat{\mu}_X, \widehat{\sigma}_X)$ is the normal distribution in which the mean and standard deviation are replaced with the sample mean and standard deviation. $\mathbf{D}_L$ measures the distance between the empirical distribution function and the normal distribution fitted to the data. Large realizations of $\mathbf{D}_L$ indicate that H_0 should be rejected. Conover (see Section 6.1 and Table 15 in [88]) provides tables with thresholds for rejection as a function of sample size and significance level. Stephens [356] offers approximate formulae for the same purpose.

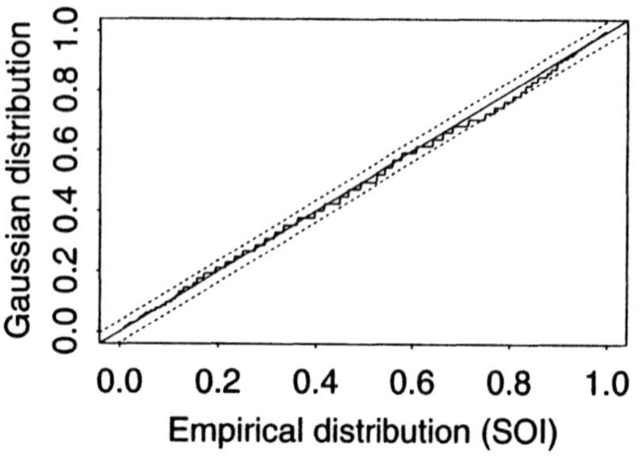

Figure 5.2: *The empirical distribution function (5.3) of the SOI plotted against the cumulative distribution function of the standard normal random variables. Points are expected to lie approximately on the $y = x$ line if the SOI is normal. The lines parallel to $y = x$ are thresholds that, if crossed, indicate that H_0: 'sample is normal' should be rejected at the 5% significance level (see [4.1.10]). The test may not be reliable because the sampling assumptions are not satisfied by the SOI.*

Stephens [356] and Pearson and Hartley [307] describe how to adjust several goodness-of-fit tests, including the Kolmogorov–Smirnov test, when sample sizes are small and when it is necessary to estimate the parameters of the distribution specified in H_0.

Figure 5.2 redisplays the empirical distribution function of the SO index as $(\widehat{F}_X(x), F^*_{\mathcal{N}}(x))$ pairs where $F^*_{\mathcal{N}}(x)$ is the normal distribution function with mean and variance estimated from the SO data. These points are expected to more or less lie on the $\widehat{F}_X(x) = F^*_{\mathcal{N}}(x)$ line when the fit is good (i.e., when H_0 is true). Note that the placement of the thresholds parallel to the $\widehat{F}_X(x) = F^*_{\mathcal{N}}(x)$ line is correct only if the iid assumption holds for the SOI, which is known not to be true. The results of the test can therefore not be taken literally.

5.2.4 Probability Plots. Subsection [3.1.3] discusses the format of a probability plot that is similar to Figure 5.2, but more useful for determining whether $F_Y = F_X$. A probability plot depicts the graph of the function $y \to F_X^{-1}[F_Y(y)]$, where F_Y is some prescribed, possibly hypothesized, distribution and F_X is the distribution of the data. The graph is plotted linearly in y but the horizontal axis is labelled with the probabilities $F_Y(y)$ (see Figure 3.2).

A probability plot may be derived from a finite sample by plotting points $(F_Y^{-1}(\widetilde{F}_X(\mathbf{x}_i)), \mathbf{x}_i)$ where $\widetilde{F}_X$ is an estimator of the distribution function. Since $\widehat{F}_X(\mathbf{x}_n) = 1$ we can *not* use $\widetilde{F}_X = \widehat{F}_X$. Otherwise the scatter plot would include the point $(\infty, \mathbf{x}_n)$. Alternative estimators are

$$\begin{aligned}\widetilde{F}_X(x) &= \frac{|\{\mathbf{X}_k : \mathbf{X}_k \leq x\}|}{n+1} \\ &= \frac{n}{n+1}\widehat{F}_X(x)\end{aligned}$$

and

$$\begin{aligned}\widetilde{F}_X(x) &= \frac{|\{\mathbf{X}_k : \mathbf{X}_k \leq x\}| - 0.5}{n} \\ &= \widehat{F}_X(x) - \frac{0.5}{n}\end{aligned} \tag{5.4}$$

so that the points to be plotted are $(F_Y^{-1}(\frac{i}{n+1}), \mathbf{x}_i)$ or $(F_Y^{-1}(\frac{i-0.5}{n}), \mathbf{x}_i)$. Equation (5.4) is used in [8.3.13].

5.2.5 Estimating the First Moment. The first moment $\mu^{(1)} = \mu$ of a real-valued random variable $\mathbf{X}$ with probability density function f_X is the expected value of $\mathbf{X}$, $\mathcal{E}(\mathbf{X})$, given by

$$\mu = \int_{-\infty}^{\infty} x f_X(x)\, dx. \tag{5.5}$$

We identified the sample mean (4.2)

$$\widehat{\mu} = \overline{\mathbf{X}} = \frac{1}{n}\sum_{k=1}^{n} \mathbf{X}_k \tag{5.6}$$

as a reasonable estimator of μ in [4.3.2] because its expectation is μ and its variance goes to zero as the sample size n increases. However, the relationship between (5.5) and (5.6) is not immediately obvious.

A heuristic argument that links the two expressions is as follows. First, let $\mathbf{X}_{(i|n)}$, $i = 1, \dots, n$, be the order statistics of sample $\{\mathbf{X}_1, \dots, \mathbf{X}_n\}$ (see [2.6.9]). Then equation (5.5) can be rewritten as

$$\mu = \int_{-\infty}^{(\mathbf{X}_{(1|n)}+\mathbf{X}_{(2|n)})/2} x f_X(x)\, dx \qquad (5.7)$$
$$+ \sum_{i=2}^{n-1} \int_{(\mathbf{X}_{(i-1|n)}+\mathbf{X}_{(i|n)})/2}^{(\mathbf{X}_{(i|n)}+\mathbf{X}_{(i+1|n)})/2} x f_X(x)\, dx$$
$$+ \int_{(\mathbf{X}_{(n-1|n)}+\mathbf{X}_{(n|n)})/2}^{\infty} x f_X(x)\, dx.$$

Now, in the ith sub-integral, we approximate the integrand $x f_X(x)$ with $\mathbf{X}_{(i|n)} f_X(x)$. Thus, the ith sub-integral, for $i = 2, \dots, n-1$, is approximated as

$$\mathbf{X}_{(i|n)} \int_{(\mathbf{X}_{(i-1|n)}+\mathbf{X}_{(i|n)})/2}^{(\mathbf{X}_{(i|n)}+\mathbf{X}_{(i+1|n)})/2} f_X(x)\, dx \qquad (5.8)$$
$$= \mathbf{X}_{(i|n)} \left[F_X\left(\frac{\mathbf{X}_{(i|n)} + \mathbf{X}_{(i+1|n)}}{2}\right) - F_X\left(\frac{\mathbf{X}_{(i-1|n)} + \mathbf{X}_{(i|n)}}{2}\right)\right].$$

Similarly, the first sub-integral is approximated as

$$\mathbf{X}_{(1|n)} \left[F_X\left(\frac{\mathbf{X}_{(1|n)} + \mathbf{X}_{(2|n)}}{2}\right) - 0\right] \qquad (5.9)$$

and the nth sub-integral is approximated as

$$\mathbf{X}_{(n|n)} \left[1 - F_X\left(\frac{\mathbf{X}_{(n-1|n)} + \mathbf{X}_{(n|n)}}{2}\right)\right]. \qquad (5.10)$$

The next step is to approximate the true distribution function F_X with its estimator (5.3). Note that each of the cumulative distribution function differences in (5.8)–(5.10) straddles one of the 'steps' in (5.3). Thus, each of these differences is equal to $1/n$ and the ith sub-integral in (5.7) is further approximated as $\frac{1}{n}\mathbf{X}_{(i|n)}$. Finally we obtain

$$\mu \approx \sum_{i=1}^{n} \frac{1}{n} \mathbf{X}_{(i|n)}$$
$$= \frac{1}{n} \sum_{i=1}^{n} \mathbf{X}_i = \widehat{\mu}.$$

5.2.6 Estimating the Second and Higher Moments. Useful estimators for the jth moment $\mu^{(j)} = \int_{-\infty}^{\infty} x^j f_X(x)\, dx$ can be defined, in a manner similar to that of the first moment, as

$$\widehat{\mu^{(j)}} = \frac{1}{n} \sum_{k=1}^{n} \mathbf{X}_k^j.$$

For the second central moment, the variance, we have

$$\widehat{\sigma}^2 = \frac{1}{n} \sum_{k=1}^{n} (\mathbf{X}_k - \widehat{\mu})^2. \qquad (5.11)$$

Note that the estimator (5.11) differs from the sample variance (4.5) by a factor of $n/(n-1)$. We return to this point in [5.3.7].

The same rules that apply to moments apply to the estimated moments as well. For example, for $\widehat{\sigma}^2$ as given in (5.11),

$$\widehat{\sigma}^2 = \widehat{\mu^{(2)}} - \left(\widehat{\mu^{(1)}}\right)^2.$$

5.2.7 Mean Vectors, Covariances, and Correlations. The univariate estimators of the mean and variance defined above are easily extended to apply to samples of n iid random vectors $\{\vec{\mathbf{X}}_1, \dots, \vec{\mathbf{X}}_n\}$ distributed as the random vector $\vec{\mathbf{X}}$. The mean vector is estimated as

$$\widehat{\vec{\mu}} = \frac{1}{n} \sum_{i=1}^{n} \vec{\mathbf{X}}_i \qquad (5.12)$$

and, in analogy to the sample variance, the covariance matrix Σ (see [2.8.7] and (2.32)) may be estimated with the *sample covariance matrix* as

$$\widehat{C} = \frac{1}{n-1} \sum_{i=1}^{n} (\vec{\mathbf{X}}_i - \widehat{\vec{\mu}})(\vec{\mathbf{X}}_i - \widehat{\vec{\mu}})^{\mathrm{T}}. \qquad (5.13)$$

As with the variance, we can also define an estimator $\widehat{\Sigma}$, expressed in terms of the moments of the sample, and obtained by dividing the sum of products in (5.13) by n rather than $n-1$:

$$\widehat{\Sigma} = \frac{1}{n} \sum_{i=1}^{n} (\vec{\mathbf{X}}_i - \widehat{\vec{\mu}})(\vec{\mathbf{X}}_i - \widehat{\vec{\mu}})^{\mathrm{T}}.$$

When we want to clarify that the estimated covariance matrix refers to the random vector $\vec{\mathbf{X}}$, we add subscripts to matrices $\widehat{C}$ or $\widehat{\Sigma}$. The elements of the estimated covariance matrix $\widehat{\Sigma}$, denoted $\widehat{\sigma}_{jk}$, are given by

$$\widehat{\sigma}_{jk} = \frac{1}{n} \sum_{i=1}^{n} (\mathbf{X}_{i;j} - \widehat{\mu}_j)(\mathbf{X}_{i;k} - \widehat{\mu}_k), \qquad (5.14)$$

where $\mathbf{X}_{i;j}$ represents the jth component of the ith random vector $\vec{\mathbf{X}}_i$. Similarly $\widehat{\mu}_j$ is the jth component of the estimated mean vector $\widehat{\vec{\mu}}$.

It may happen, in practice, that there are missing values in some of the n sample vectors $\vec{\mathbf{x}}_1, \dots, \vec{\mathbf{x}}_n$. Then the summations in (5.12) and (5.14) are taken only over the non-missing values and the sums are divided not by n but by the number of terms in the sum. Theoretical results concerning properties of the estimators may not extend smoothly when there are gaps in the data.

The correlation between the jth and kth elements of $\vec{\mathbf{X}}$ is

$$\rho_{ij} = \frac{\sigma_{jk}}{\sqrt{\sigma_{jj}\sigma_{kk}}},$$

where σ_{jk} is the covariance between $\mathbf{X}_j$ and $\mathbf{X}_k$, and σ_{jj} and σ_{kk} are the corresponding variances (see [2.8.7] and (2.33)).

This correlation is estimated with the *sample correlation*

$$\widehat{\rho}_{jk} = \frac{\widehat{\sigma}_{jk}}{\sqrt{\widehat{\sigma}_{jj}\widehat{\sigma}_{kk}}}. \tag{5.15}$$

5.2.8 Estimating L-Moments. Recall that L-moments (see [2.6.9] and (2.20)–(2.24)) are the expected values of linear combinations of order statistics of samples that are the same size as the order of the L-moment. For example, the third L-moment is the expected value of a linear combination of the order statistics of a sample of size three. The natural way to estimate an L-moment [183] is with a *U statistic* (first described by Hoeffding [178]). That is, if the third L-moment is to be estimated, then, at least conceptually, all possible sub-samples of size three are selected from the full sample, the linear combination is computed, as for the expected order statistics, from the order statistics of each sub-sample, and these linear combinations are averaged. Hosking [183] uses combinatorial arguments to show that the jth L-moment can be estimated as

$$\widehat{\lambda^{(j)}} = \sum_{l=0}^{j-1}(-1)^{j-l-1}\binom{j-1}{l}\binom{j+l-1}{l}b_l \tag{5.16}$$

where

$$b_l = \frac{1}{n}\sum_{i=1}^{n}\frac{(i-1)(i-2)\cdots(i-l)}{(n-1)(n-2)\cdots(n-l)}\mathbf{X}_{(i|n)}.$$

5.3 Properties of Estimators

5.3.1 Estimator Selection Criterion. Chapter 4 mentions that a good estimator will produce estimates $\widehat{\alpha}$ in the neighbourhood of the true parameter value α. A mathematically concise definition of 'in the neighbourhood' is obtained by defining a 'distance' such as the *mean squared error*

$$M(\widehat{\alpha};\alpha) = \mathcal{E}\big((\widehat{\alpha}-\alpha)^2\big). \tag{5.17}$$

The mean squared error allows us to compare two estimators. In particular, we have the following definition about the relative efficiency of estimators:

Let $\widehat{\alpha}$ and $\widetilde{\alpha}$ be two competing estimators of a parameter α. Then $\widehat{\alpha}$ is said to be a more efficient *estimator of α than $\widetilde{\alpha}$ if $M(\widehat{\alpha};\alpha) < M(\widetilde{\alpha};\alpha)$ for all possible values of α.*

Estimators that have mean squared error less than or equal to that of all other estimators of α are obviously desirable. However, other properties, such as *unbiasedness* (defined in [5.3.3]) are also desirable. In [5.3.7] we show that the mean squared error may be written as the sum of the mean squared *bias* and the variance of the estimator. Because lack of bias is often very desirable, the search for efficient estimators is often restricted to unbiased estimators. Thus, statisticians often search for *minimum variance unbiased* estimators. The search is often further restricted to estimators that can be expressed as linear combinations of the random variables that make up the sample.

We will continue to discuss the bias and variance of a variety of estimators after formally defining bias.

5.3.2 Definition: Bias. *Let α be a parameter of the distribution of random variable $\mathbf{X}$ and let $\widehat{\alpha}$ be an estimator of this parameter. Then the* bias *of estimator $\widehat{\alpha}$ is its expected, or mean, error, which is given by*

$$B(\widehat{\alpha}) = \mathcal{E}(\widehat{\alpha}) - \alpha.$$

Positive bias indicates that $\widehat{\alpha}$ overestimates α, on average, when the experiment that generates the sample is repeated several times. Similarly, negative bias indicates that $\widehat{\alpha}$ underestimates α, on average. An estimator that has no bias is said to be *unbiased.*

Positive bias does not imply that all realizations of $\widehat{\alpha}$ are greater than α, although that could be true if $B(\widehat{\alpha})$ is large compared with the variability of $\widehat{\alpha}$. Also, unless we know something about the distribution of $\widehat{\alpha}$, we can not say what proportion of realizations of $\widehat{\alpha}$ will be greater than α. For example, if $\widehat{\alpha}$ is positively biased and distributed symmetrically about $\mathcal{E}(\widehat{\alpha})$, then we can say that more than 50% of all estimates will be larger than α. However, if the distribution of $\widehat{\alpha}$ is skewed, then we can make this statement only if we know that the *median*[4] value of $\widehat{\alpha}$ is greater than α. Similar comments apply if $\widehat{\alpha}$ is negatively biased.

It is highly desirable to have estimators with little or no bias, but, as we will see below, it may be necessary to balance small bias against other desirable properties.

5.3.3 The Bias of Some Estimators.

We now derive the bias of some frequently used estimators. The propositions to be proved appear in italics.

The empirical distribution function $\widehat{F}_X$ (5.3) has zero bias as an estimator of the cumulative distribution function F_X. That is,

$$B(\widehat{F}_X) = 0. \tag{5.18}$$

To prove this, recall that $n\widehat{F}_X(y)$ is the number of random variables $\mathbf{X}_k$ in the sample such that $\mathbf{X}_k < y$. As usual, all random variables are assumed to be independent and identically distributed. Since the random variables are identically distributed, $P(\mathbf{X}_k \le y) = F_X(y)$. Thus, using independence, we see that the integer-valued random variable $n\widehat{F}_X(y)$ has the binomial distribution $\mathcal{B}(n, F_X(y))$. Therefore $\mathcal{E}(n\widehat{F}_X(y)) = nF_X(y)$ for all y. This proves (5.18).

The sample mean $\widehat{\mu}$ (5.6) is an unbiased estimator of μ. That is,

$$B(\widehat{\mu}) = 0. \tag{5.19}$$

The proof of (5.19) is straightforward:

$$\mathcal{E}(\widehat{\mu}) = \frac{1}{n}\sum_k \mathcal{E}(\mathbf{X}_k) = \frac{1}{n}n\mathcal{E}(\mathbf{X}) = \mu.$$

The sample variance $\mathbf{S}^2$ (4.5) is an unbiased estimator of σ^2, while $\widehat{\sigma}^2$ (5.11), is a biased estimator of σ^2. The bias of the latter is given by[5]

$$B(\widehat{\sigma}^2) = \frac{1}{n}\sigma^2. \tag{5.20}$$

The bias of $\mathbf{S}^2$ and $\widehat{\sigma}^2$ is derived as follows. First, note that

$$\begin{aligned}\sum_{i=1}^{n}(\mathbf{X}_i - \widehat{\mu})^2 &= \sum_{i=1}^{n}(\mathbf{X}_i - \mu - \widehat{\mu} + \mu)^2 \\ &= \sum_{i=1}^{n}(\mathbf{X}_i - \mu)^2 - n(\widehat{\mu} - \mu)^2.\end{aligned}$$

Then

$$\begin{aligned}\mathcal{E}(\widehat{\sigma}^2) &= \frac{1}{n}\mathcal{E}\big(\textstyle\sum_{k=1}^{n}(\mathbf{x}_k - \widehat{\mu})^2\big) \\ &= \frac{1}{n}\mathcal{E}\big(\textstyle\sum_{k=1}^{n}(\mathbf{x}_k - \mu)^2\big) - \mathcal{E}\big((\widehat{\mu} - \mu)^2\big) \\ &= \frac{1}{n}\sum_{k=1}^{n}\sigma^2 - \mathrm{Var}(\widehat{\mu}) \qquad (5.21)\\ &= \sigma^2 - \mathrm{Var}(\widehat{\mu}). \qquad (5.22)\end{aligned}$$

The step that results in (5.21) requires the 'identically distributed' assumption. We will show below that $\mathrm{Var}(\widehat{\mu}) = \frac{1}{n}\sigma^2$ if the random variables in the sample are also independent.[6] Thus, (5.20) is proven. The unbiasedness of $\mathbf{S}^2$ follows from the relationship $\mathbf{S}^2 = (\frac{n}{n-1})\widehat{\sigma}^2$.

Similar results are obtained for the multivariate mean and the sample covariance matrix:

$$\begin{aligned}B(\widehat{\vec{\mu}}) &= 0\\ B(\widehat{C}) &= 0\\ B(\widehat{\Sigma}) &= \frac{1}{n}\Sigma.\end{aligned}$$

The uncertainty of the estimator of the mean vector is easily characterized as

$$\mathrm{Cov}\big(\widehat{\vec{\mu}}, \widehat{\vec{\mu}}\big) = \frac{1}{n}\Sigma,$$

but the uncertainty of the estimator of the covariance matrix Σ is not easily characterized

[4]The *median* of a random variable $\mathbf{X}$ is a value $\mathbf{x}_{0.5}$ such that $P(\mathbf{X} \le \mathbf{x}_{0.5}) \le 0.5$ and $P(\mathbf{X} \ge \mathbf{x}_{0.5}) \ge 0.5$ (see [2.6.4]). If the distribution of $\mathbf{X}$ is symmetric about the mean $\mu = \mathcal{E}(\mathbf{X})$ (i.e., $f_X(x - \mu) = f_X(x + \mu)$ for all $x \ge 0$), then $\mathbf{x}_{0.5} = \mu$. If $\mathbf{X}$ is skewed, with a large tail to the right, $\mathbf{x}_{0.5} < \mu$, and $\mathbf{x}_{0.5} > \mu$ if $\mathbf{X}$ is skewed with a large tail on the left.

[5]It is assumed here that the sample consists of iid random variables. Both estimators are, in general, biased if the independence assumption is replaced by the more general assumption that the sample is obtained from a stationary, ergodic stochastic process.

[6]The bias is caused by the $\mathrm{Var}(\widehat{\mu})$ term in (5.22). This term can be considerably greater than σ^2/n when the independence assumption is replaced by the stationary and ergodic assumption. Then the 'memory' within the sample tends to inflate the variance of $\widehat{\mu}$ (see Section 6.6).

because it involves all of the fourth moments of $\vec{\mathbf{X}}$. This is possible using the Wishart distribution when $\vec{\mathbf{X}}$ is multivariate normal [2.8.9] (see [197] [147]).

5.3.4 Asymptotically Unbiased Estimators.

We have shown that the empirical distribution function $\widehat{F}_X$, the sample mean $\widehat{\mu} = \overline{\mathbf{X}}$, and the sample variance $\mathbf{S}^2$ are all unbiased estimators of the distribution function, of the mean, and of the variance, respectively, when the sample consists of iid random variables. On the other hand, $\widehat{\sigma}^2$ (5.11) is a biased estimator of the variance. Here the bias disappears as sample size increases. Indeed,

$$\lim_{n\to\infty} B(\widehat{\sigma}^2) = 0.$$

Estimators with this property are said to be *asymptotically unbiased.*

Many biased estimators are asymptotically unbiased, for example, the estimator of the correlation coefficient ρ (5.15) or the estimator of the L-moments (5.16).

5.3.5 Variances of Some Estimators.

We derive here the expression for the variance of the sample mean used in [5.3.3] as well as some other results. Again we assume that the sample consists of n independent and identically distributed random variables.

The variance of the empirical distribution function $\widehat{F}_X$ (5.3) at point x is given by

$$\mathrm{Var}\big(\widehat{F}_X(x)\big) = \frac{1}{n} F_X(x)(1 - F_X(x)). \tag{5.23}$$

The proof of (5.18) shows that $n\widehat{F}_X(x) \sim \mathcal{B}(n, F_X(x))$. Therefore, using (2.9), we obtain $\mathrm{Var}\big(n\widehat{F}_X(x)\big) = nF_X(x)\,(1 - F_X(x))$, proving (5.23).

The variance of the sample mean $\widehat{\mu}$ (5.6) is given by

$$\mathrm{Var}(\widehat{\mu}) = \frac{1}{n}\sigma^2. \tag{5.24}$$

To demonstrate this we first note that

$$\begin{aligned}
&\mathrm{Var}(\widehat{\mu}) \\
&= \mathcal{E}\Big(\big(\tfrac{1}{n}\textstyle\sum_{k=1}^n \mathbf{x}_k\big)^2 - \mu^2\Big) \\
&= \frac{1}{n^2}\sum_{k,j=1}^{n} \mathcal{E}\big(\mathbf{x}_k\mathbf{x}_j - \mu^2\big) \\
&= \frac{1}{n^2}\sum_{k,j=1}^{n} \mathcal{E}\big((\mathbf{x}_k - \mu)(\mathbf{x}_j - \mu)\big).
\end{aligned}$$

Now, using independence, all the expectations in the last expression vanish except those where $k = j$. Consequently

$$\mathrm{Var}(\widehat{\mu}) = \frac{1}{n^2}\sum_k \sigma^2 = \frac{1}{n}\sigma^2.$$

The variance of $\widehat{\sigma}^2$ (5.11) is given by

$$\begin{aligned}
\mathrm{Var}(\widehat{\sigma}^2) &= \frac{1}{n}(\gamma^* - \sigma^4) \\
&\quad - \frac{2}{n^2}(\gamma^* - 2\sigma^4) + \frac{1}{n^3}(\gamma^* - 3\sigma^4),
\end{aligned} \tag{5.25}$$

where $\gamma^* = \mathcal{E}\big((\mathbf{X} - \mu)^4\big)$ is the fourth central moment.[7] *The variance of $\mathbf{S}^2$ is $n^2/(n-1)^2$ times the variance of $\widehat{\sigma}^2$.*

The proof of this result is lengthy but elementary (see [325]).

When the sample consists of iid normal random variables, the variance of the sample variance $\mathbf{S}^2$ and the biased variance estimator $\widehat{\sigma}^2$ are

$$\mathrm{Var}(\widehat{\sigma}^2) = \frac{2(n-1)}{n^2}\sigma^4 \tag{5.26}$$

$$\mathrm{Var}(S^2) = \frac{2}{n-1}\sigma^4. \tag{5.27}$$

For normal random variables, $\gamma_2 = 0$, so (5.26) and (5.27) are a direct consequence of (5.25).

It can be shown that the estimator (5.15) of the correlation coefficient ρ has asymptotic variance equal to $(1 - \rho_{ij}^2)/n$, meaning that

$$\lim_{n\to\infty} \mathrm{Var}(\widehat{\rho}_{ij}) = \frac{1 - \rho_{ij}^2}{n}.$$

We describe the uncertainty of this estimator when samples are finite in [8.2.3].

Hosking provides an expression for the asymptotic covariance matrix of the L-moment estimator (5.16), but this expression is difficult to use because it depends upon the form of the distribution of the elements of the sample.

5.3.6 Consistency.

Another desirable property of an estimator is that it be *consistent.*

An estimator $\widehat{\alpha}$ is 'consistent' if its mean squared error (5.17) goes to zero with increasing sample size. That is, if

$$\lim_{n\to\infty} M(\widehat{\alpha}; \alpha) = 0.$$

All of the estimators discussed in [5.3.3]–[5.3.5] can be shown to be consistent using the following proposition.

[7] The fourth central moment is related to the kurtosis via $\gamma_2 = \gamma^*/(\sigma^4 - 3)$ (see (2.19)).

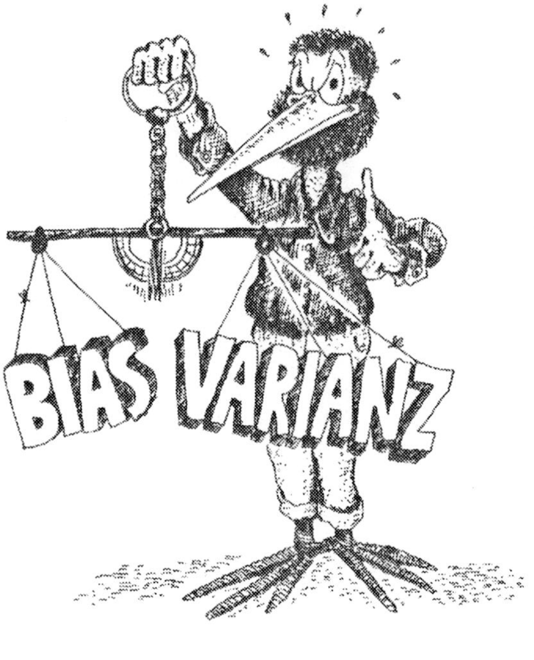

Figure 5.3: *Bias* **and** *variance contribute to the expected mean squared error.*

The mean squared error of an estimator $\widehat{\alpha}$ is the sum of its squared bias and its variance (see Figure 5.3). That is,

$$M(\widehat{\alpha};\alpha) = [B(\widehat{\alpha})]^2 + \mathrm{Var}(\widehat{\alpha}). \tag{5.28}$$

The proof is easy to demonstrate.

$$\begin{aligned} M(\widehat{\alpha};\alpha) &= \mathcal{E}\big((\widehat{\alpha}-\alpha)^2\big) \\ &= \mathcal{E}\big((\widehat{\alpha}-\mathcal{E}(\widehat{\alpha}) - (\alpha - \mathcal{E}(\widehat{\alpha})))^2\big) \\ &= \mathcal{E}\big((\widehat{\alpha}-\mathcal{E}(\widehat{\alpha}))^2\big) + \big(\alpha-\mathcal{E}(\widehat{\alpha})\big)^2 \\ &\quad - 2\big(\alpha-\mathcal{E}(\widehat{\alpha})\big)\mathcal{E}\big(\widehat{\alpha}-\mathcal{E}(\widehat{\alpha})\big). \end{aligned}$$

The cross-product term in the last expression is zero, so (5.28) follows. Therefore, any asymptotically unbiased estimator with variance that is asymptotically zero is consistent.

5.3.7 Bias Correction and the Jackknife. We showed in [5.3.3] that $\widehat{\sigma}^2$ is a biased estimator of σ^2 with $B(\widehat{\sigma}^2) = \frac{1}{n}\sigma^2$. We also showed that the sample variance $\mathbf{S}^2$ corrects this bias by multiplying the estimator $\widehat{\sigma}^2$ by $n/(n-1)$.

Many bias corrections are of the above form, that is, a bias correction is often made by scaling $\widehat{\alpha}$, a biased estimator of α, by a constant $c(n)$ so that the resulting estimator $\widetilde{\alpha} = \widehat{\alpha}/c(n)$ is an unbiased estimator of α. Biases and the corresponding bias corrections come in a variety of forms, however, so there is no general rule about the form of these corrections.

The consequences of bias correction are interesting even in this limited context, that is, where a scale correction will make an estimator unbiased. In particular, the 'improved' $\widetilde{\alpha}$ may not always be more efficient than the original $\widehat{\alpha}$. If the scaling factor $c(n) > 1$, then $\widetilde{\alpha}$ is more efficient than $\widehat{\alpha}$ because both components of the expected mean square error, the squared bias, and the variance, have been reduced. On the other hand, if $c(n) < 1$, the bias is reduced but the variance is enhanced. Thus, it is generally advised that the 'improved' estimator be accepted with caution.

The scaling factor that turns biased $\widehat{\sigma}^2$ into the unbiased $\mathbf{S}^2$ is $c(n) = (n-1)/n < 1$. The mean squared error for the unbiased estimator $\mathbf{S}^2$ is

$$M(S^2;\sigma^2) = \mathrm{Var}\big(S^2\big) = \frac{2}{n-1}\sigma^4,$$

while that for the biased estimator $\widehat{\sigma}^2$ is

$$\begin{aligned} M(\widehat{\sigma}^2;\sigma^2) &= \frac{1}{n^2}\sigma^4 + \frac{2(n-1)}{n^2}\sigma^4 \\ &= \frac{2n-1}{n^2}\sigma^4. \end{aligned}$$

Since

$$\frac{2n-1}{n^2} < \frac{2}{n-1},$$

we see that the biased estimator $\widehat{\sigma}^2$ is slightly more efficient than the unbiased estimator $\mathbf{S}^2$. We will see shortly that the biased estimator is also the *maximum likelihood estimator* of σ^2.

An empirical approach frequently used to find bias corrections is called the *jackknife* (see Efron [111] or Quenouille [326]). The idea is that the estimator is computed from the full sample, then recomputed n times, leaving a different observation out each time. These estimators are denoted $\widehat{\alpha}$ and $\widehat{\alpha}_{(i)}$, where the subscript (i) indicates that $\widehat{\alpha}_{(i)}$ is computed with $\mathbf{X}_i$ removed from the sample. The *jackknife bias correction*, which is subtracted from $\widehat{\alpha}$, is then given by

$$\widehat{\alpha}_B = (n-1)(\widehat{\alpha}_{(\cdot)} - \widehat{\alpha}),$$

where

$$\widehat{\alpha}_{(\cdot)} = \frac{1}{n}\sum_{i=1}^{n}\widehat{\alpha}_{(i)}.$$

The *jackknifed* estimator, $\widetilde{\alpha} = \widehat{\alpha} - \widehat{\alpha}_B$, can often be re-expressed in the form $\widetilde{\alpha} = \widehat{\alpha}/c(n)$.

It can be shown, with some algebraic manipulation, that the jackknifed bias correction for $\widehat{\sigma}^2$ is

$$\widehat{\sigma}^2_B = -\frac{1}{n(n-1)}\sum_{i=1}^{n}(\mathbf{X}_i - \overline{\mathbf{X}})^2.$$

Therefore the jackknifed estimator of σ^2 is

$$\widetilde{\sigma}^2 = \widehat{\sigma}^2 - \widehat{\sigma}_B^2 = \mathbf{S}^2.$$

A jackknifing approach can also be used to estimate variance of an estimator $\widehat{\alpha}$. Tukey [374] suggested that the variance of $\widehat{\alpha}$, say $\sigma_{\widehat{\alpha}}^2$, could be estimated with

$$\widehat{\sigma}_{\widehat{\alpha}}^2 = \frac{n-1}{n}\sum_{i=1}^{n}(\widehat{\alpha}_{(i)} - \widehat{\alpha}_{(\cdot)})^2.$$

Efron [111] explains why this works. The jackknife estimator of the variance of the sample mean is

$$\widehat{\sigma}_{\widehat{\mu}}^2 = \frac{1}{n}\mathbf{S}^2,$$

which is also the estimator obtained when we replace σ^2 with $\mathbf{S}^2$ in (5.24).

5.3.8 Maximum Likelihood Method. The estimators introduced in this section have been arbitrary so far. One systematic approach to obtaining estimators is the *Method of Maximum Likelihood*, introduced by R.A. Fisher [119, 120] in the 1920s.

The Maximum Likelihood Estimator of the Parameter of the Binomial Distribution. The idea is most easily conveyed through an example. For simplicity, suppose that our sample consists of n iid Bernoulli random variables $\{\mathbf{X}_1, \ldots, \mathbf{X}_n\}$ [2.4.2], which take values 0 or 1 with probabilities $1 - p$ and p, respectively. The problem is to estimate p.

The probability of observing a particular set of realizations $\{\mathbf{x}_1, \ldots, \mathbf{x}_n\}$ is

$$P(\mathbf{X}_1 = \mathbf{x}_1, \ldots, \mathbf{X}_n = \mathbf{x}_n) = p^h(1-p)^{n-h} \tag{5.29}$$

where $\mathbf{h} = \sum_{i=1}^{n}\mathbf{x}_i$. Therefore, we see that the useful information about p is carried not by the individual random variables $\mathbf{X}_i$ but by their sum

$$\mathbf{H} = \sum_{i=1}^{n}\mathbf{X}_i.$$

We come to this conclusion because (5.29) has the same value regardless of the order in which the contributions to $\mathbf{h}$ (i.e., the 0s and 1s) were observed. Thus our estimator should be based on the statistic $\mathbf{H}$.

The probability distribution of $\mathbf{H}$ is the binomial distribution (2.7)

$$f_H(h; p) = \binom{n}{h}p^h(1-p)^{n-h}. \tag{5.30}$$

Now suppose that we have observed $\mathbf{H} = \mathbf{h}$. The *likelihood* of observing $\mathbf{h}$ for a particular value of the parameter p is given by the *likelihood function*

$$L_H(p) = f_H(h; p). \tag{5.31}$$

The likelihood function is identical to the probability distribution of our statistic $\mathbf{H}$ except that it is now viewed as a function of the parameter p.

The *maximum likelihood estimator* (MLE) of p is now obtained by determining the value of parameter p for which the observed value $\mathbf{h}$ of $\mathbf{H}$ is most likely. That is, given $\mathbf{H} = \mathbf{h}$, (5.31) is maximized with respect to p.

It is often easier to maximize the *log-likelihood function*

$$l_H(p) = \ln(L_H(p)),$$

which is defined as the natural log of the likelihood function. For this example the log-likelihood is given by

$$l_H(p) = \ln\binom{n}{h} + h\ln(p) + (n-h)\ln(1-p). \tag{5.32}$$

We maximize (5.32) by taking the derivative of $l_H(p)$ with respect to p and solving the equation obtained by setting the derivative to zero. In the present example there will be only one solution to this equation. However, there may be many solutions in general, and it is necessary to select the solution that produces the overall maximum of l (or, equivalently, L).

Taking the partial derivative of (5.32) and setting it to zero, we obtain

$$\frac{\partial l_H(p)}{\partial p} = \frac{h}{p} - \frac{n-h}{1-p} = 0. \tag{5.33}$$

The unique solution of (5.33) is $p' = h/n$. The corresponding MLE of p, written in random variable form, is $\widehat{p} = \mathbf{H}/n$. Thus, we have discovered that here the estimator we would intuitively use to estimate p is also its maximum likelihood estimator.

The Maximum Likelihood Estimator in General. We will continue to assume that our sample consists of n iid random variables, $\{\mathbf{X}_1, \ldots, \mathbf{X}_n\}$, all distributed as random variable $\mathbf{X}$. For convenience we will assume that they are continuous, and refer to probability density functions rather than probability distributions. However, everything here can be repeated with probability distributions simply by replacing all occurrences of density functions with probability distributions.

Let $f_X(x;\vec{\alpha})$ be the density function of $\mathbf{X}$, where $\vec{\alpha}$ is a vector containing the parameters of the distribution of $\mathbf{X}$. The joint probability density function for the random vector $(\mathbf{X}_1,\dots,\mathbf{X}_n)^{\mathrm{T}}$ is

$$f_{X_1\dots X_n}(x_1,\dots,x_n;\vec{\alpha}) = \prod_{i=1}^n f_X(x_i;\vec{\alpha})$$

(see (2.12)). Suppose we have observed $\mathbf{X}_i = \mathbf{x}_i,\ i=1,\dots,n$. Then the *likelihood function* for the unknown parameters $\vec{\alpha}$ is

$$L_{X_1\dots X_n}(\vec{\alpha}) = \prod_{i=1}^n f_X(x_i;\vec{\alpha}), \tag{5.34}$$

and the corresponding *log-likelihood function* is given by

$$l_{X_1\dots X_n}(\vec{\alpha}) = \sum_{i=1}^n \ln(f_X(x_i;\vec{\alpha})). \tag{5.35}$$

The *maximum likelihood estimator* $\widehat{\vec{\alpha}}$ of $\vec{\alpha}$ is found by maximizing (5.34) or (5.35) with respect to $\vec{\alpha}$.

The Appeal of Maximum Likelihood Estimators. There are several good reasons to use maximum likelihood estimators. First, as we have noted, the method of maximum likelihood provides a systematic way to search for estimators. Second, MLEs tend to have pleasing asymptotic properties. They can be shown to be consistent and asymptotically normal under fairly general conditions (see, e.g., Cox and Hinkley [92], Section 9.2). The asymptotic normality can, in turn, be used to construct asymptotic confidence regions.[8]

5.3.9 Maximum Likelihood Estimators of the Mean and the Variance of a Normal Random Variable. We derive the MLEs of the mean and the variance of a normal distribution $\mathcal{N}(\mu,\sigma^2)$ from a sample of n iid normal random variables using (5.35). The natural log of the normal density function is given by

$$\ln(f_X(x;\mu,\sigma^2)) = -\frac{1}{2}\ln(2\pi\sigma^2) - \frac{(\mathbf{x}-\mu)^2}{2\sigma^2}.$$

Consequently, the log-likelihood function is given by

$$l_{X_1\dots X_n}(\mu,\sigma^2) = -\frac{n}{2}\ln(2\pi\sigma^2) - \sum_{i=1}^n \frac{(\mathbf{x}_i-\mu)^2}{2\sigma^2}.$$

[8] That is, confidence regions that attain the specified coverage, say 95%, as the sample becomes large.

Differentiation yields

$$\frac{\partial l_{X_1\dots X_n}(\mu,\sigma^2)}{\partial\mu} = \sum_{i=1}^n \frac{\mathbf{x}_i-\mu}{\sigma^2} \tag{5.36}$$

$$\frac{\partial l_{X_1\dots X_n}(\mu,\sigma^2)}{\partial\sigma^2} = -\frac{n}{2\sigma^2} + \sum_{i=1}^n \frac{(\mathbf{x}_i-\mu)^2}{2\sigma^4}. \tag{5.37}$$

We obtain the MLE of the mean by setting (5.36) to zero, to obtain

$$\mu' = \frac{1}{n}\sum_{i=1}^n \mathbf{x}_i. \tag{5.38}$$

Re-expressing (5.38) in the random variable form, we find that the sample mean $\widehat{\mu} = \overline{\mathbf{X}}$ (see [4.3.1], [5.3.3] and [5.3.5]) is the maximum likelihood estimator of the mean.

Similarly, setting (5.37) to zero, we obtain

$$\sigma^{2'} = \frac{1}{n}\sum_{i=1}^n (\mathbf{x}_i-\mu)^2.$$

Then, replacing μ with its MLE, and rewriting the resulting expression in random variable form, we obtain

$$\widehat{\sigma}^2 = \frac{1}{n}\sum_{i=1}^n (\mathbf{X}_i-\widehat{\mu})^2$$

as the maximum likelihood estimator of the variance. Thus, we see that the MLE of the variance is the biased estimator introduced in [5.2.6].

5.3.10 MLEs of Related Estimators. The following theorem (see, for example, Pugachev [325]) extends the utility of a maximum likelihood estimator:

Consider a random vector $\vec{\mathbf{X}}$ with two parameters $\vec{\alpha}$ and $\vec{\beta}$, related to each other through $g(\vec{\alpha}) = \vec{\beta}$ and $g^{-1}(\vec{\beta}) = \vec{\alpha}$, where g and g^{-1} are both continuous. If $\widehat{\vec{\alpha}}$ is an MLE of $\vec{\alpha}$, then $\widehat{\vec{\beta}} = g(\widehat{\vec{\alpha}})$ is an MLE of $\vec{\beta}$. Similarly, if $\widehat{\vec{\beta}}$ is an MLE of $\vec{\beta}$, then $\widehat{\vec{\alpha}} = g^{-1}(\widehat{\vec{q}}\,)$ is an MLE of $\vec{\alpha}$.

There are various applications of this theorem. For example, suppose $\vec{\mathbf{X}}$ is a normal random vector with covariance matrix Σ. Let $\{\lambda_1,\dots,\lambda_n\}$ be the eigenvalues of Σ and let $\{\vec{e}^{\,1},\dots,\vec{e}^{\,n}\}$ be the corresponding eigenvectors (see Chapter 13). Both the covariance matrix (corresponding to $\vec{\alpha}$ in the theorem above) and its eigenvalues and eigenvectors (corresponding to $\vec{\beta}$) are parameter vectors of $\vec{\mathbf{X}}$. Moreover, there is a continuous, one-to-one relationship between these two representations of the covariance structure of $\vec{\mathbf{X}}$. Therefore, since the covariance estimator $\widehat{\mathbf{\Sigma}}$ in [5.2.7] is the maximum likelihood estimator of $\mathbf{\Sigma}$, it follows from the theorem that the eigenvalues and eigenvectors of $\widehat{\mathbf{\Sigma}}$ are MLEs of the eigenvalues and eigenvectors of $\mathbf{\Sigma}$.

5.4 Interval Estimators

5.4.1 What are Confidence Intervals? So far we have dealt with *point estimates*, that is, prescriptions that describe how to use the information in a sample to estimate a specific parameter of a random variable. We were sometimes able to make statements about the statistical properties of the estimators in repeated sampling, such as their mean squared errors, their biases and their variances.

In the following we deal with *interval estimation*, that is, the estimation of intervals or regions that will cover the unknown, but fixed, parameter with a given probability.

Statisticians often use the word *coverage* when discussing confidence intervals since the location of parameter α is fixed on the real line. A $\tilde{p} \times 100\%$ confidence interval for α is constructed from two statistics $\widehat{\alpha}_L$ and $\widehat{\alpha}_U, \widehat{\alpha}_L < \widehat{\alpha}_U$, such that

$$P\big((\widehat{\alpha}_L, \widehat{\alpha}_U) \ni \alpha\big) = \tilde{p}. \tag{5.39}$$

We use the symbol $\ni$ to mean that the set on the left covers the point on the right in (5.39). The confidence level $\tilde{p}$ is chosen to be relatively large (e.g., $\tilde{p} = 0.95$). The upper and lower limits of the confidence interval are random variables; they are functions of the n random variables $\mathbf{X}_1, \ldots, \mathbf{X}_n$ that represent the sampling mechanism. Thus, the interval varies in length and location on the real line. The interval is constructed so that it will *cover* the fixed point α on the real line $\tilde{p} \times 100\%$ of the time. That is, $\tilde{p} \times 100\%$ of the realizations of the confidence interval will lie on top of point α. Figure 5.4 illustrates this concept.

Many authors use the word 'contain' in the context of confidence intervals, that is, they state that the confidence interval will contain the unknown parameter $\tilde{p} \times 100\%$ of the time. We have found this language to be a great source of confusion because it somehow implies that the parameter α is random. Rather, it is the endpoints of the confidence interval that are random; they vary from one realization of the sample to the next. Note that, conditional upon a particular sample, everything about the confidence interval is fixed (both the endpoints and parameter α) and,

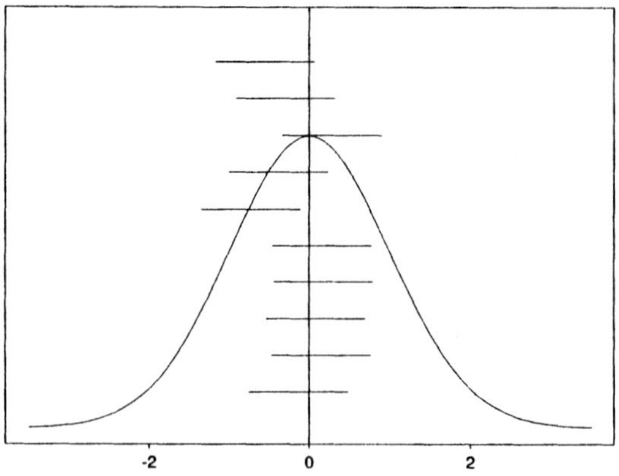

Figure 5.4: *Ten realizations of a 95% confidence interval for unknown parameter α. On average, 19 out of 20 intervals will cover α. In this example, $\alpha = 0$. The curve shows the density function of the sampled random variable.*

thus, no probabilitist interpretation can be given to the interval. Rather, the interval is interpreted as reporting a range of parameter values that are strongly consistent with the realized sample (i.e., this is a range of possible parameters for which the likelihood function [5.3.8] is large). The *confidence level* indicates the average behaviour of the reporting procedure, but it does not, and can not, give a probabilitist interpretation to any one realization of the confidence interval.

5.4.2 Confidence Interval for a Random Variable—Optional.[9] While the discussion to this point has focused on the probability that a random interval $(\widehat{\alpha}_L, \widehat{\alpha}_U)$, defined as a function of random variables $\mathbf{X}_1, \ldots, \mathbf{X}_n$, covers a fixed parameter α, our thinking need not be restricted to fixed targets.

Consider an experiment in which $n + 1$ observations are obtained in such a way that they can be represented by $n + 1$ iid random variables $\mathbf{X}_1, \ldots, \mathbf{X}_n, \mathbf{X}_{n+1}$. Suppose that there is an interval between the time the first n observations are obtained and the time the $(n+1)$th observation becomes available. Then we might be interested in using the information in the first n observations to predict an interval

$$(\mathbf{X}_L[\mathbf{X}_1, \ldots, \mathbf{X}_n], \mathbf{X}_U[\mathbf{X}_1, \ldots, \mathbf{X}_n])$$

[9] This type of interval estimator is suitable when a regression equation is used to specify the value of an unknown dependent variable (see Chapter 8). A typical application in climatology and meteorology is a statistical forecast improvement procedure in which forecasts from a numerical weather forecast are enhanced using regression equations.

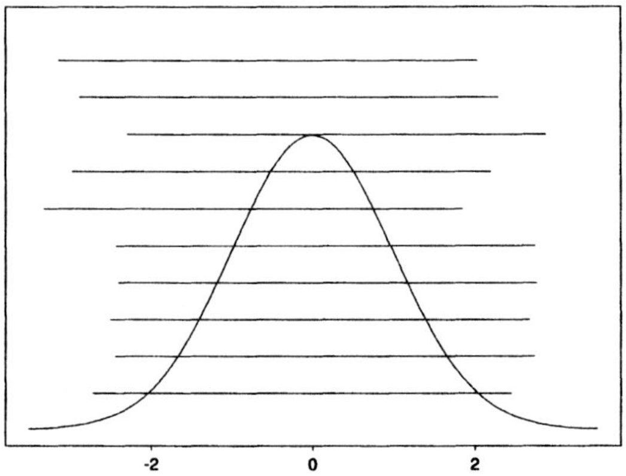

Figure 5.5: *Ten realizations of a 95% confidence interval for a random variable* $\mathbf{X}$. *On average, 19 out of 20 intervals will cover the next realization of* $\mathbf{X}$. *The curve shows the density function of* $\mathbf{X}$.

that will cover $\mathbf{X}_{n+1}$ $\tilde{p} \times 100\%$ of the time. This is a confidence interval for a random variable (see Figure 5.5). The random intervals are now wider than they were in Figure 5.4 because they need to be able to cover a moving, rather than fixed, target.

Again note that the confidence level refers to the average behaviour of the interval

$$(\mathbf{X}_L[\mathbf{X}_1, \ldots, \mathbf{X}_n], \mathbf{X}_U[\mathbf{X}_1, \ldots, \mathbf{X}_n])$$

in relation to the unknown random variable $\mathbf{X}_{n+1}$. The interval is constructed so that in repeated sampling of $\mathbf{X}_1, \ldots, \mathbf{X}_n, \mathbf{X}_{n+1}$ the probability of coverage is

$$P(\mathbf{X}_L[\mathbf{X}_1, \ldots, \mathbf{X}_n] < \mathbf{X}_{n+1} < \mathbf{X}_U[\mathbf{X}_1, \ldots, \mathbf{X}_n]) = \tilde{p}.$$

Note, that if we condition on the observed values $\mathbf{x}_1, \ldots, \mathbf{x}_n$ *of* $\mathbf{X}_1, \ldots, \mathbf{X}_n$ *and continue to think of* $\mathbf{X}_{n+1}$ *as random, then the coverage of the interval is no longer exactly* $\tilde{p}$. *However, in most practical applications the coverage will be close to* $\tilde{p}$ *because n will be relatively large. That is, we do not expect the upper and lower bounds of the interval to move a great deal due to variation in* $\mathbf{X}_1, \ldots, \mathbf{X}_n$.

5.4.3 Constructing Confidence Intervals. In general, a *confidence region* is defined indirectly as a set $\Theta_{\tilde{p}}(\mathbf{X}_1, \ldots, \mathbf{X}_n)$ such that

$$P\left(\Theta_{\tilde{p}}(\mathbf{X}_1, \ldots, \mathbf{X}_n) \ni \mathbf{A}\right) = \tilde{p}. \qquad (5.40)$$

That is, $\Theta_{\tilde{p}}(\mathbf{X}_1, \ldots, \mathbf{X}_n)$ is constructed so that it covers $\mathbf{A}$ $\tilde{p} \times 100\%$ of the time in repeated sampling. Depending on the situation, $\mathbf{A}$ denotes either a fixed parameter or a random variable. The definition of $\Theta_{\tilde{p}}(\mathbf{X}_1, \ldots, \mathbf{X}_n)$ depends on the assumed statistical model (e.g., the sample can be represented by iid normal random variables), the nature of the target (i.e., either a parameter or a random variable), and the confidence level $\tilde{p}$.

For the moment we limit ourselves to univariate problems (and thus intervals) instead of the more general multivariate problems (which require the use of multi-dimensional confidence regions). Multivariate problems arise in the context of regression analysis (see Chapter 8), for example.

As with point estimators, there are various ways to derive interval estimators. The only condition that *must* be satisfied is (5.40). Other reasonable requirements are that the set $\Theta_{\tilde{p}}(\mathbf{X}_1, \ldots, \mathbf{X}_n)$ has minimum size, on average, and that it is compact. The latter implies, in the univariate case, that confidence regions can only be intervals.

If the target is a parameter, the general procedure is as follows. We start with an efficient estimator $\widehat{\alpha}$ of parameter α. We then derive the distribution of $\widehat{\alpha}$. This distribution will depend on α somehow. There will generally be a way to transform $\widehat{\alpha}$ so that the distribution of the transformed variable no longer depends on α. For example, if α is a *location parameter* such as a mean, then the distribution of $\mathbf{Z} = \widehat{\alpha} - \alpha$ will not depend upon α. Similarly, if α is a *scale parameter* such as a variance, then the distribution of $\Psi = \widehat{\alpha}/\alpha$ will not depend on α. The distribution of the transformed variable is then used to construct the confidence interval.

For a location parameter we find critical values $\mathbf{z}_L$ and $\mathbf{z}_U$ so that $P(\mathbf{z}_L \le \mathbf{Z}) = 1 - \tilde{p}/2$ and $P(\mathbf{Z} \ge \mathbf{z}_U) = 1 - \tilde{p}/2$. Therefore, in repeated sampling,

$$\begin{aligned} \tilde{p} &= P(\mathbf{z}_L < \mathbf{Z} < \mathbf{z}_U) \\ &= P(\mathbf{z}_L < \widehat{\alpha} - \alpha < \mathbf{z}_U) \\ &= P(-\mathbf{z}_U < \alpha - \widehat{\alpha} < -\mathbf{z}_L) \\ &= P(\widehat{\alpha} - \mathbf{z}_U < \alpha < \widehat{\alpha} - \mathbf{z}_L). \end{aligned} \qquad (5.41)$$

Thus, the $\tilde{p} \times 100\%$ confidence interval for location parameter α has the form $\widehat{\alpha} - \mathbf{z}_U < \alpha < \widehat{\alpha} - \mathbf{z}_L$. Note that it is centred on estimator $\widehat{\alpha}$ and that it excludes equal proportions of the upper and lower tails of the distribution of $\widehat{\alpha}$.

For a scale parameter, we find critical values Ψ_L and Ψ_U so that $P(\Psi_L \le \Psi) = 1 - \tilde{p}/2$ and $P(\Psi \ge \Psi_U) = 1 - \tilde{p}/2$. Both critical values will be positive because we are dealing with a scale parameter. Also, for large values of $\tilde{p}$, Ψ_L will be less than 1 and Ψ_U will be greater than 1. We expect that in repeated sampling

$$\begin{aligned}\tilde{p} &= P(\Psi_L < \Psi < \Psi_U) \\ &= P\Big(\Psi_L < \frac{\widehat{\alpha}}{\alpha} < \Psi_U\Big) \\ &= P\Big(\frac{1}{\Psi_U} < \frac{\alpha}{\widehat{\alpha}} < \frac{1}{\Psi_L}\Big) \\ &= P\Big(\frac{\widehat{\alpha}}{\Psi_U} < \alpha < \frac{\widehat{\alpha}}{\Psi_L}\Big). \end{aligned} \quad (5.42)$$

Thus, the $\tilde{p} \times 100\%$ confidence interval for the scale parameter α has the form $\frac{\widehat{\alpha}}{\Psi_U} < \alpha < \frac{\widehat{\alpha}}{\Psi_L}$. This will generally be an asymmetric interval about $\widehat{\alpha}$ because the sampling distributions of scale parameters are usually skewed. None the less, the interval has been constructed to exclude equal portions of the lower and upper tails of the distribution of $\widehat{\alpha}$.

If the target is a random variable, a model is needed to predict the value of the unknown random variable from the observed random variables $\mathbf{X}_1, \dots, \mathbf{X}_n$. Often, the model has the form $\mathbf{X}_i = \alpha + \mathbf{E}_i, i = 1, \dots, n+1$, where α is a location parameter and the errors $\mathbf{E}_i$ are iid with mean zero. The approach is to estimate the location parameter and then predict $\mathbf{X}_{n+1}$ as $\widehat{\mathbf{X}}_{n+1} = \widehat{\alpha} + \widehat{\mathbf{E}}_{n+1}$. Because the errors are iid we can only predict $\widehat{\mathbf{E}}_{n+1} = 0$. Thus, the prediction error is $\mathbf{A}_{pred} = \mathbf{X}_{n+1} - \widehat{\alpha} = (\alpha - \widehat{\alpha}) + \mathbf{E}_{n+1}$. The next step is to find the distribution of the prediction error, and then to find critical values $\mathbf{A}_L$ and $\mathbf{A}_U$ such that $P\big(\mathbf{A}_L \le \mathbf{A}_{pred}\big) = 1 - \tilde{p}/2$ and $P\big(\mathbf{A}_{pred} \le \mathbf{A}_U\big) = 1 - \tilde{p}/2$. We expect that in repeated sampling

$$\begin{aligned}\tilde{p} &= P\big(\mathbf{A}_L < \mathbf{A}_{pred} < \mathbf{A}_U\big) \\ &= P\,(\mathbf{A}_L < \mathbf{X}_{n+1} - \widehat{\alpha} < \mathbf{A}_U) \\ &= P\,(\widehat{\alpha} + \mathbf{A}_L < \mathbf{X}_{n+1} < \widehat{\alpha} + \mathbf{A}_U).\end{aligned}$$

The confidence interval has structure similar to that of α, but is substantially wider because the critical values $\mathbf{A}_L$ and $\mathbf{A}_U$ account for sampling variation in both $\widehat{\alpha}$ and $\mathbf{X}_{n+1}$.

These confidence intervals may depend upon yet more parameters. For example, the limits of a confidence interval for a location parameter may depend upon the value of a scale parameter. Such parameters are called *nuisance parameters* (see also [4.1.7]). The only solution is to estimate the nuisance parameter and then reformulate the confidence interval to account for the sampling variability of the nuisance parameter estimator.

Examples of confidence intervals for location and scale parameters are described below.

5.4.4 Confidence Intervals for the Mean. Let $\mathbf{X}_1, \dots, \mathbf{X}_n$ represent a sample of iid normal random variables with mean μ and variance σ^2. Then

$$\mathbf{Z} = \sqrt{n}(\overline{\mathbf{X}} - \mu)/\sigma \quad (5.43)$$

has the standard normal distribution $\mathcal{N}(0, 1)$ (see [4.3.3]). The quantiles of this distribution are tabulated in Appendix D. Using the Appendix, we find critical values $\mathbf{z}_l$ and $\mathbf{z}_U$ for a given confidence level such that[10]

$$\tilde{p} = P(\mathbf{z}_L < \mathbf{Z} < \mathbf{z}_U).$$

The shortest $\tilde{p} \times 100\%$ confidence interval is obtained by choosing $\mathbf{z}_U$ so that $P(\mathbf{Z} < \mathbf{z}_U) = 0.5 + \tilde{p}/2$ and selecting $\mathbf{z}_L = -\mathbf{z}_U$. Then substituting (5.43) for $\mathbf{Z}$ and manipulating as in (5.41), we find that

$$\tilde{p} = P\Big(\overline{\mathbf{X}} - \mathbf{z}_U\frac{\sigma}{\sqrt{n}} < \mu < \overline{\mathbf{X}} + \mathbf{z}_U\frac{\sigma}{\sqrt{n}}\Big).$$

Thus, when σ^2 is known, the $\tilde{p} \times 100\%$ confidence interval for μ is

$$\Big(\overline{\mathbf{X}} - \mathbf{z}_U\frac{\sigma}{\sqrt{n}},\ \overline{\mathbf{X}} + \mathbf{z}_U\frac{\sigma}{\sqrt{n}}\Big). \quad (5.44)$$

We still express the distance between $\overline{\mathbf{X}}$ and μ in dimensionless units as in (5.43), but we replace σ with the estimator $\mathbf{S}$. The resulting *t statistic*,

$$\mathbf{T} = \sqrt{n}(\overline{\mathbf{X}} - \mu)/\mathbf{S},$$

has a *t distribution* with $n - 1$ degrees of freedom (see [2.7.9] and [4.3.3]). Proceeding as above, we find that, when σ is unknown, the $\tilde{p} \times 100\%$ confidence interval for μ is

$$\Big(\overline{\mathbf{X}} - t_U\frac{\mathbf{S}}{\sqrt{n}},\ \overline{\mathbf{X}} + t_U\frac{\mathbf{S}}{\sqrt{n}}\Big), \quad (5.45)$$

where t_U is the $0.5 + \tilde{p}/2$ quantile of the $t(n-1)$ distribution (see Appendix F).[11]

Be aware that the coverage of intervals (5.44) and (5.45) deviates from the nominal $\tilde{p} \times 100\%$ level when one or more of the assumptions we have made is violated. For example, serial correlations within the sample will tend to reduce the coverage of these intervals (see Chapter 4, [5.1.2] and [6.6.7–9]).

[10] For example, $\mathbf{z}_U = 1.96$ (1.63) for $\tilde{p} = 0.95$ (0.90).

[11] For example, $t_U = 2.776$ (2.132) for $\tilde{p} = 0.95$ (0.90) when $\mathbf{T}$ has 4 degrees of freedom.

5.4.5 Confidence Intervals for the Variance. Again, let $\mathbf{X}_1, \dots, \mathbf{X}_n$ represent a sample of iid $\mathcal{N}(\mu, \sigma^2)$ random variables. As described in [5.4.3], confidence intervals for scale parameters, such as σ^2, are constructed by first expressing an estimator of σ^2 in dimensionless units. Here we use

$$\Psi = (n-1)\mathbf{S}^2/\sigma^2,$$

which is $\chi^2(n-1)$ distributed (see [2.7.8]). Upper and lower tail critical values, Ψ_U and Ψ_L, of the χ^2 distribution are tabulated in Appendix E. These values are chosen so that $P(\Psi < \Psi_L) = 0.5 - \tilde{p}/2$ and $P(\Psi < \Psi_U) = 0.5 + \tilde{p}/2$.[12] Following the derivation in (5.42), we see that the $\tilde{p} \times 100\%$ confidence interval for σ^2 is

$$\left(\frac{(n-1)\mathbf{S}^2}{\Psi_U}, \frac{(n-1)\mathbf{S}^2}{\Psi_L}\right). \qquad (5.46)$$

This interval contains the point estimator $\mathbf{S}^2$, but unlike the confidence interval for the mean, it is not located at its centre. As with the mean, the coverage of (5.46) is sensitive to departures from the assumptions.

5.5 Bootstrapping

5.5.1 Concept. The interval estimation methods of the previous section use a fully parametric model to express the uncertainty of the corresponding point estimator. That is, all elements of the assumed statistical model are required to derive the confidence interval. However, it is often not possible to make a distributional assumption, or a distributional assumption can be made but derivation of a confidence interval is mathematically intractable. The *bootstrap* [111] provides a solution in both instances.

Suppose we assume only that the sample can be represented by iid random variables $\mathbf{X}_1, \dots, \mathbf{X}_n$. Each has the same distribution function, $F_X(x)$, but its form is not known. If we did know the distribution we could easily write down the joint density function of the random vector $\vec{\mathbf{X}} = (\mathbf{X}_1, \dots, \mathbf{X}_n)^{\mathrm{T}}$, and with luck derive the distribution of parameter estimator $\widehat{\alpha}(\vec{\mathbf{X}})$.[13] To keep the discussion simple, assume that α is a location parameter and that the distribution of $\widehat{\alpha}(\vec{\mathbf{X}}) - \alpha$ is free of nuisance parameters (see [4.1.7], [5.4.3]). Then we can find a confidence interval for α simply by finding the lower and upper tail critical values of $\widehat{\alpha}(\vec{\mathbf{X}}) - \alpha$. The bootstrap procedure solves the problem of the missing distribution function $F_X(x)$ by replacing it with a consistent estimator, the empirical distribution function $\widehat{F}_X(x)$ (see [5.2.2]). Then, following the same steps outlined above, we arrive at an estimate of the distribution of $\widehat{\alpha}(\vec{\mathbf{X}}) - \alpha$ that converges to the true distribution as the sample size increases. The estimated distribution can be used to obtain an approximate confidence interval for p or an estimate of the variance of $\widehat{\alpha}(\vec{\mathbf{X}})$.

The steps that produce *bootstrapped* confidence intervals or variance estimates can sometimes be performed analytically (see, e.g., Efron [111]). In general, though, the mathematics are intractable, and Monte Carlo simulation is used instead. The steps are as follows.

1 Generate a random sample $\mathbf{y}_1, \dots, \mathbf{y}_n$ from the population that has distribution function $\widehat{F}_X(x)$.[14] This can be done by using a random number generator to simulate a sample $\mathbf{u}_1, \dots, \mathbf{u}_n$ from the $\mathcal{U}(0, 1)$ distribution and then solving $\widehat{F}_X(\mathbf{y}_j) = \mathbf{u}_j$ for each $j = 1, \dots, n$.

2 Evaluate $\widehat{\alpha}$ for the realized sample.

3 Repeat steps 1 and 2 a large number of times.

The resulting sample of realizations of $\widehat{\alpha}$ can be used to estimate properties of the distribution of $\widehat{\alpha}$ such as its variance or its quantiles. The $(1-\tilde{p})/2$ and $(1+\tilde{p})/2$ quantiles are the lower and upper bounds of the bootstrapped $\tilde{p} \times 100\%$ confidence interval for α. The inferences made with bootstrapping procedures are approximate because the distribution of the parameter estimate is derived from an estimated distribution function. There may also be additional uncertainty if only a small number of bootstrap samples are generated. Inferences made with the bootstrap are asymptotically *exact*[15] provided that $\widehat{F}_X(x)$ is a consistent estimator of $F_X(x)$.

---

[12] For example, $\Psi_U = 12.8$ (11.1) and $\Psi_L = 1.24$ (1.64) for $\tilde{p} = 0.95$ (0.90) when Ψ has 5 degrees of freedom.

[13] We write $\widehat{\alpha}(\vec{\mathbf{X}})$ instead of just $\widehat{\alpha}$ to emphasize that $\widehat{\alpha}$ is a random variable whose distribution is derived from that of $\vec{\mathbf{X}}$.

[14] In the ordinary bootstrap, $\widehat{F}_X(x)$ is the empirical distribution function. However, other estimators of the distribution function can also be used. For example, $\widehat{F}_X(x)$ could be a parametric form in which the unknown parameters are replaced with efficient estimators. In this case the procedure is known as the 'parametric' bootstrap.

[15] That is, the true convergence of confidence intervals and true significance levels of tests will approach the specified values when samples become large.

5.5.2 Ordinary Bootstrap. The bootstrap is an example of a *resampling procedure*. When $\widehat{F}_X(x)$ is given by the empirical distribution function (5.3), step 1 above is equivalent to taking a sample of size n, *with replacement*, from the n observations $\mathbf{x}_1, \ldots, \mathbf{x}_n$.[16]

To see that this is so, consider again step 2 above. When $\widehat{F}_X$ is a smooth function, there will be a unique $\mathbf{y}_j$ for every $\mathbf{u}_j$. However, when $\widehat{F}_X$ is a step function such as the empirical distribution function, a range of $\mathbf{u}$ values can produce the same $\mathbf{y}_j$; the resulting sample may therefore contain a given $\mathbf{y}_j$ more than once. In particular, the empirical distribution function has n steps of equal height that completely partition the interval $(0, 1)$. Thus it follows that $\mathbf{y}_j$ will be equal to $\mathbf{x}_i$ for some i, and that every member of the sample $\{\mathbf{x}_1, \ldots, \mathbf{x}_n\}$ has the same probability of selection, that is, random resampling with replacement.

Each sample produced by the procedure described above is called a *bootstrap sample*. When $\widehat{F}_X$ is the empirical distribution function, 2^n different samples can be generated. Consequently, the bootstrapped estimate of the distribution of $\widehat{\alpha}(\vec{\mathbf{x}})$ will be quite coarse when n is small. However, the 'resolution' of the estimator quickly increases with increasing n. Even for moderate sample sizes, the cost of evaluating $\widehat{\alpha}(\vec{\mathbf{x}})$ for all possible samples becomes prohibitive (and is generally not necessary). Satisfactory bootstrapped variance estimates can often be made with as few as 100 bootstrap samples. A somewhat larger number of samples is required to produce good confidence intervals since these require estimates of quantiles in the tails of the $\widehat{\alpha}$ distribution.

There are some problems for which bootstrap estimators can be derived analytically. For example, the bootstrap estimator of σ^2 is $\widehat{\sigma}^2 = \frac{n-1}{n}\mathbf{S}^2$ (see [5.2.6] and [5.3.3]).

5.5.3 Moving Blocks Bootstrap. As with all the other estimators discussed in this chapter, bootstrapped estimators are vulnerable to the effects of departures from the sampling assumptions. Zwiers [442] illustrates what can happen when serial correlation is ignored. The difficulty arises because the resampling procedure does not preserve the temporal dependence of the observations in the sample; the resampling done in the ordinary bootstrap produces samples of independent observations regardless of the dependencies that may exist within the original sample.

A simple adaptation that accounts for short-term dependence is called the *moving blocks bootstrap* (see Künsch [235], Liu and Singh [254], and also Leger, Politis, and Romano [248]). Instead of resampling individual observations, blocks of l consecutive observations are resampled, thus preserving much of the dependent structure in the observations. In general, the block length should be related to the 'memory,' or persistence, of the process that has been sampled, with longer block lengths used when the process is more persistent. Wilks [423] points out that care is required to choose l appropriately. Blocks that are too long will result in confidence intervals with coverage greater than the nominal $\tilde{p}$, and vice versa. Theoretical work [235, 254] also shows that the block length l should increase with sample size n in such a way that l/n tends to zero as n approaches infinity.

Wilks [423] describes the use of the moving block bootstrap when constructing confidence intervals for the difference of two means from data that are serially correlated. He gives simple expressions for the block length that can be used when data come from *auto-regressions of order 1 or 2* (Chapter 10). Wang and Zwiers [414] applied the moving blocks bootstrap to GCM simulated precipitation.

[16]That is, the elements of the sample are obtained one at a time by drawing an observation at random, noting its value, and returning it to the pool of observations.

Part II

Confirmation and Analysis

Overview

In Part II we address the problem of determining the correctness of a certain statistical model[1] in the light of empirical evidence. To make sure that the assessment is fair, the model must be tested with information that is gathered independently of that which is used to formulate the model. The standard method that is used is called statistical *hypothesis testing*. We deal with this concept in some length in Chapter 6 (see also [1.2.7] and [4.1.7–11]).[2] Examples of applications in climate research are presented in Chapter 7.

The application of hypothesis testing in climate research is fraught with problems that are not always encountered in other fields.

- In climate research it is rarely possible to perform real *independent experiments* (see Navarra [289]) with the observed climate system. There is usually only one observational record, which is analysed again and again until the processes of building and testing hypotheses are hardly separable. Dynamical climate models often provide a way out of this dilemma. Hypotheses that are formed by analysing the observed record can frequently be tested by running independent experiments with GCMs. However, even these experiments are not completely independent of the observed record since GCMs rely heavily on parameterizations that have been tuned with the observed record.

  Even though fully independent tests are not possible, testing is often useful as an interpretational aid because it helps quantify unusual aspects of the data. On the other hand, we need to be wary of indiscriminate testing because it sometimes allows unusual quirks to draw our attention away from physically significant aspects of our data.

- Almost all data in climate research have spatial and temporal correlations, which is most useful since it allows us to infer the space–time state of the atmosphere and the ocean from a limited number of observations (cf. [1.2.2]). However this correlation causes difficulties in testing problems since most standard statistical techniques assume that the data are realizations of independent random variables.

  Because of these difficulties, the use of statistical tests in a *cookbook* manner is particularly dangerous. Tests can become very unreliable when the statistical model implicit in the test procedure does not properly account for properties such as spatial or temporal correlation.

  The problems caused by the indiscriminate use of recipes are compounded when obscure sophisticated techniques are used. It is fashionable to surprise the community with miraculous new techniques, even though the statistical model implicit in the method is often not understood.

Hypothesis testing is carried out by formulating two propositions: the *null* hypothesis that is to be tested, and the *alternative* hypothesis, which usually encompasses a range of possibilities that may be true if the null hypothesis is false. The alternative hypothesis indirectly influences the test because it affects the interpretation of the evidence against the null hypothesis. The null hypothesis is rejected if the evidence against it is strong enough; it is not rejected when the evidence is weak, but this does not imply rejection of the alternative. We then continue to entertain the possibility that either of the hypotheses is true.

Null hypotheses are typically of the type $A = B$, and in climate research the alternative $A \neq B$ is usually correct. Often, though, the difference between A and B is small and physically irrelevant. Statistical tests can not be used to detect the difference between physically significant and insignificant differences. The strength of the evidence against the null hypothesis, and thus for the detection of a 'statistically significant' difference, depends on the amount of evidence, that is, the number of independent samples. As the sample size increases so do the chances of detecting $A \neq B$. With the very large sample sizes that can be constructed with GCMs, almost every physically irrelevant difference can achieve statistically significant status.

[1]To be more precise: an attempt is made to determine whether the model is *incorrect*; absolute *correctness* can not be determined statistically.

[2]There are two approaches to statistical decision making. We use the *frequentist* approach, since it is more common in climatology than the *Bayesian* approach (see, e.g., Gelman et al. [139]).

Obviously, this is not satisfactory. We introduce *recurrence analysis* (see Sections 6.9–6.10) as an alternative for assessing the strength of the difference $A - B$. This technique produces estimates of the degree of separation between A and B that are independent of the sample size.

6 The Statistical Test of a Hypothesis

6.0.0 Summary. In this chapter we introduce the ideas behind the art of testing statistical hypotheses (Section 6.1). The general concepts are described, terminology is introduced, and several elementary examples are discussed. We also examine some philosophical questions about testing and some extensions to cases in which it is difficult to build the statistical models needed for testing.

The *significance level, power, bias, efficiency,* and *robustness* of statistical tests are discussed in Section 6.2. The application of Monte Carlo simulation in problem testing is discussed in Section 6.3, and in Section 6.4 we examine how hypotheses are formulated and explore some of the limitations of statistical testing. The spatial correlation structure of the atmosphere often impacts testing problems. Strategies for coping with and using this structure are discussed in Section 6.5. A number of tests of the null hypothesis of equal means and variances are discussed in Sections 6.6 and 6.7. Tests designed to provide a *global* interpretation for a field of *local* decisions, called *field significance tests*, are presented in Section 6.8. Univariate and multivariate *recurrence analysis* are discussed in Sections 6.9 and 6.10.

6.1 The Concept of Statistical Tests

6.1.1 Introduction. Since we should now be somewhat comfortable with the ideas underlying hypothesis testing (see [1.2.7], [4.1.7–11], and the preamble to this part of the book), we only briefly characterize the testing paradigm here.

Statistical hypothesis testing is a formalized process that uses the information in a sample to decide whether or not to reject H_0, the *null hypothesis*. The evidence is judged in the context of a statistical model in such a way that the risk of falsely rejecting H_0 is known. A second proposition, the alternative hypothesis H_a, generally describes the range of possibilities that may be true when H_0 is false. The alternative hypothesis affects the decision making process by altering the way the evidence in the sample is judged.

A hypothesis testing process can only have two outcomes: either H_0 is rejected or it is not rejected. The former does not imply acceptance of H_a—it simply means that we have fairly strong evidence that H_0 is false. Failure to reject H_0 simply means that the evidence in the sample is not inconsistent with H_0.

6.1.2 The Ingredients of a Test. We need two objects to perform a statistical test: the object to be examined—a set of observations that, for convenience, we collect in a single vector $\vec{\mathbf{x}}$—and a rule that determines whether to reject the null hypothesis or not. This rule usually takes the form '*reject* H_0 *if* $S(\vec{\mathbf{x}}) > \kappa_{\tilde{p}}$,' where S is a predetermined function that measures the evidence against H_0, and $\kappa_{\tilde{p}}$ is a threshold value for S beyond which we are willing to risk making the reject decision.

The rule is defined in three steps.

First, we regard the set of observations $\vec{\mathbf{x}}$ as a realization of a random vector $\vec{\mathbf{X}}$. The latter represents the ensemble of values that $\vec{\mathbf{x}}$ is able to take, when H_0 is true, under infinite replication of the 'experiment' that produced the set of observations. A statistical model is built for the experiment by representing the likelihood of observing a particular realization in this ensemble with a probability distribution $f_{\vec{X}}$.

Second, we specify the *significance level*, the probability of rejecting the null hypothesis when it is true, at which the test is to be conducted. The choice of the significance level affects the power, or sensitivity, of the test. Thus the consequences of falsely rejecting H_0 should be balanced against the consequences of failing to reject H_0 when H_0 is false. In Section 6.2 we present this idea in more concrete terms.

Finally, the chosen significance level, the alternative hypothesis, and the statistical model are used jointly to derive the decision making criterion for the test. This is usually expressed in terms of a *test statistic* and a range of values of that statistic, or *non-rejection region*,[1] that is consistent with the null hypothesis.

6.2 The Structure and Terminology of a Test

6.2.1 Risk and Power. The general mathematical setup is derived from the three components described above. A statistical model is developed to describe the stochastic characteristics of the observations and the way in which they were obtained, provided that H_0 is true. This model is expressed in terms of a random vector $\vec{\mathbf{X}}$ and its probability distribution. Then a probability $\tilde{p} \in [0;1]$ and a domain $\Theta(\tilde{p})$ are chosen so that $\tilde{p} \times 100\%$ of all realizations of $\vec{\mathbf{X}}$ fall inside $\Theta(\tilde{p})$, that is,

$$P\big(\vec{\mathbf{X}} \in \Theta(\tilde{p})\big) = \tilde{p}. \tag{6.1}$$

The null hypothesis H_0 is rejected if $\tilde{\vec{x}} \notin \Theta(\tilde{p})$. The probability of rejecting H_0 when it is actually true is $1 - \tilde{p}$. This probability, the *risk* of false rejection, is called the *significance level* of the statistical test.

The probability $\tilde{p}$ is chosen to be large, typically 95% or 99%, so that the *non-rejection region* $\Theta(\tilde{p})$ contains the realizations of $\vec{\mathbf{X}}$ most likely to occur when H_0 is true. Only the $(1 - \tilde{p}) \times 100\%$ of realizations that are unusual, and therefore constitute evidence contrary to H_0, are excluded from $\Theta(\tilde{p})$.

The probability of rejecting H_0 when H_0 is false is the *power* of the test. While we would like the power to be large, it is sometimes small, often when the alternative hypothesis describes a probability distribution similar to that described by H_0. Then $P\big(\vec{\mathbf{X}} \notin \Theta(\tilde{p})\big)$ under H_a will be close to that under H_0.

Two types of decision making errors can occur in the testing process. First, H_0 can be rejected when it is true. This is referred to as a *type I error*. The probability of a type I error, $1 - \tilde{p}$, is equal to the *significance level*.

The significance level is chosen by the user of the test. However, reducing the likelihood of a type I error comes at the cost of increasing the likelihood of the *type II error*: the failure to reject H_0 when it is false. The probability of a type II error is $1 - power$. Thus, reduced significance level comes at the cost of decreased power. Ultimately, the user must choose $\tilde{p}$ to balance the risk of a type I error with the costs of a type II error.

6.2.2 The Non-rejection Region When an Alternative Hypothesis is not Specified. To conduct a test it is necessary to derive the non-rejection region $\Theta(\tilde{p})$. Intuitively, it should contain all events except those that are unusual under the null hypothesis *and* consistent with the alternative hypothesis. We will assume for now that $H_a = \neg H_0$. In this context the non-rejection region contains all events except those that are unusual under H_0.

In particular, if the observations are realizations of continuous random variables, then the non-rejection region will cover all possible realizations $\vec{\mathbf{x}}$ for which the density function $f(\vec{\mathbf{x}})$ under the null hypothesis is larger than some threshold $\alpha_{\tilde{p}}$, that is,

$$\Theta(\tilde{p}) = \{\vec{x} : f(\vec{x}) \geq \alpha_{\tilde{p}}\}. \tag{6.2}$$

In many applications the derivation of $\Theta(\tilde{p})$ is facilitated by assuming that the sampling procedure and stochastic characteristics of the observations are such that $\vec{\mathbf{X}} \sim \mathcal{N}(\vec{\mu}, \Sigma)$. Then the outer surface of $\Theta(\tilde{p})$ is given by $f(\vec{x}) = \alpha_{\tilde{p}}$, an ellipsoidal surface defined by

$$\mathcal{D}^2(\vec{x}) = (\vec{x} - \vec{\mu})^{\mathrm{T}} \Sigma^{-1} (\vec{x} - \vec{\mu}) = \kappa_{\tilde{p}}.$$

The domain $\Theta(\tilde{p}) = \{\vec{x} : \mathcal{D}^2(\vec{x}) \leq \kappa_{\tilde{p}}\}$ is the interior of the ellipsoid. Thus the statement $\vec{\mathbf{x}} \notin \Theta(\tilde{p})$ is equivalent to $\mathcal{D}^2(\vec{\mathbf{x}}) > \kappa_{\tilde{p}}$, and the test statistic is $\mathcal{D}^2$.

When H_0 is true, the random variable $\mathcal{D}^2(\vec{\mathbf{X}})$ has a χ^2 distribution with m degrees of freedom [2.7.8], where m is the dimension of $\vec{\mathbf{X}}$. Therefore it is easy to determine $\kappa_{\tilde{p}}$ so that the test operates at the appropriate significance level. The non-rejection region is sketched in Figure 6.1 for $m = 1$ and $m = 2$.

In the univariate case, $\vec{\mathbf{X}} = \mathbf{X}$ and the matrix Σ degenerates to the scalar σ^2. The surface of the ellipsoid $(\mathbf{x} - \mu)^{\mathrm{T}} \Sigma^{-1} (\mathbf{x} - \mu) = \kappa_{\tilde{p}}$ is given by the equation $(\mathbf{x} - \mu)^2/\sigma^2 = \kappa_{\tilde{p}}$. Only two points satisfy this equation, so the ellipsoid $\Theta(\tilde{p})$ degenerates to an interval that has two points as its 'surface' (Figure 6.1a). The null hypothesis is rejected whenever an observation $\mathbf{x}$ lies outside the interval; it is not rejected when an observation $\mathbf{x}$ falls inside the interval.

The isolines of a bivariate normal density function f are plotted in Figure 6.1b (with $\Sigma =$

[1] This is admittedly an awkward expression. The term 'acceptance region' is sometimes used instead, but this expression is imprecise as it implies that we might be able to actively support the validity of the null hypothesis. Instead we just do not reject the null hypothesis—so 'non-rejection' is the correct word.

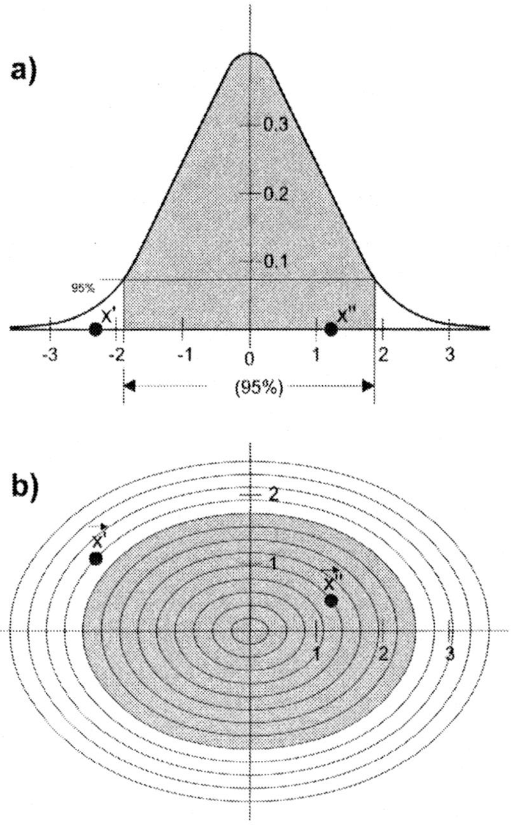

Figure 6.1: *Schematic diagrams illustrating the domains for which the null hypothesis '$\vec{\mathbf{x}}$ is drawn from $\vec{\mathbf{X}}$' is accepted. The shaded area represents the non-rejection region $\Theta(95\%) = \{\vec{x} : f(\vec{x}) \geq \alpha_{95\%}\}$ (a) univariate distribution; (b) bivariate distribution. The points $\mathbf{x}'$ and $\vec{\mathbf{x}}'$ are examples of realizations of the sampling process that provide evidence contrary to the null hypothesis, whereas the realizations $\mathbf{x}''$ and $\vec{\mathbf{x}}''$ are consistent with the null hypothesis [396].*

diag(1, 2)). The maximum of f is located in the centre of the diagram, and the region bounded by the $\Theta(95\%)$-ellipsoid is shaded. In both cases, the observation $\vec{\mathbf{x}}'$ leads to the rejection of the null hypothesis H_0, whereas $\vec{\mathbf{x}}''$ leads to the conclusion that the observations are consistent with the null hypothesis.

6.2.3 The Non-rejection Region When H_a is Specified. The choice of the non-rejection region may be constrained in various ways when an alternative hypothesis is specified. The region must satisfy (6.1) to ensure that the test operates at the selected significance level, but it need not necessarily satisfy (6.2), which was derived under the assumption that the alternative hypothesis is the complement of the null hypothesis, that is, $H_a = \neg H_0$. This particular choice of alternative hypothesis dictates that all 'unusual' values of $\vec{\mathbf{X}}$ represent evidence contrary to H_0. However, we often have prior knowledge about the expected kind of departure from the null hypothesis. An example: if we summarize the response of the climate system to a doubling of CO_2 with the global mean (near-surface) temperature and the global mean precipitation, then we anticipate an increase in temperature, but we might be uncertain about the sign of the change in precipitation. This prior knowledge, which is expressed as the alternative hypothesis, results in a non-rejection region that is constrained in some way.

Consider again the simple examples of the previous subsection. Figure 6.1 illustrates non-rejection regions when H_a is the complement of H_0. However, suppose that we anticipate, as in the climate change example above, that the mean of $\mathbf{X}_1$ will be greater than zero if H_0 is false (we use the subscript '1' to indicate the first element of $\vec{\mathbf{X}}$). Then a reasonable non-rejection region that accounts for H_a is given by $\Theta(\tilde{p}) = \{\vec{x} : f(\vec{x}) \geq \alpha_p \text{ and } x_1 \geq 0\} \cap \{\vec{x} : f(0, x_2) \geq \alpha_p \text{ and } x_1 \leq 0\}$, where $\alpha_{\tilde{p}}$ is chosen to satisfy (6.1). The alternative hypothesis has modified the 'rules of evidence' by instructing the test not to treat unusually large negative values of $\mathbf{x}_1$ as evidence inconsistent with H_0. The change in the non-rejection region is illustrated in Figure 6.2. This change reduces the magnitude of $\vec{\mathbf{X}}$ realizations needed on the right hand side of the $\mathbf{x}_1 = 0$ plane to reject H_0. Hence the power of the test is increased against alternatives for which $\mathcal{E}(\mathbf{X}_1)$ is positive.

6.2.4 Efficiency. A test may not be *efficient* even if it operates at the selected significance level, that is, the constraint (6.1) is satisfied. For example, one might choose the non-rejection region $\Theta(\tilde{p}) = \{\vec{x} : f(\vec{x}) \leq \alpha_{\tilde{p}}\}$. This would lead to the rejection of the null hypothesis for realizations of $\vec{\mathbf{X}}$ that are close to 'normal' and hence nearest the null hypothesis. Although this is a test of H_0, it is clearly an absurd one. One could also choose to ignore the data by tossing a coin that comes up heads $(1 - \tilde{p}) \times 100\%$ of the time. Generally speaking, inefficient low-power tests are avoided if the non-rejection region satisfies (6.1) and contains the outcomes $\vec{\mathbf{x}}$ that are most likely to occur under H_0. Technical details of the construction of optimal tests can be found in standard texts on mathematical statistics such as [335] or [92].

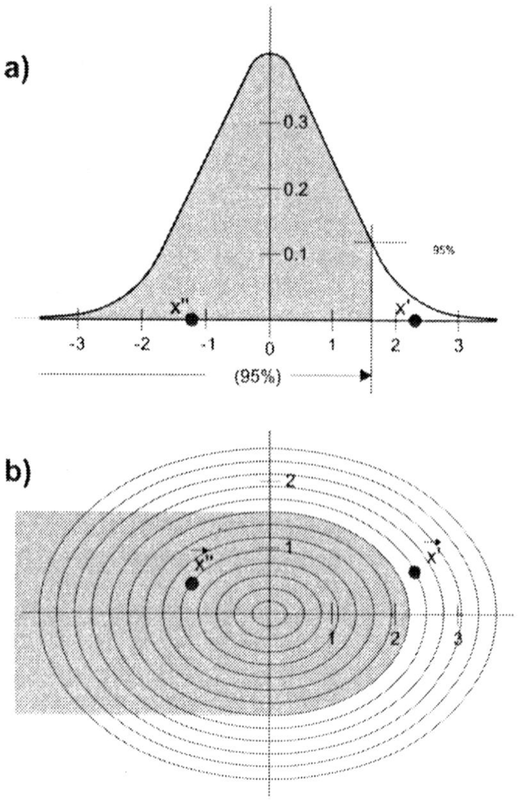

Figure 6.2: *Same as Figure 6.1 but for a one-sided test. The non-rejection region is described in the text.*

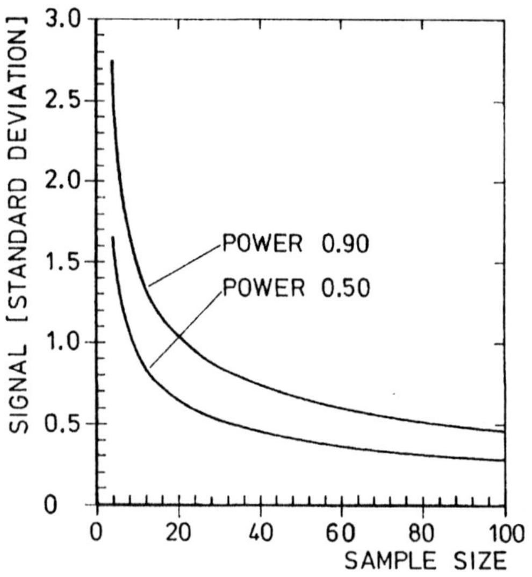

Figure 6.3: *Signal-strength* $\delta = \frac{\mu_Y - \mu_X}{\sigma}$ *for which* H_0: $\mu_Y = \mu_X$ *is rejected with probability* 50% *or* 90% *at the* 5% *significance level, shown as a function of n, the number of realizations of each* **X** *and* **Y**. *It is assumed that* $\mathbf{X} \sim \mathcal{N}(\mu_X, \sigma)$ *and* $\mathbf{Y} \sim \mathcal{N}(\mu_Y, \sigma)$. *[404]*

6.2.5 Statistical and Physical Significance. Suppose we wish to test the null hypothesis, H_0: $\mu_X = \mu_Y$, that the means of two random variables are equal. This can be accomplished by collecting a sample from both populations and computing a confidence interval for the difference of means, $\mu_Y - \mu_X$, similar to (5.45). The null hypothesis is rejected at the 5% significance level when the hypothesized value for $\mu_Y - \mu_X$, 0, is not covered by the 95% confidence interval.

Zero will lie outside just about every realization of the confidence interval when the two populations are well separated, regardless of the size of the sample, since there is probably a large, physically significant difference between the populations. On the other hand, suppose that the true difference of means is small and of little physical consequence, and that the populations have heavy overlap. Zero will often be inside the confidence intervals when the sample size is small. However, the width of the confidence interval decreases with increasing sample size. Given large enough samples, zero will again lie outside most realizations of the confidence interval. Thus, even though the difference between μ_X and μ_Y is physically insignificant, we will judge it to be statistically significant given large enough samples (i.e., resources).

This is illustrated in Figure 6.3, which shows the minimum strength of the difference of means *signal* $\mu_X - \mu_Y$ for which an ordinary t test (see [6.6.1]) will reject H_0: $\mu_X = \mu_Y$ with probability 50% or 90%. These power curves are shown as a function of sample size under the assumptions that both populations have the same variance σ^2 and size n. The figure shows, for example, that if $\mu_X - \mu_Y = 0.5\sigma$, then samples of approximately $n = 24$ observations are needed to detect the signal with a probability of 50%. Eighty-eight observations are needed in each sample to increase the power to 90%. The size of signal that can be detected with a given level of reliability tends to zero as $\mathcal{O}(1/\sqrt{n})$.

Another way to illustrate these ideas is shown in Figure 6.4, where we see the density functions of a control and an experimental random variable (solid and dashed curves labelled $n = 1$) and corresponding sampling distributions of the means for samples of 10 and 40. The population means differ by one standard deviation. The two density functions overlap considerably; a

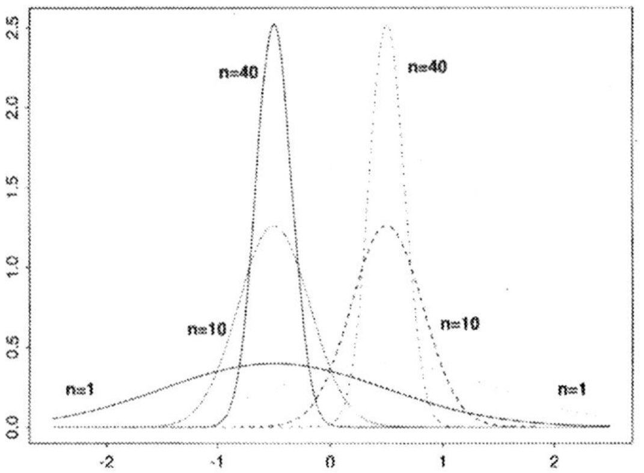

Figure 6.4: *The solid curves display the distribution of the mean of samples of size* $n = 1, 10,$ *and* 40 *taken from a* $\mathcal{N}(-0.5, 1)$ *population. The dashed curves show the same distributions for the* $\mathcal{N}(+0.5, 1)$ *population. Note that the overlap is very large when* $n = 1$, *and virtually nonexistent when* $n = 40$.

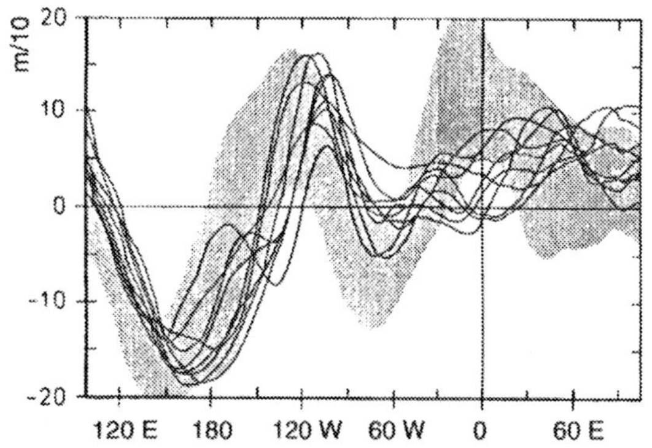

Figure 6.5: *Zonal distribution of the meridionally averaged (30° N–60° N) eddy component of January mean 500 hPa height in decametres. Shaded: the observed univariate 95% confidence band at each longitude. Curves: 10 individual states simulated with a General Circulation Model [397].*

large portion of experimental states can occur under control conditions and vice versa. However, as the sample size increases, the spread of the density functions of the sample means decreases, and eventually there is virtually no overlap. Under these circumstances the control and the experimental random variables can be distinguished with almost perfect reliability. Thus, given a large enough sample, it will be possible to state with confidence that the experimental and control random variables cluster around different means.

Thus the likelihood of rejection of the null hypothesis depends not only on the strength of the signal but also on the amount of available data. We must therefore be careful to distinguish between *statistical* and *physical* significance. We return to this point when we introduce *recurrence analysis* in Sections 6.9–6.10.

6.2.6 Example: AGCM Validation. One application of statistical tests occurs in the validation of the climate simulated by an Atmospheric General Circulation Model (AGCM). The assessment is performed by comparing individual fields $\vec{\mathbf{y}}$ generated by the AGCM with a statistical model $\vec{\mathbf{X}}$ that is fitted to an ensemble of fields obtained from the observed climate.

In the following example (see [397]) the observed random vector of interest is $\vec{\mathbf{X}}$ = {*meridionally averaged* (30°–60° N) *eddy component of January mean* 500 hPa *height*} and we let $\vec{\mathbf{Y}}$ be the corresponding random vector that is simulated by the AGCM. The null hypothesis is that $\vec{\mathbf{X}}$ and $\vec{\mathbf{Y}}$ have the same distributions. In the absence of prior knowledge about the AGCM's biases, we take the alternative hypothesis to be the complement of the null and use the non-rejection region $\Theta(95\%) = \{\vec{x} : f(\vec{x}) \geq \alpha_{95\%}\}$. We find that 6 of the 10 AGCM realizations $\vec{\mathbf{y}}$ lie outside $\Theta(95\%)$, so we reject the null hypothesis that the model simulates the observed climate.

The 10 $\vec{\mathbf{y}}$ curves are displayed in Figure 6.5 together with the *univariate* 95% *confidence band* (i.e., the univariate $\Theta(\tilde{p})$ at each longitude; shaded). Some of the simulated fields are fairly realistic but most have severe distortions. We return to this example in Section 7.1.

6.2.7 Example: Sign Test. Suppose $\mathbf{X}_1, \ldots, \mathbf{X}_m$ are iid random variables that represent a sample from a population $\mathbf{X}$, and that we want to decide whether or not $\mathcal{E}(\mathbf{X})$ has a particular value a. That is, we want to test

$$\text{H}_0\text{: } \mathcal{E}(\mathbf{X}) = a. \tag{6.3}$$

The following is a simple *non-parametric* solution (see [4.2.2]).

Assume that $\mathbf{X}$ has a *symmetrical* distribution, that is, that there exists a constant b such that $f(b - \mathbf{x}) = f(b + \mathbf{x})$ for all $\mathbf{x}$. Then (6.3) is equivalent to H_0: $b = a$.

Now consider the test statistic

$$n(\mathbf{X}_1, \ldots, \mathbf{X}_m) = \text{number of } \mathbf{X}_j \geq a. \tag{6.4}$$

Since we have assumed independence, we can think of $\mathbf{N} = n(\mathbf{X}_1, \ldots, \mathbf{X}_m)$ as the number of heads in m tosses of a coin where the probability of a head on the jth toss is $p_j = P(\mathbf{X}_j \geq a)$. When H_0 is correct, $p_j = 0.5$ and thus $\mathbf{N}$ has the binomial distribution: $\mathbf{N} \sim \mathcal{B}(m, 0.5)$. If $\mathbf{n}$ is the actual number of observations $\mathbf{x}_j$ for which $\mathbf{x}_j \geq a$, then the probability of observing $\mathbf{N} \geq \mathbf{n}$ is given by

$$P(\mathbf{N} \geq \mathbf{n}|\mathrm{H}_0) = \sum_{n \geq \mathbf{n}} \frac{m!}{n!(m-n)!} 0.5^m. \tag{6.5}$$

We reject H_0 when $\mathbf{N}$ is unusually large in the context of H_0, i.e., when $P(\mathbf{N} \geq \mathbf{n}|\mathrm{H}_0)$ is small (e.g., 5% or 1%).

We illustrate the sign test with an example from AMIP, the Atmospheric Model Intercomparison Project (see Gates [137]).

AMIP established a benchmark 10-year climate simulation experiment that was performed by a large number of modelling groups. One feature of these experiments is that the monthly mean SSTs and sea-ice extents observed between January 1979 and December 1988 were prescribed as time varying lower boundary conditions. Therefore, since AMIP simulations experience the same 'forcing' at the lower boundary as the real atmosphere, it is natural to compare the variability in the AMIP simulations with that in observations.

In particular, suppose that we want to test the null hypothesis, H_0, that the spatial variability of the December, January, February (DJF) mean 500 hPa height (ϕ_{500}) that is simulated by model X is the same as that contained in the US National Meteorological Center (NMC) global ϕ_{500} analyses. The table below gives measures of spatial variability computed from the analyses and AMIP simulations performed with two climate models.

	Spatial variance of DJF mean ϕ_{500} in m^2		
Year	NMC analyses	Model A	Model B
79/80	451	471	205
80/81	837	209	221
81/82	598	521	373
82/83	979	988	419
83/84	555	234	334
84/85	713	331	265
85/86	598	217	291
86/87	448	487	351
87/88	270	448	582

The analysed observations contain more spatial variability than does Model A in $\mathbf{n} = 5$ of nine DJF seasons. Using (6.5) we find that the probability of observing $\mathbf{n} \geq 5$ under H_0 is $(126+84+36+9+1)(0.5)^9 = 0.5$. Thus we cannot conclude that the spatial variability of the DJF climate simulated by Model A is significantly different from that which is observed. On the other hand, $\mathbf{n} = 8$ for Model B, and $P(\mathbf{N} \geq 8) = (9+1)^9 = 0.0195$. Thus the null hypothesis can be rejected for Model B at about the 2% significance level.

Not all of the assumptions required by the sign test are satisfied in this example. The measure of spatial variability we used, $\langle(\phi_{500} - \langle\phi_{500}\rangle)^2\rangle$ where $\langle\cdot\rangle$ denotes global average, is not likely to be exactly symmetrically distributed, although a Central Limit Theorem [2.7.5] type of argument can be used to show that its distribution is close to the normal distribution. Also, the spatial variability is not likely to be identically distributed in all years since it is strongly affected by ENSO (see [1.2.3]). Both of these departures from the assumptions will have some effect on the significance level and power of the test.

6.2.8 Sufficient Statistics. The decisions in the previous example [6.2.7] were made on the basis of a statistic that is a function of the pairs of variance differences, not the variances themselves. It is obvious that such reductions of data are necessary, but how do statisticians choose the statistic that results in the most effective test? In this example the hypothesis concerns the value of a parameter of the binomial distribution. The nine random variables that represent the variance differences may be transformed into nine other random variables such that distribution of one of the random variables, say $\mathbf{S}$, depends upon the unknown binomial parameter and the remaining eight of the random variables have a joint distribution that depends only upon the value of $\mathbf{S}$. If such a transformation exists, then $\mathbf{S}$ is said to be a *sufficient statistic* for the unknown parameter because it contains all the information that can be found in the sample about the unknown parameter. Sufficient statistics are therefore very good test statistics.

6.3 Monte Carlo Simulation

6.3.1 General. The analytical procedures mentioned above, as well as other theoretical methods used to derive the distributions of test statistics, often result in intractable mathematical problems.

The Monte Carlo method is often used when this happens.[2] The idea is to simulate the statistical model on a computer under the assumption that H_0 is true. The computer is used to generate a large number of realizations of the test statistic, say **S**, which in turn are used to construct an empirical estimate of the distribution of **S** under H_0. Finally, the estimated distribution is used to determine the critical value $\kappa_{\tilde{p}}$ just as its analytical counterpart would be used if it were available.

The Monte Carlo method is a powerful tool because it substantially increases the range of problems that will yield to statistical reasoning. As with all powerful tools, there are also a number of pitfalls to be avoided. Although the Monte Carlo approach can be applied to any statistic, heuristically derived statistics may not be efficient and can result in misleading inferences. For example, the *invariance principle* [4.3.3], which requires that the same inference be made under all linear transformations of the data, may be violated.

6.3.2 Example. The Monte Carlo method was used to study the relationship between the appearance of tropical storms in the Southwest Pacific and the phase of the tropical Madden-and-Julian Oscillation (MJO) [399]. The latter is a stochastic oscillation that affects the intensity of convection in the tropical West Pacific. Intensified convection may, in turn, be associated with increased tropical cyclogenesis and vice versa.

We therefore consider the null hypothesis: 'H_0: *the frequency of tropical storms in the West Pacific is independent of the phase of the Madden-and-Julian Oscillation.*' To test this hypothesis we need an objective measure of the phase of the MJO. One such measure is given by the oscillation's 'POP index' [15.2.3]. The observed phases can then be classified into one of eight 45° sectors. Each tropical cyclone is assigned to the sector corresponding to the phase of the MJO on the day of genesis. Then, if $\mathbf{F}_k$, $k = 1, \ldots, 8$, is the frequency of storms in sector k, the null hypothesis may be re-expressed as

$$H_0\colon \zeta_k = 1/8, \qquad (6.6)$$

where $\zeta_k = \mathcal{E}(\mathbf{F}_k)$.

A reasonable alternative hypothesis H_a is

$$H_a\colon \max_j \left[\sum_{k=j}^{j+3} (\zeta_k - \zeta_{k+4}) \right] > 0,$$

[2]The ideas discussed here are closely related to the bootstrapping ideas discussed in Section 5.5.

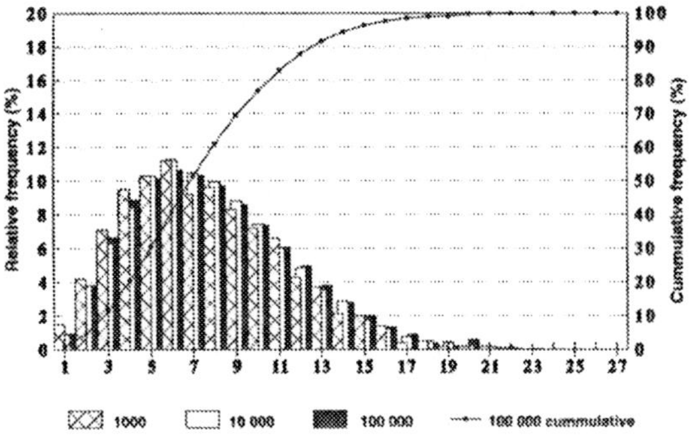

Figure 6.6: *Monte Carlo simulation of the probability function $f_S(j)$ of (6.7) with n = 51 cases. The functions are derived from 1000, 10 000, and 100 000 trials. The distribution function $F_S(j)$, estimated from 100 000 trials is also shown.*

with the convention $\zeta_k = \zeta_{k-8}$ if $k > 8$. This alternative was chosen because it was anticipated that the ζ_k will vary smoothly with k if H_0 is false in such a way that phases on one half of the circle are preferred over those in opposite sectors. A natural test statistic for this setup is

$$\mathbf{S} = \max_j \left[\sum_{k=j}^{j+3} (\mathbf{F}_k - \mathbf{F}_{k+4}) \right]. \qquad (6.7)$$

S is a discrete random variable that takes values between zero and n, the total number of storms observed. In this example, 51 storms were observed in a five year period.

To make an inference about (6.6) we need to determine the probability distribution $f_S(j)$ of **S** given that H_0 is true. This was done with the Monte Carlo method by repeatedly:

- generating n independent realizations $\mathbf{x}_1, \ldots, \mathbf{x}_n$ from the discrete uniform distribution on the set of integers $\{1, \ldots, 8\}$ [2.4.4],
- computing the frequencies $\mathbf{f}_1, \ldots, \mathbf{f}_8$, and
- finally obtaining a realization of **S** by substituting the realized frequencies into (6.7).

By doing this often, the probabilities $P(\mathbf{S} = j)$ for $j = 1, \ldots, n$ can be estimated.

Estimates based on 1000, 10 000, and 100 000 samples are shown in Figure 6.6. The three estimates are very similar. The differences arise from sampling variations: slightly different estimates of the true probability function are

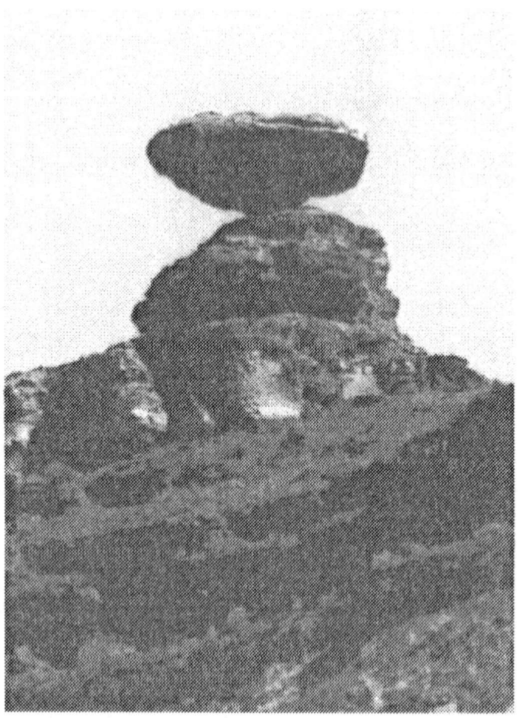

Figure 6.7: *The Mexican Hat at the border between Utah and Arizona—is this rock naturally formed? [3]*

obtained each time the Monte Carlo procedure is repeated. The estimate obtained from the 100 000 trial sample, of course, has less uncertainty than that obtained from the 1000 trial sample.

The observed set of 51 storms is distributed on the eight classes as follows: $\mathbf{f}_{1,\dots,8} = 3, 9, 16, 6, 3, 4, 2, 8$, which results in $\mathbf{s} = 19$. The corresponding critical value is $\kappa(5\%) = 14$ (derived from 100 000 trials; see the distribution function F_S in Figure 6.6). Hence we reject the null hypothesis that the occurrence of tropical cyclones in the Southwest Pacific is independent of the phase of the MJO.

6.4 On Establishing Statistical Significance

6.4.1 Independence of the Null Hypothesis. A rock formation called the *Mexican Hat* (Figure 6.7), near the border between Arizona and Utah, consists of a very large boulder perched precariously on a rocky outcrop. It is instructive to think briefly about whether we can use statistical methods to test the null hypothesis that this rock formation has natural origins. To gather information with which to test this hypothesis we might

1 randomly select a (large) sample of rock formations that have not been altered by humans, and

2 count the number of rock formations arranged as the Mexican Hat.

Let us assume that no other Mexican Hat-like formations are found. Humans have traversed most of the rocky desert of the world at one time or another and it would appear that the Mexican Hat is unique in the collective experience of these travellers. Therefore, the chances of finding another Mexican Hat among, say, one million randomly selected rocks, are nil. Thus we may reject the null hypothesis at a small significance level, and give credence to the explanation given in Figure 6.8.

Obviously we can generalize this example to include many different null hypotheses of the type 'rare event is common.'

The problem with these null hypotheses is that they were derived from the same data used to conduct the test. We already know from previous exploration that the Mexican Hat is unique, and its rarity leads us to conjecture that it is unnatural. Unfortunately, statistical methodology can not take us any farther in this instance unless we are willing to wait a very long time so that tectonic processes can generate a new independent realization of the surface of the earth.

6.4.2 More on the Role of Statistical Inference. The Mexican Hat is a pretty obvious example—but there are many similar examples in climate research journals. There are even instances in which peer reviewers have requested that authors perform statistical tests as outlined above. One example concerns the *Labitzke and van Loon hypothesis* [238] about the relationship between the 11-year solar cycle and the atmospheric circulation in the stratosphere and the troposphere.[3] They found, using about 30 years of data, that the North Pole winter mean 30 hPa temperature is only weakly correlated

[3]The original draft of [238] did not contain statistical inferences about the relationship between atmospheric circulation and solar activity. However, reviewers of that article demanded a statistical test even though there are really only two ways to verify the Labitzke and van Loon hypothesis. These are a) develop a physical hypothesis that can be verified by numerical experimentation, and b) wait a few decades so that additional independent data can be collected for a confirmatory statistical test of the hypothesis (cf. [4.1.2]).

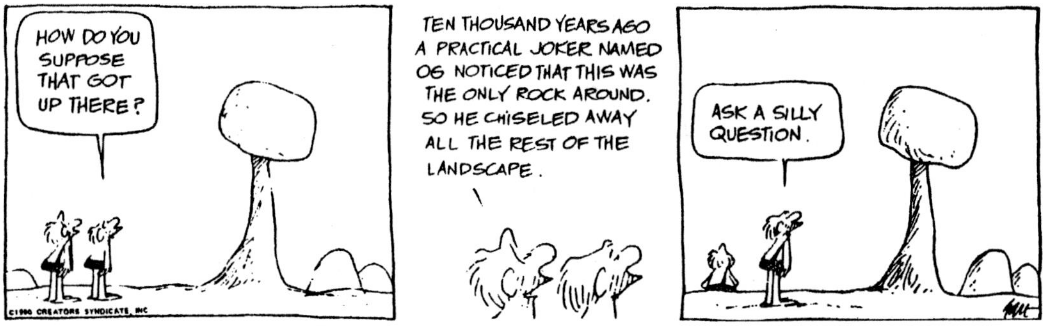

Figure 6.8: *Creation of the Mexican Hat: Null hypothesis correctly rejected!*

with solar activity. The observed correlation was 0.14 (Figure 6.9, top). The apparent strength of the relationship was much stronger when the data were stratified according to the phase of the Quasi-Biennial Oscillation (QBO; Veryard and Ebdon [382], Dunkerton [106]): A high *positive* correlation of 0.76 was obtained for the winters in which the QBO was in its west phase (Figure 6.9, middle), and a negative correlation of -0.45 when the QBO was in its east phase (Figure 6.9, bottom). The similarity of the middle and bottom curves in Figure 6.9 is certainly as remarkable as the Mexican Hat.

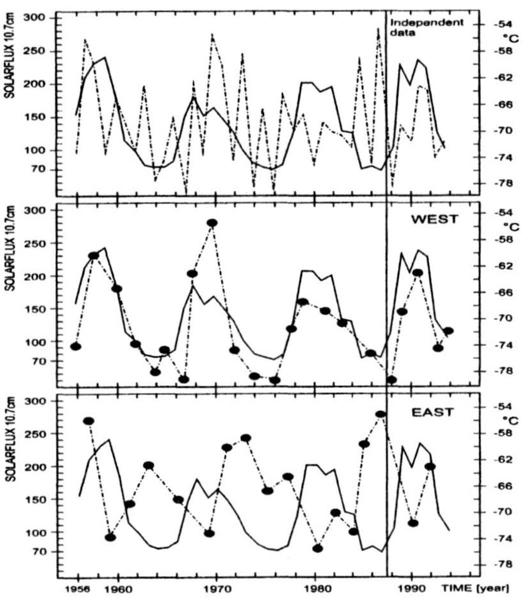

Figure 6.9: *Time series of January/February mean solar activity (solid curve) and 30 hPa temperature at the North Pole (broken curve). Top: all winters. Middle: winters when the QBO is in its west phase. Bottom: winters when the QBO is in its east phase. From Labitzke and van Loon [238].*

6.4.3 What if Confirmatory Analysis is not Possible? Although it is frequently not possible to make confirmatory statistical inferences once an exploratory analysis has suggested questions, methods of statistical inference, such as testing, are valuable. They serve to underline the unusual quantitatively and thus help us to focus on unusual aspects of the data. But the statistical test can not be viewed as an objective and unbiased judge of the null hypothesis under these circumstances.

6.4.4 What Constitutes Independent Data? Confirmatory analysis, as discussed in [6.4.1], requires additional independent data. Independence is the essential point here; it is generally not sufficient to have additional data from independent *sources*. For example, workers sometimes claim that they use independent data when they use station data to *derive* a hypothesis and grid point data from the same or a similar period to *confirm* the hypothesis. While it is certainly valuable to analyse both data sets to make sure that the hypothesis does not come about as a result of, for example, systematic biases in an ensemble of analyses fields, the two data sets are strongly correlated.

This observation limits any *confirmatory* statistical analysis with observed (atmospheric or other geophysical) data. Truly independent confirmatory analyses can only be performed with observations in the future because we can only collect the necessary independent information in the future. One alternative is to carefully construct a sensitivity experiment with a GCM to test the question. This avoids waiting, and often gives the experimenter opportunities to control or eliminate extraneous sources of variability that obscure the effects of interest in observations. Another alternative is to divide the observations into *learning* and *validation* data sets. The latter is set aside and reserved for confirmatory analysis of questions that arise from exploratory analysis of the former.

6.5 Multivariate Problems

6.5.0 Overview. The spatial covariance characteristics of the climate system have a profound effect on the analysis of just about any climate quantity that is distributed in space. Subsection 6.5.1 describes a prototypical problem in which we might want to use a multivariate test or multiple univariate tests. In both cases it is necessary to be aware of the relevant spatial covariance structure to interpret the results correctly. In subsection 6.5.2 we discuss the interpretation of multiple univariate tests, conducted, for example, at each grid point of a GCM. Another approach is to conduct a multivariate test on the entire field [6.5.3].

However, we often have information that can be used to sharpen the alternative hypothesis and therefore improve the efficiency of the multivariate test. The impact of ignoring this information is discussed in subsection 6.5.4. The prior information is expressed as a set of 'guess patterns' [6.5.6] and it is used by projecting the observed fields onto the space spanned by the guess patterns, therefore reducing the dimension of the multivariate testing problem. There are also practical considerations that motivate the dimension reduction [6.5.5]. Even after dimension reduction, it may be possible to further increase the sensitivity of the test by searching for a pattern in the space spanned by the guess patterns that optimizes the signal-to-noise ratio [6.5.7]. Finally, it is sometimes possible to develop a hierarchy of nested sets of guess patterns, and this inevitably leads to a step-wise testing procedure [6.5.8].

6.5.1 GCM Experiments. Analyses of GCM experiments are usually multivariate in nature simply because such models produce fields, such as monthly mean 500 hPa height fields, as output. GCM experiments are either *sensitivity experiments* or *simulations* of the present or a past climate of Earth or another planet.

A typical sensitivity study will consist of two climate simulations. One run, labelled the *control run* is conducted under 'normal' conditions, and the other, the *experimental run*, is conducted with, for example, the anomalous boundary conditions or a modified parameterization of a sub-grid scale physical process. Statistical tests are often used to determine whether the changes affect the distribution of climatic states simulated by the model. Since distributional changes alter the moments (such as mean and variance, see [2.6.7]), a basic problem is to test H_0: $\mu_{control} = \mu_{experiment}$, that is, the null hypothesis that the changes do not affect the mean state of the simulated climate. Examples are given in Section 7.2, where we compare two simulated climates, and Section 7.1, where a simulation is compared with the observed climate.

6.5.2 The Effect of Spatial Correlation on Multiple Univariate Tests. The simplest approach to comparing the mean states of the climates simulated in a pair of GCM experiments is to conduct a univariate difference of means test [6.6.1] at every grid point. This is called the *local test* approach because a *local null hypothesis* is tested at each grid point.

There can, however, be difficulty with the global interpretation of the results of a collection of local tests.

Assume, for the moment, that the *treatment* applied to the experimental simulation has no effect on the simulated mean state. Then the local equality of means hypothesis is true everywhere. The *global null hypothesis* that corresponds to the collection of local hypotheses is that the mean fields are equal. Now suppose that the local null hypothesis is tested at each of m grid points at the 5% significance level. Under the global null hypothesis we expect that roughly 5% of the local test decisions will be reject decisions. Each test is analogous to the toss of a fair 20-sided die that has 19 black faces and 1 white face. The white face will come up 5% of the time on average, but the proportion of white faces observed varies between replications of an m-roll die-rolling experiment. In the same way there is variability in the number of reject decisions that will be made in any one replication of the climate simulation experiment.

If decisions made at adjacent grid points are independent of each other, then the 20-sided die model can be used to predict the probability distribution of the number of reject decisions under the global null hypothesis. In fact, the probability of making reject decisions at k or more grid points is given by the binomial distribution that has cumulative distribution function $F_m(k) = \sum_{i=k}^{m} \mathcal{B}(m, 5\%)(i)$. For example, if the local test is conducted at $m = 768$ grid points, the probability of obtaining more than 48 local rejections under the global null hypothesis is 5%. Thus, in this example with independent grid points, a reasonable *global test* is to reject the global null hypothesis if the local reject decision is made at the 5%-significance level at 49 or more grid points.

In the real world, decisions made at adjacent grid points are *not* independent because meteorological fields are spatially correlated. Thus the binomial distribution does not provide the appropriate null distribution for the number of local reject decisions.

This was demonstrated in an experiment in which seven independent integrations were conducted with a simplified GCM [384]. Each integration produced one monthly mean field. The runs were identical except for small variations in their initial conditions. Because small-scale errors quickly cascade to all resolved spatial scales in AGCMs, this produced a set of seven independent realizations of the same geophysical process.

The set of $K = 7$ simulated monthly mean fields was arbitrarily split up into two sets, the first i and the last $K-i$. The first set was used to estimate the statistical parameters of the simulated geophysical process. Each realization in the second set of fields was tested at each grid point to see if it belonged to the population represented by the first set. The local rejection rate was subsequently calculated. On average, the reject decision was made 5.2% of the time, nearly the nominal 5% rate specified by the null hypothesis. However, there are instances in which the rate of incorrect decision was as high as 10%. We would expect reject rates to vary between 3.4% and 6.6% in the absence of spatial correlation. Thus it appears that spatial correlation affects the variability of the proportion of reject decisions.

The effect of spatial correlation is illustrated in Figure 6.10 where we see one field of erroneous rejections. Note that erroneous rejections do not occur at isolated points. Rather, the spatial correlation structure results in pools of reject decisions. On average these pools will occupy 5% of the map. Map to map variation in the area covered by the pools depends on the average size of the pools, which in turn is determined by the spatial correlation structure of the field. The map to map variation is smallest when the 'pools' degenerate to isolated points that are not spatially correlated.

6.5.3 Multivariate Tests of the Mean.

There are at least two ways to test the global null hypothesis of the equality of mean fields. One is to find the correct distribution for the number of false local rejections under the null hypothesis. Livezey and Chen [257] have suggested methods that are widely used [6.8.1–3]. Another is to use multivariate techniques such as the Hotelling test or a permutation test [6.6.4–7].

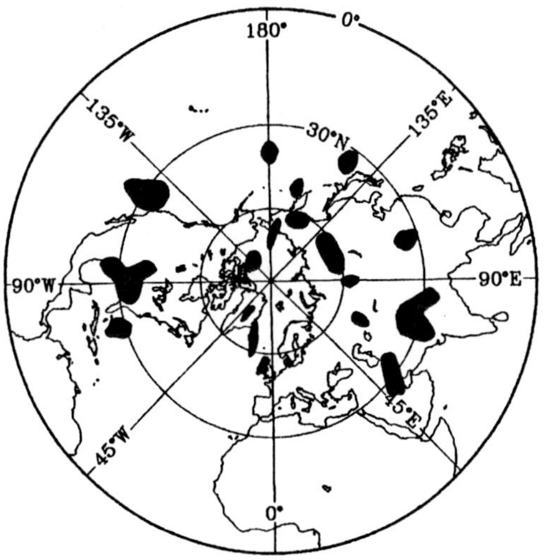

Figure 6.10: *The spatial distribution of false rejections of local null hypotheses in a Monte Carlo experiment [384].*

The multivariate method induces strategic and technical problems related to the dimension of the observed climate fields. We discuss these in the next two subsections.

6.5.4 Strategic Problems.

The strategic problem arises because the signal induced by the experimental 'treatment' may not be present in all components of the observed field. Often it resides in a low-dimensional subspace spanned by only a few vectors. The total m-dimensional space that contains the climate realizations may be represented as a sum of two spaces Ω_S and Ω_N with dimensions m_S and m_N respectively, where $m_S + m_N = m$. The signal is confined to Ω_S. Both Ω_S and Ω_N contain variations due to random fluctuations. A multivariate test of the equality of means hypothesis (i.e., the signal is absent) will be more powerful if is restricted to Ω_S because the *signal-to-noise* ratio in Ω_S is greater than it is in the full space $\Omega_S \cup \Omega_N$.

This is demonstrated in the following example. Let $\vec{\mathbf{X}}$ be an m-dimensional normal random vector with mean $\vec{\mu} = (0, \ldots, 0)^{\mathrm{T}}$ and covariance matrix

$\Sigma = \mathcal{I}$.[4] Let $\vec{\mathbf{Y}}$ be another m-dimensional normal random vector defined by $\vec{\mathbf{Y}} = \vec{\mathbf{X}} + \vec{a}$ where $\vec{a} = (2, 0, \ldots, 0)$ and let $\vec{\mathbf{y}}$ be a realization of $\vec{\mathbf{Y}}$. We want to test the null hypothesis, H_0, that $\vec{\mathbf{y}}$ belongs to the population defined by $\vec{\mathbf{X}}$. The Mahalanobis test statistic

$$\mathcal{D}^2(\vec{y}) = (\vec{y} - \vec{\mu})^{\mathrm{T}}\Sigma^{-1}(\vec{y} - \vec{\mu}) = \sum_{i=1}^{m} y_i^2 \tag{6.8}$$

has a χ^2 distribution with m degrees of freedom under H_0. Its expected value under the alternative hypothesis, which is true by construction, is $\mathcal{E}(\mathcal{D}^2) = 2^2 + m$. These expected values, and corresponding 5% significance level values for the test statistic under H_0, are:

m	$\mathcal{E}(\mathcal{D}^2)$ under H_a	$\chi^m_{5\%}$ under H_0
1	5	3.8
2	6	6.0
3	7	7.8
4	8	9.5

We see that for $m = 1$ the expected Mahalanobis distance is larger than the critical value; usually the null hypothesis will correctly be rejected. However, as more components that contain only noise are included, the chances of detecting the signal deteriorate.

6.5.5 Practical Problems. A practical problem arises in multivariate difference of means tests because the covariance matrix is generally not known. The problem was avoided in the previous example because Σ was specified. Consequently, we were able to use $\mathcal{D}^2(\vec{\mathbf{Y}})$ (6.8) as the test statistic. In most problems, though, Σ must be estimated. One implication is that we must base the test on the Hotelling T^2 statistic, which is the counterpart to $\mathcal{D}^2$ if Σ is replaced with the sample covariance matrix. To compute T^2 we must be able to invert the sample covariance matrix, which means that we need to have a sample of $n = m + 1$ realizations of the climate represented by $\vec{\mathbf{X}}$. However, in most climate applications, there are many more spatial degrees of freedom than observations (i.e., $n \ll m$). Then, reducing the number of spatial degrees of freedom by restricting the test to a subspace that is thought to contain the signal of interest is also a practical expedient.

[4] $\mathcal{I}$ denotes the $m \times m$ identity matrix.

6.5.6 Guess Patterns. The spatial degrees of freedom may be reduced by approximating the full m-dimensional fields $\vec{\mathbf{X}}$ as a linear combination of a set of $\tilde{m}$ patterns $\vec{p}^{\,i}$, as

$$\vec{\mathbf{X}} \approx \sum_{i=1}^{\tilde{m}} \alpha_i \vec{p}^{\,i}. \tag{6.9}$$

The coefficients α_i are usually fitted by a least square approximation (see Chapter 8). The *guess patterns* $\vec{p}^{\,i}$ should be specified independently of the outcome of the experiment.

There are various ways to obtain guess patterns.

1 Patterns known to yield efficient approximations of the analysed fields $\vec{\mathbf{X}}$: examples are Empirical Orthogonal Functions (EOFs; see Chapter 13) or, in case of a spherical geometry, surface spherical harmonics.

2 Problem-related patterns: patterns that were found as signals in similar but independent GCM experiments or patterns that were diagnosed from similar observations.

3 Physically based patterns: patterns that were derived by means of simplified theory that is appropriate to the hypothesis the experiment is designed to test.

It is often more profitable to invest in choices 2 and 3, which provide patterns with a physical basis, rather than to try to improve the power of the statistical tests. These choices also provide confirmation that the physical reasoning that leads to the experimental design and choice of patterns is correct. For example, if empirical guess patterns are derived from observations on the basis of physical reasoning (choice 2) and the null hypothesis that their 'experimental' treatment does not induce a climate signal is rejected, then there is statistical confirmation that the GCM has reproduced these aspects of the observed climate. If dynamically derived patterns are used (choice 3), rejection is an indication that the simplified theory behind the guess patterns operates within the GCM, at least to a first order of approximation. Examples are presented in Sections 6.9, 6.10 and Chapter 7.

6.5.7 Optimizing the Signal-to-Noise Ratio. Hasselmann [166, 168] suggested the following interesting way to construct an *optimal* guess pattern $\vec{p}^{\,o}$ from a given guess pattern $\vec{p}$.

Let $\vec{\mathbf{X}}$ be a random vector of dimension m with covariance matrix Σ and expectation $\vec{\mu}_X$. Let $\vec{\mathbf{Y}}$

be another m-dimensional random vector with the same covariance matrix and expectation $\vec{\mu}_Y \neq \vec{\mu}_X$. Next, let $\vec{p}$ be a guess pattern representing the anticipated form of the true signal $\Delta = \vec{\mu}_Y - \vec{\mu}_X$. This pattern will not point in exactly the same direction as Δ, but we will act as if $\vec{p}$ were the true signal. Then the challenge is to find an *optimal* guess pattern $\vec{p}^{\,o}$ that maximizes the likelihood of signal detection.

To do so we consider the *signal-to-noise ratio*

$$r = \frac{\langle \vec{p}\,, \vec{p}^{\,o}\rangle^2}{\mathrm{Var}\Big(\langle \vec{\mathbf{Y}} - \vec{\mathbf{X}}, \vec{p}^{\,o}\rangle\Big)}, \tag{6.10}$$

where $\langle \cdot, \cdot \rangle$ denotes the scalar, or dot, product of two vectors. The numerator in (6.10) is the strength of the (anticipated) signal in the direction of the optimal guess pattern $\vec{p}^{\,o}$. The denominator is the variance of the noise, $\vec{\mathbf{Y}} - \vec{\mathbf{X}}$, in the direction of $\vec{p}^{\,o}$. When r is large, the likelihood of rejecting the null hypothesis H_0: $\mu_Y - \mu_X = 0$, and thus detecting a nonzero signal Δ in the direction of $\vec{p}$, is also large.

We now specify $\vec{p}^{\,o}$. Because r does not depend on $\|\vec{p}^{\,o}\|$ we may constrain $\vec{p}^{\,o}$ so that

$$\langle \vec{p}\,, \vec{p}^{\,o}\rangle^2 = 1. \tag{6.11}$$

Then r may be maximized by minimizing the denominator of (6.10),

$$\mathrm{Var}\Big(\langle \vec{\mathbf{Y}} - \vec{\mathbf{X}}, \vec{p}^{\,o}\rangle\Big) = 2(\vec{p}^{\,o})^{\mathrm{T}} \mathbf{\Sigma} \vec{p}^{\,o}. \tag{6.12}$$

The guess pattern that minimizes (6.12) satisfies

$$\frac{d}{d\vec{p}^{\,o}}\Big[2(\vec{p}^{\,o})^{\mathrm{T}}\mathbf{\Sigma}\vec{p}^{\,o} - \nu\Big(\langle \vec{p}\,, \vec{p}^{\,o}\rangle^2 - 1\Big)\Big] = 0, \tag{6.13}$$

where ν is a Lagrange multiplier used to enforce the constraint (6.11). Note that any solution of (6.13) satisfies (see, e.g., Graybill [148])

$$2\mathbf{\Sigma}\vec{p}^{\,o} = \nu\langle \vec{p}\,, \vec{p}^{\,o}\rangle \vec{p}\,. \tag{6.14}$$

Thus the only solution $\vec{p}^{\,o}$ of (6.13) is

$$\vec{p}^{\,o} = \frac{1}{2}\nu\mathbf{\Sigma}^{-1}\vec{p}\,,$$

with $\nu = 2(\vec{p}^{\,\mathrm{T}}\mathbf{\Sigma}^{-1}\vec{p}\,)^{-1}$.

When $\mathbf{\Sigma} = \mathrm{diag}(\sigma_1^2, \ldots, \sigma_m^2)$, that is, $\mathbf{\Sigma}$ is diagonal, the ith component of $\vec{p}^{\,o}$ is expressed in terms of the ith component of $\vec{p}$ as $p_i^o = p_i/\sigma_i$. That is, the original guess pattern is rotated towards directions with small values of σ_i, directions that have little 'noise' relative to the signal.

An example of an application of this optimization procedure (Hegerl et al. [172]) is given in some detail in Section 7.4. Other applications include Bell [37, 39], Mikolajewicz, Maier-Reimer, and Barnett [277] and Hannoschöck and Frankignoul [161].

6.5.8 Hierarchies. When an extended set of guess patterns is available, step-wise test procedures are also possible within the multivariate testing paradigm discussed in this section.

For example, suppose a set of guess patterns contains a subset of patterns $\Xi = \{\vec{p}^{\,i} : i \in I\}$ that are physically derived (choices 2 and 3 in [6.5.6]). Here I is a collection of indices. We call the low-dimensional space Ω_Ξ, which is spanned by Ξ, the 'signal space.' The space spanned by the full collection of guess patterns is then given by the full set of patterns that are likely to contain the sought after signal $\Omega = \Omega_\Xi \cup \Omega_\Xi^\perp$ where $\Omega_\Xi^\perp$ is the space spanned by the guess patterns that are not contained in Ξ. The full response, say $\vec{\mathbf{Z}} = \vec{\mathbf{Y}} - \vec{\mathbf{X}}$, is then written as $\vec{\mathbf{Z}} = \vec{\mathbf{Z}}_\Xi + \vec{\mathbf{Z}}_\Xi^\perp$. The components *parallel* and *perpendicular* to the signal space, $\vec{\mathbf{Z}}_\Xi$ and $\vec{\mathbf{Z}}_\Xi^\perp$, are then tested. The parallel component is projected on the problem-specific guess patterns contained in Ξ and the perpendicular component is tested using problem-independent guess patterns (choice 1 in [6.5.6]) such as EOFs.

An example is given in Section 7.2.

The approach discussed above imposes a simple ordering on a set of guess patterns: the full set of patterns that are likely to contain the sought after signal and a smaller subset of patterns derived from problem-specific reasoning. A hierarchical approach to testing would involve conducting a test in the space spanned by the problem-specific patterns, and then, if a signal is detected, conducting a second test in the full space. Of course, this approach is not limited to two levels; a hierarchy of nested vector spaces could be constructed by scaling arguments, for example. A sequence of tests could then be conducted [22], either in order of increasing or decreasing dimension, to isolate the region on the supposed response space (the space spanned by the full set of guess patterns) that contains the signal (see Section 7.3).

6.6 Tests of the Mean

6.6.1 The Difference of Means Test. The t test, also known as Student's t test, is a parametric test of the null hypothesis that two univariate random variables **X** and **Y** have equal means, that is,

$$\mathrm{H_0}: \mathcal{E}(\mathbf{X}) = \mathcal{E}(\mathbf{Y}) \text{ or } \mu_X = \mu_Y. \tag{6.15}$$

The statistical model required to conduct the test is built by making three assumptions [454]. The first is a *sampling* assumption that every realization of **X** or **Y** occurs independently of all other realizations. The second and third are *distributional assumptions*: first, that the distribution that generates realizations of **X** (or **Y**) is the same for each observation in the **X** (or **Y**) sample and, second, that the distributions are normal[5] and have equal variance σ^2. The t test is moderately robust against departures from the normal distribution, particularly if relatively large samples of both random variables are available. However, the test is not robust against departures from the sampling assumption (see [454] and [6.6.6]) or against large departures from the assumption that all realizations in a sample come from the same distribution.

The optimal test statistic, within the constraints of the statistical model implied by the three assumptions, is conceptually different from that used for the sign test.[6] The difference of means is estimated and then scaled by an estimate of its own standard deviation, making it dimensionless.

The optimal test statistic is given by

$$t = \frac{\widehat{\mu}_X - \widehat{\mu}_Y}{S_p\sqrt{\frac{1}{n_X} + \frac{1}{n_Y}}}, \tag{6.16}$$

where n_X and n_Y indicate the size of the **X** and **Y** samples respectively, $\widehat{\mu}_X$ and $\widehat{\mu}_Y$ are the sample means of $\{\mathbf{x}_1, \ldots, \mathbf{x}_{n_X}\}$ and $\{\mathbf{y}_1, \ldots, \mathbf{y}_{n_Y}\}$, and S_p is the pooled estimate of the common standard deviation

$$S_p^2 = \frac{\sum_{i=1}^{n_X}(\mathbf{x}_i - \widehat{\mu}_X)^2 + \sum_{i=1}^{n_Y}(\mathbf{y}_i - \widehat{\mu}_Y)^2}{n_X + n_Y - 2}. \tag{6.17}$$

Under the null hypothesis (6.16) has a t distribution with $n_X + n_Y - 2$ degrees of freedom [2.7.9].[7] This is fortunate because it means that the reference distribution under the null hypothesis does not depend upon either the unknown common population mean $\mu = \mu_X = \mu_Y$ or standard deviation $\sigma = \sigma_X = \sigma_Y$. Consequently, only a small number of reference distributions, indexed by $n_X + n_Y - 2$, are required. Critical values for this family of distributions are tabulated in Appendix F.

6.6.2 Components of the t Statistic. It is useful to take a slight diversion to dissect (6.16) and better understand why it has the t distribution under the null hypothesis.

A random variable **T** has the t distribution with m degrees of freedom, written $\mathbf{T} \sim t(m)$ [2.7.9], when

$$\mathbf{T} = \frac{\mathbf{A}}{\sqrt{\mathbf{B}/m}}, \tag{6.18}$$

where **A** is a standard normal random variable, $\mathbf{A} \sim \mathcal{N}(0,1)$, and **B** is a χ^2 random variable with m degrees of freedom, $\mathbf{B} \sim \chi^2(m)$, that is independent of **A**. Under the null hypothesis of equality of means we find that

$$\mathbf{A} = \frac{\widehat{\mu}_X - \widehat{\mu}_Y}{\sigma\sqrt{1/n_X + 1/n_y}} \sim \mathcal{N}(0,1),$$

$$\mathbf{B} = \frac{n_X + n_Y - 2}{\sigma^2} S_p^2 \sim \chi^2(n_X + n_Y - 2),$$

and that **A** and **B** are independent. By substituting these quantities into (6.18) we see that the test statistic for the difference of means test (6.16) is $\mathbf{T} \sim t(n_X + n_Y - 2)$.

6.6.3 When the Variance is Known. The t test discussed above has been derived assuming that the variance is unknown. When the variance is known, its square root may be substituted directly for S_p in (6.16). The resulting *z-statistic* has the standard normal distribution $\mathcal{N}(0,1)$ under the null hypothesis. Critical values may be obtained from Appendix D.

[5]The test is said to be *parametric* because it concerns parameters (the means μ_X and μ_Y) of a specific distribution (the normal distribution). A non-parametric version of the test (see [6.6.11]) would focus on the expected values of **X** and **Y** and would use less specific information about the distribution of these random variables to construct the statistical model needed to conduct the test.

[6]The sign test is an example of a non-parametric test. The Mann–Whitney test [6.6.11] is another example of a non-parametric test.

[7]The term *degrees of freedom* has geometrical roots. The random variable T, of which t is a realization, is a function of deviations $\mathbf{x}_i - \widehat{\mu}_X, i = 1, \ldots, n_X$ and $\mathbf{y}_j - \widehat{\mu}_Y, j = 1, \ldots, n_Y$. When these $n_X + n_Y$ random deviations are organized into an $(n_X + n_Y)$-dimensional random vector, we find that the random vector is confined to an $(n_X + n_Y - 2)$-dimensional vector space. This happens because the n_X **X** deviations must sum to zero as must the n_Y **Y** deviations. A derivation of this distribution of (6.16) may be found in, among others, [280] or [272].

6.6.4 Relaxing the Assumptions. The difference of means test described above operates as expected (e.g., the risk of false rejection is equal to that specified) only if the assumptions are fulfilled. In the following subsections we discuss methods that can be used when:

- the variances of **X** and **Y** are unequal, $\sigma_X \neq \sigma_Y$ (see [6.6.5]),
- the observations are paired in such a manner that pairs $(\mathbf{x}_i, \mathbf{y}_i)$ are independent realizations of a random vector $(\mathbf{X}, \mathbf{Y})^{\mathrm{T}}$ that has dependent components (see [6.6.6]),
- the observations are *auto-correlated* (see [6.6.7,8]).

6.6.5 Unequal Variances. We suppose now that the sampling and distributional assumptions of [6.6.1] continue to hold except that $\mathrm{Var}(\mathbf{X}) \neq \mathrm{Var}(\mathbf{Y})$.[8] Under these circumstances only some of the ingredients that lead to the t distribution as reference distribution are obtainable. The natural estimator of the true difference of means is still $\widehat{\mu}_X - \widehat{\mu}_Y$. This is a normal random variable with mean $\mu_X - \mu_Y$ and variance $\sigma_X^2/n_X + \sigma_Y^2/n_Y$. The variance is estimated by $S_X^2/n_X + S_Y^2/n_Y$ with S_X^2 and S_Y^2 defined as usual by $S_X^2 = \frac{1}{n_x-1}\sum_{i=1}^{n_X}(\mathbf{x}_i - \widehat{\mu}_X)^2$. Thus the difference of means is expressed in dimensionless units as

$$t = \frac{\widehat{\mu}_X - \widehat{\mu}_Y}{\sqrt{S_X^2/n_X + S_Y^2/n_Y}}. \qquad (6.19)$$

The square of the denominator can be shown to be statistically independent of the numerator but it does *not* have a distribution proportional to the χ^2 distribution. Therefore the test statistic does not have a t distribution under the null hypothesis.

The accepted solution to this problem, which is known in the statistical literature as the *Behrens–Fisher problem*, is to approximate the distribution of this statistic with a t distribution whose degrees of freedom are estimated from the data. The formula used to determine the approximating t distribution is obtained by comparing the first and second moments of $S_X^2/n_X + S_Y^2/n_Y$ with those of the χ^2 distribution. The resulting formula for the approximating number of degrees of freedom is

$$df = \frac{(S_X^2/n_X + S_Y^2/n_Y)^2}{\frac{(S_X^2/n_X)^2}{n_X-1} + \frac{(S_Y^2/n_Y)^2}{n_Y-1}}. \qquad (6.20)$$

[8]When the equality of the two variances is uncertain, one might resort to an F test for the equality of variances (Section 6.7).

Hypothesis (6.15) is tested by comparing the t-value computed using (6.19) with the critical values of the t distribution with df degrees of freedom, where df is computed with (6.20). This recipe constitutes a test that operates at an actual significance level close, but not exactly equal, to the level specified by the user.

6.6.6 The Paired Difference Test. Not all experimental designs lead to pairs of samples that are independent of each other. For example, one may conduct an experiment consisting of a series of five-day simulations with an AGCM to study the effects of a particular cloud parameterization. Suppose that two parameterizations are chosen, and that pairs of five-day runs are conducted from the same initial conditions. The initial conditions are selected randomly from a much longer run of the same AGCM, and the total liquid water content of the atmosphere is computed at the end of each five-day integration.

Because the integrations are short, one can imagine that the pairs of liquid water fields obtained from each set of initial conditions are not independent of each other. Thus the difference of means tests discussed above are not appropriate for testing the null hypothesis that the change in parameterization has not affected the total liquid water content of the atmosphere. The statistical model used with these tests relies upon the independence of all observations.

The solution to this problem is to compute the difference fields and test the null hypothesis that the mean difference is zero using a one sample t test. It is reasonable to assume that the observed differences are independent of one another because the initial conditions were chosen randomly. The distributional assumptions are that the differences have a normal distribution and that all the differences come from the same distribution. The former may not be true, even approximately, because moisture related variables, such as total liquid water, often exhibit strongly skewed distributions. However, let us continue to assume that the differences are normally distributed for the purposes of this discussion. The second distributional assumption, that the differences are identically distributed, may not hold if we failed to account for other sources of variation, such as the annual cycle, in our experimental design. To avoid such problems, the choice of initial conditions should be constrained to one season or calendar month, and one time of day.

Let $\mathbf{d}_i$ represent the ith realization of the change in total liquid water $\mathbf{D}$. The null hypothesis to be tested is H_0: $\mu_D = 0$. The optimal test statistic for this problem is

$$t = \frac{\widehat{\mu}_D}{S_D/\sqrt{n}}, \tag{6.21}$$

where n is the size of the sample of differences, $\widehat{\mu}_D = \sum_{i=1}^{n} \mathbf{d}_i/n$ is the mean difference, and $S_D^2 = \sum_{i=1}^{n}(\mathbf{d}_i - \widehat{\mu}_D)^2/(n-1)$ is the sample variance of the observed differences. This statistic has a t distribution with $n-1$ degrees of freedom under the null hypothesis.[9] Thus the paired difference test is conducted by computing the differences, then computing (6.21) with the sample moments, obtaining the appropriate critical value from Appendix F and finally comparing t with the critical value to make a decision.

The paired difference test is an example of a *one-sample t* test. One-sample tests are used to test hypotheses of the form H_0: $\mu_X = c$ where c is a constant that is chosen *a priori*. These tests are performed by computing

$$t = \frac{\widehat{\mu}_X - c}{S_X/\sqrt{n}} \tag{6.22}$$

and comparing with critical values for $t(n-1)$.

6.6.7 Auto-Correlation. As noted in [6.6.1], the t test is not robust against departures from the independence assumption. In particular, meteorological *time series* are generally *auto-correlated* if the time increment between observations is not too large. Under these circumstances, a t test such as that based on (6.16) becomes *liberal*, that is, it rejects the null hypothesis when it is true more frequently than indicated by the significance level.

Intuitively, observations taken in an auto-correlated sequence vary less quickly than observations obtained completely at random. An auto-correlated series therefore contains less information about the population mean than a completely random sequence of the same length. Consequently, the standard error of $\widehat{\mu}_X - \widehat{\mu}_Y$ is larger for auto-correlated data than for independent observations. However, the denominators of t statistics, such as (6.16), estimate the standard deviation of $\widehat{\mu}_X - \widehat{\mu}_Y$ under the independence assumption. Therefore the denominator in (6.16) underestimates the variability of $\widehat{\mu}_X - \widehat{\mu}_Y$ with the consequence that the absolute value of t tends to be too large.

Resolution of this problem is non-trivial [454]. Heuristic arguments, such as that given above, lead to a t test in which the denominator of the t statistic is inflated by a factor related to the time scales at which the time series varies. The resulting statistic, detailed below, is compared with critical values from a t distribution with an estimated number of degrees of freedom. This approach, while not exact, has the advantages that it is easy to use, easy to understand, and *asymptotically* optimal (i.e., it becomes optimal as the sample size becomes infinitely large). It can be used safely when samples are relatively large, as defined below. When samples are not large the 'Table-Look-Up' test [6.6.9] should be employed.

The large sample difference of means test is developed heuristically as follows. We assume that the memory of the observed time series is finite so that the full samples $\{\mathbf{X}_1, \dots, \mathbf{X}_{n_X}\}$ and $\{\mathbf{Y}_1, \dots, \mathbf{Y}_{n_Y}\}$ contain subsets of independent observations. For example, suppose that $\{\mathbf{x}_1, \dots, \mathbf{x}_{100}\}$ is a time series of 100 daily surface temperature anomalies. Consecutive observations are certainly highly correlated, but any two observations separated by 10 days or more are nearly independent. Thus the sample contains a subset of at least 11 roughly independent observations. However, we do not throw away the other 89 observations. Instead, we attempt to estimate the information content of the entire sample by deriving an *equivalent sample size*.

The measure of information used in the difference of means problem is one over the variance of the sample mean. Thus the smaller the variance of the sample mean, the more information the sample contains about the unknown population mean. The equivalent sample size n'_X is defined as the number of independent random variables that are needed to provide the same amount of information about μ_X as the sample of dependent random variables $\{\mathbf{X}_1, \dots, \mathbf{X}_{n_X}\}$. Equivalent sample size n'_Y is defined analogously.[10] We anticipate that $n'_X < n_X$ and $n'_Y < n_Y$ when observations are auto-correlated.[11]

This paradigm leads us to estimators $\widehat{n}_X$ and $\widehat{n}_Y$, which replace n'_X and n'_Y in the ordinary difference of means tests with equal (see [6.6.1]) or unequal

[9] There are $n-1$ degrees of freedom because the deviations $\mathbf{d}_i - \widehat{\mu}_D$ are elements of an n-dimensional random vector that is constrained to vary within an $(n-1)$-dimensional vector space.

[10] Note that the definition of the equivalent sample size depends upon the parameter that is being tested and the way is which information is measured. The equivalent sample sizes for an equality of variance test, for example, are different from those for the equality of means tests. The measure of information used here, the inverse of the variance of the sample mean, is called *Fisher's information* (see [92]).

[11] Strictly speaking, this happens when time series are persistent, that is, when adjacent anomalies have the same sign. It is possible to have $n'_X > n_X$ and $n'_Y > n_Y$ when adjacent anomalies tend to have opposite sign.

(see [6.6.5]) variances. When the samples are large enough, t statistics computed in this way with (6.16) as

$$t' = \frac{\widehat{\mu}_X - \widehat{\mu}_Y}{S_p\sqrt{\frac{1}{\widehat{n}'_X} + \frac{1}{\widehat{n}'_Y}}} \tag{6.23}$$

or with (6.19) as

$$t' = \frac{\widehat{\mu}_X - \widehat{\mu}_Y}{\sqrt{\frac{S_X^2}{\widehat{n}'_X} + \frac{S_Y^2}{\widehat{n}'_Y}}} \tag{6.24}$$

can be compared with critical values from the $t(\widehat{n}'_X + \widehat{n}'_Y - 2)$ or $t(\widehat{df})$ distribution respectively where, in the latter case, $\widehat{df}$ is computed with (6.20) by substituting the equivalent sample size estimates for the sample sizes themselves.

There are two problems left:

- estimating the equivalent sample size (see [6.6.8]), and
- determining whether the t' in (6.23) or (6.24) is distributed as a Student's t random variable under the null hypothesis. When n'_X and n'_Y are small ($n'_X + n'_Y < 30$ if t' is computed with (6.23); $n'_X < 30$ and $n'_Y < 30$ if t' is computed with (6.24)), the distribution of t' deviates markedly from *any* t distribution [363]. Thus the t test can not be used with small equivalent sample sizes. An alternative is described in [6.6.9].

While the discussion above has focused on the difference of means test, the same considerations apply to one-sample t tests such as the paired difference test (cf. [6.6.6]).

6.6.8 The Definition and Estimation of the Equivalent Sample Size. Let us assume that the data are given with constant time steps δ, such that the ith sample $\mathbf{x}_i$ is taken at time $t = i\delta$. Then the variance of the sample mean is

$$\mathrm{Var}(\overline{\mathbf{X}}) = \sigma^2/n'_X, \tag{6.25}$$

where

$$n'_X = \frac{n_X}{1 + \sum_{k=1}^{n_X-1}\left(1 - \frac{k}{n_X}\right)\rho_X(k)}, \tag{6.26}$$

(see Section 17.1) and $\rho_X(k)$ is the *auto-correlation* function

$$\rho_X(k) = \frac{1}{\sigma^2}\mathrm{Cov}(\mathbf{X}_i, \mathbf{X}_{i+k})$$

(see Section 11.1). We will drop the subscript 'X' for notational convenience in the rest of this subsection.

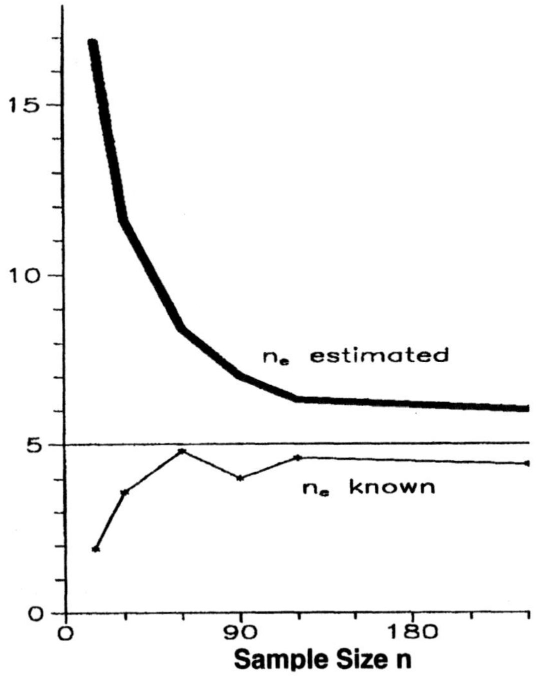

Figure 6.11: *The reject rate percentage of the one-sample t test when the observations are auto-correlated (see text). The 'equivalent sample size' n' is given by (6.26) (thin curve) and is estimated with (6.26) (thick curve).*

We conducted a Monte Carlo experiment (cf. [6.3.1]) with a one-sample t test to examine how well it works when the equivalent sample size n' is determined by (6.26). *Time series* of length $n = 15, 30, 60\ldots$ were generated by an *auto-regressive process of first order* with $\alpha = 0.6$ (see Chapter 10). Such processes have auto-correlation functions of the form $\rho(k) = \alpha^{|k|}$. If we insert the equivalent sample size, as defined by (6.26), into (6.22) and use a significance level of 5%, we observe fewer rejections of the true null hypothesis 'H_0: $\mu = 0$' (Figure 6.11) than expected. The deviation from the nominal 5% level is considerable when n is less than 30. This happens because the distribution of t' is not well approximated by the distribution of $t(n')$ under H_0.

Estimates of n' can be obtained either from physical reasoning or by means of a statistical estimator. Estimates based on physical reasoning should state lower bounds for n' because optimistic estimates will result in t'-values that are frequently too large and, consequently, cause more frequent rejection of the null hypothesis when it is true than indicated by the significance level.

Statistical estimators of n' use estimates of the auto-correlation function $\rho(k)$ in combination with

(6.26). Various reasonable estimators of n' [363, 454] result in t tests that tend to reject H_0 more frequently than specified by the significance level. Estimation is discussed further in [17.1.3].

The Monte Carlo experiment described above was repeated using this best estimator $\widehat{n}'$ in place of the known equivalent sample size n' (6.26). Figure 6.11 shows that the test now rejects the true null hypothesis more frequently than specified by the significance level of 5%. We therefore suggest that the t test *not* be used with *equivalent sample sizes smaller than 30*. Instead, we advise the use of 'Table-Look-Up test,' described next, in such predicaments.

6.6.9 The 'Table-Look-Up Test.' The 'Table-Look-Up test' [454] is a small sample alternative to the conventional t test that avoids the difficulties of estimating an equivalent sample size while remaining as efficient as the optimal asymptotic test when equivalent sample sizes are large.[12]

The Table-Look-Up test procedure is as follows.

- The paired difference (or one sample) case: to test 'H_0: $\mu = \mu_0$' using a sample of size n_X compute

$$t = \frac{(\bar{\mathbf{x}} - \mu_0)}{S_X/\sqrt{n_X}}, \tag{6.27}$$

where $\bar{\mathbf{x}}$ is the sample mean and S_X^2 is the sample variance.

Compute the sample lag-1 correlation coefficient $\widehat{\alpha}_X$ using

$$\widehat{\alpha}_X = \frac{\sum_{i=1}^{n_X} \mathbf{x}_i'\mathbf{x}_{i-1}'}{(n_x - 1)S_X^2} \tag{6.28}$$

where $\mathbf{x}_i' = \mathbf{x}_i - \widehat{\mu}_X$. Use Appendix H to determine the critical value for t that is appropriate for a sample of size n with lag-1 correlation coefficient $\widehat{\alpha}_X$.

- The two sample case (assuming $\sigma_X = \sigma_Y$ and that lag-1 correlation $\alpha_X = \alpha_Y$): to test 'H_0: $\mu_y = \mu_x$' using **X** and **Y** samples of size n_X and n_Y respectively, compute

$$t = \frac{\bar{\mathbf{x}} - \bar{\mathbf{y}}}{S_p\sqrt{\frac{1}{n_X} + \frac{1}{n_Y}}}, \tag{6.29}$$

where $\bar{\mathbf{x}}$ and $\bar{\mathbf{y}}$ are sample means and S_p^2 is the pooled sample variance (6.17). Compute the pooled sample lag-1 correlation coefficient $\widehat{\alpha}$ using

$$\widehat{\alpha} = \frac{\sum_{i=2}^{n_X} \mathbf{x}_i'\mathbf{x}_{i-1}' + \sum_{i=2}^{n_Y} \mathbf{y}_i'\mathbf{y}_{i-1}'}{(n_X + n_Y - 2)S_p^2}, \tag{6.30}$$

where $\mathbf{x}_i' = \mathbf{x}_i - \widehat{\mu}_X$ and $\mathbf{y}_i' = \mathbf{y}_i - \widehat{\mu}_Y$. Use Appendix H to determine the critical value of t that is appropriate for a sample of size $n_X + n_Y$, which has a lag-1 correlation coefficient $\widehat{\alpha}$.

[12] The Table-Look-Up test assumes that the sample(s) comes from auto-regressive processes of order 1 (Chapter 10). Departures from this assumption will compromise the test. Wilks [423] suggests an alternative approach for situations when the assumption does not hold.

6.6.10 The Hotelling T^2 test. The multivariate version of the t test, which is used to test the null hypothesis

$$H_0: \vec{\mu}_X = \vec{\mu}_Y, \tag{6.31}$$

is called the Hotelling T^2 test. The assumptions implicit in this parametric test are identical to those required for the t test except that they apply to vector, rather than scalar, realizations of an experiment. It is necessary to make the sampling assumption that the realizations of the m-dimensional random vectors $\vec{\mathbf{X}}$ and $\vec{\mathbf{Y}}$ occur independently of each other. It is also necessary to make similar distributional assumptions: that all observations in a sample come from the same distribution and that those distributions are multivariate normal. In addition, we also assume that both $\vec{\mathbf{X}}$ and $\vec{\mathbf{Y}}$ have the same covariance matrix Σ, so that $\vec{\mathbf{X}} \sim \mathcal{N}(\vec{\mu}_X, \Sigma)$ and $\vec{\mathbf{Y}} \sim \mathcal{N}(\vec{\mu}_Y, \Sigma)$.

The covariance matrix $\Sigma = (\sigma_{ij})$ is generally not known and must be estimated from the data in a manner analogous to (6.17):

$$\widehat{\sigma}_{ij} = \frac{\sum_k^{n_X} \mathbf{x}_{ik}'\mathbf{x}_{jk}' + \sum_k^{n_Y} \mathbf{y}_{ik}'\mathbf{y}_{jk}'}{n_X + n_Y - 2}, \tag{6.32}$$

where $\mathbf{x}_{jk}' = \mathbf{x}_{jk} - \widehat{\mu}_{Xj}$ and $\mathbf{y}_{jk}' = \mathbf{y}_{jk} - \widehat{\mu}_{Yj}$.

The optimal test statistic is given by

$$T^2 = \frac{n_X + n_Y - m - 1}{m(n_X + n_Y - 2)}\left(\frac{1}{n_X} + \frac{1}{n_Y}\right) \times (\widehat{\vec{\mu}}_X - \widehat{\vec{\mu}}_Y)^{\mathrm{T}}\widehat{\Sigma}^{-1}(\widehat{\vec{\mu}}_X - \widehat{\vec{\mu}}_Y). \tag{6.33}$$

This statistic measures the distance in m space between the sample mean vectors $\widehat{\vec{\mu}}_X$ and $\widehat{\vec{\mu}}_Y$ in dimensionless units. Note the similarity to the t statistic (6.16). In fact, when $m = 1$, $T^2 = t^2$, ensuring that both the Hotelling T^2 test and the t test will make the same decision. Also note that T^2 is a scaled version of the Mahalanobis distance

(6.8) that is computed with an estimate of the covariance matrix.

T^2 has the F distribution with $(m, n_X + n_Y - m - 1)$ degrees of freedom [280] when H_0 is true.[13] Thus the Hotelling test is conducted by comparing T^2 with critical values from this distribution. Critical F values may be found in Appendix G.

When the covariance matrix $\mathbf{\Sigma}$ is known, the Hotelling test reduces to the χ^2 test (see [6.7.2]). The test statistic is then given by

$$C^2 = \frac{n_X + n_Y}{n_X n_Y}(\widehat{\vec{\mu}}_X - \widehat{\vec{\mu}}_Y)^{\mathrm{T}} \mathbf{\Sigma}^{-1} (\widehat{\vec{\mu}}_X - \widehat{\vec{\mu}}_Y), \tag{6.34}$$

and is compared with the critical values of the χ^2 distribution with m degrees of freedom.[14] Again, note the scalar case analogy. When $m = 1$, C^2 reduces to $\mathbf{Z}^2$ with $\mathbf{Z} \sim \mathcal{N}(0, 1)$. Also note that C^2 is a scaled version of the Mahalanobis distance $\mathcal{D}^2$ (6.8). Critical χ^2 values may be found in Appendix E.

6.6.11 The Mann–Whitney Test. Sometimes it is not possible to make all the assumptions required for a parametric test, so it may be desirable to use a non-parametric test that can be applied under a less restrictive set of assumptions.

The Mann–Whitney test (cf. [4.1.8]) is an example of a non-parametric test of H_0: $\mu_X = \mu_Y$. The same sampling assumption is required as in the t test and it is also necessary to assume that all observations in a sample come from the same distribution, but the distributional assumption itself is relaxed. Rather than specifying a particular functional form (e.g., the normal distribution), the Mann–Whitney test requires that the density functions of $\mathbf{X} - \mathcal{E}(\mathbf{X})$ and $\mathbf{Y} - \mathcal{E}(\mathbf{Y})$ be identical.

With these assumptions, the distribution of any function of the $n_X + n_Y$ observations $\mathbf{x}_1, \ldots, \mathbf{x}_{n_X}, \mathbf{y}_1, \ldots, \mathbf{y}_{n_Y}$ is independent of the ordering of the samples under H_0. The Mann–Whitney test exploits this fact by examining the positions of the $\mathbf{X}$ observations when the combined sample is sorted in increasing order.

The samples are fully separated when $\mathbf{X}_k > \mathbf{Y}_j$, or vice versa, for all $k = 1, \ldots, n_X$ and $j = 1, \ldots, n_Y$. Combinatorial arguments show that the combined sample can be partitioned into two groups of size n_X and n_Y in $\frac{(n_X+n_Y)!}{n_X! n_Y!}$ ways. Thus the probability of observing fully separated samples under H_0 such that all observations in the $\mathbf{X}$ sample are greater than all observations in the $\mathbf{Y}$ sample is $\frac{n_X! n_Y!}{(n_X+n_Y)!}$. Similarly the probability that $\mathbf{x}_k > \mathbf{y}_j$ for all j and all but one $k = 1, \ldots, n_X$ is $n_Y / \frac{(n_X+n_Y)!}{n_X! n_Y!}$.

These examples indicate that it makes sense to define a test statistic based on the ordering of the combined sample. To do so we introduce the concept of *ranks* in the joint sample

$$\vec{\mathbf{z}} = (\mathbf{x}_1, \ldots, \mathbf{x}_{n_x}, \mathbf{y}_1, \ldots, \mathbf{y}_{n_Y})^{\mathrm{T}}. \tag{6.35}$$

Now let R_1 be the rank of $\mathbf{x}_1$ in $\vec{\mathbf{z}}$; that is, if $\mathbf{x}_1$ is the ith smallest observation in $\vec{\mathbf{z}}$, then we set $R_1 = i$. Define $R_2, \ldots, R_{n_x+n_y}$ similarly.[15] The test statistic is then defined to be the rank sum of all $\mathbf{X}$ observations,

$$S = \sum_{i=1}^{n_X} R_i. \tag{6.36}$$

The distribution of S, under H_0, is obtained through combinatorial arguments [88]. Critical values $\kappa_{1-\tilde{p}}$ are tabulated in Appendix I. For large samples sizes, approximate critical values for tests at the $(1 - \tilde{p}) \times 100\%$ significance level are given by [88] as

$$\kappa_{\tilde{p}} = \frac{n_X(n_X + n_Y + 1)}{2} - Z_{\tilde{p}} \sqrt{\frac{n_X n_Y (n_X + n_Y + 1)}{12}}, \tag{6.37}$$

where $Z_{\tilde{p}}$ is the $\tilde{p}$-quantile of the standard normal distribution (Appendix D). A two-sided test of H_0: $\mu_x = \mu_y$ versus H_a: $\mu_x \neq \mu_y$ is performed at the $(1 - \tilde{p}) \times 100\%$ significance level by rejecting H_0 when $S < \kappa_{(1-\tilde{p})/2}$ or $S > S_{max} - \kappa_{(1-\tilde{p})/2}$, where $S_{max} = n_x(n_x + 2n_y + 1)/2$ is the largest possible value that S can take. A one-sided test of H_0: $\mu_x \geq \mu_y$ versus H_a: $\mu_x < \mu_y$ is performed by rejecting H_0 when $S < \kappa_{(1-\tilde{p})}$.

The added flexibility of the Mann–Whitney test compared with its conventional parametric counterpart, the t test, comes at the cost of slightly reduced efficiency when the observations are normally distributed. The *asymptotic relative*

[13] The derivation of the distribution of T^2 follows that of t closely. The statistic can be written as the ratio of two independent quadratic forms that each have the χ^2 distribution under the H_0. It follows that T^2 has an F distribution because the latter is characterized as a ratio of χ^2 random variables [2.7.10].

[14] Note the analogy with T^2. Here the statistic consists of a single quadratic form.

[15] Of course, the ranks can be defined equally well in ascending order so that the largest value receives the rank 1, etc.

efficiency[16] of the Mann–Whitney test is 0.955 when the data are normally distributed. That means that, asymptotically, the t test is able to achieve the same power as the Mann–Whitney test using only 95.5% of the observations needed by the latter. However, this disadvantage disappears for some distributions other than the normal distribution. The asymptotic relative efficiency is 1.0 when the data come from the uniform distribution and it is 1.5 if the data have the double exponential distribution, indicating that the t test requires 1.5 times as many observations.

6.6.12 A Permutation Test. The following test of H_0: $\mu_X = \mu_Y$, first proposed by Pitman [314, 315, 316], can be applied to univariate as well as multivariate problems. It also allows us to relax the distributional assumption somewhat further than the Mann–Whitney test allows. We will need the standard sampling assumption (i.e., independence), the assumption that observations are identically distributed within samples, and a third assumption that distributions differ only with respect to their expectations, if they differ at all. Note that the sampling assumption is crucial. In particular, the permutation test performs very poorly when observations are serially correlated [442].

Let us first consider the univariate case. As in the Mann–Whitney test, let $\vec{\mathbf{z}}$ be the vector of all $\mathbf{X}$ and $\mathbf{Y}$ observations: $\vec{\mathbf{z}} = (\mathbf{x}_1, \ldots, \mathbf{x}_{n_X}, \mathbf{y}_1, \ldots, \mathbf{y}_{n_Y})^T$. Under the null hypothesis, the distributions of $\mathbf{X}$ and $\mathbf{Y}$ are identical and thus any statistic S of $\vec{\mathbf{Z}}$ has a distribution that is independent of the ordering of the components of $\vec{\mathbf{Z}}$. That is, if π is a random permutation of $\{1, \ldots, n_X + n_Y\}$, then $S(\vec{\mathbf{Z}})$ has the same distribution as $S(\vec{\mathbf{Z}}_\pi)$. Consequently, any arrangement $\vec{\mathbf{z}}_\pi$ of the observed $\vec{\mathbf{z}}$ is as likely under the null hypothesis as any other. Hence the probability that the observed test statistic $S(\vec{\mathbf{z}})$ takes a value in the upper fifth percentile of values that can be taken by $S(\vec{\mathbf{z}}_\pi)$ is exactly 5% under the null hypothesis.

In contrast, ordering becomes important under the alternative hypothesis, where possible values of $S(\vec{z}_\pi)$ obtained via permutation are not equally likely. The unpermuted vector precisely divides the observations according to their population of origin, and consequently $S(\vec{\mathbf{z}})$ should lie at the extremes of the collection of $S(\vec{\mathbf{z}}_\pi)$ values.[17]

A test is therefore constructed by comparing $S(\vec{\mathbf{z}})$ with the ensemble of values obtained by evaluating $S(\vec{\mathbf{z}}_\pi)$ for all permutations π. If the collection of permutations is very large, the distribution of $S(\vec{\mathbf{z}}_\pi)$ may be estimated by randomly selecting a subset of permutations. For most applications a subset containing 1000 permutations will do.

To express the test mathematically, let Π be the set of all permutations π. Then compute (or estimate if Π is large)

$$H = \frac{|\{\pi \in \Pi : S(\vec{\mathbf{z}}_\pi) > S(\vec{\mathbf{z}})\}|}{|\Pi|}, \tag{6.38}$$

where $|A|$ denotes the number of entries in a set A. Since H is an estimate of the probability of observing a more extreme value of the test statistic under the null hypothesis, we may reject H_0 if H is less than the specified significance level.

The permutation test approach is easily extended to multivariate problems [397]. One approach is to define a multivariate test statistic S' in terms of univariate test statistics S_j, $j = 1, \ldots, m$, as

$$S' = \sum_{j=1}^{m} |S_j|. \tag{6.39}$$

The same procedure as outlined above is then applied to S' instead of S. One should exercise some caution with this expedient. For example, the multivariate test that is obtained is not always *invariant* [4.3.3] under linear transformation of the m-dimensional field.

One drawback of the permutation test is that it is not supported by a rich statistical theory. We do know that permutation tests are asymptotically as efficient as their parametric counterparts [274], but we must rely on Monte Carlo methods to obtain information about the small sample properties of the test in specific situations.

6.7 Test of Variances

6.7.1 Overview. Until now our focus has been on tests about the first moments (i.e., means) of scalar and vector random variables. We briefly describe a few ways in which to test hypotheses about the second central moments (i.e., variances) of scalar random variables in this section. Tests

[16] The efficiency of two tests is measured by comparing the sample sizes needed to achieve the same power at the same significance level against the same alternative. The sample size ratio often becomes independent of power, significance level, and the particular alternative as one of the sample sizes tends to infinity. When this happens, the limiting sample size ratio is called the *asymptotic relative efficiency* (ARE). See Conover [88] for more details.

[17] At least, this should be true if S efficiently estimates a monotone function of the difference between the two populations.

about the second central moments of random vectors (i.e., covariance matrices) are beyond the scope of this book.[18]

6.7.2 The χ^2 Test. Suppose $\mathbf{X}_1, \ldots, \mathbf{X}_n$ are iid random variables that represent a sample of size n from the normal distribution. Then $\mathbf{C}^2 = (n-1)S_X^2/\sigma_X^2$ has the $\boldsymbol{\chi}^2(n-1)$ distribution (cf. [2.7.8]).

The null hypothesis H_0: $\sigma_X^2 = \sigma_o^2$ can then be tested at the $(1-\tilde{p})$ significance level by computing $\mathbf{C}^2 = (n-1)S_X^2/\sigma_o^2$ and making decisions as follows.

- H_a: $\sigma_X^2 < \sigma_o^2$: reject H_0 when $\mathbf{C}^2$ is less than the $(1-\tilde{p})$-quantile of the $\boldsymbol{\chi}^2(n-1)$ distribution. The χ^2 distribution is partially tabulated in Appendix E. For example, when $n = 10$, we would reject H_0 at the 5% significance level when $\mathbf{C}^2$ is less than 3.33. The non-rejection region is $[3.3, \infty)$.

- H_a: $\sigma_X^2 \neq \sigma_o^2$: reject H_0 when $\mathbf{C}^2$ is less than the $((1-\tilde{p})/2)$-quantile of the $\boldsymbol{\chi}^2(n-1)$ distribution, or greater than its $((1+\tilde{p})/2)$-quantile. When $n = 10$, the non-rejection region for the 5% significance level test is $[2.70, 19.0]$.

- H_a: $\sigma_X^2 > \sigma_o^2$: reject H_0 when $\mathbf{C}^2$ is greater than the $\tilde{p}$-quantile of the $\boldsymbol{\chi}^2(n-1)$ distribution. When $n = 10$, the non-rejection region for the 5% significance level is $[0, 16.9]$.

The χ^2 test is *more* sensitive to departures from the normal distribution assumption than the tests of the mean discussed in the previous section. This sensitivity arises because $\mathbf{C}^2$ is a sum of squared deviations. Data that are not completely normal tend to have at least some deviations from the sample mean that are larger than would be observed in a completely normal sample. Because these deviations are squared, they have a very large effect on the value of $\mathbf{C}^2$. Inferences are consequently unreliable.

6.7.3 The F Test. The one sample χ^2 test of the previous subsection has relatively limited applications. On the other hand, there are many problems in which it is necessary to decide whether two samples came from populations with equal variances. For example, this is needed when selecting a test for the equality of means (see [6.6.1] and [6.6.5]). There are also a myriad of climate analysis problems in which we want to compare variances. For example, we may want compare the variability of two simulated climates on some time scale, the variability of the observed climate with that of a simulated climate, or the variability under different climatic regimes (e.g., warm versus cold ENSO events).

The standard procedure for testing H_0: $\sigma_X^2 = \sigma_Y^2$ is the *F test*. It can be applied when we have two independent samples $\mathbf{X}_1, \ldots, \mathbf{X}_{n_X}$ and $\mathbf{Y}_1, \ldots, \mathbf{Y}_{n_Y}$, each consisting of iid normal random variables. Then

$$\mathbf{F} = \frac{S_X^2}{S_Y^2} \tag{6.40}$$

has the $F(n_X-1, n_Y-1)$ distribution under the null hypothesis [2.7.10]. Critical values of the F distribution are tabulated in Appendix G. The test is performed at the $(1-\tilde{p}) \times 100\%$ significance level as follows.

- H_a: $\sigma_X^2 > \sigma_Y^2$: reject H_0 when $\mathbf{f}$ is greater than the $\tilde{p}$-quantile of the $F(n_X-1, n_Y-1)$ distribution. For example, when $n_X = 9$ and $n_Y = 10$, the non-rejection region for a test conducted at the 10% significance level is $[0, 2.47]$.

- H_a: $\sigma_X^2 \neq \sigma_Y^2$: reject H_0 when $\mathbf{f}$ is less than the $(1-\tilde{p})/2$-quantile of the $F(n_X-1, n_Y-1)$ distribution, or greater than its $((1+\tilde{p})/2)$-quantile. Note that most tables do not list the lower tail quantiles of the F distribution, because when $F \sim F(n_X-1, n_Y-1)$, then $\frac{1}{F} \sim F(n_Y-1, n_X-1)$. Thus the $((1-\tilde{p})/2)$-quantile of $F(n_X-1, n_Y-1)$ is 1 over the $((1+\tilde{p})/2)$-quantile of $F(n_Y-1, n_X-1)$. When $n_X = 9$ and $n_Y = 10$, the non-rejection region for a 10% significance level test is $[0.295, 3.23]$.

Just as for the χ^2 test, the F test is sensitive to departures from the normal distribution. Also, it is not robust against outlying observations caused by, for example, observational or data management errors. It is therefore useful to have a non-parametric alternative even if the relative efficiency of the test is low when data are normal. A non-parametric test is discussed in the next subsection.

The F test also does not perform as expected when there is dependence within the samples.

[18] Interested readers can find entry points to literature on this subject in, for example, Graybill [147], Johnson and Wichern [197], Morrison [281], or Seber [342].

If the samples are time series, spectral analysis methods (see Section 12.3) can be used to describe the variability in the samples as functions of time scale. F tests can then be used to compare variability within the samples at various time scales.

Finally, the F test is not particularly powerful. For example, to reject H_0 reliably when $\sigma_X^2 = 2\sigma_Y^2$, say with power 95% in a 5% significance level test, requires samples of size $n_X = n_Y \approx 100$. Since power can always be increased somewhat at the cost of greater risk of false rejection, F tests are often performed at the 10% significance level whereas t tests are usually performed at the 5% or 1% significance levels.

6.7.4 A Non-parametric Test of Dispersion. There are several simple non-parametric tests of equality of variance.[19] We will describe two of them here. In both cases, the standard sampling assumptions are required. That is, it must be possible to represent the samples by iid random variables, and the samples must be independent of each other. It is also necessary to assume that the two populations have the same distribution when they are standardized by subtracting the mean and dividing by the standard deviation.

The first test is performed by converting both samples into absolute deviations from the respective sample means:

$$\mathbf{u}_i = |\mathbf{x}_i - \overline{\mathbf{x}}|, \quad i = 1, \ldots, n_X \text{ and}$$
$$\mathbf{v}_j = |\mathbf{y}_j - \overline{\mathbf{y}}|, \quad j = 1, \ldots, n_Y.$$

The combined samples of absolute deviations $\mathbf{u}_1, \ldots, \mathbf{u}_{n_X}, \mathbf{v}_1, \ldots, \mathbf{v}_{n_Y}$ are then assigned ranks, as in the Mann–Whitney test [6.6.11]. The sum of the ranks

$$S = \sum_{i=1}^{n_X} R_i \tag{6.41}$$

is used as the test statistic. Critical values are the same as for the Mann–Whitney test (see Appendix I). This is an approximate test when samples are small because ranked entities, the absolute deviations, are not quite independent of one another.[20]

The idea behind this simple test is that the deviations in one sample will tend to be smaller than deviations in the other when H_0 is false, resulting in either unusually small or large rank sums (6.41). It is clear that this test can never be as powerful as the Mann–Whitney test because the two samples of absolute deviations can never be completely separated. Regardless of the variance, both samples are likely to have some small deviations near zero.

One way to improve the power of this test is to focus more attention on the largest absolute deviations. The second test, the squared-ranks test, does this by using

$$T = \sum_{i=1}^{n_X} R_i^2 \tag{6.42}$$

as the test statistic instead of (6.41). Decisions are made at the $(1 - \tilde{p}) \times 100\%$ significance level by using the critical values in Appendix J as follows.

- H_0: $\sigma_X^2 < \sigma_Y^2$: reject when T is unusually small, that is, when T is less than the $(1 - \tilde{p})$-quantile of T. When $n_X = 7$, $n_Y = 8$, $(1 - \tilde{p}) = 0.05$, we would reject when $T < 426$.
- H_0: $\sigma_X^2 \neq \sigma_Y^2$: reject when T is less than the $((1 - \tilde{p})/2)$-quantile of T, or greater than the $((1 + \tilde{p})/2)$-quantile. When $n_X = 7, n_Y = 8$, and $(1 - \tilde{p}) = 0.05$, reject when $T < 384$ or $T > 935$.
- H_a: $\sigma_X^2 > \sigma_Y^2$: reject when T is greater than the $\tilde{p}$-quantile of T. When $n_X = 7$, $n_Y = 8$ and $(1 - \tilde{p}) = 0.05$, reject when $T > 896$.

When n_X or n_Y is greater than 10, the $(1 - \tilde{p})$-quantile of T can be approximated by

$$T_{(1-\tilde{p})} = \frac{n_Y(N+1)(2N+1)}{6} - Z_{\tilde{p}}\sqrt{\frac{n_X n_Y (N+1)(2N+1)(8N+1)}{180}}, \tag{6.43}$$

where $N = n_X + n_Y$ and $Z_{\tilde{p}}$ is the $\tilde{p}$-quantile of the standard normal distribution (Appendix D). Note that, as with the first non-parametric test of the variance, this test is also an approximate test when samples are small.

Even with the improved power, the squared-ranks test is inefficient when the data are really normal. Conover [88] notes that the test has asymptotic relative efficiency 0.76 in this case (i.e., the F test with samples of size 760 will be as efficient as the squared-ranks test is with samples of size 1000). On the other hand, when the data are actually distributed as the double exponential distribution (a wide-tailed asymmetric distribution that peaks sharply at the mean), the asymptotic relative efficiency is 1.08.

[19] Strictly speaking, these are tests of *dispersion* because they are designed to look for differences in the spread of the samples.

[20] The deviations within a sample are dependent because they sum to zero.

6.8 Field Significance Tests

6.8.1 Constructing Field Significance Tests from Local Tests. We discussed the use of a field of *local* test decisions for making a *global* decision about a global null hypothesis in [6.5.2]. We reconsider this problem here in more generality.

The *global null hypothesis* is H_0^G: 'all *local* null hypotheses are correct.' We assume that all local tests are conducted at the $(1-\tilde{p})$ significance level. The alternative hypothesis is that 'at least one local null hypothesis is incorrect.' Note that we must specify two significance levels: $(1-\tilde{p})$, the significance level of the local test; and $(1-\tilde{\tilde{p}})$, the significance level of the global test. We will see that $\tilde{\tilde{p}}$ can be chosen independently of $\tilde{p}$. However, the power of the global test is not independent of the power of the local tests.

Let $\vec{\mathbf{D}}$ be an m-dimensional random vector of binary random variables $\mathbf{D}_i$ that take values 0 or 1. Each of these random variables represents the result of a local test. These binary random variables are identically distributed with $P(\mathbf{D}_i = 1) = 1-\tilde{p}$ and $P(\mathbf{D}_i = 0) = \tilde{p}$ under the global null hypothesis.

Now let test statistic S be the number of local rejections, or formally $S = \vec{\mathbf{D}}^{\mathrm{T}}\vec{\mathbf{D}}$. Under the global null hypothesis, $S \sim \mathcal{B}(m, 1-\tilde{p})$ if local test decisions are made independently of one another [6.5.2]; unfortunately, this usually doesn't happen.

Livezey and Chen [257] suggested several solutions to this problem. One approach is to reduce the number of degrees of freedom [6.8.2] (similar to the modification of the t test when the data are serially correlated [6.6.3]). Another is to use a series of Monte Carlo experiments to simulate the statistical properties of the random variables that enter the local decisions [6.8.3].

6.8.2 Reduced Number of Spatial Degrees of Freedom. In many applications the local decisions are made on a regular grid so that each point has approximately the same number of points in its immediate neighbourhood. The observations used to test a local hypothesis at a grid point are often strongly correlated with those used at nearby neighbours and roughly independent of those at distant grid points. Then it may be possible to select a subset of grid points so that the observations at these grid points are mutually independent.

Suppose there are m grid points in total and that the size of the subset is m^*. Let $\vec{\mathbf{D}}$ be the vector of all decisions and let $\vec{\mathbf{D}}^*$ be the vector of decisions at the subset of points. The relative frequency of rejections of local null hypotheses will, on average, be about the same in $\vec{\mathbf{D}}$ and in $\vec{\mathbf{D}}^*$. That is,

$$\vec{\mathbf{D}}^{*\mathrm{T}}\vec{\mathbf{D}}^*/m^* \approx \vec{\mathbf{D}}^{\mathrm{T}}\vec{\mathbf{D}}/m.$$

Independence ensures that

$$\vec{\mathbf{D}}^{*\mathrm{T}}\vec{\mathbf{D}}^*/m^* \sim \mathcal{B}(m^*, \tilde{p}).$$

Thus the challenge is to select m^* in such a way that the distribution of $\vec{\mathbf{D}}^{\mathrm{T}}\vec{\mathbf{D}}/m$ is approximately that of $\vec{\mathbf{D}}^{*\mathrm{T}}\vec{\mathbf{D}}^*/m^*$.

One way to determine m^* is to use physical reasoning. Usually this approach will lead to only vague estimates, but often this approach does yield upper limits on m^*. Another approach is to compute the minimum m^* for which H_0^G can be rejected at the $(1-\tilde{\tilde{p}})$ significance level. Clearly, if $m^* > m$, the global null hypothesis H_0^G cannot be rejected. See [6.8.4].

6.8.3 Livezey and Chen's Example. Livezey and Chen [257] describe an analysis of the relationship between the Southern Oscillation, as represented by an SO index, and the Northern Hemisphere extratropical circulation, given by gridded 700 hPa height fields poleward of 20° N. Correlations between the winter (DJF) mean SO index and corresponding winter mean height anomalies were estimated at $m = 936$ grid points. The local null hypothesis H_0^j that the true correlation at grid point j is zero was tested at the $(1-\tilde{p}) = 5\%$ level at each of the 936 grid points using a method that accounts for serial correlation. The local null hypothesis was rejected at 11.4% of grid points—that is, $\vec{\mathbf{d}}^{\mathrm{T}}\vec{\mathbf{d}}/m = 0.114$. This is substantially larger than the 5% frequency that would be expected if all local null hypotheses were correct.

Figure 6.12a illustrates the rejection frequency $\varphi = \vec{\mathbf{d}}^{\mathrm{T}}\vec{\mathbf{d}}/m$ required to reject the global null hypothesis at a global 5% significance level as a function of the number of independent spatial degrees of freedom m^*. The rejection frequency φ is given by

$$\min_{\varphi} \sum_{j=\varphi\cdot m^*}^{m^*} \mathcal{B}(m^*, 1-\tilde{p})(j) \geq (1-\tilde{\tilde{p}}).$$

We see that a local rejection rate of $\varphi = 11.4\%$ supports rejection of the global null hypothesis in fields that have $m^* = 52$ or more spatial degrees of freedom. However, seasonal mean 700 hPa height is a very smooth field with very large spatial

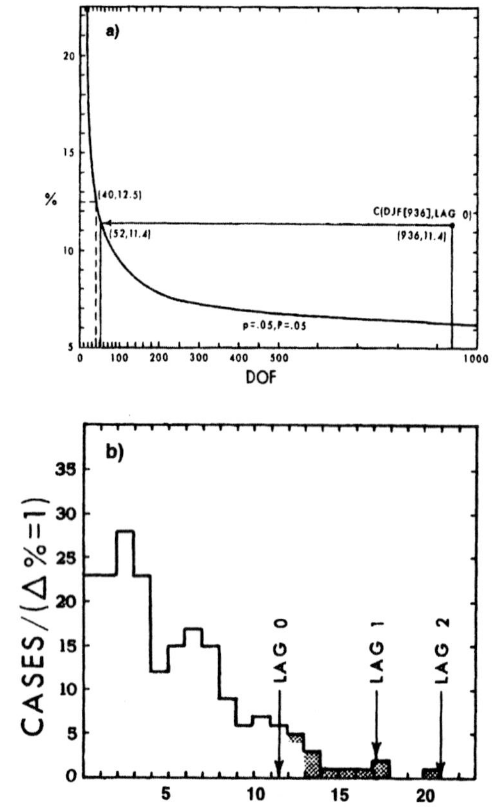

Figure 6.12:
a) Estimated percentage of rejected local null hypotheses required to reject the global null hypothesis (that all local null hypotheses are valid) at the 5% level. From Livezey and Chen [257].
b) Livezey and Chen's example [257]. Monte Carlo estimate (200 trials) of the rate φ of erroneous rejections of local null hypothesis when the global null hypothesis is true. The hatched area marks the 10 largest random S statistics so that the critical value $\kappa_{\tilde{p}}$ is 12.5%. The value to be tested, $S = 11.4\%$, is marked by the lag-0 *arrow.*

covariance structures, so it is unlikely that this field contains as many as 52 spatial degrees of freedom. Hence there is insufficient evidence to reject the global null hypothesis.

Livezey and Chen [257] also describe an attempt to use Monte Carlo methods to estimate the distribution of $S = \vec{\mathbf{D}}^{\mathrm{T}}\vec{\mathbf{D}}/m$ under the global null hypothesis. The authors conducted the Monte Carlo experiment by replacing the SO index time series with a random ('white noise') time series. This ensured that all local correlations were zero. The authors did not simulate the 700 hPa height fields. Thus, the reference distribution they obtained is conditional upon the observed sequence of height fields. The process of simulating the SO index and computing the test statistic S was repeated 200 times. The resulting distribution function is shown in Figure 6.12b. Note that 5% of all randomly generated S statistics are greater than 12.5%. Thus we again find that the global null hypothesis can not be rejected at the 5% level.

6.9 Univariate Recurrence Analysis

6.9.0 Motivation. The t test was introduced in Section 6.6 to test the null hypothesis, H_0: $\mu_X = \mu_Y$, that a pair of univariate random variables $\mathbf{X}$ and $\mathbf{Y}$ have equal means. The power of the test depends upon two factors. It increases when the 'signal' $\mu_Y - \mu_X$ increases, and when the sample sizes n_X and n_Y increase. This is illustrated in Figure 6.3, where we displayed the signals $\delta = (\mu_X - \mu_Y)/\sigma$ for which a test conducted at the 5% significance level has power 50% and 90% given sample sizes $n_X = n_Y = n$. Note that the probability of rejecting H_0 is 90% when $\delta = 0.5$ and $n = 100$, but that it is less than 50% when $n = 20$.

More generally we find, for all significance levels $(1 - \tilde{p})$ and all signals $\delta \neq 0$, that the probability of rejecting H_0 converges to 1 as $n \to \infty$. Thus, paradoxically, poor scientists are less likely to detect physically insignificant differences than rich scientists (see [6.2.5]).

One solution to this problem is to use scientific knowledge to identify the size of signal that is not physically significant and then to derive a test that rejects H_0 only when there is evidence of a larger signal. This is the idea behind *recurrence analysis.* We introduce the univariate concept [404] in this section, and the multivariate generalization [452] in Section 6.10.

Applications of the recurrence analysis include [141, 175, 223, 404, 452].

6.9.1 Definition. Two random variables $\mathbf{X}$ and $\mathbf{Y}$ are said to be (q, p)-recurrent if

$$P(\mathbf{Y} > X_q) = p, \tag{6.44}$$

where X_q is the qth quantile of the random variable $\mathbf{X}$.

In many climate modelling applications $\mathbf{X}$ represents the *control climate* and $\mathbf{Y}$ represents a climate disturbed by anomalous boundary conditions or modified parameterizations of sub-grid scale processes. The word *recurrence* refers to the probability p of observing $\mathbf{Y} > X_q$. The strength

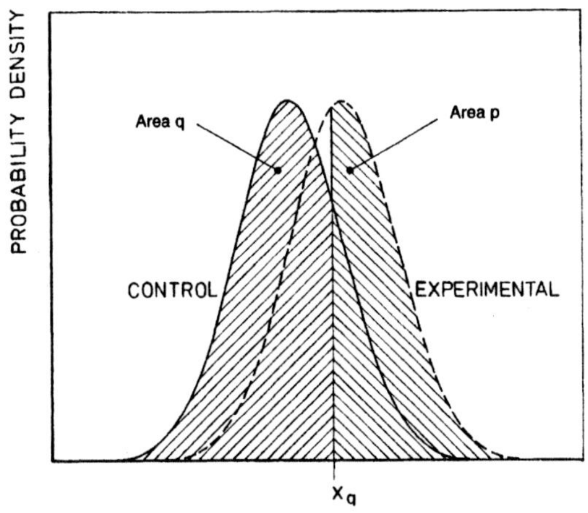

Figure 6.13: *Definition of (q, p)-recurrence: the 'control' represents the random variable* **X** *and the 'experimental' the random variable* **Y**. *The size of the area hatched to the left is q, so that $P(\mathbf{X} < X_q) = q$. The size of the area the hatched to the right is p, and $P(\mathbf{Y} > X_q) = p$ [404].*

of the effect of the anomalous boundary conditions or modified parameterizations is measured against the reference value q.

In many applications $q = 50\%$ so that the reference level is the mean of **X**. In that case we simply speak of p-recurrence.

6.9.2 Illustration. The idea of the (q, p)-recurrence is illustrated in Figure 6.13, where X_q represents a point on the right hand tail of f_X. By definition, the proportion of **X** realizations that are less than X_q is q. This point also represents a point on the left hand tail of f_Y and, according to (6.44), the proportion of **Y** realizations that are greater than X_q is p. Thus the definition states that two random variables **X** and **Y** are (q, p)-recurrent if there is a point between f_X and f_Y such that proportion q of all **X** realizations lie to the left of X_q and proportion p of all **Y** realizations lie to the right of this point. If p and q are close to 1, then the two random variables are almost perfectly separated. On the other hand, if the distributions are symmetrical and $p = q = 0.5$, then the means are equal.

6.9.3 Classification. Another way to understand the idea of (q, p)-recurrence is to think of a classification problem. Let us assume that we have a pair of random variables **X** and **Y** that are (q, p)-recurrent, and a realization **z** that is drawn from either **X** or **Y**. We want to determine which population **z** was actually drawn from. Furthermore we want to know the probability of making an incorrect decision.

The decision algorithm is:

- '**z** is drawn from **X**' if $\mathbf{z} < X_q$
- '**z** is drawn from **Y**' if $\mathbf{z} \geq X_q$

If **z** is really drawn from **X** then $P(\mathbf{z} < X_q) = q$ so that the probability of a correct decision is q in this case. On the other hand, if **z** is really drawn from **Y**, then by (6.44) $P(\mathbf{z} > X_q) = p$ so that the probability of a correct decision is p. The probabilities of incorrect decisions are $1 - q$ and $1 - p$, respectively.

6.9.4 The Murray Valley Encephalitis Example. Before we discuss mathematical aspects of recurrence, we present a concrete example of a recurrence analysis.

Between 1915 and 1984 there were seven outbreaks of *Murray Valley encephalitis* (MVE) in the Murray Valley in southeast Australia. The prevalence of MVE virus depends on the abundance of mosquitos, which in turn depends on climate. Nicholls [292] studied the relationship between the appearance of MVE and the state of the Southern Oscillation (see [1.2.2]), and found that annual mean sea-level pressure at Darwin was unusually low in all seven MVE years.

The frequency histograms of annually averaged Darwin pressure in MVE and non-MVE years are plotted in Figure 6.14. The random variables **X** (Darwin pressure conditional on the presence of MVE), and **Y** (Darwin pressure conditional on the absence of MVE) are highly recurrent, with $p = 95\%$ and $q = 86\%$. Clearly, the estimates of p and q might change drastically when the sample size increases, but the main conclusions, that the two distributions are very well separated and that the probability of misclassification is small, are not likely to change.[21]

6.9.5 Non-uniqueness of the Numbers p and q. The point of separation X_q in Figure 6.13 may be shifted; thus (q, p)-recurrence is equivalent to (q', p')-recurrence for an infinite number of pairs (q', p'). In particular, there is always one number p'' so that (q, p)-recurrence is equivalent to (p'', p'')-recurrence.

[21]Note, however, that the relationship between the SO and MVE outbreaks has changed since the discovery of the link because precautionary measures are now taken to control outbreaks when the SO index is low.

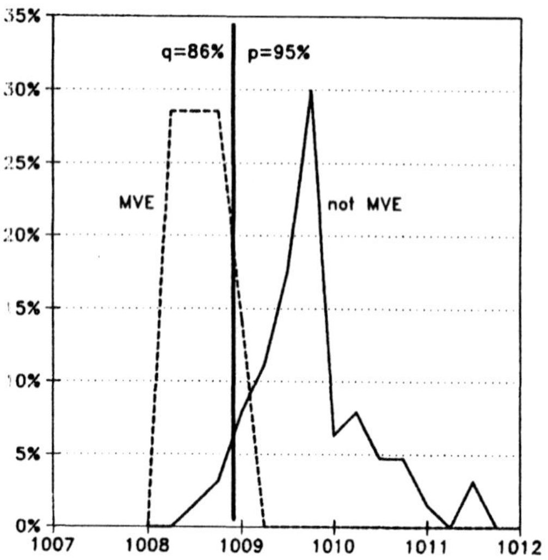

Figure 6.14: *Frequency distribution of annually averaged (March to February) Darwin sea-level pressure for seven years when Murray Valley encephalitis (MVE) was reported and for 63 years when no cases of MVE were reported. The two distributions are estimated to be* (86%, 95%)*-recurrent.*

If random variables $\mathbf{X}$ and $\mathbf{Y}$ have identical symmetrical distributions except for their means, then (q, p)-recurrence is equivalent to (p, q)-recurrence.

6.9.6 ***p*-recurrence.** Recall that p-recurrence is synonymous with $(0.5, p)$-recurrence. Suppose now that both random variables $\mathbf{X}$ and $\mathbf{Y}$ are normally distributed with means μ_X and μ_Y and a common standard deviation σ. If $\mathbf{X}$ and $\mathbf{Y}$ are p-recurrent (with $p \geq 0.50$ so that $\mu_X < \mu_Y$), then

$$\begin{aligned} p &= P(\mathbf{Y} > \mu_X) \\ &= 1 - F_{\mathcal{N}}\Big(\frac{\mu_X - \mu_Y}{\sigma}\Big) \\ &= F_{\mathcal{N}}\Big(\frac{\mu_Y - \mu_X}{\sigma}\Big), \end{aligned} \tag{6.45}$$

where $F_{\mathcal{N}}$ is the distribution function of the standard normal distribution $\mathcal{N}(0, 1)$. Thus the difference between $\mathbf{X}$ and $\mathbf{Y}$ is p-recurrent when

$$\frac{\mu_Y - \mu_X}{\sigma} = Z_p, \tag{6.46}$$

where $Z_p = F_{\mathcal{N}}^{-1}(p)$ (see Appendix D).

A reasonable estimator of p-recurrence can be obtained from (6.45) by replacing μ_X, μ_Y, and σ with the corresponding estimators $\overline{\mathbf{X}}, \overline{\mathbf{Y}}$, and S_p, where S_p^2 is the pooled sample variance (6.17). Then

$$\hat{p} = F_{\mathcal{N}}\Big(\frac{\overline{\mathbf{Y}} - \overline{\mathbf{X}}}{S_p}\Big). \tag{6.47}$$

6.9.7 Testing for (q, p)-recurrence. To test that the response to experimental conditions is at least (q, p)-recurrent, we assume that we have n_X realizations $\mathbf{x}_1, \ldots, \mathbf{x}_{n_X}$ of the control state $\mathbf{X}$, n_Y realizations $\mathbf{y}_1, \ldots, \mathbf{y}_{n_Y}$ of the experimental state $\mathbf{Y}$, and that all realizations are mutually statistically independent. The null hypothesis is that $\mathbf{X}$ and $\mathbf{Y}$ are *less* than (q, p)-recurrent, that is

$$\mathrm{H}_0\colon P\big(\mathbf{Y} > X_q\big) < p. \tag{6.48}$$

Two classes of tests are suggested in [404]: one is a *parametric test* based on the assumption of normality and the other is a *non-parametric permutation test*. We present the parametric test in the next subsection.

6.9.8 A Parametric Test. To construct a parametric test we adopt a statistical model for the random variables $\mathbf{X}$ and $\mathbf{Y}$, namely that both random variables are normally distributed with identical variances σ^2. Using (6.46), we formulate the null hypothesis H_0 that the response is less than p-recurrent as

$$\mathrm{H}_0\colon \frac{\mu_Y - \mu_X}{\sigma} < Z_p. \tag{6.49}$$

If the null hypothesis is valid, the standard t-statistic (6.15) has a *non-central t distribution* (see Pearson and Hartley [307]) with $n_X + n_Y - 2$ degrees of freedom and a *non-centrality parameter* Δ such that

$$\Delta < \frac{Z_p}{\sqrt{\frac{1}{n_X} + \frac{1}{n_Y}}}. \tag{6.50}$$

Therefore, to test H_0 we compute the usual t-statistic (6.16)

$$t = \frac{\bar{\mathbf{Y}} - \bar{\mathbf{X}}}{S_p\sqrt{\frac{1}{n_X} + \frac{1}{n_Y}}}.$$

If $1 - \tilde{p}$ is the acceptable risk of erroneously rejecting the null hypothesis, this t-value is compared with the $\tilde{p}$ percentile, $t_{n_X+n_Y-2,\Delta,\tilde{p}}$, of the non-central t distribution with $(n_X + n_Y - 2)$

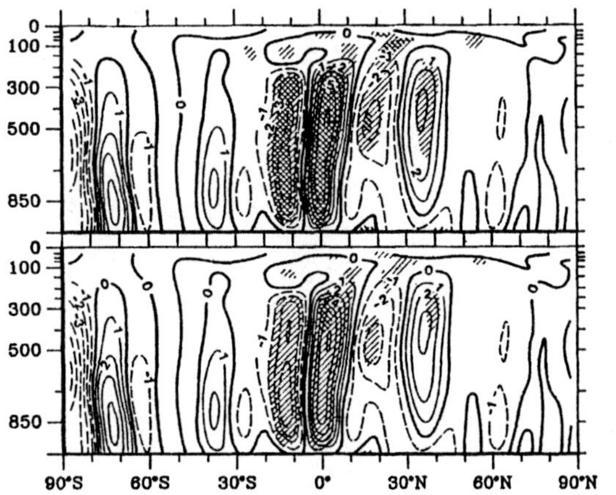

Figure 6.15: *The cross-section of monthly mean zonally averaged vertical 'velocity'* $[\omega]$ *in a paired AGCM experiment on the effect of the anomalous SST conditions in June 1988. The contours lines show the 11-sample difference between the 'June 1988 SST anomaly' run and the 'control' run [398].*
Top: The points for which the null hypothesis of equal means can be rejected with a standard t test [6.6.1] at the 5% (light shading) or 1% (dark shading) significance level.
Bottom: Points at which the univariate estimate of $(0.5, p)$*-recurrence (6.47) is less than 20% or greater than 80% are shaded.*

degrees of freedom and non-centrality parameter Δ (6.50). These percentiles are given in [307] and also in some statistical software libraries (e.g., IMSL [193]). For large sample sizes, the percentiles can be approximated by

$$t_{n_X+n_Y-2,\Delta,\tilde{p}} = \Delta + Z_{\tilde{p}}. \tag{6.51}$$

6.9.9 (p, p)-recurrence. The multivariate generalization of the concept of recurrence in Section 6.10 requires (p, p)-recurrence. Under the conditions of [6.9.6], that is, both distributions are normal with the same variance, (p, p)-recurrence is equivalent to

$$\frac{\mu_Y - \mu_X}{\sigma} \geq 2Z_p. \tag{6.52}$$

To test the null hypothesis that **Y** and **X** are less than (p, p)-recurrent, we proceed as in [6.9.8] except that the non-centrality parameter Δ in (6.50) is replaced by

$$\Delta = \frac{2Z_p}{\sqrt{\frac{1}{n_X} + \frac{1}{n_Y}}}. \tag{6.53}$$

6.9.10 A Univariate Analysis: The Effect of Cold Equatorial Pacific SSTs on the Zonally Averaged Atmospheric Circulation. In June 1988, cold surface waters were observed in the Eastern and Central Equatorial Pacific. This event attracted interest in the scientific community because of its timing (northern summer) and strength (these were the coldest June conditions in the last 60 years). A numerical experiment was performed to quantify the effect of such anomalous lower boundary conditions on the atmospheric circulation. Two 11-month perpetual July simulations were performed: once with standard sea-surface temperatures and once with the anomalous June 1988 SST distribution superimposed (von Storch et al. [398]).

Monthly mean cross-sections of the zonally averaged vertical 'velocity' $[\omega]$ obtained in the two simulations were compared with univariate recurrence analysis. The difference between the mean $[\omega]$ cross-sections is shown in Figure 6.15. Shading in the upper panel shows where the difference of means is significantly different from zero at the 5% (light) and 1% (dark) levels. Clearly there is very strong evidence of change in the mean Hadley circulation. On the other hand, the lower panel in Figure 6.15 shows that two $[\omega]$ distributions overlap substantially, even in the tropics. Regions are shaded where the response is more than 80%-recurrent or less than 20%-recurrent. There were no locations at which the response to the anomalous SSTs was more than 95%-recurrent or less than 5%-recurrent, indicating that the anomalous SST does not excite a response strong enough to eliminate the overlap between the two density functions.

The physical message of the lower panel of Figure 6.15 is that the inclusion of the anomalous tropical SST markedly modifies the *Hadley cell* but that the atmospheric circulation poleward of, say, 20° latitude is not affected by the anomalous forcing.

In this case the upper panel gives roughly the same message; there is not much difference between locations where there are significant differences (upper panel, Figure 6.15) and where there is substantial recurrence. However, when samples are larger, the estimated recurrence generally gives a clearer indication of physically significant responses than the local significance test, since the rate of rejection in the latter is sensitive to sample size.

6.10 Multivariate Recurrence Analysis

6.10.1 Motivation. We described univariate recurrence analysis as a classification problem in [6.9.3]. Specifically, if a realization $\mathbf{z}$ is drawn randomly from $\mathbf{X}$ or $\mathbf{Y}$, then the probability of incorrectly determining the origin of $\mathbf{z}$ is $1 - p$ when $\mathbf{X}$ and $\mathbf{Y}$ are (p, p)-recurrent.

Figure 6.16 illustrates two bivariate normal distributions $\vec{\mathbf{X}}$ and $\vec{\mathbf{Y}}$ with overlapping density functions. We want to quantify this overlap in the *multivariate recurrence analysis*, so we divide the full two-dimensional plane into two disjoint sets Θ_X and Θ_Y so that

$$\begin{aligned} P(\vec{\mathbf{X}} \in \Theta_Y) &= P(\vec{\mathbf{Y}} \in \Theta_X) = 1 - p \\ P(\vec{\mathbf{X}} \in \Theta_X) &= P(\vec{\mathbf{Y}} \in \Theta_Y) = p. \end{aligned} \tag{6.54}$$

The probability of a misclassification is then $1 - p$.

The sets Θ_X and Θ_Y are easily found when $\vec{\mathbf{X}}$ and $\vec{\mathbf{Y}}$ are multivariate normal [2.8.9] and have the same covariance matrix $\boldsymbol{\Sigma}$. The solution in our bivariate example is sketched in Figure 6.16 (bottom); Θ_X lies above the straight line and Θ_Y below. In this example, $p = 87.6\%$. In general, when $\vec{\mathbf{X}}$ and $\vec{\mathbf{Y}}$ are of dimension m, Θ_X and Θ_Y are separated by an $(m-1)$-dimensional hyper-plane.

We now sketch the basic ideas of multivariate recurrence analysis. A more involved discussion of this approach can be found in Zwiers and von Storch [452]. An application can be found in Hense et al. [175].

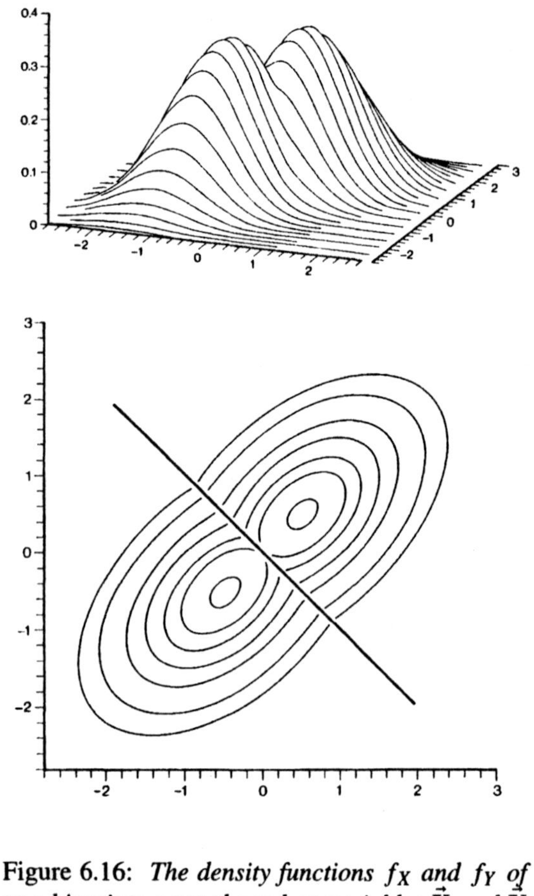

Figure 6.16: *The density functions f_X and f_Y of two bivariate normal random variables $\vec{\mathbf{X}}$ and $\vec{\mathbf{Y}}$ that differ only in their mean values.*
Top: Three-dimensional representation of $\max(f_X, f_Y)$.
Bottom: Contour lines of constant densities $\max(f_X, f_Y)$ *in the two-dimensional plane. The straight line separates the full spaces into the two subsets Θ_X and Θ_Y so probability of misclassification is $1 - p = 12.4\%$.*
From Zwiers and von Storch [452].

6.10.2 The Discrimination Function and the Probability of Misclassification. The line (or more generally, hyper-plane) in Figure 6.16 (bottom) that defines the sets Θ_X and Θ_Y is given by $\vec{\mathbf{z}} = W^{-1}(0)$ where $W(\cdot)$ is the *discrimination function*[22]

$$\begin{aligned} W(\vec{\mathbf{z}}) &= \vec{\mathbf{z}}^{\mathrm{T}}\boldsymbol{\Sigma}^{-1}(\vec{\mu}_X - \vec{\mu}_Y) \\ &\quad - \frac{1}{2}(\vec{\mu}_X - \vec{\mu}_Y)^{\mathrm{T}}\boldsymbol{\Sigma}^{-1}(\vec{\mu}_X - \vec{\mu}_Y). \end{aligned} \tag{6.55}$$

The sets Θ_X and Θ_Y are then given by

$$\begin{aligned} \Theta_X &= W^{-1}([0, +\infty)) \\ \Theta_Y &= W^{-1}((-\infty, 0)). \end{aligned} \tag{6.56}$$

[22] The discrimination function is used in *multiple discriminant analysis* (see Anderson [12], for example).

The discriminating function is used to identify the source of $\vec{\mathbf{z}}$ when it is drawn randomly from either $\vec{\mathbf{X}}$ or $\vec{\mathbf{Y}}$. When $W(\vec{\mathbf{z}}) \geq 0$, $\vec{\mathbf{z}}$ is classified as being drawn from $\vec{\mathbf{X}}$ and vice versa when $W(\vec{\mathbf{z}})$ is negative. The probability of correctly classifying $\vec{\mathbf{Z}}$ is

$$P(\vec{\mathbf{Y}} \in \Theta_Y) = P(\vec{\mathbf{X}} \in \Theta_X) = p$$

where p is given by

$$p = F_{\mathcal{N}}(\mathcal{D}/2), \tag{6.57}$$

and $\mathcal{D}$ is the Mahalanobis distance [6.5.4],

$$\mathcal{D}^2 = (\vec{\mu}_X - \vec{\mu}_Y)^{\mathrm{T}}\boldsymbol{\Sigma}^{-1}(\vec{\mu}_X - \vec{\mu}_Y). \tag{6.58}$$

$\mathcal{D}$ is a dimensionless measure of the distance between the means of $\vec{\mathbf{X}}$ and $\vec{\mathbf{Y}}$.

6.10.3 Definition of Multivariate (p, p)-Recurrence. *Let $\vec{\mathbf{X}}$ and $\vec{\mathbf{Y}}$ be independent multivariate random vectors with identical covariance matrices $\Sigma_X = \Sigma_Y = \Sigma$ and mean vectors $\vec{\mu}_X$ and $\vec{\mu}_Y$ that are separated by Mahalanobis distance $\mathcal{D}$ (6.58). Then the difference between $\vec{\mathbf{X}}$ and $\vec{\mathbf{Y}}$ is said to be (p, p)*-recurrent *when $p = F_{\mathcal{N}}(\mathcal{D}/2)$ (6.57).*

In contrast with the univariate definition, the definition above is restricted to multivariate normal distributions with identical covariance matrices. The concept is not easily extended to other multivariate settings because, in general, derivation of the surface that separates Θ_X from Θ_Y becomes intractible. For the same reason, (q, p)-recurrence with $p \neq q$ is also not defined.

6.10.4 Estimation of the Level of (p, p)-recurrence. Zwiers and von Storch [452] considered several estimators of the level of (p, p)-recurrence and found that an estimator originally proposed by Okamoto [299, 300] worked well. Hense et al. [175] suggested the following modified form of this estimator for p:

$$\hat{p} = 1 - \mathrm{erf}\Big(-\frac{\mathcal{D}_S}{2}\Big) + \frac{f_{\mathcal{N}}(-\mathcal{D}_S/2)}{n_X + n_Y - 2}\Big(\frac{\mathcal{D}_S}{16}\Big(\frac{(n_X + n_Y - 1)^2 - 1}{n_X n_Y}\Big) - \frac{m-1}{4\mathcal{D}_S}\Big(\frac{(n_X - 3n_Y)(n_X + n_Y - 2)}{n_X n_Y} - \mathcal{D}_S^2\Big)\Big), \tag{6.59}$$

where $\mathcal{D}_S$ is the 'shrunken' Mahalanobis distance

$$\mathcal{D}_S^2 = \frac{n_X + n_y - m - 3}{n_X + n_Y - 2}\mathcal{D}_{\hat{\Sigma}} \tag{6.60}$$

$$\mathcal{D}_{\hat{\Sigma}} = (\overline{\mathbf{Y}} - \overline{\mathbf{X}})^{\mathrm{T}}\hat{\Sigma}(\overline{\mathbf{X}} - \overline{\mathbf{Y}}), \tag{6.61}$$

and $f_{\mathcal{N}}$ is the standard normal density function.

6.10.5 Testing for (p, p)-recurrence. A parametric test of (p, p)-recurrence can be constructed following the ideas of the parametric test of univariate recurrence in [6.9.9]. The null hypothesis H_0 is again '$\vec{\mathbf{X}}$ and $\vec{\mathbf{Y}}$ are less than (p, p)-recurrent.' H_0 can be tested with Hotelling T^2 statistic [6.6.10]:

$$T^2 = \frac{n_X + n_Y - m - 1}{m(n_X + n_Y - 2)}\mathcal{D}_{\hat{\Sigma}}. \tag{6.62}$$

Under H_0 T^2 has a non-central F distribution (see, e.g., [307]) with m and $n_X + n_Y - m - 1$ degrees of freedom and non-centrality parameter

$$\Delta = \frac{n_X n_Y}{n_X + n_Y}\mathcal{D}, \tag{6.63}$$

where $\mathcal{D} = 2F_{\mathcal{N}}^{-1}(p)$ (6.57).

7 Analysis of Atmospheric Circulation Problems

7.0.0 Summary. In this chapter we present examples of hypothesis tests in the contexts of confirming, or validating, Atmospheric General Circulation Models (AGCMs) (Section 7.1, see also [1.2.7]) and the analysis of paired sensitivity experiments (Section 7.2, see also [1.2.7]). Similar applications in the literature include [105, 132, 134, 135, 161, 393]. See also Frankignoul's review of the topic [130], and the *recurrence analysis* examples presented in Sections 6.9 and 6.10. An application of the Hotelling test is described in Section 7.3 and an example of the anthropogenic CO_2 signal is discussed in Section 7.4.

7.1 Validating a General Circulation Model

7.1.1 The Problem. Climate models in general, and AGCMs specifically, are mathematical representations of the climate that are built from first principles. On short time scales they simulate the day-to-day variations in the weather, ideally in such a way that the statistics of the observed climate are reproduced when the model is run for a long period of time. A careful strategy is needed to determine, even partly, whether a model has achieved this goal. The problem is complex because, in principle, we would need to compare the statistics of a state vector that characterizes all aspects of the thermo- and hydrodynamics of the atmosphere. The statistics should include time averaged fields of various variables at various levels, and temporal and spatial cross-covariances of different variables on different scales.

It would be difficult, but not impossible, to characterize the simulated climate in this way. On the other hand, it simply cannot be done for the observed climate because our observations are far from complete. In reality, model validation efforts must be restricted to an incomplete state vector that represents only a few variables of interest.

In the following example the state vector is only a single variable: the zonal distribution of geopotential height at 500 hPa in the Northern Hemisphere extratropics. The comparison is often performed with a statistical test of the null hypothesis that the observed and simulated vectors have the same distribution.[1] Thus, as we noted in [6.9.1], given large enough samples we will eventually discover that the simulated climate is 'significantly'[2] different from that which is observed because no model is perfect.[3]

That is, a fully satisfactory 'verification' or 'validation' is impossible with the hypothesis testing paradigm. Are there more satisfying ways to prove the 'correctness' of a model? Oreskes et al. [301] argue that a positive answer can be given only if the model describes a closed sub-system of the full system, that is, a sub-system with completely known 'external' forcings. The atmosphere and the climate system, as a whole, are not closed systems but open to various external factors, such as variations in solar radiation, volcanic eruptions, or the Milankovicz cycle. Even if these external factors were known in detail, the part of the climate system represented by an AGCM cannot be viewed as a closed sub-system because the atmosphere loses energy and moisture into other parts of the system.

Sometimes, a possible alternative to the 'hypothesis testing' strategy is to use the models as forecasting instruments, then assess their ability to predict atmospheric variations (see Chapter 18) correctly. Unfortunately, this approach is applicable only in cases when there

[1] The test may concentrate on a specific aspect of the distributions, such as the means (Section 6.6) or variances (Section 6.7), or it may be concerned with the whole distribution ([5.2.3], [5.3.3] and Sections 6.9 and 6.10)

[2] Statistically, not necessarily physically, significant.

[3] One of the unavoidable errors is due to space-time truncation that determines the modelled sub-space. The part of the phase space that is disregarded by the truncation affects the real system also in the resolved part of its phase space, but has no impact on the model's phase space.

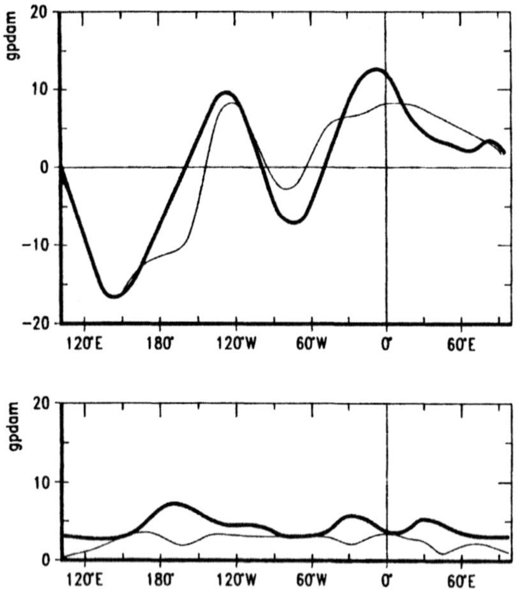

Figure 7.1: *Sample mean (top) and standard deviation (bottom) of the* 30° *N–*60° *N meridional average of 500 hPa height simulated in a GCM (light lines), and derived from observations (heavy lines) (cf. Figure 6.5) [397].*

is predictive skill, as in case of short-term forecasts or in case of externally induced anomalies (as the injection of volcanic aerosols). Also, it is often impractical because we lack independent observed data on the time scales of interest.

Regardless of the validation strategy used, it is always possible that the model verifies correctly for the wrong reasons. For example, Zwiers and Hamilton [447] showed that CCC GCMI[4] simulates the semi-diurnal thermal tide very realistically. However, the tide in the observed atmosphere is excited primarily by solar heating of the stratosphere at levels well above the model's 10 hPa 'lid.' The lid apparently allows standing oscillations to develop from the weak solar heating that takes place below 10 hPa.

In summary, there are strong limitations to statistical model validation. Neither the testing nor prediction approaches are fully satisfactory. When we do satisfy ourselves that some aspect of the distribution of the simulated climate matches that which is observed, it then becomes necessary to confirm that the same physical mechanisms operate in both.

[4]The first GCM of the Canadian Climate Centre [53].

7.1.2 Example: Extratropical Geopotential Height at 500 hPa. We return to an example first described in [6.2.6], which dealt with January mean 500 hPa heights, meridionally averaged between 30° N and 60° N. In [6.2.6] we asked whether the individual zonal distributions of height, $\vec{\mathbf{X}}$, simulated by a GCM were distributed similarly to those observed. Here we use the same model output to test the null hypothesis that the means and variances of the simulated zonal distribution are equal to those of the observations [397].

The permutation test [6.6.12] is used with statistic (6.36) to test the null hypothesis that the means of the simulated and observed climates are equal. The assumptions needed in this case are (i) the observed and simulated samples can be represented by iid random vectors $\vec{\mathbf{X}}_1, \ldots, \vec{\mathbf{X}}_{n_X}$ and $\vec{\mathbf{Y}}_1, \ldots, \vec{\mathbf{Y}}_{n_Y}$, (ii) the samples are mutually independent, and (iii) the variances of the simulated meridional means are equal to those of the observed means. Assumption (i) may be violated for the observations since low frequency interactions between the ocean and the atmosphere, such as the Southern Oscillation, may result in weak dependence between consecutive January meridional means. This should not cause major problems with the test procedure. Departures from the third assumption are more obvious, but fortunately Monte Carlo experiments have shown that this violation does not lead to strong biases in the risk of incorrectly rejecting the null hypothesis.

The result of the test is that the equality of means hypothesis can be rejected at a significance level of less than 5%. This isn't at all surprising since we saw in [6.2.6] that six out of ten simulated monthly means were not likely to have been observed in the real climate. The observed and simulated sample means are shown in the upper panel of Figure 7.1.

The permutation test can also be used to test the null hypothesis of equal standard deviations, but the data must first be centred. That is, observations $\vec{\mathbf{x}}_i$ and $\vec{\mathbf{y}}_j$ are replaced with the corresponding deviations $\vec{\mathbf{x}}_i - \bar{\vec{\mathbf{x}}}$ and $\vec{\mathbf{y}}_j - \bar{\vec{\mathbf{y}}}$. Figure 7.1 (lower panel) shows the model underestimates the natural variability of the considered parameter.[5]

The major conclusions of this study [397] were that the GCM suffered from systematic errors in the mean distribution of 500 hPa height in the extratropics and that the *interannual* variability of monthly means was significantly underestimated.

[5]This is a malady shared by many models. See, for example, Zwiers et al. [449]

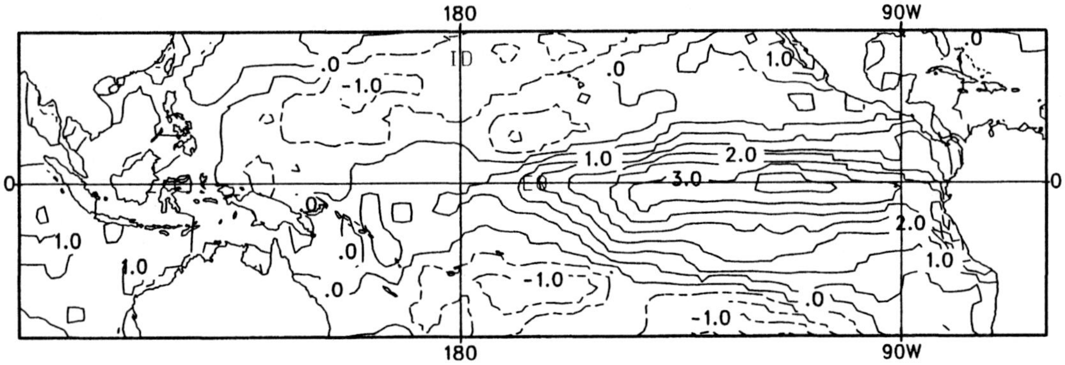

Figure 7.2: *The DJF 1982/83 mean SST anomaly relative to the 1948–94 DJF mean in the tropical Pacific. The SSTs are from the Hadley Centre GISST data set. Parker et al. [302]. Courtesy V. Kharin.*

7.2 Analysis of a GCM Sensitivity Experiment

7.2.1 Experimental Set-up. GCMs are very important 'lab tools' in climate research because they can be used to perform controlled experiments that determine the atmospheric response to variations in one external factor (say factor X) while all other external factors are held fixed. Two sets of experiments are usually performed: one set with unaltered *normal*, or *control*, conditions, and another set with *anomalous* conditions, in which a specific external factor is changed, such as the sea-surface temperature in a certain area, the atmospheric load of aerosol, or the formulation of a parameterization of the cloud-radiation interaction. More complicated experimental designs can be constructed to examine the combined effects of variations in more than one external factor (see Chapter 9; examples include Gough and Welch [145] and Chapman et al. [79]).

The evaluation of such experiments may be done by formulating and testing the null hypothesis: 'the change in factor X has no effect on the state of the (modelled) atmosphere.' Again, we need to keep the limitation of the testing paradigm in mind. Rejection implies that the response is *statistically* significant; physical insight is required to ascertain that it is *physically* significant as well. Non-rejection may indicate that the change has no effect, or simply that the experiment is too small,[6] and therefore that the signal remains hidden in the noise. The remainder of this section presents details of one of these paired sensitivity experiments [386].

[6]That is, the simulated sample contains too few independent realizations.

7.2.2 Example: The Effect of ENSO Sea-surface Temperature Anomalies on the Extratropical Atmospheric Flow. The El Niño/Southern Oscillation (ENSO) phenomenon is considered to be the strongest climate variation on time scales of a few years (for further details refer to the short description in [1.2.2]). A significant feature of this phenomenon is the appearance of anomalous sea-surface temperatures on large spatial scales in the tropics, which affects the overlying convective activity in the atmosphere. The DJF mean SST anomaly for the 1982/83 ENSO event is shown in Figure 7.2. It is not immediately obvious how anomalous temperatures at the lower boundary might affect the overall circulation of the atmosphere. A large body of literature has been published on this subject, describing approaches that range from theoretical considerations [185] to numerical experiments [48, 51, 95, 146, 221, 244, 288].

The experiment was conducted by integrating an atmospheric GCM twice under conditions that were identical apart from the sea-surface temperature distribution in the equatorial Pacific. One integration, the 'control' run, used climatological SST. In the other integration, an exaggerated El Niño SST anomaly was superimposed onto the climatological SST. The anomaly has a maximum of about 4 °C and is centred in the equatorial Central and Eastern Pacific. Both integrations were performed in the perpetual January mode,[7] with a

[7]The expressions 'perpetual' and 'permanent January mode' refer to GCM experiments in which the solar radiation and the lower boundary conditions, such as SST, are kept constant to fixed January conditions—a design that saves computer time *and* produces many iid samples. This experimental set-up introduces some systematic errors, mostly related to hydrological processes such as the accumulation of snow, when compared with runs done with a regular annual cycle (see Zwiers and Boer [446]).

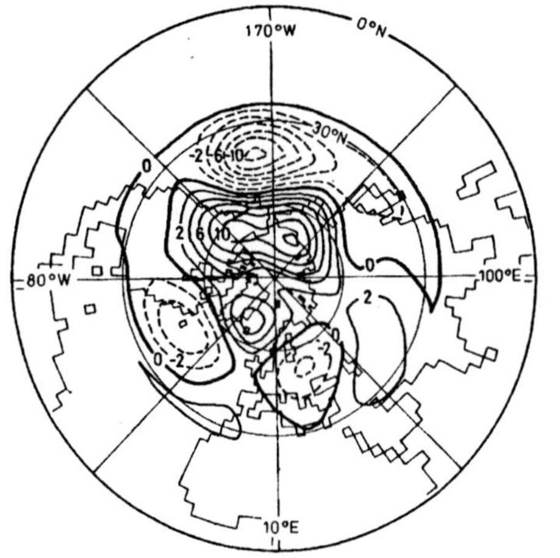

Figure 7.3: *GCM experiment on the extratropical atmospheric response to tropical El Niño sea-surface temperature anomalies. The variable shown is the 500 hPa height 'full signal,' that is, the 'El Niño minus control' difference field derived from all samples. Units: dkm. From von Storch [386].*

spin-up period of 200 days and a sampling period of 1200 days.[8]

Here, we consider the effect of the anomalous boundary conditions on the tropospheric state in the extratropical Northern Hemisphere, in particular the monthly mean 500 hPa height. The random variable $\vec{\mathbf{X}}$ is the January mean of this field for the control run and $\vec{\mathbf{Y}}$ is that for the experimental run. The null hypothesis is $\mathcal{E}(\vec{\mathbf{X}}) = \mathcal{E}(\vec{\mathbf{Y}})$. The 1200 day sampling period is subdivided as follows: the time series is broken into adjacent 40 day intervals; the first 10 days of each sub-interval are disregarded and the remaining 30 days are retained for analysis. The result is a collection of 30 roughly independent Januaries for both the control and experimental conditions. Note that a 10-day gap is sometimes not enough to ensure independence.[9]

The 'full' signal, that is, the overall 500 hPa height 'El Niño minus control' difference field, is shown in Figure 7.3. The equality of means hypothesis was tested with the permutation test [6.6.12] after projecting the data onto a set of guess patterns [6.5.6].

Three different sets of guess patterns were used.

1 *EOFs as Guess Patterns*
First, the eddy component $\vec{z}^*$ of the 30° N–60° N 500 hPa height meridional average was considered.[10] To reduce the number of spatial degrees of freedom, the first five Empirical Orthogonal Functions (EOFs, see Chapter 13) of the control experiment were used as guess patterns. The $\vec{z}^*$-field of each individual month was projected onto these guess patterns, and the Hotelling T^2 statistic (6.33) was used in combination with the permutation test to determine whether the means of the first five EOF coefficients changed significantly when the El Niño SST anomaly was imposed. The result was that it is highly unlikely ($\leq$ 1% chance) that the simulated differences between the $\vec{z}^*$-fields in the control and experimental runs were caused only by random variations. Therefore the null hypothesis was rejected.

2 *Splitting the GCM data to Obtain a Guess Pattern*
A more detailed analysis was performed on specific aspects of the full Northern Hemisphere 500 hPa height field. The experiments were integrated over a fairly long time in order to obtain a large number of samples. It is therefore possible to split the control and experimental samples into two sub-samples of equal size. The first sub-sample from each simulation was used to estimate the signal. The second pair of sub-samples was used to test the equality of means hypothesis using the estimated signal from the first pair of sub-samples as guess patterns. Since only one guess pattern is used, the number of spatial degrees of freedom is reduced to one, and a univariate difference of means test (6.29) may be used.

The difference was found to be significant at much less than the 1% significance level. The estimated signal, obtained by multiplying the guess patterns by the change in the mean coefficient (not shown) is very similar to the full signal (Figure 7.3). A test was also performed to see if there was a signal orthogonal to the guess pattern (see [6.5.4]). This was done using the EOF method described above. The null hypothesis that a

[8] A 'spin-up period' is the time needed for a model to travel through its phase space from the initial conditions to quasi-equilibrium; that is, the time needed by the model to 'forget' the initial conditions.

[9] For example, when there is 'blocking,' the memory of the atmosphere might be a few weeks.

[10] The *eddy component* of a random field $\vec{\mathbf{Z}}$ is the deviation from the zonal mean, $\vec{\mathbf{Z}}^* = \vec{\mathbf{Z}} - [\vec{\mathbf{Z}}]$, where $[\cdot]$ denotes the zonal averaging operator. Here $[\vec{\mathbf{Z}}] = \vec{\mathbf{Z}}^{\mathrm{T}}\vec{1}/m$, where m is the number of elements in $\vec{\mathbf{Z}}$ and $\vec{1}$ is the m-dimensional vector of units. See also [7.2.1].

component of the signal lies in a direction orthogonal to the guess pattern was not rejected.

3 *Observed Fields as Guess Patterns*
The GCM experiment was conducted to simulate the atmospheric response to anomalous SST conditions in the tropical Pacific. Therefore, the January mean 500 hPa height anomaly fields observed during three El Niño events (1973, 1977, and 1983) were used as guess patterns. A separate univariate test was performed with each guess pattern.

The January 1973 guess pattern successfully extracted part of the signal, although the change in pattern coefficient was negative rather than positive, as in the previous item. The estimated signal, obtained by multiplying the guess pattern with the change in its coefficient (Figure 7.4, top), had about half of the strength of the signal obtained by splitting the GCM data. The most variance was contained in a sector covering the Atlantic and Eurasia. The part of the full signal that appeared in the direction of the guess patterns was actually weaker than the components that were orthogonal to the guess pattern.

The January 1977 guess pattern successfully captured a large fraction of the GCM signal. There was strong evidence against the null hypothesis, and the strength of the projection was about 75% of the value found through splitting the GCM data. The parallel component (Figure 7.4, bottom) was very similar to the full signal (Figure 7.3). The orthogonal part of the full signal (not shown) was still significantly nonzero.

The last guess pattern, January 1983, represented the observed atmospheric response to the most intense ENSO event on record up to 1997. Analysis of observational data has shown that the January 1983 Northern Hemisphere extratropical 500 hPa height field was substantially different from 'normal' January mean height fields [385]. None the less, this field failed to capture the simulated ENSO signal when it was used as a guess pattern. In fact, the GCM output was almost orthogonal to the January 1983 500 hPa height anomaly.

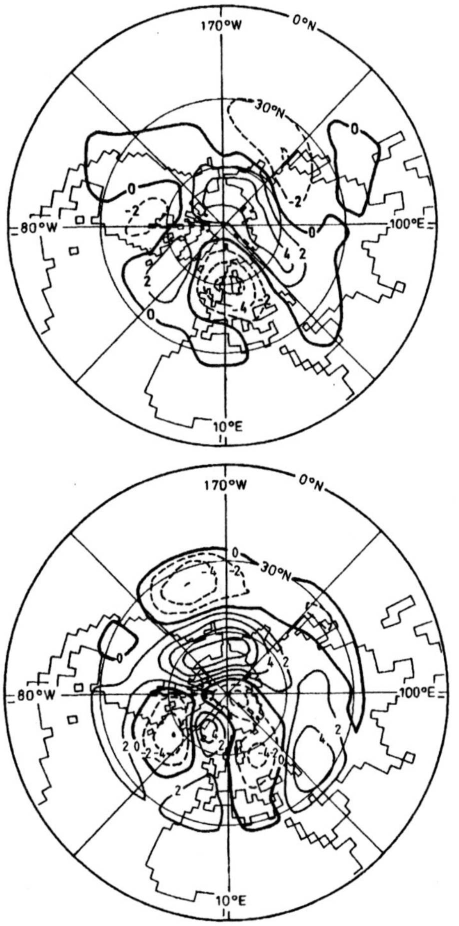

Figure 7.4: *GCM experiment on the extratropical atmospheric response to tropical El Niño SST anomalies. Statistically significant projections of the full 500 hPa height signal (Figure 7.3) on the January 1973 guess pattern (top: note that the signal is almost zero in the Pacific sector, where the El Niño related signal is expected to be strongest) and on the January 1977 guess pattern (bottom). Units: dkm. From von Storch [386].*

The major conclusion drawn from this statistical analysis [386] was that the El Niño SST anomalies excite a statistically significant response in the extratropical atmospheric circulation. The model simulated a response similar to the observed circulation anomaly from January 1977, but largely orthogonal to the observed January 1973 and 1983 anomalies.

7.3 Identification of a Signal in Observed Data

7.3.1 General. Dramatic events sometimes take place in the global environment, such as the appearance of large-scale ENSO sea-surface temperature anomalies of 1982/83 (Figure 7.2) or the injection of large amounts of aerosols into the stratosphere by an erupting volcano such as the Pinatubo in 1992 (see, e.g., McCormick, Thomason, and Trepte [268] or Pudykiewicz and Dastoor [324]). The large events can be viewed as natural sensitivity experiments, so it is of interest to know whether the state of the atmosphere during, and after, the event is different from that of the undisturbed, 'normal' climate. The observations that represent the normal climate are regarded as independent realizations of a 'control' climate state vector $\vec{\mathbf{X}}$. The observation taken during the event of interest is labelled $\vec{\mathbf{y}}_1$, and the null hypothesis: '$\vec{\mathbf{y}}_1$ *is drawn from* $\vec{\mathbf{X}}$' is examined. If the null hypothesis is rejected, $\vec{\mathbf{y}}_1 - \bar{\vec{\mathbf{x}}}$ is regarded as an informative, but uncertain, estimate of the effect of the event.[11]

7.3.2 The 1982/83 El Niño and its Impact on the Extratropical Circulation. In Section 7.2 we described an analysis of a simulated response to a prescribed tropical Pacific SST anomaly. In this subsection we describe the analysis of observed response.

Hense [174] examined monthly anomalies of Northern Hemisphere stream function for the period January 1982 to September 1983, a period containing the largest ENSO event on record (until 1997). The monthly anomalies used in the study were obtained by subtracting the 1967–81 mean appropriate to the month from each monthly mean in the 21-month study period. The covariance structure varies with the time of year, so the statistical analysis is done separately for each calendar month.

As in [6.1.3], the null hypothesis for each of the 21 months from January 1982 to September 1983 is that the respective monthly anomaly $\vec{\mathbf{y}}$ is drawn from the random variable $\vec{\mathbf{X}}$, where $\vec{\mathbf{X}}$ represents the 'normal' monthly mean stream function distribution appropriate to the month in which $\vec{\mathbf{y}}$ is observed. The $\vec{\mathbf{X}}$ sample for a given month of the year is taken to be the 15 monthly mean stream function fields observed for that month between 1967 and 1981. The null hypothesis is tested with Hotelling T^2 [6.6.10], which means that a number of assumptions are made implicitly. Specifically, it is assumed that the monthly mean 500 hPa stream function is multivariate normal, that the realizations for a given month of the year are independent, that the realizations during the 1967–81 period all come from random vectors with the same distribution, and that the covariance structure during the 1982/83 ENSO was the same as that during the preceding 15 years. Clearly these assumptions are not all satisfied. None the less, the analysis based on this model is useful, even if it is not fully precise.

The data were available on a 576 point grid. The guess patterns used in this study are the *surface spherical harmonics*, written as $P_j^m(\phi)\cos(m\lambda)$ and $P_j^m(\phi)\sin(m\lambda)$, where ϕ is latitude, λ is longitude, P_j^m is the corresponding *associated Legendre polynomial*, for $j = 0, 1, \ldots, \infty$ and $m = 0, \ldots, j$ [15]. The surface spherical harmonics are orthonormal[12] functions. The index j specifies the spatial scale, that is, any two surface spherical harmonics with the same index j share the same spatial scale whereas a larger j indicates a smaller scale. Only functions with odd 'two-dimensional wavenumbers' $m + j$ are needed to represent a hemispherical field. There is only one function for each (m, j) combination when $m = 0$, but there are two functions, one displaced zonally $\frac{\pi}{2m}$ radians relative to the other, when m is nonzero. The cosine form of the $(1, 1)$ and $(1, 2)$ spherical harmonics are shown in Figure 7.5.

A hierarchy [6.5.8] was chosen as shown in Figure 7.6: the hierarchy with $K = 1$ element contains only the function $P_1^0(\phi)$; the hierarchy with $K = 3$ contains that and the functions $P_2^1(\phi)\cos(\lambda)$ and $P_2^1(\phi)\sin(\lambda)$ as well, and so on. The hierarchy does not contain an element with $K = 2$ guess patterns.

The projection of the full signal $\vec{\mathbf{y}} - \bar{\vec{\mathbf{x}}}$ onto a subset of K guess patterns represents a truncated signal. The *optimal signal* is identified as the truncated signal that goes with the K for which the evidence against the equality of means hypothesis is the strongest. Barnett et al. [22] call this *selection rule C*.

The following results were obtained:

- *Results for November 1982* are shown in the bottom panel of Figure 7.6. The statistic that is displayed for each level K of the hierarchy is a scaled version of Hotelling T^2 (6.33) that

[11] If more than one event is examined, the observations from the events are regarded as samples of another random variable $\vec{\mathbf{Y}}$ and the null hypothesis is H_0: $\mathcal{E}(\vec{\mathbf{X}}) = \mathcal{E}(\vec{\mathbf{Y}})$. If H_0 is rejected, the difference $\bar{\vec{\mathbf{y}}} - \bar{\vec{\mathbf{x}}}$ is understood to be an estimate of the mean response of the climate system to the external events.

[12] *Orthonormal* means that the scalar product of any two non-identical surface spherical harmonics is zero, and that of a spherical harmonic with itself is one. In fact, $\frac{1}{2\pi^2}\int_0^{2\pi}\int_0^{\pi} P_j^m(\phi)(\cos(m\lambda) + i\sin(m\lambda))P_k^n(\phi)(\cos(n\lambda) - i\sin(n\lambda))\,d\phi\,d\lambda = \delta_{mn}\delta_{jk}$, where δ_{il} is one if $i = l$ and zero otherwise.

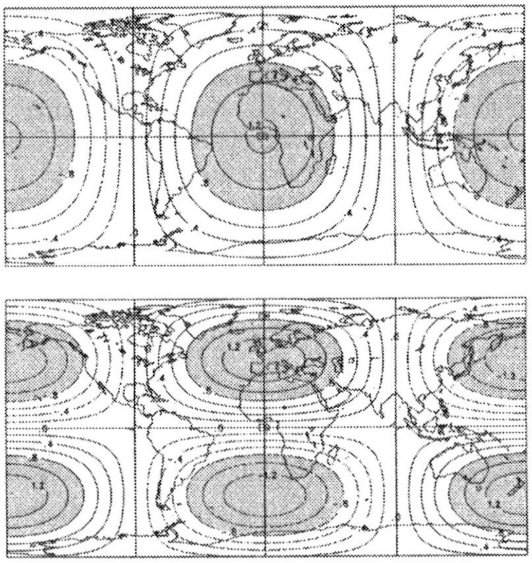

Figure 7.5: *Two surface spherical harmonics. The upper panel represents a larger spatial scale than the lower panel.*
Top: $P_1^1(\phi)\cos(\lambda)$. *Bottom:* $P_2^1(\phi)\cos(\lambda)$.

is given by

$$\widehat{\mathcal{D}}^2(\vec{\mathbf{y}} - \bar{\vec{\mathbf{x}}}) = \frac{K(n_X + n_Y - 2)}{n_X + n_Y - K - 1} \times \left(\frac{1}{n_X} + \frac{1}{n_Y}\right)^{-1} T^2,$$

where $n_X = 15$ and $n_Y = 1$. The critical values are those of the T^2 statistic, that is, they are upper tail quantiles of the $F(K, n_X + n_Y - K - 1)$ distribution, also scaled by the same factor.

The null hypothesis can be rejected at the 5% significance level for $K = 3, \ldots, 8$. The evidence against H_0 is strongest for $K = 3$. The first conclusion is that there is a significant signal in the data. The second conclusion is that the projection of the full signal on the three first guess patterns, $P_1^0(\phi)$, $P_2^1(\phi)\cos(\lambda)$, and $P_2^1(\phi)\sin(\lambda)$, yields the optimal model in the hierarchy.

- *Results for All Months.* The hierarchal testing procedure was repeated in each of the 21 months from January 1982 to September 1983. The null hypothesis was rejected at the 5% significance level or less for at least one member of the hierarchy in every month from July 1982 until September 1983. The optimal signal was often found in the $K = 3$ hierarchy. The strongest signals, in terms of significance, were found from September 1982 to June 1983.

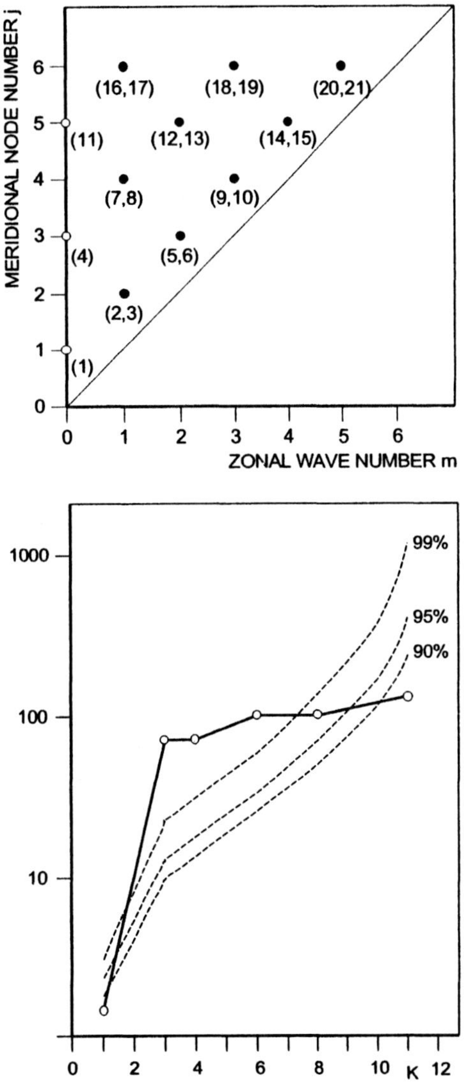

Figure 7.6: *Analysis of extratropical 500 hPa height during the 1982/83 El Niño event [174].*
Top: Hierarchy in the set of surface spherical harmonics functions, used as guess patterns in Section 7.3.
Bottom: Results for November 1982.

A total of 21 tests were conducted and the null hypothesis was rejected in 15. We would expect only one or two rejections to occur at the 5% significance level if H_0 was correct throughout the 21-month period, and if the 21 decisions are statistically independent of one another (which they are not). Assuming that we have made the equivalent of seven independent decisions, the probability of making the reject decision under H_0 15 or more times in the 21 tests is well below 1% [6.8.2].

The major conclusion of this study [174] is that the Northern Hemisphere extratropical circulation during the 1982–83 El Niño was substantially different from the circulation during the preceding 15 years.

7.4 Detecting the 'CO_2 Signal'

7.4.1 A Perspective on Global Warming. The prospect of man changing the world's climate by modifying the chemical composition of the atmosphere was first discussed by Arrhenius [16] in 1896. He argued that a change in the atmospheric concentration of radiatively active gases, such as a carbon dioxide, will cause a change in the physical state of the atmosphere in general and the near-surface temperature of the globe in particular. Arrhenius's result was mostly of academic interest for many decades, but since the late 1970s it has become one of the top environmental topics. The scientific challenge was, and is, to determine whether the changing composition of the atmosphere will result in *physically* or *socially significant* climate changes.

Early climate model experiments indicated large effects, which were not matched by the observational record. These were 'equilibrium' experiments designed to estimate the effect of doubling the atmosphere's CO_2 concentration; they were typically performed with AGCMs that were coupled to thermodynamic models of sea ice and the upper (i.e., mixed layer) part of the global oceans.[13]

More recent simulations[14] have used *coupled climate system models* that incorporate an AGCM, a dynamical ocean model, sometimes a dynamical sea-ice component, and the effects of tropospheric aerosols.[15] The greenhouse gas and tropospheric aerosol concentration in these experiments is changed gradually in time to reflect the effects of human activities on the environment. These sophisticated simulations, performed at a number of institutions, appear to agree broadly with observed climate change and also agree broadly on the size and distribution of future climate change. None the less, these simulations are only plausible scenarios for the future since many aspects of the simulated system, such as the low-frequency variability of the oceans and the role of clouds in regulating climate, are still poorly understood.

7.4.2 Methodological Considerations. As in the preceding section, where we dealt with the 'signal' excited during an episode with large tropical sea-surface temperature anomalies, the statistical 'climate change detection' problem consists of evaluating one event, say the latest record of the global distribution of near-surface temperature, in the context of the natural variability of near-surface temperature. The problem is to determine whether the recent warming is consistent with the variations of temperature due to internal, and thus undisturbed, dynamics.

The main methodological obstacle is the lack of observations that sample the 'control' regime. Most of the available instrumental record consists of surface observations taken during the last century or so. This record may be contaminated by the greenhouse gas signal but, more importantly, it is not large enough to provide us with a reliable estimate of the natural variability of the climate on the time scales on which the climate change is expected to occur. In the next subsection we summarize the approach to this problem developed by Hegerl et al. [172].[16]

7.4.3 A Detection Strategy. The first problem in developing a 'detection strategy' that aims to identify the 'greenhouse signal' is to choose which variable to exploit (such as sea-level pressure, near-surface temperature, the vertical distribution of moisture in the atmosphere, etc.). Whatever the variable, it should satisfy the following criteria.

- There should be a long historical record of the variable, containing observations with wide spatial coverage that are made in a consistent manner throughout the year. The only data in the instrumental record

[13] An equilibrium climate change experiment is described by Boer et al. [52].

[14] See Gates et al. [138] and Kattenberg et al. [215] for an overview.

[15] Such as SO_4, which reflects sunlight and therefore cools the climate, and black carbon, which absorbs sunlight and therefore warms the climate.

[16] There is an extensive literature on climate change detection. Some additional important entry points to the recent literature include Barnett and Schlesinger [23], Bell [38, 39], Hasselmann [168], Hegerl et al. [172], Karoly et al. [214], Mitchell et al. [279], North, Kim, and Shen [297], Parker et al. [303], Santer et al. [339] and Stevens and North [359]. Santer et al. [340] provide an extensive overview.

that satisfy this criterion are rainfall, sea-level pressure, near-surface temperature, and sea-level observations taken during the last hundred years or so. For example, Folland, Karl, and Vinnikov [121] use a *frozen grid* analysis to show that global annual mean near-surface temperature can be reliably estimated from about 1860 onwards.[17] Proxy data [3.1.10] hold promise for climate change detection because they cover much longer periods than the instrumental record. However, proxy data are difficult to use because the information they contain is often specific to a particular region and time of year (e.g., the growing season).

- The observational record should be homogeneous, and free of biases caused by changes in the observing network configuration, the instruments and their immediate physical environment (see Figure 1.9), and observing practices. The three atmospheric data records mentioned above have been made somewhat homogeneous by means of laborious 'homogenization' techniques (see, e.g., Jones [201], or Vincent [383]). Rainfall is the least reliable variable in this respect. Sea-level data are contaminated by land rising and sinking, among other processes (see [3.2.2]).

- The 'signal-to-noise' ratio should be large. For example, GCM experiments indicate that sea-level pressure has a much weaker signal-to-noise ratio than screen temperature. See Barnett, Schlesinger, and Jiang [24] for more details.

- The variable should be well simulated by climate models, for reasons explained below. It is felt that current models do not yet simulate precipitation or sea level well.

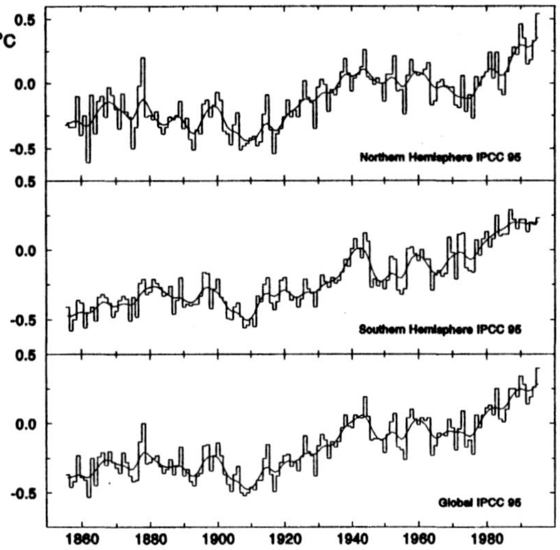

Figure 7.7: *Estimated annual mean near-surface temperature expressed as anomalies relative to the 1950–79 mean. From Nicholls et al. [294]. Courtesy P. Jones.*

Hegerl et al. therefore used the instrumental near-surface temperature record as the basis for their detection strategy. Jones et al. [202] [203] have carefully compiled a widely used gridded (5° longitude × 5° latitude) near-surface temperature data set for 1855 onwards. The coverage increases with time as more stations become available.

The 'detection question' is formulated as a statistical testing problem. The null hypothesis is that the 'trend' (Figure 7.7) found in the observational record stems from natural variability. Physical reasoning, as well as results obtained from recent 'transient' climate simulations [7.4.1], indicate that rejection of this hypothesis will be consistent with greenhouse gas induced climate warming.

The methodical problems connected with this test are as follows.

1 The state variable is a high-dimensional vector. Before performing a test, the spatial degrees of freedom have to be reduced by projecting the raw data onto a guess pattern [6.5.6]. Hegerl et al. [172] used data from a climate model, which was forced with increasing concentrations of greenhouse gases, to build a simple guess pattern: the simulated 100-year change in the near-surface temperature.

  We introduced an algorithm in [6.5.7] to increase the power of a test by 'rotating' the guess pattern towards the anticipated signal in such a way that the signal-to-noise ratio is optimized. To achieve this we must project the guess pattern, and the observations, onto a low-dimensional vector

[17] Folland et al. [121] compare global mean temperatures time series computed from a number of observing networks representing the distribution of observing stations at a number of points in time. Shen, North, and Kim [349] and Zwiers and Shen [451] use more rigorous arguments to come to the same conclusion.

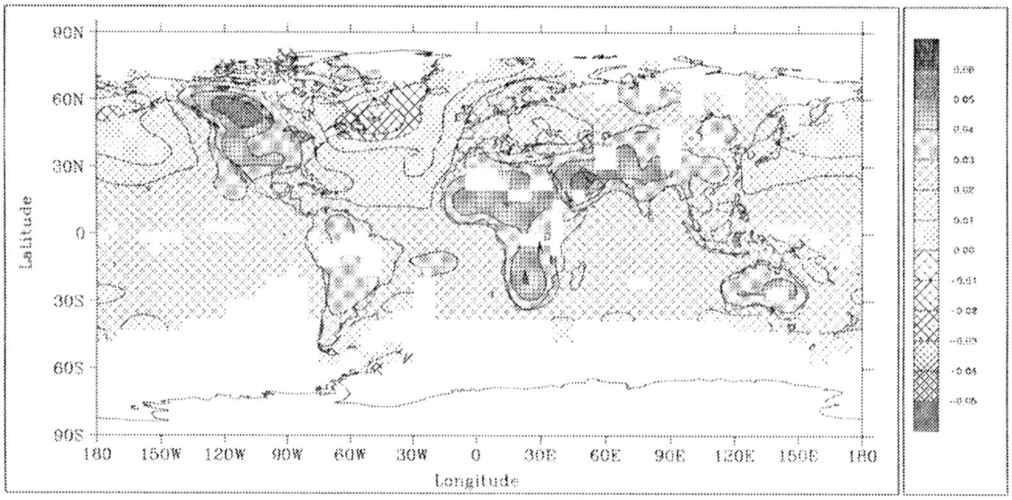

Figure 7.8: Optimized guess pattern used by Hegerl et al. [172] to 'detect' the impact of an anthropogenic greenhouse gas effect on recent 20-year near-surface temperature trends. The pattern is taken from a 'scenario' run with a climate model that was forced with continuously increasing greenhouse gas concentrations.

space that captures most of the climate's natural variability on decadal and longer time scales. We also need an accurate estimate of the covariance matrix of the natural variability in this subspace. Hegerl et al. chose to use the subspace spanned by the first 10 EOFs (see Chapter 13) of the 'transient' climate simulated by the model used to produce the guess pattern.[18] The covariance matrix Σ needed for the optimization was also estimated from this simulation.[19] If $\vec{p}$ is the raw guess pattern in the 10-dimensional subspace, then the optimized guess pattern $\vec{p}^{\,0}$ is given by $\Sigma^{-1}\vec{p}$. Furthermore, if $\vec{\mathbf{T}}_t$ represents the detection variable at time t (i.e., observed near-surface temperature projected onto the four-dimensional subspace), then the optimized detection variable is given by

$$\alpha_0(\vec{\mathbf{T}}_t) = \langle \vec{p}^{\,0}, \vec{\mathbf{T}}_t \rangle. \tag{7.1}$$

Hegerl et al. [172] performed the analysis with both the raw guess pattern and the optimized guess pattern.[20] We limit our report here to the improved results that were obtained with the optimized pattern.

2 The observed data are not complete. Data is missing sporadically in some 5° × 5° boxes, and other boxes have extended intervals of missing data. This means that the scalar product cannot be used to project the observed temperatures onto the guess pattern. Instead, the projection is determined by solving a least squares problem. Let $\vec{\mathbf{t}}_t = (\mathbf{t}(1,t), \ldots, \mathbf{t}(10,t))^{\mathrm{T}}$ be the realized projection at time t, let $\mathbf{t}_o(i,j,t)$ represent the near-surface temperature observed at time t in the (i,j)th 5° × 5° box, and let $\upsilon_k(i,j)$, for $k = 1, \ldots, 10$ represent the 10 EOFs. Also, let $A(i,j)$ be the area of the (i,j)th box. Then $\vec{\mathbf{t}}_t$ is found by minimizing

$$\sum\sum\left(\mathbf{t}_o(i,j,t) - \hat{t}_o(i,j,t)\right)^2 A(i,j),$$

where

$$\hat{t}_o(i,j,t) = \sum_{k=1}^{10} \mathbf{t}(k,t)\upsilon_k(i,j),$$

and where the double sum is taken over those grid boxes that contain data.[21] Simulation

[18] The first 10 EOFs of the transient simulation were used because they capture the guess pattern much more effectively than the EOFs of the control simulation.

[19] We treat Σ as known since it was estimated from a very long simulation.

[20] Also sometimes called a 'fingerprint.'

[21] For a more detailed representation of the problem of determining EOF coefficients in case of gappy data, refer to [13.2.8].

experiments have shown that changes in data density may cause inhomogeneities in $\alpha_0(\vec{\mathbf{t}})$. To limit this effect, Hegerl et al. used only those grid boxes for which the record from 1949 onwards was complete. Therefore the entire southern and northern polar regions and the Southern Ocean are disregarded.

Figure 7.8 shows the optimized guess pattern truncated to the area that has complete data coverage for 1949 onwards.

3 The natural variability of the optimized detection variable cannot be estimated from the observations. The observed record is contaminated by the presumed signal and the data are correlated in time so that only a few independent realizations of the 'naturally varying' state variable are available.

There are, in principle, two ways to deal with this problem. The first approach is to remove the expected climate signal from the observed record by constructing a linear model of the form

$$\begin{aligned} \mathbf{T}_t^* &= \mathbf{T}_t - \mathbf{T}_t^{CO_2} \qquad (7.2) \\ \mathbf{T}_t^{CO_2} &= \int_0^\infty S(\Delta) \ln\Big(\frac{C(t-\Delta)}{C(0)}\Big)\, d\Delta . \end{aligned}$$

Here $\mathbf{T}_t$ is the observed temperature record, $\mathbf{T}_t^{CO_2}$ is an estimate of the CO_2 induced temperature signal, and $\mathbf{T}_t^*$ is the residual. The variability of $\mathbf{T}_t^*$ is assumed to be the same as that of the undisturbed climate system. The function $C(t)$ is the atmospheric CO_2-concentration at time t, and $S(\cdot)$ is a transfer function. The variability of the detection variable is then derived from $\mathbf{T}^*$ instead of $\mathbf{T}$.[22]

One problem with this approach is that it does not eliminate the effects of serial correlation; even without the signal it is difficult to estimate the natural variability of the climate on decadal and longer time scales from the observed record.

Another problem with this approach, apart from adopting the model (7.2), is that the remaining variability also includes contributions from other external factors such as aerosol forcing caused by human pollution and volcanos. While volcanos may be considered stationary in time, the effect of pollution

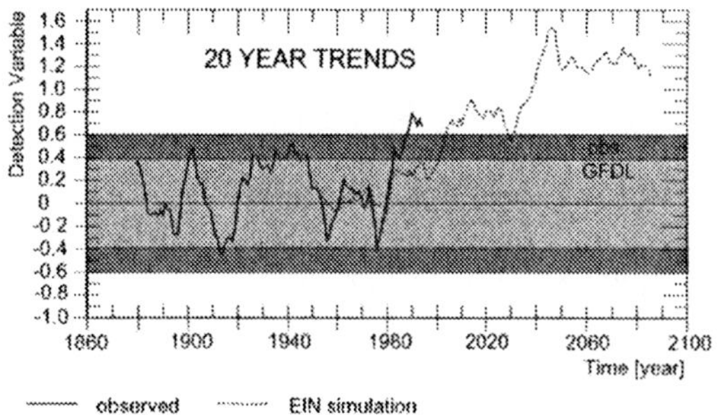

Figure 7.9: *Time evolution of 20-year trends of the optimized detection variable* $\alpha_o(\vec{\mathbf{t}}) = \langle \vec{p}^{\,o}, \vec{\mathbf{t}} \rangle$ *for near-surface temperature. Labels on the abscissa identify the last year (1879 until 1994) of each 20-year period.*
The solid line is derived from observed data since 1860. The dotted line, labelled 'EIN', is derived from 150 years of climate model output. The climate model was forced with anomalous radiative forcing corresponding to the observed 1935–85 greenhouse gas concentrations during the first 50 years of the simulation. A scenario (IPCC scenario A) was to prescribe greenhouse gas concentrations from '1985' onwards. Twenty-year trends from the simulation (dashed curve) are shown to compare the observed evolution with that anticipated by a climate model.
The narrow shaded band, labelled 'GFDL', is an estimate of the natural variability of the 20-year trend derived from a 1000-year control simulation (Manabe and Stouffer [266]). It should contain the trend coefficient 95% of the time if there is no trend. The wider band, labelled 'obs', is derived from observations after an estimate of the greenhouse gas (GHG) is removed. From Hegerl et al. [172].

is not stationary since the concentration of airborne pollutants increases substantially in the latter part of the observed record.

To cope with this problem, the null hypothesis should be reformulated to state that observed variations are consistent with natural variability originating from natural external processes as well as internal dynamical processes. The anthropogenic aerosol effect probably causes a cooling that counteracts the expected greenhouse warming; the presence of this effect in the observed data inflates the estimate of the variability and dampens the signal, diminishing the overall power of the test.

The second approach is to consider the output of 'control' climate model runs

[22] Subsection 17.5.7 also deals with the problem of removing a suspected signal from a time series.

without any external forcing so that all variability originates from internal dynamical processes. This approach has the advantage that, at least in principle, very long samples can be created without inhomogeneities in accuracy or varying spatial coverage. A major disadvantage, though, is that the models may not simulate the natural low-frequency variability correctly.

Hegerl et al. used both approaches. In two steps, 95% confidence intervals for the natural variability of 20-year trends in the optimized detection variable were constructed from both observed anomalies (7.2) and climate model output. In both cases an auto-regressive process of order 1 was fitted to the optimal detection variable.[23] Monte Carlo simulations were then performed with the fitted auto-regressive models to estimate the natural variability of 20-year trends in the optimized detection variable.[24] The test is eventually performed by comparing recent 20-year trends with the estimated 95% confidence intervals.

The result of the exercise is summarized in Figure 7.9, which shows the time evolution of 20-year trends of the optimal detection variable together with the 95% confidence intervals derived from several sources. The latest trends do indeed exceed the upper confidence limit, so we may conclude that the prevailing trend is not likely to be due to internal processes. This conclusion, of course, depends crucially on the validity of the natural variability estimates. For further reading on climate change detection and attribution see Santer et al. [340] and Zwiers [445].

[23] An auto-regressive process of order 1 (an AR(1) process) is written formally as $\mathbf{X}_t = \alpha \mathbf{X}_{t+1} + \mathbf{N}_t$, where $\mathbf{N}_t$ is a series of independent random variables (sometimes called 'white noise'). Chapters 10 and 11 explain AR(p) processes in some detail.

[24] This procedure is closely related to the bootstrap (Section 5.5).

Part III

Fitting Statistical Models

Overview

In this part of the book we introduce two classical, fully developed[1] methods of inference: 'regression' and 'analysis of variance' (ANOVA). We do not expect that there will be significant changes in the overall formulation of these techniques, but new applications and improved approaches for special cases may emerge.

Both regression and ANOVA are methods for the estimation of *parametric* models of the relationship between related random variables, or between a random variable and one or more non-random external factors. While regression techniques have been used almost from the beginning of quantitative climate research in different degrees of complexity (see, e.g., Brückner [70]), ANOVA has only recently been applied to climatic problems [441, 444].

The regression technique is introduced in detail and illustrated with several examples in Chapter 8.[2] Regression is used to describe relationships that involve variables and factors measured on a continuous scale. Examples of regression problems include modelling the trend in a time series by means of a polynomial function of time (which would be a non-random external factor), or the description of the link between two concurrent events, such as the width of a tree ring and the temperature, with the purpose of constructing 'best guesses' of temperature in ancient times when no instrumental data are available. Also, time-lagged events are linked through regression, such as the wind force in the German Bight and the water level in Hamburg several hours later. The derived model is then used for storm surge forecasts.

The reader may notice that climatologists often use the term *specify* when they refer to regressed values, as opposed to the term *forecast* commonly used by statisticians. Neither word is perfect. 'Forecast' implies that there will be error in the estimated value, but sometimes has irrelevant time connotations. 'Specify' eliminates the confusion about time but suggests that the estimate is highly accurate. However, despite its inadequacies, we use 'specify,' except when discussing projections forward in time, in which case we refer to forecasts.

The *analysis of variance* was designed by R.A. Fisher for problems arising in agriculture. In his words, ANOVA deals with 'the separation of the variance ascribable to one group of causes from the variance ascribable to other groups.' Separation of variance is also often required in climate diagnostics. A typical problem is to discriminate between the effect of internal and external processes on the global mean temperature. In that case, an internal process might be the formation and decay of storms in midlatitudes, while an external factor might be the stratospheric loading of volcanic aerosols. Another typical application treats sea-surface variability on monthly and longer time scales as an external process. In this case several independent climate simulations might be performed such that the same time series of sea-surface temperatures is prescribed in each simulation. ANOVA methods are then used to identify the simulated atmospheric variability that results from the prescribed sea-surface temperatures. The ANOVA technique is explained in detail in Chapter 9 and its merits are demonstrated with examples.

[1] By 'fully developed' we mean that for each parameter involved there is at least an asymptotic distribution theory so hypothesis tests and confidence intervals can be readily constructed.

[2] In fact, regression techniques appear throughout the book, as in Sections 14.3 and 14.4, which deal with Canonical Correlation Analysis and Redundancy Analysis, respectively.

8 Regression

8.1 Introduction

8.1.1 Outline. We start by describing methods used to estimate and make inferences about correlation coefficients. Then, we describe some of the ideas that underly regression analysis, methods in which the mean of a *response* (or dependent) variable is described in terms of a simple function of one or more *predictors* (or independent variables). The models we consider are said to be *linear* because they are linear in their unknown parameters. We describe a variety of inferential methods and model diagnostics, and consider the robustness of the estimators of the model parameters.

A simple example is a naive model of climate change in which global annual mean temperature increases, on average, logarithmically with CO_2 concentration:

$$T_{year}^{globe} = a_0 + a_1 \ln(c_{CO_2}) + \epsilon_{year}. \qquad (8.1)$$

We know that global annual mean temperature is subject to fluctuation induced by a variety of physical processes whose collective effect results in apparently stochastic behaviour. On the other hand, CO_2 concentration appears to have only a minor stochastic component, at least on interannual time scales, and can therefore be considered to be deterministic to a first approximation. The model proposes that global annual mean temperature, denoted T_{year}^{globe}, is trending upwards approximately logarithmically as the CO_2 concentration, denoted c_{CO_2}, increases. It also proposes that T_{year}^{globe} has a stochastic component, which is represented by the noise process $\{\epsilon_{year}\}$. There are two free parameters, a_0 and a_1, that must be estimated from the data. This is something that is often (although not always best) done using the method of least squares. Here least squares estimation of the parameters is simple because the model is linear in its parameters. If inferences are to be made about the parameters (e.g., tests of hypothesis or construction of confidence intervals), then it is required that (8.1) also include some sort of assumption about the characteristics of the stochastic component, typically that this component behaves as normally distributed white noise. Other less restrictive assumptions are possible, but they may require the use of more sophisticated inference methods than those described in this chapter.

After introducing simple linear models, our discussion of regression goes on to consider multivariate linear models and methods for model selection. We close the chapter with two short sections on model selection and some other related topics, including nonlinear regression models. It is worth repeating that statisticians distinguish between linear and nonlinear models on the basis of the model's parameters, not on how the predictors enter the model.

An example of a simple nonlinear model, which may be better suited than (8.1) to the example above, is

$$T_{year}^{globe} = b_0 + b_1 \ln(c_{CO_2 year} + b_2) + \epsilon_{year}.$$

Note that this model is nonlinear in b_2.

8.1.2 The Statistical Setting. Most of the discussion in this chapter takes place in the context of normal random variables, not because other types of data are uncommon, but because it is relatively easy to introduce concepts in this framework. Nevertheless, note that departures from assumptions can affect the reliability of some statistical analyses quite drastically.

8.1.3 Example: ENSO Indices. This example was considered briefly in [1.2.2] and [2.8.8]. Wright [426] described a tropical Pacific sea-surface temperature index that captures information about ENSO that is very similar to the information captured by the classical Southern Oscillation Index (SOI) based on the difference between mean sea-level pressure at Darwin and Tahiti. Wright's index is based on SSTs observed east of the date line and roughly between 5° N and 10° S. A scatter plot of the monthly mean values of these indices for 1933–84 inclusive is

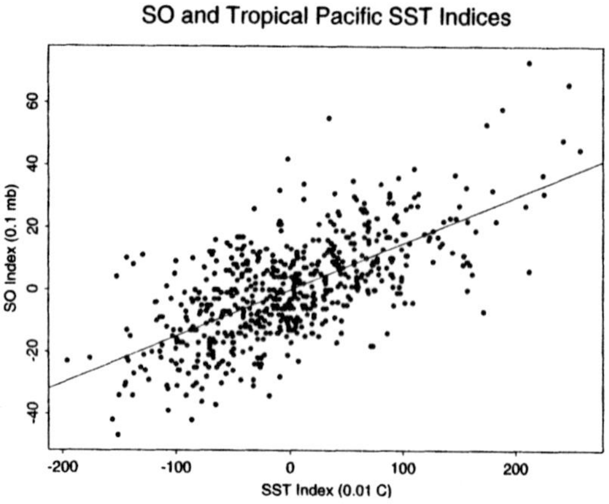

Figure 8.1: *Scatter plot of monthly values of the SO index versus the SST index for 1933–84 inclusive. Units: 0.1 mb (SOI),* 0.01 °*C (SST Index).*

shown in Figure 8.1, and their corresponding time evolutions are shown in Figure 1.4. Both diagrams show the strong tendency for the two indices to co-vary; when the SOI is large and positive, the tropical Pacific SSTs east of the date line also tend to be large and positive. We return to this example in Sections 8.2 and 8.3.

8.1.4 Example: Radiative Transfer Parameterization in a GCM. AGCMs use *parameterizations* to describe the effect of unresolved sub-grid scale processes in terms of larger resolved scale quantities [6.6.6]. One such process is the transmission of short wave radiation (i.e., light) through the atmosphere to the land surface, where this energy is either reflected or converted into other forms (such as latent and sensible heat). The propagation of light through the atmosphere at a specific location is strongly affected by factors such as the three-dimensional structure of the cloud field and the distribution of other materials, such as aerosols that may reflect, refract, or absorb light.

AGCMs need to know the grid box average of light energy incident upon the ground (or passing though an atmospheric layer). Radiation transfer codes used in AGCMs estimate these averages from other grid scale parameters that are simulated by the model.

Barker [17] describes a radiative transfer parameterization that requires the mean ($\overline{\tau}$) and standard deviation (σ_τ) of cloud optical depth τ within the grid box as input.[1] In contrast, the cloud parameterizations used in GCMs can estimate $\overline{\tau}$ and A_c, the fraction of the grid box that is cloud covered, but they are not able to estimate σ_τ. However, it turns out that the mean log cloud optical depth $\overline{\ln \tau}$ is closely related to σ_τ. Thus the radiative transfer calculation can be performed once estimates of $\overline{\tau}$ and $\overline{\ln \tau}$ are available. The latter can be obtained from $\overline{\tau}$ and A_c by means of a simple regression model.

We use satellite data described by Barker, Wielicki, and Parker [18] in Section 8.4 to examine the observed relationship between $\overline{\ln \tau}$ and corresponding ($\overline{\tau}$, A_c) pairs. The data consists of 45 estimates of ($\overline{\ln \tau}$, $\overline{\tau}$, A_c) that were derived from 45 ocean images taken by the Landsat satellite. Each image covers an area of about 3400 km^2. Figure 8.2 shows three of these images, and Figure 8.3 shows the derived data. Note that the relationship between $\overline{\tau}$ and $\overline{\ln \tau}$ is curvilinear (Figure 8.3, left). Also, note that, even though there are a substantial number of scenes that are fully covered (i.e., $A_c = 1$), this does not preclude variability of $\overline{\ln \tau}$.

[1] Optical depth is a measure of opacity.

8.2 Correlation

8.2.1 Covariance. The covariance between two random variables **X** and **Y** is defined as

$$\mathrm{Cov}(\mathbf{X}, \mathbf{Y}) = \mathcal{E}((\mathbf{X} - \mu_X)(\mathbf{Y} - \mu_Y)), \qquad (8.2)$$

where μ_X and μ_Y are the mean values of **X** and **Y** respectively. (See also Section 2.8.)

Climatologists often interpret covariances involving winds as transports [311]. For example, Figure 8.4 displays the meridional *transient eddy transport* of zonally averaged zonal momentum, as simulated by a GCM in the December, January, February (DJF) season. The 'eddy component' of any variable, here the wind, is the deviation from the spatial mean, here the zonal mean. A significant part of the variability in this component stems from cyclones or 'eddies.' The 'transient' part of the wind statistic is the variability around the time mean (the 'stationary' component). The transient eddy transport is the zonally averaged covariance between the space–time variable part of, for instance, the zonal and meridional wind.

The following notation is often used by climatologists. The eddy and transient components of a field are indicated by superscripts '*' and ''' respectively. The time mean (equivalent to the sample mean in this context) is denoted by an overbar, and square brackets denote the zonal average. With this notation the meridional transient eddy

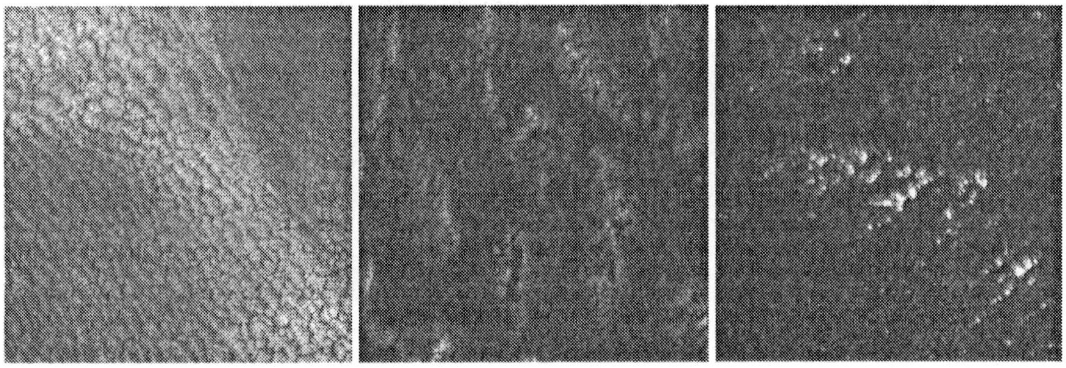

Figure 8.2: *Optical depth inferred from three* 0.83 μm *Landsat images. The brightest pixels in these images correspond to an optical depth of about 20. From Barker et al. [18].*
Left: Scene A3. Overcast stratocumulus, $A_c = 1.000$ *and* $\overline{\tau} = 11.868$.
Middle: Scene B2. Broken stratocumulus, $A_c = 0.644$ *and* $\overline{\tau} = 3.438$.
Right: Scene C14. Scattered cumulus, $A_c = 0.291$ *and* $\overline{\tau} = 3.741$.

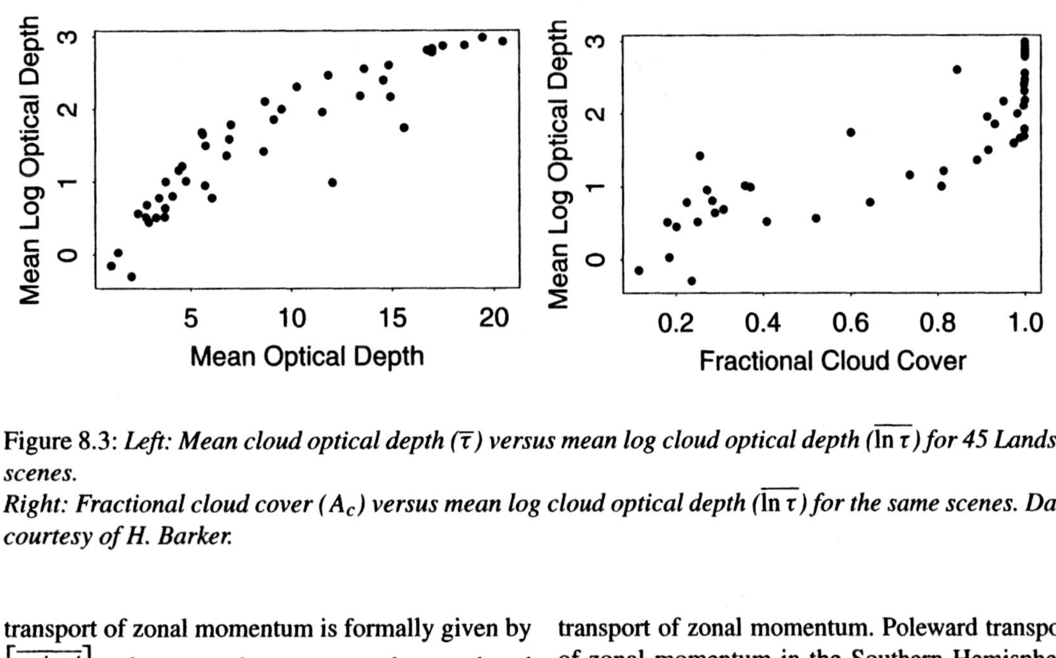

Figure 8.3: *Left: Mean cloud optical depth* ($\overline{\tau}$) *versus mean log cloud optical depth* ($\overline{\ln \tau}$) *for 45 Landsat scenes.*
Right: Fractional cloud cover (A_c) *versus mean log cloud optical depth* ($\overline{\ln \tau}$) *for the same scenes. Data courtesy of H. Barker.*

transport of zonal momentum is formally given by $\left[\overline{u^{*\prime} v^{*\prime}}\right]$, where u and v represent the zonal and meridional wind components.[2]

When $\left[\overline{u^{*\prime} v^{*\prime}}\right] > 0$ in the Northern Hemisphere, as in Figure 8.4, then easterly u anomalies ($u^{*\prime} > 0$) are usually connected with northerly v anomalies ($v^{*\prime} > 0$), and westerly anomalies ($u^{*\prime} < 0$) with southerly anomalies ($v^{*\prime} < 0$). The distribution in Figure 8.4 represents a northward (poleward) transport of zonal momentum. Poleward transport of zonal momentum in the Southern Hemisphere is indicated by negative covariances. Figure 8.4 illustrates that the transient eddies are a powerful agent for exporting zonal momentum from the tropical and subtropical latitudes polewards in both hemispheres.

8.2.2 The Correlation Coefficient. The correlation coefficient is given by

$$\rho_{XY} = \frac{\mathcal{E}((\mathbf{X} - \mu_X)(\mathbf{Y} - \mu_Y))}{\sigma_X \sigma_Y},$$

where $\sigma_X = \sqrt{\mathrm{Var}(\mathbf{X})}$ and σ_Y is defined analogously. Note that ρ_{XY} takes values in the range $[-1, 1]$.

[2] The complete decomposition of the *total transport* is $[uv] = [\overline{u^{*\prime} v^{*\prime}}] + [\bar{u}^* \bar{v}^*] + \overline{[u]'[v]'} + [\bar{u}\bar{v}]$. The first two terms represent the transport by transient and stationary eddies, and the last two terms the transports by the transient and stationary cells. For maps and further details, see Peixoto and Oort [311].

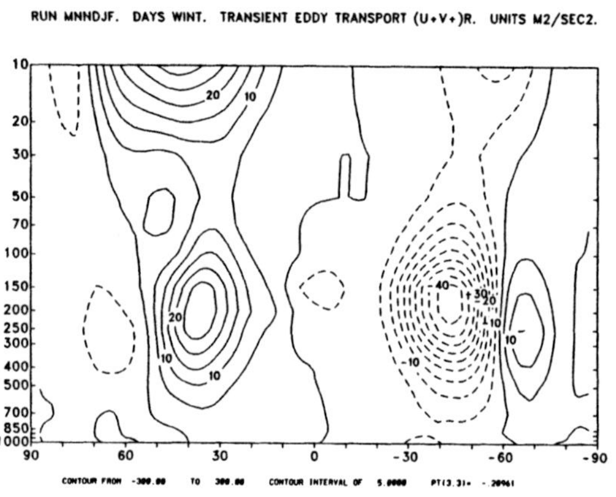

Figure 8.4: *Zonally averaged covariance between the 'transient' eddy components of the zonal and meridional wind = 'meridional transient eddy transport of zonally averaged zonal momentum' during DJF simulated by a GCM. Units:* $\mathrm{m}^2/\mathrm{s}^2$.

As noted in Section 2.8, the correlation coefficient measures the tendency of $\mathbf{X}$ and $\mathbf{Y}$ to co-vary (see Example [2.8.12] and Figure 2.10); the greater $|\rho|$, the greater the ability of $\mathbf{X}$ to specify $\mathbf{Y}$.

Suppose that $\mathbf{X}$ and $\mathbf{Y}$ are bivariate normally distributed with means μ_X and μ_Y, variances σ_X^2 and σ_Y^2, and correlation coefficient ρ_{XY}. Their joint density function is given by (2.35). Suppose also that only $\mathbf{X}$ is observable and we want to find a function, say $g(\mathbf{X})$, that specifies the value of $\mathbf{Y}$ as accurately as possible on average. A reasonable measure of accuracy is the mean squared error, given by

$$\mathcal{E}((\mathbf{Y} - g(\mathbf{X}))^2). \tag{8.3}$$

It can be shown that

$$g(\mathbf{X}) = \mu_Y + \frac{\sigma_Y}{\sigma_X}\rho_{XY}(\mathbf{X} - \mu_X)$$

minimizes (8.3) when g is linear in $\mathbf{X}$ and that the mean squared error is $\sigma_Y^2(1-\rho_{XY}^2)$. To reduce the mean squared error to less than 50% of the variance of $\mathbf{Y}$, it is necessary that $|\rho_{XY}| > 1/\sqrt{2}$. That is, $\mathbf{X}$ *represents* at least 50% of the variance of $\mathbf{Y}$ when $|\rho_{XY}| > 1/\sqrt{2}$. To reduce the root mean squared error to less than 50% of the standard deviation of $\mathbf{Y}$ it is necessary that $|\rho_{XY}| > \sqrt{3}/2 \approx 0.87$.

Using the estimated correlation $\widehat{\rho} = 0.667$ between Wright's [426] monthly SST index and the monthly SOI [8.1.4] we estimate that the mean square error of the SO index is 58% of its variance when using the monthly mean SST anomaly to specify the SOI, or the root mean square error is 76% of the standard deviation. This is in general agreement with the level of scatter displayed in Figure 8.1.

Note that the mean squared error is zero when $\rho_{XY} = 1$; that is, $\mathbf{Y} = \mu_Y + \frac{\sigma_Y}{\sigma_X}(\mathbf{X} - \mu_x)$ with probability 1 when $\rho_{XY} = 1$. Also, note that zero correlation is generally not the same as independence (except when $\mathbf{X}$ and $\mathbf{Y}$ are normally distributed, then $\mathbf{X}$ and $\mathbf{Y}$ are independent if and only if $\rho_{XY} = 0$; see [2.8.14]).

8.2.3 Making Inferences about Correlations. When the sample $\{(\mathbf{X}_i, \mathbf{Y}_i)^{\mathrm{T}} : i = 1, \ldots, n\}$ consists of independent, identically distributed random vectors of length two, a good estimator of the correlation coefficient ρ_{XY} is

$$\widehat{\rho}_{XY} = \frac{\sum_{i=1}^{n}(\mathbf{X}_i - \overline{\mathbf{X}})(\mathbf{Y}_i - \overline{\mathbf{Y}})}{\sqrt{\sum_{i=1}^{n}(\mathbf{X}_i - \overline{\mathbf{X}})^2 \sum_{i=1}^{n}(\mathbf{Y}_i - \overline{\mathbf{Y}})^2}}. \tag{8.4}$$

This is the maximum likelihood estimator [5.3.8] when $(\mathbf{X}, \mathbf{Y})$ is bivariate normally distributed. Furthermore, (8.4) is asymptotically normally distributed with mean ρ_{XY} and variance $(1 - \rho_{XY}^2)^2/n$. However, because $\widehat{\rho}_{XY}$ converges slowly to its asymptotic distribution, this result is generally not used to make inferences about ρ_{XY}. Instead, inferences are based on Fisher's z-transform,

$$z = \frac{1}{2}\ln\left(\frac{1+\widehat{\rho}_{XY}}{1-\widehat{\rho}_{XY}}\right), \tag{8.5}$$

which converges quickly to the normal distribution $\mathcal{N}\left(\frac{1}{2}\log\left(\frac{1+\rho_{XY}}{1-\rho_{XY}}\right), \frac{1}{n-3}\right)$ when ρ_{XY} is nonzero. It is then easily demonstrated that an approximate $\tilde{p} \times 100\%$ confidence interval for ρ_{XY} is given by

$$(\tanh(z_L),\ \tanh(z_U)), \tag{8.6}$$

where

$$z_L = z - Z_{(1+\tilde{p})/2}/\sqrt{n-3}$$
$$z_U = z + Z_{(1+\tilde{p})/2}/\sqrt{n-3},$$

and $Z_{(1+\tilde{p})/2}$ is the $(1+\tilde{p})/2$-quantile of the standard normal distribution (see Appendix D). David [100] (see also Pearson and Hartley [308]) gives tables for exact confidence intervals for ρ_{XY}.

In the SOI example $\widehat{\rho}_{SST,SOI} = 0.667$ and thus $z = 0.805$. For $(1+\tilde{p}/2) = 0.05$, $Z_{(1+\tilde{p})/2} = 1.96$, so that $Z_L = 0.805 - 1.96/\sqrt{621} = 0.727$, assuming that each of the 52×12 months in the index series are independent. This latter assumption is, of course, invalid, but it serves our pedagogical purposes at this point. Similarly, $Z_U = 0.884$. Finally, from (8.6) we obtain $(0.621, 0.708)$ as the 95% confidence interval for $\rho_{SST,SOI}$. This interval is almost symmetric about $\widehat{\rho}_{SST,SOI}$ because the sample size is large; it will be less symmetric for smaller samples. Note also that this confidence interval is probably too narrow because it does not account for dependence within the data.

An approximate test of H_0: $\rho_{XY} = 0$ can be performed by computing

$$\mathbf{T} = |\widehat{\rho}_{XY}|\sqrt{\frac{n-2}{1-\widehat{\rho}_{XY}^2}} \tag{8.7}$$

and comparing $\mathbf{T}$ with critical values from the t distribution with $n-2$ degrees of freedom (see Appendix F). The type of test, one sided or two sided, is determined by the form of the alternative hypothesis.

Confidence interval (8.6) and test (8.7) both require the normal assumption. A non-parametric approach based on ranks can be used when the observations are thought not to be normal. The sample $\{(\mathbf{X}_i, \mathbf{Y}_i): i = 1, \dots, n\}$ is replaced by the corresponding sample of ranks $\{(\mathbf{R}_{\mathbf{X}_i}, \mathbf{R}_{\mathbf{Y}_i}): i = 1, \dots, n\}$ where $\mathbf{R}_{\mathbf{X}_i}$ is the rank of $\mathbf{X}_i$ amongst the $\mathbf{X}$s and $\mathbf{R}_{\mathbf{Y}_i}$ is defined similarly.[3] The dependence between $\mathbf{X}$ and $\mathbf{Y}$ is then estimated with the *Spearman rank correlation coefficient* $\widehat{\rho}_{XY}^S$

$$\widehat{\rho}_{XY}^S = \frac{\sum_{i=1}^n \mathbf{R}_{X_i}\mathbf{R}_{Y_i} - N}{\sqrt{\left(\sum_{i=1}^n \mathbf{R}_{X_i}^2 - N\right)\left(\sum_{i=1}^n \mathbf{R}_{Y_i}^2 - N\right)}} \tag{8.8}$$

where

$$N = n\left(\frac{n+1}{2}\right)^2.$$

This is just the ordinary sample correlation coefficient[4] (8.4) of the ranks. Note that $-1 \le \widehat{\rho}_{XY}^S \le 1$, that $\widehat{\rho}_{XY}^S = +1$ when the rank orders of the two random variables are equal, and that $\widehat{\rho}_{XY}^S = -1$ when the two rank orders are the reverse of each other. Small sample critical values for testing H_0: $\rho_{XY} = 0$ with $\widehat{\rho}_{XY}^S$ are given in Appendix K. Approximate large sample (i.e., $n > 30$) critical values for testing H_0 against H_a: $\rho_{XY} \neq 0$ at the $(1-\tilde{p}) \times 100\%$ significance level are given by $\pm Z_{(1+\tilde{p})/2}\sqrt{n-1}$ where $Z_{(1+\tilde{p})/2}$ is the $((1+\tilde{p})/2)$-quantile of the standard normal distribution. Critical values for one-sided tests are obtained analogously.

In contrast to tests of the mean (see Section 6.6), inference about the correlation coefficient seems to be relatively weakly affected by serial correlation, at least when correlations are small [442]. A resampling scheme that further reduces the impact of serial correlation on inferences made about the correlation coefficient is described by [110].

[3] If there are ties, the tied observations are assigned the corresponding average rank.

[4] Also known as *Pearson's r*.

8.2.4 More Interpretations of Correlation. The correlation coefficient can also be interpreted as a measure of the proportion of the variance of one variable, say $\mathbf{Y}$, that can be represented by constructing a linear model of the dependence of the mean of $\mathbf{Y}$ upon $\mathbf{X}$. Assume that $(\mathbf{X}, \mathbf{Y})$ are bivariate normally distributed with joint density function $f_{XY}(x, y)$ given by (2.35). We factor $f_{XY}(x, y)$ into the product of the density function of $\mathbf{Y}$ conditional upon $\mathbf{X} = x$ and the marginal density function of $\mathbf{X}$ (see Sections 2.5 and 2.8) to obtain

$$\begin{aligned} f_{Y|X=x}(y|X=x) &= \frac{f_{XY}(x,y)}{f_X(x)} \\ &= \frac{\exp\left(-(y-\mu_{Y|X=x})^2/2\sigma_Y^2(1-\rho_{XY}^2)\right)}{\sqrt{2\pi\sigma_Y^2(1-\rho_{XY}^2)}} \end{aligned}$$

where

$$\mu_{Y|X=x} = \mu_Y - \rho_{XY}\frac{\sigma_Y}{\sigma_X}(\mu_X - x).$$

The variance of $\mathbf{Y}$ conditional upon $\mathbf{X} = x$ is $\sigma_Y^2(1-\rho_{XY}^2)$, the same factor discovered in [8.2.2] when we considered $\mathbf{X}$ as a predictor of $\mathbf{Y}$. The conditional variance does not depend upon the specific realized value of $\mathbf{X}$. The mean of $\mathbf{Y}$ varies linearly with the realized value of $\mathbf{X}$ when ρ_{XY} is nonzero. Note that the mean of one of the pair of variables is completely determined by the realized value of the other. The squared correlation coefficient only tells us the proportion of the variance of $\mathbf{Y}$ that is attributable to knowledge of the conditional mean.

Yet another way to view the relationship between $\mathbf{Y}$ and $\mathbf{X}$ is to write $\mathbf{Y}$ in the form

$$\mathbf{Y} = a_0 + a_1\mathbf{X} + \mathbf{E}, \tag{8.9}$$

where $\mathbf{E}$ is independent of $\mathbf{X}$. In geometrical terms, a realization of the pair $(\mathbf{X}, a_0 + a_1\mathbf{X})$ randomly selects a point on one of the axes of the ellipse depicted in Figure 2.10, and $\mathbf{Y}$ is subsequently determined by deviating vertically from the chosen point. By computing means and variances we obtain

$$\begin{aligned}\sigma_E^2 &= \sigma_Y^2(1-\rho_{XY}^2)\\ a_1 &= \rho_{XY}\frac{\sigma_Y}{\sigma_X}\\ a_0 &= \mu_Y - \rho_{XY}\frac{\sigma_Y}{\sigma_X}\mu_X.\end{aligned}$$

The purpose of regression analysis, discussed in the next section, is to diagnose relationships such as (8.9) between a response (or dependent) variable and one or more factors (or independent) variables. As the derivation above showed, the language used in many statistics textbooks can be misleading. If the factors that affect the mean of the response variable are determined externally to the studied system, either by an experimenter (as in a doubled CO_2 experiment conducted with a GCM) or by nature (e.g., by altering the climate's external forcing through the effects of volcanos), then words such as *dependent* and *independent* or *response* and *factor* can be used to describe relationships between variables. However, often in climatology both $\mathbf{X}$ and $\mathbf{Y}$ are responses of the climate system to some other unobserved factor. Then regression analysis can be used to document the relationship between the means of $\mathbf{X}$ and $\mathbf{Y}$, but it would be inappropriate to use language that implies causality.

8.3 Fitting and Diagnosing Simple Regression Models

Our purpose here is to describe the anatomy of a *simple linear regression* in which it is postulated that the conditional mean of a response variable $\mathbf{Y}$ depends linearly upon a random factor $\mathbf{X}$ (the arguments in the next few subsections work equally well if this factor is deterministic). Suppose that we have n pairs of observations $\{(\mathbf{x}_i, \mathbf{y}_i)\colon i = 1, \ldots, n\}$, each representing the realizations of a corresponding random variable pair $(\mathbf{X}_i, \mathbf{Y}_i)$, all pairs being independent and identically bivariate normally distributed.

8.3.1 Least Squares Estimate of a Simple Linear Regression. Assume that the conditional means satisfy

$$\mu_{\mathbf{Y}_i|\mathbf{X}=\mathbf{x}_i} = a_0 + a_1\mathbf{x}_i$$

so that conditional upon $\mathbf{X}_i = \mathbf{x}_i$, the ith response can be represented as a random variable $\mathbf{Y}_i$ such that

$$\mathbf{Y}_i = a_0 + a_1\mathbf{x}_i + \mathbf{E}_i. \tag{8.10}$$

Following on from the discussion in Section 8.2, the random variables $\mathbf{E}_i$ must be independent normal random variables with mean zero and variance

$$\sigma_E^2 = \sigma_Y^2(1-\rho_{XY}^2). \tag{8.11}$$

The corresponding representation for the realized value of $\mathbf{Y}_i$ is

$$\mathbf{y}_i = a_0 + a_1\mathbf{x}_i + \mathbf{e}_i,$$

where $\mathbf{e}_i$ represents the realized value of $\mathbf{E}_i$. If we have estimates $\widehat{a}_0$ and $\widehat{a}_1$ of the unknown coefficients a_0 and a_1, estimates of the realized errors (which are generally called *residuals*) are given by

$$\widehat{\mathbf{e}}_i = \mathbf{y}_i - \widehat{a}_0 - \widehat{a}_1\mathbf{x}_i. \tag{8.12}$$

A reasonable strategy for estimating a_0 and a_1 is to minimize some measure of the size of the estimated errors $\widehat{\mathbf{e}}_i$. While many metrics can be used, the sum of squared errors $\sum_{i=1}^n \widehat{\mathbf{e}}_i^2$ is the most common. The resulting estimators of a_0 and a_1 are called *least squares* estimators. We will see later that least squares estimators have some potential pitfalls that may not always make them the best choice. However, they are prominent in the normal setup because of the tractability of their distributional derivation, ease of interpretation, and optimality within this particular restricted parametric framework.

The least squares estimators of a_0 and a_1 are obtained as follows. The sum of squared errors is

$$\mathcal{SSE} = \sum_{i=1}^n (\mathbf{y}_i - \widehat{a}_0 - \widehat{a}_1\mathbf{x}_i)^2. \tag{8.13}$$

Taking partial derivatives with respect to the unknown parameters $\widehat{a}_0$ and $\widehat{a}_1$ and setting these to zero yields the normal equations

$$\sum_{i=1}^n (\mathbf{y}_i - \widehat{a}_0 - \widehat{a}_1\mathbf{x}_i) = 0 \tag{8.14}$$

$$\sum_{i=1}^n (\mathbf{y}_i - \widehat{a}_0 - \widehat{a}_1\mathbf{x}_i)\mathbf{x}_i = 0. \tag{8.15}$$

The normal equations have solutions

$$\widehat{a}_0 = \overline{\mathbf{y}} - \widehat{a}_1\overline{\mathbf{x}} \tag{8.16}$$

$$\widehat{a}_1 = \frac{\sum_{i=1}^n \mathbf{x}_i\mathbf{y}_i - n\overline{\mathbf{x}}\,\overline{\mathbf{y}}}{\sum_{i=1}^n \mathbf{x}_i^2 - n\overline{\mathbf{x}}^2}. \tag{8.17}$$

As will be shown in [8.3.20], an unbiased estimate of σ_E^2 (8.11) is given by

$$\widehat{\sigma}_E^2 = \frac{\mathcal{SSE}}{n-2}. \tag{8.18}$$

Returning to our SO example [8.1.3], the parameter estimates obtained using (8.16)–(8.18) are $\widehat{a}_0 = -0.09$, $\widehat{a}_1 = 0.15$, and $\widehat{\sigma}_E = 12.2$. The fitted line is shown as the upwards sloping line that passes through the cloud of points in Figure 8.1, and $\widehat{\sigma}_E$ is an estimate of the standard deviation of the vertical scatter about the fitted line. Note that the eye is not always a good judge of where the least squares line should be placed; our initial impression of Figure 8.1 is that the slope of the fitted line is not steep enough.

8.3.2 Partitioning Variance. The slope estimate (8.17) is often written

$$\widehat{a}_1 = \frac{S_{XY}}{S_{XX}},$$

where

$$\begin{aligned} S_{XY} &= \sum_{i=1}^{n}(\mathbf{x}_i - \overline{\mathbf{x}})(\mathbf{y}_i - \overline{\mathbf{y}}) \\ &= \sum_{i=1}^{n}\mathbf{x}_i\mathbf{y}_i - n\overline{\mathbf{x}}\,\overline{\mathbf{y}} \end{aligned}$$

and

$$\begin{aligned} S_{XX} &= \sum_{i=1}^{n}(\mathbf{x}_i - \overline{\mathbf{x}})^2 \\ &= \sum_{i=1}^{n}\mathbf{x}_i^2 - n\overline{\mathbf{x}}^2. \end{aligned}$$

The sum of squared errors can be expressed similarly as

$$\mathcal{SSE} = S_{YY} - \widehat{a}_1 S_{XY},$$

where

$$\begin{aligned} S_{YY} &= \sum_{i=1}^{n}(\mathbf{y}_i - \overline{\mathbf{y}})^2 \\ &= \sum_{i=1}^{n}\mathbf{y}_i^2 - n\overline{\mathbf{y}}^2. \end{aligned}$$

S_{YY} is often called the *total sum of squares* and denoted $\mathcal{SST}$. Be aware of the potential confusion here between the common climatological practice of referring to sea-surface temperature as SST and the equally common statistical practice of referring to the total sum of squares as $\mathcal{SST}$. The quantity

$$\mathcal{SSR} = \widehat{a}_1 S_{XY}$$

is often called the *sum of squares due to regression* and denoted $\mathcal{SSR}$. It is easily verified that $\mathcal{SSR} = \sum_{i=1}^{n}(\widehat{a}_0 + \widehat{a}_1\mathbf{x}_i - \overline{\mathbf{y}})^2$. The least squares fitting process thus provides a partition of the total variability into a component that is attributed to the fitted line ($\mathcal{SSR}$) and a component that is due to departures from that line ($\mathcal{SSE}$). That is,

$$\mathcal{SST} = \mathcal{SSR} + \mathcal{SSE}. \tag{8.19}$$

In the SOI example, this partitioning of the total sum of squares is

Source	Sum of squares
Regression ($\mathcal{SSR}$)	74 463.2
Error ($\mathcal{SSE}$)	92 738.1
Total ($\mathcal{SST}$)	167 201.3

8.3.3 Coefficient of Multiple Determination. An immediately available diagnostic of the ability of the fitted line to explain variation in the data is the *coefficient of multiple determination*, denoted R^2, given by

$$R^2 = \mathcal{SSR}/\mathcal{SST}. \tag{8.20}$$

The use of the phrase *coefficient of determination* to describe this number seems natural enough because it is a measure of the extent to which $\mathbf{X}$ determines $\mathbf{Y}$. The adjective *multiple* is added because in multiple regression (Section 8.4) this number is a measure of the extent to which all variables on the right hand side of the regression equation determine $\mathbf{Y}$. While a useful diagnostic, it is just one of several tools which should be used to assess the utility and goodness-of-fit of a model. R^2 is discussed further in [8.3.12]. Additional diagnostic tools are discussed in [8.3.13,14,16,18] and [8.4.11].

In our SOI example, $R^2 = 0.445$, meaning that somewhat less then one-half of the total variability in the SO index is represented by the SST index. This is clearly in agreement with Figure 8.1 where we see quite a bit of scatter about the fitted line.

8.3.4 The Relationship Between Least Squares and Maximum Likelihood Estimators. When the random variables $\mathbf{E}_i$ (8.10) are independent and identically normally distributed, it is easy to demonstrate that the least squares estimators are also maximum likelihood estimators. Under these conditions, the log-likelihood function $l(a_0, a_1|\mathbf{x}_i, \mathbf{y}_i)$, for $i = 1, \ldots, n$, is given by

$$\begin{aligned} -2l(a_0, a_1|\mathbf{x}_i, \mathbf{y}_i) &= n\log(2\pi\sigma_E^2) \\ &\quad + \frac{1}{\sigma_E^2}\sum_{i=1}^{n}(\mathbf{y}_i - a_0 - a_1\mathbf{x}_i)^2. \end{aligned}$$

The likelihood estimators are chosen to maximize the likelihood, or equivalently the log-likelihood, of the estimated errors $\mathbf{y}_i - a_0 - a_1\mathbf{x}_i$. Maximizing the log-likelihood with respect to a_0 and a_1 results in precisely the least squares estimators. This means that least squares estimators have the optimality properties of maximum likelihood estimators (Section 5.3) when the normal distributional assumption is satisfied.

8.3.5 Properties. While the estimators (8.16), (8.17), and (8.18) have been written in their realized forms, they can also be considered as random variables whose distribution is conditional on the realized values of **X**. We will briefly state the distributional properties of these estimators. The derivation of these properties is discussed in [8.3.20].

1 $\widehat{a}_0, \widehat{a}_1$, and $\widehat{\sigma}_E^2$ are unbiased estimators of a_0, a_1, and σ_E^2 respectively.

2 $\widehat{\sigma}_E^2$ is independent of $\widehat{a}_0$ and $\widehat{a}_1$.

3 $(n-2)\widehat{\sigma}_E^2/\sigma_E^2 \sim \chi^2(n-2)$.

4 $\widehat{a}_1 \sim \mathcal{N}(a_1, (\sigma_E^2/S_{XX})^2)$.

5 $\widehat{a}_0 \sim \mathcal{N}(a_0, (\sigma_E^2 \sum_{i=1}^n \mathbf{x}_i^2/(nS_{XX}))^2)$.

8.3.6 Inferential Methods. The distributional properties stated above provide a number of inferential results that are useful for interpreting a fitted regression model. Bear in mind, however, that inferences made in the following way may be compromised if the assumptions embedded in the procedures are violated. See [8.3.17] for more discussion about this.

8.3.7 A Confidence Interval for the Slope Parameter. A $\tilde{p} \times 100\%$ confidence interval for the slope of the regression line, a_1, is given by

$$\left(\widehat{a}_1 - \frac{t_{(1+\tilde{p})/2}\,\widehat{\sigma}_E}{\sqrt{S_{XX}}},\ \widehat{a}_1 + \frac{t_{(1+\tilde{p})/2}\,\widehat{\sigma}_E}{\sqrt{S_{XX}}}\right),$$

where $t_{(1+\tilde{p})/2}$ is the $((1+\tilde{p})/2)$-quantile of the t distribution with $n-2$ degrees of freedom (see Appendix F).

In our SOI example $n - 2 = 622$, $S_{XX} = 3.320 \times 10^6$ and $\widehat{\sigma}_E = 12.2$. Therefore, assuming that there is no dependence between observations (an assumption we know to be false), the 95% confidence interval for the slope of the fitted line is (0.137, 0.163). However, dependence between observations causes the actual 95% confidence interval for a_1 to be wider.

8.3.8 Tests of the Slope Parameter. The null hypothesis that a_1 has a particular value, say a_1^*, can be tested by comparing

$$\mathbf{T} = \frac{\widehat{a}_1 - a_1^*}{\widehat{\sigma}_E/\sqrt{S_{XX}}}$$

against critical values from the t distribution with $n-2$ degrees of freedom. It is often of interest to know whether or not a_1 is significantly different from zero, that is, whether or not there is a regression relationship between **X** and **Y**.

To test H_0: $a_1 = 0$ against H_a: $a_1 \neq 0$ in our SOI example, we compute

$$\begin{aligned} \mathbf{t} &= \frac{\widehat{a}_1}{\widehat{\sigma}_E/\sqrt{S_{XX}}} \\ &= \frac{0.15}{12.2/\sqrt{3.320 \times 10^6}} = 22.4\,. \end{aligned}$$

This realized value of **T** is compared with critical values from $t(622)$ and is found to be significant at much less than the 0.1% level. The effect of dependence between observations is, generally, to increase the frequency with which the null hypothesis is rejected when it is true, that is, to decrease the apparent significance level. Here it is certain that H_0 is false, but often when the evidence is more equivocal, it is important to consider the effects of dependence (see Section 6.6).

Another approach to testing whether or not a regression relationship exists is based on the observation that, when $a_1 = 0$, the regression sum of squares $\mathcal{SSR}$ is an unbiased estimator of the error variance which is distributed $\chi^2(1)$ and is independent of $\widehat{\sigma}_E^2$. (These results can be proved using methods similar to that in [8.3.20].) Since $(n-2)\widehat{\sigma}_E^2/\sigma_E^2$ is distributed $\chi^2(n-2)$, we obtain that

$$\mathbf{F} = \frac{\mathcal{SSR}}{\widehat{\sigma}_E^2} \sim F(1, n-2)$$

under the null hypothesis. Thus the test can be conducted by comparing **F** with critical values from Appendix G.

Because we have fitted a linear model that depends upon only one factor, the t and F tests are equivalent. In fact, $\mathbf{F} = \mathbf{T}^2$, and the square of a t random variable with $n-2$ df is distributed as $F(1, n-2)$. Thus identical decisions are made provided that the t test is conducted as a two-sided test.

8.3.9 Inferences About the Intercept. A $\tilde{p} \times 100\%$ confidence interval for the intercept of the regression line, a_0, has bounds given by

$$\widehat{a}_0 \pm \frac{t_{(1+\tilde{p})/2}\widehat{\sigma}_E\sqrt{\sum_{i=1}^n \mathbf{x}_i^2}}{\sqrt{S_{XX}}}$$

The null hypothesis that the intercept has a particular value, say a_0^*, can be tested by comparing

$$\mathbf{T} = \frac{\widehat{a}_0 - a_0^*}{\widehat{\sigma}_E\sqrt{\sum_{i=1}^n \mathbf{x}_i^2/nS_{XX}}}$$

also against critical values from the t distribution with $n-2$ degrees of freedom (see Appendix F). Setting $a_0^* = 0$ determines whether or not the fitted line passes through the origin.

A test of the intercept is only of pedagogical interest in the SOI example because both the SO and SST indices are expressed as departures from arbitrarily selected base period means. None the less, to test H_0: $a_0 = 0$ against H_a: $a_0 \neq 0$ we compute

$$\begin{aligned}\mathbf{t} &= \frac{\widehat{a}_0}{\widehat{\sigma}_E\sqrt{\sum_{i=1}^n \mathbf{x}_i^2/nS_{XX}}}\\ &= \frac{-0.09}{12.2\sqrt{\frac{3.322\times10^6}{624\times3.320\times10^6}}} = -0.184.\end{aligned}$$

When this value of $\mathbf{T}$ is compared with critical values of $t(622)$ we see that it is not significantly different from zero. Accounting for dependence further reduces the amplitude of $\mathbf{t}$ and therefore does not affect our inference about a_0.

8.3.10 A Confidence Interval for the Mean of the Response Variable. The conditional mean $\mu_{Y|X=x}$ of the response variable $\mathbf{Y}$ for a realization $\mathbf{x}$ of $\mathbf{X}$ is estimated from the fitted regression equation as

$$\widehat{\mu}_{Y|X=x} = \widehat{a}_0 + \widehat{a}_1\mathbf{x}.$$

By substituting for $\widehat{a}_0$ with (8.16) we obtain

$$\widehat{\mu}_{Y|X=x} = \overline{\mathbf{y}} + \widehat{a}_1(\mathbf{x} - \overline{\mathbf{x}}). \tag{8.21}$$

Computing variances, we see that

$$\sigma^2_{\widehat{\mu}_{Y|X=x}} = \sigma_E^2\Big(\frac{1}{n} + \frac{(\mathbf{x}-\overline{\mathbf{x}})^2}{S_{XX}}\Big). \tag{8.22}$$

This can be derived by first substituting (8.17) for $\widehat{a}_1$ in (8.21), then substituting the model (8.10) wherever $\mathbf{Y}_i$ appears, and finally computing

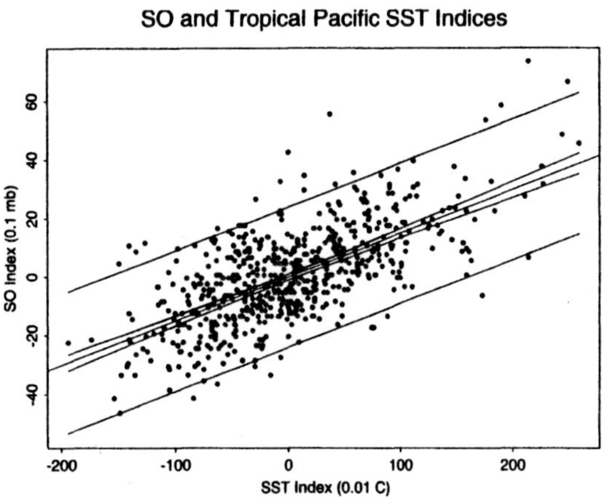

Figure 8.5: *A simple linear regression fitted by ordinary least squares to 1933–84 monthly mean SO and SST indices (see [8.1.3]). The pair of curved lines closest to the regression line indicate, at each point* $\mathbf{x}$*, the upper and lower 95% confidence bounds for the mean of the response variable* $\mu_{Y|X=x}$ *conditional upon* $\mathbf{X} = \mathbf{x}$ *(see [8.3.10]). The pair of more widely curved lines indicates, at each point* $\mathbf{x}$*, the upper and lower 95% confidence bound for the response variable* $\mathbf{Y}$ *conditional upon* $\mathbf{X} = \mathbf{x}$ *(see [8.3.11]).*

the variance of the resulting expression. A corresponding estimate is obtained by substituting $\widehat{\sigma}_E^2$ for σ_E^2. Now note that the estimate is proportional to a $\chi^2(n-2)$ random variable and that it is independent of $\widehat{\mu}_{Y|X=x}$, which is normally distributed. Taking care to scale the normally distributed and χ^2 components correctly, we finally obtain that

$$\mathbf{T} = \frac{\widehat{\mu}_{Y|X=x} - \mu_{Y|X=x}}{\widehat{\sigma}_E\sqrt{\frac{1}{n} + \frac{(\mathbf{x}-\overline{\mathbf{x}})^2}{S_{XX}}}}$$

is distributed $t(n-2)$. Thus a $\tilde{p} \times 100\%$ confidence interval for the conditional mean at $\mathbf{x}$ has bounds

$$\widehat{\mu}_{Y|X=x} \pm t_{(1+\tilde{p})/2}\widehat{\sigma}_E\sqrt{\frac{1}{n} + \frac{(\mathbf{x}-\overline{\mathbf{x}})^2}{S_{XX}}}, \tag{8.23}$$

where $t_{(1+\tilde{p}/2)}$ is the $((1+\tilde{p})/2)$-quantile of $t(n-2)$ (Appendix F).

An example of a fitted regression line and the confidence bound curves defined by (8.23) is illustrated in Figure 8.5. The pair of curves closest to the regression line illustrates a *separate* 95% confidence interval at each $\mathbf{x}$. The curves bound the vertical interval at each $\mathbf{x}$ that covers the regression line 95% of the time on average. As mentioned

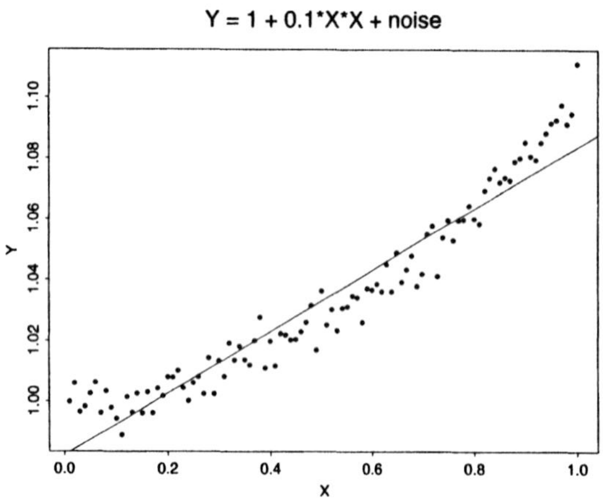

Figure 8.6: *This diagram illustrates the least squares fit of a straight line to a sample of 100 observations generated from the model* $\mathbf{Y} = 1 + 0.1\mathbf{x}^2 + \mathbf{E}$ *where* $\mathbf{E} \sim \mathcal{N}(0, 0.005^2)$. *Even though* $R^2 = 0.92$, *the model fits the data poorly.*

previously, accounting for dependence would increase the distance between the confidence bound curves.

8.3.11 A Confidence Interval for the Response Variable. While confidence interval (8.23) accounts for uncertainty in our estimate of the conditional mean, it does not indicate the range of values of the response variable that is likely for a given value $\mathbf{x}$ of $\mathbf{X}$. To solve this problem we need to interpret the fitted regression equation, when evaluated at $\mathbf{x}$, as an estimate of $\mathbf{Y}$ rather than as an estimate of the conditional mean $\mu_{Y|X=x}$. The estimation (or specification) error in this context is

$$\mu_{Y|X=x} + \mathbf{E} - \widehat{\mu}_{Y|X=x}.$$

Since $\mathbf{E}$ is independent of $\widehat{\mu}_{Y|X=x}$, we see using (8.22) that the variance of the estimation error is

$$\sigma_E^2\left(1 + \frac{1}{n} + \frac{(\mathbf{x} - \overline{\mathbf{x}})^2}{S_{XX}}\right).$$

Then, replacing σ_E^2 with the estimator $\widehat{\sigma}_E^2$ we obtain the confidence interval for $\mathbf{Y}$ with bounds

$$\widehat{\mu}_{Y|X=x} \pm t_{(1+\tilde{p})/2}\,\widehat{\sigma}_E\sqrt{1 + \frac{1}{n} + \frac{(\mathbf{x} - \overline{\mathbf{x}})^2}{S_{XX}}}$$

where $t_{(1+\tilde{p})/2}$ is the $((1 + \tilde{p})/2)$-quantile of $t(n-2)$ (Appendix F).

The wider pair of curves in Figure 8.5 (they really are very shallow hyperbolas) illustrates the confidence bounds for the response variable (the SO index) in our SOI example. Again, the exact interpretation here hinges upon the independence of observations. However, dependence has a relatively minor effect on this particular inference because the regression line itself is well estimated; only the sampling variability of the regression line is affected by dependence. Note also that in this case the curves *do not* bound the region that will simultaneously cover 95% of all possible values of the response variable.

8.3.12 Diagnostics: R^2. The inferential methods described above are based on the assumptions that the conditional mean of $\mathbf{Y}$ given $\mathbf{X} = \mathbf{x}$ is a linear function of $\mathbf{x}$ and that the errors $\mathbf{E}_i$ in model (8.10) are iid normal.

We have already seen one diagnostic (8.20)

$$R^2 = \mathcal{SSR}/\mathcal{SST}$$

associated with a fitted model. However, R^2, the proportion of variance in the response variable that is explained by the fitted model, should not be confused with the model's *goodness-of-fit*. The correct interpretation of R^2 is that it is an estimate of the model's ability to specify unrealized values of the response variable $\mathbf{Y}$.

A large R^2 does not indicate that the model fits well in a statistical sense (i.e., that inferences made with the methods above are reliable). Figure 8.6 illustrates the least squares fit of a linear regression model to data that closely approximate a quadratic. The R^2 for this fit is large ($R^2 = 0.92$) but it would not be correct to say that the fit is a good one because the deviations from the fitted line display systematic behaviour. In this case the assumption that the errors are iid normal is not satisfied and thus inferences are not likely to be reliable.

Neither does a small R^2 indicate that the model fits poorly. Figure 8.7 illustrates a least squares fit of a linear regression model to simulated data from a linear model. The R^2 for this fit is only moderately large ($R^2 = 0.51$) but the deviations from the fitted line do not show any kind of systematic behaviour. It is likely that inferences made in this case will be reliable even though the model's ability to specify $\mathbf{Y}$ from given values of $\mathbf{X}$ is low. Despite the relatively low R^2, the fitted regression line $\widehat{\mu}_{Y|X=x} = 1.0047 + 0.0972\mathbf{x}$ estimates the true conditional mean $\mu_{Y|X=x} = 1 + 0.1\mathbf{x}$ very well.

While R^2 summarizes well the extent to which the fitted line specifies the realized values $\mathbf{y}_i$ of $\mathbf{Y}$ given the corresponding values $\mathbf{x}_i$ of $\mathbf{X}$ for $i = 1, \ldots, n$, it is well recognized

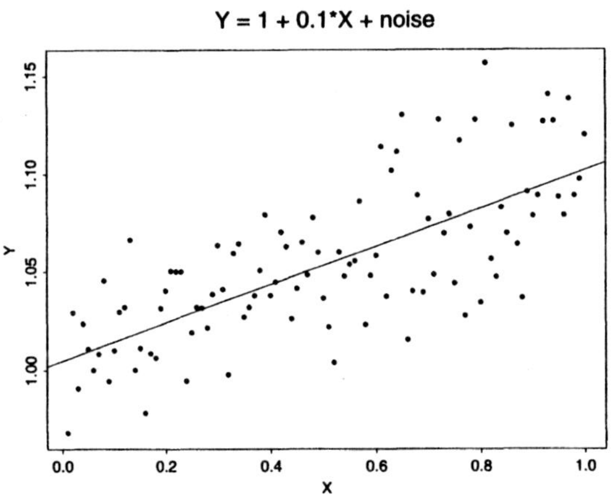

Figure 8.7: *This diagram illustrates the least squares fit of a straight line to a sample of 100 observations generated from the model* $\mathbf{Y} = 1 + 0.1\mathbf{x} + \mathbf{E}$ *where* $\mathbf{E} \sim \mathcal{N}(0, 0.025^2)$. *Even though* $R^2 = 51\%$, *the model fits the data well.*

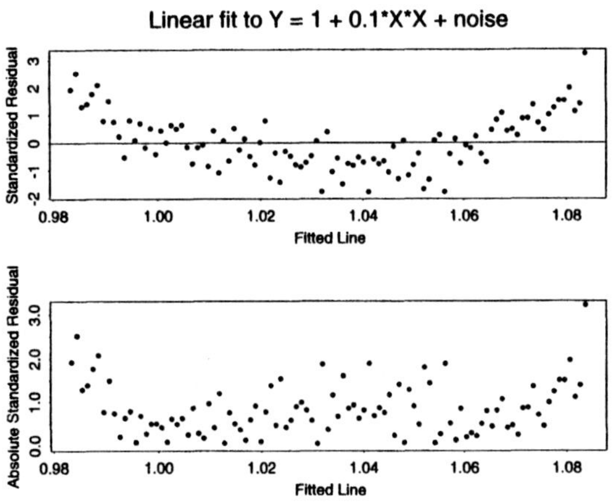

Figure 8.8: *In the upper panel the standardized residuals (departures from the fitted line divided by* $\widehat{\sigma}_E$*) are plotted as a function of the estimated conditional mean* $\widehat{\mu}_{Y|X=x}$ *for the fit displayed in Figure 8.6. The absolute values of the residuals are plotted in the lower panel.*

that R^2 is an optimistic indicator of model specification performance for unrealized values of $\mathbf{X}$ (see, e.g., Davis [101]). Climatologists and meteorologists call this phenomenon *artificial skill*. The artificial skill arises because the fitted model, as a consequence of the fitting process, has adapted itself to the data. Cross-validation (see Section 18.5) provides a more reliable means of predicting future model performance.

8.3.13 Diagnostics: Using Scatter Plots. Some fundamental tools in model diagnostics include scatter plots of the standardized residuals $\widehat{\mathbf{e}}_i/\widehat{\sigma}_E$ (see (8.18)) against the corresponding estimates of the conditional mean (8.21), and scatter plots of the absolute standardized residuals against the estimates of the conditional mean.

Figure 8.6 illustrates a violation the assumption that the conditional mean varies linearly with $\mathbf{x}$. This is revealed through systematic behaviour in standardized residuals, as displayed in Figure 8.8. This type of behaviour is generally easier to detect in displays of the standardized residuals (upper panel of Figure 8.8) than in displays of the absolute standardized residuals (lower panel of Figure 8.8). Other kinds of departures from the fitted model are easier to detect in displays of the absolute standardized residuals.

Figure 8.9 illustrates an example in which the assumption that the errors $\mathbf{E}_i$ all have common variance is violated. This is known as *heteroscedasticity*. In this case error variance appears to increase until $\mathbf{x} = 0.5$ and then decrease again beyond $\mathbf{x} = 0.5$. Heteroscedasticity is generally easier to detect in scatter plots of the absolute residuals. Heteroscedastic errors can sometimes be dealt with by transforming the data before fitting a regression model [8.6.2]. Other times it may be necessary to use *weighted* regression techniques in which the influence of a squared error in determining the fit is inversely proportional to its variance (see Section 8.6 and [104]).

Finally, Figure 8.10 results from a simulated linear regression with two inserted errant observations. Attempts to detect these observations are made by looking for *outliers*, that is, residuals that are greater in absolute value than the rest. As a general rule, residuals more than three standard deviations from the fitted line should be examined for errors in the corresponding observations of the response and factor variables. Outliers are generally easier to detect using the plot of the absolute residuals. However, they may not always be easy to detect, especially when more than one outlier is present in a sample. In this example, the data were generated using the model $\mathbf{Y} = 1 + 0.1\mathbf{x} + \mathbf{E}$, where $\mathbf{E}$ is normally distributed noise with mean zero and standard deviation 0.05, and $\mathbf{x}$ varies between 0 and 1. The error at $\mathbf{x} = 0.5$ was set to be 0.15 (3 standard deviations) and the error at $\mathbf{x} = 0.95$ was set to be -0.15. The outlier at $\mathbf{x} = 0.5$ is detected in our residual display, but

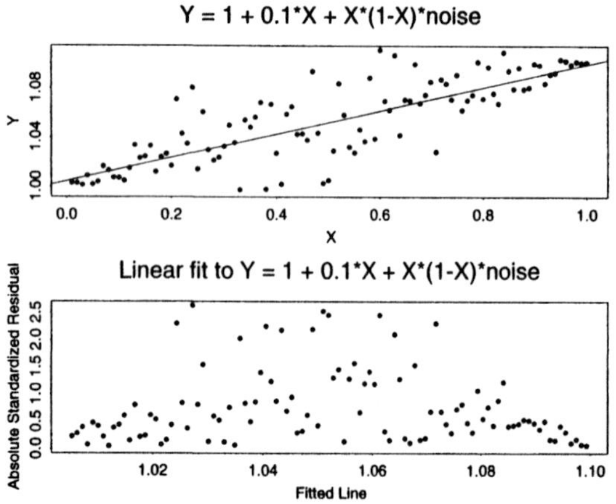

Figure 8.9: *A pair of scatter plots illustrating heteroscedasticity. The data were generated from* $\mathbf{Y} = 1 + 0.1\mathbf{x} + \mathbf{x}(1-\mathbf{x})\mathbf{E}$, *where* $\mathbf{E} \sim \mathcal{N}(0, 0.1^2)$. *The upper panel shows 100 simulated data points and the line fitted by least squares. The lower panel displays the absolute standardized residuals as a function of the fitted line.*

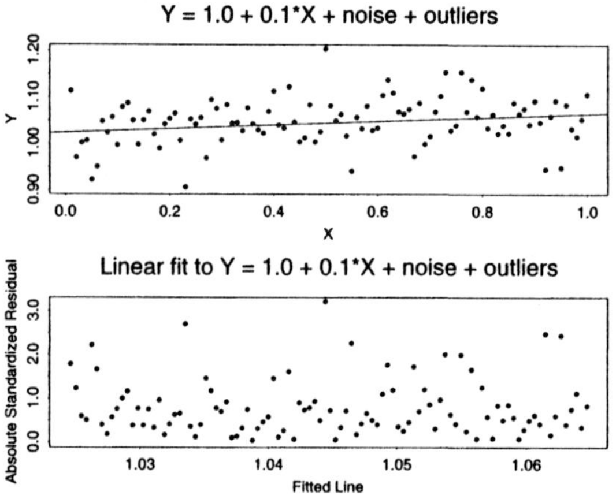

Figure 8.10: *A scatter plot illustrating data generated from* $\mathbf{Y} = 1 + 0.1\mathbf{x} + \mathbf{E}$ *where* $\mathbf{E} \sim \mathcal{N}(0, 0.05^2)$. Two outliers *have been inserted by setting the realizations of* $\mathbf{E}$ *at* $\mathbf{x} = 0.5$ *and* $\mathbf{x} = 0.95$ *to 0.15 and* -0.15 *respectively.*

that at $\mathbf{x} = 0.95$ is hidden, for reasons discussed in [8.3.18].

Studentized residuals, rather than standardized residuals, are often used in diagnostic plots. A studentized residual is obtained at point $\mathbf{x}_i$ by fitting the regression model without the data pair $(\mathbf{x}_i, \mathbf{y}_i)$, computing the difference between $\mathbf{y}_i$ and the estimate obtained from the fit, and finally dividing this deviation by the estimate of the standard error obtained from the fit. Outliers hidden in ordinary residual plots often become apparent in plots of studentized residuals because they do not affect the fit of the model used to estimate the studentized residual. Unfortunately, studentized residuals fail to identify the hidden outlier in Figure 8.10.

Diagnostic scatter plots of the residuals from the fitted regression of the SO index on Wright's SST index are displayed in Figure 8.11. No evidence of heteroscedasticity or systematic departure from the fitted line is apparent. However, three outliers can be observed, all of which are positive. Only one deviation (occurring in February 1983) corresponds to a known El Niño warm event.

8.3.14 Diagnostics: Probability Plots. As will be discussed in [8.3.15], skewness of the residuals (e.g., a tendency for there to be more residuals of one sign than another) should not immediately lead to the conclusion that all inferences about the regression model are invalid. None the less, once it has been determined that the model fits the data reasonably well, it is still useful to examine the residuals to see if there are gross departures from the normal distribution assumption, which might compromise the inferences. A useful diagnostic for this purpose is a *normal probability plot*[5] of the (ordered) standardized residuals $\widehat{\mathbf{e}}_{(i|n)}/\widehat{\sigma}_E$ against the $((i-0.5)/n)$-quantiles of the standard normal distribution.

As discussed in [3.1.3] and [4.2.2], such plots are constructed by plotting the points

$$\left(F_{\mathcal{N}}^{-1}\Big(\frac{i-0.5}{n}\Big), \frac{\widehat{\mathbf{e}}_{(i|n)}}{\widehat{\sigma}_E}\right) \quad \text{for } i = 1, \ldots, n.$$

The points will lie on an approximately straight line sloping upwards at a 45° angle when the residuals are approximately normal with variance $\widehat{\sigma}_E^2$.

The probability plot for our SOI example is shown in Figure 8.12. We see that the central body of the distribution is very close to normal. The diagram shows that the left hand tail of the distribution is slightly narrower than that of a normal distribution and the right hand tail is slightly wider. The three outliers we identified previously can be seen at the upper right hand

[5]Sometimes also called *qq plots*, or *quantile–quantile plots*.

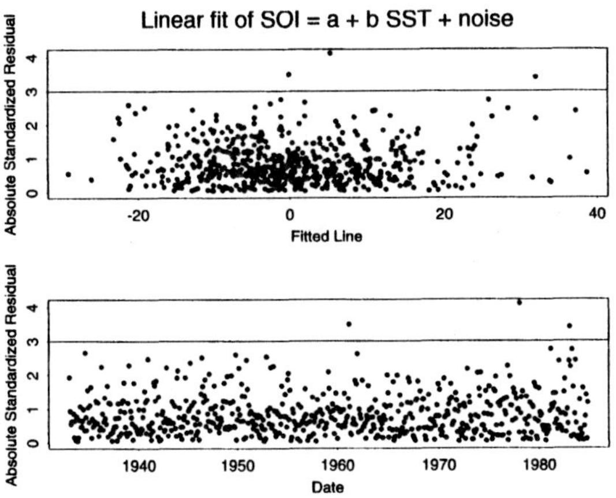

Figure 8.11: *Scatter plots illustrating the fit of the regression of the SO index on the SST index. This example is introduced in [8.1.3]. Three outliers, occurring in March 1961, February 1978 and February 1983 can be identified. In the upper panel the absolute standardized residuals are plotted against the estimated conditional mean. They are plotted against time in the lower panel.*

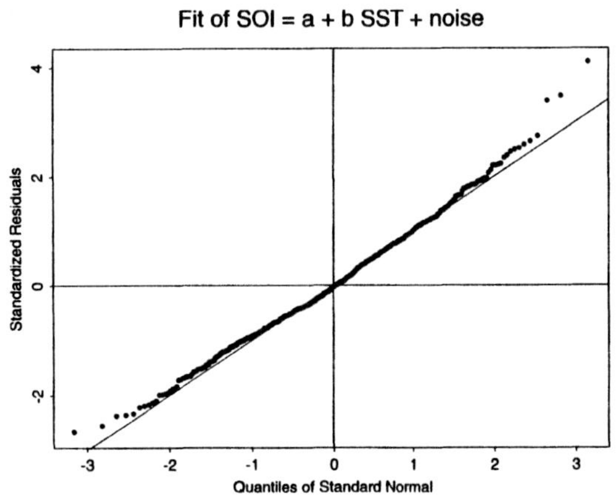

Figure 8.12: *A probability plot of the standardized quantiles of the residuals from the regression of the SO index on the SST index, against the quantiles of the standard normal distribution.*

corner of the graph. In general, these residuals are acceptably close to being normally distributed.

8.3.15 Why Use Least Squares? While we have, on occasion, warned that inferences made with least squares estimators may not be robust, their widespread use is justified for more reasons than just computational ease and the tractability of inference when errors are independent and normally distributed.

- As a consequence of the Gauss–Markov Theorem (see [147, p. 219], or [197, p. 301]), least squares estimators of linear model parameters have minimum variance amongst all unbiased linear estimators as long as the errors are independent and identically distributed with zero mean and constant finite variance. This is a relatively strong reason to use least squares estimators, despite the insistence that the estimators be *linear* (i.e., that they be expressible as linear combinations of the response variables $\mathbf{Y}_i$), because our ability to construct nonlinear estimators is limited. This property of least squares estimators does not persist if errors do not have constant variance (see, e.g., Section 8.6, and [62, pp. 352–353]).
- When the errors $\mathbf{E}_i$ are elements of a stationary time series, the least squares estimators are still, under relatively broad conditions, asymptotically the best (i.e., minimum variance) linear unbiased estimators of the regression parameters (see [323, pp. 588–595]).

However, be aware that even minor departures from the normal distribution assumption can have a detrimental effect on inferences made about the error variance.

8.3.16 Diagnostics: Serial Correlation. While the last item above reassures us that least squares estimators can be consistent when errors are dependent, it says nothing about the reliability of inferences under dependence. Unfortunately, the inference procedures outlined above are very sensitive to departures from independence (see Section 6.6; [62, p. 375]; and also [363], [442], [454]).

The Durbin–Watson statistic (see [104], [107], [108], and [109]), computed as

$$d = \frac{\sum_{i=1}^{n-1}(\widehat{\mathbf{e}}_{i+1} - \widehat{\mathbf{e}}_i)^2}{\mathcal{SSE}}, \tag{8.24}$$

is commonly used to detect serial correlation. When errors have positive serial correlation, the differences $(\widehat{\mathbf{e}}_{i+1} - \widehat{\mathbf{e}}_i)^2$ tend to be small compared with those when errors are independent. Therefore small values of d (near zero) indicate positive serial correlation. When errors are independent, we see from (8.24) that

$$\begin{aligned} d &= \frac{\sum_{i=1}^{n-1}(\widehat{\mathbf{e}}_{i+1} - \widehat{\mathbf{e}}_i)^2}{\mathcal{SSE}} \\ &= \frac{\sum_{i=1}^{n-1}(\widehat{\mathbf{e}}_{i+1}^2 + \widehat{\mathbf{e}}_i^2 - 2\widehat{\mathbf{e}}_{i+1}\widehat{\mathbf{e}}_i)}{\mathcal{SSE}} \\ &\approx \frac{\mathcal{SSE} + \mathcal{SSE} - 0}{\mathcal{SSE}} = 2. \end{aligned}$$

Hence values of d near 2 are consistent with independent errors. If the alternative hypothesis is that the errors are negatively (rather than positively) correlated, then the test statistic should be $4 - d$.

Computation of the significance of the observed d under the null hypothesis of independence is somewhat involved. Durbin and Watson give a range of critical values for samples of size $n \leq 100$. The tabulated critical values consist of pairs d_L and d_U such that H_0 can always be rejected if $\max(d,\ 4-d) < d_L$ and H_0 should not be rejected if $\min(d,\ 4-d) > d_H$. Between these limits, the determination of whether or not d is significantly different from 2 depends on the specific values $\mathbf{x}_i$, for $i = 1, \ldots, n$, taken by the independent variable. Durbin and Watson [108, 109] describe an approximation to the distribution of d based on the beta distribution that can be used with moderate to large sample sizes when the test based on the tabulated values is inconclusive or when the sample is large.

A 'rough-and-ready' approach that can be used when the samples are large is based on the observation that $d = 2(1 - \widehat{\rho}_{\hat{\epsilon}\hat{\epsilon}}(1))$, where $\widehat{\rho}_{\hat{\epsilon}\hat{\epsilon}}(1)$ is the estimated lag-1 correlation coefficient of the residuals. An approximate test can therefore be performed by comparing $\widehat{\rho}_{\hat{\epsilon}\hat{\epsilon}}(1)/\sqrt{n}$ with critical values from the standard normal distribution (Appendix D). If the null hypothesis can not be rejected with this test, then it will also not be rejected with d. On the other hand, if H_0 is rejected with this test, Durbin and Watson's approximation [108, 109] should be used to confirm that this decision will stand when the details of the independent variable (i.e., the values $\mathbf{x}_i$) are taken into account.

The value of the Durbin–Watson statistic in our SOI example is 2.057, which means that $\widehat{\rho}_{\hat{\epsilon}\hat{\epsilon}}(1) = -0.0285$. This value is not significantly different from zero.

Another approach to testing for serial correlation in the residuals is to perform a *runs test* (see, e.g., Draper and Smith [104] or Lehmann and D'Abrera [249]) to determine whether the residuals change sign less frequently (i.e., there is positive serial correlation) or more frequently (negative serial correlation) than would be expected in a sequence of independent errors. The test statistic used in the runs test, denoted $\mathbf{U}$, is the number of sign changes plus 1. Draper and Smith [104, pp. 160–161] give tabulated critical values when the number of residuals of both signs is small (≤ 10). A normal approximation can be used when samples are large. It can be shown that the mean and variance of $\mathbf{U}$ under H_0 are

$$\begin{aligned} \mu_U &= \frac{2n_1n_2}{n_1+n_2} + 1 \\ \sigma_U^2 &= \frac{2n_1n_2(2n_1n_2 - n_1 - n_2)}{(n_1+n_2)^2(n_1+n_2-1)}, \end{aligned}$$

where n_1 and n_2 are the number of positive and negative residuals. Then H_0: *no serial correlation* can be tested against H_a: *positive serial correlation* by comparing $(\mathbf{U} - \mu_U + \frac{1}{2})/\sigma_U$ against the lower tail critical values of the standard normal distribution (Appendix D). Here we are approximating a discrete distribution with a continuous distribution; so the half that is added is a *continuity correction* that accounts for this. For our SOI example, we have $n_1 = 295$ and $n_2 = 329$ so that $\mu_U = 312.17$ and $\sigma_U = 12.44$. We observe $\mathbf{u} = 307$, a value that is not significantly different from μ_U.

8.3.17 Are Least Squares Estimators Robust?

To understand the influence outliers have on least squares estimates, think about the sample mean. A positive outlier will increase the sample mean in direct proportion to the size of the outlier. In fact, there is no upper limit on the effect that can be induced on the sample mean by an outlier. On the other hand, the effect of an outlier on the sample median is bounded; once the outlier becomes the largest observation in the sample it has no further influence on the median. Thus the sample median and mean are examples of estimators that are robust and not robust, respectively.

Least squares estimators are not robust to the effects of outlying observations. Other fitting methods (see [8.3.18]), such as robust M-estimation (see, e.g., [154]) can be used, but at the expense of computer time (perhaps not such an issue these days), some loss of the rich body of inferential methods available for least squares estimators, and some loss of efficiency when errors are actually iid normally distributed.

8.3.18 Influence and Leverage: the Effects of Outliers. In regression analysis, the effect of an outlying realization of $\mathbf{Y}$ is also influenced by the value of $\mathbf{X}$. One can think of the regression line as a bar balanced on a pivot point at $(\overline{\mathbf{x}}, \overline{\mathbf{y}})$. An outlier directly above (or below) the pivot point pulls the bar up (or down) and has a relatively small effect on the fitted conditional mean. An outlier near the end of the bar has a very large influence on the fitted line.

Suppose an outlying point $(\mathbf{x}, \mathbf{y})$ is located above the fitted line and that the line passes through $(\mathbf{x}, \widehat{\mathbf{y}})$. Then a physical analogy for the outlier's effect is that it exerts an upwards force of $(\mathbf{y} - \widehat{\mathbf{y}})^2$ units on the line at a distance $\mathbf{x} - \overline{\mathbf{x}}$ units from the pivot point of the bar. The farther from the pivot point, the greater the ability of the outlier to affect the fit, that is, the greater its ability to use the line as a *lever*. Hence the term *leverage*.

We can now understand why the relatively small outlier in Figure 8.10 at $\mathbf{x} = 0.5$ is easy to detect while the relatively large outlier at $\mathbf{x} = 0.95$ is not. The outlier at $\mathbf{x} = 0.5$ exerts little influence on the fitted line. Thus the line has little opportunity to 'adapt' to this outlier, leaving the outlier plainly visible above the fitted line. The large outlier at $\mathbf{x} = 0.95$ has much greater influence on the fitted line, which 'adapts' well to this outlier, hiding its presence.

Statisticians have devised a number of sophisticated techniques for estimating the influence of an individual observation. Without going into detail, the idea behind these methods is that the influence of an individual observation can be estimated by fitting the model with, and without, that observation. The change in the fit, measured in some objective manner, determines the influence of that observation. See [41], [78], and [90] for details and methods.

Bounded influence regression (M-estimation, see [154])—of which *median absolute deviation regression* is a special case—has become a popular way to protect against the effects of influential outliers. Such techniques are now generally available in statistical packages and subroutine libraries. Two kinds of action are taken to control the effects of outliers. First, the errors $\widehat{\mathbf{e}}_i$ (8.12) are weighted (see Section 8.6) so that observations corresponding to outlying values of the factor $\mathbf{X}$ receive less weight. Second, rather than substituting the weighted errors into normal equations (8.14) and (8.15) to obtain parameter estimators, *bounded errors* are substituted into the equations. That is, the M-estimates are obtained by solving equations of the form

$$\sum_{i=1}^{n} \Psi(\widehat{\mathbf{e}}_i) = 0$$
$$\sum_{i=1}^{n} \Psi(\widehat{\mathbf{e}}_i)\mathbf{x}_i = 0,$$

where $\Psi(\cdot)$ is a function that preserves the sign of its argument but limits its magnitude. For example, Huber [190] uses

$$\Psi(t) = \begin{cases} -c, & t < c \\ t, & |t| \le c \\ c, & t > c. \end{cases}$$

8.3.19 Matrix-vector Formulation of Least Squares Estimators. We have formulated the least squares estimators for simple linear regression by basic brute force, but it is easier to form estimators and derive distributional results for multiple linear regression problems when matrix-vector notation is used.

Let $\vec{\mathbf{Y}}$ denote the n-dimensional random vector whose ith element is $\mathbf{Y}_i$. Let $\mathcal{X}$ be the $n \times 2$ matrix that has units in the first column and $\mathbf{x}_i$ as the ith element of the second column. That is,

$$\mathcal{X} = \begin{pmatrix} 1 & \mathbf{x}_1 \\ 1 & \mathbf{x}_2 \\ \vdots & \vdots \\ 1 & \mathbf{x}_n \end{pmatrix}.$$

Matrix $\mathcal{X}$ is called the *design matrix*. Let $\vec{\mathbf{E}}$ denote the n-dimensional random vector whose ith element is $\mathbf{E}_i$, and let $\vec{\mathbf{a}}$ be the two-dimensional vector whose elements are a_0 and a_1. Then the matrix-vector representation of (8.10) is

$$\vec{\mathbf{Y}} = \mathcal{X}\vec{\mathbf{a}} + \vec{\mathbf{E}}. \quad (8.25)$$

The least squares estimates are obtained by choosing $\vec{\mathbf{a}}$ so that the squared length of $\vec{\mathbf{E}}$, given by

$$\mathcal{SSE} = \vec{\mathbf{E}}^{\mathrm{T}}\vec{\mathbf{E}} = (\vec{\mathbf{Y}} - \mathcal{X}\vec{\mathbf{a}})^{\mathrm{T}}(\vec{\mathbf{Y}} - \mathcal{X}\vec{\mathbf{a}}), \quad (8.26)$$

is minimized. Differentiating with respect to $\vec{\mathbf{a}}$ (see, e.g., [148, pp. 350–360]), we obtain the normal equations

$$2\mathcal{X}^{\mathrm{T}}(\vec{\mathbf{Y}} - \mathcal{X}\vec{\mathbf{a}}) = \vec{0},$$

where $\vec{0}$ is a two-dimensional vector of zeros. The solutions of the normal equations are given by

$$\widehat{\vec{\mathbf{a}}} = (\mathcal{X}^{\mathrm{T}}\mathcal{X})^{-1}\mathcal{X}^{\mathrm{T}}\vec{\mathbf{Y}}. \quad (8.27)$$

Some simple algebra reveals that estimator (8.27) is identical to estimators (8.16) and (8.17) derived previously.

The sums of squares appearing in (8.19) are also easily re-expressed in matrix-vector form. Substituting (8.27) into (8.26) we obtain

$$\mathcal{SSE} = \vec{\mathbf{Y}}^{\mathrm{T}}(\mathcal{I} - \mathcal{X}(\mathcal{X}^{\mathrm{T}}\mathcal{X})^{-1}\mathcal{X}^{\mathrm{T}})\vec{\mathbf{Y}}, \tag{8.28}$$

where $\mathcal{I}$ denotes the $n \times n$ identity matrix. By noting that

$$\overline{\mathbf{y}} = \vec{\mathbf{Y}}^{\mathrm{T}} \begin{pmatrix} 1/n \\ 1/n \\ \vdots \\ 1/n \end{pmatrix}$$

we obtain that the sum of squares due to regression, given by $\sum_{i=1}^{n}(\widehat{\mu}_{\mathbf{Y}|\mathbf{X}=\mathbf{x}_i} - \overline{\mathbf{y}})^2$, can be expressed as

$$\mathcal{SSR} = \vec{\mathbf{Y}}^{\mathrm{T}}(\mathcal{X}(\mathcal{X}^{\mathrm{T}}\mathcal{X})^{-1}\mathcal{X}^{\mathrm{T}} - \mathcal{U})\vec{\mathbf{Y}}, \tag{8.29}$$

where $\mathcal{U}$ is an $n \times n$ matrix with each entry equal to $1/n$. The total sum of squares is given by

$$\mathcal{SST} = \vec{\mathbf{Y}}^{\mathrm{T}}(\mathcal{I} - \mathcal{U})\vec{\mathbf{Y}}. \tag{8.30}$$

8.3.20 Distributional Results. Here we briefly demonstrate how Properties 1–5 stated in [8.3.5] are obtained and provide a geometrical interpretation of the concept of degrees of freedom. These ideas generalize easily to include regression models that contain more than one factor.

Now suppose again that the errors $\mathbf{E}_i$ are iid normally distributed with mean zero. Then $\vec{\mathbf{Y}}$ has a multivariate normal distribution with mean $\mathcal{X}\vec{\mathbf{a}}$ and covariance matrix $\sigma_E^2\mathcal{I}$. It follows that $\widehat{\vec{\mathbf{a}}}$ is normally distributed with mean $\vec{\mathbf{a}}$ and covariance matrix $\sigma_E^2(\mathcal{X}^{\mathrm{T}}\mathcal{X})^{-1}$ (see Section 2.8).

Next we demonstrate that $\mathcal{SSE}/\sigma_E^2$ is independent of $\widehat{\vec{\mathbf{a}}}$ and distributed $\chi^2(n-2)$. Let $\vec{\mathbf{k}}_1$ and $\vec{\mathbf{k}}_2$ be orthonormal vectors spanning the column space of the design matrix $\mathcal{X}$. Choose $\vec{\mathbf{k}}_3, \dots, \vec{\mathbf{k}}_n$ so that $\vec{\mathbf{k}}_1, \vec{\mathbf{k}}_2, \dots, \vec{\mathbf{k}}_n$ form a complete orthonormal basis for $\mathbb{R}^n$. Let $\vec{\mathbf{Z}} = \mathcal{K}^{\mathrm{T}}\vec{\mathbf{Y}}$ where $\mathcal{K}$ is the $n \times n$ matrix that has $\vec{\mathbf{k}}_i$ as its ith column. Then, since $\mathcal{K}^{\mathrm{T}}\mathcal{K} = \mathcal{K}\mathcal{K}^{\mathrm{T}} = \mathcal{I}$, $\vec{\mathbf{Y}} = \mathcal{K}\vec{\mathbf{Z}}$. Now substituting for $\vec{\mathbf{Y}}$ in expression (8.26), we have

$$\begin{aligned} \mathcal{SSE} &= (\vec{\mathbf{Y}} - \mathcal{X}\vec{\mathbf{a}})^{\mathrm{T}}(\vec{\mathbf{Y}} - \mathcal{X}\vec{\mathbf{a}}) \\ &= (\mathcal{K}\vec{\mathbf{Z}} - \mathcal{X}\vec{\mathbf{a}})^{\mathrm{T}}(\mathcal{K}\vec{\mathbf{Z}} - \mathcal{X}\vec{\mathbf{a}}) \\ &= (\vec{\mathbf{Z}} - \mathcal{K}^{\mathrm{T}}\mathcal{X}\vec{\mathbf{a}})^{\mathrm{T}}(\vec{\mathbf{Z}} - \mathcal{K}^{\mathrm{T}}\mathcal{X}\vec{\mathbf{a}}). \end{aligned}$$

Because the first two columns of $\mathcal{K}$ span the columns of $\mathcal{X}$, we have that $\mathcal{K}^{\mathrm{T}}\mathcal{X}$ is of the form

$$\mathcal{K}^{\mathrm{T}}\mathcal{X} = \begin{pmatrix} \mathcal{X}_1^* \\ \mathcal{X}_2^* \end{pmatrix},$$

where $\mathcal{X}_1^*$ is a nonzero 2×2 matrix and $\mathcal{X}_2^*$ is the $(n-2) \times 2$ matrix of zeros. Therefore $\mathcal{SSE}$ is of the form

$$\mathcal{SSE} = (\vec{\mathbf{Z}}_1 - \mathcal{X}_1\vec{\mathbf{a}})^{\mathrm{T}}(\vec{\mathbf{Z}}_1 - \mathcal{X}_1\vec{\mathbf{a}}) + \vec{\mathbf{Z}}_2^{\mathrm{T}}\vec{\mathbf{Z}}_2,$$

where $\vec{\mathbf{Z}}_1$ consists of the first two elements of $\vec{\mathbf{Z}}$ and $\vec{\mathbf{Z}}_2$ consists of the remaining $(n-2)$ elements. Upon minimization we see that

$$\mathcal{SSE} = \vec{\mathbf{Z}}_2^{\mathrm{T}}\vec{\mathbf{Z}}_2 = \sum_{i=3}^{n} \mathbf{Z}_i.$$

Now from the matrix-vector form of the regression model we see that the elements of $\vec{\mathbf{Z}}$ are independent and have common variance σ_E^2 (the covariance matrix of both $\vec{\mathbf{Y}}$ and $\mathcal{K}\vec{\mathbf{Y}}$ is $\sigma_E^2\mathcal{I}$). Therefore $\mathcal{SSE}/\sigma_E^2$ is $\chi^2(n-2)$ distributed. Note that $n-2$ is the dimension of the sub-space *not* spanned by the columns of the design matrix. Moreover, because $\mathcal{SSE}$ depends only upon $\vec{\mathbf{Z}}_2$ and $\widehat{\vec{\mathbf{a}}}$ depends only upon $\vec{\mathbf{Z}}_1$, we see that $\mathcal{SSE}$ is independent of $\widehat{\vec{\mathbf{a}}}$.

8.4 Multiple Regression

The simple linear regression model we have examined up to this point, while enormously useful in climatology and meteorology, has severely limited flexibility. Many methods, such as the *MOS* (model output statistics) and *perfect prog* statistical forecast improvement procedures (see, for example, Klein and Glahn [226], Klein [224], Klein and Bloom [225], Brunet, Verret, and Yacowar [71]), require the use of regression models with more than one explanatory factor.

The working example we develop as we progress through the section is the cloud parameterization example introduced in [8.1.4].

8.4.1 The Multiple Regression Model. A multiple linear regression model expresses a response variable as an error term plus a mean that is conditional upon several factors. Suppose we observe a response variable $\mathbf{Y}$ and k factors denoted by $\mathbf{X}_1, \dots, \mathbf{X}_k$ that are thought to affect the expected value of $\mathbf{Y}$. These random variables are all observed n times. The result is a sample of n $(k+1)$-tuples represented by random variables $(\mathbf{Y}_i, \mathbf{X}_{1,i}, \dots, \mathbf{X}_{k,i})$ whose actual observed, or realized, values are represented by $(\mathbf{y}_i, \mathbf{x}_{1,i}, \dots, \mathbf{x}_{k,i})$, for $i = 1, \dots, n$. The multivariate version of (8.10) is given by

$$\mathbf{Y}_i = a_0 + \sum_{l=1}^{k} a_l \mathbf{x}_{li} + \mathbf{E}_i, \tag{8.31}$$

where the $\mathbf{E}_i$, for $i = 1, \ldots, n$, are iid random variables with mean zero. We usually assume that these errors are normally distributed.

This model states that the mean of $\mathbf{Y}$, conditional upon the realized values of the factors $\mathbf{X}_j$, can be expressed as a linear combination of the factors. Thus the model is linear in its parameters. However, the factors themselves can be nonlinear functions of other variables. For example, the model specifies a polynomial of order k in $\mathbf{X}$ if $\mathbf{X}_{li} = (\mathbf{X}_i)^l$.

The model we will fit to the Landsat data (cf. [8.1.4]) has the form

$$\overline{\ln \tau} = a_0 + a_1 \ln(\overline{\tau}) + a_2 A_c + \mathbf{E}. \qquad (8.32)$$

The $\ln(\overline{\tau})$ term is used to account for the curvilinear relationship between $\overline{\tau}$ and $\overline{\ln \tau}$ that is apparent in Figure 8.3 (left). See also [8.6.2].

8.4.2 Matrix-vector Representation of the Multiple Linear Regression Model. The development of least squares estimators and inferential methods for multiple regression parallels that for the simple linear regression model once the model has been expressed in matrix-vector form.

As in [8.3.19], let $\vec{\mathbf{Y}}$ represent the n-dimensional random vector whose ith element is $\mathbf{Y}_i$. Define $\vec{\mathbf{E}}$ similarly. Let the design matrix $\mathcal{X}$ be the $n \times (k+1)$ matrix given by

$$\mathcal{X} = \begin{pmatrix} 1 & \mathbf{x}_{1,1} & \ldots & \mathbf{x}_{k,1} \\ 1 & \mathbf{x}_{1,2} & \ldots & \mathbf{x}_{k,2} \\ \vdots & \vdots & & \vdots \\ 1 & \mathbf{x}_{1,n} & \ldots & \mathbf{x}_{k,n} \end{pmatrix}.$$

Let $\vec{\mathbf{a}}$ be the $(k+1)$-dimensional vector consisting of model parameters $a_0, a_1, \ldots, a_k$. With this notation, the matrix-vector representation of (8.31) is identical to that of the simple linear regression case given in (8.25), where we have $\vec{\mathbf{Y}} = \mathcal{X}\vec{\mathbf{a}} + \vec{\mathbf{E}}$.

The least squares estimator of $\vec{\mathbf{a}}$ and the variance components $\mathcal{SST}$, $\mathcal{SSR}$, and $\mathcal{SSE}$ are computed as in (8.27)–(8.30).

The degrees of freedom for the variance components are as follows:

Source	Sum of Sq.	df
Regression	$\mathcal{SSR}$	$df_R = k$
Error	$\mathcal{SSE}$	$df_E = n - k - 1$
Total	$\mathcal{SST}$	$df_T = n - 1$

When model (8.32) is fitted to the Landsat data described in [8.1.4], we obtain parameter estimates $\widehat{a}_0 = -0.747$, $\widehat{a}_1 = 0.794$, $\widehat{a}_2 = 1.039$ and $\widehat{\sigma}_E = 0.233$. The coefficient of multiple determination, R^2, is equal to 0.938, indicating that $\overline{\tau}$ and A_c jointly represent about 94% of the variability in $\overline{\ln \tau}$ in the data set. The total variability in the 45 $\overline{\ln \tau}$ values of the Landsat data set is partitioned by the fitted model as follows:

Source	Sum of Sq.	df
Regression	34.705	2
Error	2.287	42
Total	36.992	44

The methods of [8.3.20] can be used to prove the following properties, which form the basis of the inference procedures used in multiple regression:

1. $\widehat{\vec{\mathbf{a}}}$ is an unbiased estimate of $\vec{\mathbf{a}}$,
2. $\widehat{\sigma}_E^2 = \frac{\mathcal{SSE}}{df_E}$ is an unbiased estimate of σ_E^2,
3. $\widehat{\vec{\mathbf{a}}} \sim \mathcal{N}(\vec{\mathbf{a}}, \sigma_E^2(\mathcal{X}^{\mathrm{T}}\mathcal{X})^{-1})$.
4. $\widehat{\vec{\mathbf{a}}}$ is independent of $\mathcal{SSE}$.
5. $\mathcal{SSE}/\sigma_E^2 \sim \chi^2(df_E)$.

8.4.3 Multiple Regression Model Without an Intercept. Sometimes it may be desirable to force the fitted regression surface to pass through the origin. In this case coefficient a_0 in (8.31) is set to zero and the column of 1s in the design matrix is deleted. The least squares estimator is computed as before by substituting the modified design matrix into (8.27). The variance components are computed using

$$\begin{aligned} \mathcal{SSR} &= \vec{\mathbf{Y}}^{\mathrm{T}}(\mathcal{X}(\mathcal{X}^{\mathrm{T}}\mathcal{X})^{-1}\mathcal{X}^{\mathrm{T}})\vec{\mathbf{Y}} \\ \mathcal{SSE} &= \vec{\mathbf{Y}}^{\mathrm{T}}(\mathcal{I} - \mathcal{X}(\mathcal{X}^{\mathrm{T}}\mathcal{X})^{-1}\mathcal{X}^{\mathrm{T}})\vec{\mathbf{Y}} \\ \mathcal{SST} &= \vec{\mathbf{Y}}^{\mathrm{T}}\vec{\mathbf{Y}}. \end{aligned}$$

The corresponding degrees of freedom are

Source	Sum of Sq.	df
Regression	$\mathcal{SSR}$	$df_R = k$
Error	$\mathcal{SSE}$	$df_E = n - k$
Total	$\mathcal{SST}$	$df_T = n$

In particular, notice that there is one additional degree of freedom for error because it was not necessary to fit the intercept parameter.

8.4.4 A Confidence Interval for the Mean of the Response Variable. Let $\vec{\mathbf{X}}$ represent the $(k+1)$-dimensional vector $\vec{\mathbf{X}} = (1, \mathbf{X}_1, \ldots, \mathbf{X}_k)^{\mathrm{T}}$. The rows of the design matrix can be thought of as a collection of n realizations of $\vec{\mathbf{X}}$. From (8.31) we see that the expected value of $\mathbf{Y}$ conditional upon $\vec{\mathbf{X}} = \vec{\mathbf{x}}$ is given by

$$\mu_{\vec{\mathbf{Y}}|\vec{\mathbf{X}}=\vec{\mathbf{x}}} = \vec{\mathbf{x}}^{\mathrm{T}}\vec{\mathbf{a}},$$

which is estimated by

$$\widehat{\mu}_{\vec{\mathbf{Y}}|\vec{\mathbf{X}}=\vec{\mathbf{x}}} = \vec{\mathbf{x}}^{\mathrm{T}}\widehat{\vec{\mathbf{a}}}.$$

Property 3 of [8.4.2] tells us that

$$\widehat{\mu}_{\vec{\mathbf{Y}}|\vec{\mathbf{X}}=\vec{\mathbf{x}}} \sim \mathcal{N}(\vec{\mathbf{x}}^{\mathrm{T}}\vec{\mathbf{a}}, \sigma_E^2\vec{\mathbf{x}}^{\mathrm{T}}(\mathcal{X}^{\mathrm{T}}\mathcal{X})^{-1}\vec{\mathbf{x}}).$$

Using properties 4 and 5 of [8.4.2] we obtain that

$$\mathbf{T} = \frac{\widehat{\mu}_{\vec{\mathbf{Y}}|\vec{\mathbf{X}}=\vec{\mathbf{x}}} - \mu_{\vec{\mathbf{Y}}|\vec{\mathbf{X}}=\vec{\mathbf{x}}}}{\widehat{\sigma}_E\sqrt{\vec{\mathbf{x}}^{\mathrm{T}}(\mathcal{X}^{\mathrm{T}}\mathcal{X})^{-1}\vec{\mathbf{x}}}} \sim t(df_E).$$

Thus a $\tilde{p} \times 100\%$ confidence interval for the conditional mean at $\vec{\mathbf{x}}$ has bounds

$$\widehat{\mu}_{\vec{\mathbf{Y}}|\vec{\mathbf{X}}=\vec{\mathbf{x}}} \pm t_{(1+\tilde{p})/2}\widehat{\sigma}_E\sqrt{\vec{\mathbf{x}}^{\mathrm{T}}(\mathcal{X}^{\mathrm{T}}\mathcal{X})^{-1}\vec{\mathbf{x}}}, \qquad (8.33)$$

where $t_{(1+\tilde{p})/2}$ is the appropriate quantile of the t distribution with df_E degrees of freedom obtained from Appendix F. As for simple linear regression, the true response surface (a plane) will be covered by the range of hyper-surfaces described by this expression $\tilde{p} \times 100\%$ of the time.

8.4.5 A Confidence Interval for the Response Variable. As with simple linear regression, a $\tilde{p} \times 100\%$ confidence interval for the response variable $\mathbf{Y}$ at $\vec{\mathbf{X}} = \vec{\mathbf{x}}$ is obtained by adding 1 to the quantity under the radical sign in (8.33).

8.4.6 A Confidence Interval for Parameter a_l. Let $\vec{e}_l$ be the $(k+1)$-dimensional vector

$$\vec{e}_l = (\delta_{l,0}, \delta_{l,1}, \ldots, \delta_{l,k})^{\mathrm{T}} \qquad (8.34)$$

where $\delta_{lj} = 1$ if $l = j$ and $\delta_{lj} = 0$ otherwise. The $\tilde{p} \times 100\%$ confidence interval for a_l is obtained by substituting $\vec{e}_l$ for $\vec{\mathbf{x}}$ in (8.33).

The matrix $(\mathcal{X}^{\mathrm{T}}\mathcal{X})^{-1}$ for the Landsat data fitted with model (8.32) is

$$\begin{pmatrix} 0.1714 & -0.0649 & -0.0371 \\ -0.0649 & 0.0842 & -0.1409 \\ -0.0371 & -0.1409 & 0.4416 \end{pmatrix}. \qquad (8.35)$$

Therefore, the 95% confidence intervals for the estimated parameters are

Parameter	95% Confidence Interval
a_0	(−0.936, −0.557)
a_1	(0.661, 0.927)
a_2	(0.735, 1.343)

We see that the estimated value of a_2 is somewhat less certain than that of a_0 and a_1. However, we can safely infer that all three parameters are significantly different from zero. We should add the caveat that these inferences are valid only if our assumptions about the errors (i.e., that they are iid normal) hold.

The parameter estimators $\widehat{\mathbf{a}}_l$ are seldom independent because $(\mathcal{X}^{\mathrm{T}}\mathcal{X})^{-1}$ is seldom a diagonal matrix. Therefore multiple $\tilde{p} \times 100\%$ confidence intervals for, say, m different parameters *do not* constitute a joint $\tilde{p}^m \times 100\%$ confidence region for the m parameters taken as a group (see [8.4.7]).[6]

Property 3 of [8.4.2] tells us that the covariance matrix of $\widehat{\vec{a}}$ can be estimated with $\widehat{\sigma}_E^2(\mathcal{X}^{\mathrm{T}}\mathcal{X})^{-1}$. The estimates for our example are:

Correlation	$\widehat{a}_0$	$\widehat{a}_1$	$\widehat{a}_2$
$\widehat{a}_0$	1.000	−0.532	−0.135
$\widehat{a}_1$	−0.532	1.000	−0.731
$\widehat{a}_2$	−0.135	−0.731	1.000

8.4.7 Joint Confidence Regions for More Than One Parameter. A joint $\tilde{p} \times 100\%$ confidence region for p parameters $a_{l_1}, \ldots, a_{l_p}$ can be obtained as follows.

First, let $\mathcal{U}$ be the $(k+1) \times p$ matrix that has $\vec{e}_{l_j}$, where $\vec{e}_{l_j}$ is given by (8.34), in column j, for $j = 1, \ldots, p$. Then the vector $\vec{\mathbf{a}}^s = \mathcal{U}^{\mathrm{T}}\vec{\mathbf{a}}$ contains the p parameters of interest and is estimated by $\widehat{\vec{\mathbf{a}}}^s = \mathcal{U}^{\mathrm{T}}\widehat{\vec{\mathbf{a}}}$. Using Property 3 of [8.4.2] we see that the estimator has a normal distribution given by

$$\widehat{\vec{\mathbf{a}}}^s \sim \mathcal{N}(\mathcal{U}^{\mathrm{T}}\vec{\mathbf{a}}, \sigma_E^2\mathcal{U}^{\mathrm{T}}(\mathcal{X}^{\mathrm{T}}\mathcal{X})^{-1}\mathcal{U})$$

(see [2.8.9]). Now let

$$\mathcal{V} = [\mathcal{U}^{\mathrm{T}}(\mathcal{X}^{\mathrm{T}}\mathcal{X})^{-1}\mathcal{U}]^{-1/2}$$

so that $\mathcal{V}^{\mathrm{T}}\mathcal{V} = [\mathcal{U}^{\mathrm{T}}(\mathcal{X}^{\mathrm{T}}\mathcal{X})^{-1}\mathcal{U}]^{-1}$, and define $\vec{\mathbf{Z}}$ to be the p-dimensional normal random vector

$$\vec{\mathbf{Z}} = \mathcal{V}(\widehat{\vec{\mathbf{a}}}^s - \vec{\mathbf{a}}^s).$$

[6] This type of rectangular region in parameter space is also not a good way to construct a joint confidence region when estimators are independent. Construction of a confidence region should use the principle that any point in parameter space outside the confidence region should be less likely given the data than points inside the confidence region. For iid normal data this means that the boundaries of confidence regions should be ellipsoids. See [6.2.2] and Figure 6.16.

Then

$$\vec{\mathbf{Z}} \sim \mathcal{N}(\vec{0}, \sigma_E^2 \mathcal{I}),$$

where $\mathcal{I}$ is the $p \times p$ identity matrix (see [2.8.9]). Therefore

$$\frac{p}{\sigma_E^2}(\widehat{\vec{\mathbf{a}}}^s - \vec{\mathbf{a}}^s)^{\mathrm{T}} \mathcal{V}^{\mathrm{T}} \mathcal{V} (\widehat{\vec{\mathbf{a}}}^s - \vec{\mathbf{a}}^s) \sim \chi^2(p).$$

We now have the ingredients needed to construct a simultaneous confidence region for parameters $a_{l_1}, \ldots, a_{l_p}$. By Properties 4 and 5 of [8.4.2], the $\chi^2(p)$ random variable above is independent of the $\chi^2(df_E)$ random variable $df_E \frac{\mathcal{SSE}}{\sigma_E^2}$. Therefore, from (2.29), we see that

$$\frac{(\widehat{\vec{\mathbf{a}}}^s - \vec{\mathbf{a}}^{s\mathrm{T}}) \mathcal{V}^{\mathrm{T}} \mathcal{V} (\widehat{\vec{\mathbf{a}}}^s - \vec{\mathbf{a}}^s)}{\widehat{\sigma}_E^2} \sim F(p, df_E).$$

Thus the $\tilde{p} \times 100\%$ confidence region, an ellipsoid, is composed of all points in the $(k+1)$-dimensional parameter space that satisfy the inequality

$$\frac{(\widehat{\vec{\mathbf{a}}}^s - \vec{\mathbf{a}}^s)^{\mathrm{T}} \mathcal{V}^{\mathrm{T}} \mathcal{V} (\widehat{\vec{\mathbf{a}}}^s - \vec{\mathbf{a}}^s)}{\widehat{\sigma}_E^2} < F_{\tilde{p}}, \qquad (8.36)$$

where $F_{\tilde{p}}$ is the $\tilde{p}$-quantile of the F distribution with (p, df_E) df obtained from Appendix G.

Let us consider the problem of constructing a joint $\tilde{p}$ confidence region for a subset of two parameters, (a_1, a_2), in our Landsat example. Proceeding as above, we have

$$\mathcal{V}^{\mathrm{T}}\mathcal{V} = \left[\begin{pmatrix} 0 & 1 & 0 \\ 0 & 0 & 1 \end{pmatrix} (\mathcal{X}^{\mathrm{T}}\mathcal{X})^{-1} \begin{pmatrix} 0 & 0 \\ 1 & 0 \\ 0 & 1 \end{pmatrix} \right]^{-1}$$
$$= \begin{pmatrix} 25.47 & 8.13 \\ 8.13 & 4.86 \end{pmatrix}.$$

Expanding (8.36), we find that the points in the joint $\tilde{p}$ confidence region for (a_1, a_2) satisfy

$$25.47(\widehat{a}_1 - a_1)^2 + 2 \times 8.13(\widehat{a}_a - a_1)(\widehat{a}_2 - a_2) + 4.86(\widehat{a}_2 - a_2)^2 < F_{\tilde{p}}\widehat{\sigma}_E^2$$

where $F_{\tilde{p}}$ is the $\tilde{p}$-quantile of $F(2, 42)$.

The 95% confidence region computed in this way is displayed in Figure 8.13. The tilt of the ellipse reflects the correlation between $\widehat{a}_1$ and $\widehat{a}_2$. The point estimate is shown in the middle of the ellipse. The dashed lines indicate the 95% confidence intervals for a_1 and a_2 computed with (8.33). Note that the rectangular region defined by their intersection is substantially larger than the region enclosed by the ellipse.

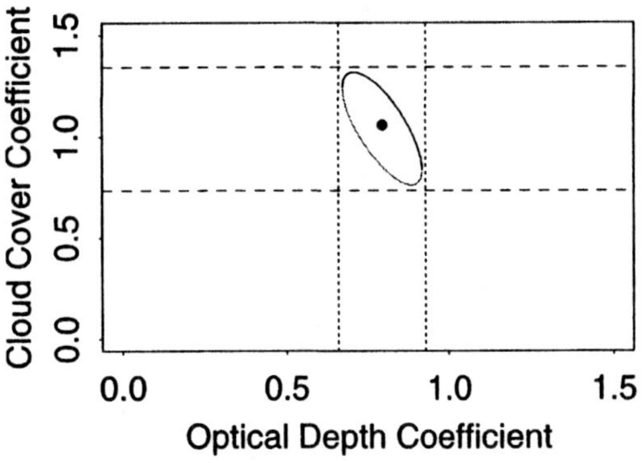

Figure 8.13: *The joint 95% confidence region for the* $\ln(\overline{\tau})$ *and* A_c *coefficients of model (8.32). The estimated coefficients are indicated by the dot. The dashed lines indicate the individual 95% confidence intervals computed as in (8.33).*

8.4.8 Is There a Regression Relationship? This question is answered by testing the null hypothesis H_0: $a_1 = \ldots = a_k = 0$. We could proceed as we did above when constructing the joint confidence region by constructing a suitable *kernel* matrix $\mathcal{V}^{\mathrm{T}}\mathcal{V}$ and then developing a test statistic of the form

$$\mathbf{F} = \frac{\widehat{\vec{\mathbf{a}}}^{\mathrm{T}} \mathcal{V}^{\mathrm{T}} \mathcal{V} \widehat{\vec{\mathbf{a}}}}{\widehat{\sigma}_E^2}, \qquad (8.37)$$

which is distributed $F(df_R, df_E)$ under H_0. However, in this case there is an easier way. It can be shown that (8.37) is also given by

$$\mathbf{F} = \frac{\mathcal{SSR}/df_R}{\mathcal{SSE}/df_E},$$

which is easily computed as a byproduct of the least squares fitting procedure. Large values of $\mathbf{F}$ are evidence contrary to H_0, so the test is conducted at the $(1-\tilde{p}) \times 100\%$ significance level by rejecting H_0 when $\mathbf{f} > F_{\tilde{p}}$, the $\tilde{p}$-quantile of $F(df_R, df_E)$.

We find $\mathbf{f} = 318.6$ in our Landsat example, a value that is significant at much less than the 0.1% level.

8.4.9 Are all Parameters in a Subset Zero? We could answer this question as well by constructing a suitable kernel $\mathcal{V}^{\mathrm{T}}\mathcal{V}$ and computing $\mathbf{F}$ as in (8.37). Again, there is an easier and more intuitively appealing answer.

Consider the following possible approach for testing H_0: $a_{l_1} = \cdots = a_{l_p} = 0$.

- Fit the full regression model including the p factors $\mathbf{X}_{l_1}, \ldots, \mathbf{X}_{l_p}$. Denote the resulting regression and sum of squared errors as $\mathcal{SSR}_F$ and $\mathcal{SSE}_F$, respectively, where the subscript F indicates that these variance components were obtained by fitting the full model.

- Fit the *restricted* regression model specified by the null hypothesis by excluding factors $\mathbf{X}_{l_1}, \ldots, \mathbf{X}_{l_p}$ from the design matrix. Denote the resulting regression sum of squares as $\mathcal{SSR}_R$.

- The increase in the regression sum of squares that is obtained by adding factors $\mathbf{X}_{l_1}, \ldots, \mathbf{X}_{l_p}$ to the restricted model is given by $\mathcal{SSR}_F - \mathcal{SSR}_R$. Under H_0, $[(\mathcal{SSR}_F - \mathcal{SSR}_R)/\sigma_E^2] \sim \chi^2(p)$ and is independent of $\mathcal{SSE}_F$. Thus, using property 5 of [8.4.2], we obtain a test statistic

$$\mathbf{F} = \frac{(\mathcal{SSR}_F - \mathcal{SSR}_R)/p}{\mathcal{SSE}_F/df_{E_F}} = \frac{(\mathcal{SSR}_F - \mathcal{SSR}_R)/p}{\widehat{\sigma}_E^2}$$

that is distributed $F(p, df_{E_F})$ under H_0. Here df_{E_F} is the degrees of freedom of the sum of squared errors for the full regression.

The test is conducted at the $(1-\tilde{p}) \times 100\%$ significance level by rejecting H_0 when $\mathbf{f} > F_{\tilde{p}}$, the $\tilde{p}$-quantile of $F(p, df_E)$.

8.4.10 Diagnostics. We have two things in mind when we think about the fit of the model. The first is, how well does the model specify values of $\mathbf{Y}$ from the factors $\mathbf{X}_l$? The coefficient of multiple determination $R^2 = \frac{\mathcal{SSR}}{\mathcal{SST}}$ (see [8.3.12]) gives a quick but somewhat optimistic answer. Use cross-validation (see Section 18.5) if it is important to obtain a good estimate of future model performance [18.5.2].

The second worry is whether or not inferences are made reliably. Implicit in the discussion to this point are the assumptions that the errors in (8.25) are iid normally distributed and that the full model adequately represents the conditional mean of $\mathbf{Y}$. Therefore the diagnostic procedures discussed in Section 8.3 should be applied to confirm that the distributional assumptions are as close to being satisfied as possible and that the inferences can be properly qualified. Scatter plots (see [8.3.13]) of residuals should be examined for evidence of outliers, heteroscedasticity, and lack-of-fit. For multiple regression, residuals should be plotted against the estimated conditional mean (i.e., the fitted model) *and* against the values of individual factors. Bear in mind that outliers (see [8.3.13]) will be more difficult to detect than in the case of simple linear regression. Use objective methods for detecting influential observations (see [8.3.18]) if at all possible. Use probability plots (see [8.3.14]) to detect departures from the assumption of a normal distribution. The general considerations of [8.3.15] apply, so we can proceed cautiously if the normal distribution assumption is in doubt. When appropriate, use the Durbin–Watson statistic (8.24) or runs test to check for dependence amongst the errors (see [8.3.16]).

We now briefly examine the fit of model (8.32) to the Landsat data set described in [8.1.4]. Figure 8.14 shows studentized residuals plotted against $\widehat{\ln \tau}$ (right). The left hand panel shows one outlier with undue influence on the fit. One effect of this outlier, the extreme point in the lower left corner of the right hand panel, is to shift the other quantiles in the probability plot upwards, thereby giving the impression that the upper tail of the error distribution may be narrower than that of the normal distribution.

Figure 8.15 shows the same diagnostics for the fit that is obtained after removing the outlier from the data set. The left hand panel shows that there may still be one or two observations that need investigation. Other diagnostics also indicate that these observations, corresponding to the two largest remaining studentized residuals, are somewhat more influential than we might like. The right hand panel shows improvement in the distributional characteristics of the residuals after removal of the outlier.

Removing the single outlier results in fairly large changes to the fitted model. There is little change in the estimated intercept (the new value of $\widehat{a}_0$ is -0.0748), but there are substantial changes in the coefficients of $\overline{\tau}$ ($\widehat{a}_1 = 0.866$) and A_c ($\widehat{a}_2 = 0.866$). Also, $\widehat{\sigma}_E$ is reduced to 0.208 and R^2 increases slightly to 95.2%, a further indication that the fit is improved.[7]

[7] The outlying observation comes from a Landsat image identified as scene $C4$ by Barker et al. (see [18, Table 2]). The image contains scattered cumulus clouds and appears to have large mean optical depth relative to its fractional cloud coverage. However, the image was taken when the solar zenith angle was 68°. Optical depth is difficult to estimate accurately in this scene because of the oblique trajectory of light incident on the clouds.

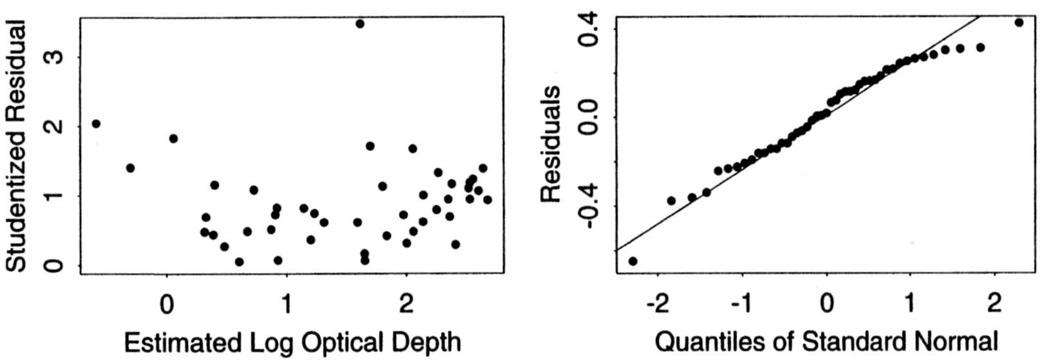

Figure 8.14: *Left: Absolute studentized residuals plotted against* $\widehat{\ln \tau}$ *for the fit of the model (8.32) to the Landsat data described in [8.1.4].*
Right: A probability plot of the ordinary residuals $\overline{\ln \tau} - \widehat{\ln \tau}$.

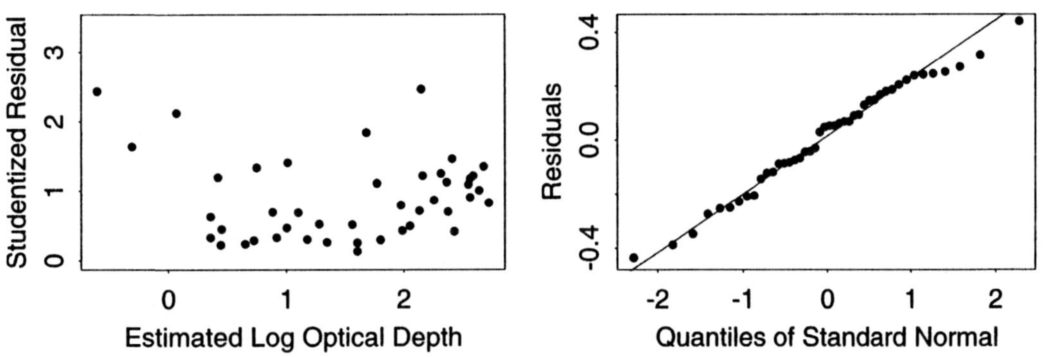

Figure 8.15: *As Figure 8.14, except these diagrams illustrate the fit that is obtained when the large outlier is removed.*
Left: Absolute studentized residuals.
Right: Probability plot of ordinary residuals.

8.4.11 Multicolinearity. We have, by now, learned to think of the factors in a multiple regression as columns in the design matrix. Two or more factors are *multicolinear* when the corresponding columns in the design matrix point in similar directions in $\mathbb{R}^n$, that is, when they are strongly correlated. Therefore, one way to look for multicolinearity is simply to study the correlation matrix of the non-constant factors. Large correlations indicate potential multicolinearity problems.

The effect of multicolinearity is to make the matrix $\mathcal{X}^{\mathrm{T}}\mathcal{X}$ nearly uninvertible, resulting in highly variable parameter estimators (see Property 3 of [8.4.2]) and making it difficult to diagnose the factors that are most important in specifying $\mathbf{Y}$.

Parameter estimates are sensitive to small variations in the data when there is multicolinearity. The sensitivity of the model is estimated from the *condition number* $\kappa(\mathcal{X})$ of the design matrix $\mathcal{X}$, which is defined as the ratio between the largest and smallest singular values of $\mathcal{X}$ (see Appendix B). A good introduction to the use of $\kappa(\mathcal{X})$ for detecting multicolinearity and strategies for coping with estimator sensitivity are contained in [78, pp. 138–144] (see also [104]). Some statistical packages, such as SPlus [36], are able to produce sensitivity estimates.

8.4.12 Ridge Regression. One way to cope with multicolinearity is to remove redundant factors from the model. However, this is not always possible or desirable for either aesthetic or visual reasons. In this case *ridge regression* (see, e.g., [104] or [420]) is an alternative. The idea is to give up the unbiased property of least squares estimation in exchange for reduced estimator uncertainty.

In ridge regression, constraints are implicitly placed on the model parameters and the least squares problem is then solved subject to those constraints. These constraints result in a modified, less variable, least squares estimator. First, note that the ordinary least squares estimator (8.27) can be written

$$\hat{\vec{\mathbf{a}}} = \mathcal{P}^{\mathrm{T}}\Lambda^{-1}\mathcal{P}\mathcal{X}^{\mathrm{T}}\vec{\mathbf{y}},$$

where the columns of $\mathcal{P}$ are the normalized eigenvectors of $\mathcal{X}^{\mathrm{T}}\mathcal{X}$ and Λ is the corresponding diagonal matrix of eigenvalues. That is, $\mathcal{X}^{\mathrm{T}}\mathcal{X} = \mathcal{P}\Lambda\mathcal{P}$. One form of a *generalized ridge regression* estimator, which conveys the general idea and the source of the term 'ridge,' is given by

$$\hat{\vec{\mathbf{a}}}_{ridge} = \mathcal{P}^{\mathrm{T}}(\Lambda + \mathcal{D})^{-1}\mathcal{P}\mathcal{X}^{\mathrm{T}}\vec{\mathbf{y}}$$

where $\mathcal{D}$ is a diagonal matrix of positive constants. The effect of inflating the eigenvalues in this way is to downplay the importance of the off-diagonal elements of $\mathcal{X}^{\mathrm{T}}\mathcal{X}$ when this matrix is inverted. The constants, are, of course not known. Ridge regression algorithms use a variety of procedures to choose appropriate constants for a given design matrix $\mathcal{X}$.

8.5 Model Selection

None of the inference methods described in Sections 8.3 and 8.4 performs reliably if factors are missing from the model. On the other hand, if the model contains unnecessary factors it will be unnecessarily complex and will specify more poorly than it could otherwise. We therefore briefly discuss methods helpful for developing *parsimonious* models. The main goal here is not so much to specify accurately or estimate a complete model, as it is to perform screening to discover which factors contribute significantly to variation in the response.

The primary screening principle we use is that a variable should not be included in a model if it does not significantly increase the regression sum of squares $\mathcal{SSR}$. A careful and systematic approach is needed because a test of an individual parameter, which asks whether a specific factor makes a significant contribution after accounting for all other factors, may hide the importance of that factor within a group of factors.

When the number of factors in a problem is small it is usually possible to choose a suitable model, as in the example above, using the tools of [8.4.9]. However, in problems with a large number of factors that are each potentially important for representing the conditional mean of the response variable, an automated procedure is needed.

8.5.1 Stepwise Regression: Introduction. Stepwise regression is the iterative application of forward selection and backward elimination steps. We first describe these procedures and then return to the subject of stepwise regression. However, we need to introduce some additional notation before delving into detail.

We use $\mathcal{SSR}_{l_1,\dots,l_p}$ to represent the sum of squares due to regression when the p factors $\mathbf{X}_{l_1},\dots,\mathbf{X}_{l_p}$ are included in the multiple regression model. Similar notation is used for the sum of squared errors. We use $\mathcal{SSR}_{l_{(p+1)}|l_1,\dots,l_p}$ to denote the increase in the regression sum of squares that comes about by adding factor $\mathbf{X}_{l_{(p+1)}}$ to the model. That is

$$\mathcal{SSR}_{l_{(p+1)}|l_1,\dots,l_p} = \mathcal{SSR}_{l_1,\dots,l_{(p+1)}} - \mathcal{SSR}_{l_1,\dots,l_p}.$$

8.5.2 Forward Selection. Before any fitting is done, a decision should be made about whether or not to include an intercept in the model. If an intercept is to be included, it should be included at all steps of the forward selection procedure. The steps are as follows.

1 Simple linear regression is performed with each factor. The factor $\mathbf{X}_{l_1}$ for which $\mathcal{SSR}_{l_1}$ is greatest is selected as the *initial factor*.

2 Search for factor $\mathbf{X}_{l_2}$, $l_2 \notin \{l_1\}$, for which the incremental regression sum of squares $\mathcal{SSR}_{l_2|\{l_1\}}$ is greatest. The notation $\{l_1\}$ denotes the list of previously selected factors and $l_2 \notin \{l_1\}$ denotes any factor not in $\{l_1\}$. This list contains only the initial factor after step 1 has been completed.

3 Test the hypothesis that inclusion of $\mathbf{X}_{l_2}$ significantly reduces the regression sum of squares by computing

$$\mathbf{F} = \frac{\mathcal{SSR}_{l_2|\{l_1\}}}{(\mathcal{SSE}_{\{l_1\},l_2})/(n' - (1 + |\{l_1\}|))}$$

where $n' = n$ or $n - 1$ depending upon whether or not the intercept is included, and $\{l_1\}$ denotes the list of previously selected factors. $\mathbf{F}$ is compared with the critical values of $F(1, n' - (1 + |\{l_1\}|))$.

4 Stop at the previous iteration if $\mathbf{X}_{l_2}$ does not significantly increase the regression sum of squares. Otherwise, include $\mathbf{X}_{l_2}$ in the model and repeat steps 2 and 3.

8.5.3 Backward Elimination. The backward elimination procedure operates similarly to the forward selection procedure.

1 Fit the full model.

2 Search for the factor that reduces the regression sum of squares by the smallest amount when it is removed from the model.

3 Conduct an F test to determine whether this factor explains a significant amount of variance in the presence of all other factors remaining in the model at this point. Remove the variable from the model if it does not contribute significant variance.

4 Repeat steps 2 and 3 until no variable can be removed from the model.

8.5.4 Stepwise Regression. The stepwise regression procedure combines forward selection with backward elimination. As forward selection progresses, factors selected early on may become redundant when related factors are selected during later steps. Therefore, in stepwise regression, backward elimination is performed after every forward selection step to remove redundant variables from the model. Forward regression and backward elimination steps are repeated until no further change can be made to the model.

8.5.5 All Subsets Regression. Another screening approach that has become feasible with increased computing power is all subsets regression. As the name suggests, the procedure fits all 2^k possible subsets of factors to the response variable. The screening statistic C_p

$$C_{p\{l_1,\dots,l_p\}} = \frac{\mathcal{SSE}_{\{l_1,\dots,l_p\}}}{\widehat{\sigma}_E^2} - (n-2p)$$

is computed for every model and a plot of points $(p, C_{p\{l_1,\dots,l_p\}})$ is produced. Note that the error variance estimate is generally obtained from the full model. A model that fits well will have a computed C_p that lies close to the $C_p = p$ line. This is therefore used as a guide for selecting models that require more careful examination (see [104] or [420] for details).

Alternatively, Akaike's information criterion (AIC) [6] could be used as the screening statistic. The idea here is to choose the model that minimizes the AIC criterion given by

$$\begin{aligned} AIC &= -2l(\widehat{\mathbf{a}}_{l_1},\dots,\widehat{\mathbf{a}}_{l_p}) + 2p \\ &= n\log(2\pi\widehat{\sigma}_E^2) + \frac{\mathcal{SSE}_{\{l_1,\dots,l_p\}}}{\widehat{\sigma}_E^2} + 2p, \end{aligned}$$

where $l(\widehat{\mathbf{a}}_{l_1},\dots,\widehat{\mathbf{a}}_{l_p})$ is the log-likelihood function (see [8.3.4]). That is, minimizing AIC is equivalent to maximizing likelihood, but penalized for the number of parameters in the model. As with C_p, we use the best available estimate of the error variance when computing AIC, the estimator of σ_E^2 obtained from the least squares fit of the full model. Note the similarity between C_p and AIC.

8.5.6 Numerical Forecast Improvement. One meteorological application for screening regression techniques is in the development of statistical procedures for improving numerical weather forecasts.[8] Improvement is required because global, and even regional, numerical forecast models do not accurately represent sub-grid scale processes. Statistical procedures attempt to exploit systematic relationships between the large-scale flow of the free atmosphere, which is both well observed and well represented by numerical forecast models, and local phenomena.

MOS procedures (see Glahn and Lowry [140] or Klein and Glahn [226]) rely upon 'specification equations' that describe statistical relationships between numerical forecasts of atmospheric conditions in the troposphere (i.e., model output) and observed variables at specific points on the surface, such as precipitation and temperature. The primary tool used is multiple linear regression. The advantage of MOS over perfect prog is that it inherently corrects for forecast model biases in both the mean and variance. A disadvantage of MOS is that the specification equations need to adapt constantly to the changing characteristics of the numerical forecast model and its associated data assimilation systems.

Perfect prog procedures (See Klein, Lewis, and Enger [227], Brunet et al. [71]) are similar to MOS procedures except that the specification equations describe simultaneous relationships between the analysed (as opposed to forecast) free atmosphere and observed variables at specific points on the surface. The resulting specification equations are more stable than the MOS equations because the

[8]Many other techniques, such as cluster analysis [163, 115], multiple discriminant analysis [267] and classification and regression trees [63] are also used. See, for example, Yacowar [435].

data used to fit the equations are less affected by periodic model changes. However, perfect prog specification equations do not account for forecast model biases. Statistical downscaling procedures (see [97, 152, 252, 403]) that link regional and local aspects of simulated climate change are a variation of perfect prog.

Screening regression is strongly affected by the artificial skill phenomenon discussed in [8.3.12] and also [18.4.7] (see, e.g., Ross [332] or Unger [377]) because these methods select a model from a set of possible models that adapts most closely to the data. Ross [332] citing Copas [91] and Miller [278] points out that using the same sample to select the model and estimate its coefficients is 'overfitting' and can lead to models that perform very poorly on independent data. It may therefore be wise to use three data sets in conjunction with screening techniques; one with which to identify the model, one with which to estimate coefficients, and one for validation.

Small data sets often make this strategy impossible to use. An alternative method for estimating the skill of the model is cross-validation, but Unger [377] demonstrates that cross-validation does not provide reliable skill estimates because of the way in which it interacts with the screening methods. He proposes the use of a method called *bi-directional retroactive real-time* (BRRT) validation instead. The idea is that a substantial subset of recent data is withheld. A screening technique is used to fit a model to the earlier data (called the base data set). This model is used to forecast the first observation in the withheld set. It is then added to the base data set and the process is repeated, thereby collecting a set of verification statistics of the same size as the withheld data set. More verification data are collected by running the same process in reverse (hence the term 'bi-directional'). Unger finds that BRRT gives reliable estimates of skill 'when the number of candidate predictors is low.'

8.6 Some Other Topics

8.6.1 Weighted Regression. The working assumption to this point has been that the errors $\mathbf{E}_i$ are normally distributed, independent, and identically distributed. That is, the vector of errors $\vec{\mathbf{E}}$ is jointly distributed $\mathcal{N}(\vec{0}, \sigma_E^2\mathcal{I})$, where $\mathcal{I}$ denotes the $n \times n$ identity matrix. We noted in [8.3.16] that departures from the independence assumption lead to difficulties. If the errors are not independent, $\vec{\mathbf{E}} \sim \mathcal{N}(\vec{0}, \Sigma_{\vec{\mathbf{E}}})$, where $\Sigma_{\vec{\mathbf{E}}}$ is a non-diagonal covariance matrix. If there are departures from the constant variance assumption (heteroscedasticity; see [8.3.13]), then although $\Sigma_{\vec{\mathbf{E}}}$ may be diagonal, the elements on the diagonal are not constant. In general, ordinary least squares estimates are less than optimal (they are no longer maximum likelihood estimates) whenever $\Sigma_{\vec{\mathbf{E}}} \neq \sigma_E^2\mathcal{I}$.

When $\Sigma_{\vec{\mathbf{E}}}$ is known, the optimality properties of ordinary least squares estimators are restored by solving the *generalized normal equations.* Instead of minimizing $(\vec{\mathbf{Y}} - \mathcal{X}\vec{\mathbf{a}})^{\mathrm{T}}(\vec{\mathbf{Y}} - \mathcal{X}\vec{\mathbf{a}})$, we choose $\vec{\mathbf{a}}$ to minimize

$$(\vec{\mathbf{Y}} - \mathcal{X}\vec{\mathbf{a}})^{\mathrm{T}}\Sigma_{\vec{\mathbf{E}}}^{-1}(\vec{\mathbf{Y}} - \mathcal{X}\vec{\mathbf{a}}). \tag{8.38}$$

The *generalized least squares estimators* are therefore given by

$$\widehat{\vec{\mathbf{a}}} = (\mathcal{X}^{\mathrm{T}}\Sigma_{\vec{\mathbf{E}}}^{-1}\mathcal{X})^{-1}\mathcal{X}^{\mathrm{T}}\Sigma_{\vec{\mathbf{E}}}^{-1}\vec{\mathbf{Y}}.$$

Weighted regression is the special case in which $\Sigma_{\vec{\mathbf{E}}}$ is diagonal. Then quadratic form (8.38) reduces to

$$\sum_{i=1}^{n} w_i^2\Big(\mathbf{Y}_i - \sum_{l=1}^{k} a_l \mathbf{x}_l\Big)^2,$$

where weight w_i is proportional to $1/\sigma_{E_i}$.

Weighted regression is an option to consider when errors are heteroscedastic, and transformation of the response variable [8.6.2] does not result in a model with a reasonable physical interpretation. Note that in order to perform weighted regression it is only necessary to know the relative sizes of the error variances, not the variances themselves. Very good prior information about the relative variances may be available from sampling or physical considerations.

8.6.2 Transformations. Transformation of variables can be used in several ways in regression analysis. First, many models that appear to be nonlinear in their parameters can easily be made linear.

- Multiplicative models, such as

  $$\mathbf{Y} = a_0\mathbf{x}_1^{a_1}\mathbf{x}_2^{a_2}\mathbf{x}_3^{a_3}\mathbf{E},$$

  can be made linear by taking logarithms to obtain

  $$\begin{aligned}\ln\mathbf{Y} &= a_0' + a_1\ln\mathbf{x}_1 + a_2\ln\mathbf{x}_2 \\ &\quad + a_3\ln\mathbf{x}_3 + \mathbf{E}'.\end{aligned}$$

  Fitting can now proceed provided appropriate assumptions can be made about $\mathbf{E}' = \ln\mathbf{E}$.

- Reciprocal models, such as

$$\mathbf{Y} = \frac{1}{a_0 + a_1\mathbf{x}_1 + \mathbf{E}},$$

can be made linear by inverting the dependent variable to obtain

$$\frac{1}{\mathbf{Y}} = a_0 + a_1\mathbf{x}_1 + \mathbf{E}.$$

- Bilinear models, such as

$$\mathbf{Y} = \frac{a_0\mathbf{x}_1}{a_1 + a_2\mathbf{x}_2 + \mathbf{E}},$$

can be made linear by cross-multiplying to obtain

$$\frac{\mathbf{x}_1}{\mathbf{Y}} = \frac{a_1}{a_0} + \frac{a_2}{a_0}\mathbf{x}_2 + \mathbf{E}',$$

or by inversion to obtain

$$\frac{1}{\mathbf{Y}} = \frac{a_1}{a_0}\frac{1}{\mathbf{x}_1} + \frac{a_2}{a_0}\frac{\mathbf{x}_2}{\mathbf{x}_1} + \frac{\mathbf{E}'}{\mathbf{x}_1},$$

where $\mathbf{E}' = \mathbf{E}/a_0$. Least squares estimators can then be obtained for the ratios a_1/a_0 and a_2/a_0. The particular form that is chosen depends upon whether or not $\mathbf{x}_1$ can become zero, and whether $\mathbf{E}/\mathbf{x}_1$ better satisfies the distributional assumptions needed to make statistical inferences about the model than $\mathbf{E}$ itself.

- Many models can be made linear in their parameters through a combination of transformations. For example, a model of the form

$$\mathbf{Y} = \frac{1}{1 + a_0\mathbf{x}_1^{a_1}\mathbf{E}}$$

can be re-expressed as

$$\ln\Big(\frac{1}{\mathbf{Y}} - 1\Big) = a_0' + a_1 \ln \mathbf{x}_1 + \mathbf{E}'.$$

Some models are intrinsically nonlinear and can not be re-expressed in a way that is linear in the parameters. For example, Xu and Randall [434] propose the following parameterization for the fraction C_S of the sky in a GCM grid box that is covered by stratiform clouds:

$$C_S = r_H^p(a - e^{-\alpha\bar{q}_e}), \tag{8.39}$$

where r_H is relative humidity, $\bar{q}_e$ is the large-scale condensate (cloud water plus ice) mixing ratio, and

$$\alpha = \alpha_0\big((1 - r_H)q^*\big)^{-\gamma}, \tag{8.40}$$

where q^* is the water vapour mixing ratio. Constants p, α_0, and γ are scalar parameters that are estimated by fitting model (8.39, 8.40) to the output from a high resolution *cloud ensemble model* (CEM); see, for example, Xu and Krueger [433]. CEMs are used in the development of cloud parameterizations because detailed observational data on cloud fields are scarce.

A second reason for using transformations in regression is to change the model so that it better satisfies the assumptions necessary to make inferences about the estimated parameters and about unobserved values of the dependent variable. For example, the heteroscedasticity displayed in Figure 8.8 can be removed by fitting the model

$$\frac{\mathbf{Y}}{\mathbf{x}(1-\mathbf{x})} = a_0 + a_1\frac{1}{1-\mathbf{x}} + \mathbf{E}'$$

instead of

$$\mathbf{Y} = a_0 + a_1\mathbf{x} + \mathbf{E}.$$

Suitable variance stabilizing transforms are found by physical reasoning and by plotting residuals against the independent variables.

8.6.3 Nonlinear Regression. Many of the ideas discussed in this chapter can be extended to the fitting and analysis of intrinsically nonlinear models such as (8.39, 8.40) provided it is possible to assume that errors are iid and normally distributed. Then a reasonable nonlinear regression model for the conditional mean of the response variable has the form

$$\mathbf{Y}_i = h(\mathbf{x}_{1,i}, \ldots, \mathbf{x}_{k,i}|\mathbf{a}_1, \ldots, \mathbf{a}_p) + \mathbf{E}_i.$$

That is, the conditional mean of the response variable is a function $h(\cdot|\cdot)$ of k factors that is known up to the value of p coefficients. Function h is nonlinear in at least some of the unknown coefficients. Parameters are estimated by using function minimization techniques (such as the method of steepest descent, see [322]) to minimize the sum of squared errors

$$\mathcal{SSE} = \sum_{i=1}^{n}(\mathbf{y}_i - h(\mathbf{x}_{1,i}, \ldots, \mathbf{x}_{k,i}|\mathbf{a}_1, \ldots, \mathbf{a}_p))^2.$$

Approximate inferences are possible by linearizing h about $\widehat{\mathbf{a}}$. See Bates and Watts [35] or Draper and Smith [104] for more details.

9 Analysis of Variance

9.1 Introduction

In this chapter we describe some methods that can be used to diagnose qualitative relationships between a quantitative response variable, that is, a variable measured on a continuous scale, and one or more factors that are classified, perhaps according to level, or perhaps only according to their presence or absence.

Our purpose is to introduce only some of the concepts of *experimental design* and *analysis of variance* (*ANOVA*). We illustrate the general patterns of analysis and thought with these methods using a couple of examples from the climate literature. Our coverage of the subject is necessarily far from complete. A more complete treatment of the topic can be found in Box, Hunter, and Hunter [59]. Cochran and Cox [87] provide a classical treatment. Anderson and McLean [13] provide a good description of ANOVA for non-specialists.

9.1.1 Terminology and Purpose of Experimental Design. The classical setting for ANOVA and experimental design methods is agricultural experiments, so much of the associated terminology has its roots in agriculture.

For example, a typical agricultural experiment might be designed to determine the effect of two factors, say, fertilizer (applied at one of three different levels) and tillage (the land is either tilled, or not tilled before seeding) on crop yield. The experiment might be conducted as a *factorial experiment* in which each possible treatment combination is applied to a separate plot of land according to an experimental design.

The simplest experimental design is a *completely randomized* design in which treatment combinations are randomly assigned to plots of land (or more generally, *experimental units*: anything to which treatments are applied). In experiments without *replication*, each treatment combination is applied exactly once. Thus in the simple agricultural example introduced here, six plots of land would be used. In experiments with replication, at least some of the treatment combinations are applied more than once.

9.1.2 Experimental Designs in Climatology. The experimental units are simulations in designed experiments conducted with General Circulation Models. Treatments applied to the simulations could be various combinations of parameterizations of sub-grid scale processes, parameter values for a given set of parameterizations (as in Gough and Welch [145]), conditions imposed at the top of the atmosphere (e.g., a rigid lid as opposed to a sponge layer) or at the lower boundary (e.g., to examine the model's systematic response to an imposed sea-surface temperature anomaly such as the standard Rasmusson and Carpenter El-Niño anomaly [330], as in Boer [51]), vertical resolutions for a model, and so on.

Unfortunately, developers of GCMs have not generally relied upon designed experiments to differentiate objectively between treatments because GCM experimentation is quite expensive. However, developers of models that are cheaper to run (such as basin scale ocean models and sea-ice models) have started to study their models objectively through the use of designed experiments. Gough and Welch [145], Chapman et al. [79], and Bowman, Sacks, and Chang [58] are examples. The Gough and Welch example is discussed in Section 9.5.

9.1.3 Isolating External Sources of Variability. A deficiency of the completely randomized design is that variation in the response variable is induced both by the treatments and by variations between experimental units. In agricultural experiments, variations might occur because the fertility is not uniform from one plot to the next. In GCM experiments, simulations might be conducted with different computers, which, owing to the peculiarities of a particular machine, leads to small differences amongst simulated climates. In the language of statisticians, the treatment effects are *confounded* with the plot effects in the completely randomized design.

The ability to detect treatment effects can be enhanced if experimental designs are constructed that reduce or eliminate external sources of variation. One such design is the *randomized complete block design*. In our pedagogical agricultural example, we could split each plot into six sub-plots, then randomly assign treatments to sub-plots with the constraint that every treatment combination appears once within every plot. Presumably fertility is relatively uniform within each plot, so all responses within a plot are subject to the same variations induced by differences in plot fertility.

An extra factor, the *block* (or plot) *effect*, is effectively introduced into the experiment. When the results of the experiment are subsequently analysed using the methods of ANOVA, we will be able to isolate variation in the data induced by the blocks from variation induced by the treatments, and therefore make better inferences about the effect of the treatments.

9.1.4 Randomized Complete Block Climate Experiments. Designed climate experiments, because of their huge cost, might have to be run on several computers, perhaps not all of the same type. Different types of machines have different schemes for representing real numbers, slightly different implementations of intrinsic functions, different numerical precisions, etc., resulting in simulated climates that are slightly, but sometimes detectably, different.

However, complete block experiments may not be feasible as there may not be sufficient computing resources available on a given machine to replicate every treatment combination. It may therefore be necessary to use another design, such as a *fractional factorial design* (see Box et al. [59]) in which only some fraction of treatment combinations is applied to the simulations conducted on each computer. The effects of some treatment combinations will be confounded with the block effect in a fractional design. The art of designing a fractional factorial experiment depends primarily on making informed choices about the effects that are likely to be small enough to be safely confounded with the block effect.

9.1.5 What is ANOVA and How is it Different from Regression Analysis? There is a very strong connection between the experimental design and the subsequent analysis of variance used to analyse the data generated by the experiment. Formally, the models fitted using ANOVA are regressions in which the factors on the right hand side of the equation are indicator variables. The choice of model is not very flexible because the indicator variables are used to identify the specific treatment and block combination that resulted in each realization of the response variable. Some terms in ANOVA models may be of little direct interest to the analyst because they are only present to account for the variation, such as between block variation, that the experiment was designed to isolate from the effects of interest.

Perhaps because of the limited flexibility in the choice of model, the estimated values of model coefficients are generally of less interest than the partitioning of variability according to its source and determining which sources contribute significantly to the variation in the data obtained from the experiment. The examples discussed in this chapter show that this is also largely true in climatological applications of ANOVA methodology. The model coefficients or, at least, the relationships between model coefficients, are only of interest after it has been determined that a factor has a significant effect on the response variable. The specific value of the coefficient is irrelevant in many problems because the factor level may not have been measured quantitatively. Even when the levels are known, values of the response variable might only be available for a few levels of a factor, making it inappropriate to attempt to diagnose systematic relationships between the factor and the mean of the response variable.

9.1.6 Applications to Climatology. In the past, it was relatively uncommon to apply ANOVA to climatological and meteorological problems. This is partly because our observational data do not lend themselves well to analysis using methods appropriate for designed experiments, and partly because the cost of properly designed climate model experiments was prohibitively high in the past, although this situation is now changing.

We will describe applications of ANOVA to the analysis of interannual variability in an experiment consisting of multiple AMIP[1]

[1] The AMIP (Atmospheric Model Intercomparison Project) encompasses most of the world's climate modelling groups (see Gates [137] for a description of the project and its goals). All participants ran a standard 10-year atmospheric simulation imposing observed 1979–88 monthly mean sea-surface temperatures and sea-ice extents at the lower boundary. Several groups, such as the Canadian Centre for Climate Modelling and Analysis, ran multiple AMIP simulations from randomly selected initial conditions.

simulations conducted with the CCC GCMII (see McFarlane et al. [270] for a description of CCC GCMII; and see Zwiers [444, 449] and Wang and Zwiers [414] for analysis of the AMIP experiments).[2] We will also describe an application of so-called *space filling experimental designs* to the problem of parameter specification in a basin scale ocean model (Gough and Welch [145]).

9.1.7 Outline. The models and methods used in *one way analysis of variance* are described in Section 9.2. These are methods suitable for use in simple experiments that intercompare the mean responses to a number of different treatments, or levels of one treatment. One way ANOVA methods are also appropriate when it is necessary to intercompare the means of two or more samples.

Both *fixed* and *random* effects models are discussed in Section 9.2. A *fixed effects* model describes the effect of a treatment as a change in the mean of the response variable. This is a deterministic response to a treatment that can be replicated from one realization of the experiment to the next. A *random effects* model describes the effect of the treatment with a random variable, a form of response that can not be replicated from one experiment to the next. Methods of inference are discussed for both types of one way model. The relationship between ANOVA and regression is described at the end of Section 9.2.

The models and methods used in *two way analysis of variance* are described in Section 9.3. These models are used to analyse experiments conducted with randomized complete block designs or completely randomized designs in which two different kinds of treatment have been applied. The discussion in this section is limited to fixed effects models.

The Canadian Centre for Climate Modelling and Analysis (CCCma) AMIP experiment is used as a working example throughout Sections 9.2 and 9.3. This experiment is analysed in more detail in Section 9.4 with a two way model containing a mixture of fixed and random effects. An additional example is discussed in Section 9.5, where we describe Gough and Welch's [145] use of space filling designs to study the sensitivity of a basin scale ocean GCM to its parameter settings.

[2]Several other analyses of *ensembles* of climate variability have recently appeared in the climate literature, including Rowell [336], Rowell and Zwiers [337], Kumar et al. [232], Folland and Rowell [123], Stern and Miyakoda [358], and Anderson and Stern [11].

We complete this section by briefly introducing the CCCma multiple AMIP simulations.

9.1.8 Example: Multiple AMIP Simulations. AMIP is a *Level 2* model intercomparison as defined by the WGNE (Working Group on Numerical Experimentation). The more primitive Level 1 intercomparisons apply common diagnostics to climate simulations as available. At Level 2, simulations are conducted under standard conditions, common diagnostics are computed, and validation is made against a common data set. Level 3 encompasses Level 2 and also requires that models use a common resolution and common subroutines.

An AMIP simulation (see Gates [137]) is a 10-year simulation conducted with an atmospheric climate model in which the monthly mean sea-surface temperatures and sea-ice boundaries are prescribed to follow the January 1979 to December 1988 observations.

The CCCma AMIP simulations were conducted with a spectral model ([270] and [52]) that operates at 'T32' horizontal resolution (approximately $3.75° \times 3.75°$), has 10 layers in the vertical, and a 20-minute time step. The first simulation, conducted on a Cray XMP, was initiated from 1 January 1979 FGGE (First GARP Global Experiment [44]) conditions. Five additional AMIP simulations, performed on a NEC SX/3, were started from previously simulated 1 January model states. These initial states were selected from the control run at two-year intervals. Analysis of the AMIP simulations begins in June of the first simulated year. That is, the first five months of each simulation is regarded as a 'spin-up' period during which the model forgets about its initial conditions, and slow (primarily land surface) processes equilibrate with the imposed lower boundary conditions. Because the atmosphere forgets its initial state very quickly, the effect of selecting different initial conditions is basically to select independent realizations of the simulated climate's path through its phase space. For all intents and purposes, these six simulations can be regarded as having been initiated from randomly selected initial states.

9.2 One Way Analysis of Variance

9.2.1 The One Way ANOVA Model. Suppose that an experiment has been conducted that results in J samples of size n represented by random variables $\mathbf{Y}_{ij}$, for $i = 1, \ldots, n$ and $j = 1, \ldots, J$. The subscript j identifies the sample, and the subscript i identifies the element of the sample. Assume that the sampling is done in such a way that all random variables are independent, normal, and have the same variance. Also assume that the means are constant within samples. That is, in sample j

$$\mathcal{E}(\mathbf{Y}_{ij}) = \mu_j$$

for all $i = 1, \ldots, n$, or equivalently that

$$\mathcal{E}(\mathbf{Y}_{ij}) = \mu + a_j$$

for all $i = 1, \ldots, n$, where μ is the overall mean given by

$$\mu = \frac{1}{J}\sum_{j=1}^{J} \mu_j,$$

and a_j is the difference

$$a_j = \mu_j - \mu$$

between the expectation of $\mathbf{Y}_{ij}$ and the overall mean. The coefficients a_j are often called *treatment effects*.

An appropriate statistical model for this type of data is

$$\mathbf{Y}_{ij} = \mu + a_j + \mathbf{E}_{ij}, \tag{9.1}$$

where the errors $\mathbf{E}_{ij}$ are iid zero mean normal random variables with variance σ_E^2 (i.e., $\mathbf{E}_{ij} \sim \mathcal{N}(0, \sigma_E^2)$) and the coefficients a_j are constrained to sum to zero.

9.2.2 Where Do the Data Come From? Data of this sort might be a result of a planned experiment that examined the effects of J treatments by applying each treatment to n experimental units. The experimenter would have made sure that the *experimental units* (e.g., people, rats, plots of land, climate simulations, etc.) were representative of the population from which they were drawn and that the treatments were applied to the experimental units in random order.

However, data of this sort might also have been obtained with somewhat less attention to experimental design. Suppose, for example, that we wish to use an ensemble of AMIP simulations to determine whether the specified sea-surface temperatures and sea-ice boundaries have an effect on the interannual variability of the simulated December, January, February (DJF) climate. The 10-year AMIP period (January 1979 to December 1988) includes nine complete DJF seasons. Each DJF season can be thought of as the result of a different treatment (the specified sea-surface temperature and sea-ice regime) applied to a different experimental unit (a year in a simulation). Because the AMIP simulations are conducted with an atmospheric model, it seems reasonable to assume that consecutive mean DJF states simulated by the model are approximately independent of each other. Thus a simulation can be thought of as the outcome of a completely randomized experiment in which each of the $J = 9$ treatments is applied once. Each simulation in an ensemble of AMIP simulations can be considered a replication of the nine treatment experiment. Because the AMIP simulations in the six member CCCma ensemble were started from randomly selected initial conditions, the replications can also be assumed to be independent of one another. Thus it appears that seasonal mean data from the CCCma AMIP experiment can be analysed using a one way ANOVA appropriate for data obtained from a replicated completely randomized design with $J = 9$ treatments and $n = 6$ replicates.

9.2.3 Partitioning Variance into Treatment and Error Components. In regression analysis (see Chapter 8) we started with a model such as (9.1), developed parameter estimators, and slowly proceeded towards an analysis of variance that partitioned the total sum of squares into regression and sum of squared errors. That approach is also useful in analysis of variance because it provides a direct means of obtaining distributional properties for confidence intervals and test statistics. However, here we use a more intuitive approach to the analysis of variance that begins with the partitioning of variability.

Before beginning, let us introduce a little notation. Let

$$\overline{\mathbf{Y}}_{\circ\circ} = \frac{1}{nJ}\sum_{i=1}^{n}\sum_{j=1}^{J} \mathbf{Y}_{ij}$$

be the mean of all the observations and let

$$\overline{\mathbf{Y}}_{\circ j} = \frac{1}{n}\sum_{i=1}^{n} \mathbf{Y}_{ij}$$

be the mean of all the observations that were the result of the jth treatment. The '$\circ$' notation indicates averaging over the missing subscript. By substituting the model (9.1) into these expressions and taking expectations, it is easily shown that $\overline{\mathbf{Y}}_{\circ\circ}$ is an unbiased estimator of μ and that $\overline{\mathbf{Y}}_{\circ j}$ is an unbiased estimator of $\mu + a_j$. Therefore $\overline{\mathbf{Y}}_{\circ j} - \overline{\mathbf{Y}}_{\circ\circ}$ is an unbiased estimator of a_j.

The total sum of squares $\mathcal{SST}$, given by

$$\mathcal{SST} = \sum_{i=1}^{n}\sum_{j=1}^{J}(\mathbf{Y}_{ij} - \overline{\mathbf{Y}}_{oo})^2,$$

can be partitioned as follows. First, subtract and add $\overline{\mathbf{Y}}_{oj}$ inside the squared difference to obtain

$$\mathcal{SST} = \sum_{i=1}^{n}\sum_{j=1}^{J}\left((\mathbf{Y}_{ij} - \overline{\mathbf{Y}}_{oj}) + (\overline{\mathbf{Y}}_{oj} - \overline{\mathbf{Y}}_{oo})\right)^2.$$

Then square and sum the individual terms to obtain

$$\begin{aligned}\mathcal{SST} &= n\sum_{j=1}^{J}(\overline{\mathbf{Y}}_{oj} - \overline{\mathbf{Y}}_{oo})^2 \\ &\quad + \sum_{i=1}^{n}\sum_{j=1}^{J}(\mathbf{Y}_{ij} - \overline{\mathbf{Y}}_{oj})^2 \\ &\quad - 2\sum_{i=1}^{n}\sum_{j=1}^{J}(\overline{\mathbf{Y}}_{oj} - \overline{\mathbf{Y}}_{oo})(\mathbf{Y}_{ij} - \overline{\mathbf{Y}}_{oj}).\end{aligned}$$

The sum of the cross-products is zero because $\sum_{i=1}^{n}(\mathbf{Y}_{ij} - \overline{\mathbf{Y}}_{oj}) = 0$ for each j. Thus we have

$$\mathcal{SST} = \mathcal{SSA} + \mathcal{SSE},$$

where

$$\mathcal{SSA} = n\sum_{j=1}^{J}(\overline{\mathbf{Y}}_{oj} - \overline{\mathbf{Y}}_{oo})^2, \qquad (9.2)$$

and

$$\mathcal{SSE} = \sum_{i=1}^{n}\sum_{j=1}^{J}(\mathbf{Y}_{ij} - \overline{\mathbf{Y}}_{oj})^2.$$

$\mathcal{SSA}$ is often referred to as the *treatment sum of squares* or the *between blocks sum of squares*. $\mathcal{SSE}$ is referred to as the *sum of squared errors* or *within blocks sum of squares*. The latter names are particularly descriptive of the calculations that were performed.

The treatment sum of squares is taken over J deviations that sum to zero, thus it has $J - 1$ degrees of freedom (df). The sum of squared errors is taken over nJ deviations such that deviations within a particular *block* (or sample) must sum to zero. That is, the sum of squared errors is taken over deviations that are subject to J constraints. Consequently, $\mathcal{SSE}$ has $(n-1)J$ df. The total sum of squares is summed over nJ deviations which are subject to only one constraint (i.e., that they sum to zero) and therefore the total sum of squares have $nJ - 1$ df. In summary, we have the following partition of the total sum of squares and degrees of freedom.

Source	Sum of Squares	df
Treatment	$\mathcal{SSA}$	$J-1$
Error	$\mathcal{SSE}$	$J(n-1)$
Total	$\mathcal{SST}$	$Jn-1$

9.2.4 Testing for a Treatment Effect. The effect of the jth treatment is represented by coefficient a_j in model (9.1). Thus the no treatment effect hypothesis can be expressed as

$$\mathrm{H}_0\colon a_1 = \cdots = a_J = 0, \qquad (9.3)$$

or, equivalently, as

$$\mathrm{H}_0\colon \sum_{j=1}^{J} a_j^2 = 0.$$

We wish to test H_0 against the alternative hypothesis that at least some of the coefficients a_j are different from zero. That is, we test H_0 against

$$\mathrm{H}_a\colon \sum_{j=1}^{J} a_j^2 > 0.$$

We have already noted that $\overline{\mathbf{Y}}_{oj} - \overline{\mathbf{Y}}_{oo}$ is an unbiased estimator of a_j, so it would seem reasonable that a test of H_0 should be based on $\mathcal{SSA}$, since it is proportional to the sum of squared coefficient estimates. Therefore let us examine the treatment sum of squares $\mathcal{SSA}$, given in (9.2), more closely.

Substituting the model (9.1) into (9.2) we obtain

$$\begin{aligned}\mathcal{SSA} &= n\sum_{j=1}^{J}(\mu + a_j + \overline{\mathbf{E}}_{oj} - (\mu + \overline{\mathbf{E}}_{oo}))^2 \\ &= n\sum_{j=1}^{J}a_j^2 + n\sum_{j=1}^{J}(\overline{\mathbf{E}}_{oj} - \overline{\mathbf{E}}_{oo})^2. \qquad (9.4)\end{aligned}$$

Now note that the second term in (9.4) estimates $(J-1)\sigma_E^2$. We can show this by means of (4.6) after noting the following.

1 $\overline{\mathbf{E}}_{oj}$ is the average of n iid errors that have variance σ_E^2. Therefore, using (4.4), we see that the variance of $\overline{\mathbf{E}}_{oj}$ is σ_E^2/n.

2 All errors $\mathbf{E}_{ij}$ are independent. Therefore the within block mean errors $\overline{\mathbf{E}}_{oj}$ are also independent.

It follows that the expected value of $\mathcal{SSA}$ is

$$\mathcal{E}(\mathcal{SSA}) = n\sum_{j=1}^{J}a_j^2 + (J-1)\sigma_E^2. \qquad (9.5)$$

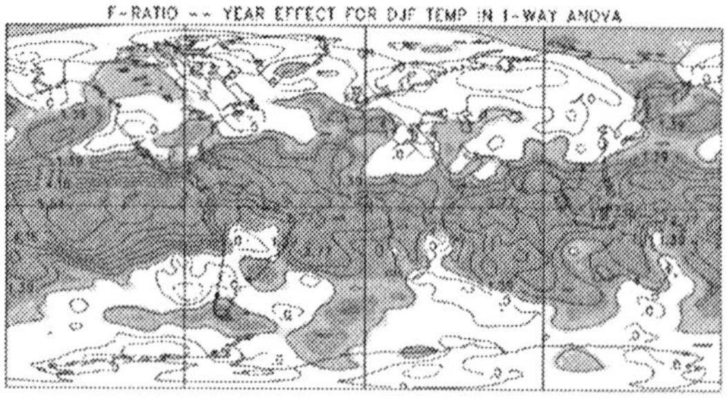

Figure 9.1: *The natural log of the F-ratios for the year effect obtained from a one way analysis of variance of DJF mean 850 hPa temperature simulated in the six member ensemble of CCCma AMIP simulations. The shading indicates ratios that are significantly greater than 1 at the 10% significance level.*

Equation (9.5) shows that $\mathcal{SSA}/(J-1)$ estimates σ_E^2 when H_0 is true, and that it estimates a number larger than σ_E^2 when H_0 is false. It may therefore be possible to construct a test of H_0 if another statistic can be found that estimates only σ_E^2 regardless of whether or not H_0 is true. An argument similar to the one we just completed shows that $\mathcal{SSE}/((n-1)J)$ has this property. Hence

$$\mathbf{F} = \frac{\mathcal{SSA}/(J-1)}{\mathcal{SSE}/(J(n-1))} \tag{9.6}$$

may be a suitable statistic for testing H_0.

In order to use $\mathbf{F}$ in a test we must find its distribution under the null hypothesis. Methods like those of [8.3.20] can be used to demonstrate that

- $\mathcal{SSA}/\sigma_E^2 \sim \chi^2(J-1)$, under H_0,
- $\mathcal{SSE}/\sigma_E^2 \sim \chi^2((n-1)J)$, and
- $\mathcal{SSA}$ is independent of $\mathcal{SSE}$.

Therefore, using [2.7.10], we find that

$$\mathbf{F} \sim F(J-1, (n-1)J)$$

under H_0. Thus we finally obtain the result that H_0 can be tested at the $(1-\tilde{p})$ significance level by comparing $\mathbf{F}$ computed from (9.6) against the $\tilde{p}$-quantile of $F(J-1,(n-1)J)$ obtained from Appendix G.

9.2.5 Application of a One Way Fixed Effects Model to the CCCma AMIP Experiment. The results of the one way analysis of variance of DJF mean 850 hPa temperature conducted with the six member ensemble of CCCma AMIP simulations are shown in Figure 9.1. In this case the variance components and F-ratio were computed at every point on the model's grid.[3] The F-ratio (9.6) is plotted on a log scale in such a way that a one contour increment indicates a factor of two increase in $\mathbf{f}$. The no treatment effect hypothesis is rejected at the 10% significance level over 65.7% of the globe. Experience with fields that have spatial covariance structure similar to that of 850 hPa temperature indicates that this rejection rate is certainly field significant (see Section 6.8).

Note that very large F-ratios (i.e., $\mathbf{f} > 8$ or $\ln(\mathbf{f}) > 2.77$) cover the entire tropical Pacific and Indian Oceans. Significantly large F-ratios are also found over the North Pacific, the midlatitude North and South Atlantic, and the southern Indian Oceans.

9.2.6 The Proportion R^2 of Variance Due to Treatments. As in regression analysis (Chapter 8) it is possible to compute a *coefficient of multiple determination*

$$R^2 = \mathcal{SSA}/\mathcal{SST} \tag{9.7}$$

that diagnoses the proportion of the response variable variance that is explained by the fitted model. As with regression, this is a somewhat optimistic estimate of the ability of the model to specify the response given the treatment.

An adjustment that attempts to reduce the tendency for R^2 to be optimistic is derived as follows. The expected value of the total sum of squares is

$$\mathcal{E}(\mathcal{SST}) = n\sum_{j=1}^{J} a_j^2 + (nJ-1)\sigma_E^2.$$

Therefore the proportion of the expected total sum of squares that is due to the treatments is

$$\frac{n\sum_{j=1}^{J} a_j^2}{n\sum_{j=1}^{J} a_j^2 + (nJ-1)\sigma_E^2}. \tag{9.8}$$

However, (9.5) shows that the numerator of (9.7) is a biased estimator of the numerator of (9.8). We therefore adjust $\mathcal{SSA}$ in (9.7) so that it becomes an

[3] Often, a pattern analysis approach (see Chapters 13-16) provides richer and more insightful results. A pattern analysis technique is used to obtain patterns representing the dominant modes of variation. The fields are then projected onto these patterns. The *loadings*, or pattern coefficients, are subsequently analysed in an ANOVA.

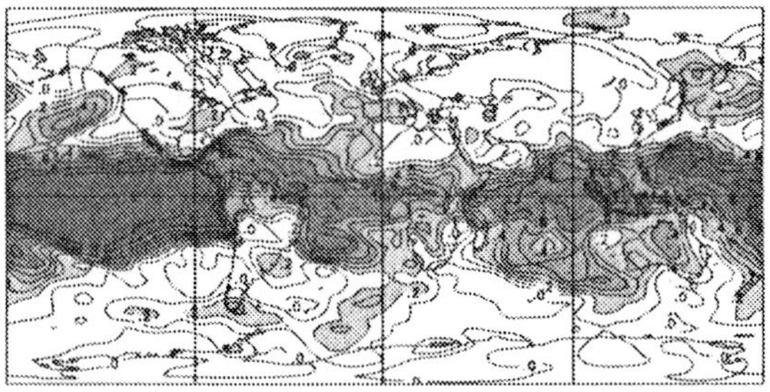

Figure 9.2: *The adjusted proportion R_a^2 of the total (i.e., interannual plus intersimulation) variance of DJF mean 850 hPa temperature that is explained by the imposed lower boundary conditions in the six member CCCma ensemble of AMIP simulations. Shading indicates values of R_a^2 greater than 0.2.*

unbiased estimator of the numerator in (9.8). The resulting adjusted R^2 is

$$R_a^2 = \frac{\mathcal{SSA} - \frac{(J-1)}{J(n-1)}\mathcal{SSE}}{\mathcal{SST}}.$$

Note that sampling variability occasionally causes R_a^2 to be negative.

9.2.7 AMIP Example: Adjusted R^2. The spatial distribution of R_a^2 for our AMIP example is illustrated in Figure 9.2. Notice that R_a^2 is large primarily over the tropical oceans. Note also that there is a one-to-one correspondence between R_a^2 and **F**. In fact, we may write

$$R_a^2 = \frac{\mathbf{F} - 1}{\mathbf{F} + \frac{J(n-1)}{J-1}} \text{ and } \mathbf{F} = \frac{1 + \frac{J(n-1)}{J-1} R_a^2}{1 - R_a^2}.$$

Thus both statistics convey the same information, and critical values of **F** are easily expressed as critical values of R_a^2. None the less, the messages conveyed by Figures 9.1 and 9.2 are not the same. The latter gives a much clearer picture of the physical relevance of the response to the forcing imposed by the bottom boundary conditions.

9.2.8 A One Way Random Effects Model. The one way model given by (9.1) and discussed above regards the treatment effects a_j, for $j = 1, \dots, J$, as *fixed* (non-random) effects that can be replicated from one experiment to the next. However, it is easy to conceive of experiments in which the response to the treatments is random and therefore can not be replicated from one experiment to the next. Treatments that have this property add variability to the response variable rather than changing its mean. Their effect is modelled using the *random effects* version of (9.1), which is given by

$$\mathbf{Y}_{ij} = \mu + \mathbf{A}_j + \mathbf{E}_{ij},$$

where the errors are iid $\mathcal{N}(0, \sigma_E^2)$ and the 'random effects' $\mathbf{A}_j$ are iid $\mathcal{N}(0, \sigma_A^2)$. Random variables $\mathbf{A}_j$ are assumed to be independent of the errors. With these assumptions we see that

$$\mathbf{Y}_{ij} \sim \mathcal{N}(0, \sigma_A^2 + \sigma_E^2).$$

Rather than testing that the treatment changes the mean of the response variable, we are now interested in testing the null hypothesis that the treatments do not induce between block (or between sample) variability, that is,

$$\text{H}_0\text{: } \sigma_A^2 = 0. \tag{9.9}$$

The statistic (9.6), used to test (9.3) in the fixed effects case, is also used to test (9.9) in the random effects case. The statistic also has the same distribution under the null hypothesis.

The differences between the fixed and random effects cases lie only in the interpretation of the model and the treatment sum of squares. The model tells us only that the treatments may increase interblock (or intersample) variability. The treatment sum of squares is an estimator of this variability. In fact,

$$\mathcal{E}(\mathcal{SSA}/(J-1)) = n\sigma_A^2 + \sigma_E^2. \tag{9.10}$$

9.2.9 R^2 for Random Effects Models. When random effects are assumed, we see from (9.10) that the variance of the random treatment effect can be estimated as [336]

$$\widehat{\sigma}_A^2 = \frac{\mathcal{SSA}/(J-1) - \mathcal{SSE}/(J(n-1))}{n}.$$

The proportion of variance of the response variable that is caused by the treatment effects is therefore estimated as

$$R'^2 = \frac{\widehat{\sigma}_A^2}{\widehat{\sigma}_A^2 + \widehat{\sigma}_E^2} = \frac{\mathcal{SSA} - \frac{(J-1)}{J(n-1)}\mathcal{SSE}}{\mathcal{SST} - \mathcal{SSE}/J}.$$

Note again that sampling variability may result in negative estimates of $\widehat{\sigma}_A^2$, and hence R'^2. Also, there is again a one-to-one relationship between R'^2 and **F**. In this case

$$R'^2 = \frac{\mathbf{F} - 1}{\mathbf{F} + (n-1)} \text{ and } \mathbf{F} = \frac{1 + (n-1)R'^2}{a - R'^2}.$$

While the form of R'^2 is similar to that of the adjusted coefficient of determination R_a^2, the interpretation is quite different because specification is impossible in the random effects setup. R'^2 simply estimates the proportion of variance that is induced by the 'treatment' variations.

9.2.10 Unequal Sample Sizes. Although experiments may be planned so that all treatments are replicated the same number of times, an experiment often yields samples of unequal size. Also, we must often adapt analysis of variance techniques to data that were not originally gathered for ANOVA purposes. We therefore briefly consider one way models with unequal sample sizes:

$$\mathbf{Y}_{ij} = \mu + a_j + \mathbf{E}_{ij} \quad \text{for } i = 1, \dots, n_j, \text{ and } j = 1, \dots, J.$$

As usual, we assume that errors $\mathbf{E}_{ij}$ are iid $\mathcal{N}(0, \sigma_E^2)$. The treatment effects can be either fixed or random. The number of replicates subjected to treatment j is denoted n_j.

The total sum of squares can still be partitioned into treatment and error components as in [9.2.3]. We have

$$\mathcal{SST} = \sum_{j=1}^{J}\sum_{i=1}^{n_j}(\mathbf{Y}_{ij} - \overline{\mathbf{Y}}_{\circ\circ})^2$$

$$\mathcal{SSA} = \sum_{j=1}^{J} n_j(\overline{\mathbf{Y}}_{\circ j} - \overline{\mathbf{Y}}_{\circ\circ})^2$$

$$\mathcal{SSE} = \sum_{j=1}^{J}\sum_{i=1}^{n_j}(\mathbf{Y}_{ij} - \overline{\mathbf{Y}}_{\circ j})^2.$$

As in the equal sample size case, $\mathcal{SSA}$ and $\mathcal{SSE}$ are statistically independent, and

$$\mathcal{SSE}/\sigma_E^2 \sim \chi^2(N - J),$$

where

$$N = \sum_{j=1}^{J} n_j.$$

A difficulty, however, is that $\mathcal{SSA}/\sigma_E^2$ is *not* distributed $\chi^2(J-1)$ under the null hypothesis that there is no treatment effect, either fixed or random. This violation of the usual distributional theory occurs because $\mathcal{SSA}$ can not be rewritten as a sum of $(J-1)$ squared normal random variables that all have the same variance. In this case the block mean errors $\overline{\mathbf{E}}_{\circ j}$ are independent, zero mean normal random variables with variance σ_E^2/n_j.

Consequently, the F test conducted by comparing

$$\mathbf{F} = \frac{\mathcal{SSA}/(J-1)}{\mathcal{SSE}/(N-J)}$$

against critical values from $F(J-1, N-J)$ is *approximate* rather than exact; the exact significance level of the test will be somewhat different from the specified significance level. Another consequence of unequal sample sizes is that the *power* of the test (recall [6.2.1]) is determined primarily by the size of the smallest sample. Thus, even when the same total number of experimental units are used, experiments with unequal sample sizes are generally less efficient than experiments with equal sample sizes. However, if variations in sample size are not enormous and all other assumptions implicit in the analysis are satisfied, the loss of power and precision usually do not pose a serious problem.

9.2.11 Relationships Between Treatments. We now return to the fixed effects model of (9.1). The only inferential consideration so far has been whether the treatment effects a_j are jointly zero. However, once this hypothesis has been rejected one would like to extract additional information from the data. Tools that can be used for this purpose are called *linear contrasts*.

9.2.12 Linear Contrasts. Linear contrasts are used to test hypotheses about specific relationships between treatment means that may have arisen from physical considerations. For example, the AMIP period included the strongest El Niño event on record (1982/83) and a relatively weak El Niño event (1986/87). Thus we might ask, within the confines of our one way setup, whether the mean anomalous response to 1982/83 lower boundary conditions is similar to the response to the 1986/87 lower boundary conditions.

These kinds of questions can be asked using linear contrasts. Tests of simple contrasts, which compare only two treatments or samples, are similar to the tests employed in *composite analysis* (see Section 17.3). However, the tests of contrasts may be more powerful than tests of composite differences because the test of the contrast uses more information about within sample variability.

A *linear contrast* is any linear combination of the treatment (or sample) means

$$w_c = \sum_{j=1}^{J} c_j \mu_j$$

for which $\sum_{j=1}^{J} c_j = 0$. Questions such as that discussed above are expressed as null hypotheses about linear contrasts:

$$\mathrm{H}_0\colon \sum_{j=1}^{J} c_j \mu_j = 0. \tag{9.11}$$

In the AMIP example we might set $c_j = 0$ for all j except 1982/83, for which we might choose $c_{82/83} = 1$, and 1986/87, for which we might choose $c_{86/87} = -1$. This contrast would satisfy the requirement that the coefficients sum to zero, and the null hypothesis would read 'the mean response in 1982/83 is equal to that in 1986/87.'

9.2.13 Testing Linear Contrasts. The test of the linear contrast is constructed in the now familiar fashion. First, we construct an estimator of the contrast

$$\widehat{w}_c = \sum_{j=1}^{J} c_j \mathbf{Y}_{\circ j}. \tag{9.12}$$

We substitute the model (9.1) into (9.12), and compute the expectation of $\widehat{w}_c^2$. We learn that

$$\mathcal{E}(\widehat{w}_c^2) = \Big(\sum_{j=1}^{J} c_j a_j\Big)^2 + \sigma_E^2 \sum_{j=1}^{J} \frac{c_j^2}{n_j}.$$

This suggests that a suitable test of (9.11) is based on

$$\mathbf{F} = \frac{\widehat{w}_c^2}{\frac{\mathcal{SSE}}{N-J}\sum_{j=1}^{J}\frac{c_j^2}{n_j}},$$

and that H_0 should be rejected when $\mathbf{F}$ is unusually large. Next we show that $\widehat{w}_c$ and $\mathcal{SSE}$ are independent. Then we argue that $\widehat{w}_c$ is normal because it is a linear combination of normal random variables. Also, the mean of $\widehat{w}_c$ is zero under the null hypothesis, and therefore the numerator of $\mathbf{F}$, when properly scaled, is distributed $\chi^2(1)$ under H_0. Finally, we conclude that $\mathbf{F} \sim F(1, N-J)$ under H_0. Thus the test is conducted at the $(1-\tilde{p})$ significance level by comparing the computed $\mathbf{f}$ with the $\tilde{p}$-quantile of $F(1, N-J)$ (see Appendix G).

Note that the test of the linear contrast adapts itself correctly to account for unequal sample sizes, but the test for the treatment effect does not.

Note also that if two contrasts, say c_j and d_j, for $j = 1, \ldots, J$, are *orthogonal*, meaning that

$$\sum_{j=1}^{J} \frac{c_j d_j}{n_j} = 0,$$

then the resulting tests are statistically independent.

Finally note that $J-1$ orthonormal contrasts could be used to partition the treatment sum of squares into $(J-1)$ independent components, each with one degree of freedom, and each independent of the sum of squared errors $\mathcal{SSE}$.

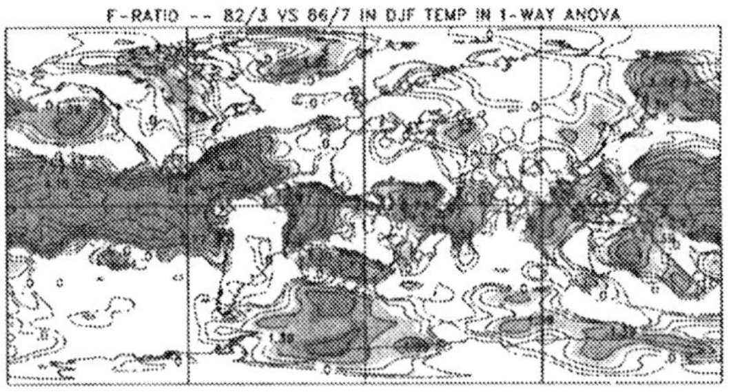

Figure 9.3: *The natural log of the F-ratios for the contrast comparing 1982/83 DJF 850 hPa temperature with 1986/87 DJF 850 hPa temperature in the CCCma six run ensemble of AMIP simulations. The shading indicates ratios that are significantly greater than 1 at the 10% significance level.*

9.2.14 The Response of the CCC GCMII to the 1982/83 El Niño Using the Method of Linear Contrasts. The F-ratios comparing the mean response to the 1982/83 and 1986/87 boundary conditions are shown in Figure 9.3. $\mathbf{F}$ is significantly greater than 1 over 34.9% of the globe. The diagram shows that there are substantial differences in the atmospheric response to the two warm events in the tropical Pacific, the North Pacific, and the South Atlantic. On the other hand, the response to the two warm events is similar over Africa and the Indian Ocean during DJF. Larger differences evolve in subsequent seasons reflecting the difference in the phasing of these two events.[4]

9.2.15 Diagnostics. We have not concerned ourselves much, to this point, with diagnostics of the fitted model. Many of the diagnostics discussed in connection with regression models (see [8.3.13] and [8.3.14]) are useful here as well. In particular, scatter plots of the residuals as a function of the treatment are useful for detecting outliers

[4]The five-month running mean SO index reached a minimum in January of 1983 and again in March or April of 1987.

and changes in error variance between treatments. Changes in variability from one treatment to the next can also be conveniently tested with *Bartlett's test*.

9.2.16 Bartlett's Test. Suppose we have J samples (or treatments) of possibly unequal sizes $n_1, \ldots, n_J$ and we wish to test the null hypothesis that all errors, either in the fixed or random effects models, have the same variance. The alternative is that at least one sample or treatment has a variance that is different. That is, we wish to test

$$\text{H}_0\text{: } \sigma^2_{E_j} = \sigma^2_E \text{ for all } j = 1, \ldots, J$$

against the alternative that the variances are not all equal. Here we use $\sigma^2_{E_j}$ to denote the variance of the random variables that represent sample or treatment j. Let $S_1^2, \ldots, S_J^2$ be the corresponding sample variances of the errors; that is,

$$S_j^2 = \sum_{i=1}^{n_j} (\mathbf{Y}_{ij} - \overline{\mathbf{Y}}_{\circ j})^2/(n_j - 1),$$

and let S_p^2 be the pooled estimate of the variances given by

$$S_p^2 = \frac{\sum_{j=1}^{J}(n_j - 1)S_j^2}{N - J} = \frac{\mathcal{SSE}}{N - J},$$

where $N = \sum_{j=1}^{J} n_j$.

With this notation, Bartlett's statistic is given by

$$\mathbf{B} = \frac{\mathbf{Q}}{h},$$

where

$$\mathbf{Q} = (N - J)\ln(S_p^2) - \sum_{j=1}^{J}(n_j - 1)\ln(S_j^2),$$

and

$$h = 1 + \frac{1}{3(J-1)}\left(\sum_{j=1}^{J}\frac{1}{(n_j - 1)} - \frac{1}{(N - J)}\right).$$

Statistic $\mathbf{B}$ is approximately distributed $\chi^2(J-1)$ under H_0. Large values of $\mathbf{B}$ are interpreted as evidence that H_0 is false. Therefore the test is conducted at the $(1 - \tilde{p})$ significance level by comparing the realized value of $\mathbf{B}$ against the $\tilde{p}$-quantiles of $\chi^2(J-1)$ (see Appendix E).

9.2.17 Equivalent Representation of a One Way ANOVA Model as a Regression Model. It may be useful at this point to make the connection between ANOVA and regression models. We can write model $\mathbf{Y}_{ij} = \mu + a_j + \mathbf{E}_{ij}$ from (9.1) in matrix vector form as follows. Let $\vec{\mathbf{Y}}$ be the N-dimensional random vector constructed by concatenating the J n_j-dimensional vectors $(\mathbf{Y}_{1,j}, \mathbf{Y}_{2,j}, \ldots, \mathbf{Y}_{n_j,j})^{\mathrm{T}}$, and define $\vec{\mathbf{E}}$ similarly. Let $\vec{\mathbf{A}}$ be the $(k+1)$-dimensional vector of parameters $(\mu, a_1, \ldots, a_J)^{\mathrm{T}}$. Then, (9.1) can be expressed as

$$\vec{\mathbf{Y}} = \mathcal{X}\vec{\mathbf{A}} + \vec{\mathbf{E}},$$

where $\mathcal{X}$ is the $N \times (J+1)$ *design matrix* given by

$$\mathcal{X} = \begin{pmatrix} \left.\begin{matrix} 1 & 1 & 0 & \ldots & 0 \\ \vdots & \vdots & \vdots & & \vdots \\ 1 & 1 & 0 & \ldots & 0 \end{matrix}\right\} & n_1 \text{ rows} \\ \left.\begin{matrix} 1 & 0 & 1 & \ldots & 0 \\ \vdots & \vdots & \vdots & & \vdots \\ 1 & 0 & 1 & \ldots & 0 \end{matrix}\right\} & n_2 \text{ rows} \\ \begin{matrix} \vdots & \vdots & \vdots & & \vdots \end{matrix} & \\ \left.\begin{matrix} 1 & 0 & 0 & \ldots & 1 \\ \vdots & \vdots & \vdots & & \vdots \\ 1 & 0 & 0 & \ldots & 1 \end{matrix}\right\} & n_k \text{ rows} \end{pmatrix}$$

The normal equations that provide the least squares estimators of $\vec{\mathbf{A}}$ are given by

$$\mathcal{X}^{\mathrm{T}}\mathcal{X}\vec{\mathbf{A}} = \mathcal{X}\vec{\mathbf{Y}}. \tag{9.13}$$

These equations have solutions given by

$$\widehat{\vec{\mathbf{a}}} = (\mathcal{X}^{\mathrm{T}}\mathcal{X})^{-}\mathcal{X}^{\mathrm{T}}\vec{\mathbf{Y}},$$

where $(\mathcal{X}^{\mathrm{T}}\mathcal{X})^{-}$ denotes the *generalized* inverse of $\mathcal{X}^{\mathrm{T}}\mathcal{X}$ (see Graybill [148]). The generalized inverse is required because $\mathcal{X}^{\mathrm{T}}\mathcal{X}$ is a non-invertible matrix (the first column of $\mathcal{X}$ is the sum of the remaining J columns). In fact, solution (9.13) defines a one-dimensional subspace of the parameter space $\mathbb{R}^{(J+1)}$ such that every point in the subspace minimizes the sum of squared errors. We select the solution of interest by imposing the constraint $\sum_{j=1}^{J}\widehat{\mathbf{a}}_j = 0$. These solutions are given by

$$\begin{aligned} \widehat{\mu} &= \overline{\mathbf{Y}}_{\circ\circ} \\ \widehat{\mathbf{a}}_j &= \overline{\mathbf{Y}}_{\circ j} - \overline{\mathbf{Y}}_{\circ\circ}, \text{ for } j = 1, \ldots, J. \end{aligned}$$

It is easily shown that the regression sum of squares is equal to the treatment sum of squares derived above, and that the test of the null hypothesis that there is not a regression relationship [8.4.8] is equivalent to the test that there is not a treatment effect [9.2.4].

9.3 Two Way Analysis of Variance

We now extend the model discussed in Section 9.2 so that it is possible to account for the effects of two treatments (if a completely randomized design has been used) or the effects of a treatment and a block (if a randomized block design has been used).

The example we wish to keep in mind is the CCCma AMIP experiment. Recall that we have nine DJF seasons in each simulation, each of which is subjected to a different 'treatment' (i.e., the sea-surface temperature and sea-ice regime). The experiment is replicated six times in six different simulations started from randomly chosen initial conditions. We can think of the six simulations as blocks.

9.3.1 The Two Way ANOVA Model—Introduction. Suppose an experiment was conducted that resulted in one outcome per treatment or treatment/block combination. (The language we use refers to treatments and blocks because that coincides most closely with our example.) Suppose that I different treatments were used, and that these were applied in random order to I experimental units in J blocks. We represent the resulting IJ outcomes of the experiment with random variables $\mathbf{Y}_{ij}$, for $i = 1, \ldots, I$, and $j = 1, \ldots, J$, which we assume to be independent and normally distributed.

A *fixed effects* model for data of this sort is the *two way model without interaction* given by

$$\mathbf{Y}_{ij} = \mu + a_i + b_j + \mathbf{E}_{ij}. \qquad (9.14)$$

The parameters are subject to the constraints

$$\sum_{i=1}^{I} a_i = 0 \text{ and } \sum_{j=1}^{J} b_j = 0.$$

The errors are assumed to be iid $\mathcal{N}(0, \sigma_E^2)$.

An important, and limiting, aspect of this model is that the treatment and block effects are assumed to be *additive*. This assumption may not be correct, but we can not determine this with the limited number of data that are available.

To test the additivity assumption it is necessary to have data from a replicated experiment. If a completely randomized design is used, every treatment combination must be used more than once. If a blocked design is used, each treatment must be used within each block more than once.

The outcome of a replicated experiment is represented by random variables $\mathbf{Y}_{ijl}$, for $i = 1, \ldots, I$, $j = 1, \ldots, J$, and $l = 1, \ldots, n_{ij}$, which we again assume to be independent and normally distributed.

A fixed effects *two way model with interaction* is given by

$$\mathbf{Y}_{ijl} = \mu + a_i + b_j + c_{ij} + \mathbf{E}_{ijl}. \qquad (9.15)$$

The parameters are subject to the constraints

$$\sum_{i=1}^{I} a_i = \sum_{j=1}^{J} b_j = 0, \text{ and}$$
$$\sum_{j=1}^{J} c_{ij} = \sum_{i=1}^{I} c_{ij} = 0 \text{ for all } i \text{ and } j,$$

and the errors are assumed to be iid $\mathcal{N}(0, \sigma_E^2)$.

In both experiments with and without replication, it is possible to construct models with some or all of the effects treated as random effects. As in [9.2.8], the test statistics used to test for block and treatment effects are identical to the fixed effects case, but the interpretation of the tests is quite different. There are also differences in the calculation of variance proportions.

We do not discuss random effects models in this section, but a two way model with a combination of fixed and random effects is discussed in detail in Section 9.4 in the context of this chapter's working example.

9.3.2 Two Way Model Without Interaction. In the setup of (9.14), the total sum of squares is partitioned into treatment, block, and sum of squared errors as

$$\mathcal{SST} = \mathcal{SSA} + \mathcal{SSB} + \mathcal{SSE},$$

where

$$\mathcal{SST} = \sum_{i=1}^{I}\sum_{j=1}^{J} (\mathbf{Y}_{ij} - \overline{\mathbf{Y}}_{\circ\circ})^2,$$

$$\mathcal{SSA} = J \sum_{i=1}^{I} (\overline{\mathbf{Y}}_{i\circ} - \overline{\mathbf{Y}}_{\circ\circ})^2,$$

$$\mathcal{SSB} = I \sum_{j=1}^{J} (\overline{\mathbf{Y}}_{\circ j} - \overline{\mathbf{Y}}_{\circ\circ})^2, \qquad (9.16)$$

$$\mathcal{SSE} = \sum_{i=1}^{I}\sum_{j=1}^{J} (\mathbf{Y}_{ij} - \overline{\mathbf{Y}}_{i\circ} - \overline{\mathbf{Y}}_{\circ j} + \overline{\mathbf{Y}}_{\circ\circ})^2.$$

Using methods similar to those in [9.2.4], the following can be shown.

1 $\mathcal{E}(\mathcal{SSA}) = J \sum_{i=1}^{I} a_i^2 + (I-1)\sigma_E^2$.

2 If H_0: $a_1 = \cdots = a_I = 0$ is true, then $\mathcal{SSA}/\sigma_E^2 \sim \chi^2(I-1)$.

3 $\mathcal{E}(\mathcal{SSB}) = I\sum_{j=1}^{J} b_j^2 + (J-1)\sigma_E^2$.

4 If H_0: $b_1 = \cdots = b_J = 0$ is true, then $\mathcal{SSB}/\sigma_E^2 \sim \chi^2(J-1)$.

5 $\mathcal{E}(\mathcal{SSE}) = (I-1)(J-1)\sigma_E^2$.

6 $\mathcal{SSE}/\sigma_E^2 \sim \chi^2((I-1)(J-1))$.

7 $\mathcal{SSA}$, $\mathcal{SSB}$, and $\mathcal{SSE}$ are independent.

It follows from items 1, 2, and 5–7 that the null hypothesis of no treatment effect, that is,

$$H_0: a_1 = \cdots = a_I = 0,$$

can be tested against the alternative hypothesis that there is a treatment effect by comparing

$$\mathbf{F} = \frac{\mathcal{SSA}/(I-1)}{\mathcal{SSE}/((I-1)(J-1))}$$

with $F(I-1, (I-1)(J-1))$ critical values (see Appendix G). Similarly, items 3–7 are used to show that the no block effect null hypothesis, that is,

$$H_0: b_1 = \cdots = b_J = 0 \qquad (9.17)$$

can be tested against the alternative hypothesis that there is a block effect by comparing

$$\mathbf{F} = \frac{\mathcal{SSB}/(J-1)}{\mathcal{SSE}/((I-1)(J-1))}$$

with $F(J-1, (I-1)(J-1))$ critical values (see Appendix G).

One possible reason for testing for a block effect is to determine whether or not the block sum of squares can be pooled with the sum of squared errors. If this can be done, that is, if (9.17) is not rejected, then the between blocks variation can be used to improve the estimate of error variance and hence increase the power of the test for treatment effects. In this case we compute

$$\mathbf{F} = \frac{\mathcal{SSA}/(I-1)}{(\mathcal{SSB}+\mathcal{SSE})/(I(J-1))}$$

and compare with critical values from $F(I-1, I(J-1))$. It is easily shown that this test is equivalent to the test for treatment effects in the one way model with fixed effects [9.2.4].

The interaction terms in (9.15) are confounded with error when the experiment is not replicated. Then the mean sum of squared errors, $\mathcal{SSE}/((I-1)(J-1))$, is inflated by the interaction terms; it estimates a number greater than σ_E^2. The effect is to reduce the power of the tests described above.

The linear contrast methodology described in [9.2.12] and [9.2.13] naturally extends to the two way case and is not detailed here. Both the treatment and block sums of squares can be partitioned into independent components if needed.

Diagnostic opportunities for the two way model without interaction are relatively limited because of the relatively large number of fitted parameters compared with the number of degrees of freedom available for error. None the less, scatter plots of estimated errors, plotted by treatment and block, can be useful for identifying observations with large influence.

9.3.3 Two Way ANOVA of the CCCma Multiple AMIP Experiment. We now use the two way model with $I = 9$ treatments and $J = 6$ blocks. Because there is only one replication per treatment/block combination, there are $(I-1)(J-1) = 40$ df for error.

The F-ratios for the boundary forced effect on 850 hPa DJF temperature (not shown) are very similar to those computed using the one way model. The small reduction in the number of degrees of freedom available for error results in a test that is slightly less powerful than in the one way case. However, the estimate of error variability is not contaminated by the confounding block effect. The result is that the test for the sea-surface temperature effect on 850 hPa DJF temperature rejects the null hypothesis at the 10% significance level over a slightly larger area (66.1% of the globe).

The F-ratios for the block effect on 850 hPa DJF temperature are shown in Figure 9.4. The F-ratio exceeds the 10% critical value for $F(5, 40)$ over about 13.1% of the globe. Previous experience with field significance tests (see Section 6.8) conducted with fields with comparable spatial covariance structure suggests that this rate is not significantly greater than 10%. However, the same test conducted with 500 hPa DJF geopotential (not shown) resulted in a rejection rate of 23%, which is likely field significant. Therefore, while a block, or *run*, effect is difficult to detect in lower tropospheric temperature, it appears to be detectable in the integrated temperature of the lower half of the atmosphere.

The CCCma experiments were actually conducted on two computers. One 10-year simulation was conducted on a Cray-XMP while the remaining five were conducted on a NEC SX/3.

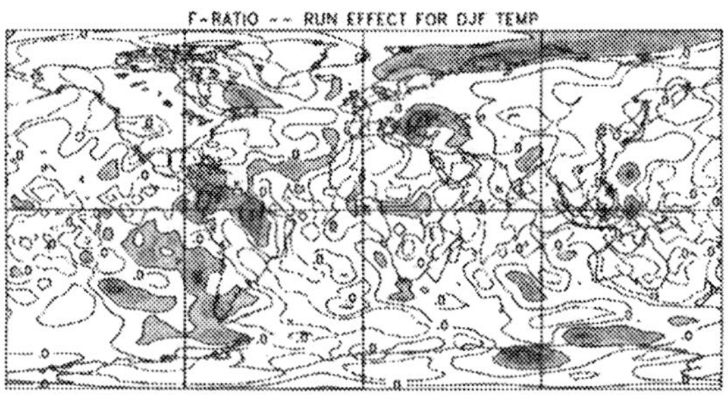

Figure 9.4: *The natural log of the F-ratios for the block or run effect in the CCCma AMIP experiment. Each contour indicates a doubling of the F-ratio. The shading indicates ratios that are significantly greater than 1 at the 10% significance level.*

We label the Cray 'block' as block number 1. The hypothesis that the block effect for the Cray was equal to that for the NEC was tested with the contrast $c = (1, -0.2, -0.2, -0.2, -0.2, -0.2)$. Specifically, the null hypothesis

$$\mathrm{H}_0\colon \sum_{j=1}^{J} c_j b_j = 0 \qquad (9.18)$$

was tested against the alternative that the contrast is nonzero. The contrast was estimated by computing

$$\widehat{w}_c = \sum_{j=1}^{J} c_j \overline{\mathbf{Y}}_{\circ j}.$$

Under (9.18), the squared contrast has expectation

$$\mathcal{E}(\widehat{w}_c^2) = \left(\sum_{j=1}^{J} c_j b_j\right)^2 + \frac{\sigma_E^2}{I}\sum_{j=1}^{J} c_j^2.$$

Therefore (9.18) can be tested by comparing

$$\mathbf{F} = \frac{\widehat{w}_c^2}{\frac{\mathcal{SSE}}{I(I-1)(J-1)}\sum_{j=1}^{J} c_j^2}$$

with critical values from $F(1, (I-1)(J-1))$. We obtained a rejection rate of 18.1% when (9.18) was tested in DJF 850 hPa temperature at the 10% significance level (not shown).

We can test whether there is significant inter-run variation that is orthogonal to contrast (9.18) by computing

$$\mathbf{F} = \frac{\frac{\mathcal{SSB} - I\widehat{w}_c^2/\sum_{j=1}^{J} c_j^2}{J-2}}{\frac{\mathcal{SSE}}{(I-1)(J-1)}}$$

and comparing $\mathbf{F}$ with critical values from $F(J-2, (I-1)(J-1))$. The null hypothesis that there is additional inter-run variation not explained by the computer change is rejected at the 10% significance level over 9.4% of the globe.

The run effect is observed much more strongly in June, July, August (JJA) 500 hPa geopotential for which (9.18) is rejected over 52% of the globe (primarily in the tropics).

The differences between the Cray and NEC simulations were not primarily due to the differences between machines (see Zwiers [449]).[5] It turns out, however, that the change in machine type coincided with a change in the source of initialization data. CCCma's initialization procedure diagnoses the atmospheric mass from the initialization data. The model subsequently conserves that mass for the duration of the simulation. The resulting atmospheric mass for the Cray simulation is equivalent to a global mean surface pressure of 985.01 hPa. In contrast, the masses diagnosed from the initial conditions used for the NEC simulations varied between 984.55 and 984.58 hPa. This difference between the Cray and NEC simulations, approximately 0.44 hPa, corresponds to a change in 500 hPa geopotential height in the tropics of about 3.5 m. The large, and unexpected, block effect described above is primarily the result of the change in the source of initialization data. This example illustrates that it is difficult to design an experiment so that it excludes unwanted external variability, since such variability often arrives from unanticipated sources.

9.3.4 Two Way Model with Interaction. We now briefly consider the two way fixed effects model with interaction given by (9.15) in the case in which each treatment or treatment/block combination is replicated n times. The calculation of the variance components is easily extended to the case in which each combination is not replicated equally. However, the tests for treatment, block, and interaction are then only approximate (see [9.2.10]) if the corresponding sum of squares has more than one df.

In the setup of (9.15) the total sum of squares is partitioned into four independent components for treatment, block, interaction, and error, as follows:

$$\mathcal{SST} = \mathcal{SSA} + \mathcal{SSB} + \mathcal{SSI} + \mathcal{SSE}, \qquad (9.19)$$

[5] Differences in the way in which the two machines represented floating point numbers did lead to surface elevation changes at three locations on the latitude row just north of the equator, but these were not judged to be the cause of large-scale effects in the tropical climate.

where

$$\mathcal{SST} = \sum_{i=1}^{I}\sum_{j=1}^{J}\sum_{l=1}^{n}(\mathbf{Y}_{ijl} - \overline{\mathbf{Y}}_{\circ\circ\circ})^2, \tag{9.20}$$

$$\mathcal{SSA} = nJ\sum_{i=1}^{I}(\overline{\mathbf{Y}}_{i\circ\circ} - \overline{\mathbf{Y}}_{\circ\circ\circ})^2, \tag{9.21}$$

$$\mathcal{SSB} = nI\sum_{j=1}^{J}(\overline{\mathbf{Y}}_{\circ j\circ} - \overline{\mathbf{Y}}_{\circ\circ\circ})^2, \tag{9.22}$$

$$\mathcal{SSI} = n\sum_{i=1}^{I}\sum_{j=1}^{J}(\overline{\mathbf{Y}}_{ij\circ} - \overline{\mathbf{Y}}_{i\circ\circ} - \overline{\mathbf{Y}}_{\circ j\circ} + \overline{\mathbf{Y}}_{\circ\circ\circ})^2, \tag{9.23}$$

$$\mathcal{SSE} = \sum_{i=1}^{I}\sum_{j=1}^{J}\sum_{l=1}^{n}(\mathbf{Y}_{ijl} - \overline{\mathbf{Y}}_{ij\circ})^2. \tag{9.24}$$

Assuming fixed effects and iid $\mathcal{N}(0, \sigma_E^2)$ errors, the following can be shown.

- $\mathcal{E}(\mathcal{SSA}) = nJ\sum_{i=1}^{I} a_i^2 + (I-1)\sigma_E^2$.
- If H$_0$: $a_1 = \cdots = a_I = 0$ is true, then $\mathcal{SSA}/\sigma_E^2 \sim \chi^2(I-1)$.
- $\mathcal{E}(\mathcal{SSB}) = nI\sum_{j=1}^{J} b_j^2 + (J-1)\sigma_E^2$.
- If H$_0$: $b_1 = \cdots = b_J = 0$ is true, then $\mathcal{SSB}/\sigma_E^2 \sim \chi^2(J-1)$.
- $\mathcal{E}(\mathcal{SSI}) = n\sum_{i=1}^{I}\sum_{j=1}^{J} c_{ij}^2 + (I-1)(J-1)\sigma_E^2$.
- If H$_0$: $c_{1,1} = \cdots = c_{IJ} = 0$ is true, then $\mathcal{SSI}/\sigma_E^2 \sim \chi^2((I-1)(J-1))$.
- $\mathcal{E}(\mathcal{SSE}) = IJ(n-1)\sigma_E^2$.
- $\mathcal{SSE}/\sigma_E^2 \sim \chi^2(IJ(n-1))$.
- $\mathcal{SSA}$, $\mathcal{SSB}$, $\mathcal{SSI}$, and $\mathcal{SSE}$ are independent.

Tests for treatment, block, and interaction effects as well as tests of linear contrasts among treatments, blocks, and interactions follow in the usual way.

As in [9.3.2], the power of the test for treatment effects can be enhanced if the block and/or interaction sums of squares can be pooled with the sum of squared errors. For example, if the null hypothesis that there is no block/treatment interaction is accepted, then an improved estimator of σ_E^2 that has $nIJ - (I + J - 1)$ df instead of $(n-1)IJ$ df is given by

$$\hat{\hat{\sigma}}_E^2 = \frac{\mathcal{SSI} + \mathcal{SSE}}{nIJ - (I + J - 1)}.$$

The effect of pooling interaction and sum of squared errors (when it can be done) is particularly dramatic if the number of replicates is small.

9.4 Two Way ANOVA with Mixed Effects of the CCCma AMIP Experiment

We continue the analysis of [9.3.3] by introducing a two way model with interaction terms and a mixture of fixed and random effects.

The data we use are monthly means of 850 hPa temperature for December, January, and February from which the annual cycle common to all six simulations has been removed (see Zwiers [444] for details of the procedure used). For each DJF season we regard the three monthly means obtained for the season as three replicates of the treatment (i.e., sea-surface temperature) and block (i.e., simulation) combination that corresponds to that season. Although these replicates are not quite independent of one another, we operate, for now, as if they were.

9.4.1 The Model. The model we use to represent this data is a two way model with interaction in which some effects are fixed and others are random. The model is given by

$$\mathbf{Y}_{ijl} = \mu + a_i + \mathbf{B}_j + \mathbf{C}_{ij} + \mathbf{E}_{ijl}, \tag{9.25}$$

where $i = 1979, \ldots, 1987$ indicates the year of the December month in each DJF season, $j = 1, \ldots, 6$ indicates the member of the ensemble of simulations, $l = 1, 2, 3$ indicates the 'replicate' (i.e., December, January or February).

We treat the year effects a_i as fixed effects because every simulation was forced with the same sea-surface temperature and sea-ice record as dictated by the AMIP protocol (see Gates [137]). A fixed mean response to a given sea-surface temperature and sea-ice regime is anticipated in each simulation. This is not to say that each simulation is identical, since low-frequency variations from internal sources ensures that the simulations are different. However, the fixed sea-surface temperature and sea-ice signal are assumed to induce the same amount of interannual variability in each simulation.

The block effects $\mathbf{B}_j$ are treated as random effects and assumed to be independently distributed $\mathcal{N}(b_j, \sigma_B^2)$ where the fixed parts of the block effects, b_j, are constrained to sum to zero. That is, we represent the block effect as $\mathbf{B}_j = b_j + \mathbf{B}_j^*$ where the $\mathbf{B}_j^*$s are iid $\mathcal{N}(0, \sigma_B^2)$. The idea is that the fixed part of the block effect represents variation in the simulation configuration (such as the source of initialization data) and the random part represents excess intersimulation variability caused by the particular choice of initial conditions. Variations in initial conditions might cause CCC GCMII to produce simulations that occupy distinctly different parts of the model's phase space if the model has more than one stable regime. Rejection of H$_0$: $\sigma_B^2 = 0$ might be evidence of this. However, except for the possibilities of this sort of chaotic behaviour and computing glitches, we do not expect block effects to contribute significantly to total variability. We will see below that it is possible to separate the fixed and random components of the block effects in model (9.25) provided additional assumptions are made about the structure of the fixed components.

The interaction effects $\mathbf{C}_{ij}$ are treated as pure random effects and are assumed to be iid $\mathcal{N}(0, \sigma_C^2)$ random variables that are independent of the block effects. The interaction effects represent interannual variations that are not common to all runs. That is, this term in (9.25) represents the effects of slow processes in the climate system that do not evolve the same way in every simulation. For example, CCC GCMII contains a simple land surface processes model (see McFarlane et al. [270]). The evolution of the soil moisture field in this land surface model will certainly be affected by the prescribed evolution of sea-surface temperature and sea ice, but it will not be completely determined by these forcings. Therefore about 30% of the lower boundary of the simulated climate evolves differently from one simulation to the next. The effects of these variations in the lower boundary over land, and other slow variations generated internally by the GCM, are not common to all simulations and will therefore be reflected in the interaction term.

The noise terms $\mathbf{E}_{ijl}$ represent the effects of intra-seasonal variations caused by processes (such as daily weather) that operate on shorter than interannual time scales (see Zwiers [444] and the discussion of potential predictability in Section 17.2). We assume that the errors are identically distributed $\mathcal{N}(0, \sigma_E^2)$ and that they are independent of the block and interaction effects.

There are certainly problems with this last assumption that should make us cautious about the subsequent inferences we make. For example, our assumptions imply that the amount of variability at high frequencies is not affected by either the imposed sea-surface temperature and sea-ice regime or by the state of slowly varying internal processes. Also note that we were careful *not* to make the assumption that the errors are independent, because they are actually weakly correlated within seasons. We therefore assume only that errors $\mathbf{E}_{ijl}$ and $\mathbf{E}_{i'j'l'}$ are independent for $(i, j) \neq (i', j')$. Errors $\mathbf{E}_{ijl}$ and $\mathbf{E}_{ijl'}$ for $l \neq l'$ are not assumed to be independent.

9.4.2 Partition of the Total Sum of Squares. With all these assumptions, we are able to partition the total sum of squares into treatment, block, interaction, and error components as in [9.3.4] (see (9.19)–(9.24)). Because model (9.25) has mixed effects and some dependence amongst errors, the interpretation of the variance components is somewhat different from that in [9.3.4]. By taking expectations and making arguments such as those in [9.2.4] and [9.2.6] we obtain the following.

1 $\mathcal{E}(\mathcal{SSA}) = nJ\sum_{i=1}^{I} a_i^2 + n(I-1)(\sigma_{AB}^2 + \sigma_{\bar{E}_{ij\circ}}^2)$.

2 If H$_0$: $a_1 = \cdots = a_I = 0$ is true, then

$$\frac{\mathcal{SSA}}{n(\sigma_C^2 + \sigma_{\bar{E}_{ij\circ}}^2)} \sim \chi^2(I-1).$$

3 $\mathcal{E}(\mathcal{SSB}) = nI\sum_{j=1}^{J} b_j^2 + nI(J-1)\sigma_B^2 + n(J-1)(\sigma_C^2 + \sigma_{\bar{E}_{ij\circ}}^2)$.

4 If H$_0$: $b_1 = \cdots = b_J = \sigma_B^2 = 0$ is true, then

$$\frac{\mathcal{SSB}}{n(\sigma_C^2 + \sigma_{\bar{E}_{ij\circ}}^2)} \sim \chi^2(J-1).$$

5 $\mathcal{E}(\mathcal{SSI}) = n(I-1)(J-1)(\sigma_{AB}^2 + \sigma_{\bar{E}_{ij\circ}}^2)$.

6 If H$_0$: $\sigma_C^2 = 0$ is true, then

$$\frac{\mathcal{SSI}}{n(\sigma_C^2 + \sigma_{\bar{E}_{ij\circ}}^2)} \sim \chi^2((I-1)(J-1)).$$

7 $\mathcal{SSA}$, $\mathcal{SSB}$, $\mathcal{SSI}$, and $\mathcal{SSE}$ are independent.

Here $\sigma_{\bar{E}_{ij\circ}}^2$ indicates the variance of the seasonal mean error.

9.4.3 Variance of the Seasonal Mean Error. In our specific application, in which $n = 3$, the variance of the seasonal mean error is

$$\sigma^2_{\overline{E}_{ij\circ}} = \mathrm{Var}\Big(\frac{\mathbf{E}_{i,j,1} + \mathbf{E}_{i,j,2} + \mathbf{E}_{i,j,3}}{3}\Big) = T_0\,\sigma_E^2/3,$$

where $\sigma_E^2/3$ is the variance of the mean of three iid errors and T_0 is a factor that reflects how the dependence between the errors inflates the variance. T_0 is called the *decorrelation time* (see Sections 17.1 and 17.2 for a detailed discussion of the decorrelation time and its estimation). In this case it is easily shown that

$$T_0 = 1 + \frac{2}{3}(2\rho_1 + \rho_2), \tag{9.26}$$

where ρ_1 is the correlation between errors in adjacent months, that is,

$$\rho_1 = \mathrm{Cor}\big(\mathbf{E}_{i,j,1}, \mathbf{E}_{i,j,2}\big) = \mathrm{Cor}\big(\mathbf{E}_{i,j,2}, \mathbf{E}_{i,j,3}\big),$$

and ρ_2 is the correlation between errors separated by a month, that is,

$$\rho_2 = \mathrm{Cor}\big(\mathbf{E}_{i,j,1}, \mathbf{E}_{i,j,3}\big).$$

We will analyse the effect of the correlated errors shortly.

9.4.4 Distribution of the Variance Components. However, we first illustrate how items 1–7 in [9.4.2] are obtained by considering items 1 and 2 in detail.

Recall from (9.21) that

$$\mathcal{SSA} = nJ\sum_{i=1}^{I}(\overline{\mathbf{Y}}_{i\circ\circ} - \overline{\mathbf{Y}}_{\circ\circ\circ})^2.$$

The χ^2 assertion (item 2) is verified by using arguments similar to those in [8.3.20] to show that (9.21) can be rewritten as a sum of $I - 1$ squared independent normal random variables with mean zero. Hence $\mathcal{SSA}$, when scaled by the variance of these normal random variables, is distributed $\chi^2(I-1)$.

The scaling variance (item 1) is obtained as follows. Using model (9.25) we see that

$$\begin{aligned}\overline{\mathbf{Y}}_{i\circ\circ} &= \mu + a_i + \overline{\mathbf{B}}_\circ + \overline{\mathbf{C}}_{i\circ} + \overline{\mathbf{E}}_{i\circ\circ}\\ \overline{\mathbf{Y}}_{\circ\circ\circ} &= \mu + \overline{\mathbf{B}}_\circ + \overline{\mathbf{C}}_{\circ\circ} + \overline{\mathbf{E}}_{\circ\circ\circ}\end{aligned}$$

where the over-bar and $\circ$ notation have the usual meaning, given in [9.2.3]. Taking differences, we see that

$$(\overline{\mathbf{Y}}_{i\circ\circ} - \overline{\mathbf{Y}}_{\circ\circ\circ}) = a_i + (\overline{\mathbf{C}}_{i\circ} - \overline{\mathbf{C}}_{\circ\circ}) + (\overline{\mathbf{E}}_{i\circ\circ} - \overline{\mathbf{E}}_{\circ\circ\circ}).$$

Squaring and summing, we obtain

$$\begin{aligned}\mathcal{SSA} &= nJ\sum_{i=1}^{I}a_i^2 + nJ\sum_{i=1}^{I}(\overline{\mathbf{C}}_{i\circ} - \overline{\mathbf{C}}_{\circ\circ})^2\\ &\quad + nJ\sum_{i=1}^{I}(\overline{\mathbf{E}}_{i\circ\circ} - \overline{\mathbf{E}}_{\circ\circ\circ})^2\\ &\quad + \text{cross-terms}.\end{aligned}$$

When taking expectations, we see that the expected values of the cross-terms in this expression are zero (some cross-terms are products of independent, zero mean random variables; others are products between constants and zero mean random variables). Therefore, the expected value of $\mathcal{SSA}$ reduces to

$$\begin{aligned}\mathcal{E}(\mathcal{SSA}) &= nJ\sum_{i=1}^{I}a_i^2\\ &\quad + nJ\mathcal{E}\Big(\textstyle\sum_{i=1}^{I}(\overline{\mathbf{C}}_{i\circ} - \overline{\mathbf{C}}_{\circ\circ})^2\Big)\\ &\quad + nJ\mathcal{E}\Big(\textstyle\sum_{i=1}^{I}(\overline{\mathbf{E}}_{i\circ\circ} - \overline{\mathbf{E}}^2_{\circ\circ\circ})\Big).\end{aligned}$$

Therefore, using (4.5) and (4.6), we see that

$$\begin{aligned}\mathcal{E}(\mathcal{SSA}) &= nJ\sum_{i=1}^{I}a_i^2 + nJ(I-1)\sigma_C^2\\ &\quad + nJ(I-1)\sigma^2_{\overline{E}_{i\circ\circ}}.\end{aligned}$$

Finally, we note that $\sigma^2_{\overline{E}_{i\circ\circ}} = \sigma^2_{\overline{E}_{ij\circ}}/J$. Assertion 1 follows.

9.4.5 Testing the Year Effect: Potential Predictability from External Sources. Items 1–7 in [9.4.2] provide us with sufficient information to construct tests about year and block effects. As in [9.2.5] and [9.3.3], a test of

$$\mathrm{H}_0\colon a_1 = \cdots = a_I = 0 \tag{9.27}$$

determines whether there is a detectable signal attributable to the external boundary forcing. If so, the climate may be predictable on seasonal time scales because we believe the lower boundary conditions (i.e., sea-surface temperature and sea-ice extent) to be predictable on these time scales due to the much large thermal inertia of the upper ocean and cryosphere.

From items 1, 2, and 5–7 in [9.4.2] we see that hypothesis (9.27) is tested against the alternative that some of the year effects are nonzero by comparing

$$\mathbf{F} = \frac{\mathcal{SSA}/(I-1)}{\mathcal{SSI}/((I-1)(J-1))} \tag{9.28}$$

with critical values of $F(I-1,(I-1)(J-1))$. Note that this F-ratio was also used to test this hypothesis in the two way model without interaction that was applied to the seasonal means in [9.3.2] and [9.3.3]. The numerical values of the ratios are also identical because only seasonal means are used in the calculation of (9.28).

As reported in [9.3.3], there is a significant sea-surface temperature effect. These effects are *potentially predictable* (see Section 17.2 and also [9.4.7–11]). Hindcast experiments (see Zwiers [444]) demonstrate that in this case potential predictability is *actual* predictability.

9.4.6 Testing the Block Effect. Using items 3–7 in [9.4.2] we can construct a test of the null hypothesis that there is not a block effect

$$\text{H}_0\colon b_1 = \cdots = b_J = \sigma_B^2 = 0 \tag{9.29}$$

against the alternative hypothesis that there is a block effect. This particular form of the null hypothesis comes about because we assumed, in [9.4.1], that the block (i.e., simulation) effect has both a fixed and a random component. That is, we assumed that $\mathbf{B}_j \sim \mathcal{N}(b_j, \sigma_B^2)$ with the constraint that $\sum_{j=1}^J b_j = 0$. The fixed and random components are confounded in our experimental design, so it is not possible to construct separate tests about σ_B and the b_js without making further assumptions about the fixed parts of the block effect.

Hypothesis (9.29) is tested by comparing

$$\mathbf{F} = \frac{\mathcal{SSB}/(J-1)}{\mathcal{SSI}/((I-1)(J-1))}$$

against $F(J-1,(I-1)(J-1))$ critical values. Again, the test is identical to that for block effects reported in [9.3.3]. Figure 9.4 showed weak evidence for a block effect, which appears to be associated with a change in computing hardware part way through the experiment.

Further dissection of the block effect is possible if we assume that only the computer type and source of initial data affect the fixed part of the block effect (i.e., if we assume $b_2 = \cdots = b_J$). Then, using linear contrasts, $\mathcal{SSB}$ (9.24) can be partitioned into statistically independent components as:

$$\mathcal{SSB} = \mathcal{SSB}_{\mathcal{F}} + \mathcal{SSB}_{\mathcal{R}}$$

where

$$\mathcal{SSB}_{\mathcal{F}} = nI\frac{J-1}{J}\Big(\bar{\mathbf{Y}}_{1\circ\circ} - \frac{1}{J-1}\sum_{j=2}^{J}\bar{\mathbf{Y}}_{j\circ\circ}\Big)^2$$

$$\mathcal{SSB}_{\mathcal{R}} = nI\sum_{j=2}^{J}\Big(\bar{\mathbf{Y}}_{j\circ\circ} - \frac{1}{J-1}\sum_{j=2}^{J}\bar{\mathbf{Y}}_{j\circ\circ}\Big)^2.$$

$\mathcal{SSB}_{\mathcal{F}}$ is proportional to the squared difference between the mean state simulated in the Cray and the mean state simulated in the five NEC simulations, and $\mathcal{SSB}_{\mathcal{R}}$ can be recognized as a scaled estimate of the intersimulation variance that is computed from those simulations that are assumed to have the same configuration effects.

Taking expectations, we can show that

$$\mathcal{E}(\mathcal{SSB}_{\mathcal{R}}) = n(J-2)\Big(I\sigma_B^2 + \sigma_C^2 + \sigma_{\bar{E}_{ij\circ}}^2\Big)$$

and, using now familiar arguments, we can demonstrate that $\text{H}_0\colon \sigma_B^2 = 0$ can be tested by comparing

$$\mathbf{F} = \frac{\mathcal{SSB}_{\mathcal{R}}/(J-2)}{\mathcal{SSI}/((I-1)(J-1))}$$

with $F(J-2,(I-1)(J-1))$ critical values. No evidence was found to suggest that $\sigma_B^2 > 0$ in the CCCma ensemble of AMIP simulations.

Again, taking expectations, it can be shown that

$$\mathcal{E}(\mathcal{SSB}_{\mathcal{F}}) = n\bigg(I\Big(b_1 - \frac{1}{J-1}\sum_{j=2}^{J} b_j\Big)^2 + \frac{J-1}{J}\Big(I\sigma_B^2 + \sigma_C^2 + \sigma_{\bar{E}_{ij\circ}}^2\Big)\bigg).$$

Thus, if $\text{H}_0\colon \sigma_B^2 = 0$ has not been rejected, the null hypothesis that there is not a configuration effect ($\text{H}_0\colon b_1 = \frac{b_2+\cdots+b_J}{J-1}$) can be tested by comparing

$$\mathbf{F} = \frac{J\mathcal{SSB}_{\mathcal{F}}/(J-1)}{\mathcal{SSI}/((I-1)(J-1))}$$

with critical values from $F(1,(I-1)(J-1))$. When there is evidence that $\sigma_B^2 > 0$, the no configuration effect hypothesis should be tested by comparing

$$\mathbf{F} = \frac{J\mathcal{SSB}_{\mathcal{F}}/(J-1)}{\mathcal{SSB}_{\mathcal{R}}/(J-2)}$$

with critical values from $F(1, J-2)$. As noted previously, there were significant configuration effects in the CCCma AMIP ensemble.

9.4.7 Testing the Interaction Effect: Potential Predictability from Internal Sources. The interaction effects in this experiment are particularly interesting because they represent slow, and hence potentially predictable, processes in the simulated climate of CCC GCMII that are internal to the climate system. An earlier investigation with the predecessor model to CCC GCMII (see Zwiers [440]) found evidence for such variations in a simulated climate when the sea-surface temperatures and sea-ice boundaries follow a fixed annual cycle.

It will be necessary to account for the effects of dependence within seasons to test the null hypothesis

$$\mathrm{H}_0\colon \sigma_C^2 = 0 \tag{9.30}$$

that there are no interaction effects. It is easily shown that the expected value of the sum of squared errors in our application is given by

$$\mathcal{E}(\mathcal{SSE}) = IJ(3 - T_0)\sigma_E^2,$$

where T_0 is given by (9.26). This is smaller than the expected value of $\mathcal{SSE}$ when errors are fully independent and not a convenient quantity to use in a test of (9.30). Item 5 in [9.4.2] indicates that a suitable test statistic should be of the form

$$\mathbf{F} = \frac{\mathcal{SSI}/((I-1)(J-1))}{\widehat{n\sigma}^2_{\bar{E}_{ij\circ}}}$$

where $\widehat{n\sigma}^2_{\bar{E}_{ij\circ}}$ is an estimator of $n\sigma^2_{\bar{E}_{ij\circ}} = T_0\sigma_E^2$.

The distribution of $\mathbf{F}$ under (9.30) is most easily found if $\widehat{n\sigma}^2_{\bar{E}_{ij\circ}}$ is also independent of $\mathcal{SSI}$ and distributed as a χ^2 random variable because $\mathbf{F}$ will then be F distributed under H_0.

9.4.8 A Rough and Ready Interaction Test.

Two solutions are available to the problem of testing for interaction effects in the presence of within season dependence.

A rough and ready solution is based on the argument that the correlation within seasons is small, and that it is negligible if monthly means are separated by at least a month. We could therefore drop the middle month in each season when computing $\mathcal{SSE}$ and adjust the degrees of freedom for error accordingly. That is, we compute

$$\mathcal{SSE}^* = \sum_{i=1}^{I}\sum_{j=1}^{J}\left(\mathbf{Y}_{i,j,1} - \frac{\mathbf{Y}_{i,j,1}+\mathbf{Y}_{i,j,3}}{2}\right)^2 + \left(\mathbf{Y}_{i,j,3} - \frac{\mathbf{Y}_{i,j,1}+\mathbf{Y}_{i,j,3}}{2}\right)^2.$$

Each of the IJ terms in this sum consists of the sum of two squared deviations that are constrained to add to zero. Thus each term contributes only 1 df for a total of IJ df. The effect of within season dependence can then be ignored and a test of (9.30) can be conducted by comparing

$$\mathbf{F} = \frac{\mathcal{SSI}/((I-1)(J-1))}{\mathcal{SSE}^*/(IJ)}$$

against $F((I-1)(J-1), IJ)$ critical values. Note that (9.27) and (9.29) can still be tested with the full data set.

Application of the 'rough and ready' method to 850 hPa temperature from the six simulation CCCma AMIP experiment demonstrates weak evidence for interaction effects (the null hypothesis is rejected over 14% of the globe). What makes the result interesting is that most of these rejections occur over land. They are apparently related to land surface processes that evolve differently from simulation to simulation. We return to the interaction effects in this experiment in [9.4.11].

9.4.9 A More Refined Test for Interaction Effects.

The 'rough and ready test' is not entirely satisfactory for a couple of reasons. An aesthetic objection is that the problem of within season dependence has been avoided rather than solved. More troubling is the loss of one-third of the data available for estimating error variability. We therefore embark on a path that results in full use of the data.

Our goal is to find factors C and n^* such that

A. $C\times\mathcal{SSE}/(n^*IJ)$ is an approximately unbiased estimator of $T_0\sigma_E^2$,

B. $C\times\mathcal{SSE}/(T_0\sigma_E^2)$ is approximately distributed $\chi^2(n^*IJ)$, and

C. $C\times\mathcal{SSE}/(n^*IJ)$ is independent of variance components $\mathcal{SSA}$, $\mathcal{SSB}$, and $\mathcal{SSI}$.

As with T_0, factors C and n^* are implicitly functions of the within season dependence.

Once these results are obtained, it is possible to test (9.30) by comparing

$$\mathbf{F} = \frac{\mathcal{SSI}/((I-1)(J-1))}{C\mathcal{SSE}/(n^*IJ)} \tag{9.31}$$

with $F((I-1)(J-1), n^*IJ)$ critical values.

Our first step in developing a test like (9.30) is to note that, in our application, $\mathcal{SSE}$ contains IJ statistically independent terms of the form

$$\mathbf{S}_{ij} = \sum_{l=1}^{3}(\mathbf{E}_{ijl} - \bar{\mathbf{E}}_{ij\circ})^2.$$

We find an approximating distribution for each individual $\mathbf{S}_{ij}$. We then use this result together with independence arguments to obtain items **A–C** in order to test (9.30) using (9.31).

Zwiers [444], using a method similar to that outlined in [8.3.20], shows that $\mathbf{S}_{ij}$ can be written

$$\mathbf{S}_{ij} = \mathbf{Z}_1^2 + \mathbf{Z}_2^2,$$

where $\mathbf{Z}_1$ and $\mathbf{Z}_2$ are independent zero mean normal random variables with variances $\sigma_E^2\lambda_1$ and $\sigma_E^2\lambda_2$, respectively. Parameters λ_1 and λ_2, which characterize the within season dependence, are the nonzero eigenvalues of the matrix $\mathcal{A}^T\mathcal{R}\mathcal{A}$ where $\mathcal{R}_{ij} = \rho_{|i-j|}$, and

$$\mathcal{A} = \frac{1}{3}\begin{pmatrix} 2 & -1 & -1 \\ -1 & 2 & -1 \\ -1 & -1 & 2 \end{pmatrix}.$$

Here $\rho_0 = 1$, ρ_1 is the correlation between $\mathbf{E}_{i,j,l}$ and $\mathbf{E}_{i,j,l+1}$ for $l = 1, 2$, and ρ_2 is the correlation between $\mathbf{E}_{i,j,1}$ and $\mathbf{E}_{i,j,3}$. The eigenvalues are given by

$$\begin{aligned}\lambda_1 &= 1 - \rho_2 \\ \lambda_2 &= 1 - \frac{4}{3}\rho_1 + \frac{1}{3}\rho_2.\end{aligned}$$

Because $\mathbf{Z}_1$ and $\mathbf{Z}_2$ do not generally have equal variances, the exact distribution of $\mathbf{S}_{ij}$ is difficult to find. In fact, the exact distribution can neither be expressed analytically nor tabulated efficiently. We therefore need to find an approximating distribution.

It is reasonable to select the χ^2 distribution as the approximating distribution because $\mathbf{S}_{ij} \sim \chi^2(2)$ when $\mathbf{Z}_1$ and $\mathbf{Z}_2$ have equal variances (i.e., when $\lambda_1 = \lambda_2 = 1$) and $\mathbf{S}_{ij} \sim \chi^2(1)$ when one of the eigenvalues is zero.[6] A χ^2 distribution with a fractional number of degrees of freedom somewhere between these two extremes should therefore work well. Thus we need to find a constant c and equivalent degrees of freedom n^* such that $c\chi^2(n^*)$ approximates the distribution of $\mathbf{S}_{ij}$. We do this by matching the mean and variance of $\mathbf{S}_{ij}$ with that of a $c\chi^2(n^*)$ random variable.

If Ξ is a $c\chi^2(n^*)$ random variable, then the mean and variance of Ξ are given by

$$\begin{aligned}\mathcal{E}(\Xi) &= cn^* \\ \mathrm{Var}(\Xi) &= 2c^2n^*.\end{aligned}$$

The mean and variance of $\mathbf{S}_{ij}$ are given by

$$\begin{aligned}\mathcal{E}(\mathbf{S}_{ij}) &= \sigma_E^2(\lambda_1 + \lambda_2) \\ &= 2\sigma_E^2(1 - \frac{2}{3}\rho_1 - \frac{1}{3}\rho_2) \\ \mathrm{Var}(\mathbf{S}_{ij}) &= 2\sigma_E^4(\lambda_1^2 + \lambda_2^2) \\ &= 4\sigma_E^4(1 - \frac{4}{3}\rho_1 + \frac{8}{9}\rho_1^2 - \frac{2}{3}\rho_2 \\ &\quad - \frac{4}{9}\rho_1\rho_2 + \frac{5}{9}\rho_2^2).\end{aligned}$$

[6] Recall that if $\mathbf{Z}_1, \ldots, \mathbf{Z}_n$ are iid $\mathcal{N}(0, \sigma^2)$, then $(\sum_{i=1}^n \mathbf{Z}_i^2)/\sigma^2 \sim \chi^2(n)$.

Equating means and variances and solving for c and n^* yields

$$\begin{aligned}c &= \sigma_E^2\frac{\lambda_1^2 + \lambda_2^2}{\lambda_1 + \lambda_2} \\ n^* &= \frac{(\lambda_1 + \lambda_2)^2}{\lambda_1^2 + \lambda_2^2},\end{aligned}$$

which, after substitution for λ_1 and λ_2, yields

$$c = \frac{9 - 12\rho_1 + 8\rho_1^2 - 6\rho_2 - 4\rho_1\rho_2 + 5\rho_2^2}{9 - 6\rho_1 - 3\rho_2} \tag{9.32}$$

$$n^* = \frac{2(3 - 2\rho_1 - \rho_2)^2}{9 - 12\rho_1 + 8\rho_1^2 - 6\rho_2 - 4\rho_1\rho_2 + 5\rho_2^2}. \tag{9.33}$$

We can check our work by testing these expressions when within season errors are iid; that is, when $\rho_1 = \rho_2 = 0$. We see we get the right answers, $c = 1$ and $n^* = 2$, by substituting $\rho_1 = \rho_2 = 0$ into (9.32) and (9.33). When $1 > \rho_1 > \rho_2 \geq 0$, we see that $c \leq 1$ (as expected, because $\lambda_1 \leq 1$ and $\lambda_2 \leq 1$) and $n^* \leq 2$.

Because the components $\mathbf{S}_{ij}$ of $\mathcal{SSE}$ are independent, (9.32) and (9.33) provide us with the result that

$$\mathcal{SSE}/c \sim \chi^2(n^*IJ).$$

Therefore the constant C required by items **A–C** above is given by

$$\begin{aligned}C &= \frac{T_0\sigma_E^2}{c} \\ &= \frac{(3 + 4\rho_1 + 2\rho_2)(3 - 2\rho_1 - \rho_2)}{9 - 12\rho_1 + 8\rho_1^2 - 6\rho_2 - 4\rho_1\rho_2 + 5\rho_2^2}.\end{aligned} \tag{9.34}$$

In summary, we account for within season dependence in our test of H_0 (9.30) by computing $\mathbf{F}$ as in (9.31), and comparing with $F((I-1)(J-1), n^*IJ)$ critical values. The 'shrinkage factor' C is given by (9.34). The 'equivalent degrees of freedom' for the denominator are n^*IJ, where n^* is given by (9.33).

9.4.10 Estimating Within Season Dependence. We need to know the within season correlations ρ_1 and ρ_2 to perform the test derived above. Since we do not know them, they must be estimated, and we must be careful to do this in such a way that items **A–C** are not seriously compromised.

Unfortunately, ρ_1 and ρ_2 can not be estimated directly from the monthly data because, in this context, the usual estimator has extremely large bias and variability. Instead, ρ_1 and ρ_2 are obtained by fitting a *parametric time series* model (see Chapter 10) to the daily data after they have been adjusted for the annual cycle, and then inferring ρ_1 and ρ_2 from the fitted model.

Because the parameters of the fitted time series model are estimated from a very large number of days of data (4860 in case of the CCCma AMIP experiment), they have very little sampling variability. Consequently, the derived estimates of ρ_1 and ρ_2 also have very little sampling variability, and therefore items **A–C** will not be seriously compromised provided that the fitted time series model fits the daily data well (see Zwiers [444] for discussion).

The particular time series model used is the *auto-regressive model of order 1* (10.3). With this model it is assumed that day-to-day variations within a season behave as *red noise* (see Sections 10.3 and 17.2). If we let $\{\mathbf{W}_{ijt} : t = 1, \ldots, 90\}$ represent the daily weather within season i of simulation j after removal of the annual cycle, then the red noise assumption states that

$$\mathrm{Cor}(\mathbf{W}_{i,j,t_1}, \mathbf{W}_{i,j,t_2}) = \rho^{|t_2-t_1|}, \qquad (9.35)$$

where ρ is the correlation between, say, 850 hPa temperature on adjacent days.

The monthly means, which are the object of our study, are given by

$$\mathbf{Y}_{ijl} = \sum_{t=(l-1)30+1}^{l30} \mathbf{W}_{ijt}/30. \qquad (9.36)$$

Using (9.35) and (9.36) we obtain, after some simplification, that

$$\mathrm{Var}(\mathbf{Y}_{ijl}) = \frac{\sigma_W^2}{30}\Big(1 + 2\sum_{\tau=1}^{29}(1 - \frac{\tau}{30})\rho^\tau\Big) \qquad (9.37)$$

and

$$\mathrm{Cov}(\mathbf{Y}_{i,j,l}, \mathbf{Y}_{i,j,(l+k)}) = \frac{\sigma_W^2\rho^{30(k-1)}}{30} \times \Big(\rho^{30} + \sum_{\tau=1}^{29}\Big(\frac{\tau}{30} + \Big(1 - \frac{\tau}{30}\Big)\rho^{30}\Big)\rho^\tau\Big). \qquad (9.38)$$

Further simplification yields that, for $\rho < 0.9$,

$$\rho_1 \approx \frac{\rho}{30^2(1-\rho)^2}$$
$$\rho_2 \approx \frac{\rho^{31}}{30^2(1-\rho)^2}.$$

It is reasonable to assume that $\rho_2 = 0$, except when ρ is large ($\rho > 0.9$).

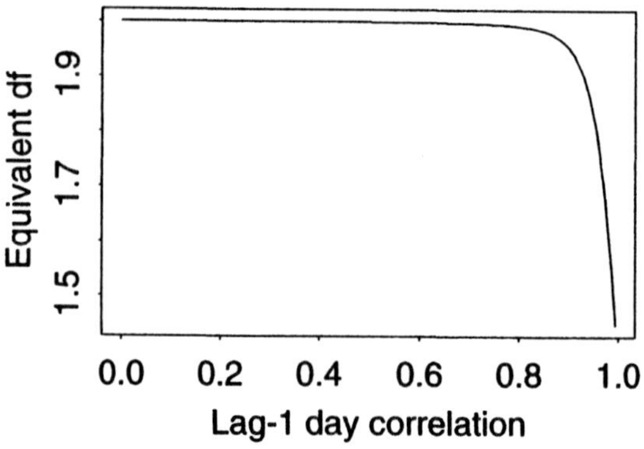

Figure 9.5: *Equivalent degrees of freedom n^* displayed as a function of the lag-1 day correlation when within season variations behave as red noise.*

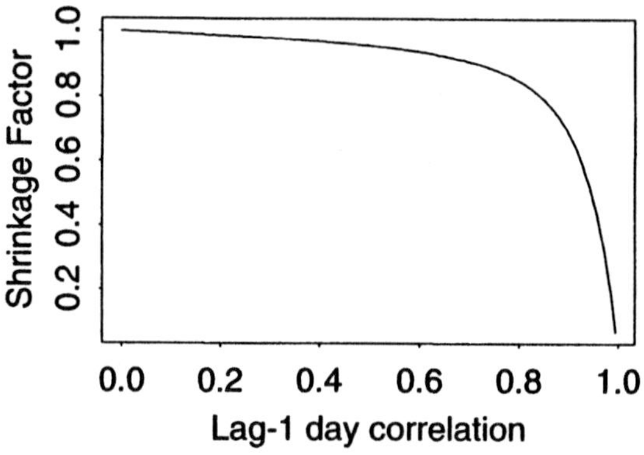

Figure 9.6: *Shrinkage factor for the unadjusted F-ratio for interaction effects.*

We substituted the exact expressions (9.37) and (9.38) into (9.33) and evaluated n^* as a function of ρ (see Figure 9.5). We see that $n^* > 1.95$ for $\rho < 0.9$. This was expected because $n^* = 2$ in the absence of terms affected by ρ_2, which becomes important only when ρ is very large. Hence, the degrees of freedom of the test for interaction effects need only be adjusted if day-to-day dependence is very strong.

We also substituted the exact expressions (9.37) and (9.38) into (9.34). The fraction $1/C$, used to shrink the unadjusted F-ratio for interaction effects, is illustrated in Figure 9.6. The shrinkage factor decreases slowly with increasing ρ when ρ is small, and drops very quickly as ρ approaches 1. When $\rho = 0.9$, it is necessary to shrink the unadjusted F-ratio for interaction effects by a factor of approximately 32%.

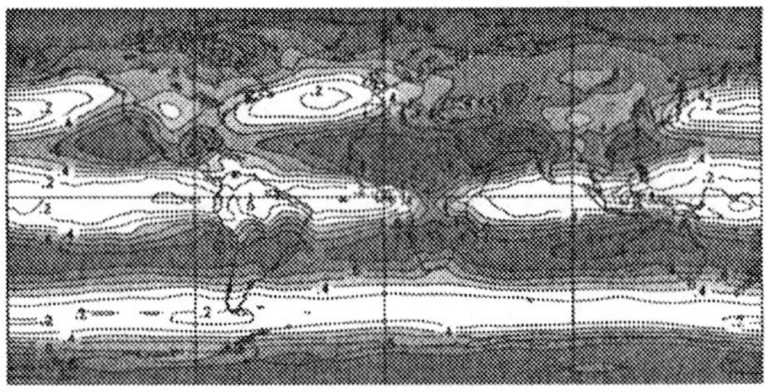

Figure 9.7: *Lag-1 day correlation for 850 hPa DJF temperature in the CCCma six member AMIP ensemble. Correlations greater than 0.4 are shaded.*

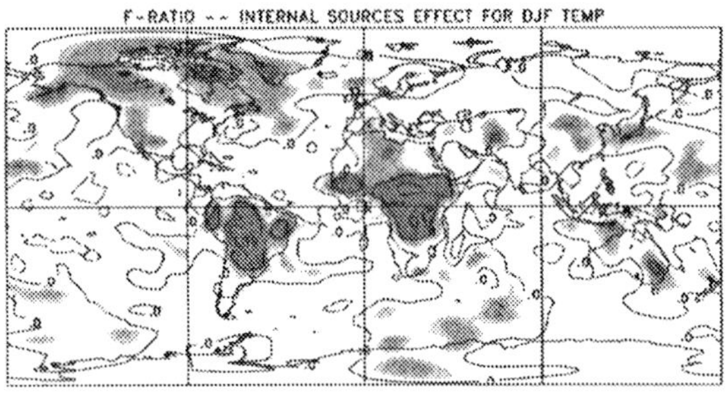

Figure 9.8: *The natural log of the F-ratios for the interaction effect for 850 hPa temperature in the CCCma AMIP experiment using the variance component adjustment method. Each contour indicates a doubling of the F-ratio. The shading indicates ratios which are significantly greater than 1 at the 10% significance level.*

9.4.11 Results for the CCCma AMIP Experiment. Estimates of lag-1 day correlation ρ for DJF 850 hPa temperature computed from the CCCma AMIP simulations using (9.37) and (9.38) are shown in Figure 9.7. We see that the simulated lower tropospheric temperature is generally most persistent on a day-to-day time scale where there is subsidence, and least persistent in the tropics and in the extratropical storm tracks. Estimated lag-1 day correlations range between $\widehat{\rho} = 0.0765$ and $\widehat{\rho} = 0.891$. Corresponding values for C (9.34) range between 1.005 and 1.409, and those for n^* range between $n^* = 2$ and $n^* = 1.96$. The varying amounts of dependence result in substantial spatial variation in the adjustment to the F-ratio but almost no spatial variation in the degrees of freedom of the F test for interaction effects.

The adjusted F-ratios (9.31) required to test H_0 (9.30) are displayed in Figure 9.8. The null hypothesis of the absence of the interaction effect is rejected over 17.5% of the globe at the 10% significance level. Experience suggests that this rate of rejection is field significant. The structure of this field of F-ratios is very similar to that obtained with the 'rough-and-ready' test, but the rate of rejection is higher because all of the data are used, rather than only two-thirds.

Figure 9.8 illustrates that the interaction effects are confined primarily to locations over land. As noted in [9.4.9], this suggests that land surface properties do not evolve identically in each AMIP simulation. The effects of slow variations in soil moisture and surface albedo are apparently detectable in the temperature of the lower troposphere. These effects do not appear to be detectable in the mean flow of the atmosphere as represented by 500 hPa geopotential. In this case, the no interaction effect hypothesis is rejected at the 10% significance level over only 12.4% of the globe in DJF and there does not appear to be a preferred location for the significantly large F-ratios.

9.5 Tuning a Basin Scale Ocean Model

9.5.1 Tuning an Ocean Model. We now briefly describe a designed experiment of a different sort. As discussed previously, geophysical models use parameterizations to describe sub-grid scale processes (see [6.6.6]). The sensitivity of such a model to a small number of parameters can be explored systematically with designed experiments provided individual runs of the model can be made at reasonable computational cost. Even today, this constraint places fairly tight bounds on the complexity of models that can be studied in this way and ingenuity is required to develop experimental designs that adequately explore parameter space.

Gough and Welch [145] describe a study of an *isopycnal* mixing parameterization in an ocean general circulation model[7] (OGCM) that has seven adjustable parameters (diapycnal and isopycnal diffusivity, vertical and horizontal eddy

[7] *Isopycnal* parameterizations represent mixing processes on surfaces of constant density (isopycnals) and their perpendiculars (diapycnals). Conventional parameterizations (as in Bryan [72] and Cox [93]) represent these processes on surfaces of constant height (horizontal levels) and their perpendiculars.

viscosity, horizontal background eddy diffusivity, maximum allowable isopycnal slope, and peak wind stress). Had they used a standard factorial design (see [9.1.1]) with, say, three different values of each parameter, it would have been necessary to integrate the model $3^7 = 2187$ times. Instead, they used a design called a *random Latin hypercube* (McKay, Conover, and Beckman [271]) that enabled them to adequately explore the model's parameter space with just 51 runs.[8] All runs were 1500 years long and were started with the ocean at rest.

The design employed by Gough and Welch exploits the fact that OGCMs are fully deterministic and converge to a steady state at long times, given a particular set of parameter values and no random forcing. Thus the experimental outcomes do not contain random noise in the conventional sense. This means that stochastic variation can be introduced into the response by means of the parameter settings, and subsequently that statistical methods similar to multivariate regression analysis (see Section 8.4) can be used to relate model response to the settings (see Gough and Welch [145, p. 782]).

The initial experiment performed by Gough and Welch consisted of 26 simulations with parameter settings selected as follows. A range of values was identified for each parameter, which was divided into 25 equal length intervals. The 26 values that delineate the boundaries of the intervals were recorded. The first combination of parameter settings was obtained by randomly selecting one value from each of the seven sets of 26 values. The second combination of parameter settings is obtained by randomly selecting a value from each of the 25 remaining values, and so on. The result is a random Latin hypercube design with seven treatments and 26 levels (values) of each treatment, combined at random in such a way that every level of every treatment occurs once in the 26 combinations of parameter settings. The objective is to obtain uniform (but necessarily sparse) coverage of the parameter space. One indicator of success in this regard is low correlation between the selected values of pairs of OGCM parameters. The objective is not always achieved with the randomization procedure because large correlations can occur by chance. Iman and Conover [192] describe a method for transforming a given random Latin hypercube into one with better correlation properties. Gough and Welch used this method iteratively to improve their experimental design.

A difficulty encountered by Gough and Welch is that the parameter space for which the OGCM converges to a steady state is not a hypercube (i.e., a seven-dimensional rectangle). In fact, four of the 26 runs displayed explosive behaviour, and one evolved to an 'unconverged' oscillatory solution. The regression-like analysis methods alluded to above were applied to the 21 successful runs to estimate the relationship between the parameters and the response, but the information that the experiment yielded was not considered sufficient to ensure accuracy. Twenty-five additional simulations were thus performed using parameter settings selected to be distant from the original 26 settings and also distant from one another; 15 of these converged to a steady state.

The final collection of 36 simulations successfully captured most of the dependence between the model's steady state circulation and the seven adjustable parameters. The resulting systematic description of the dependence between model outputs and parameter settings makes it easier to tune the model to reproduce an observed circulation feature. Gough and Welch were also able to study the interaction between pairs of parameters. For example, they found that diapycnal eddy diffusivity modifies the effect that the maximum allowable isopycnal slope has on the number of ocean points at which convection occurs. They thus demonstrated that this is a highly effective means of systematically exploring an unknown parameter space.

---

[8] Similar studies have been performed with an ice model [79] and a simplified atmospheric model [58].

Part IV

Time Series

Overview

In this part we deal with *time series analysis*, that is, the statistical description of stochastic processes and the use of sample time series for the identification of properties and the estimation of parameters. The motivation for our non-conventional development of the subject is explained in Section 10.1.

We introduce the concept of a stochastic process and its realizations, called time series, in Chapter 10. Special emphasis is placed upon auto-regressive processes since they may be interpreted as discretized linear differential equations with random forcing. At this stage we do not concern ourselves with the tools needed to characterize such processes, namely the covariance function and the spectrum. Instead we use a non-conventional non-parametric characterization, based on the frequency distribution of *run length*, that is, the duration of excursions above or below the mean. It allows us to intuitively examine characteristic properties of stochastic processes, such as memory or quasi-oscillatory behaviour, without using more complex mathematical tools such as the Fourier transform. Also, we differentiate between the variability caused by the internal dynamics of the process and that caused by the driving noise.

The conventional parametric characterization of a stochastic process, in terms of the auto- or cross-covariance function and the spectrum, is introduced in Chapter 11. While the concept of the covariance function poses no special problems, that of the spectrum is more difficult. The spectrum is often taken literally as the decomposition of a stochastic process into oscillations at a set of fixed frequencies. This interpretation is only appropriate in certain limited circumstances when there are good physical reasons to believe that the time series contains only a finite number of regular oscillatory signals. In general, though, the process will also contain noise, in which case the spectrum can not be interpreted as glibly. For example, the white noise process does not contain regular or oscillatory features; thus the interpretation of its spectrum as the decomposition of the white noise into equally important *oscillatory* components is misleading.

This part of the book is completed with Chapter 12, in which we describe techniques for inferring information about the true covariance function and spectrum.

10 Time Series and Stochastic Processes

10.1 General Discussion

10.1.1 The Role of Noise. This part of the book deals with stochastic processes and their realizations, time series. We begin with a general discussion of some of the basic ideas and pitfalls. The language and terminology we use is necessarily vague; more precise definitions will follow later in this chapter and in Chapters 11 and 12.

A time series $\mathbf{X}_t$ often consists of two components, a dynamically determined component D_t and a stochastic component $\mathbf{N}_t$, such that

$$\mathbf{X}_t = D_t + \mathbf{N}_t.$$

Sometimes the time evolution of D_t is independent of the presence of the stochastic component $\mathbf{N}_t$; in such cases the evolution of D_t is *deterministic*.[1] Examples are externally forced oscillations such as the tides or the annual cycle. At other times the dynamically determined part depends on the random component. Such processes become deterministic when the stochastic component is absent. When the stochastic component (or *noise*) is present, typical features, such as damped oscillations, are masked and therefore not clearly detectable. One goal of time series analysis is to detect and describe the characteristics of the *dynamical* component when the stochastic component is present.

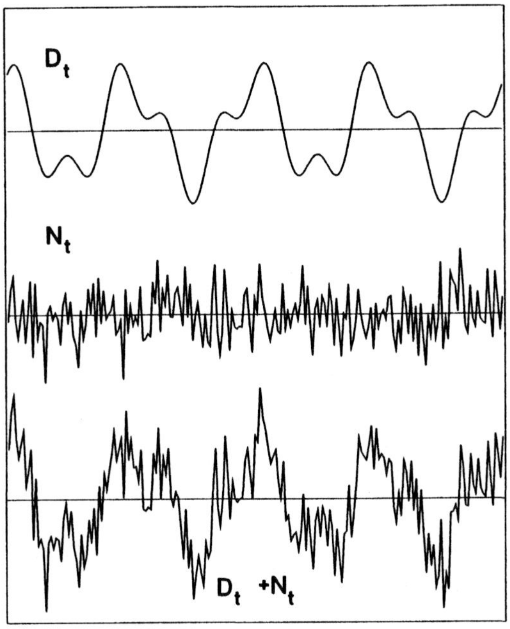

Figure 10.1: *A realization of a process* $\mathbf{X}_t = D_t + \mathbf{N}_t$ *in which the dynamical component* D_t *is not affected by the stochastic component* $\mathbf{N}_t$.
Top: A dynamical component D_t *made up of two oscillations.*
Middle: A 'white noise' component $\mathbf{n}_t$.
Bottom: The sum of both components.

Figures 10.1 and 10.2 illustrate these concepts. Figure 10.1 displays a purely deterministic oscillation D_t, a realization of a white noise process $\mathbf{n}_t$, and the sum $D_t+\mathbf{n}_t$. The addition of the noise introduces some uncertainty, but it does not modify the period or phase of the oscillations. In contrast, Figure 10.2 illustrates a damped system in which $D_t = \alpha \mathbf{x}_{t-1}$. Without noise ($\mathbf{N}_t = 0$), any nonzero value decays to zero in a characteristic time. The addition of noise transforms this decay into a stationary sequence of episodes (i.e., *runs*) during which D_t is continuously positive or negative. The distribution of the length of these excursions is a characteristic of such processes. When the dynamical component generates cyclical features in the absence of noise, pieces of such cyclical features will also be present when the noise is turned on. However, the 'period' will fluctuate, often around the period of D_t when noise is absent, and the phase will vary unpredictably. We refer to this as *quasi-oscillatory behaviour*.

The two types of stochastic processes differ with respect to their *predictability*. Here, we say a system is predictable at lead time τ if the conditional distribution of $\mathbf{X}_{t+\tau}$ given D_t is different from the unconditional distribution

[1] We depart slightly from our standard notation by using D_t, the dynamical component, to represent both the deterministic and stochastic forms.

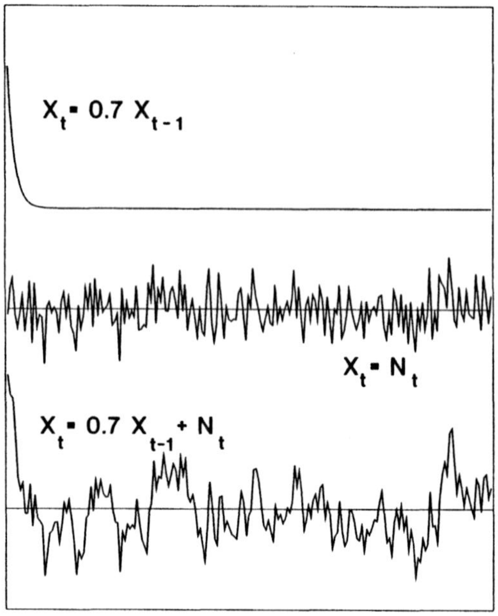

Figure 10.2: *A realization of a process* $\mathbf{X}_t = D_t + \mathbf{N}_t$ *for which the dynamical component* $D_t = 0.7\mathbf{X}_{t-1}$ *is affected by the stochastic component* $\mathbf{N}_t$.
Top: Evolution of the dynamical component $\mathbf{X}_t = D_t$ *from an arbitrary initial value when noise is absent.*
Middle: A 'white noise' component $\mathbf{N}_t$.
Bottom: Evolution from an arbitrary initial value when noise is present. The noise is the same as that used in Figure 10.1.

of $\mathbf{X}_{t+\tau}$. In that sense, the case in which the dynamical component D_t evolves independently of the stochastic component exhibits unlimited predictability. For example, the mean temperature in Hamburg in the winter of 3130 will be lower than the mean temperature in summer of that year. However, the system is inherently *unpredictable* beyond a certain time lag when the evolution of the dynamical part depends on the noise.[2]

10.1.2 The Probabilistic Structure of Time Series. We consider, for the moment, processes in which the dynamical state is determined by the history of the noise. To fully describe the stochastic, or probabilistic, structure of such a process it is necessary to specify joint density functions $f(\mathbf{X}_{t_1}, \mathbf{X}_{t_2}, \ldots, \mathbf{X}_{t_N})$ for an arbitrary number N of observing times and arbitrary times $t_1, \ldots, t_N$. This is generally not practical. Instead, the most important aspects of this probabilistic structure are described with either the *auto-covariance function* or, equivalently, the *spectrum.* Both descriptions require that we make a stationarity assumption of some sort about the stochastic process, that is, we need to assume that the statistical properties of the process are not time dependent.

The spectrum is the Fourier transform (see Appendix C) of the auto-covariance function. While both functions contain the same information, the spectrum is often more useful than the auto-covariance function for inferring the nature of the dynamical part of the process. In particular, the presence of multiple quasi-oscillatory components in a process causes *peaks* in the spectrum. The frequency at which a peak occurs often corresponds to that of a periodicity in the deterministic component of the process, and the width of the peak is representative of the damping rate.

The truth of this is difficult to deduce when the spectrum is defined as the Fourier transform of the auto-covariance function. Therefore conventional approaches for introducing the spectrum use another avenue. They often start with a representation of a stochastic process as the inverse Fourier transform of a random complex valued function (or *measure*) that is defined in the frequency domain rather than the time domain (i.e., the so-called Wiener spectral representation of a stochastic process [229, 422]). The spectrum is then defined as the expectation of the squared modulus of the random spectral measure and, finally, the auto-covariance function is shown to be the inverse Fourier transform of the spectrum.

A difficulty with the conventional approach, however, is that the dynamical aspects of the studied process are obscured. Hence, here we use a non-conventional time domain characterization of stochastic processes. We return to more conventional approaches in Chapters 11 and 12.

Another difficulty with the conventional approach concerns the way in which the spectrum is estimated from a time series. Suppose that the stochastic process is observed at times $t = 0, 1, \ldots, T$ and, for convenience, that T is even. Most spectral estimators use the Fourier expansion

$$\mathbf{x}_t = \sum_{k=-T/2}^{T/2} \mathbf{a}_k e^{-i2\pi kt}$$

to represent the observed time series. When this approximation is inverted, a *line spectrum* $|\mathbf{a}_k|^2$, for $k = 0, \pm 1, \ldots, T/2$ is obtained that can be

[2]Note that this statement is not related to ideas concerning chaos or nonlinear dynamics in general.

interpreted as a raw estimator of the spectrum. This raw estimator is not generally very useful, as is easily demonstrated by calculating it for a white noise time series. The true spectrum is flat ('white') but the raw estimate exhibits many large peaks, which are not manifestations of the 'dynamics' of the white noise process. In fact, when the calculation is repeated for another realization of the white noise process, peaks appear at entirely different frequencies.[3]

The mathematical inconsistency is that the trigonometric expansion is defined only for finite time series and periodic infinite time series, but stochastic processes are neither finite nor periodic. Thus, the expansion does not converge as the length of the time series increases. Note also that a line spectrum is a discrete object, defined for frequencies $0, 1/T, 2/T, \ldots, 1/2$. The spectrum of the sampled stochastic process, on the other hand, is continuous on the interval $[0, 1/2]$.

However, this approach can still be used to construct consistent estimates of the spectrum, provided it is done carefully. These are powerful methods when properly applied, but misleading conclusions about the spectrum are frequently obtained when they are used naively.

10.1.3 Overview. In this chapter we first introduce the concepts of *characteristic times* and *stochastic processes* (Section 10.2). *Auto-regressive* processes are the most widely used type of stochastic process in climate research, since they may be seen as approximations of ordinary linear differential equations subject to stochastic forcing (Section 10.3). As such they represent an important special case of Hasselmann's 'Stochastic Climate Models' (Section 10.4; [165]).

[3]This observation, and the realization that the spectral analysis of a stochastic time series can not be done by simply extending the time series periodically, are relatively recent developments. Indeed, at the turn of the twentieth century there was a frenzy of efforts to detect periodicities in all kinds of data, particular weather-related data, at almost all possible frequencies. Various climate forecast schemes were built on this futile approach, some of which can still be found in the literature.

The search for regular weather cycles resulted in a 1936 monograph that contained a four and half page list, entitled 'Empirical periods derived from the examination of long series of observations by arithmetic manipulation or by inspection,' describing supposed periodicities varying from 1 to 260 years in length (Shaw [347], pp. 320–325). In the light of our present understanding of the climate system, this search seems rather absurd, but modesty is advised. Modern workers also often use allegedly 'powerful,' poorly understood techniques in order to obtain 'interesting' results. Future climate researchers will probably find some of our present activities just as absurd and amusing as the search for periodicities.

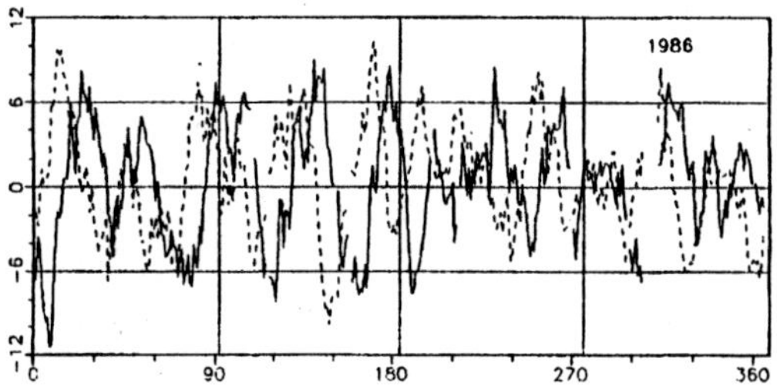

Figure 10.3: *A two-dimensional representation of the MJO for 1986 [388].*

In Section 10.5 we deal with two concepts of lesser importance in climate research, namely the large class of linear processes called *auto-regressive moving average processes* and a special class of nonlinear processes called *regime-dependent auto-regressive processes.*

10.2 Basic Definitions and Examples

10.2.1 Introduction: Characteristic Times. A time series is a finite sequence of real or complex numbers or vectors that are ordered by an index t and understood to be a realization of part of a stochastic process. The index usually represents time but could also represent some other non-stochastic variable that imposes order on the process, such as distance along a transect or depth in an ice core. Figure 10.3 shows a pair of real time series that jointly form a (bivariate) index of the so-called Madden-and-Julian Oscillation (MJO; [388], see [1.2.3], [15.2.4]). Both time series exhibit the typical features of a process in which the dynamical component is affected by noise. In particular, the time series lack any strict regularity; unlike time series of, for example, tidal sea level, prediction at long lead times appears to be impossible.

Despite the absence of strict periodicities, the two time series do exhibit some regularities. For example, the series exhibit 'memory' in the sense that, if a series is positive, it will tend to stay positive for some time. That is, $P(\mathbf{X}_{t+\tau} > 0 | \mathbf{X}_t > 0) > 0.5$ for small values of τ. However, for sufficiently large 'lags' τ, we find that knowledge of the sign of $\mathbf{X}_t$ does not inform us about the sign of $\mathbf{X}_{t+\tau}$. Thus,

$$P\left(\mathbf{X}_{t+\tau'} > 0 | \mathbf{X}_t > 0\right) = 0.5, \qquad (10.1)$$

for all τ' greater than some limit τ. The smallest τ satisfying (10.1), labelled τ_M, is a *characteristic time* that represents the time after which there is no forecast skill;[4] τ_M is a measure of the 'memory' of the stochastic process. Inspection of Figure 10.3 indicates that τ_M is at least 10–20 days for both time series.

There are various other ways to define characteristic times, and [10.3.7] shows that τ_M is not particularly useful in many applications. Another time scale is the average waiting time between successive local minima or maxima. By this measure, it would appear that both time series in Figure 10.3 exhibit *quasi-periodicity* of about 40 days. Note that, even though the quasi-periodicities occur on a similar time scale, they are shifted relative to each other. In the words of *spectral analysis*, the two time series vary *coherently* and are approximately 90° *out-of-phase* on the time scale of the quasi-periodicity.

Two important goals of time series analysis are *to identify characteristic time scales in stochastic processes, and to determine whether two time series share common information.*

In the following we consider exclusively time series samples in discrete time. Also, for the sake of brevity, the time step between two consecutive data is arbitrarily set to 1.

10.2.2 Stochastic Processes. We have, so far, used the expression 'time series' rather informally. Time series may be seen as randomly selected *finite* sections of infinitely long sequences of random numbers. In that sense, a time series is a *random sample* of a *stochastic process*, an ordered set of random variables $\mathbf{X}_t$ indexed with an integer t (which usually represents time).

In general, the state $\mathbf{X}_t$ of the process at any specific time t depends on the state of the process at *all* other 'times' s. In particular, for any pair of 'times' (t, s), there is a bivariate density function f_{ts} such that

$$P(\mathbf{X}_t \in [a, b] \text{ and } \mathbf{X}_s \in [c, d]) = \int_a^b \int_c^d f_{ts}(x, y)\, dx\, dy. \tag{10.2}$$

The marginal density functions derived from f_{ts} (see [2.8.3]) are, of course, the density functions of $\mathbf{X}_t$ and $\mathbf{X}_s$, given by

$$f_t(x) = \int_{-\infty}^{\infty} f_{ts}(x, y)\, dy$$
$$f_s(y) = \int_{-\infty}^{\infty} f_{ts}(x, y)\, dx.$$

[4] Note that the direction of the inequalities in (10.1) does not affect the definition of τ_M.

Random variables $\mathbf{X}_t$ and $\mathbf{X}_s$ are usually dependent. This does not prevent the estimation of process parameters, but it does compromise the various interval estimation approaches discussed in Section 5.4 because the dependence violates the fundamental 'iid' assumption. Similarly, most hypothesis testing procedures described in Chapter 6 no longer operate as specified when the data are *serially correlated* or otherwise dependent.

10.2.3 Example: White Noise. *White noise*, an infinite sequence of zero mean iid normal random variables, is the simplest example of a stochastic process. Such processes contain no memory by construction, that is, for every t, element $\mathbf{X}_t$ is independent of every other element in the process. A realization of a white noise process is shown in Figure 10.1.

The characteristic time $\tau_M = 1$, since for any nonzero τ (10.2)

$$P(\mathbf{X}_{t+\tau} > 0 | \mathbf{X}_t > 0) = \frac{\int_0^\infty \int_0^\infty f_{ts}(x, y)\, dx\, dy}{\int_0^\infty f_t(x)\, dx} = \frac{\int_0^\infty f_\mathcal{N}(x)\, dx \times \int_0^\infty f_\mathcal{N}(x)\, dx}{\int_0^\infty f_\mathcal{N}(y)\, dy} = 0.5.$$

The probability of observing a *run* (i.e., a sequence of consecutive $\mathbf{x}_s$s of the same sign) of length $\mathbf{L}$ beginning at an arbitrary time t is obtained from an independence argument. Runs are observed when $-\mathbf{X}_{t-1}, \mathbf{X}_t, \ldots, \mathbf{X}_{t+L-1}$ and $-\mathbf{X}_{t+L}$ all have the same sign. Therefore, since two signs are possible,

$$P(\mathbf{L} = L) = 2 \times 2^{-(L+2)} = 2^{-(L+1)}. \tag{10.3}$$

Note that

$$P(\mathbf{L} = 0) = 1 - P(\mathbf{L} > 0) = 1/2.$$

That is, there is probability $1/2$ that a run does not begin at time t. The probability of observing a run of length $\mathbf{L} = L$, given that a run begins at time t, is 2^{-L}. Thus the probability that a run beginning at a given time will become exactly $L = 3$ time units in length is $2^{-3} = 0.125$. The probability that the run will last at least three time steps is $\sum_{L=3}^{\infty} 2^{-L} = 0.25$. The corresponding probabilities for $L = 10$ are only 0.01 and 0.02.

10.2.4 Definition: Stationary Processes. A stochastic process $\{\mathbf{X}_t : t \in \mathbb{Z}\}$ is said to be *stationary* if all stochastic properties are independent of index t.

It follows that if $\{\mathbf{X}_t\}$ is stationary, then:

1 $\mathbf{X}_t$ has the same distribution function F for all t, and

2 for all t and s, the parameters of the joint distribution function of $\mathbf{X}_t$ and $\mathbf{X}_s$ depend only on $|t - s|$.

10.2.5 Weakly Stationary Processes. For most purposes, the assumption of strict stationarity can usually be replaced with the less stringent assumption that the process is *weakly stationary*, in which case

- the mean of the process, $\mathcal{E}(\mathbf{X}_t)$, is independent of time, that is, the mean is constant, and
- the second moments $\mathcal{E}(\mathbf{X}_s\mathbf{X}_t)$ are a function only of the time difference $|t - s|$.

A consequence of the last condition is that the variance of the process, $\text{Var}(\mathbf{X}_t)$, does not change with time.

The two conditions required for weak stationarity are less restrictive than the conditions enumerated in [10.2.4], and are often sufficient for the methods used in climate research. Even so, the weaker assumptions are often difficult to verify. Provided there are not contradictory dynamical arguments, it is generally assumed that the process is weakly stationary.

10.2.6 Weakly Cyclo-stationary Processes. The assumption that a process is stationary, or weakly stationary, is clearly too restrictive to represent many climatological processes accurately. Often we know that stochastic properties are linked to an externally enforced deterministic cycle, such as the annual cycle, the diurnal cycle, or the Milankovitch cycles. When we deal with variations on time scales of months and years, the annual cycle is important. For time scales of hours and days the diurnal cycle is important. For variations on time scales of thousands to hundreds of thousands of years, the Milankovitch cycle will affect the data significantly. We therefore consider processes with the following properties.

1 The mean is a function of the time within the external cycle, that is, $\mathcal{E}(\mathbf{X}_t) = \mu_{t|_m}$, where $t|_m = t \bmod L$ and L is the length of the external cycle measured in units of observing intervals.

2 $\mathcal{E}\big((\mathbf{X}_t - \mu_{t|_m})(\mathbf{X}_s - \mu_{s|_m})\big)$, the central second moment, is a function only of the time difference $|t - s|$ and the phase $t|_m$ of the external cycle.

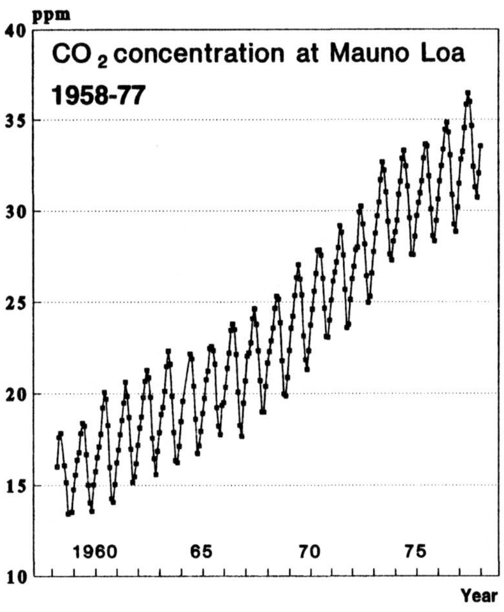

Figure 10.4: *1958–77 time series of monthly mean atmospheric CO_2 concentration measured at the Mauna Loa Observatory in Hawaii.*

We refer to processes with such properties as *weakly cyclo-stationary processes*. It follows from the second condition that the variance is also a function of the time within the externally determined cycle. Cyclo-stationary behaviour can be seen in Figures 1.7, 1.8, and 10.4. Huang and North [189] and Huang, Cho, and North [188] describe cyclo-stationary processes and cyclo-stationary spectral analysis in detail.

The conditions for weak cyclo-stationarity parallel those for ordinary weak stationarity, except that the parameters of interest are indexed by the phase of the external cycle. Statistical inference problems that can be solved for weakly stationary processes can generally also be solved for weakly cyclo-stationary processes. However, the utility of these models is strongly constrained by the very large demands they place on the data sets used for parameter estimation. Cyclo-stationary models generally have many more parameters than their stationary counterparts and all of these parameters must be estimated from the available data.

10.2.7 Examples. Suppose $\mathbf{X}_t$ is a stationary process. If a linear *trend* is added, the resulting process $\mathbf{Y}_t = \mathbf{X}_t + \alpha t$ is no longer stationary: its distribution function, $F_{Y_t}(y) = f_X(y - \alpha t)$, depends on t.

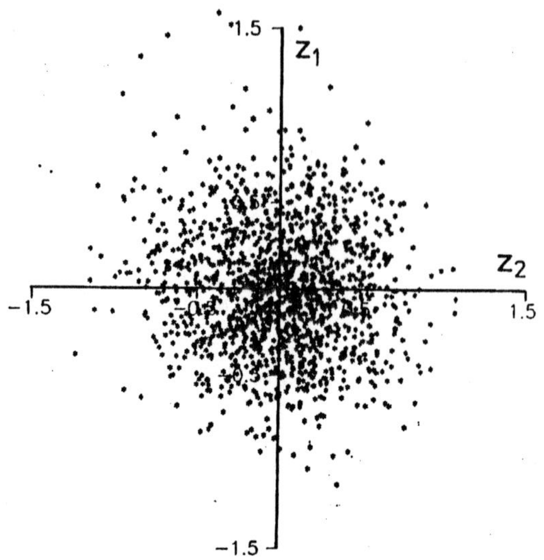

Figure 10.5: *Scatter diagram of the bivariate MJO index, in units of standard deviations. A sub-segment of the full time series is shown in Figure 10.3 [388].*

If an oscillation is added to stationary process $\mathbf{X}_t$, the resulting process $\mathbf{Y}_t = \mathbf{X}_t + \cos(\omega t)$ is *cyclo-stationary* and has distribution function $F_{Y_t}(y) = F_X(y - \cos(\omega t))$.

A time series that exhibits both a trend and a cyclo-stationary component is the famous CO_2 concentration curve measured at Mauna Loa in Hawaii. A segment of this series is shown in Figure 10.4. Note that the trend is not strictly linear; both the rate of change and the amplitude of the annual cycle increase with time. The maximum CO_2 concentrations occur during northern winter, and the minima during northern summer.

The time series displayed in Figure 10.3 are approximately stationary with means near zero and nearly equal variances. A *scatter diagram* illustrating the joint variation of the two time series in units of standard deviations is plotted in Figure 10.5. The time series appear to be jointly normal. In particular, note that the points are scattered symmetrically about the origin with maximum density near the origin.

10.2.8 Example: A Random Walk and the Long-range Transport of Pollutants. If $\mathbf{Z}_t$ is white noise, then $\mathbf{X}_t$, given by

$$\mathbf{X}_t = \sum_{j=1}^{t} \mathbf{Z}_j, \tag{10.4}$$

is a non-stationary process. The first moment of $\mathbf{X}_t$ is independent of time, but the variance increases with time. In fact,

$$\mathcal{E}(\mathbf{X}_t) = \mathcal{E}\Big(\sum_{j=1}^{t} \mathbf{Z}_j\Big) = \sum_{j=1}^{t} \mathcal{E}(\mathbf{Z}_j) = 0,$$

and

$$\begin{aligned}
\mathrm{Var}(\mathbf{X}_t) &= \mathcal{E}\Big(\big(\sum_{j=1}^{t} \mathbf{Z}_j\big)^2\Big) \\
&= \sum_{j,k=1}^{t} \mathcal{E}(\mathbf{Z}_j\mathbf{Z}_k) = \sum_{j=1}^{t} \mathcal{E}\big(\mathbf{Z}_j^2\big) \\
&= t\sigma_Z^2.
\end{aligned}$$

This stochastic process, a *random walk*, is stationary with respect to the mean, but non-stationary with respect to the variance.

This process describes the path of a particle that experiences random displacements. If a large number of such particles are considered, the centre of gravity will not move, that is, $\mathcal{E}(\mathbf{X}_t) = 0$, but the scatter increases continuously. Thus the random walk is sometimes a useful *stochastic model* for describing the transport of atmospheric or oceanic tracers.

The movement of a particle, perhaps emitted from a smoke stack, is determined by the deterministic flow U and many small unpredictable displacements. If the particle is located at $\mathbf{R}(t)$ at time t, then its location at time $t+1$ is given by $\mathbf{R}(t+1) = \mathbf{R}(t) + U + \mathbf{Z}_t$, where $\mathbf{Z}_t$ represents white noise, and its location at time $t+l$ is given by $\mathbf{R}(t+l) = \mathbf{R}(t) + lU + \sum_{s=t}^{t+l-1} \mathbf{Z}_s$. If many particles are 'emitted' and transported this way, the time evolution of the concentration may be modelled in three dimensions.

The result of such a simulation is shown in Figure 10.6. The left hand panel displays a 24-hour forecast of the 1000 hPa height field over Western Europe. Note the cyclonic around the low over the coast of Norway. A pollutant, SO_2, was injected into the simulated atmosphere at a constant rate from a point source in east England. The right hand panel displays the simulated SO_2 distribution at the end of the 24-hour period. Evidence of both deterministic advection processes and random diffusive processes can be seen.

10.2.9 Ergodicity. Unfortunately, stationarity, or weak stationarity, alone is not enough to ensure that the moments of a process can be estimated from a single time series. Koopmans [229] elegantly illustrates this with the following example.

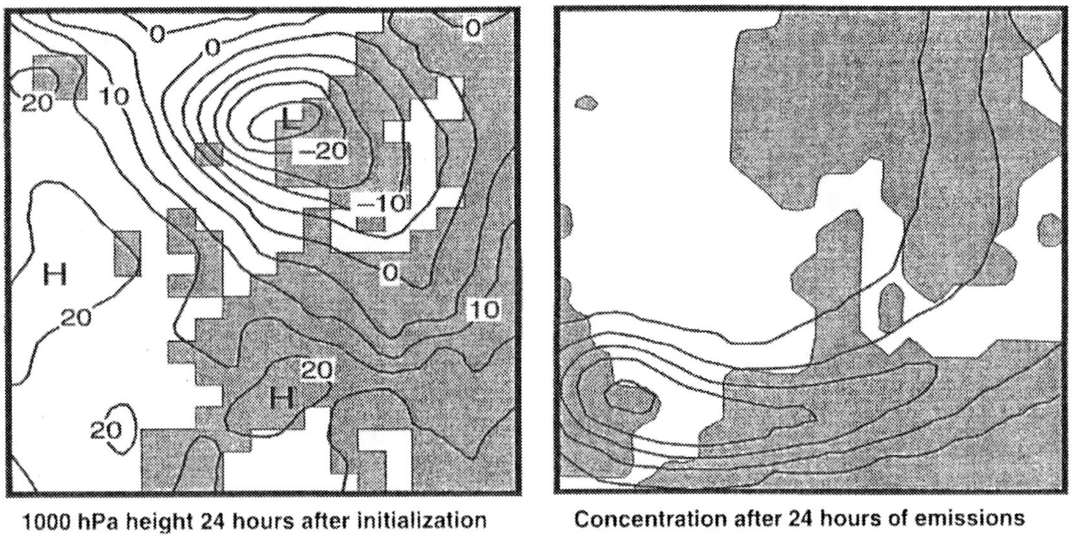

Figure 10.6: *Example of a simulation of long-range transport of air pollutants.*
Left: Simulated 1000 hPa height field 24 hours after model initialization.
Right: Distribution of pollutant continuously emitted in east England after 24 hours.
From Lehmhaus et al. [250].

Consider a stochastic process $\mathbf{X}_t$ such that each realization is constant in time. That is, suppose $\mathbf{x}_t = \mathbf{a}$, where $\mathbf{a}$ is a realization of an ordinary random variable $\mathbf{A}$. Every realization of $\mathbf{X}_t$ is thus a line parallel to the time-axis. It is easily shown that the process $\mathbf{X}_t$ is weakly stationary; the mean and variance of the process, which are equal to $\mathcal{E}(\mathbf{A})$ and $\mathrm{Var}(\mathbf{A})$, respectively, are independent of time and all covariances $\mathrm{Cov}(\mathbf{X}_t, \mathbf{X}_s)$ are also equal to $\mathrm{Var}(\mathbf{A})$ and hence independent of time. However, the usual estimator of the process mean, $\frac{1}{n}\sum_{t=1}^{n}\mathbf{X}_t = \frac{1}{n}\sum_{t=1}^{n}\mathbf{A} = \mathbf{A}$, does not converge to the process mean, $\mathcal{E}(\mathbf{A})$, as the length of the averaging interval increases. Since the individual realizations of the process do not contain any variability, a single realization of this process does not provide sufficient information about the process to construct consistent estimators of process parameters.

Stochastic processes must be *ergodic* as well as stationary in order to ensure that individual realizations of the process contain sufficient information to produce consistent parameter estimates. A technical description of ergodicity is beyond the scope of this book (see, e.g., Brockwell and Davis [68], Koopmans [229] or Hannan [157]). However, in loose terms, ergodicity ensures that the time series varies quickly enough in time that increasing amounts of information about process parameters can be obtained by extending the time series.[5] Clearly this does not happen in Koopmans's example. However, ergodicity is not generally a problem in climate research.

10.3 Auto-regressive Processes

10.3.0 General. We will explore the properties of *auto-regressive processes* in some detail in this section. The collection of all weakly stationary auto-regressive models forms a general purpose class of parametric stochastic process models. This class is not complete but, given any weakly stationary ergodic process $\{\mathbf{X}_t\}$, it is possible to find an auto-regressive process $\{\mathbf{Y}_t\}$ that approximates $\{\mathbf{X}_t\}$ arbitrarily closely.

Auto-regressive processes are popular in climate research, mainly because they represent discretized versions of ordinary differential equations [10.3.1]. Conventional auto-regressive processes operate with constant coefficients and generate weakly stationary time series. By allowing the coefficients to vary periodically, the resulting time series become weakly cyclo-stationary.

[5] Another way of describing an ergodic process is to say that it does not have excessively long memory. Thus the ergodic property is often expressed in terms of a 'mixing condition' that involves the rate of decay of the auto-covariance function with increasing lag. A typical mixing condition specifies that the auto-covariance function should be absolutely summable.

Such processes are called *seasonal auto-regressive processes* [10.3.8]. The name 'auto-regressive' indicates that the process evolves by regressing past values towards the mean and then adding noise.

The plan for the remainder of the section is as follows. An ordinary auto-regressive (AR) process is defined in [10.3.1] and its mean and variance are derived in [10.3.2]. Some specific AR processes are examined in [10.3.3,4], and the conditions under which an AR process is stationary are discussed in [10.3.5]. As noted above, AR processes can be thought of as discretized differential equations. We show, in [10.3.6], the effect that the 'dynamics' of these processes have on their time evolution. Next, we introduce the notion in [10.3.7] that these processes have a 'memory' that can be described in general terms by a characteristic time. We generalize the AR processes so that seasonal behaviour is also accounted for in [10.3.8,9], and the concept is extended to multivariate processes in [10.3.10].

Looking ahead, we will take a short excursion into stochastic climate modelling in Section 10.4, but will then return to the subject of parametric stochastic models in Section 10.5 where we will see that the class of AR models is one of three more or less equivalent classes of models.

10.3.1 Definition: Auto-regressive Processes. The dynamics of many physical processes can be approximated by first- or second-order ordinary linear differential equations, for example,

$$a_2\frac{d^2x(t)}{dt^2} + a_1\frac{dx(t)}{dt} + a_0x(t) = z(t),$$

where z is some external forcing function. Standard time discretization yields

$$\begin{aligned} a_2(x_t + x_{t-2} - 2x_{t-1}) \\ + a_1(x_t - x_{t-1}) + a_0x_t = z_t, \end{aligned}$$

or

$$x_t = \alpha_1 x_{t-1} + \alpha_2 x_{t-2} + z'_t. \qquad (10.5)$$

where

$$\begin{aligned} \alpha_1 &= \frac{a_1 + 2a_2}{a_0 + a_1 + a_2} \\ \alpha_2 &= -\frac{a_2}{a_0 + a_1 + a_2} \\ z'_t &= \frac{1}{a_0 + a_1 + a_2} z_t. \end{aligned}$$

If z_t is a white noise process, then (10.5) defines a second-order auto-regressive or *AR(2) process.*

An auto-regressive process of order p, or an *AR(p) process*, is generally defined as follows:

$\{\mathbf{X}_t : t \in \mathbb{Z}\}$ is an auto-regressive process of order p if there exist real constants α_k, $k = 0, \dots, p$, with $\alpha_p \neq 0$ and a white noise process $\{\mathbf{Z}_t : t \in \mathbb{Z}\}$ such that

$$\mathbf{X}_t = \alpha_0 + \sum_{k=1}^{p} \alpha_k \mathbf{X}_{t-k} + \mathbf{Z}_t. \qquad (10.6)$$

The most frequently encountered AR processes are of first or second order; an AR(0) process is white noise. Note that $\mathbf{X}_t$ is independent of the part of $\{\mathbf{Z}_t\}$ that is in the future, but that it is dependent upon the parts of the noise process that are in the present and the past.

10.3.2 Mean and Variance of an AR(p) Process. Taking expectations on both sides of (10.6) we see that

$$\mathcal{E}(\mathbf{X}_t) = \frac{\alpha_0}{1 - \sum_{k=1}^{p} \alpha_k}. \qquad (10.7)$$

If we set $\mu = \mathcal{E}(\mathbf{X}_t)$, then (10.6) may be rewritten as

$$\mathbf{X}_t - \mu = \sum_{k=1}^{p} \alpha_k (\mathbf{X}_{t-k} - \mu) + \mathbf{Z}_t. \qquad (10.8)$$

The variance of $\mathbf{X}_t$ is obtained by multiplying both sides of (10.8) with $\mathbf{X}_t - \mu$, and again taking expectations on both sides of the equation. We see that

$$\begin{aligned} \mathrm{Var}(\mathbf{X}_t) &= \sum_{k=1}^{p} \alpha_k \mathcal{E}((\mathbf{X}_t - \mu)(\mathbf{X}_{t-k} - \mu)) \\ &\quad + \mathcal{E}((\mathbf{X}_t - \mu)\mathbf{Z}_t) \\ &= \sum_{k=1}^{p} \alpha_k \rho_k \mathrm{Var}(\mathbf{X}_t) + \mathrm{Var}(\mathbf{Z}_t), \end{aligned}$$

where

$$\rho_k = \frac{\mathcal{E}((\mathbf{X}_{t-k} - \mu)(\mathbf{X}_t - \mu))}{\mathrm{Var}(\mathbf{X}_t)}.$$

Thus,

$$\mathrm{Var}(\mathbf{X}_t) = \frac{\mathrm{Var}(\mathbf{Z}_t)}{1 - \sum_{k=1}^{p} \alpha_k \rho_k}. \qquad (10.9)$$

The function ρ_k, $k = 0, \pm 1, \dots$ is known as the *auto-correlation function* (see Chapter 11).

We assume in the following, for convenience, that $\alpha_0 = 0$ so that $\mathcal{E}(\mathbf{X}_t) = \mu = 0$.

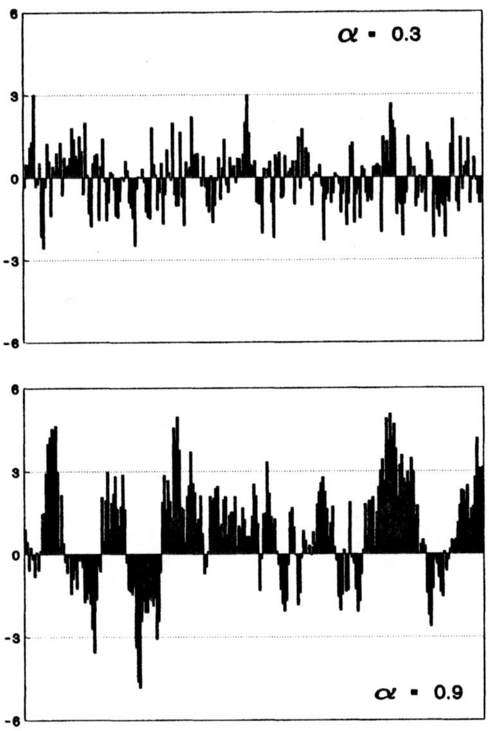

Figure 10.7: *240 time step realizations of AR(1) processes with* $\alpha_1 = 0.3$ *(top) and* 0.9 *(bottom). Both processes are forced by unit variance normally distributed white noise.*

10.3.3 AR(1) Processes. AR(1) processes may be understood as discretized first-order differential equations. Such systems have only one degree of freedom and are unable to oscillate when the damping coefficient is positive. A nonzero value $\mathbf{x}_t$ at time t tends to be damped with an average damping rate of α_1 per time step.[6] Obviously the system can only be stationary if $\alpha_1 < 1$.[7] Figure 10.7 shows realizations of AR(1) processes with $\alpha_1 = 0.3$ and 0.9. The upper time series is very noisy and usually changes sign within just a few time steps; the lower one has markedly longer 'memory' and tends to keep the same sign for 10 and more consecutive time steps.

What is the variance of an AR(1) process? Because of the independence of $\mathbf{X}_{t-1}$ and the driving noise $\mathbf{Z}_t$ we find that

$$\rho_1 = \frac{\mathcal{E}(\mathbf{X}_{t-1}\mathbf{X}_t)}{\mathrm{Var}(\mathbf{X}_t)} = \alpha_1$$

[6]Specifically, $\mathcal{E}(\mathbf{X}_{t+\ell}|\mathbf{X}_t = \mathbf{x}_t) = \alpha_1^\ell \mathbf{x}_t$.

[7]The realizations of $\{\mathbf{X}_t\}$ grow explosively when $\alpha_1 > 1$, and the process with $\alpha_1 = 1$ behaves as a random walk (see [10.2.8]).

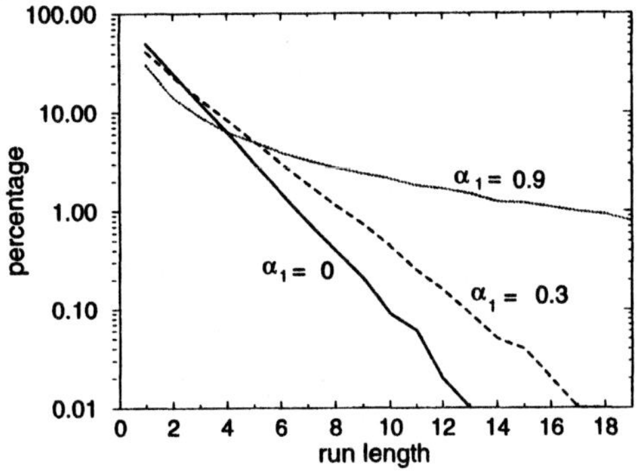

Figure 10.8: *The frequency distribution of the run length* **L** *as derived from* 100 000 *time step random realizations of three AR(1) processes* $\mathbf{X}_t$ *with different process parameters* α_1.
50 095 *runs were found for* $\alpha_1 = 0$,
40 280 *runs for* $\alpha_1 = 0.3$,
14 375 *runs for* $\alpha_1 = 0.9$.
The horizontal axis indicates the run length L.

and thus, using (10.9),

$$\mathrm{Var}(\mathbf{X}_t) = \frac{\sigma_z^2}{1-\alpha_1^2}. \tag{10.10}$$

Thus, the variance of the process is a linear function of the variance σ_z^2 of the 'input' noise $\mathbf{Z}_t$ and a nonlinear function of the memory parameter α_1. For processes with small memory, that is, $\alpha_1 \approx 0$, the variance of $\mathbf{X}_t$ is almost equal to the variance of $\mathbf{Z}_t$. When $\alpha_1 > 0$, $\mathrm{Var}(\mathbf{X}_t) > \mathrm{Var}(\mathbf{Z}_t)$, and when α_1 is almost 1, the variance of $\mathbf{X}_t$ becomes very large. The variance of (10.9) is not defined when $\alpha_1 = 1$. Figure 10.7 neatly demonstrates that the variance of an AR(1) process increases with the process parameter α_1.

Now recall the run length random variable **L**, discussed in [10.2.3]. We were able to derive the distribution of **L** analytically for white noise (i.e., $\alpha_1 = 0$). The derivation can not be repeated when $\alpha_1 \neq 0$ because then elements of the process are serially correlated. We therefore estimated the distribution of **L** with a Monte Carlo experiment (see Section 6.3). The experiment was conducted by generating a time series of length 100 000 from an AR(1) process. The runs of length $\mathbf{L} = L$ were counted for each $L > 0$. The result of this exercise is shown in Figure 10.8.

When $\alpha_1 = 0$, the Monte Carlo result agrees well with the analytical result (10.3) for $L \leq 10$. For larger run lengths, the relative uncertainty of the estimate becomes large because so few runs are observed. The frequency of short runs (e.g.,

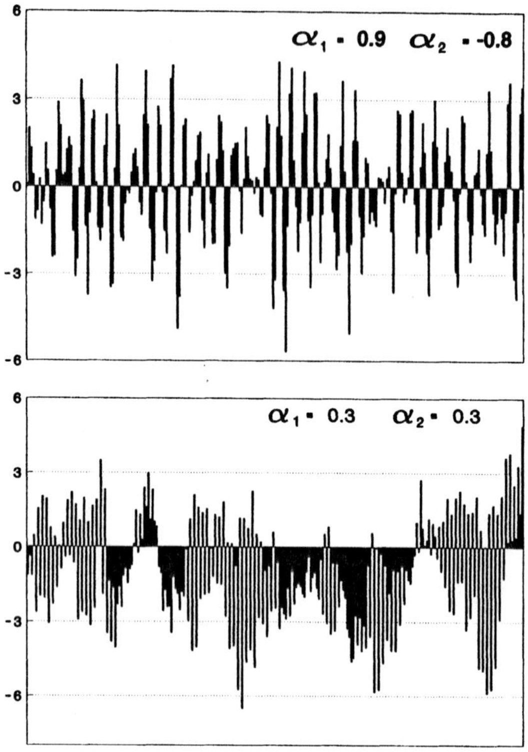

Figure 10.9: *240 time step realizations of an AR(2) process with* $\alpha_1 = 0.9$ *and* $\alpha_2 = -0.8$ *and with* $\alpha_1 = \alpha_2 = 0.3$.

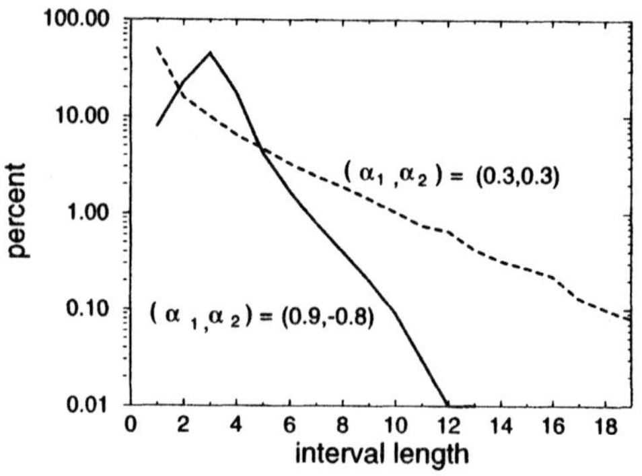

Figure 10.10: *The frequency distribution of the run length* **L** *as derived from* 100 000 *time step random realizations of two AR(2) processes* $\mathbf{X}_t$ *with different process parameters* (α_1, α_2). *There were* 35 684 *intervals for* $(\alpha_1, \alpha_2) = (0.3, 0.3)$, *and* 33 326 *runs for* $(\alpha_1, \alpha_2) = (0.9, -0.8)$. *The horizontal axis indicates the run length L.*

$L = 1$) decreases with increasing α_1, while the frequency of longer runs increases. For instance, in the white noise case we expect one run in 1000 will be of length 10. In contrast, when $\alpha_1 = 0.3$, about four runs in 1000 are of length 10, and when $\alpha_1 = 0.9$, this number increases to 20.

10.3.4 AR(2) Processes. AR(2) processes, which represent discretized second-order linear differential equations (see [10.3.1]), have two degrees of freedom and can oscillate with one preferred frequency (see also [11.1.8]). Finite segments of realizations of two AR(2) processes are shown in Figure 10.9. The time series with $(\alpha_1, \alpha_2) = (0.9, -0.8)$ exhibits clear quasi-periodic behaviour with a period of about six time steps. The other time series, with $(\alpha_1, \alpha_2) = (0.3, 0.3)$, has behaviour comparable to that of an AR(1) process with large memory. The diagram hints that there may be a longer quasi-periodicity, say of the order of 150 or more time steps. However, we will see later that the (0.3, 0.3) process does not generate periodicities of any kind.

When we repeated the Monte Carlo experiment described above for the (0.9, −0.8) AR(2) process, we observed 33 355 runs in a 100 000 time unit simulation. The relative frequency distribution of **L** that was obtained is shown in Figure 10.10. Note that the $L = 1$ category is not the most frequent. Instead, runs of length $L = 3$, comprising 44% of all runs, are most common. This is consistent with our perception that this process has a quasi-periodicity of about six time units in length.

If the (0.9, −0.8) AR(2) process is truly quasi-oscillatory with a period of approximately six time steps, we should expect to frequently observe runs of approximately three time units in length. We therefore counted the number of times that a run of length, say, L_2 adjoined a run of length L_1. The results are given in Table 10.1. Note that two consecutive runs tend to have joint length $L_1 + L_2 = 6$ more often than would be expected by chance. On the other hand, pairs of intervals with $L_1 + L_2 = 4, 5$ or more than 7 are under-represented. Any two neighbouring intervals must have different signs, by the definition of **L**, so that the $(L_1, L_2) = (2, 4)$ and $(3, 3)$ combinations represent 'quasi-oscillatory' events in the time series.

The time series generated with the parameter combination $(\alpha_1, \alpha_2) = (0.3, 0.3)$ exhibits a strange pattern of extended intervals with continuous sign reversals and prolonged persistence. The reason for this pattern will become clear in [10.3.6].

L_2			L_1		
1	1	2	3	4	5
1	571	1248	1523	917	334
	351	*23*	*−875*	*−31*	*108*
2		1428	6418	3304	741
		−282	*−280*	*650*	*109*
3			7341	5048	1081
			783	*−144*	*−153*
4				846	418
				−182	*−70*
5					59
					0

Table 10.1: *Absolute frequency with which a run of length L_1 is preceded or followed by a run of length L_2 in a 100 000 time unit simulation of an AR(2) process with $(\alpha_1, \alpha_2) = (0.9, -0.8)$. The entries in italics display the deviation from the expected cell frequency computed under the assumption that consecutive run lengths are independent.*

10.3.5 Stationarity of AR Processes. The conditions under which the AR processes of definition [10.3.1] are stationary are not immediately obvious. Clearly, AR processes can be non-stationary. An AR(1) process with $\alpha_1 = 2$ and $\mu = 0$ initiated from a random variable $\mathbf{X}_0$ that has finite variance is stationary with respect to the mean but non-stationary with respect to variance. In this case we note that, for $t > 0$,

$$\mathbf{X}_t = 2^t\mathbf{X}_0 + \sum_{i=1}^{t} 2^{t-i}\mathbf{Z}_{t-i+1}$$

and therefore that

$$\begin{aligned} \mathcal{E}(\mathbf{X}_t) &= 2^t\mathcal{E}(\mathbf{X}_0) = 0 \text{ and} \\ \mathrm{Var}(\mathbf{X}_t) &= 4^t\mathrm{Var}(\mathbf{X}_0) + \sum_{i=1}^{t} 4^{t-i}\mathrm{Var}(Z) \\ &= \frac{4^t}{3}\Big(1 - \frac{1}{4^t}\Big). \end{aligned}$$

Thus the variance of this process grows at an exponential rate.

The stationarity of an AR(p) process depends entirely on the dynamical AR coefficients α_k, $k \neq 0$. In fact,

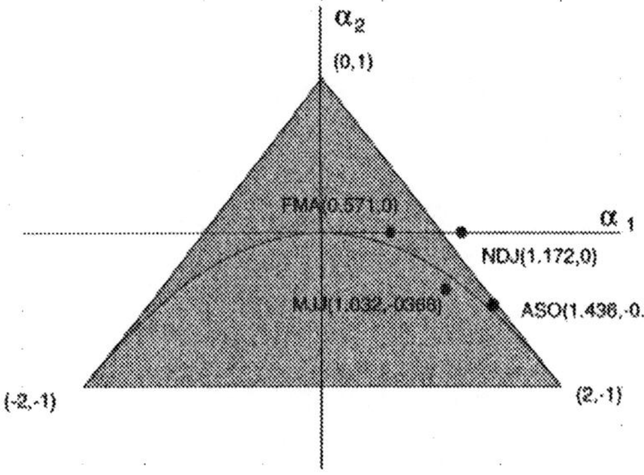

Figure 10.11: *The triangle identifies the range of parameters for which an AR(2) process is stationary. The four points represent the parameters of a seasonal AR(2) process used to represent the SO index (see [10.3.7]). Processes with parameters below the curve defined by $\alpha_1^2 + 4\alpha_2 = 0$ have quasi-oscillatory behaviour (see [10.3.6]).*

An AR(p) process with AR coefficients α_k, for $k = 1, \ldots, p$, is stationary if and only if all roots of the characteristic polynomial

$$\mathrm{p}(y) = 1 - \sum_{k=1}^{p} \alpha_k y^k \tag{10.11}$$

lie outside the circle $|y| = 1$.

Note that (10.11) has p roots y_j, some of which are real and others of which may appear in complex conjugate pairs.

Thus the stationarity condition for an AR(1) process is simply

$$|\alpha_1| < 1. \tag{10.12}$$

Stationarity conditions are somewhat more involved for an AR(2) process, where it is necessary that

$$\begin{aligned} \alpha_2 + \alpha_1 &< 1 \\ \alpha_2 - \alpha_1 &< 1 \\ |\alpha_2| &< 1. \end{aligned} \tag{10.13}$$

The region of *admissible process parameters* defined by (10.13) consists of points (α_1, α_2) in the two-dimensional plane that also lie in the triangle depicted in Figure 10.11.

10.3.6 More about the Characteristic Polynomial. Equation (10.11) has interesting implications. Let y_j, for $j = 1, \ldots, p$, be the roots of the characteristic polynomial $\mathrm{p}(y)$. Given a fixed j, set $X_{t-k,j} = y_j^k$ for $k = 1, \ldots, p$. Substitute these values into (10.6), disregard the noise term, and recall that we have assumed that $\alpha_0 = 0$. Then, using (10.11), we see that $X_t = y_j^0 = 1$.[8] That is, each root y_j identifies a set of 'typical initial conditions' $\mathcal{IC}_j = (X_{t-1,j}, \ldots, X_{t-p,j})$ that lead to $X_t = 1$ when the noise $\mathbf{Z}_t$ is disregarded. Since these 'initial conditions' are linearly independent, any set of states $(X_{t-1}, \ldots, X_{t-p})$ can be represented as a linear combination $\sum_{j=1}^{p} \beta_j \mathcal{IC}_j$ of the initial states. In the absence of noise, the future evolution of these states will be

$$X_{t+\tau} = \sum_{j=1}^{p} \beta_j y_j^{-\tau}. \qquad (10.14)$$

Note that some of the $X_{t-k,j}$s may be complex and therefore will appear in conjugate complex pairs. When this is true, the corresponding coefficients β_j will also appear as complex conjugate pairs.

When X_t is an AR(1) process and the noise is absent, $X_{t-1} = 1/\alpha_1$ is the only initial condition that leads to $X_t = 1$ in one time step.

In the case of an AR(2) process, the roots of the characteristic polynomial (10.11) are

$$y_j = \frac{-\alpha_1 - (-1)^j \sqrt{\alpha_1^2 + 4\alpha_2}}{2\alpha_2}, \quad j = 1, 2.$$

The roots are either both real or they are complex conjugates.

Both roots are *real* when $\alpha_1^2 > -4\alpha_2$. The AR(2) process with $(\alpha_1, \alpha_2) = (0.3, 0.3)$ belongs to this category. Its characteristic polynomial has roots $y_1 = 1.39$ and $y_2 = -2.39$, and 'typical initial conditions,' which lead to $X_t = 1$, are $\mathcal{IC}_1 = (X_{t-2,1}, X_{t-1,1}) = (1.93, 1.39)$ and $\mathcal{IC}_2 = (X_{t-2,2}, X_{t-1,2}) = (5.71, -2.39)$.

The first 'mode,' which is initiated by $\mathcal{IC}_1$, has a damping rate of $X_{t-1,1}/X_{t-2,1} = X_{t,1}/X_{t-1,1} = 1/y_1 = 0.72$. The time development initiated by such an initial state is that of an exponential decay with constant sign.

The second mode has a damping rate $1/|y_2| = 0.42$ and a clear tendency for perpetual sign reversals.

These two modes underlie the 'strange pattern' of variation seen in Figure 10.9. There are some periods when the process undergoes continual sign reversals, and others when the system retains the same sign. Change between the two regimes is instigated by the noise $\mathbf{Z}_t$.

[8] Note that now we are neither dealing with the stochastic process $\mathbf{X}_t$ nor with a random realization $\mathbf{x}_t$. We therefore use the notation X_t.

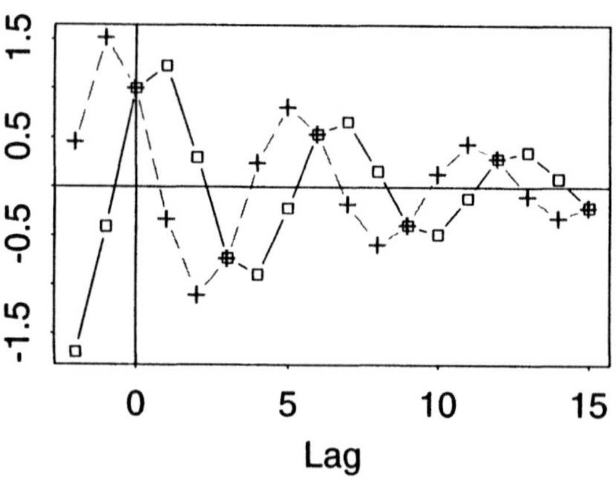

Figure 10.12: *Initial conditions at times X_{t-2} and X_{t-1} which lead an AR(2) process, with parameters 0.9 and −0.8, to $X_t = 1$, and their future development $X_{t+\tau}$ in the absence of noise.*

The roots of the characteristic polynomial of an AR(2) process are complex when $\alpha_1^2 < -4\alpha_2$, and can therefore be written in the form

$$y_j = r \cdot \exp\big(-(-1)^j i\phi\big), \quad j = 1, 2. \qquad (10.15)$$

It is easily shown that $r = 1.11$ and $\phi = \frac{\pi}{3}$ when $(\alpha_1, \alpha_2) = (0.9, -0.8)$. Since the process parameters are real, (10.11) may be rewritten as

$$\begin{aligned} 0 &= 1 - (\alpha_1 \mathrm{Re}(y) + \alpha_2 \mathrm{Re}(y^2)) \\ &= 1 - (\alpha_1 r \cos(\phi) + \alpha_2 r^2 \cos(2\phi)) \end{aligned}$$

and

$$\begin{aligned} 0 &= \alpha_1 \mathrm{Im}(y) + \alpha_2 \mathrm{Im}(y^2) \\ &= \alpha_1 r \sin(\phi) + \alpha_2 r^2 \sin(2\phi) \end{aligned}$$

so that the two sets of 'typical initial conditions' that evolve into $X_t = 1$ are

$$\mathcal{IC}_j = (X_{t-2,j}, X_{t-1,j})$$

with

$$X_{t-2,j} = r^2(\cos(2\phi) - (-1)^j \sin(2\phi))$$

and

$$X_{t-1,j} = r(\cos(\phi) - (-1)^j \sin(\phi)).$$

Thus (10.14) determines the future states as

$$X_{t+\tau,1} = r^{-\tau}(\cos(\tau\phi) + \sin(\tau\phi))$$

and

$$X_{t+\tau,2} = r^{-\tau}(\cos(\tau\phi) - \sin(\tau\phi)).$$

The two sets of initial conditions (labelled '1' and '2') and the future evolution of the process without the noise $\mathbf{Z}_t$ are plotted in Figure 10.12. We see that the process generates damped oscillations with a period of $\frac{\pi}{\phi} = 6$ time steps for arbitrary nonzero initial conditions. The initial conditions serve only to determine the phase and amplitude of the oscillation.

The region of admissible process parameters (10.13) for a stationary AR(2) process (see Figure 10.11) can be split into two sub-regions. An upper area, delimited by $\alpha_1^2 + 4\alpha_2 > 0$, indexes AR(2) processes whose characteristic polynomials have two real solutions and thus consist of two non-oscillatory damped modes. The rest of the parameter space, delimited by (10.13) and the constraint $\alpha_1^2 + 4\alpha_2 < 0$, indexes processes with characteristic polynomials that have a pair of conjugate roots, and thus one quasi-oscillatory mode.

10.3.7 Characteristic Time. What is the characteristic time (10.1) of an AR(p) process? According to (10.1), we must find a lag τ_M such that auto-correlations $\rho_{X_t,X_{t+\tau}}$ vanish for lags $\tau \geq \tau_M$. In the case of an AR(1) process with $\mu = 0$ we find

$$\begin{aligned}\rho_{X_t,X_{t+\tau}} &= \frac{\mathcal{E}(\mathbf{X}_t\mathbf{X}_{t+\tau})}{\mathrm{Var}(\mathbf{X}_t)} \\ &= \frac{\alpha_1^\tau \mathcal{E}(\mathbf{X}_t\mathbf{X}_t)}{\mathrm{Var}(\mathbf{X}_t)} \\ &= \alpha_1^\tau \neq 0, \qquad (10.16)\end{aligned}$$

for *all* lags τ. Thus $\tau_M = \infty$ for an AR(1) process. This statement holds for all AR processes. Thus definition (10.1) is not useful for such processes. We suggest an alternative definition in Section 17.1.

10.3.8 Seasonal AR Processes. The 'stationary' AR(p) process defined by (10.6) can be easily generalized to *seasonal* or *cyclo-stationary* [10.2.5] AR(p) processes. However, before giving a definition we need to establish some notation. First, we assume that there exists an external deterministic 'cycle' that is indexed by time $\tau = 1, \ldots, N$. This index may count months within a year or hours in a day. We then express an arbitrary time as a pair (t, τ), where t counts repetitions of the external cycle, so that $(t, \tau + N) \equiv (t + 1, \tau)$. Then, $\{\mathbf{X}_{t\tau}: t \in \mathbb{Z}, \tau = 1, \ldots, N\}$ is said to be a *cyclo-stationary AR(p) process* if

1. there are constants $\alpha_{k\tau}$, $k = 0, 1, \ldots, p$ such that $\alpha_{k,\tau+N} = \alpha_{k\tau}$ for all τ and $\alpha_{p\tau} \neq 0$ for some τ,
2. there is a sequence of independent, zero mean random variables $\{\mathbf{Z}_{t\tau} : t \in \mathbb{Z}, \tau = 1, \ldots, N\}$ that have variance $\sigma_{Z\tau}^2$ which depends only on τ and such that the sequence $\{\mathbf{Z}_{t\tau}/\sigma_{Z\tau} : t \in \mathbb{Z}, \tau = 1, \ldots, N\}$ behaves as white noise, and
3. $\mathbf{X}_{t\tau}$ satisfies the difference equation

$$\mathbf{X}_{t,\tau} = \alpha_{0,\tau} + \sum_{k=1}^{p} \alpha_{k\tau}\mathbf{X}_{t,\tau-k} + \mathbf{Z}_{t\tau} \qquad (10.17)$$

for all (t, τ).

Such processes are able to exhibit cycles of length N of the mean, the variance, and the auto-covariance function.

Suppose, now, that a process satisfying (10.17) is weakly cyclo-stationary. This means that the process parameters are constrained in such a way that all means, variances, and covariances exist. This constraint, together with (10.17), is sufficient to ensure that the mean and variance are only a function of τ and that the auto-covariance function is only a function of the absolute time difference and the location in the external cycle. With these assumptions it is possible to derive the 'seasonal cycle' of mean, variance and auto-covariance.

To illustrate, consider the calculation of the annual cycle of the mean. We apply the expectation operator $\mathcal{E}(\cdot)$ to (10.17) for all τ to obtain

$$\mu_\tau = \alpha_{0,\tau} + \sum_{k=1}^{p} \alpha_{k\tau}\mu_{\tau-k}. \qquad (10.18)$$

This is a closed linear system since both μ_τ and $\alpha_{k\tau}$ are periodic in τ with period N. It can therefore be re-expressed in matrix-vector form and solved using standard techniques.

Calculation of the seasonal cycle of the variance is more complicated. First, the past states $\mathbf{X}_{t,\tau-1}$, $\mathbf{X}_{t,\tau-2}, \ldots$ in (10.18) are replaced with linear combinations of previous states by recursive application of (10.17). This recursion yields an infinite series (an 'infinite moving average process'; see [10.5.2])

$$\mathbf{X}_{t\tau} = \beta_{0,\tau} + \sum_{j=1}^{\infty} \beta_{j\tau}\mathbf{Z}_{t,\tau-j+1}. \qquad (10.19)$$

The βs are functions of the seasonal AR(p) parameters and the cyclo-stationarity conditions alluded to above ensure that this sum converges in a suitable manner. The noise contributions

$\mathbf{Z}_{t\tau}$ have zero expectation and are mutually independent so that

$$\mathcal{E}(\mathbf{X}_{t\tau}) = \beta_{0,\tau} \qquad (10.20)$$

$$\text{Var}(\mathbf{X}_{t\tau}) = \sum_{j=1}^{\infty} \beta_{j\tau}^2 \sigma_{Z,\tau-j+1}^{\tau-j+1}.$$

10.3.9 Example: A Seasonal Model of the SST Index of the Southern Oscillation. A seasonal AR(2) process can be used to model the SST index of the Southern Oscillation [453]. A segment of the full monthly time series is shown in Figure 1.4 (dashed curve). The model was fitted to seasonal means so that one 'seasonal cycle' comprises $N =$ 4 time steps, namely FMA, MJJ, ASO and NDJ.

The estimated process parameters $\widehat{\alpha}_{k\tau}$ and the standard deviation of the driving noise $\widehat{\sigma}_{Z\tau}$, which fit the data best, are:

Season τ	$\widehat{\alpha}_{0,\tau}$	$\widehat{\alpha}_{1,\tau}$	$\widehat{\alpha}_{2,\tau}$	$\widehat{\sigma}_{Z,\tau}$
FMA	0.39	0.571	0	0.332
MJJ	−0.17	1.032	−0.368	0.374
ASO	2.55	1.436	−0.471	0.362
NDJ	3.56	1.172	0	0.271

When we examine the four sub-models for FMA, MJJ, ASO, and NDJ separately using (10.13) to determine whether they satisfy the stationarity condition of an AR(2) process, we find that the FMA, MJJ, and ASO processes satisfy the condition but that the NDJ process lies outside the 'admissible' triangle of Figure 10.11. The transition from NDJ to FMA, with $\widehat{\alpha}_{1,FMA} =$ 0.571, is connected with substantial damping. On the other hand, the step from ASO to NDJ, with $\widehat{\alpha}_{1,NDJ} = 1.172$, is associated with amplification of the process. Despite this, the full process is cyclo-stationary.

The estimated annual cycle of the means, $\widehat{\mu}_{X\tau}$, and standard deviations, $\widehat{\sigma}_{X\tau}$, derived from the fitted model are displayed in the following table:[9]

Season τ	$\widehat{\mu}_{X\tau}$ (°C)	$\widehat{\sigma}_{X\tau}$ (°C)
FMA	0.058	0.621
MJJ	0.033	0.554
ASO	0.046	0.743
NDJ	0.091	0.911

The overall mean value, as well as the expected values for the four seasons, are slightly positive. The standard deviation varies strongly with the season. Maximum variability occurs in the season with the largest SST anomalies (NDJ); weakest variability occurs in northern summer (MJJ). Note that the NDJ variance is 2.7 times greater than the MJJ variance.

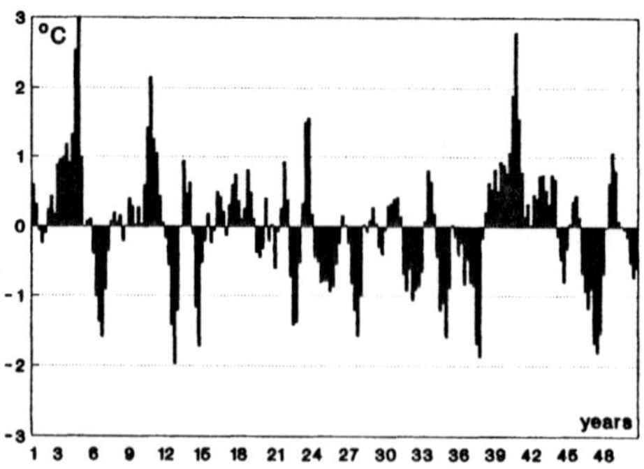

Figure 10.13: *A 50-year random realization of the seasonal AR(2) process which models the SST index of the SO. Compare with Figure 1.4.*

A simulated 200 time step realization of the fitted process is displayed in Figure 10.13. The character of the time series is similar to that of the original displayed in Figure 1.4. It resembles the output of an ordinary AR(2) process with frequent occurrences of positive (or negative) anomalies extending over four and more seasons. The preference for maxima to occur in NDJ distinguishes the fitted process from an ordinary AR(2) process. A non-seasonal process does not have a preferred season for generating extremes. This preference is indeed a characteristic feature of the SO.

10.3.10 Bivariate and Multivariate AR Processes. The 'univariate' definition (10.6) or (10.8) of an AR process can be easily generalized to a multivariate setting. A sequence of ℓ-dimensional random vectors $\{\vec{\mathbf{X}}_t : t \in \mathbb{Z}\}$ is said to be a *multivariate AR(p) process* if $\vec{\mathbf{X}}_t$ satisfies a vector difference equation of the form

$$\vec{\mathbf{X}}_t = \mathcal{A}_0 + \sum_{k=1}^{p} \mathcal{A}_k \vec{\mathbf{X}}_{t-k} + \vec{\mathbf{Z}}_t \qquad (10.21)$$

for all t where

1. $\mathcal{A}_0$ is an ℓ-dimensional vector of constants,
2. $\mathcal{A}_k$, for $k = 1, \dots, p$, are $\ell \times \ell$ matrices of constants such that $\mathcal{A}_p \neq 0$, and
3. $\{\vec{\mathbf{Z}}_t : t \in \mathbb{Z}\}$ is a sequence of iid zero mean ℓ-dimensional random vectors.

[9] The estimated means are different from zero because the seasonal AR process was fitted to anomalies computed relative to a reference period that was shorter than the full record.

Bivariate AR(p) processes that describe the joint evolution of two processes and multivariate AR(1) processes are of particular interest. For example, a multivariate AR(1) process (i.e., $\mathcal{A}_i = 0$ for $i \geq 2$) is fitted in *Principal Oscillation Pattern* analysis (see Chapter 15).

10.4 Stochastic Climate Models

10.4.1 Historic Excursion. What are the physical processes that excite slow climate variations such as the Ice Ages, the Medieval Warm Time, or the Little Ice Age? The early scientific mainstream opinion was that such variability stems exclusively from external forcings, such as variations in the Earth's orbital parameters. It was argued that the weather fluctuations were irrelevant because their influence would diminish through the process of time integration [see 10.4.2]. That is, short-term statistical forcing was not believed to affect the dynamics of systems that respond slowly to such forcing. Hasselmann ([165]; see [10.4.3]) was apparently the first to recognize the inconsistency of this concept. He demonstrated that low-frequency variability in systems like the climate could simply be the integrated response of a linear (or nonlinear) system forced by short-term variations, such as those of the macroturbulent atmospheric flow at midlatitudes. The success of this proposal is demonstrated in [10.4.3] and possible generalizations are briefly mentioned in [10.4.4].

10.4.2 Statistical Dynamical Models. The purpose of *Statistical Dynamical Models* (SDM) is to describe the behaviour of a 'climate variable' y_t that varies on time scales τ_Y and has dynamics that are described by a differential equation of the form

$$\frac{dy}{dt} = V(y, x) + f. \tag{10.22}$$

Here x_t is another climate variable that varies on a much shorter time scale τ_X. Generally, V is some nonlinear function of y_t and x_t, and f represents external forcing.

Now let A_τ be an operator that averages a climate variable over the time scale τ. Because $\tau_X \ll \tau_y$, there is a time scale τ^* such that

$$\begin{aligned} A_{\tau^*}(x) &\approx \text{constant} \\ \frac{dA_{\tau^*}(y)}{dt} &\approx \frac{dy}{dt}. \end{aligned}$$

Thus (10.22) may be reformulated as

$$\frac{dy}{dt} = V^*(y) + f^*. \tag{10.23}$$

The modified operator V^* includes the effect of averaging and, in particular, the constant contribution from the 'fast' component x. The modified forcing f^* represents the slow component of the forcing.

Equation (10.23) is a 'dynamical' model because the dynamics are explicitly accounted for by the function V^*. It is also called a 'statistical' model because the averaging operator has embedded the moments of the noisy component x into function V^*. However, this nomenclature is somewhat misleading since (10.23) does not contain random components, but rather describes the deterministic evolution of the moments of a random variable. Equation (10.23) is fully deterministic and may, at least in principle, be solved if adequate initial conditions and forcing functions are available. Consequently, the study of climate variability is reduced to the analysis of the structure of the forcing functions. The system (10.23) can generate many complicated modes of variation if it is nonlinear. To understand such a system it is necessary to identify a subspace of the full phase space that contains the relevant nonlinear dynamics.[10]

10.4.3 Stochastic Climate Models. Neither the search for external forcing functions nor the search for nonlinear sub-systems has been convincingly successful in explaining the observed variability in the climate system. Hasselmann [165] suggested a third mechanism for generating low-frequency variations in the system described by (10.22). This concept, *Stochastic Climate Modelling*, is now used widely.

Suppose the forcing f in (10.22) is zero and consider the evolution of the system from an initial value.

Early on, for $0 \leq t < \tau_Y$, one may assume that $V(y_t, x_t) \approx V(y_0, x_t)$ so that V acts only in response to random variable $\mathbf{X}_t$. During this time period

$$\frac{d\mathbf{Y}_t}{dt} = V(y_0, x_t) \tag{10.24}$$

behaves as a stochastic process, say $\mathbf{Z}_t$. Since $\mathbf{X}_t$ varies on time scales $\tau_X \ll \tau_Y$, the derived

[10]This is easier said than done. One possibility is to fit *Principal Interaction Patterns* (see [15.1.6] and Hasselmann [167]) to observed or simulated data. Regardless of the method used, the investigator must have a clear understanding of the dynamics of the studied process.

process $\mathbf{Z}_t$ also varies on short time scales. After discretization of (10.24) we find

$$\mathbf{Y}_{t+1} = \alpha \mathbf{Y}_t + \mathbf{Z}_t \qquad (10.25)$$

with $\alpha = 1$. Equation (10.25) describes a random walk when $\mathbf{Z}_t$ is a white noise process [10.2.6]. Thus, the system gains energy and the excursions grow, even if, in an ensemble sense, the mean solution is constant.

Later, when $t \geq \tau_Y$, the operator V does depend on $\mathbf{Y}_t$. Since the trajectories of the system are bounded, a *negative feedback* mechanism must be invoked. An approximation of the form

$$V(\mathbf{Y}_t, \mathbf{X}_t) \approx -\beta \mathbf{Y}_t + \mathbf{Z}_t \qquad (10.26)$$

is often suitable. This leaves (10.25) unchanged except that $\alpha = 1 - \beta$. Equation (10.25) now describes an AR(1) process. The stationarity condition $\alpha < 1$ is obtained for sufficiently small time steps.

We now return to (10.22) with $f = 0$, except we consider a system that varies around an equilibrium state. If we assume that the disturbances are small, then the nonlinear operator V can be linearized as

$$V(x, y) = v_x x + v_y y \qquad (10.27)$$

so that we again arrive at (10.25) with $\mathbf{Z}_t = v_x \mathbf{X}_t$.

In both of these cases, the full nonlinear system can be approximated by a stationary AR process as long as there is negative feedback. Section 10.3 shows that such systems possess substantial low-frequency variations that are not related to (deterministic) internal nonlinear dynamics or to (also deterministic) external forcing. Instead, the system is fully random: it is entirely driven by the short-term fluctuating noise $\mathbf{X}_t$.

10.4.4 Examples. Frankignoul, in two reviews [129, 131], summarizes a number of applications in which dynamical systems have been modelled explicitly as stochastic climate models. Such systems include the sea-surface temperature at midlatitudes, and Arctic and Antarctic sea ice and soil moisture.

For the midlatitude sea-surface temperature (SST) the variable y is the SST and the variables x that vary on short time scales are the air–sea heat flux and the wind stress (Frankignoul and Hasselmann [133]). The characteristic times are $\tau_{SST} \approx 6$ months $\gg \tau_x \approx 8$ days. Similarly, for Arctic sea ice extent (Lemke [251]), the low-frequency variable is the sea-ice extent and the short time scale variable represents *weather noise.*

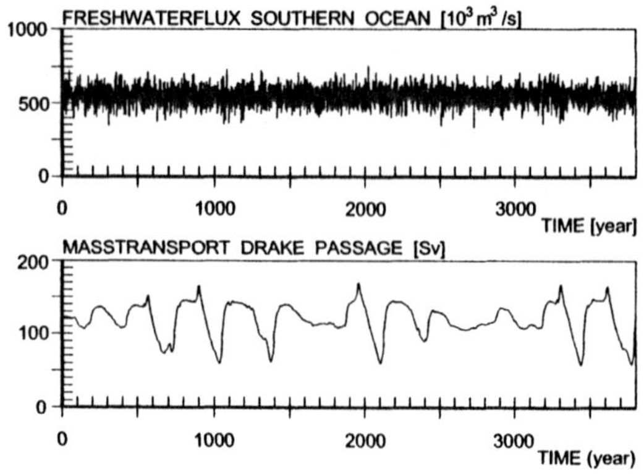

Figure 10.14: *Result of an extended Ocean General Circulation Model experiment forced with white noise freshwater fluxes.*
Top: Net freshwater flux into the Southern Ocean.
Bottom: Mass transport through the Drake Passage.
From Mikolajewicz and Maier-Reimer [276].

Mikolajewicz and Maier-Reimer [276] provide a particularly convincing example without explicitly fitting a simple stochastic climate model. They ran an Ocean General Circulation Model with upper boundary forcing consisting of constant wind stress, and heat and freshwater fluxes. Additional freshwater flux anomalies with characteristic time $\tau_x \sim 1$ were also added (Figure 10.14, top). These additional anomalies were white in time and almost white in space. The 'response,' characterized by the mass transport through the Drake Passage, is dominated by low-frequency variations with typical times $\tau_y > 100$ years (Figure 10.14, bottom).[11] Subsequent research has shown that this result is at least partly an artifact of the model and its boundary conditions. None the less, this example effectively demonstrates that the dynamics of a physical system can turn short-term stochastic forcing into low-frequency climate variability.

Stochastic Climate Models can not be used to reproduce a physical system in detail. Nevertheless, they are instrumental in the understanding of the dynamics that prevail in complex general circulation models or observations.

10.4.5 Generalizations. The main purpose of the stochastic climate model is to explain fundamental dynamics from a zero-order approximation. Examples from various aspects of the climate system support the general concept that

[11] See also Weaver and Hughes [418].

short-term variations are a significant source of low-frequency variability, although, of course, the dynamics may be more complicated. The operator V may have preferred time scales, nonlinearities, complex feedbacks and resonances, requiring approximations other than (10.26) or (10.27). However, the principle will still be valid. Also, multivariate systems may be considered—we present various examples of multivariate systems that are successfully represented by multivariate AR(1) processes when we introduce the *Principal Oscillation Patterns* in Chapter 15.

10.5 Moving Average Processes and Regime-dependent AR Processes

10.5.1 Overview. This section deals with some topics that, up to now, have been only marginally relevant to climate research applications. Some readers might find it convenient to skip directly to Chapter 11.

Auto-regressive processes are part of a larger class of processes known as *auto-regressive moving average processes* or *ARMA processes.* These models, first made popular by Box and Jenkins [60], are widely used in some parts of geophysical science. We discuss them here for completeness. We also briefly discuss *regime-dependent auto-regressive processes,* which are nonlinear generalizations of the seasonal AR processes.

We begin by defining a moving average process.

10.5.2 Definition: Moving Average Processes. Moving average processes are a special class of stochastic processes that have finite memory τ_M. Such models represent physical systems that integrate the effects of only the last m encounters with a random forcing mechanism. A process $\mathbf{X}_t$ is said to be a *moving average process* of order q, or equivalently, an MA(q) process, if

$$\mathbf{X}_t = \mu_X + \mathbf{Z}_t + \sum_{l=1}^{q} \beta_l \mathbf{Z}_{t-l} \qquad (10.28)$$

where

1 μ_X is the mean of the process,

2 $\beta_1, \ldots, \beta_q$ are constants such that $\beta_q \neq 0$, and

3 $\{\mathbf{Z}_t : t \in \mathbb{Z}\}$ is a white noise process.

A moving average process is stationary with mean μ_X and variance $\mathrm{Var}(\mathbf{X}_t) = \mathrm{Var}(\mathbf{Z}_t)(1 + \sum_{l=1}^{q} \beta_l^2)$.

10.5.3 Infinite Moving Averages and Auto-Regressions. It is useful, for technical reasons, to be able to discuss infinite moving averages. A process $\mathbf{X}_t$ is said to be an *infinite moving average process* if

$$\mathbf{X}_t = \mu_X + \mathbf{Z}_t + \sum_{l=1}^{\infty} \beta_l \mathbf{Z}_{t-l} \qquad (10.29)$$

where

1 μ_X is the mean of the process,

2 $\{\beta_j : j = 1, 2, \ldots\}$ is a sequence of coefficients such that $\sum_{j=1}^{\infty} |\beta_j| < \infty$, and

3 $\{\mathbf{Z}_t : t \in \mathbb{Z}\}$ is a white noise process.

Infinite auto-regressions are defined similarly. A process $\mathbf{X}_t$ is said to be an *infinite auto-regressive process* if

$$\mathbf{X}_t = \alpha_0 + \sum_{k=1}^{\infty} \alpha_k \mathbf{X}_t + \mathbf{Z}_t \qquad (10.30)$$

where

1 $\{\alpha_k : k = 0, 1, \ldots\}$ is a sequence of coefficients such that $\sum_{k=0}^{\infty} |\alpha_k| < \infty$, and

2 $\{\mathbf{Z}_t : t \in \mathbb{Z}\}$ is a white noise process.

10.5.4 Examples. Figure 10.15 shows finite samples of two MA(q) processes with $q = 2$ and 10, respectively, $\mu_X = 0$, and $\mathrm{Var}(\mathbf{Z}_t) = 1$. We have set all coefficients $\beta_l = 1$ so that these MA(q) processes are running sums of length $q + 1$ of a white noise process. The variance of the MA(q) process is $q + 1$. The longer the summing interval for the 'forcing' process $\mathbf{Z}_t$, the longer the memory and the longer the typical excursions of the 'responding' process $\mathbf{X}_t$ from the mean.

What are the characteristic times τ_M (10.1) for the MA(q) processes in Figure 10.15? Note that

$$\mathcal{E}(\mathbf{X}_t \mathbf{X}_{t+\tau}) = \sum_{l,m=0}^{q} \beta_l \beta_m \mathcal{E}\big(\mathbf{Z}_{t+i} \mathbf{Z}_{t+\tau+j}\big)$$
$$= \begin{cases} \sum_{l=0}^{q} \beta_l \beta_{l-\tau} \mathrm{Var}(\mathbf{Z}) & |\tau| \leq q \\ 0 & |\tau| > q. \end{cases}$$

Therefore, since we have implicitly assumed that $\mathbf{Z}_t$ (and hence $\mathbf{X}_t$) is normally distributed, it follows that $P(\mathbf{X}_{t+\tau} > 0 | \mathbf{x}_t > 0) = 0.5$ for all $\tau \geq q + 1$. Hence the characteristic time (10.1) of an MA(q) process is $\tau_M = q + 1$.

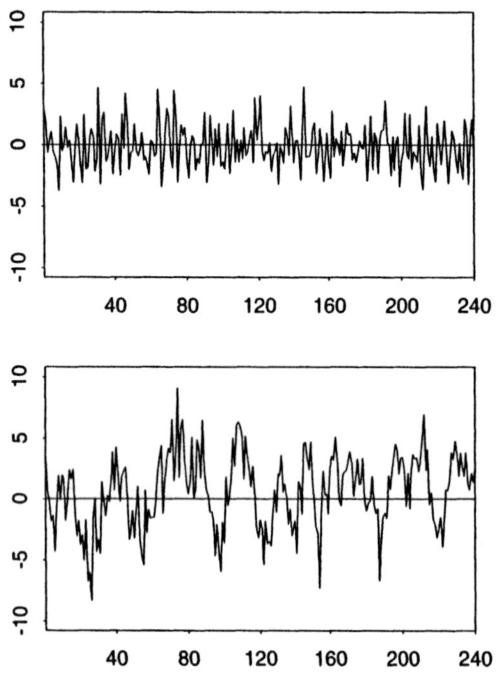

Figure 10.15: *Top: A 240 time step realization of an MA(q) process with* $q = 2$, $\mu_X = 0$, *and* $\beta_l = 1$, *for* $l = 1, \ldots, q$.
Bottom: As top, except $q = 10$.

10.5.5 Auto-regressive Moving Average Processes. An auto-regressive moving average (ARMA) process of order (p, q) [60] is simply an auto-regressive process of order p (10.6) that is forced by a zero mean moving average process of order q (10.28) instead of by white noise.

An ARMA(p, q) process is formally defined as follows: $\mathbf{X}_t$ is said to be an *auto-regressive moving average process of order* (p, q) if

$$(\mathbf{X}_t - \mu_X) - \sum_{i=1}^{p} \alpha_i \mathbf{X}_{t-i} = \mathbf{Z}_t + \sum_{j=1}^{q} \beta_j \mathbf{Z}_{t-j} \qquad (10.31)$$

where

1. μ_X is the mean of the process,
2. $\alpha_1, \ldots, \alpha_p$ and $\beta_1, \ldots, \beta_q$ are constants such that $\alpha_p \neq 0$ and $\beta_q \neq 0$, and
3. $\{\mathbf{Z}_t : t \in \mathbb{Z}\}$ is a white noise process.

There is substantial overlap between the classes of moving average, auto-regressive, and ARMA models. In particular, it can be shown that any weakly stationary ergodic process can be approximated arbitrarily closely by any of the three types of models. However, the ARMA models can approximate the behaviour of a given weakly stationary ergodic process to a specified level of accuracy with fewer parameters that can a pure AR or MA model. That is, they are more *parsimonious* than their AR or MA counterparts.

The parsimony of the ARMA models is of some practical significance when fitting models to a finite data set because fewer parameters need to be estimated from a limited data resource. However, this comes at the cost of developing dynamical models that are forced by stochastic processes with memory. This may be desirable if specific knowledge that can be used to choose the memory of the forcing (i.e., order of the moving average) appropriately is at hand. However, in the absence of such knowledge, the analyst risks obscuring the true dynamical nature of the process under study by resorting to the more parsimonious statistical model.

10.5.6 Invertible Linear Processes. All of the models described in this section can be represented formally in terms of a *backward shift operator* B that acts on the time index of the stochastic process. The operator B is defined so that

$$B[X_t] = X_{t-1}. \qquad (10.32)$$

AR, MA, and ARMA processes can all formally be written in terms of the back shift operator. Specifically, we define the auto-regressive operator $\phi(B)$ as the polynomial

$$\phi(B) = \alpha_0 - \sum_{i=1}^{p} \alpha_i B^i \qquad (10.33)$$

and we define the moving average operator $\theta(B)$ as the polynomial

$$\theta(B) = 1 + \sum_{j=1}^{q} \beta_j B^j. \qquad (10.34)$$

AR, MA, and ARMA processes are then formally stochastic processes that satisfy equations of the form

$$\begin{aligned} \phi(B)\mathbf{X}_t &= \mathbf{Z}_t & \text{(AR)} \qquad (10.35) \\ \mathbf{X}_t &= \theta(B)\mathbf{Z}_t & \text{(MA)} \qquad (10.36) \\ \phi(B)\mathbf{X}_t &= \theta(B)\mathbf{Z}_t & \text{(ARMA)}, \qquad (10.37) \end{aligned}$$

where $\{\mathbf{Z}_t : t \in \mathbb{Z}\}$ is a white noise process.

This formality is introduced to provide the tools needed to briefly explore the connections between AR and MA models.

Consider an MA process represented with the polynomial backshift operator $\theta(B)$ as in (10.36). Suppose now that there exists a power series

$$\theta^{-1}(B) = 1 - \sum_{i=1}^{\infty} \beta_i' B^i$$

such that the power series $\theta(B)\theta^{-1}(B)$ converges to 1 for B in some region in the complex plane that contains the unit circle. That is, all roots of the MA backshift operator $\theta(B)$ must lie outside the unit circle. Then, the MA process can be 'inverted' to produce an infinite auto-regressive process

$$\theta^{-1}(B)\mathbf{X}_t' = \mathbf{Z}_t \tag{10.38}$$

or, equivalently

$$\mathbf{X}_t' - \sum_{i=1}^{\infty} \beta_i' \mathbf{X}_{t-i}' = \mathbf{Z}_t. \tag{10.39}$$

Given that the invertibility condition is satisfied, the process defined by (10.39) is stochastically indistinguishable from the process that satisfies (10.36). Such a process is called an *invertible MA process*.

Note that the invertibility condition for MA processes is analogous to the stationarity condition for AR processes; both conditions can be expressed in terms of the roots of the corresponding backshift operator. As we have just argued, when the MA backshift operator is invertible, the process can be represented as an infinite AR process. On the other hand, when the AR operator has all its roots outside the unit circle, the process is stationary and the AR operator can be inverted so that the process can be represented as an infinite moving average.

A stationary AR process can therefore be approximated with arbitrary precision by truncating its infinite MA representation at some suitable point. Similarly, an invertible MA process can be well approximated by a high order AR process. Also, it is obvious that stationary and invertible ARMA processes can be closely approximated by either a high order AR or a high order MA process simply by inverting and truncating the appropriate backshift operator.

10.5.7 Regime-dependent Auto-regressive Processes. *Regime-dependent auto-regressive processes*, or 'RAMs,' are nonlinear AR processes introduced into climate research by Zwiers and von Storch [453].

The idea is that the dynamics of a stochastic process $\mathbf{X}_t$ are controlled by an external process Y. The RAM has the form

$$\mathbf{X}_t = \alpha_{0,k} + \sum_{i=1}^{p} \alpha_{ik}\mathbf{X}_{t-j} + \mathbf{Z}_{tk}, \tag{10.40}$$

where $k = 1, \ldots, K$ identifies one of K regimes. Within each regime the process behaves as an AR process of some order no greater than p. The dynamics in each regime are forced by their own white noise process. The choice of regime k at any given time t depends on the external state variable $Y(t)$. The regime k is set to l when $Y(t) \in [T_{l-1}, T_l]$. The 'thresholds' are chosen as part of the model fitting process. In principle, other nonlinear dependencies of k on $Y(t)$ could be specified, but the above formulation is piecewise linear, which makes the estimation easier.

A RAM was used to model the SST index of the Southern Oscillation [453]. Two external factors were analysed, namely the intensity of the Indian monsoon, with $K = 2$, and the strength of the Southwest Pacific circulation, with $K = 3$. It was found that the probability of a warm or cold event of the Southern Oscillation did indeed seem to depend on the state of the external variable $Y(t)$.

11 Parameters of Univariate and Bivariate Time Series

Time series analysis deals with the estimation of the characteristic properties and times of stochastic processes. This can be achieved either in the *time domain* by studying the *auto-covariance function*, or in the *frequency domain* by studying the *spectrum*. This chapter introduces both approaches.[1]

11.1 The Auto-covariance Function

11.1.0 Complex and Real Time Series. Note that, even though the auto-covariance and auto-correlation functions of both real and complex-valued time series are defined below, in this chapter we generally limit ourselves to real time series.

11.1.1 Definition. *Let $\mathbf{X}_t$ be a real or complex-valued stationary process with mean μ. Then*

$$\begin{aligned}\gamma(\tau) &= \mathcal{E}\big((\mathbf{X}_t-\mu)(\mathbf{X}_{t+\tau}-\mu)^*\big)\\ &= \mathrm{Cov}(\mathbf{X}_t,\mathbf{X}_{t+\tau})\end{aligned}$$

is called the auto-covariance function *of $\mathbf{X}_t$, and the normalized function,*

$$\rho(\tau) = \frac{\gamma(\tau)}{\gamma(0)}$$

is called the auto-correlation function *of $\mathbf{X}_t$*. The argument τ is called the *lag*. Note that the auto-correlation and auto-covariance functions have the same shape but that they differ in their units; the covariance $\gamma(\tau)$ is expressed in the units of $\mathbf{X}_t^2$ while the correlation $\rho(\tau)$ is expressed in dimensionless units. When required for clarity, we will identify the auto-covariance and auto-correlation functions of process $\mathbf{X}_t$ as γ_{xx} and ρ_{xx}, respectively.

[1] We recommend [60, 49, 68], and [195] for further reading about the technical aspects of this subject.

11.1.2 Auto-correlation and Persistence Forecast. The auto-correlation function can be interpreted as an indication of the skilfulness of the *persistence forecast* of $\mathbf{X}_{t+\tau}$ that is constructed when an observation $\mathbf{x}_t$ is 'persisted' τ time steps into the future. In this context $\rho(\tau)$ is the correlation between the forecast made at time t and the verifying realization that is obtained lag τ time steps later. The proportion of variance 'explained' by the persistence forecast is $\rho^2(\tau)$.

As we saw in Chapter 10, a slowly varying time series, that is, one with relatively long memory, tends to retain anomalies of the same sign for several time steps. Persistence forecasts made for such a process are likely to be more successful than those made for a process with short memory. Thus we anticipate, and are soon able to show, that the auto-correlation function of a long memory process decays to zero more slowly than that of a short memory process.

11.1.3 Examples. The auto-correlation function of the Southern Oscillation Index, which is shown in Figure 1.3 in [1.2.2], is positive for lags shorter than 12 months and oscillates irregularly around zero at longer lags. We will see later that these irregular variations at large lags are typical of auto-correlation function estimates. They are probably the result of sampling variability and the true auto-correlation function is likely to be zero at large lags. Only the first part of the curve, in which the correlation function estimates lie beyond those levels that can be induced solely by sampling variation, is of interest. Figure 1.3 shows us that once a positive (or negative) SOI anomaly has developed it will, on average, persist for up to 12 months.

The interpretation is similar if $\mathbf{X}_t$ is a complex-valued process. For convenience, assume that $\mathbf{X}_t$ has mean zero. Note that we may express the auto-covariance function in polar coordinates as

$$\gamma(\tau) = \mathcal{E}\big(\mathbf{X}_t\mathbf{X}_{t+\tau}^*\big) = r(\tau)e^{i\phi(\tau)}$$

where the amplitude $r(\tau)$ and phase $\phi(\tau)$ are functions of the lag τ. Thus the product $\mathbf{x}_t\mathbf{x}^*_{t+\tau}$ of two realizations τ time steps apart, averaged over many times t, will equal $r(\tau)e^{i\phi(\tau)}$. This tells us that, on average, a real $\mathbf{x}_t$ is followed τ time steps later by a complex $\mathbf{x}_{t+\tau}$ centred on $r(\tau)\,(\cos(\phi(\tau))\mathbf{x}_t - i\sin(\phi(\tau))\mathbf{x}_t)$. That is the persistent part of $\mathbf{X}_t$ follows a damped rotation in the complex plane.

This behaviour is often seen in climate data. An example is the estimated auto-correlation function of the bivariate MJO index (Figure 10.3) that is shown in Figure 11.1. Since $\mathrm{Re}(\hat{\rho}(\tau))$ is approximately zero at about lag-10 days, we estimate that this bivariate index will rotate 90° to the right in about 10 days on average. Similarly, it will rotate about 180° in 22 days, and 270° in 37 days. The estimated auto-correlation function is certainly contaminated by sampling variation after about day 20 (see Section 12.1).

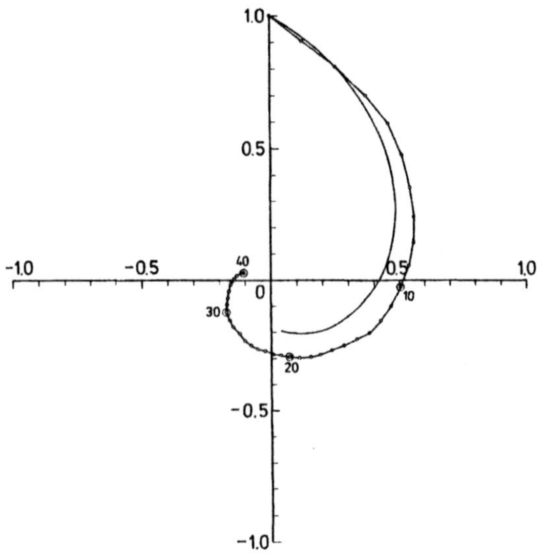

Figure 11.1: *The auto-correlation function of a complex index of the Madden-and-Julian Oscillation. The dots represent the estimated auto-correlation function. The continuous line displays the theoretical auto-correlation function of a fitted complex AR(1) process. The real part of the auto-correlation function is represented by the vertical axis, and the imaginary part by the horizontal axis. From von Storch and Baumhefner [388].*

11.1.4 Properties of the Auto-correlation Function. We note that the auto-correlation function is symmetric about the origin,

$$\rho(\tau) = \rho(-\tau),$$

and that it does not take values outside the interval [-1,1] (if $\mathbf{X}_t$ is real) or outside the unit circle (if $\mathbf{X}_t$ is complex). That is,

$$|\rho(\tau)| \le 1.$$

11.1.5 The Auto-correlation Function of White Noise. Because the elements of white noise are independent, it immediately follows that the auto-correlation function is

$$\rho(\tau) = \begin{cases} 1 & \text{if } \tau = 0 \\ 0 & \text{otherwise.} \end{cases}$$

11.1.6 The Yule–Walker Equations for an AR(p) Process. If we multiply a zero mean AR(p) process $\mathbf{X}_t$ (10.6) by $\mathbf{X}_{t-\tau}$, for $\tau = 1, \ldots, p$,

$$\mathbf{X}_t\mathbf{X}_{t-\tau} = \sum_{i=1}^{p} \alpha_i \mathbf{X}_{t-i}\mathbf{X}_{t-\tau} + \mathbf{Z}_t\mathbf{X}_{t-\tau}, \qquad (11.1)$$

and take expectations, we obtain a system of equations

$$\Sigma_p \vec{\alpha}_p = \vec{\gamma}_p \qquad (11.2)$$

that are known as the *Yule–Walker equations*. The equation relates the auto-covariances

$$\vec{\gamma}_p = \big(\gamma(1), \gamma(2), \ldots, \gamma(p)\big)^{\mathrm{T}}$$

at lags $\tau = 1, \ldots, p$ to the process parameters

$$\vec{\alpha}_p = (\alpha_1, \alpha_2, \ldots, \alpha_p)^{\mathrm{T}}$$

and the auto-covariances $\gamma(\tau)$ at lags $\tau = 0, \ldots, p-1$ through the $p \times p$ matrix

$$\Sigma_p = \begin{pmatrix} \gamma(0) & \gamma(1) & \ldots & \gamma(p-1) \\ \gamma(1) & \gamma(0) & \ldots & \gamma(p-2) \\ \vdots & \vdots & \ddots & \vdots \\ \gamma(p-1) & \gamma(p-2) & \ldots & \gamma(0) \end{pmatrix}.$$

This system of equations has two applications.

First, if $\gamma(0), \ldots, \gamma(p)$ are known (or have been estimated from a time series), the parameters of the AR(p) process can be determined (or estimated) by solving (11.2) for $\vec{\alpha}_p$. Once the parameters have been estimated, both the auto-covariance function for lags $\tau > p$ [11.1.7] and the spectrum (Section 11.2) of the unknown process can be estimated by the corresponding characterizations of the fitted AR(p) process.

Second, if $\vec{\alpha}_p$ is known, then (11.2) can be recast as a linear equation with unknowns $\gamma(1), \ldots, \gamma(p)$, given the variance of the process $\gamma(0)$. Thus the Yule–Walker equations can be used to derive the first $p+1$ elements $1, \rho(1), \ldots, \rho(p)$ of the auto-correlation function. The full auto-covariance or auto-correlation function can now be derived by recursively extending equations (11.2). This is done by evaluating equation (11.1) for $\tau \geq p$ and taking expectations to obtain

$$\gamma(\tau) = \sum_{k=1}^{p} \alpha_k \gamma(k-\tau)$$

and

$$\rho(\tau) = \sum_{k=1}^{p} \alpha_k \rho(k-\tau). \tag{11.3}$$

11.1.7 Auto-covariance and Auto-correlation Functions of Some Low-order AR(p) Processes.

- $p = 1$:
  The Yule–Walker equation (11.2) for an AR(1) process is

  $$\alpha_1 \gamma(0) = \gamma(1).$$

  Hence $\rho(1) = \alpha_1$. Applying (11.3) recursively we see that

  $$\rho(\tau) = \alpha_1^{|\tau|}. \tag{11.4}$$

- $p = 2$:
  The Yule–Walker equations (11.2) for an AR(2) process are

  $$\begin{aligned} \alpha_1\gamma(0) + \alpha_2\gamma(1) &= \gamma(1) \\ \alpha_1\gamma(1) + \alpha_2\gamma(0) &= \gamma(2). \end{aligned}$$

  Using the first equation, we see that

  $$\rho(1) = \frac{\alpha_1}{1-\alpha_2}. \tag{11.5}$$

  Recursion (11.3) can be used to extend the auto-correlation function to higher lags. For example, the auto-correlation at lag-2 is

  $$\rho(2) = \frac{\alpha_1^2 - \alpha_2^2 + \alpha_2}{1-\alpha_2}.$$

- $p = 3$:
  The Yule–Walker equations (11.2) for an AR(3) process are

  $$\begin{aligned} \alpha_1\gamma(0) + \alpha_2\gamma(1) + \alpha_3\gamma(2) &= \gamma(1) \\ \alpha_1\gamma(1) + \alpha_2\gamma(0) + \alpha_3\gamma(1) &= \gamma(2) \\ \alpha_1\gamma(2) + \alpha_2\gamma(1) + \alpha_3\gamma(0) &= \gamma(3). \end{aligned}$$

  Using the first two equations, we obtain

  $$\begin{aligned} \rho(1) &= \frac{\alpha_1 + \alpha_2\alpha_3}{1-\alpha_2-\alpha_1\alpha_3-\alpha_3^2} \\ \rho(2) &= \frac{(\alpha_1+\alpha_3)\alpha_1 + (1-\alpha_2)\alpha_2}{1-\alpha_2-\alpha_1\alpha_3-\alpha_3^2}. \end{aligned}$$

  Recursion relationship (11.3) can again be used to extend $\rho(\tau)$ to longer lags.

- $p \geq 4$:
  The calculations required at higher orders become increasingly laborious, but no more complex.

Note that the auto-covariance function can be obtained by using (10.9) to compute the variance $\text{Var}(\mathbf{X}_t)$ and then applying

$$\gamma(\tau) = \text{Var}(\mathbf{X}_t)\,\rho(\tau).$$

11.1.8 Examples. We will now discuss the auto-correlation functions of the processes that were used as examples in [10.3.2]. Recall that there are two AR(1) processes with $\alpha_1 = 0.3$ and 0.9, and two AR(2) processes with $(\alpha_1, \alpha_2) = (0.9, -0.8)$ and $(0.3, 0.3)$. Sample realizations of these processes are shown in Figures 10.7 and 10.9.

The auto-correlation functions of the AR(1) processes (Figure 11.2a) decrease monotonically. The value of the auto-correlation function for the $\alpha_1 = 0.3$ process is less than 0.5 for all nonzero lags; thus the persistence forecast is able to forecast less than 25% of process variance at any lag. When $\alpha_1 = 0.9$, it takes five time steps for the skill to fall below 25%; this process is much more persistent than the $\alpha_1 = 0.3$ process. This is consistent with the analysis of the distributions of the run length $\mathbf{L}$ in [10.3.3].

The auto-correlation functions of the AR(2) processes are shown in Figure 11.2b. The first two auto-correlations of the $(\alpha_1, \alpha_2) = (0.3, 0.3)$ process are $\rho(1) = \rho(2) = 0.43$, and those for the $(\alpha_1, \alpha_2) = (0.9, -0.8)$ process are $\rho(1) = 0.5$, $\rho(2) = -0.35$. In the $(0.3, 0.3)$ case the auto-correlation function is always positive and has a pattern similar to that of an AR(1) process. The $(0.9, -0.8)$ case reveals considerably more structure. The main feature is a damped 'periodicity' of about six time steps in length. This result is also consistent with the run length analysis in [10.3.3].

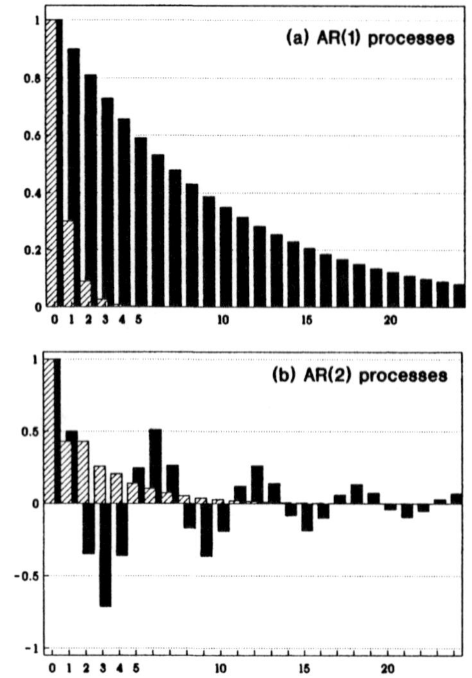

Figure 11.2: *Auto-correlation functions of auto-regressive processes.*
a) Two AR(1) processes with $\alpha_1 = 0.3$ *(hatched bars) and 0.9 (solid bars).*
b) Two AR(2) processes with $(\alpha_1, \alpha_2) = (0.3,\ 0.3)$ *(hatched bars) and* $(0.9,\ -0.8)$ *(solid bars).*

11.1.9 The General Form of the Auto-correlation Function of an AR(p) Process. The auto-correlation function of a weakly stationary AR(p) process can be expressed as

$$\rho(\tau) = \sum_{k=1}^{p} a_k y_k^{-|\tau|} \tag{11.6}$$

for all τ, where y_k, $k = 1, \ldots, p$, are the roots of the characteristic polynomial (10.11), $\phi(B) = 1 - \sum_{k=1}^{p} \alpha_k B^k$ (see, e.g., [195, 60]). Since the characteristic polynomial can be factored as a product of linear and quadratic functions, the roots y_k are either real or come in complex conjugate pairs. The constants a_k can be derived from the process parameters $\vec{\alpha}_p$. When y_k is real, the corresponding coefficient a_k is also real, and when y_k and y_l are complex conjugates, the corresponding coefficients a_k and a_l are also complex conjugates.

Regardless of whether the roots are real or complex, the weak stationarity assumption ensures that $|y_k| > 1$ for all k (see [10.3.5]). Thus each real root contributes a component to the auto-correlation function (11.6) that decays exponentially. Similarly, each pair of complex conjugate roots contributes an exponentially damped oscillation.

We now consider some specific cases.

First, suppose $\mathbf{X}_t$ is a weakly stationary AR(1) process. The characteristic polynomial is $\phi(B) = 1 - \alpha_1 B$ and the only root is $y_1 = (\alpha_1)^{-1}$. Note that $|y_1| > 1$ since $|\alpha_1| < 1$. Thus the auto-correlation function (11.6) consists of a single term $\rho(\tau) = a_1(\alpha_1)^{\tau}$ that decays exponentially. The constant $a_1 = 1$.

Now suppose $\mathbf{X}_t$ is an AR(2) process. We saw in [10.3.6] there are two types of AR(2) processes; one has a pair of decaying modes, the other has a single damped oscillatory mode. The first occurs when $\alpha_1^2 + 4\alpha_2 > 0$, in which case (10.11) has real roots y_1 and y_2, and the auto-correlation function (11.6) is the sum of two terms that decay exponentially.

The (0.3, 0.3) process (see [10.3.5]) belongs to this class. The roots of its characteristic polynomial are $y_1 = 1.39$ and $y_2 = -2.39$. The y_1-mode has a monotonically decaying auto-correlation function $a_1(y_1^{-1})^{|\tau|} = a_1(0.72)^{|\tau|}$. The y_2-mode has auto-correlation function $a_2(y_2^{-1})^{|\tau|} = a_2(-0.42)^{|\tau|}$, which decays even more quickly but has alternating sign.

The constants a_1 and a_2 can be calculated from (11.5) and (11.6). Since

$$\begin{aligned} \rho(0) &= 1 = a_1 + a_2 \\ \rho(1) &= \frac{\alpha_1}{1-\alpha_2} = a_1 y_1^{-1} + a_2 y_2^{-1}, \end{aligned}$$

it follows that

$$a_1 = \frac{\rho(1) - y_2^{-1}}{y_1^{-1} - y_2^{-1}}, \quad a_2 = \frac{y_1^{-1} - \rho(1)}{y_1^{-1} - y_2^{-1}}. \tag{11.7}$$

In this example, $a_1 = 0.74$ and $a_2 = 0.26$.

When $\alpha_1^2 + 4\alpha_2 < 0$, equation (10.11) has a pair of complex conjugate roots, $y_1 = y_2^* = y$. Consequently, for positive τ, equation (11.6) reduces to

$$\rho(\tau) = a_1 y^{-\tau} + a_2^*(y^*)^{-\tau},$$

where $a_1 = a_2^* = a$. If we write $y = re^{i\phi}$, this may be rewritten as

$$\rho(\tau) = \frac{2\,\mathrm{Re}(a)\cos(\tau\phi) - 2\,\mathrm{Im}(a)\sin(\tau\phi)}{r^{\tau}}. \tag{11.8}$$

To determine the complex constant a we first evaluate (11.8) at $\tau = 0$ and obtain $\mathrm{Re}(a) = 1/2$. We then evaluate (11.8) at $\tau = 1$ and obtain

$$\rho(1) = \frac{\cos(\phi) - 2\,\mathrm{Im}(a)\sin(\phi)}{r}$$

so that

$$\mathrm{Im}(a) = \frac{\cos(\phi) - r\,\rho(1)}{2\,\sin(\phi)} \tag{11.9}$$

where $\rho(1)$ is given by (11.5). Finally, we see that the auto-correlation function (11.8) may be rewritten as

$$\rho(\tau) = \frac{\sqrt{1+4\,\mathrm{Im}(a)^2}}{r^\tau}\cos(\tau\phi + \psi)$$

with $\tan(\psi) = 2\,\mathrm{Im}(a)$. Note that $r = 1.12$ and $\phi \approx \pi/3$ in the $(0.9, -0.8)$ example, so that $a = 0.5 + i\,0.032$ and $\psi \approx -\pi/50$.

In general, the auto-correlation function of an AR(p) process is the sum of decaying exponentials (one for every real root of the characteristic polynomial) and damped oscillations (one for every pair of complex conjugate roots of the characteristic polynomial). Thus, the general auto-correlation function has the form

$$\rho(\tau) = \sum_i \frac{a_i}{y_i^\tau} + \sum_k a_k \frac{\cos(\tau\phi_k + \psi_k)}{r_k^\tau}. \tag{11.10}$$

We use this property in [11.2.7] when we discuss the general form and interpretation of the spectrum of an AR(p) process.

11.1.10 Uniqueness of the AR(p) Approximation to an Arbitrary Stationary Process. The following theorem is useful when fitting an AR(p) process to an observed time series.

Let $\mathbf{X}_t$ be a stationary process with auto-correlation function ρ. For each $p \geq 0$ there a unique AR(p) process $\mathbf{A}_{pt}$ with auto-correlation function ρ_p such that

$$\rho_p(\tau) = \rho(\tau) \quad \textit{for all } |\tau| \leq p. \tag{11.11}$$

The parameters $\vec{\alpha}_p = (\alpha_{p,1}, \ldots, \alpha_{p,p})$ of the approximating process of order p are recursively related to those of the approximating process of order $p-1$ by

$$\alpha_{p,k} = \alpha_{(p-1),k} - \alpha_{p,p}\alpha_{(p-1),(p-k)} \qquad k = 1, \ldots, p-1 \tag{11.12}$$

where

$$\alpha_{p,p} = \frac{\rho(p) - \sum_{k=1}^{p-1}\alpha_{(p-1),k}\rho(p-k)}{1 - \sum_{k=1}^{p-1}\alpha_{(p-1),(p-k)}\rho(p-k)}. \tag{11.13}$$

The recursion is started by setting $\alpha_{1,1} = \rho(1)$. A proof can be found in Appendix M.

11.1.11 The Partial Auto-correlation Function. When $\mathbf{X}_t$ is a normal process, $\alpha_{\tau,\tau}$ is called the *partial auto-correlation coefficient* between $\mathbf{X}_t$ and $\mathbf{X}_{t-\tau}$ (see [60]). A useful property of the *partial auto-correlation function* is that $\alpha_{\tau,\tau}$ becomes zero for $\tau > p$ when $\mathbf{X}_t$ is an AR(p) process. Thus an estimate of $\alpha_{\tau,\tau}$ is often plotted as a diagnostic to help identify the order of an AR process.

11.1.12 What is the Partial Auto-correlation Coefficient? Details. In technical terms, the partial auto-correlation coefficient $\alpha_{p,p}$ is the correlation between $\mathbf{X}_t$ and $\mathbf{X}_{t-p}$ when $\mathbf{X}_{t-1}, \ldots, \mathbf{X}_{t-p+1}$ are held fixed. When $\mathbf{X}_t$ is a stationary normal process,

$$\alpha_{p,p} = \frac{\mathrm{Cov}\big(\mathbf{X}_t, \mathbf{X}_{t-p} | \vec{\mathbf{G}}_t = \vec{\mathbf{g}}_t\big)}{\sigma_f \sigma_b}$$

where

$$\begin{aligned}\sigma_f^2 &= \mathrm{Var}\big(\mathbf{X}_t | \vec{\mathbf{G}}_t = \vec{\mathbf{g}}_t\big)\\ \sigma_b^2 &= \mathrm{Var}\big(\mathbf{X}_{t-p} | \vec{\mathbf{G}}_t = \vec{\mathbf{g}}_t\big),\end{aligned}$$

and where, for notational convenience, $\vec{\mathbf{G}}_t$ is the $(p-1)$-dimensional random vector $\vec{\mathbf{G}}_t = (\mathbf{X}_{t-1}, \ldots, \mathbf{X}_{t-p+1})^{\mathrm{T}}$. The value of this correlation does not depend upon the specific realization $\mathbf{x}_{t-1}, \ldots, \mathbf{x}_{t-p+1}$ of $\mathbf{X}_{t-1}, \ldots, \mathbf{X}_{t-p+1}$.[2]

The easiest way to understand the partial correlation coefficient is by means of an example. Therefore suppose $\mathbf{X}_t$ is a zero mean normal AR(1) process with parameter α_1. For an arbitrary time t, let $\mathbf{Y}_1 = \mathbf{X}_{t+1}$, $\mathbf{Y}_2 = \mathbf{X}_t$ and $\mathbf{Y}_3 = \mathbf{X}_{t-1}$. These random variables have variance-covariance matrix

$$\mathbf{\Sigma}_{1,2,3} = \sigma_X^2 \begin{pmatrix} 1 & \alpha_1 & \alpha_1^2 \\ \alpha_1 & 1 & \alpha_1 \\ \alpha_1^2 & \alpha_1 & 1 \end{pmatrix}$$

which has inverse

$$\mathbf{\Sigma}_{1,2,3}^{-1} = \frac{1}{\sigma_X^2(1-\alpha_1^2)} \begin{pmatrix} 1 & -\alpha_1 & 0 \\ -\alpha_1 & 1+\alpha_1^2 & -\alpha_1 \\ 0 & -\alpha_1 & 1 \end{pmatrix}$$

Substituting into (2.34), we obtain the joint density function for these three random variables:

$$f_{1,2,3}(y_1, y_2, y_3) = \frac{1}{(2\pi\sigma_X^2)^{3/2}} \times e^{-\frac{y_1^2 + (1+\alpha_1^2)y_2^2 + y_3^2 - 2\alpha_1 y_1 y_2 - 2\alpha_1 y_2 y_3}{2(1-\alpha_1^2)\sigma_X^2}}$$

[2] In general, when the process is not normal, the value of $\alpha_{p,p}$ does depend upon the specific realization.

To understand the $\alpha_{2,2}$ partial correlation coefficient, we now derive the joint density function of $\mathbf{Y}_1$ and $\mathbf{Y}_3$ conditional upon $\mathbf{Y}_2$. Recall from [2.8.6] that

$$f_{1,3|2}(y_1, y_3|y_2) = \frac{f_{1,2,3}(y_1, y_2, y_3)}{f_2(y_2)}.$$

But

$$f_2(y_2) = \frac{1}{(2\pi\sigma_X^2)^{1/2}} e^{-y_2^2/2\sigma_X^2},$$

therefore

$$\begin{aligned} f_{1,3|2}(y_1, y_3|y_2) &= \frac{1}{2\pi\sigma_X^2} \\ &\times e^{-\frac{y_1^2+(1+\alpha_1^2)y_2^2+y_3^2-2\alpha_1 y_1 y_2-2\alpha_1 y_2 y_3-(1-\alpha_1^2)y_2^2}{2(1-\alpha_1^2)\sigma_X^2}} \\ &= \frac{1}{2\pi\sigma_X^2} e^{-\frac{(y_1-\alpha_1 y_2)^2+(y_3-\alpha_1 y_2)^2}{2\sigma_X^2(1-\alpha_1^2)}} \\ &= \frac{1}{(2\pi\sigma_X^2)^{1/2}} e^{-\frac{(y_1-\alpha_1 y_2)^2}{2\sigma_X^2(1-\alpha_1^2)}} \\ &\quad \times \frac{1}{(2\pi\sigma_X^2)^{1/2}} e^{-\frac{(y_3-\alpha_1 y_2)^2}{2\sigma_X^2(1-\alpha_1^2)}} \qquad (11.14) \\ &= f_{1|2}(y_1|y_2) f_{3|2}(y_3|y_2). \end{aligned}$$

Thus $\mathbf{Y}_1$ and $\mathbf{Y}_3$ are conditionally independent [2.8.5], since the joint conditional density function can be factored as the product of marginal conditional density functions. Hence the conditional correlation between $\mathbf{Y}_1$ and $\mathbf{Y}_3$ is also zero, which is exactly what we obtain for $\alpha_{2,2}$ if we solve (11.13) and (11.12) recursively.

Equation (11.14) is the key to understanding the true meaning here. Since $\mathbf{X}_t$ is an AR(1) process, $\alpha_1 \mathbf{Y}_2 = \alpha_1 \mathbf{X}_t$ is the best one-step ahead forecast of $\mathbf{X}_{t+1}$. Similarly, $\alpha_1 \mathbf{X}_t$ is the best one-step back 'forecast' of $\mathbf{X}_{t-1}$.[3] Equation (11.14) shows that $f_{1,3|2}$ is the joint distribution of the one-step ahead and one-step back forecast errors. If the process was actually AR of order $p > 1$, the error of the one-step ahead forecast made only with $\mathbf{X}_t$ would still depend upon $\mathbf{X}_{t-1}$, and that of the one-step back forecast made only with $\mathbf{X}_t$ would still depend upon $\mathbf{X}_{t+1}$. Since $\mathbf{X}_{t-1}$ and $\mathbf{X}_{t+1}$ are dependent, the errors would also be dependent and factorization (11.14) would not be possible.

In general, $\alpha_{\tau,\tau}$ is the correlation between the error of a one-step ahead forecast of $\mathbf{X}_t$ made with $\mathbf{X}_{t-1}, \ldots, \mathbf{X}_{t-\tau+1}$ and the error of a one-step back forecast of $\mathbf{X}_{t-\tau}$ made with the same random variables. When $\mathbf{X}_t$ is AR(p) and normal, these errors become independent for lags $\tau > p$.

11.1.13 Auto-covariance Functions of Filtered Series. An operator that replaces a process $\mathbf{X}_t$ with the process

$$\mathbf{Y}_t = \sum_{k=-\infty}^{\infty} a_k \mathbf{X}_{t+k},$$

where $\sum_{k=-\infty}^{\infty} |a_k| < \infty$, is called a *linear filter*. Filters are used to remove, or isolate, variation on certain time scales from a process (see Section 17.5). The auto-covariance function of the filtered process is

$$\gamma_{yy}(\tau) = \sum_{k,l=-\infty}^{\infty} a_k a_l^* \gamma_{xx}(\tau + k - l). \qquad (11.15)$$

11.2 The Spectrum

11.2.0 General. The variance of a time series $\{X_1, X_2, \ldots, X_T\}$ of *finite* length may be attributed to different time scales by expanding it into a finite series of trigonometric functions[4] (cf. Equation (C.1))

$$X_t = A_0 + \sum_{k=1}^{(T-1)/2} \left(a_k \cos\frac{2\pi kt}{T} + b_k \sin\frac{2\pi kt}{T} \right). \qquad (11.16)$$

Equation (11.16) distributes the variance in the time series

$$\frac{1}{T}\sum_{t=1}^{T}(X_t - \overline{X})^2 = \frac{1}{2}\sum_{k=1}^{(T-1)/2}\left(a_k^2 + b_k^2\right) \qquad (11.17)$$

to the periodic components in the expansion shown in (11.16). The elements $(a_k^2 + b_k^2)$ are collectively referred to as the *periodogram* of the finite time series $\{X_1, \ldots, X_T\}$ when they are multiplied by $T/4$ (cf. [12.3.1]).

Unfortunately, it is not readily apparent that the expansion in (11.16) is related to the spectrum of an *infinite* time series or a stationary process, although this is true. We will see below that the spectrum is a continuous function of frequency. In contrast, the periodogram is always discrete.

It is important to note that our purpose in this chapter is to describe the spectrum as a characteristic of a stochastic process (hence the use of the word 'parameter' in the title). Spectral estimation is dealt with in Chapter 12.

[3] If $\mathbf{X}_t$ is an AR(1) process with parameter α_1 then both $\mathbf{X}_t - \alpha_1\mathbf{X}_{t-1}$ and $\mathbf{X}_t - \alpha_1\mathbf{X}_{t+1}$ are white noise processes. To confirm that $\mathbf{N}_t = \mathbf{X}_t - \alpha_1\mathbf{X}_{t+1}$ is a white noise process, show that $\mathcal{E}(\mathbf{N}_t\mathbf{N}_{t+\tau}) = 0$ for all $\tau \neq 0$.

[4] We have assumed, for mathematical convenience, that T is odd. The expansion is slightly more complex when T is even.

11.2.1 Definition of the Spectrum. *Let* $\mathbf{X}_t$ *be an ergodic weakly stationary stochastic process with auto-covariance function* $\gamma(\tau)$, $\tau = 0, \pm 1, \ldots$. *Then the* spectrum *(or* power spectrum*)* Γ *of* $\mathbf{X}_t$ *is the Fourier transform*[5] $\mathcal{F}$ *of the auto-covariance function* γ*. That is*

$$\Gamma(\omega) = \mathcal{F}\{\gamma\}(\omega) \qquad (11.18)$$
$$= \sum_{\tau=-\infty}^{\infty} \gamma(\tau) e^{-2\pi i \tau\omega}$$

for all $\omega \in [-1/2, 1/2]$.

Note that since γ is an even function of τ,

$$\Gamma(\omega) = \gamma(0) + 2\sum_{\tau=1}^{\infty} \gamma(\tau)\cos(2\pi\tau\omega).$$

Note also that the spectrum and the auto-covariance function are *parameters* of the stochastic process $\mathbf{X}_t$. When the process parameters are known (not estimated from data), the spectrum is well-defined and not contaminated by any uncertainty.

As is our practice with the auto-covariance and auto-correlation functions, we will use the notation Γ_{xx} to identify Γ as the spectrum of $\mathbf{X}_t$ when required by the context.

11.2.2 Properties.

1 The spectrum of a real-valued process is symmetric. That is

$$\Gamma(-\omega) = \Gamma(\omega).$$

2 The spectrum is continuous and differentiable everywhere in the interval $[-1/2, 1/2]$. Consequently

3 $\frac{d}{d\omega}\Gamma(\omega)|_{\omega=0} = 0.$

4 The auto-covariance function can be reconstructed from the spectrum by using the inverse Fourier transform (C.6) to obtain

$$\gamma(\tau) = \int_{-\frac{1}{2}}^{\frac{1}{2}} \Gamma(\omega) e^{2i\pi\omega\tau}\, d\omega.$$

5 The spectrum describes the distribution of variance across time scales. In particular,

$$\mathrm{Var}(\mathbf{X}_t) = \gamma(0) = 2\int_0^{\frac{1}{2}} \Gamma(\omega)\, d\omega. \qquad (11.19)$$

6 The spectrum is a linear function of the auto-covariance function. That is, if γ is decomposed into two functions, $\gamma(\tau) = \alpha_1\gamma_1(\tau) + \alpha_2\gamma_2(\tau)$, then

$$\Gamma(\omega) = \alpha_1\Gamma_1(\omega) + \alpha_2\Gamma_2(\omega)$$

where $\Gamma_i = \mathcal{F}\{\gamma_i\}$.

11.2.3 Theorem: The Spectra of AR(p) and MA(q) Processes.

1 *The spectrum of an AR(p) process with process parameters* $\{\alpha_1, \ldots, \alpha_p\}$ *and noise variance* $Var(\mathbf{Z}_t) = \sigma_Z^2$ *is*

$$\Gamma(\omega) = \frac{\sigma_Z^2}{|1 - \sum_{k=1}^p \alpha_k e^{-2\pi i k\omega}|^2}. \qquad (11.20)$$

2 *The spectrum of an MA(q) process with process parameters* $\{\beta_1, \ldots, \beta_q\}$ *and noise variance* $Var(\mathbf{Z}_t) = \sigma_Z^2$ *is*

$$\Gamma(\omega) = \sigma_Z^2 |1 + \sum_{l=1}^{q} \beta_l e^{-2\pi i l\omega}|^2. \qquad (11.21)$$

Proofs can be found in standard textbooks such as [195] or [60].

11.2.4 The Spectrum of a White Noise Process. The spectrum of a white noise process $\mathbf{Z}_t$ is easily computed from (11.21). Since $\gamma(0) = \sigma_Z^2$ and $\gamma(\tau) = 0$ for nonzero τ, the spectrum is independent of ω. That is

$$\Gamma_Z(\omega) = \sigma_Z^2 \text{ for all } \omega \in [-1/2, 1/2]. \qquad (11.22)$$

The spectrum is drawn as a horizontal line, indicating that no time scale of variation is preferred, hence the allusion to white light. This agrees with the analysis of the run length $\mathbf{L}$ discussed in [10.3.3].

11.2.5 The Spectrum of an AR(1) Process. The power spectrum of an AR(1) process with lag-1 correlation coefficient α_1 is

$$\Gamma(\omega) = \frac{\sigma_Z^2}{|1 - \alpha_1 e^{-2\pi i\omega}|^2}$$
$$= \frac{\sigma_Z^2}{1 + \alpha_1^2 - 2\alpha_1\cos(2\pi\omega)}. \qquad (11.23)$$

[5] Note the specific mathematical character of the discrete Fourier transform. It operates on the set of infinite, summable, real-valued series and generates complex-valued functions that are defined on the real interval $[-1/2, 1/2]$. See Appendix C. For more reading about the Fourier transform see standard textbooks, such as [195].

This spectrum has no extremes in the interior of the interval [0, 1/2] because, everywhere inside the interval, the derivative

$$\frac{d}{d\omega}\Gamma(\omega) = -2\alpha_1\Gamma(\omega)^2\sin(2\pi\omega) \neq 0.$$

The sign of the derivative is determined by α_1. Thus the spectrum has a minimum at one end of the interval [0, 1/2] and a maximum at the other end.

When $\alpha_1 > 0$, the 'spectral peak' is located at frequency $\omega = 0$. Such processes are often referred to as *red noise processes.*

AR(1) processes with $\alpha_1 < 0$, which are sometimes called *blue noise processes,* are of little practical importance in climate research because they tend to change sign every time step. In most climate research contexts, the observed process evolves continuously. Thus α_1 will be positive given a sufficiently small time step.[6]

Figure 11.3 shows the spectra of the two 'red' AR(1) processes that were discussed in [10.3.3] and [11.1.7]. The $\alpha_1 = 0.9$ spectrum (right hand axis in Figure 11.3a) is more energetic on long time scales (ω^{-1} greater than approximately seven time steps) than the $\alpha_1 = 0.3$ spectrum (left hand axis in Figure 11.3a). At short time scales the $\alpha_1 = 0.3$ process is somewhat more energetic. This interpretation is consistent with the finding that AR(1) processes with large lag-1 correlation coefficients generate more long runs than those with small lag-1 correlation coefficients, and vice versa (see Figure 10.8).

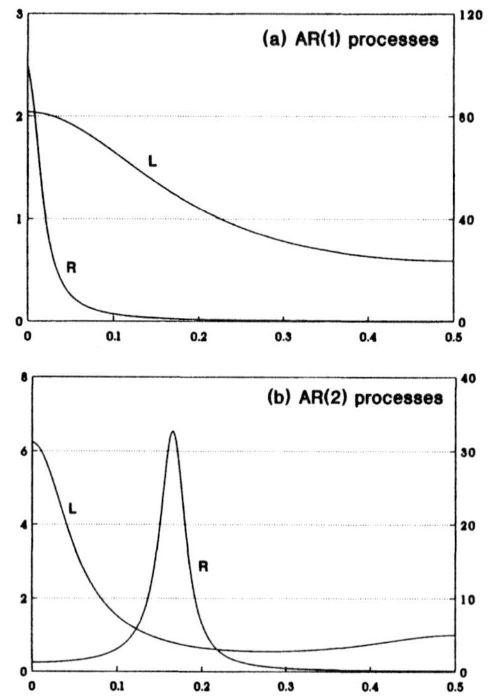

Figure 11.3: *Power spectra of various AR processes. The left hand axis applies to spectra labelled 'L' and the right hand axis applies to those labelled 'R.'*
a) AR(1) processes with $\alpha_1 = 0.3$ *(L) and* $\alpha_1 = 0.9$ *(R),*
b) AR(2) processes with $(\alpha_1, \alpha_2) = (0.3, 0.3)$ *(L) and* $(\alpha_1, \alpha_2) = (0.9, -0.8)$ *(R).*

11.2.6 The Spectrum of an AR(2) process.

The power spectrum of an AR(2) process with parameters (α_1, α_2) (11.20) is given by

$$\Gamma(\omega) = \frac{\sigma_Z^2}{1 + \alpha_1^2 + \alpha_2^2 - 2g(\omega)}$$

where

$$g(\omega) = \alpha_1(1-\alpha_2)\cos(2\pi\omega) + \alpha_2\cos(4\pi\omega).$$

Depending upon the parameters, this spectrum can have a minimum or a maximum in the interior of the interval [0, 1/2]. Figure 11.3b displays spectra of both types.

When its derivative is zero, $\Gamma(\omega)$ has a maximum or minimum, and we note that $\Gamma'(\omega) = 0$ whenever $g'(\omega) = 0$. By using the identity $\sin(4\pi\omega) = 2\sin(2\pi\omega)\cos(2\pi\omega)$, we find that

$$\begin{aligned} g'(\omega) &= -2\pi\alpha_1(1-\alpha_2)\sin(2\pi\omega) \\ &\quad - 4\pi\alpha_2\sin(4\pi\omega) \\ &= (-2\pi)\sin(2\pi\omega) \\ &\quad \times \big(\alpha_1(1-\alpha_2) + 4\alpha_2\cos(2\pi\omega)\big). \end{aligned}$$

Since $\sin(2\pi\omega) \neq 0$ for all $\omega \in (0, 1/2)$, $\Gamma'(\omega) = 0$ when

$$\cos(2\pi\omega) = -\alpha_1(1-\alpha_2)/(4\alpha_2). \qquad (11.24)$$

This last equation has a solution $\omega \in (0, 1/2)$ when $|\alpha_1(1-\alpha_2)| < 4|\alpha_2|$. This solution represents a spectral maximum when $\alpha_2 < 0$ and a spectral minimum when $\alpha_2 > 0$.

Equation (11.24) has solutions $\omega \in (0, 1/2)$ for both spectra shown in Figure 11.3b. When $\vec{\alpha} = (0.3, 0.3)$, a minimum is located at $\omega \approx 0.28$. When $\vec{\alpha} = (0.9, -0.8)$ a maximum occurs at $\omega \approx 0.17$.

[6]There are exceptions to this statement. For example, annual layer thickness in ice cores can be modelled as 'blue noise' (see, for example, Fisher et al. [118]).

11.2.7 The Spectrum of a Linearly Filtered Process. We described the auto-covariance function of a linearly filtered process $\mathbf{Y}_t = \sum_k a_k \mathbf{X}_{t+k}$ in [11.1.13]. The spectrum of such a process is (see (C.15, 11.15)):

$$\Gamma_{yy} = |\mathcal{F}\{a\}|^2 \Gamma_{xx}.$$

11.2.8 Interpretation: General. Literal interpretation of equations (11.16, 11.17) leads to the incorrect notion that all weakly stationary stochastic processes can be represented as a combination of a finite number of oscillating signals with random amplitude and phase.

However, a special class of weakly stationary processes that behave in just this way can be constructed. An example of this type of process is sea level measured at a given tide gauge. These measurements contain a tide signal made up of a (practically) finite number of astronomically forced modes that is overlaid by irregular variations excited by weather.

Such a process has infinitely long memory and is *not ergodic*. Its auto-covariance function does not go to zero for increasing lag but instead becomes periodic at long lags. Therefore, the Fourier transform of its auto-covariance function does not exist and the process has no auto-spectrum in the sense of definition [11.2.1]. A different type of characteristic spectrum must be defined for these processes, namely a discrete line spectrum.[7] We will discuss this type of process in the next subsection.

Ergodic weakly stationary processes have finite memory and summable auto-covariance functions with defined Fourier transforms. Most time series encountered in climate research are, to a reasonable approximation, of this type, at least after deterministic cycles such as the annual cycle or the diurnal cycle have been subtracted. We will discuss the interpretation of spectra of such processes in [11.2.10].

The two concepts of the power and line spectra can be formally unified by defining a generalized Fourier transform. The discrete part of the spectrum is then represented by Dirac δ-functions, functions that are infinitely large at the frequencies of the oscillations they represent and zero everywhere else. By suitably generalizing the definition of integration, the δ-function can be given an intensity such that the integral over the δ-function is equal to the variance of the oscillation.

[7]Note that the expression *spectrum* is used for a large variety of mathematical objects. Examples include the *eigenvalue spectrum* of an EOF analysis, the power spectrum, and the line spectrum discussed here. Climatologists also use *spatial* spectra that describe the distribution of energy to different spatial scales. A common characteristic of these spectra is that they are expressed as functions of a discrete or continuous set of indices that are ordered on the basis of time scale (in case of the power spectrum), relevance (eigenvalue spectrum), or other meaningful criteria.

11.2.9 Interpretation: Periodic Weakly Stationary Processes.[8] Suppose a periodic stochastic process $\mathbf{X}_t$ can be represented as

$$\mathbf{X}_t = \sum_{j=-n}^{n} \mathbf{Z}_j e^{2\pi i \omega_j t} + \mathbf{N}_t, \qquad (11.25)$$

where $\omega_j = 1/T_j$, $j = -n, \ldots, n$, $T_j \in \mathbb{Z}$ are fixed frequencies, $\mathbf{Z}_j$, $j = -n, \ldots, n$, are complex random variables, and $\mathbf{N}_t$ represents a noise term that is independent of the $\mathbf{Z}_j$. For simplicity we assume that $\mathbf{N}_t$ is white in time, but this assumption is easily generalized.

What conditions must be placed on the frequencies ω_j and random variables $\mathbf{Z}_j$ to ensure that $\mathbf{X}_t$ is real valued and weakly stationary?

To ensure that $\mathbf{X}_t$ is real for all $t = 0, \pm 1, \ldots$, the frequencies ω_j must be symmetric about zero (i.e., $\omega_{-j} = \omega_j$), and for every $j = 1, \ldots, n$, random variables $\mathbf{Z}_{-j}$ and $\mathbf{Z}_j$ must be complex conjugates.

Two conditions must be satisfied to ensure weak stationarity. First, the mean of the process,

$$\mu_X = \mathcal{E}(\mathbf{X}_t) = \sum_{j=-n}^{n} \mathcal{E}(\mathbf{Z}_j) e^{2\pi i \omega_j t} + \mathcal{E}(\mathbf{N}_t)$$

should be independent of time. This means that random variables $\mathbf{Z}_j$, $j = 1, \ldots, n$, must have mean zero.

Second, the auto-covariance function $\mathcal{E}(\mathbf{X}_{t+\tau}\mathbf{X}_t)$ must be a function of τ alone. The auto-covariance function is given by

$$\begin{aligned}\gamma(\tau) &= \mathcal{E}(\mathbf{X}_{t+\tau}\mathbf{X}_t) \\ &= \delta_{0,\tau}\,\sigma_N^2 + \sum_{j=-n}^{n} \mathcal{E}\big(|\mathbf{Z}_j|^2\big) e^{2\pi i \omega_j \tau} \\ &\quad + \sum_{j=-n}^{n} \sum_{k \neq j} \big(\mathcal{E}\big(\mathbf{Z}_j \mathbf{Z}_k^*\big) e^{2\pi i \omega_j \tau}\big) \\ &\qquad \times e^{2\pi i (\omega_j - \omega_k) t}\end{aligned}$$

for all $t = 0, \pm 1, \ldots$, where $\delta_{0,\tau} = 1$ if $\tau = 0$, and zero otherwise. Since the left hand side is constant for all t, it follows that the random variables $\mathbf{Z}_j$, $j = 0, 1, \ldots, n$, must be uncorrelated.

[8]Following Koopmans [229].

Consequently, periodic weakly stationary processes (11.25) have periodic auto-covariance functions of the form

$$\gamma(\tau) = \sum_{j=-n}^{n} \mathcal{E}(|\mathbf{Z}_j|^2)e^{2\pi i\omega_j\tau}$$
$$= \text{Var}(\mathbf{Z}_0)+4\sum_{j=1}^{n}\mathcal{E}(|\mathbf{Z}_j|^2)\cos(2\pi\omega_j\tau).$$

for $|\tau| \geq 1$.

Hence, only very special weakly stationary stochastic processes can be represented as a finite sum of discrete signals. In contrast with the ARMA processes described in Sections 10.3 and 10.5, these processes have periodic auto-covariance functions for which $\lim_{\tau\to\infty}\sum_{\ell=0}^{\tau}\gamma(\ell)^2 = \infty$. Therefore, the summation (11.18) does not converge and the spectrum does not exist. In conceptual terms: the system has infinitely long memory.

Even though the power spectrum does not exist, we can define a spectrum that distributes the variance of $\mathbf{N}_t$ with time scale in the usual way and adds a specific amount of variance, $\mathcal{E}(|Z_j|^2)$, at the discrete frequencies ω_j, $j = -k, \ldots, k$. The discrete part of this spectrum is called a *line spectrum*. As noted above, this kind of spectrum can be given a density function interpretation by resorting to Dirac δ-functions.

11.2.10 Interpretation: Ergodic Weakly Stationary Processes. Processes with limited memory, that is,

$$\lim_{\tau\to\infty}\sum_{\ell=0}^{\tau}\gamma(\ell)^2 < \infty,$$

can not be periodic in the sense of (11.25). A specific amount of variability can not be attributed to a specific frequency, otherwise we would again have a process with a discrete periodic component and infinite memory. Instead, variance is attributed to time scale ranges or frequency *intervals*. Given two frequencies $0 \leq \omega_1 < \omega_2 \leq 1/2$, we interpret

$$\int_{\omega_1}^{\omega_2}\Gamma(\omega)\,d\omega$$

as the variability generated by the process in the time scale range $(1/\omega_2, 1/\omega_1)$.

11.2.11 Interpretation: Spectra of AR Processes. In general, a peak in a spectrum indicates only that more variability is concentrated at time scales near that of the peak than at other time scales. Peaks in the spectra of ergodic, weakly stationary processes do not reflect the presence of an oscillatory component in the system.

However, peaks in the spectra of AR processes do indicate the presence of damped eigen-oscillations in the system with eigenfrequencies close to that of the peak.

To understand this, we return to (10.14),

$$X_{t+\tau} = \sum_{j=1}^{p}\beta_j y_j^{-\tau},$$

which describes the evolution of an AR(p) process $\mathbf{X}_t$ from a given state $X_{t-1}, \ldots, X_{t-p}$ when the noise is turned off. The constants β_j depend upon the process parameters $\alpha_1, \ldots, \alpha_p$ and the initial state $X_{t-1}, \ldots, X_{t-p}$. The constants y_j are the roots of the characteristic polynomial (10.11)

$$1 - \sum_{k=1}^{p}\alpha_k B^k. \tag{11.26}$$

Since there are p such roots, there are p sets of initial states $\mathcal{I}_k = \{X_{t-1,k}, \ldots, X_{t-p,k}\}$, $k = 1, \ldots, p$, for which $\beta_j = \delta_{jk}$.[9] It can be shown that the initial states are given by

$$X_{t-\tau,k} = y_k^{\tau-1},\ \tau = 1, \ldots, p.$$

Hence each set of states $\mathcal{I}_k$ represents a finite segment of a time series that is either a damped oscillation (if there are a pair of complex conjugate roots y_k) or simply decays exponentially (if y_k is a real root). (See Figure 10.12.)

Equation (10.14) shows that the evolution of the system in the absence of noise is determined by the mixture of 'initial states' $\mathcal{I}_k$. In particular, if the initial state is one of the decaying exponential, or damped oscillatory, states $\mathcal{I}_k$, it will stay in that state and continue to display the same behaviour in the future. In that sense, the roots of the characteristic polynomial represent eigensolutions of the system. However, since noise is continually added to the system, we see variability on all time scales. The eigenmodes of the system determine the way in which the variability in the input noise evolves into the future. When a sequences of states evolves that is close to one of the eigenmodes of the system, that mode tends to persist more strongly than other sequences of states. These preferences are, in turn, reflected in the tendency for there to be more variance in some parts of the spectrum than others.

[9] For details, refer to [10.3.5,6].

We saw in [11.1.9] that the auto-covariance function of the AR(p) process may be written as (11.10)

$$\rho(\tau) = \sum_i \frac{a_i}{y_i^\tau} + \sum_k a_k \frac{\cos(\tau\phi_k + \psi_k)}{r_k^\tau}$$

where the first sum is taken over the real roots and the second is taken over the complex roots. The complex roots y_k are expressed in polar coordinates as $y_k = r_k e^{i\phi_k}$. Thus, the auto-covariance function is a sum of auto-covariance functions of AR(1) and AR(2) processes, which correspond to the initial states $\mathcal{I}_j$ discussed above.

The spectrum of the AR(p) process is then the Fourier transform of the sum of auto-covariance functions, or, because of the linearity of the Fourier transform, the sum of autospectra of AR(1) and AR(2) processes. Thus, any peak in the spectrum of the AR(p) process must originate from a peak in an AR(2) spectrum, and we have seen that such peaks just correspond to first-order approximations of the eigen-oscillations of the AR(2) process (cf. [10.3.5,6]).

Things are relatively clear in this context because we have complete knowledge about the process to guide us in the interpretation of the spectrum. However, interpretation is much more difficult when spectra are estimated from finite time series. The estimates are uncertain because they are affected by sampling variability. They are also affected by the properties of the estimator itself and the way in which those properties (such as bias) are affected by the true spectrum. So we must attempt to interpret a noisy version of the true spectrum that is viewed through 'rose coloured glasses.' Moreover, in practice cases this usually must be done without complete knowledge of the nature of the process that generated the spectrum.

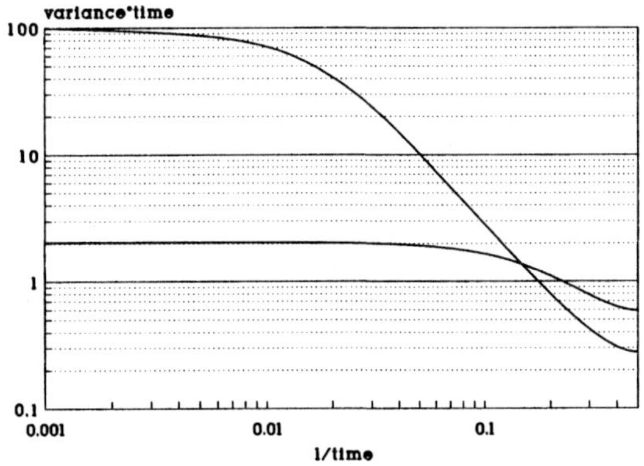

Figure 11.4: *Power spectra of the AR(1) processes shown in Figure 11.3 displayed in log-log format. Note that the derivative is zero at the origin.*

11.2.12 Units. Suppose that the process $\mathbf{X}_t$ is expressed in, say, units of A, and the time increment τ in units of δ.

- The auto-covariance function is given in units of A^2.
- To have appropriate units in (11.19) we multiply by a constant factor carrying the unit $1/\delta$ in the definition of the Fourier transform $\mathcal{F}$ so that the spectrum is expressed in units of $A^2\delta$.
- The frequency ω is expressed in units of $1/\delta$.

If, for example, we consider a process that is given in metres, with a time step of months, then the spectral, or variance, density is expressed in units of $m^2 \times$ month, and the frequency in month^{-1}. A peak at $\omega = 0.2$/month represents a period of $1/\omega = 5$ months.

11.2.13 Plotting Formats. An important practical aspect of spectral analysis concerns the format in which spectra are displayed. So far, we have used the plain format with the frequency ω as the abscissa and the spectral density as the ordinate.

The log-log presentation, in which the logarithm of the frequency is plotted against the logarithm of the spectral density, is another common display format. Spectra displayed in this way look rather different. This can be seen from Figure 11.4, which shows the same AR(1) spectra as Figure 11.3 but in log-log format. Note that both spectra become 'white' for frequencies close to the origin.

An advantage of this format for theoreticians is that certain power laws, such as $\Gamma(\omega) \sim \omega^{-k}$, appear as straight lines with a slope of $-k$. A disadvantage with this format is that the area under the curve as it is perceived by the eye is no longer proportional to the variance. Also, the frequency range that contains most of the variation is not always readily identified.

Another alternative is to plot $\omega\Gamma(\omega)$ on a log-log scale so that the units on the ordinate are independent of time.

In any case, it is advisable to clarify the plotting format and units of a spectrum before making physical interpretations. Alleged inconsistencies are sometimes entirely due to the use of different display formats.

11.3 The Cross-covariance Function

11.3.1 Definition. *Let* $(\mathbf{X}_t, \mathbf{Y}_t)$ *represent a pair of stochastic processes that are jointly weakly stationary. Then the* cross-covariance function γ_{xy} *is given by*

$$\gamma_{xy}(\tau) = \mathcal{E}\big((\mathbf{X}_t - \mu_X)(\mathbf{Y}_{t+\tau} - \mu_Y)^*\big)$$

where μ_x *is the mean of* $\mathbf{X}_t$ *and* μ_y *is the mean of* $\mathbf{Y}_t$.

Note that if $\mathbf{X}_t = \mathbf{Y}_t$, then the cross-covariance function is simply the auto-covariance function γ_{xx}.

The *cross-correlation function* ρ_{xy} is the normalized cross-covariance function

$$\rho_{xy}(\tau) = \frac{\gamma_{xy}(\tau)}{\sigma_X \sigma_Y}, \qquad (11.27)$$

where σ_X and σ_Y are the standard deviations $\sqrt{\gamma_{xx}(0)}$ and $\sqrt{\gamma_{yy}(0)}$ of processes $\{\mathbf{X}_t\}$ and $\{\mathbf{Y}_t\}$, respectively.

11.3.2 Assumption. We list here the assumptions that are needed to ensure that the cross-correlation function exists and is absolutely summable. Specifically, we assume that $\{(\mathbf{X}_t, \mathbf{Y}_t) : t \in \mathbb{Z}\}$ is an ergodic weakly stationary bivariate process. Hence we have the following results.

- The means μ_x and μ_y are independent of time.
- The auto-covariance functions γ_{xx} and γ_{yy} depend only on the absolute time difference:

$$\mathcal{E}\big((\mathbf{X}_t - \mu_x)(\mathbf{X}_s - \mu_x)\big) = \gamma_{xx}(|t-s|)$$
$$\mathcal{E}\big((\mathbf{Y}_t - \mu_y)(\mathbf{Y}_s - \mu_y)\big) = \gamma_{yy}(|t-s|).$$

- The cross-covariance functions γ_{xy} and γ_{yx} depend only on the time difference:

$$\mathcal{E}\big((\mathbf{X}_t - \mu_x)(\mathbf{Y}_s - \mu_y)\big) = \gamma_{xy}(s-t)$$
$$\mathcal{E}\big((\mathbf{Y}_t - \mu_y)(\mathbf{X}_s - \mu_x)\big) = \gamma_{yx}(s-t).$$

  Note that

$$\begin{aligned}\gamma_{xy}(\tau) &= \mathcal{E}\big((\mathbf{X}_t - \mu_x)(\mathbf{Y}_{t+\tau} - \mu_x)\big)\\ &= \mathcal{E}\big((\mathbf{Y}_{t+\tau} - \mu_x)(\mathbf{X}_t - \mu_x)\big)\\ &= \gamma_{yx}(-\tau).\end{aligned}$$

- The process has limited memory. That is, the auto-covariance function

$$\Sigma(\tau) = \begin{pmatrix} \gamma_{xx}(\tau) & \gamma_{xy}(\tau) \\ \gamma_{yx}(-\tau) & \gamma_{yy}(\tau) \end{pmatrix}$$

  of the bivariate process satisfies a *mixing condition* such as

$$\sum_{\tau=-\infty}^{\infty} |\gamma_{ab}(\tau)| < \infty$$

  for $ab = xx, xy,$ and yy.

11.3.3 Some Simple Examples. Let us consider a few cases in which $\mathbf{Y}_t$ is a simple function of a zero mean weakly stationary process $\mathbf{X}_t$.

- Suppose $\mathbf{Y}_t$ is a multiple of $\mathbf{X}_t$,

$$\mathbf{Y}_t = \alpha \mathbf{X}_t. \qquad (11.28)$$

  Then, the cross-covariance function

$$\gamma_{xy}(\tau) = \alpha \gamma_{xx}(\tau) \qquad (11.29)$$

  is proportional to the auto-covariance function of $\mathbf{X}_t$.

- We make equation (11.28) slightly more complex by adding some independent white noise $\mathbf{Z}_t$ so that

$$\mathbf{Y}_t = \alpha \mathbf{X}_t + \mathbf{Z}_t. \qquad (11.30)$$

  The noise is assumed to be independent of $\mathbf{X}_{t+\tau}$ for all lags τ. Then the auto-covariance function of $\mathbf{Y}$ and the cross-covariance function of $\mathbf{X}$ and $\mathbf{Y}$ are

$$\begin{aligned}\gamma_{yy}(\tau) &= \begin{cases} \alpha^2\gamma_{xx}(0) + \sigma^2 & \text{if } \tau = 0 \\ \alpha^2\gamma_{xx}(\tau) & \text{if } \tau \neq 0 \end{cases}\\ \gamma_{xy}(\tau) &= \alpha\gamma_{xx}(\tau). \end{aligned} \qquad (11.31)$$

  Thus, the addition of the noise changes the variance of process $\mathbf{Y}$ but not its auto-covariance or its cross-covariance with process $\mathbf{X}$. It does, however, change its auto-correlation and its cross-correlation with $\mathbf{X}$.

- Now suppose $\mathbf{Y}_t$ is obtained by shifting $\mathbf{X}_t$ by a fixed lag ζ,

$$\mathbf{Y}_t = \mathbf{X}_{t+\zeta}. \qquad (11.32)$$

  The resulting cross-covariance function is a shifted version of the auto-covariance function of $\mathbf{X}$,

$$\begin{aligned}\gamma_{xy}(\tau) &= \mathcal{E}\big(\mathbf{X}_t \mathbf{X}_{t+\zeta+\tau}\big)\\ &= \gamma_{xx}(\zeta + \tau). \end{aligned} \qquad (11.33)$$

- We could assume that $\mathbf{Y}_t$ is the discretized time derivative of $\mathbf{X}_t$, such that

$$\mathbf{Y}_t = \mathbf{X}_t - \mathbf{X}_{t-1} \approx \frac{d}{dt}\mathbf{X}_t \qquad (11.34)$$

Then

$$\begin{aligned} \gamma_{xy}(\tau) &= \mathcal{E}(\mathbf{X}_t\mathbf{X}_{t+\tau} - \mathbf{X}_t\mathbf{X}_{t-1+\tau}) \\ &= \gamma_{xx}(\tau) - \gamma_{xx}(\tau-1) \qquad (11.35) \end{aligned}$$

which one might loosely think of as

$$\gamma_{xy} \approx \frac{d}{d\tau}\gamma_{xx}(\tau). \qquad (11.36)$$

Similarly,

$$\begin{aligned} \gamma_{yy}(\tau) &= 2\gamma_{xx}(\tau) \\ &\quad - \big(\gamma_{xx}(\tau-1) + \gamma_{xx}(\tau+1)\big) \\ &\approx -\frac{d^2}{d\tau^2}\gamma_{xx}. \end{aligned}$$

A model such as (11.34) is often appropriate for conservative quantities. For example, the atmospheric angular momentum ($\mathbf{Y}$) has time-variability that is determined by the globally integrated torques ($\mathbf{X}$). Thus, the cross-covariance function between the atmospheric angular momentum and the torques is of the form (11.36).

- The last two examples are special cases of the situation in which $\mathbf{Y}_t$ is a (linearly) filtered version of $\mathbf{X}_t$. We showed in [11.1.13] that the auto-covariance function of process

$$F(\mathbf{X})_t = \sum_{k=-\infty}^{\infty} a_k \mathbf{X}_{t+k} \qquad (11.37)$$

is

$$\gamma_{F(x),F(x)}(\tau) = \sum_{k,l=-\infty}^{\infty} a_k a_l^* \gamma_{xx}(\tau - k + l). \qquad (11.38)$$

Similarly, the cross-covariance function of $\mathbf{X}_t$ and $F(\mathbf{X})_t$ is

$$\gamma_{x,F(x)}(\tau) = \sum_{k=-\infty}^{\infty} a_k^* \gamma_{xx}(\tau + k). \qquad (11.39)$$

Expressions (11.33) and (11.35, 11.36) are special cases of (11.38, 11.39), which may be re-expressed as

$$\begin{aligned} \gamma_{x,F(x)} &= F^{(*)}\{\gamma_{xx}\} \qquad (11.40) \\ \gamma_{F(x),x} &= F^{(r)}\{\gamma_{xx}\} \\ \gamma_{F(x),F(x)} &= F^{(r)}\{F^{(*)}\{\gamma_{xx}\}\} \qquad (11.41) \end{aligned}$$

by defining

$$F^{(*)}\{\gamma\}(\tau) = \sum_{k=-\infty}^{\infty} a_k^* \gamma(\tau + k)$$

$$F^{(r)}\{\gamma\}(\tau) = \sum_{k=-\infty}^{\infty} a_{-k} \gamma(\tau + k).$$

- Relationships (11.40, 11.41) can be generalized to two processes $\mathbf{X}_t$ and $\mathbf{Y}_t$ that are passed through two linear filters F and G with coefficients a_k and b_k, $k = -\infty, \infty$, respectively. Then

$$\gamma_{F(x),G(y)} = F^{(r)}\{G^{(*)}\{\gamma_{xy}\}\}.$$

11.3.4 Properties. We note the following.

1 The cross-covariance function is 'Hermitian.' That is,

$$\gamma_{yx}(\tau) = \gamma_{xy}^*(-\tau). \qquad (11.42)$$

2 We have

$$|\gamma_{xy}(\tau)| \le \sqrt{\gamma_{xx}(0)\gamma_{yy}(0)}. \qquad (11.43)$$

Therefore $|\rho_{xy}(\tau)| \le 1$, where $\rho_{xy}(\tau)$ is the cross-correlation function defined by equation (11.27).

3 The cross-covariance function is bi-linear. That is,

$$\gamma_{\alpha x,\beta y+z}(\tau) = \alpha\beta^*\gamma_{xy}(\tau) + \alpha\gamma_{xz}(\tau) \qquad (11.44)$$

for all processes $\mathbf{X}$, $\mathbf{Y}$, and $\mathbf{Z}$.

11.3.5 Example: SST and SLP in the Tropics and Extratropics. Frankignoul [131] estimated cross-correlation functions for monthly means of area averaged SST ($\mathbf{S}_t$) and turbulent heat flux ($\mathbf{H}_t$) for different areas of the Atlantic Ocean. Figure 11.5 shows the cross-correlation functions

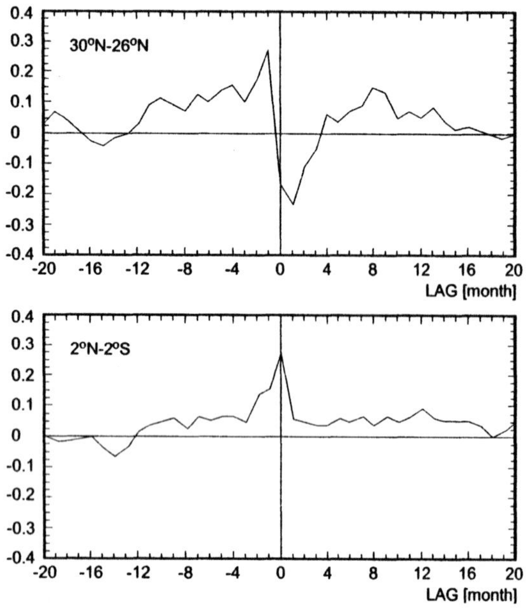

Figure 11.5: *Estimated cross-correlation functions* ρ_{hs} *for* $\mathbf{H}_t$*, the monthly mean turbulent heat flux into the atmosphere, and* $\mathbf{S}_t$*, the monthly mean sea-surface temperature (SST), averaged over different latitudinal bands in the Atlantic ocean (top: 26°–30° N, bottom: 2° S–2° N). The SST leads for negative lags* τ*. From Frankignoul [131].*

for a sub-tropical belt and the equatorial belt. The lagged correlations are small in both cases.

The cross-correlation function in the subtropics (top panel) is approximately anti-symmetric about the origin. The negative cross-correlation at $\tau = 1$ tells us that, on average, the SST is higher than normal one month *after* a negative (downward into the ocean) heat flux anomaly. Similarly, the positive cross-correlation at $\tau = -1$ indicates that a positive SST anomaly usually precedes a positive (upward into the atmosphere) heat flux anomaly. This suggests that there is a typical sequence of events of the form

$$\cdots \mathbf{H}_{t-1} < 0 \Rightarrow \mathbf{S}_t > 0 \Rightarrow$$

$$\Rightarrow \mathbf{H}_{t+1} > 0 \Rightarrow \mathbf{S}_{t+2} < 0 \cdots$$

The two quantities apparently interact with each other in such a way that an initial anomaly is damped by means of a *negative feedback* process (see [11.3.11]).

The cross-correlation function of the equatorial turbulent heat flux and SST (Figure 11.5, bottom) is more symmetric with a maximum at lag zero. The symmetry indicates that $\mathbf{H}_t$ and $\mathbf{S}_{t+\tau}$ tend to have the same sign for moderate lags τ. Such behaviour often indicates that both quantities are forced by the same external mechanism or are coupled together by a *positive feedback* mechanism (see [11.3.11]).

11.3.6 Bivariate AR(1) Processes: Notation. The next few subsections focus on the bivariate auto-regressive processes of first order. For convenience, we represent these processes in matrix-vector notation as

$$\begin{pmatrix}\mathbf{X}\\ \mathbf{Y}\end{pmatrix}_t = \mathcal{A}\begin{pmatrix}\mathbf{X}\\ \mathbf{Y}\end{pmatrix}_{t-1} + \begin{pmatrix}\mathbf{N}\\ \mathbf{M}\end{pmatrix}_t \tag{11.45}$$

where the coefficient matrix $\mathcal{A}$ is given by

$$\mathcal{A} = \begin{pmatrix}\alpha_{xx} & \alpha_{xy}\\ \alpha_{yx} & \alpha_{yy}\end{pmatrix}. \tag{11.46}$$

We also use the corresponding component-wise representation of (11.45)

$$\mathbf{X}_t = \alpha_{xx}\mathbf{X}_{t-1} + \alpha_{xy}\mathbf{Y}_{t-1} + \mathbf{N}_t \tag{11.47}$$

$$\mathbf{Y}_t = \alpha_{yx}\mathbf{X}_{t-1} + \alpha_{yy}\mathbf{Y}_{t-1} + \mathbf{M}_t, \tag{11.48}$$

where it is convenient.

The two components of the driving noise, **N** and **M**, are assumed to form a bivariate white noise process. This means that the lagged covariances and cross-covariances of $\mathbf{N}_t$ and $\mathbf{M}_t$ are zero. However, it is possible that the components of the bivariate white noise processes are correlated at zero lag (i.e., $\gamma_{nm}(0) \neq 0$).

11.3.7 Bivariate AR(1) Process: Cross-covariance Matrix. The variances $\gamma_{xx}(0)$, $\gamma_{yy}(0)$ of the components of the bivariate process and their lag zero cross-covariance $\gamma_{xy}(0)$ are obtained by solving a 3×3 system of linear equations. These equations are derived by squaring equations (11.47) and (11.48), multiplying equations (11.47, 11.48) with each other, and taking expectations to obtain

$$\mathcal{B}\begin{pmatrix}\gamma_{xx}(0)\\ \gamma_{yy}(0)\\ \gamma_{xy}(0)\end{pmatrix} = \begin{pmatrix}\gamma_{nn}(0)\\ \gamma_{mm}(0)\\ \gamma_{nm}(0)\end{pmatrix} \tag{11.49}$$

where

$$\mathcal{B} = \mathcal{I} - \begin{pmatrix}\alpha_{xx}^2 & \alpha_{xy}^2 & 2\alpha_{xx}\alpha_{xy}\\ \alpha_{yx}^2 & \alpha_{yy}^2 & 2\alpha_{yx}\alpha_{yy}\\ \alpha_{xx}\alpha_{yx} & \alpha_{xy}\alpha_{yy} & \alpha_{xx}\alpha_{yy}+\alpha_{xy}\alpha_{yx}\end{pmatrix}$$

and $\mathcal{I}$ is the 3×3 identity matrix. The *cross-covariance matrix* at nonzero lag τ,

$$\Sigma_{xy}(\tau) = \begin{pmatrix}\gamma_{xx}(\tau) & \gamma_{xy}(\tau)\\ \gamma_{yx}(\tau) & \gamma_{yy}(\tau)\end{pmatrix},$$

may be computed recursively as

$$\begin{aligned}\Sigma_{xy}(\tau) &= \mathcal{A}\Sigma_{xy}(\tau-1)\\ &= \mathcal{A}^{\tau}\Sigma_{xy}(0).\end{aligned} \qquad (11.50)$$

11.3.8 A POP[10] Example. We now consider a bivariate AR(1) process in which the coefficient matrix $\mathcal{A}$ (11.46) is a rotation matrix

$$\mathcal{A} = r\begin{pmatrix} u & -v \\ v & u \end{pmatrix} \qquad (11.51)$$

with $u^2 + v^2 = 1$ and $0 \le r \le 1$.[11] The noise components **N** and **M** are assumed to be uncorrelated and of equal variance. That is, $\gamma_{nm}(0) = 0$ and $\gamma_{nn}(0) = \gamma_{mm}(0) = \sigma_z^2$. Thus the lag zero covariance matrix for the POP coefficients $\mathbf{X}_t$ and $\mathbf{Y}_t$, which is obtained by solving (11.49), satisfies

$$\mathcal{B}\begin{pmatrix}\gamma_{xx}(0)\\ \gamma_{yy}(0)\\ \gamma_{xy}(0)\end{pmatrix} = \begin{pmatrix}\sigma_z^2\\ \sigma_z^2\\ 0\end{pmatrix} \qquad (11.52)$$

where

$$\mathcal{B} = \mathcal{I} - r^2\begin{pmatrix} u^2 & v^2 & -2uv \\ v^2 & u^2 & +2uv \\ uv & -uv & u^2 - v^2 \end{pmatrix}.$$

Since $\gamma_{xx} = \gamma_{yy}$ and $u^2 + v^2 = 1$, the solution of equation (11.52) is

$$\Sigma_{xy}(0) = \frac{\sigma_z^2}{1-r^2}\begin{pmatrix} 1 & 0 \\ 0 & 1 \end{pmatrix}.$$

To obtain the lagged cross-covariance matrix $\Sigma_{xy}(\tau)$ by means of equation (11.50), we need to calculate the powers $\mathcal{A}^{\tau}$. To do this we let η be the angle for which

$$u = \cos(2\pi\eta) \text{ and } v = \sin(2\pi\eta),$$

and then note that

$$\mathcal{A}^{\tau} = r^{|\tau|}\begin{pmatrix} \cos(2\pi\tau\eta) & -\sin(2\pi\tau\eta) \\ \sin(2\pi\tau\eta) & \cos(2\pi\tau\eta) \end{pmatrix}.$$

Then

$$\gamma_{xx}(\tau) = \gamma_{yy}(\tau) = \frac{\sigma_z^2 r^{|\tau|}\cos(2\pi\tau\eta)}{1-r^2}$$

and

$$\gamma_{xy}(\tau) = \gamma_{yx}(-\tau) = -\frac{\sigma_z^2 r^{|\tau|}\sin(2\pi\tau\eta)}{1-r^2}.$$

When η is positive (or equivalently, when v is positive), the cross-covariance $\gamma_{xy}(\tau)$ is positive for lags $0 < \tau < 1/(2\eta)$ and negative for lags $-1/(2\eta) < \tau < 0$. Thus, although the variability of the processes is uncorrelated at lag zero, the correlation becomes positive when $\mathbf{X}_t$ *leads* $\mathbf{Y}_t$ (i.e., $\tau > 0$) and negative when $\mathbf{X}_t$ *lags* $\mathbf{Y}_t$ (i.e., $\tau < 0$).

This interpretation can be verified by repeatedly applying matrix $\mathcal{A}$ to vector $(1,0)^{\mathrm{T}}$. For example, we see that

$$\mathcal{A}^{\tau}\begin{pmatrix}1\\0\end{pmatrix} = r^{\tau}\begin{pmatrix}0\\1\end{pmatrix}$$

after $\tau = 1/(4\eta)$ applications. The information that was contained in $\mathbf{X}_t$ is transferred to $\mathbf{Y}_{t+\tau}$. Thus, $\mathbf{X}_t$ *leads* $\mathbf{Y}_t$ for positive vs. Furthermore, we can interpret $1/\eta$ as a rotation 'period' and r as a damping rate. Note that $\tau = -1/\ln r$ applications of $\mathcal{A}$ to a vector of length 1 will reduce its length to $1/e$. This characteristic time is referred to as the *e-folding time.*

Auto- and cross-covariance functions for two processes with rotation time $1/\eta \approx 20$ time units are shown in Figure 11.6. In both processes, **X** leads **Y**. The functions displayed in the left panel belong to a process that is only weakly damped. Its e-folding time $-1/\ln r \approx 100$ time units. Oscillatory behaviour is clearly apparent on the 20-unit time scale. A large proportion of the information that is carried by component **X** (or **Y**) is returned to that component in approximately 20 time steps.

In contrast, the functions displayed in the right panel belong to a process that is strongly damped. Its e-folding time $-1/\ln r \approx 1.4$ time units. The peak in $\gamma_{xy}(\tau)$ that occurs for $\tau = 1, 2$ indicates that the process is attempting to convey some information from the leading component **X** to the lagging component **Y**. However, because damping is strong, not enough information is transferred to initiate oscillatory behaviour.

The characteristics of two processes with a much shorter rotation time of $1/\eta \approx 3.2$ time units are shown in Figure 11.7. Again, **X** leads **Y** in both processes. The e-folding times for these processes are 4.5 and 2.8 time units for left and right panels respectively. The main difference between the two processes is that the auto- and cross-covariance functions decay more quickly in the right panel. On average, information transfer in both processes is sufficient for oscillatory behaviour to develop.

We will revisit these four examples in [11.4.6] when we calculate the spectra of bivariate AR(1)

[10] POPs are 'Principal Oscillation Patterns' (see Chapter 15). Pairs of POP time coefficients are represented by a bivariate AR(1) process with a rotation matrix such as (11.51).

[11] Note that except for its sign, v is completely determined by u, and vice versa, i.e., $v = \pm\sqrt{1-u^2}$.

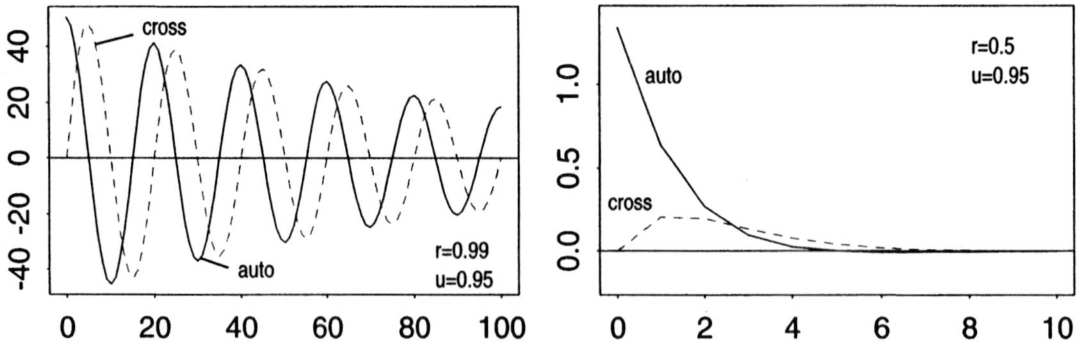

Figure 11.6: *Auto- and cross-covariance functions γ_{xx} (solid) and γ_{xy} (dashed) for two bivariate AR(1) processes with parameter matrix (11.51). The rotation time $1/\eta$ is approximately 20 time units. The e-folding times are approximately 100 time units (left) and 1.4 time units (right). The corresponding power spectra are shown in Figure 11.10.*

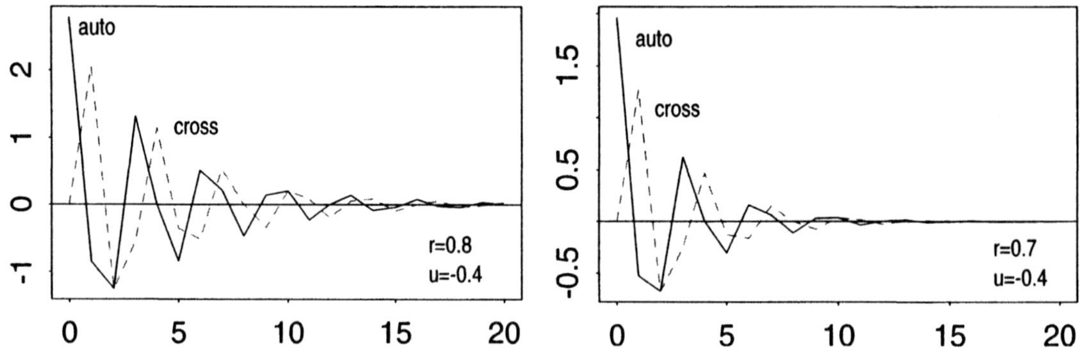

Figure 11.7: *As Figure 11.6, except the rotation time $1/\eta$ is now approximately 3.2 time units. The e-folding times are approximately 4.5 time units (left) and 2.8 time units (right). The corresponding power spectra are shown in Figure 11.10.*

processes that have rotation matrices as their parameters.

11.3.9 Example: Cross-correlation Between an AR(1) Process and its Driving Noise.

The auto-covariance function between an AR(1) process

$$\mathbf{X}_t = \alpha_1 \mathbf{X}_{t-1} + \mathbf{Z}_t \tag{11.53}$$

and its driving white noise $\mathbf{Z}_t$ can be quickly calculated by replacing (11.53) with its infinite moving average representation (see [10.5.2]) or with the mechanics developed above.

The first step in the latter approach is to represent (11.53) as bivariate AR(1) process (11.45)

$$\begin{pmatrix}\mathbf{X}\\ \mathbf{Y}\end{pmatrix}_t = \begin{pmatrix}\alpha_1 & 1\\ 0 & 0\end{pmatrix}\begin{pmatrix}\mathbf{X}\\ \mathbf{Y}\end{pmatrix}_{t-1} + \begin{pmatrix}\mathbf{N}\\ \mathbf{M}\end{pmatrix}_t,$$

where $\mathbf{N}_t = 0$ and $\mathbf{M}_t = \mathbf{Z}_{t+1}$. Then,

$$\Sigma_Z = \begin{pmatrix}0 & 0\\ 0 & \sigma_z^2\end{pmatrix},$$

where $\sigma_z^2 = \mathrm{Var}(\mathbf{Z}_t)$. Using (11.49), we see that the covariance matrix $\Sigma_{xy}(0)$ satisfies

$$\left(\mathcal{I} - \begin{pmatrix}\alpha_1^2 & 1 & 2\alpha_1\\ 0 & 0 & 0\\ 0 & 0 & 0\end{pmatrix}\right)\begin{pmatrix}\gamma_{xx}(0)\\ \gamma_{yy}(0)\\ \gamma_{xy}(0)\end{pmatrix} = \begin{pmatrix}0\\ \sigma_z^2\\ 0\end{pmatrix}$$

so that

$$\Sigma_{xy}(0) = \begin{pmatrix}\sigma_x^2 & 0\\ 0 & \sigma_z^2\end{pmatrix},$$

where $\sigma_x^2 = \sigma_z^2/(1-\alpha_1^2)$. Next, we compute,

$$\mathcal{A}^\tau = \begin{pmatrix}\alpha_1^\tau & \alpha_1^{\tau-1}\\ 0 & 0\end{pmatrix}$$

and finally use (11.50) to find that for positive τ

$$\gamma_{xx}(\tau) = \alpha_1^\tau \sigma_x^2 = \sigma_z^2 \alpha_1^\tau/(1-\alpha_1^2) \qquad (11.54)$$

$$\gamma_{xm}(\tau) = \alpha_1^{\tau-1}\sigma_z^2, \qquad (11.55)$$

$$\gamma_{mx}(\tau) = 0. \qquad (11.56)$$

Equation (11.54) is the auto-covariance function of AR(1) process (11.53) and it holds for all τ. However, note that, since $\mathbf{M}_t$ was defined as $\mathbf{Z}_{t+1}$, equations (11.55, 11.56) describe the cross-covariance function between the AR(1) process and its driving noise one time step in the future. Thus the cross-covariance function of $\mathbf{X}_t$ and $\mathbf{Z}_t$ are given by

$$\gamma_{xz}(\tau) = \alpha_1^\tau \sigma_z^2 \text{ for } \tau \geq 0, \qquad (11.57)$$

$$\gamma_{xz}(\tau) = \gamma_{zx}(-\tau) = 0 \text{ for } \tau < 0. \qquad (11.58)$$

Note that $\gamma_{xz}(\tau)$ is highly non-symmetric. It is nonzero for all *non-negative* lags τ, that is, the current $\mathbf{X}_t$ value 'remembers' the preceding and present noise with a memory that dims exponentially. On the other hand, $\gamma_{xz}(\tau)$ is zero for all negative lags. Hence $\mathbf{X}_t$ 'knows' nothing about future noise.[12]

11.3.10 Pacific SST and SLP. The following example, which is taken from Frankignoul and Hasselmann [133], illustrates that (11.57, 11.58) can be of some practical use.

Frankignoul and Hasselmann considered two indices which are representative of the large-scale monthly variability of sea-surface temperature (SST) and sea-level air pressure (SLP) in the North Pacific. The cross-correlation function $\rho_{SST,SLP}$ estimated from monthly mean data is non-symmetrical with values that are essentially zero for negative lags. Correlations for lags between zero and about six months are positive (Figure 11.8; closed dots connected by thin line segments). A stochastic climate model,[13] which is slightly more complex than the simple AR(1) process (11.53) with $\alpha_1 = 5/6$, was also used to estimate $\rho_{SST,SLP}$. The resulting cross-correlation function (Figure 11.8; open dots connected by heavy line segments) is similar to that computed from the observations, and has structure similar to that predicted by (11.57, 11.58). To a first-order approximation, the North Pacific SST may be seen as an integrated response to atmospheric forcing which is independent of the SST variability.

[12]Some authors also use the term *innovations* to describe the noise processes that force AR processes.

[13]This model is derived from a one-dimensional mixed layer ocean model. See [10.4.3] for more discussion on stochastic climate models.

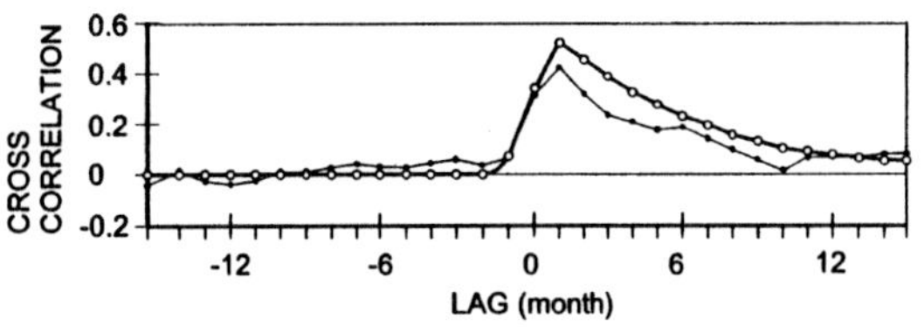

Figure 11.8: *Estimated cross-correlation functions $\rho_{SST,SLP}$ between two monthly indices of the dominant variability of SST and SLP over the North Pacific. One estimate (closed dots connected by a thin line) is estimated from data. A second estimate (open dots connected by a heavy line) is obtained from a stochastic climate model. The SST leads for negative lags. From Frankignoul and Hasselmann [133].*

11.3.11 The Effect of Feedbacks. The continuous version of an AR(1) process is a first-order differential equation of the form[14]

$$\frac{\partial \mathbf{X}_t}{\partial t} = -\lambda \mathbf{X}_t + \mathbf{Z}_t. \qquad (11.59)$$

Unfortunately, equation (11.59) is of limited physical interest because the 'forcing' $\mathbf{Z}_t$ acts on $\mathbf{X}_t$ without feedback. Frankignoul [128] added such feedbacks by replacing (11.59) with a system of two equations

$$\frac{\partial \mathbf{X}_t}{\partial t} = -\lambda_o \mathbf{X}_t + \mathbf{Z}_t + \mathbf{N}_t^x \qquad (11.60)$$

$$\mathbf{Z}_t = \lambda_a \mathbf{X}_t + \mathbf{N}_t^z \qquad (11.61)$$

with two white noise forcing terms $\mathbf{N}_t^x$ and $\mathbf{N}_t^z$. For example, we could think of variable $\mathbf{X}_t$ as SST and variable $\mathbf{Z}_t$ as the turbulent heat flux into the ocean (as in [11.3.4]). Then the change in SST is influenced by its current state (i.e., memory, represented by the parameter $\lambda_o > 0$), by the instantaneous heat flux forcing, and by some random variations. The heat flux, on the other hand, depends on the current SST and random noise induced by the turbulent flow of the atmosphere.

The cross-correlation function between $\mathbf{X}_t$ and $\mathbf{Z}_t$ depends on the value of the 'feedback' parameter λ_a. The following can be shown (cf. [128]).

[14]We will avoid mathematical questions such as the definition of continuous white noise. We use the continuous representation for reasons of convenience, and to clarify the underlying physics. In practice, the derivatives are replaced by finite differences, and the problem of how to define continuous noise, for example, disappears. A good introduction can be found in Koopmans [229].

- There is no feedback when $\lambda_a = 0$. In this case we get the result developed in [11.3.9] and discussed in [11.3.10]. Cross-correlations between SST and heat flux are zero for negative lags and positive for lags $\tau \geq 0$.

- There is a *negative feedback* when $\lambda_a < 0$. When the SST anomaly is positive ($\mathbf{X}_t > 0$), the anomalous heat flux is negative so that the SST-tendency becomes negative (i.e., $\frac{\partial \mathbf{X}_t}{\partial t} < 0$) on average. The cross-correlation function is anti-symmetric in this case.

- There is a *positive feedback* when $\lambda_a > 0$, and the cross-correlation function between $\mathbf{X}_t$ and $\mathbf{Z}_t$ is positive everywhere with a maximum near lag zero.

Thus the cross-covariance functions in Figure 11.5 suggest that, in the extratropics, the heat flux (**Z**) drives the SST (**X**), which in turn exerts a (small) negative feedback on the heat flux.[15] In the tropics the shape of the cross-correlation function suggests that there is weak positive feedback.

11.3.12 Example: Ekman Veering at the Bottom of the Ocean. Kundu [233] describes an interesting application of a *complex* cross-correlation (at lag zero). Theoretical arguments based on the Ekman theory for boundary layers predict that the currents near the bottom of the ocean will veer counter-clockwise in the Northern Hemisphere (i.e., a current close to the bottom of the ocean will be directed somewhat more to the left than a current above).

Kundu [233] used a two-month long time series of current data collected off the Oregon (USA) coast to search for observational evidence supporting the theory. Data from two current meters moored 5 m and 20 m above the bottom was first filtered to eliminate the effects of tidal and inertial modes. The 'veering angle' was then estimated from these data using three approaches.

1 The currents were averaged and the angle spanned by the mean currents 5 m and 20 m above the bottom was computed. The problem with this approach is that the mean can be strongly influenced by a few large events in the time series.

2 The angle between the currents was computed at every observing time. These angles were subsequently averaged. The disadvantage of this approach is that weak currents are associated with highly variable angles. Thus the weak current events substantially increase the uncertainty of the veering angle estimate.

3 The complex 'correlation'[16] between the two complex random variables $\mathbf{X}_t = \mathbf{U}_{5m}(t) + i\,\mathbf{V}_{5m}(t)$ and $\mathbf{Y}_t = \mathbf{U}_{20m}(t) + i\,\mathbf{V}_{20m}(t)$ was estimated. For simplicity we assume $\mathcal{E}(\mathbf{X}_t) = \mathcal{E}(\mathbf{Y}_t) = 0$. Then, the complex correlation is

$$\rho = \frac{\mathcal{E}(\mathbf{X}_t \mathbf{Y}_t^*)}{\sigma_x \sigma_y},$$

where $\sigma_x^2 = \mathcal{E}(\mathbf{X}_t \mathbf{X}_t^*)$ and σ_y^2 is defined similarly. The correlation ρ is then written in polar coordinates as

$$\rho = e^{i\xi} \frac{R}{\sigma_x \sigma_y},$$

where $e^{i\xi} R$ is the complex covariance $\mathcal{E}(\mathbf{X}_t \mathbf{Y}_t^*)$. The angle ξ is used as an estimate of the veering angle; R, σ_x, σ_y, and ξ are estimated from the finite sample in the usual manner by forming sums. Note that estimate of the complex cross-covariance can be written in the form

$$\widehat{e^{i\xi} R} = \sum_j e^{i\xi_j} R_j, \tag{11.62}$$

where R_j is the product $X_j Y_j^*$ expressed in polar coordinate form. Thus, the veering estimate obtained from (11.62) can be interpreted as the mean of all observed angles weighted by the strength of the instantaneous flow.

Kundu [233] obtained veering estimates of 3° from the angle spanned by the mean currents, 7° from the average angle, and 6° from the complex correlation.

11.4 The Cross-spectrum

11.4.0 General. The purpose of cross-spectral analysis is to learn how the variability of two time series is interrelated in the *spectral domain*—that is, to determine the time scales on which variability is related as well as the characteristics of that covariation. Conceptually, we could split a

[15] Note, however, that the similarity of cross-correlation functions is not *proof* that the proposed statistical model, say (11.60, 11.61), is correct.

[16] Kundu did not really calculate the correlation; he did not subtract the mean values.

pair of time series into slowly and quickly varying parts, say

$$\begin{aligned} \mathbf{X}_t &= \mathbf{X}_t^f + \mathbf{X}_t^s \\ \mathbf{Y}_t &= \mathbf{Y}_t^f + \mathbf{Y}_t^s, \end{aligned}$$

where f denotes the *fast* components, and s the *slow* components. We want to know, for example, whether the slow components of $\mathbf{X}_t$ and $\mathbf{Y}_t$ vary together in some way. If at a certain time t there is a 'slow positive (negative) bump,' is there a characteristic time lag τ, such that, on average, there will also be a 'slow positive (negative) bump' in $\mathbf{Y}_{t+\tau}$? If so, the two slow components vary 'coherently' with a 'phase lag' of τ/τ_s, where τ_s is the time scale of the slow variability.

Just as with spectral analysis [11.2.1], our purpose in the next several subsections is to refine these concepts in such a way that the nature of the covariability of a process can be examined over a continuum of time scales.

11.4.1 Definition: The Cross-spectrum. *Let $\mathbf{X}_t$ and $\mathbf{Y}_t$ be two weakly stationary stochastic processes with covariance functions γ_{xx} and γ_{yy}, and a cross-covariance function γ_{xy}. Then the* cross-spectrum Γ_{xy} *is defined as the Fourier transform of γ_{xy}:*

$$\begin{aligned} \Gamma_{xy}(\omega) &= \mathcal{F}\{\gamma_{xy}\}(\omega) \\ &= \sum_{\tau=-\infty}^{\infty} \gamma_{xy}(\tau) e^{-2\pi i \tau\omega} \qquad (11.63) \end{aligned}$$

for all $\omega \in [-1/2, 1/2]$.

The cross-spectrum is generally a complex-valued function since the cross-covariance function is, in general, neither strictly symmetric nor anti-symmetric.

The cross-spectrum can be represented in a number of ways.

1 The cross-spectrum can be decomposed into its real and imaginary parts as

$$\Gamma_{xy}(\omega) = \Lambda_{xy}(\omega) + i\,\Psi_{xy}(\omega).$$

The real and imaginary parts Λ_{xy} and Ψ_{xy} are called the *co-spectrum* and *quadrature spectrum*[17] respectively.

[17] Note that we define the quadrature spectrum as the positive imaginary part of the cross-spectrum. It is also sometimes defined as the negative imaginary part. This choice is arbitrary, but may cause a great deal of confusion in the definition of the frequency–wavenumber spectra, for example (see Section 11.5).

2 The cross-spectrum can be written in polar coordinates as

$$\Gamma_{xy}(\omega) = A_{xy}(\omega)\, e^{i\,\Phi_{xy}(\omega)}.$$

Then A_{xy} and Φ_{xy} are called the *amplitude spectrum* and *phase spectrum* respectively. The amplitude spectrum is given by

$$A_{xy}(\omega) = \left(\Lambda_{xy}(\omega)^2 + \Psi_{xy}(\omega)^2\right)^{1/2}.$$

The phase spectrum is given in three parts:

$$\Phi_{xy}(\omega) = \tan^{-1}\left(\Psi_{xy}(\omega)/\Lambda_{xy}(\omega)\right) \qquad (11.64)$$

when $\Psi_{xy}(\omega) \neq 0$ and $\Lambda_{xy}(\omega) \neq 0$,

$$\Phi_{xy}(\omega) = \begin{cases} 0 & \text{if } \Lambda_{xy}(\omega) > 0 \\ \pm\pi & \text{if } \Lambda_{xy}(\omega) < 0 \end{cases} \qquad (11.65)$$

when $\Psi_{xy}(\omega) = 0$, and

$$\Phi_{xy}(\omega) = \begin{cases} \pi/2 & \text{if } \Psi_{xy}(\omega) > 0 \\ -\pi/2 & \text{if } \Psi_{xy}(\omega) < 0 \end{cases} \qquad (11.66)$$

when $\Lambda_{xy}(\omega) = 0$.

3 The (squared) *coherency spectrum*

$$\kappa_{xy}(\omega) = \frac{A_{xy}^2(\omega)}{\Gamma_{xx}(\omega)\,\Gamma_{yy}(\omega)} \qquad (11.67)$$

expresses the amplitude spectrum in dimensionless units. It is formally similar to a conventional (squared) correlation coefficient.

11.4.2 Some Properties of the Cross-spectrum.

1 The cross-spectrum is bilinear. That is, for jointly weakly stationary processes $\mathbf{X}_t$, $\mathbf{Y}_t$, and $\mathbf{Z}_t$, and arbitrary constants α and β,

$$\Gamma_{\alpha x, \beta y + z}(\omega) = \alpha\beta^*\Gamma_{xy}(\omega) + \alpha\Gamma_{xz}(\omega).$$

This follows from the linearity of the Fourier transformation and the bilinearity of the cross-covariance function (11.44).

2 The cross-covariance function can be recovered from the cross-spectrum by inverting the Fourier transform (11.63)

$$\gamma_{xy}(\tau) = \int_{-\frac{1}{2}}^{\frac{1}{2}} \Gamma_{xy}(\omega) e^{2i\pi\tau\omega}\, d\omega.$$

3 It can be shown that

$$0 \leq \kappa_{xy}(\omega) \leq 1.$$

11.4.3 Properties of the Cross-spectrum of Real Processes. Let $\mathbf{X}_t$ and $\mathbf{Y}_t$ be a pair of real-valued stochastic processes that are jointly weakly stationary. Then the following additional properties hold.

1 The co-spectrum is the Fourier transform of the symmetric part of the cross-covariance function, $\gamma^s_{xy}(\tau)$, and the quadrature spectrum is the Fourier transform of the anti-symmetric part of the cross-covariance function, $\gamma^a_{xy}(\tau)$. That is,

$$\Lambda_{xy}(\omega) = \gamma_{xy}(0) + 2\sum_{\tau=1}^{\infty}\gamma^s_{xy}(\tau)\cos(2\pi\tau\omega) \tag{11.68}$$
$$\Psi_{xy}(\omega) = -2\sum_{\tau=1}^{\infty}\gamma^a_{xy}(\tau)\sin(2\pi\tau\omega).$$

The symmetric and anti-symmetric parts of the cross-covariance function are given by

$$\gamma^s_{xy}(\tau) = \frac{1}{2}\big(\gamma_{xy}(\tau) + \gamma_{xy}(-\tau)\big)$$
$$\gamma^a_{xy}(\tau) = \frac{1}{2}\big(\gamma_{xy}(\tau) - \gamma_{xy}(-\tau)\big).$$

2 Therefore, the co-spectrum is symmetric

$$\Lambda_{xy}(\omega) = \Lambda_{xy}(-\omega)$$

and the quadrature spectrum is anti-symmetric

$$\Psi_{xy}(\omega) = -\Psi_{xy}(-\omega).$$

3 When the cross-covariance function is symmetric (i.e., $\gamma_{xy} = \gamma_{yx}$), the quadrature and phase spectra are zero for all ω.

When the cross-covariance function is anti-symmetric (i.e., $\gamma_{xy} = -\gamma_{yx}$), the co-spectrum vanishes and the phase spectrum is $\Phi_{xy}(\omega) = -\frac{\pi}{2}\operatorname{sgn}(\Psi_{xy}(\omega))$, where $\operatorname{sgn}(\cdot)$ is the sign function.

4 The amplitude spectrum is positive and symmetric, and the phase spectrum is anti-symmetric, that is,

$$A_{xy}(\omega) = A_{xy}(-\omega) \ge 0 \tag{11.69}$$
$$\Phi_{xy}(\omega) = -\Phi_{xy}(-\omega).$$

5 It follows from (11.67) and (11.69) that the coherency spectrum is symmetric,

$$\kappa_{xy}(\omega) = \kappa_{xy}(-\omega).$$

6 Since $\gamma_{yx}(\tau) = \gamma_{xy}(-\tau)$ (equation (11.42)), we have

$$\Gamma_{yx}(\omega) = \Gamma^*_{xy}(\omega), \tag{11.70}$$
$$\Lambda_{yx}(\omega) = \Lambda_{xy}(\omega)$$
$$\Psi_{yx}(\omega) = -\Psi_{xy}(\omega)$$
$$A_{yx}(\omega) = A_{xy}(\omega)$$
$$\Phi_{yx}(\omega) = -\Phi_{xy}(\omega)$$
$$\kappa_{yx}(\omega) = \kappa_{yx}(\omega).$$

Thus it is sufficient to consider, and to plot, spectra only for positive ω, if the processes are real.

11.4.4 Some Simple Examples. We described the cross-covariance functions of a number of simple processes in [11.3.3]. We present the cross-spectra of these processes here.

- $\mathbf{Y}_t = \alpha\mathbf{X}_t$. From (C.7) and because $\gamma_{\alpha x,x} = \alpha\gamma_{xx}$ (see (11.29)), the cross-spectrum is a simple function of the spectrum of $\mathbf{X}$:

$$\Gamma_{xy}(\omega) = \alpha\Gamma_{xx}(\omega) \tag{11.71}$$
$$\Gamma_{yy}(\omega) = \alpha^2\Gamma_{xx}(\omega)$$
$$\Lambda_{xy}(\omega) = \alpha\Gamma_{xx}(\omega)$$
$$\Psi_{xy}(\omega) = 0$$
$$A_{xy}(\omega) = \alpha\Gamma_{xx}(\omega)$$
$$\Phi_{xy}(\omega) = 0$$
$$\kappa_{xy}(\omega) = 1.$$

These are intuitively reasonable results. All events in the two time series occur synchronously, thus the phase spectrum is zero everywhere and the coherency spectrum is one for all ω.

- Recall that we also considered the slightly more complex case in which $\mathbf{Y}$ is composed of a scaled version of $\mathbf{X}$ plus white noise $\mathbf{Z}$, as

$$\mathbf{Y}_t = \alpha\mathbf{X}_t + \mathbf{Z}_t.$$

Equations (11.30) and (11.31) show that the cross-, co-, quadrature, amplitude, and phase spectra are unaffected by the added noise. However, the power spectrum of $\mathbf{Y}$, and therefore the coherency spectrum, do change. Specifically,

$$\Gamma_{yy}(\omega) = \alpha^2\Gamma_{xx}(\omega) + \sigma_Z^2$$
$$\kappa_{xy}(\omega) = \frac{\alpha^2\Gamma_{xx}(\omega)}{\sigma_Z^2 + \alpha^2\Gamma_{xx}(\omega)} < 1.$$

The coherency is now less than 1 at all time scales, indicating that knowledge of the sequence of the events in **X** is no longer enough to completely specify the sequence of events in **Y**. The impact of the noise is small if its variance is small relative to that of $\alpha\mathbf{X}_t$ (and vice versa).

- When we shifted $\mathbf{X}_t$ by a fixed lag ζ so that

$$\mathbf{Y}_t = \mathbf{X}_{t+\zeta}$$

we found (11.32) that $\gamma_{xy}(\tau) = \gamma_{xx}(\zeta+\tau)$. Using (C.8), we find that

$$\begin{aligned}
\Gamma_{xy}(\omega) &= e^{i2\pi\zeta\omega}\Gamma_{xx}(\omega) \qquad (11.72)\\
\Gamma_{yy}(\omega) &= \Gamma_{xx}(\omega)\\
\Lambda_{xy}(\omega) &= \cos(2\pi\zeta\omega)\Gamma_{xx}(\omega)\\
\Psi_{xy}(\omega) &= \sin(2\pi\zeta\omega)\Gamma_{xx}(\omega)\\
A_{xy}(\omega) &= \Gamma_{xx}(\omega)\\
\Phi_{xy}(\omega) &= 2\pi\zeta\omega\\
\kappa_{xy}(\omega) &= 1.
\end{aligned}$$

When we shift $\mathbf{X}_t$ a fixed number of lags we obtain the same coherency spectrum as when $\mathbf{X}_t$ is simply scaled. It is 1 for all time scales meaning that the sequence of events in **Y** is completely determined by **X**. In contrast, the phase spectrum has changed from being zero for all ω to a linear function of ω. This type of linear dependency is characteristic of shifts that are independent of the time scale.

Note that if the process **X** *lags* the process **Y**(i.e., if $\zeta > 0$), then the phase spectrum Φ_{xy} is *positive* for positive frequencies.[18]

- We also considered the first difference

$$\mathbf{Y}_t = \mathbf{X}_t - \mathbf{X}_{t-1}$$

that approximates a discretized time derivative. Recall from (11.35) that

$$\begin{aligned}
\gamma_{xy}(\tau) &= \gamma_{xx}(\tau) - \gamma_{xx}(\tau-1)\\
\gamma_{yy}(\tau) &= 2\gamma_{xx}(\tau)\\
&\quad - (\gamma_{xx}(\tau-1) + \gamma_{xx}(\tau+1).
\end{aligned}$$

[18]The definition of the phase is arbitrary to some extent, and thus some care is needed. We say that **Y** *leads* **X** when certain 'events' in **Y** are followed by similar events in **X** at a later time (i.e., $\mathbf{X}_{t+\zeta} \approx \mathbf{Y}_t$). With this definition, the phase difference Φ_{yx} is positive. At the same time **X** *lags* **Y**, $\mathbf{X}_t \approx \mathbf{Y}_{t-\zeta}$, and the phase difference Φ_{xy} is negative.

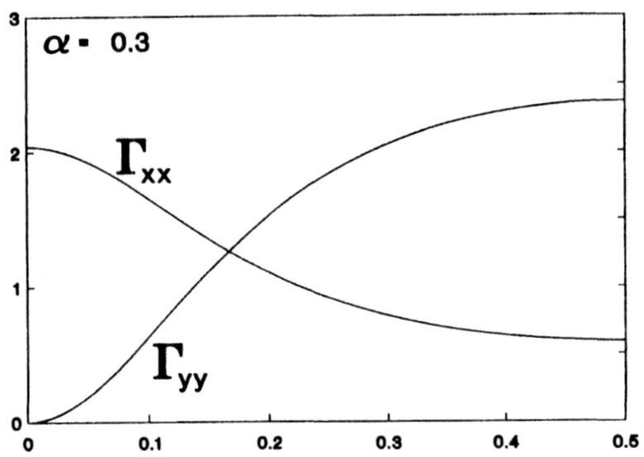

Figure 11.9: *Spectra Γ_{xx} and Γ_{yy} of an AR(1) process $\mathbf{X}_t$ with $\alpha = 0.3$ and the differenced process $\mathbf{Y}_t = \mathbf{X}_t - \mathbf{X}_{t-1}$. Note that the differencing acts as a high-pass filter.*

Thus, again using (C.8),

$$\begin{aligned}
\Gamma_{xy}(\omega) &= (1 - e^{-2\pi i\omega})\Gamma_{xx}(\omega) \qquad (11.73)\\
\Gamma_{yy}(\omega) &= 2(1-\cos(2\pi\omega))\Gamma_{xx}(\omega)\\
\Lambda_{xy}(\omega) &= (1-\cos(2\pi\omega))\Gamma_{xx}(\omega)\\
\Psi_{xy}(\omega) &= \sin(2\pi\omega)\Gamma_{xx}(\omega)\\
A^2_{xy}(\omega) &= 2(1-\cos(2\pi\omega))\Gamma_{xx}(\omega)^2\\
&= \Gamma_{xx}(\omega)\Gamma_{yy}(\omega)\\
\Phi_{xy}(\omega) &= \tan^{-1}\left(\frac{\sin(2\pi\omega)}{1-\cos(2\pi\omega)}\right)\\
&= \tan^{-1}(\cot(\pi\omega))\\
&= \pi\left(\omega - \frac{1}{2}\right) \le 0 \text{ for } \omega \ge 0\\
\kappa_{xy}(\omega) &= 1 \text{ for } \omega \ne 0.
\end{aligned}$$

Several things can be noted here.

i) The coherency is 1 at all time scales except 0. This is reasonable since integration can undo differentiation up to a constant.

ii) The spectrum of the differenced process **Y** has more short time scale variability than the spectrum of the original process **X**. Indeed, differencing acts as a high-pass filter that dampens long time scale variability and eliminates the time mean ($\Gamma_{yy}(0) = 0$). For example, Figure 11.9 displays the spectrum of an AR(1) process $\mathbf{X}_t$ with $\alpha = 0.3$ and that of the differenced process $\mathbf{Y}_t = \mathbf{X}_t - \mathbf{X}_{t-1}$. The **X**-spectrum is 'red' with a maximum at zero frequency whereas the **Y**-spectrum is 'blue' with a maximum at frequency 1/2.

iii) 'Physical reasoning' suggests that the forcing should lead the response[19] in the sense that the phase lag Φ_{yx} between the 'forcing' **Y** and the 'response' **X** is $\pi/2$. This is approximately the case for the long time scales near $\omega = 0$, since $\Phi_{xy}(0) = -\pi/2$. The phase converges towards zero on shorter time scales. This effect occurs because the time derivative is only approximated by the time difference, and the accuracy of this approximation increases with the time scale.

- Now consider again process (11.37)

$$F(\mathbf{X})_t = \sum_{k=-\infty}^{\infty} a_k \mathbf{X}_{t+k}.$$

which is obtained by passing a weakly stationary stochastic process through a linear filter. The cross-covariance function of the two processes $F(\mathbf{X})_t$ and $\mathbf{X}_t$ is (11.39)

$$\gamma_{x,F(x)}(\tau) = \sum_{k=-\infty}^{\infty} a_k^* \gamma_{xx}(\tau + k).$$

The cross-spectra are then (cf. (C.17))

$$\begin{aligned} \Gamma_{x,F(x)}(\omega) &= \mathcal{F}\{a\}^*(\omega)\Gamma_{xx}(\omega) \\ \Gamma_{F(x),x}(\omega) &= \mathcal{F}\{a\}(\omega)\Gamma_{xx}(\omega), \end{aligned} \qquad (11.74)$$

where $\mathcal{F}\{a\}(\omega)$ is the Fourier transform of the sequence of filter coefficients $\{a_k : k \in \mathbb{Z}\}$. The examples discussed above and in [11.4.3] can all be cast in a linear filter format. In particular, note the following.

i) When $\mathbf{Y}_t = \alpha \mathbf{X}_t$, the sequence of filter coefficients are $a_0 = \alpha$ and $a_k = 0$ for $k \neq 0$. Thus $\mathcal{F}\{a\}^*(\omega) = \sum_k a_k^* e^{2\pi i k\omega} = \alpha$ for all ω, and hence $\Gamma_{xy}(\omega) = \alpha\Gamma_{xx}(\omega)$.

ii) When $\mathbf{Y}_t = \alpha \mathbf{X}_{t+\zeta}$, the filter is determined by $a_k = 0$ for all $k \neq \zeta$, and $a_\zeta = 1$. The complex conjugate of the Fourier transform of this series is $\mathcal{F}\{a\}^*(\omega) = e^{2\pi i\tau\omega}$ (cf. (11.72)).

iii) When $\mathbf{Y}_t = \mathbf{X}_t - \mathbf{X}_{t-1}$, $a_0 = 1$, $a_{-1} = -1$, and $a_k = 0$ for all $k \neq 0, -1$. The complex conjugate of the Fourier transform of this filter is $\mathcal{F}\{a\}^*(\omega) = 1 - e^{-2\pi i\omega}$ (cf. (11.73)).

[19] The 'physical' argument is as follows. Suppose $dX/dt = Y$ where $Y = A\cos(\omega t)$. Then $X = A/\omega \cos(\omega t + \Phi_{xy})$ where $\Phi_{xy} = -\frac{\pi}{2}$.

- Finally, the cross-spectrum of two filtered processes $F(\mathbf{X}_t)$ and $G(\mathbf{Y}_t)$ is given by

$$\Gamma_{F(x)G(y)} = \mathcal{F}\{a\}\Gamma_{xy}\mathcal{F}\{b\}^*. \qquad (11.75)$$

11.4.5 The Spectrum of a Multivariate AR(p) Process. The following general representation of the spectrum of a multivariate AR(p) process will be useful when describing the spectra of a bivariate AR process. Let $\vec{\mathbf{X}}_t$ be a weakly stationary ℓ-dimensional AR(p) process

$$\vec{\mathbf{X}}_t = \sum_{k=1}^{p} \mathcal{A}_k \vec{\mathbf{X}}_{t-k} + \vec{\mathbf{Z}}_t.$$

The $\ell \times \ell$ *spectral matrix* $\Gamma_{\vec{x}\vec{x}}(\omega)$ of $\vec{\mathbf{X}}_t$ is constructed by placing the power spectra of the elements of $\vec{\mathbf{X}}_t$ on the diagonal and by placing cross-spectra in the off-diagonal positions. Note that for any two elements $\mathbf{X}_{it}$ and $\mathbf{X}_{jt}$ of $\vec{\mathbf{X}}_t$, $\Gamma_{x_i x_j}(\omega) = \Gamma^*_{x_j x_i}(\omega)$ (see equation (11.70)). Thus the matrix function $\Gamma_{\vec{x}\vec{x}}$ is Hermitian. It can be shown (see Jenkins and Watts [195, p. 474]) that

$$\Gamma_{\vec{x}\vec{x}}(\omega) = \mathcal{B}(\omega)^{-1}\Gamma_{\vec{z}\vec{z}}(\omega)\left(\mathcal{B}(\omega)^*\right)^{-1} \qquad (11.76)$$

where

$$\mathcal{B}(\omega) = \mathcal{I} - \sum_{k=1}^{p} \mathcal{A}_k e^{ik2\pi\omega}$$

is the characteristic polynomial of the process evaluated at $e^{i2\pi\omega}$, $\mathcal{I}$ is the $\ell \times \ell$ identity matrix, and $\Gamma_{\vec{z}\vec{z}}(\omega) = \Sigma_{\vec{z}}$ is the spectral matrix of the multivariate white noise process that drives $\vec{\mathbf{X}}_t$. The '$*$' is used to denote the conjugate transpose operation. We evaluate (11.76) for a bivariate AR(1) process in the next subsection.

11.4.6 Cross-spectrum of a Bivariate AR(1) Process. We assume, in the following, that the bivariate AR(1) process $(\mathbf{X}_t, \mathbf{Y}_t)^{\mathrm{T}}$ (11.45) has been transformed to coordinates in which the variance covariance matrix of the driving noise $\vec{\mathbf{Z}}_t$ has the form

$$\Sigma_{\vec{z}} = \sigma^2 \begin{pmatrix} 1 & 0 \\ 0 & b \end{pmatrix}.$$

For AR(1) processes, matrix function $\mathcal{B}(\omega)$ is given by

$$\mathcal{B}(\omega) = \mathcal{I} - \mathcal{A}\zeta$$

where $\zeta = e^{i2\pi\omega}$. Thus, from (11.76), we see that the spectral matrix of the process is given by

$$\Gamma_{\vec{x}\vec{x}}(\omega) = \frac{\sigma^2}{D^2}(\mathcal{I} - \mathcal{A}_j\zeta)\begin{pmatrix} 1 & 0 \\ 0 & b \end{pmatrix}(\mathcal{I} - \mathcal{A}_j\zeta)^* \qquad (11.77)$$

where D is the modulus of the determinant of $\mathcal{B}$ (i.e., $D = |\det(\mathcal{B})|$), and $\mathcal{A}_j$ is the adjoint of the coefficient matrix

$$\mathcal{A}_j = \begin{pmatrix} \alpha_{yy} & -\alpha_{xy} \\ -\alpha_{yx} & \alpha_{xx} \end{pmatrix}.$$

After some manipulation, we find

$$\begin{aligned} D^2 = 1 &+ (\alpha_{xx} + \alpha_{yy})^2 \\ &+ (\alpha_{xx}\alpha_{yy} - \alpha_{xy}\alpha_{yx})^2 - 2(\alpha_{xx} + \alpha_{yy}) \\ &\times (1 + \alpha_{xx}\alpha_{yy} - \alpha_{xy}\alpha_{yx})\cos(2\pi\omega) \\ &+ 2(\alpha_{xx}\alpha_{yy} - \alpha_{xy}\alpha_{yx})\cos(4\pi\omega). \end{aligned}$$

The spectra are consequently derived from equation (11.77) as

$$\begin{aligned} \Gamma_{xx}(\omega) &= \frac{\sigma^2}{D^2}\big((1 - \alpha_{yy}\zeta)(1 - \alpha_{yy}\zeta^*) \\ &\qquad + b\alpha_{xy}^2\zeta\zeta^*\big) \\ &= \frac{\sigma^2}{D^2}\big(1 + \alpha_{yy}^2 + b\alpha_{xy}^2 \\ &\qquad - 2\alpha_{yy}\cos(2\pi\omega)\big), \\ \Gamma_{yy}(\omega) &= \frac{\sigma^2}{D^2}\big(b + b\alpha_{xx}^2 + \alpha_{yx}^2 \\ &\qquad - 2b\alpha_{xx}\cos(2\pi\omega)\big) \end{aligned}$$

and

$$\begin{aligned} \Gamma_{xy}(\omega) &= \frac{\sigma^2}{D^2}\big(\alpha_{yx}\zeta^*(1 - \alpha_{yy}\zeta) \\ &\qquad + b(1 - \alpha_{xx}\zeta^*)\alpha_{xy}\zeta\big) \\ &= \frac{\sigma^2}{D^2}\big(\alpha_{yx}\zeta^* + b\alpha_{xy}\zeta \\ &\qquad - (\alpha_{yx}\alpha_{yy} + b\alpha_{xx}\alpha_{xy})\zeta\zeta^*\big) \\ &= \Lambda_{xy}(\omega) + i\,\Psi_{xy}(\omega), \end{aligned}$$

where the co-spectrum and quadrature spectrum are given by

$$\begin{aligned} \Lambda_{xy}(\omega) &= \frac{\sigma^2}{D^2}\big((b\alpha_{xy} + \alpha_{yx})\cos(2\pi\omega) \\ &\qquad - (\alpha_{yx}\alpha_{yy} + b\alpha_{xx}\alpha_{xy})\big) \\ \Psi_{xy}(\omega) &= \frac{\sigma^2}{D^2}(b\alpha_{xy} - \alpha_{yx})\sin(2\pi\omega). \end{aligned}$$

Note that, in all of these expressions, D^2 is a function of ω.

11.4.7 Cross-spectra of Some Special AR(1) Processes. The spectra described above are easily computed for a number of special AR(1) processes, three of which are described by Luksch, von Storch, and Hayashi [262]. These models are briefly described here, and we revisit them in [11.5.5], [11.5.8], and [11.5.11].

In the first of Luksch's examples, the two components of the AR(1) process are not connected. Also, one process is red, and the other is white. That is,

$$\mathcal{A} = \begin{pmatrix} \alpha & 0 \\ 0 & 0 \end{pmatrix}.$$

Then $D^2 = 1 + \alpha^2 - 2\alpha\cos(2\pi\omega)$ and the spectra are

$$\begin{aligned} \Gamma_{xx}(\omega) &= \sigma^2/D^2 \\ &= \frac{\sigma^2}{1 + \alpha^2 - 2\alpha\cos(2\pi\omega)} \qquad (11.78) \\ \Gamma_{yy}(\omega) &= \frac{\sigma^2}{D^2}b\left(1 + \alpha^2 - 2\alpha\cos(2\pi\omega)\right) \\ &= \sigma^2 b. \qquad (11.79) \end{aligned}$$

All other spectra, such as the cross-spectrum and the coherency spectrum, are zero. The results (11.78), (11.79) are, of course, identical to (11.22), (11.23).

Luksch's second example features two independent AR(1) processes with the same parameter, that is,

$$\mathcal{A} = \begin{pmatrix} \alpha & 0 \\ 0 & \alpha \end{pmatrix},$$

and with noise forcing of equal variance (i.e., $b = 1$). Then

$$\begin{aligned} D^2 &= 1 + 4\alpha^2 + \alpha^4 - 4(1 + \alpha^2)\alpha\cos(2\pi\omega) \\ &\qquad + 2\alpha^2\cos(4\pi\omega) \\ &= \left(1 + \alpha^2 - 2\alpha\cos(2\pi\omega)\right)^2 \end{aligned}$$

and

$$\begin{aligned} \Gamma_{xx}(\omega) &= \frac{\sigma^2}{D^2}\left(1 + \alpha^2 - 2\alpha\cos(2\pi\omega)\right) \\ &= \frac{\sigma^2}{1 + \alpha^2 - 2\alpha\cos(2\pi\omega)} \\ &= \Gamma_{yy}(\omega). \end{aligned}$$

This result is identical to (11.23) since the bivariate process considered here is composed of two independent but identical AR(1) processes.

11.4.8 Cross-spectra for the POP Process. Luksch's third example is an AR(1) process with a rotational parameter matrix and with noise components of equal variance so that $b = 1$. Recall that a 2×2 rotational parameter matrix has the form

$$\mathcal{A} = r\begin{pmatrix} u & -v \\ v & u \end{pmatrix} \tag{11.80}$$

where $u^2 + v^2 = 1$ and $0 < r < 1$.

When matrix (11.80) is applied to a vector $\vec{a}$, it rotates that vector through η radians into $\vec{b} = \mathcal{A}\vec{a}$, where $\cos(2\pi\eta) = u$. The rotated vector is returned to its initial direction by applying the matrix (11.80) $T = 1/\eta$ times. When the 'damping' and the noise are switched off (i.e., $r = 1$ and $\sigma = 0$), the system oscillates with period

$$T = 2\pi/\cos^{-1}(u)$$

or, equivalently, frequency $\eta = 1/T$. Note that $\eta < 1/4$ (and $T > 4$) when u is positive and that $\eta > 1/4$ (and $T < 4$) when u is negative. The direction of rotation is determined by v (see [11.4.9]).

The auto- and cross-covariance functions for this process are given in [11.3.8]. The spectra are given by

$$\begin{aligned}
\Gamma_{xx}(\omega) &= \frac{\sigma^2}{D^2}\big(1 + r^2 - 2ru\cos(2\pi\omega)\big) \\
&= \Gamma_{yy}(\omega) \\
\Lambda_{xy}(\omega) &= 0 \\
\Psi_{xy}(\omega) &= -2rv\sigma^2\sin(2\pi\omega)/D^2 \\
A_{xy}(\omega) &= |\Psi_{xy}(\omega)| \\
\Phi_{xy}(\omega) &= \begin{cases} -\pi/2 & \text{if } v < 0 \\ \pi/2 & \text{if } v > 0 \\ \text{undefined} & \text{if } v = 0 \end{cases} \\
\kappa_{xy}(\omega) &= \left(\frac{2rv\sin(2\pi\omega)}{1 + r^2 - 2ru\cos(2\pi\omega)}\right)^2,
\end{aligned} \tag{11.81}$$

where

$$\begin{aligned}
D^2 &= 1 + 4r^2u^2 + r^4 - 4ru(1 + r^2)\cos(2\pi\omega) \\
&\quad + 2r^2\cos(4\pi\omega).
\end{aligned} \tag{11.82}$$

The coherency spectrum has a maximum at

$$\omega_0 = \frac{1}{2\pi}\cos^{-1}\left(\frac{2ru}{1 + r^2}\right).$$

The frequency with maximum coherency approximates the oscillation frequency η. These frequencies coincide exactly only when u is zero. In general, they are different because $2r/(1+r^2) < 1$.

In practice, however, the oscillation frequency η is often diagnosed as the frequency with maximum coherency ω_0. For the POP-case we find

$$\begin{aligned}
\omega_0 &> \eta \qquad \text{for } u > 0 \ \ (\eta < 1/4) \\
\omega_0 &< \eta \qquad \text{for } u < 0 \ \ (\eta > 1/4).
\end{aligned}$$

That is, the coherency maximum underestimates the 'deterministic period' T when the deterministic frequency is low (i.e., $\eta < 1/4$) and it overestimates T for high deterministic frequencies. The discrepancy between the deterministic period and the frequency of maximum coherency increases as the 'damping' coefficient r decreases. In the limit as r tends to zero, the maximum of the coherency spectrum (which is also decreasing in magnitude) converges towards $1/4$ independently of the value of u.

Power and coherency spectra are shown in Figure 11.10 for processes with a number of combinations of r and u. Coherency spectra (dashed curves) and power spectra (solid curves) are displayed for processes with oscillation frequencies $\eta = 0.050$ ($u = 0.95$; top row) and $\eta = 0.315$ ($u = -0.4$; bottom row). The location of the deterministic period η is indicated by the vertical bar at $\eta = (2\pi)^{-1}\cos^{-1}(u)$. The same examples were discussed in [11.3.8].

Damping is almost absent in the $r = 0.99$, $u = 0.95$ case. The power spectrum has a pronounced peak at $\eta = \cos(u)/2\pi$ and the coherency spectrum peaks at about the same frequency. Both processes have maximum 'energy' and vary coherently at the $T = 2\pi/\cos^{-1}(u)$ time scale. The second example has the same u and thus has the same 'period' as the first case, namely $T \approx 20$. However, much more damping occurs with $r = 0.5$. Neither spectrum has a maximum at $2\pi/T$. Instead the power spectrum is red with a maximum at zero frequency, and the coherency spectrum peaks, with a very small maximum, at about 0.1. The strong damping almost obliterates the connection between the two components of the process.

The lower two panels display spectra for two processes with a deterministic time scale $T \approx 3.2$ and slightly different damping rates. We see that the power spectra are substantially affected by the change in damping between the two processes but that there are only subtle differences between the coherency spectra. They show that the components of these processes tend to vary coherently on a wide range of time scales. Their maxima coincide well with η in both cases, although agreement is slightly better for the process with the lower damping rate ($r = 0.8$).

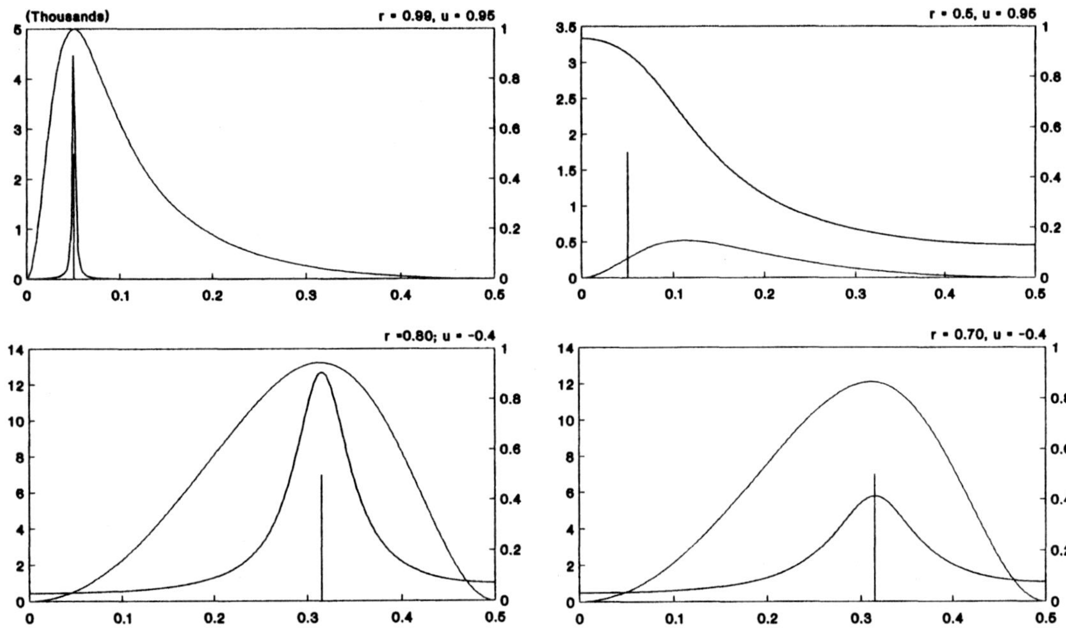

Figure 11.10: *Power spectra $\Gamma_{xx} = \Gamma_{yy}$ (heavy line, left hand axis) and coherency spectra κ_{xy} (light line, right hand axis) of four bivariate AR(1) processes with parameter matrices $\mathcal{A}$ given by (11.80). See text for details.*

11.4.9 The POP Process: The Role of v. Parameter v in the rotation matrix $\mathcal{A}$ (11.80) determines the direction of rotation. Suppose, for convenience, that there is no damping. When $v > 0$, repeated application of $\mathcal{A}$ smoothly transforms the initial state $(x, y) = (1, 0)$ into the state $(x, y) = (0, 1)$ in a quarter of a period T. Continued application of $\mathcal{A}$ then transforms $(x, y) = (0, 1)$ into $(x, y) = (-1, 0)$ in the next quarter period. Thus, for positive numbers v, the system tends to create sequences in the $(\mathbf{X},\mathbf{Y})$-space of the form

$$\begin{aligned}\cdots \rightarrow \begin{pmatrix}1\\0\end{pmatrix} \rightarrow \begin{pmatrix}0\\1\end{pmatrix} \rightarrow \begin{pmatrix}-1\\0\end{pmatrix} \\ \rightarrow \begin{pmatrix}0\\-1\end{pmatrix} \rightarrow \begin{pmatrix}1\\0\end{pmatrix} \rightarrow \begin{pmatrix}0\\1\end{pmatrix} \rightarrow \cdots\end{aligned} \tag{11.83}$$

The sign of v does not affect the auto- and coherency spectra shown in Figure 11.10. However, the phase spectrum is affected. When v is positive the phase spectrum (11.81) is *positive* (for positive frequencies ω), which is consistent with the interpretation that $\mathbf{X}$ *leads* $\mathbf{Y}$.

The opposite interpretation holds when $v < 0$: $\mathbf{X}$ *lags* $\mathbf{Y}$ and the characteristic sequences are

$$\begin{aligned}\cdots \rightarrow \begin{pmatrix}0\\1\end{pmatrix} \rightarrow \begin{pmatrix}1\\0\end{pmatrix} \rightarrow \begin{pmatrix}0\\-1\end{pmatrix} \\ \rightarrow \begin{pmatrix}-1\\0\end{pmatrix} \rightarrow \begin{pmatrix}0\\1\end{pmatrix} \rightarrow \begin{pmatrix}1\\0\end{pmatrix} \rightarrow \cdots\end{aligned} \tag{11.84}$$

These ideas resurface in the next section when we deal with *eastward* and *westward travelling waves*.

11.5 Frequency–Wavenumber Analysis

11.5.1 Introduction. Wave-like processes play an important role in the dynamics of geophysical fluids. Physical processes exhibit standing waves with maxima, minima, and nodes at fixed locations, propagating waves with wave crests that move in space, and mixed forms that are a combination of the two.

Waves are often readily described with trigonometric functions such that at any given time t the wave field $f(x, t)$ can be expanded into sines and cosines as

$$\begin{aligned}f(x, t) = \sum_{k=0}^{\infty} \Big(& c_{kt} \cos\Big(\frac{2\pi k x}{L}\Big) \\ & + s_{kt} \sin\Big(\frac{2\pi k x}{L}\Big)\Big)\end{aligned} \tag{11.85}$$

where L is some reference length such as the circumference of Earth at a given latitude. The coefficients c_{kt} and s_{kt} are given by

$$c_{kt} = \begin{cases} \int_0^L f(x,t)\,dx & k=0 \\ 2\int_0^L f(x,t)\cos(\frac{2\pi kx}{L})\,dx & k>0 \end{cases}$$

$$s_{kt} = \begin{cases} 0 & k=0 \\ 2\int_0^L f(x,t)\sin(\frac{2\pi kx}{L})\,dx & k>0. \end{cases}$$

Index k is known as the *wavenumber*. The time-dependent coefficients c_{kt} and s_{kt} sometimes oscillate with a period that is conditional upon the wavenumber k (e.g., Rossby waves). Functions that relate the variation of the period with the wavenumber are commonly referred to as *dispersion relations* because they relate a spatial scale, namely L/k, to a time scale.

It is useful to look for dispersion relationships in observed data, either to support a dynamical theory that predicts dispersion relationships or as a diagnostic that may ultimately lead to the detection of wave-like dynamics.

Frequency–wavenumber analysis, or *space–time spectral analysis*, is a tool that can be used to diagnose possible relationships between spatial and time scales. The original concept, developed by Deland [102] and Kao [211, 212], assumed that the wave field evolved in a deterministic way. Hayashi [169, 170] and Pratt [319] adapted the method by accounting for the stochastic nature of the analysed fields.

There are many examples of applications of the frequency–wavenumber analysis. For example, Hayashi and Golder [171] studied the Madden-and-Julian Oscillation with this tool. Also, many workers, including Fraedrich and coworkers [124, 126] and Speth, Madden, and others [353, 354, 419], have analysed the frequency-wavenumber spectrum of the extratropical height field.

11.5.2 The Four Steps. Frequency-wavenumber analysis is performed in four steps.

1 The field of interest is expanded into a series of sine and cosine functions (11.85).

   The field (e.g., an atmospheric process on a latitude circle), is assumed to be spatially periodic.

2 The bivariate time series, composed of the time-dependent sine and cosine coefficients c_{kt} and s_{kt}, is assumed to be a random realization of a bivariate stochastic process.

   The cross-spectrum of this process is estimated.

3 The cross-spectrum is separated into components representing the variance of eastward and westward travelling waves.

   The methods used to perform the separation are derived using heuristic arguments.[20] The total variance is assumed to consist of only 'eastward' and 'westward travelling' variance. The total variance is split up into equal contributions from eastward and westward 'travelling waves' when the processes are generated by white noise, or by non-propagating features (see examples in [11.5.5]). This seems reasonable when there are standing features that can be thought of as the sum of coherent waves that propagate in opposite directions. However, one might be skeptical about applying this approach to stochastic processes since white noise, for example, does not contain 'travelling waves.' We can live with these ambiguities in the scientific lexicon if the limitations are asserted and understood. However, use of this slang without also presenting the caveats leaves plenty of opportunity to misinterpret results.

4 Additional heuristic arguments are used to assign a part of the overall variance to standing waves.

   Pratt [319] interprets the modulus of the difference between westward and eastward travelling wave variance as 'propagating variance' and labels the remainder as 'standing wave' variance. With this interpretation, the standing variance comprises all truly standing waves plus all random fluctuations. Depending upon the sign of the difference between the eastward and westward travelling wave variance, the propagating variance is interpreted as being either purely 'eastward' or 'westward' variance.

   Hayashi [170] attributes the coherent part of the eastward and westward travelling variance to standing waves. The incoherent part is interpreted as eastward or westward propagating variance. Thus the total variance is split up into three compartments: standing waves, eastward propagating waves, and westward propagating waves. The propagating variance is described by a two-sided

[20] The adjective 'heuristic' describes an argument that is not rigorously logical or complete and may be supported by ad-hoc assumptions.

spectrum and the standing variance by a one-sided spectrum.

Unfortunately, Hayashi's partitioning of variance from two sources, the sine and cosine coefficient time series, into three components is not well-defined mathematically. It is even possible to obtain negative variances with this partitioning under some conditions. There is probably no universal method of partitioning space–time variance into standing and propagating components.

The expression *space–time spectral analysis* indicates that the method requires a spectral decomposition of the process in space, that is, the calculation of the Fourier coefficients c_{kt} and s_{kt}, and a spectral decomposition of the temporal co-variability of these coefficients. However, the analysis is far from being symmetric in terms of space and time. The spatial decomposition is only geometrical in nature; there are no sampling problems. The temporal decomposition, on the other hand, is heavily loaded with sampling problems and non-trivial assumptions, as we will see in the following.

11.5.3 The Total Variance of the Waves. We assume that the space–time stochastic process has been been decomposed, by means of (11.85), into sine and cosine coefficient stochastic processes $\mathbf{C}_{kt}$ and $\mathbf{S}_{kt}$. For convenience we will use index k only when necessary for clarity. The bivariate process formed by the sine and cosine coefficients is denoted

$$\vec{\mathbf{X}}_t = (\mathbf{C}_t, \mathbf{S}_t)^{\mathrm{T}}. \qquad (11.86)$$

We also assume, for convenience, that the means of the sine and cosine coefficient processes are zero (i.e., $\mathcal{E}(\mathbf{C}_t) = \mathcal{E}(\mathbf{S}_t) = 0$). Then, the total variance of the space–time stochastic process $\mathbf{F}(x,t)$ at spatial wavenumber k, say σ_T^2, is

$$\begin{aligned}\sigma_T^2 &= \int_0^L \mathrm{Var}\big(\mathbf{C}_t \cos(\tfrac{2\pi kx}{L}) + \mathbf{S}_t \sin(\tfrac{2\pi kx}{L})\big)\,dx \\ &= \mathcal{E}(\mathbf{C}_t^2)\int_0^L \cos^2(\tfrac{2\pi kx}{L})\,dx \\ &\quad + \mathcal{E}(\mathbf{S}_t^2)\int_0^L \sin^2(\tfrac{2\pi kx}{L})\,dx \\ &\quad + 2\mathcal{E}(\mathbf{C}_t\mathbf{S}_t)\int_0^L \cos(\tfrac{2\pi kx}{L})\sin(\tfrac{2\pi kx}{L})\,dx \\ &= \frac{\mathrm{Var}(\mathbf{C}_t) + \mathrm{Var}(\mathbf{S}_t)}{2}. \end{aligned} \qquad (11.87)$$

Using (11.19), we see that the total variance[21] at wavenumber k can be re-expressed as

$$\begin{aligned}\sigma_T^2 &= \frac{\mathrm{Var}(\mathbf{C}_t) + \mathrm{Var}(\mathbf{S}_t)}{2} \\ &= \frac{\gamma_{cc}(0) + \gamma_{ss}(0)}{2} \\ &= \int_0^{\frac{1}{2}} \Gamma_{cc}(\omega) + \Gamma_{ss}(\omega)\,d\omega. \end{aligned} \qquad (11.88)$$

11.5.4 The Variance of Eastward and Westward Propagating Waves. The next step is to split the total variance given by (11.87) into the contributions from eastward and westward travelling waves, so that

$$\sigma_T^2 = \sigma_E^2 + \sigma_W^2,$$

where σ_E^2 and σ_W^2 represent the components that propagate eastward (E) and westward (W) respectively.

Formally, we write

$$\begin{aligned}\sigma_E^2 &= \tfrac{1}{2}\sigma_T^2 + R \\ \sigma_W^2 &= \tfrac{1}{2}\sigma_T^2 - R,\end{aligned} \qquad (11.89)$$

where R is currently unknown. What properties should R have?

1 The westward and eastward variance should be non-negative. Thus

$$|R| \le \frac{1}{2}\sigma_T^2.$$

2 If the bivariate $\vec{\mathbf{X}}_t$ (11.86) contains no noise and consists of a single, undamped, eastward travelling wave, then all of σ_T^2 should be attributed to the 'eastward' compartment. That is, we would have $R = \frac{1}{2}\sigma_T^2$.

3 If $\vec{\mathbf{X}}_t$ is such that the sequences of eastward travelling waves are randomly overlaid by noise, then only part of σ_T^2 should be attributed to the eastward variance. The remaining 'unaccounted' for variance should be distributed evenly between the eastward and westward component. In this case $R < \frac{1}{2}\sigma_T^2$.

4 If the components of $\vec{\mathbf{X}}_t$ vary in an unrelated manner at all time scales, then there is no preference for a direction and $R = 0$.

[21] The total variance in the field at wavenumber k is half of the sum of the variances of the coefficients because of the way in which coefficients $\mathbf{C}_t$ and $\mathbf{S}_t$ are defined.

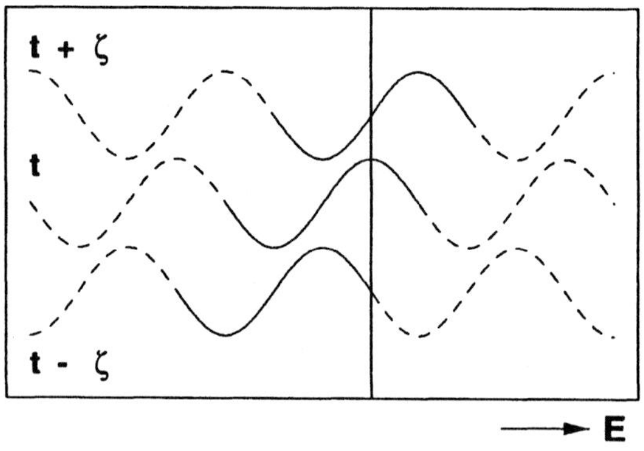

Figure 11.11: *A schematic diagram illustrating one wavelength of a wave that travels eastward 1/4 of a wavelength every ζ time units.*

One quantity that satisfies all of these requirements is the integral of the quadrature spectrum over negative frequencies,

$$R = \int_{-\frac{1}{2}}^{0} \Psi_{cs}(\omega)\, d\omega. \tag{11.90}$$

To motivate this choice for R, let us consider what an eastward travelling wave is. Suppose that the field $f(x,t)$ consists of a pure cosine pattern at time t and that, a short time τ later, the cosine pattern has been slightly damped and a weak sine pattern has been added. The effect of adding the sine pattern is that the crest of the wave moves from the $x = 0$ location to some point to the right of $x = 0$. If our conceptual diagram is oriented with north at the top of the page, the wave will have moved eastward during the interval. Eventually, after a quarter of a period, the cosine pattern is replaced by a sine pattern, and after half a period the wave crest will be replaced by a trough.

When the process is stochastic, we can no longer assume that the eastward movement is strictly uniform or that a well-defined period exists. However, if the waves tend to travel eastward, a wave crest (in the form of a pure cosine; see middle curve in Figure 11.11) will, on average, be replaced by a pure sine after a characteristic time ζ (upper curve in Figure 11.11). When there are eastward propagating waves, a large positive cosine coefficient at a given time t (the middle curve, Figure 11.11, which is represented by $\mathbf{c}_t = 1$ and $\mathbf{s}_t = 0$) will tend to be followed by a large positive sine coefficient ζ time units latter (upper curve, represented by $\mathbf{c}_{t+\zeta} = 0$ and $\mathbf{s}_{t+\zeta} = 1$), and will tend to have been preceded by a large negative sine coefficient ζ time units earlier (lower curve, represented by $\mathbf{c}_{t-\zeta} = 0$ and $\mathbf{s}_{t-\zeta} = -1$). Thus, for small values of τ, $\gamma_{cs}(\tau)$ is positive and $\gamma_{cs}(-\tau)$ is negative. For a sufficiently well-behaved process, the quadrature spectrum $\Psi_{cs}(\omega)$ will be positive for most negative ωs (cf. (11.68)) when there are eastward travelling waves. Therefore R is positive and σ_E^2 is greater than σ_W^2 (cf. (11.89)).

Similarly, when there are westward travelling waves, negative sine functions will tend to replace cosines so that R is negative and the westward travelling variance is larger than the eastward travelling variance.

If the sine and cosine coefficient processes are unrelated, then the quadrature spectrum is zero. Thus equation (11.90) satisfies the requirements listed above. The concept can be extended so that the propagating variance can be attributed to specific frequency ranges. To do that, we define the *frequency–wavenumber spectrum* as

$$\Gamma_{cs}^{fw}(\omega) = \frac{\Gamma_{cc}(\omega) + \Gamma_{ss}(\omega)}{2} - \Psi_{cs}(\omega). \tag{11.91}$$

This is a two-sided spectrum with different densities for negative and positive frequencies. Since

$$\begin{aligned}
&\int_0^{\frac{1}{2}} \Gamma_{cs}^{fw}(\omega)\, d\omega \\
&= \frac{1}{2}\int_0^{\frac{1}{2}} \frac{\Gamma_{cc}(\omega) + \Gamma_{ss}(\omega)}{2}\, d\omega - \int_0^{\frac{1}{2}} \Psi_{cs}(\omega)\, d\omega \\
&= \frac{1}{2}\sigma_T^2 + \int_{-\frac{1}{2}}^{0} \Psi_{cs}(\omega)\, d\omega \\
&= \frac{1}{2}\sigma_T^2 + R
\end{aligned}$$

it follows from equations (11.89, 11.90) that

$$\sigma_E^2 = \int_0^{\frac{1}{2}} \Gamma_{cs}^{fw}(\omega)\, d\omega$$

$$\sigma_W^2 = \int_{-\frac{1}{2}}^{0} \Gamma_{cs}^{fw}(\omega)\, d\omega.$$

Thus, negative frequencies represent westward travelling waves, and positive frequencies eastward travelling waves.[22]

[22]Note that this convention depends upon the definitions of i) the sign of the quadrature spectrum, ii) the variance of eastward and westward travelling waves, iii) R, and iv) the frequency–wavenumber spectrum. They are to some extent arbitrary. Negative frequencies are associated with westward propagation for the particular definitions used here. An advantage of this convention is that the eastward travelling wave variance appears on the right hand side of diagrams, and westward travelling wave variance on the left (cf. Figure 11.12).

11.5.5 Examples. We return to the three bivariate AR(1) examples that were discussed in [11.4.7,8]. The general representation used here is

$$\begin{pmatrix}\mathbf{C}\\ \mathbf{S}\end{pmatrix}_t = \mathcal{A}\begin{pmatrix}\mathbf{C}\\ \mathbf{S}\end{pmatrix}_{t-1} + \sigma\begin{pmatrix}1 & 0\\ 0 & \sqrt{b}\end{pmatrix}\begin{pmatrix}\mathbf{Z}_{ct}\\ \mathbf{Z}_{st}\end{pmatrix}.$$

The first example has a simple AR(1) cosine process and a white noise sine process. Thus

$$\mathcal{A} = \begin{pmatrix}\alpha & 0\\ 0 & 0\end{pmatrix}.$$

We calculated the spectra needed to determine the frequency–wavenumber spectrum in [11.4.7]. The power spectrum of **C** is a red spectrum (11.78), and that of **S** is white (11.79). The quadrature spectrum is zero so the frequency–wavenumber spectrum,

$$\Gamma_{cs}^{fw}(\omega) = \frac{\sigma^2}{2}\left(\frac{1}{1+\alpha^2-2\alpha\cos(2\pi\omega)} + b\right), \tag{11.92}$$

is symmetric. Equal variance is attributed to eastward and westward travelling features.

The second example has two unrelated red noise processes with the same parameter α and forcing noise of the same variance. That is

$$\mathcal{A} = \begin{pmatrix}\alpha & 0\\ 0 & \alpha\end{pmatrix}$$

and $b = 1$. The spectra needed to determine the frequency–wavenumber spectrum were derived in [11.4.7]. Since the quadrature spectrum is zero, the frequency–wavenumber spectrum

$$\Gamma_{cs}^{fw}(\omega) = \frac{1}{2}(\Gamma_{cc}(\omega) + \Gamma_{ss}(\omega)) = \Gamma_{cc}(\omega)$$

is again symmetric. A preferred direction wave propagation is not indicated.

The third example, with a rotational parameter matrix

$$\mathcal{A} = r\begin{pmatrix}u & -v\\ v & u\end{pmatrix},$$

is more interesting. In this case the frequency–wavenumber spectrum

$$\begin{aligned}\Gamma_{cs}^{fw}(\omega) &= \Gamma_{cc(\omega)} - \Psi_{cs}(\omega)\\ &= \frac{\sigma^2}{D^2}(1+r^2 \\ &\quad - 2r\,(u\cos(2\pi\omega) - v\sin(2\pi\omega)))\end{aligned} \tag{11.93}$$

is not symmetric.[23]

[23] See [11.4.8] for the power and quadrature spectra. D^2 is given by (11.82).

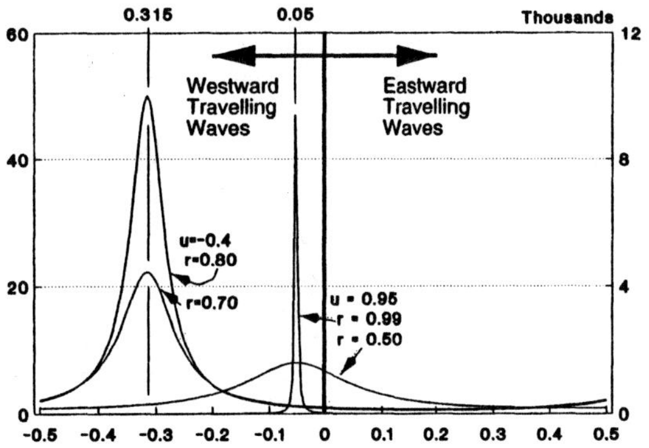

Figure 11.12: *Frequency–wavenumber spectra Γ_{cs}^{fw} of a bivariate process with a rotational parameter matrix. The same examples are shown as in Figure 11.10, namely $u = 0.95$ and $r = 0.99/0.50$, and $u = -0.4$ and $r = 0.8/0.7$. It is assumed that v is negative. The 'theoretical peak frequencies' $\eta = \cos^{-1}(u) = 0.315$ and 0.05 are marked by thin vertical lines. Variance at positive frequencies is interpreted as coming from eastward travelling waves, and variance at negative frequencies from westward travelling waves. The right axis measures the spiky spectrum, whereas the left axis is valid for the other three spectra.*

Depending upon the sign of v, this process exhibits a smooth transition either from a cosine pattern to a sine pattern (eastward travelling waves), or from a cosine pattern to an inverse sine pattern (westward travelling waves). The transition tends to occur in $\tau = \frac{1}{8\pi}\cos^{-1}(u)$ time steps. Sequence (11.83) shows that the eastward motion (i.e., to the right) occurs when v is positive. Conversely, negative v is associated with sequence (11.84) and westward motion.

Figure 11.12 shows frequency–wavenumber spectra for examples with the combinations of periods, represented by u, and damping rates r considered previously. This time we assume v is negative, so we can expect westward propagation. Indeed, we see that, except for the strongly damped process ($r = 0.5$), most variance is ascribed to westward travelling waves. The 'theoretical' peak frequencies $\eta = \frac{1}{2\pi}\cos^{-1}(u)$ are associated with maxima in the spectra.

11.5.6 Example: A Two-sided Spectrum of Travelling Wave Variance. Fraedrich and Böttger [124] studied five years of daily Northern Hemisphere winter 500 hPa geostrophic meridional wind data that was derived from daily

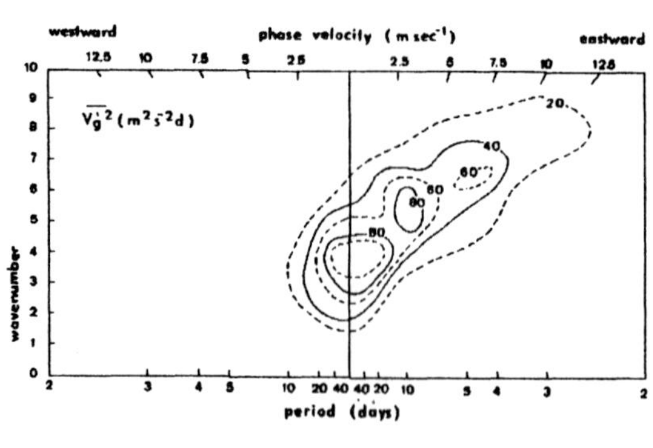

Figure 11.13: *Two-sided spectra of travelling wave variance of daily meridional geostrophic wind during winter at 50° N. The vertical axis represents the zonal wavenumbers* $k = 0, 1, \ldots, 10$. *The time scale is given on the bottom axis. The top axis gives the associated phase velocity (see text).*
From Fraedrich and Böttger [124].

geopotential analyses. They first calculated the cosine and sine coefficients of the meridional wind at 50° N for spatial wavenumbers k, $k = 0, \ldots, 10$. Two-sided spectra of the travelling wave variance (11.91) (Figure 11.13) were estimated from the cosine and sine coefficient time series.[24]

A few words are required about the presentation in Figure 11.13. First, continuous contours are used for clarity even though the frequency–wavenumber spectrum is discrete in k. Also, for each wavenumber k, period $1/\omega$ can be interpreted as a *phase velocity* that expresses the rate at which the wave crest moves. The lines of constant phase velocities are indicated by tick marks on the upper axis.

Most of the variance is attributed to zonal wavenumbers $k = 2, \ldots, 7$. The variance at large scales (wavenumbers $k = 2, 3$) is divided equally between eastward and westward travelling waves. Almost all smaller scale variability that is characteristic for baroclinic dynamics ($k \geq 5$) is attributed to eastward travelling waves. A variance maximum occurs along a line that corresponds well with the theoretical dispersion line for Rossby waves at 50° N in a zonal mean flow of about 15 m/s [56].

11.5.7 Pratt's Definition of Standing Wave Variance Spectra.

Pratt [319] tries to discriminate between 'propagating' and 'standing' wave variance by arguing that a standing wave is the sum of two waves of equal variance that propagate in opposite directions. Motivated by this reasoning, Pratt partitions the total variance σ_T^2 into two symmetric spectra: one describing the distribution of standing variance with time scale, the other describing propagating variance. Since the spectra are symmetric, they are defined so that the total variance is ascribed to positive frequencies.

The *standing wave variance spectrum* is defined as

$$\Gamma_{cs}^{st}(\omega) = 2\,\min\big(\Gamma_{cs}^{fw}(\omega), \Gamma_{cs}^{fw}(-\omega)\big). \tag{11.94}$$

Note that Γ_{cs}^{st} is symmetric in ω.

The *propagating wave variance spectrum* is defined as

$$\Gamma_{cs}^{pro}(\omega) = \max\big(\Gamma_{cs}^{p}(\omega), \Gamma_{cs}^{p}(-\omega)\big) \tag{11.95}$$

where

$$\Gamma_{cs}^{p}(\omega) = \Gamma_{cs}^{fw}(\omega) - \frac{1}{2}\Gamma_{cs}^{st}(\omega).$$

This spectrum is also symmetric.

We can express these spectra in terms of the power and quadrature spectra of the cosine and sine coefficient processes by substituting definition (11.91) of the frequency–wavenumber spectrum into (11.94) and (11.95). We find that the standing and propagating variance spectra can be expressed as

$$\Gamma_{cs}^{st}(\omega) = \Gamma_{cc}(\omega) + \Gamma_{ss}(\omega) - 2|\Psi_{cs}(\omega)| \tag{11.96}$$

and

$$\Gamma_{cs}^{pro}(\omega) = 2|\Psi_{cs}(\omega)| \tag{11.97}$$

respectively.

Pratt uses the frequency–wavenumber spectrum to label the propagating variance as eastward or westward. If the frequency–wavenumber spectrum assigns more variance to eastward than westward travelling waves at a given frequency, the propagating variance at that frequency is labelled 'eastward,' and vice versa. Equivalently, if the quadrature spectrum at a given frequency is positive, then the propagating variance in the neighbourhood of that frequency is identified as eastward.

The *total standing wave variance* and the *total propagating wave variance* are the integrals over all positive frequencies of the standing and propagating variance spectra:

$$\sigma_{st}^2 = \int_0^{\frac{1}{2}} \Gamma_{cs}^{st}(\omega)\,d\omega$$

$$\sigma_{pro}^2 = \int_0^{\frac{1}{2}} \Gamma_{cs}^{pro}(\omega)\,d\omega.$$

[24] These spectral estimates are subject to uncertainty from a number of sources, whose effects we ignore for the moment. Spectral estimation is discussed in some detail in Sections 12.3 and 12.5.

By substituting equations (11.96) and (11.97) into these expressions, we see that the sum of the standing and propagating wave variance is the total variance (11.88)

$$\sigma_T^2 = \sigma_{st}^2 + \sigma_{pro}^2.$$

While Pratt's partitioning of variability is intuitively pleasing, we should remember that it is based on heuristic arguments. Therefore, as with other aspects of the language used in frequency–wavenumber analysis, the terms 'standing and propagating wave variance' are an ambiguous description of equations (11.94) and (11.95). Literal interpretation of these quantities as *the* standing and propagating variance spectra can be misleading.

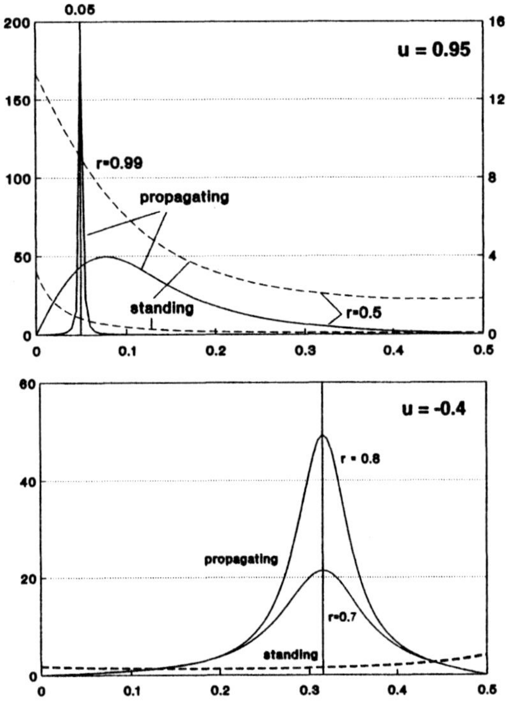

Figure 11.14: *Standing (dashed) and propagating (solid) variance spectra for a bivariate process with rotational parameter matrix (cf. Figure 11.12). The 'theoretical' peak frequency* $\eta = \frac{1}{2\pi}\cos^{-1}(u)$ *is represented by a vertical line.*
Top: $u = 0.95; r = 0.99$ *(left hand axis, the propagating wave spectrum is scaled by a factor of 0.01) and* $r = 0.5$ *(right hand axis).*
Bottom: $u = -0.4; r = 0.8$, *and* $r = 0.7$. *The standing wave spectra can not be distinguished in this diagram.*

11.5.8 Examples of Pratt's Decomposition. Recall again the three bivariate AR(1) examples developed by Ute Luksch (see [11.4.7,8] and [11.5.5]). Since the quadrature spectrum is zero in the first two examples, Pratt's formalism attributes all variance to standing waves. This clearly makes sense in the first example because the cosine coefficient varies dynamically and the sine coefficient is white noise. Interpretation is a little more difficult in the second example where cosine and sine coefficients are independent, identically structured AR(1) processes.

The propagating and standing wave variance spectra for Luksch's third example, in which the bivariate AR(1) processes have rotational parameter matrices, are shown in Figure 11.14. In all cases, most of the variance is attributed to the propagating variance and only a small portion is designated as standing variance. Except for the $u = 0.95$, $r = 0.5$ process, the peaks in the propagating spectra correspond well with the theoretical rotation rate $2\pi/\eta$ (η is indicated by the vertical line in the diagrams). As with the coherency spectrum (see Figure 11.10), the peak in the propagating spectrum when $u = 0.95$, $r = 0.5$ occurs at a frequency greater than η.

Note that negative values of v were used to compute Figure 11.14. Negative v results in negative quadrature spectra (for positive ωs) so that all propagating variance is attributed to westward travelling waves.

11.5.9 Example [11.5.6] Revisited. Fraedrich and Böttger [124] applied Pratt's formalism to time series of daily analysed 500 hPa geopotential height during winter along the 50° N latitude circle. Most of the standing wave variance was estimated to occur on time scales longer than 10 days with maximum variance for small wavenumbers (Figure 11.15, top). The bulk of the variance due to propagating waves, on the other hand, was attributed to time scales of less than 10 days and baroclinic spatial scales. Almost all variance was attributed to eastward propagating variance (Figure 11.15, bottom). Fraedrich and Böttger were able to relate dynamically the three spectral maxima in Figure 11.13 to standing waves or eastward propagating waves.

11.5.10 Hayashi's Definition of Standing Wave Variance. An alternative to Pratt's approach was offered by Hayashi [170], who defines a non-symmetric spectrum for propagating variance, and a symmetric spectrum for standing variance.

In this formalism, the standing wave spectrum is defined to be

$$\Gamma_{cs}^{st}(\omega) = C(\omega)\sqrt{\Gamma_{cs}^{fw}(\omega)\Gamma_{cs}^{fw}(-\omega)}$$

with 'coherency'

$$C^2(\omega) = \frac{\left(\Gamma_{cc}(\omega) - \Gamma_{ss}(\omega)\right)^2 + 4(\Lambda_{cs}(\omega))^2}{\left(\Gamma_{cc}(\omega) + \Gamma_{ss}(\omega)\right)^2 - 4(\Psi_{cs}(\omega))^2}. \tag{11.98}$$

The propagating variance is then defined as the remainder

$$\Gamma_{cs}^{pro}(\omega) = \Gamma_{cs}^{fw}(\omega) - \frac{1}{2}\Gamma_{cs}^{st}(|\omega|). \tag{11.99}$$

The total standing wave variance is defined as the integral of the standing wave spectrum over the positive frequencies:

$$\sigma_{st}^2 = \int_0^{\frac{1}{2}} \Gamma_{cs}^{st}(\omega)\, d\omega \tag{11.100}$$

In contrast, the total propagating wave variance is defined as the integral of the propagating wave spectrum over all frequencies:

$$\sigma_{pro}^2 = \int_{-\frac{1}{2}}^{\frac{1}{2}} \Gamma_{cs}^{pro}(\omega)\, d\omega. \tag{11.101}$$

Hayashi also devised a method for ascribing a (spatial) phase to the standing wave variance at frequency ω. For wavenumber k, the position of the train of crests and troughs of this standing wavenumber relative to the origin is given by

$$\phi_{cs}^k(\omega) = \frac{1}{2k}\tan^{-1}\left(\frac{2\Lambda_{cs}(\omega)}{\Gamma_{cc}(\omega) - \Gamma_{ss}(\omega)}\right)$$

when both the numerator and the denominator are nonzero. When both are zero, the 'coherency' $C(\omega)$ (11.98) and the standing wave variance are also zero, so the phase is meaningless. When the denominator is zero (i.e., $\Gamma_{cc}(\omega) = \Gamma_{ss}(\omega)$) the phase is given by

$$\phi_{cs}^k(\omega) = \begin{cases} \frac{\pi}{4k} & \text{if } \Lambda_{xy}(\omega) > 0 \\ -\frac{\pi}{4k} & \text{if } \Lambda_{xy}(\omega) < 0 \end{cases}$$

and when the numerator is zero (i.e., when $\Lambda_{xy}(\omega) = 0$), it is given by

$$\phi_{cs}^k(\omega) = \begin{cases} 0 & \text{if } \Gamma_{cc}(\omega) > \Gamma_{ss}(\omega) \\ \frac{\pi}{2k} & \text{if } \Gamma_{cc}(\omega) < \Gamma_{ss}(\omega). \end{cases}$$

See [170] for details.

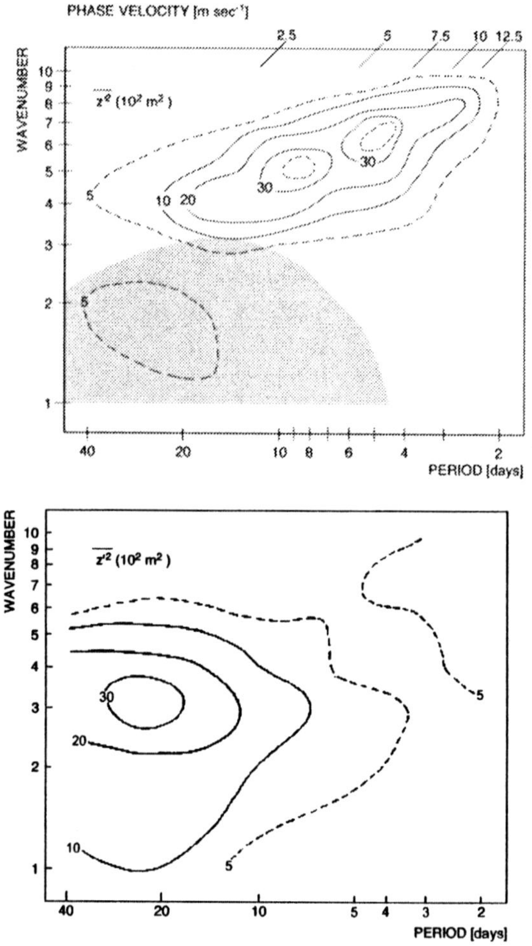

Figure 11.15: *One-sided spectra of standing and propagating wave variance (Pratt's definition) of 500 hPa geopotential height during winter at 50° N plotted as a function of zonal wavenumbers $k = 1, \ldots, 10$ in the vertical and log 'periods' in the horizontal.*
Top: Propagating wave variance. Shading indicates westward propagation.
Bottom: Standing wave variance.
From Fraedrich and Böttger [124].

We will see in the example below that Hayashi's method generally gives useful results. However, from a strictly mathematical point of view, the formalism is a not entirely satisfactory because it is sometimes possible to obtain negative propagating spectral densities (11.99).

11.5.11 Luksch's Examples Revisited. To illustrate Hayashi's formalism we return once more to the three examples described in [11.4.7,8], [11.5.5], and [11.5.8].

In the *first example*, the cosine coefficient evolves as an AR(1) process with parameter α and forcing with variance σ^2, and the sine coefficient evolves as white noise with variance $b\sigma^2$. Then (cf. (11.92))

$$\begin{aligned}\Gamma_{cs}^{fw}(\omega) &= \frac{\sigma^2}{2}\Big(\frac{1}{D^2}+b\Big)\\ C(\omega) &= \frac{|1-bD^2|}{1+bD^2},\end{aligned}$$

with $D^2 = 1+\alpha^2-2\alpha\cos(2\pi\omega)$. D^2 is symmetric in ω so that

$$\Gamma_{cs}^{st}(\omega) = \Gamma_{cs}^{fw}(\omega)C(\omega) = \frac{\sigma^2}{2\,D^2}|1-bD^2|.$$

The distribution of the total variance to the standing wave, eastward, and westward propagating components depends on the ratio $\Gamma_{cc}(\omega)/\Gamma_{ss}(\omega) = bD^2$.

When $\Gamma_{cc}(\omega) = \sigma^2/D^2$ is greater than $\Gamma_{ss}(\omega) = \sigma^2 b$, that is, if bD^2 is less than 1, then

$$\begin{aligned}\Gamma_{cs}^{st}(\omega) &= \frac{\sigma^2}{2\,D^2}(1-bD^2)\\ \Gamma_{cs}^{pro}(\omega) &= b\sigma^2\\ \phi_{cs}^{k}(\omega) &= 0.\end{aligned}$$

The cosine series dominates the sine series in this case, so setting the phase of the standing waves to zero is reasonable. In the limit, when $b = 0$, all the variance is attributed to standing wave variance. Also, note that when the standing and propagating spectra are integrated, as in equations (11.100, 11.101), the total propagating and standing variance sums to σ_T^2. In the opposite case, with the cosine coefficient spectrum smaller than the sine coefficient spectrum, that is, bD^2 is greater than 1, we obtain

$$\begin{aligned}\Gamma_{cs}^{st}(\omega) &= \frac{\sigma^2}{D^2}(bD^2-1)\\ \Gamma_{cs}^{pro}(\omega) &= \frac{\sigma^2}{D^2}\\ \phi_{cs}^{k}(\omega) &= \frac{\pi}{2}.\end{aligned}$$

This time the sine coefficient tends to be greater than the cosine coefficient, and the standing wave's crest or trough is correctly placed at $\pi/2$.

If the parameters of the process are such that the cosine and sine spectra are equal at some frequency ω_0, that is, ω_0 is a solution of $bD^2 = 1$, then the standing wave spectral density becomes zero at ω_0 and $\Gamma_{cs}^{pro}(\omega_0) = \Gamma_{cs}^{fw}(\omega_0) = b\sigma^2$.

One final point for the example is that in all three scenarios just discussed, the propagating variance spectrum is an even function of frequency. Thus, in keeping with the nature of the parameter matrix $\mathcal{A}$, variance has no preferred direction of propagation.

The *second example* consisted of cosine and sine coefficient processes generated by two independent AR(1) processes with identical AR parameter and variance. Thus, the two spectra Γ_{cc} and Γ_{ss} are equal and the 'coherency' $C(\omega)$ vanishes so that the total variance is distributed equally among the westward and eastward propagating waves at all frequencies, as

$$\Gamma_{cs}^{pro}(\omega) = \Gamma_{cs}^{fw}(\omega). \tag{11.102}$$

The *third example* used a bivariate AR(1) process with a rotation matrix as its parameter matrix $\mathcal{A}$. In the setup considered in [11.4.7], [11.5.5], and [11.5.8], the white noise forcing parameter b was set to 1, resulting in cosine and sine coefficient processes of the same variance. In this case all of the variance is again attributed to propagating waves (11.102).

When b is not 1, the variances of the cosine and sine coefficient processes are not equal and part of the joint variance is attributed to standing wave variance. In this case the frequency–wavenumber spectrum (11.93) is given by (see Luksch et al. [262] for details)

$$\begin{aligned}\Gamma_{cs}^{fw}(\omega) &= \frac{(1+b)\sigma^2}{2D^2}\\ &\quad\times\Big(1+r^2-2r(u\cos(2\pi\omega)-v\sin(2\pi\omega))\Big),\end{aligned}$$

and the squared 'coherency' is

$$\begin{aligned}C(\omega)^2 &= \Big(\frac{1-b}{1+b}\Big)^2\\ &\quad\times\frac{\big(E-(rv)^2\big)^2+4\big(rv\cos(2\pi\omega)-r^2uv\big)^2}{\big(E+(rv)^2\big)^2-4\big(rv\sin(2\pi\omega)\big)^2}\end{aligned}$$

where $E = 1+(ru)^2-2ru\cos(2\pi\omega)$. The phase (for $b \neq 1$) is

$$\phi_{cs}^{k}(\omega) = \frac{1}{2k}\tan^{-1}\Big(\frac{rv\cos(2\pi\omega)-r^2uv}{E-(rv)^2}\Big).$$

In this case, the positiveness of the propagating variance densities (11.99) is no longer guaranteed. For example, setting $b = 0.1, ru = 0.5, rv = -0.8$ results in a small negative variance density for the travelling waves at frequency $\omega = -0.17$.

In summary, Hayashi's formalism does generally yield reasonable results, but, as the previous example illustrates, caution is advised.

12 Estimating Covariance Functions and Spectra

12.0.0 Overview. The purpose of this chapter is to describe some of the methods used to estimate the second moments, the auto- and cross-covariance functions, and the power and cross-spectra, of the weakly stationary ergodic processes that were described in the previous two chapters. It is not our intention to be exhaustive, but rather to introduce some of the concepts associated with the estimation problem. We leave it to the reader to explore these concepts further in the sources that we cite.

12.0.1 Parametric and Non-parametric Approaches. We will take one of two approaches when inferring the properties of stochastic processes from limited observational evidence.

Parametric estimators assume that the observed process is generated by a member of a specific class of processes, such as the class of auto-regressive processes (AR processes; see Section 10.3). Some parametric estimation techniques further restrict the type of process considered by adding distributional assumptions. For example, it is often assumed that a process is normal, meaning that all joint distributions of arbitrary numbers of elements $\mathbf{X}_{t_1}, \ldots, \mathbf{X}_{t_n}$ are multivariate normal. The parameters of such a process are estimated by finding the member of the class of models that best fits the observational evidence. The fitting methods, such as the method of moments, least squares, or maximum likelihood estimation, are the same as those used in other branches of statistics. Once a model has been fitted, estimates of the auto-covariance function and power spectrum are obtained simply by deriving them from the fitted process.

The fitting of auto-regressive models to observed time series is discussed in Section 12.2. Auto-regressive and maximum entropy spectral estimation are briefly discussed in [12.3.19].

Non-parametric estimators make fewer assumptions about the generating process. In fact, the methods generally used in time series analysis assume only ergodicity and weak stationarity. Methods described in this chapter, aside from methods that specifically assume a time-domain model, are non-parametric.

Note that 'non-parametric' tends to have an interpretation in time series analysis that is different from that in other areas of statistics. In other areas of statistics, non-parametric inference methods often use exact distributional results that are obtained through heavy reliance on sampling assumptions, such as the assumption that the observations are realizations of a collection of independent and identically distributed random variables. Time series statisticians must replace the independence assumption with something considerably weaker (e.g., weak stationarity and ergodicity) and therefore can generally only appeal to asymptotic theory when making inferences about the characteristics of a stochastic process.

12.0.2 Outline. The second moments of an ergodic weakly stationary process have equivalent representations in the time (the auto-correlation function) and frequency (the spectral density functions) domains. We describe non-parametric and parametric approaches to the estimation of the auto-correlation function of a univariate process in Sections 12.1 and 12.2, respectively. Estimation of the corresponding spectral density function is described in Section 12.3. In this case, most of our effort is devoted to the non-parametric approach (see [12.3.1–20]; we discuss the parametric approach briefly in [12.3.21]) because the non-parametric estimators can be coupled with an effective asymptotic theory to make reliable inferences about the spectrum. Similar tools are not available with the parametric approach to spectral estimation. The ideas discussed in Sections 12.1 and 12.3 are extended to multivariate processes in Section 12.4, where we briefly describe a non-parametric estimator of the cross-correlation function, and Section 12.5, where we describe non-parametric estimators of the cross-spectral density functions.

12.1 Non-parametric Estimation of the Auto-correlation Function

12.1.0 Outline. We begin by describing the usual non-parametric product-moment estimator of the auto-correlation function in [12.1.1]. The bias and variance of this estimator are examined in [12.1.2], some examples are considered in [12.1.3,4], and a simple test of the null hypothesis that the observed process is white is described in [12.1.5]. The partial auto-correlation function, which is useful when fitting parametric models to time series, is briefly described in [12.1.6,7].

Throughout this chapter we use the notation $\mathbf{x}_1, \ldots, \mathbf{x}_T$ to represent a sample obtained by observing a single realization of an ergodic, weakly stationary, stochastic process at T consecutive times $t_0, t_0+1, \ldots, t_0+T-1$, beginning at some arbitrary time t_0. The corresponding random variables will be denoted by $\mathbf{X}_1, \ldots, \mathbf{X}_T$. We will also use the notation $\mathbf{x}'_t = \mathbf{x}_t - \overline{\mathbf{x}}$, $t = 1, \ldots, T$, to represent the time series of deviations from the sample mean $\overline{\mathbf{x}} = \frac{1}{T}\sum_{t=1}^{T} \mathbf{x}_t$, and we will write $\mathbf{X}'_t$, $t = 1, \ldots, T$, and $\overline{\mathbf{X}}$ to represent the corresponding random variables.

12.1.1 Non-parametric Estimator. A non-parametric estimator of the auto-correlation function $\rho(\tau)$ is given by

$$r(\tau) = c(\tau)/c(0) \tag{12.1}$$

where $c(\tau)$ is the sample auto-covariance function

$$c(\tau) = \frac{1}{T}\sum_{t=|\tau|+1}^{T} \mathbf{X}'_{t-|\tau|}\mathbf{X}'_t. \tag{12.2}$$

The sample auto-covariance function is set to zero for $|\tau| \geq T$.

12.1.2 Properties of the Non-parametric Estimator. Kendall (see Section 7.7 of [220]) shows that estimator (12.1) can have substantial bias. In particular, if $\mathbf{X}_t$ is a white noise process, the bias is

$$B\big(r(\tau)\big) \approx \frac{-1}{T},$$

and when $\mathbf{X}_t$ is an AR(1) process with lag-1 correlation coefficient α_1,

$$B\big(r(1)\big) \approx -\frac{1}{T}(1+4\alpha_1)$$

$$B\big(r(\tau)\big) \approx -\frac{1}{T} \times \Big(\frac{1+\alpha_1}{1-\alpha_1}(1-\alpha_1^{|\tau|}) + 3|\tau|\alpha_1^{|\tau|}\Big) \quad |\tau|>1 \tag{12.3}$$

Equation (12.1) is sometimes inflated by the factor $T/(T-|\tau|)$ to adjust for bias, but this is not generally considered helpful because it also inflates the variability of the estimator (recall [5.3.7] and Figure 5.3).

Bartlett [31], working under the assumption that $\mathbf{X}_t$ is a stationary normal process, derived a general asymptotic result about the variability of $r(\tau)$ that is useful for interpreting the sample auto-covariance function. He showed that

$$\mathrm{Var}\big(r(\tau)\big) \approx \frac{1}{T}\sum_{\ell=-\infty}^{\infty}\Big(\rho^2(\ell) + \rho(\ell+\tau)\rho(\ell-\tau) - 4\rho(\tau)\rho(\ell)\rho(\ell-\tau) + 2\rho^2(\ell)\rho^2(\tau)\Big).$$

Thus, if there exists a p such that $\rho(\tau)$ is zero for τ greater than p, then

$$\mathrm{Var}\big(r(\tau)\big) \approx \frac{1}{T}\Big(1 + 2\sum_{\ell=1}^{p}\rho^2(\ell)\Big) \tag{12.4}$$

for τ greater than p. This result can be used to conduct a rough and ready test of the null hypothesis that $\rho(\tau) = 0$ at each lag τ as follows.

1 Assume that $\rho(\ell)$ is zero for $\ell \geq \tau$.

2 Substitute $r(\ell)$, $1 \leq \ell < \tau$, into approximation (12.4) to obtain $\widehat{\sigma}^2_{r(\tau)}$, an estimate of the variance of $r(\tau)$.

3 Compare $Z = r(\tau)/\widehat{\sigma}_{r(\tau)}$ with the critical values of the standard normal distribution (Appendix D).

Summation (12.4) is usually truncated at a 'reasonable' number of lags, say 20–25.

We emphasize that this test is based on asymptotic theory and thus is not exact. Also, the user needs to be aware of the effects of 'multiplicity.' When the test is conducted at the 5% significance level, rejection of the null hypothesis should be expected at 5% of lags tested, even when it is true at all lags. None the less, this test does offer some guidance in the interpretation of the auto-correlation function. Statistical packages sometimes compute $\widehat{\sigma}^2_{r(\tau)}$ for every $\tau > 0$ and display the approximate critical values $\pm 2\widehat{\sigma}^2_{r(\tau)}$ on a graph of $r(\tau)$.

Bloomfield [49] points out a disadvantage of analysing the correlation structure of a time series in the time domain (as opposed to in the spectral domain, see Section 12.3): the estimated auto-correlation function has complex correlation structure of its own. Bartlett [31] derives the asymptotic covariance between auto-correlation function estimates at different lags and Box and Jenkins [60] use this result to show that

$$\mathrm{Cov}(r(\tau), r(\tau+\delta)) \approx \frac{1}{T}\sum_{\ell=-\infty}^{\infty} \rho(\ell)\rho(\ell+\delta).$$

If we have a process that is, for example, AR(1) with parameter $\alpha_1 > 0$, this approximation gives

$$\mathrm{Cor}(r(\tau), r(\tau+\delta)) \approx \alpha_1^{\delta}$$

at large lags τ. That is, the correlations between the auto-correlation function estimates are roughly similar to those of the process itself. Consequently, when the process is persistent, the estimated auto-correlation function will vary slowly around zero even when the real auto-correlation function has decayed away to zero, and we need to be careful to avoid over-interpreting the estimated auto-correlation function.

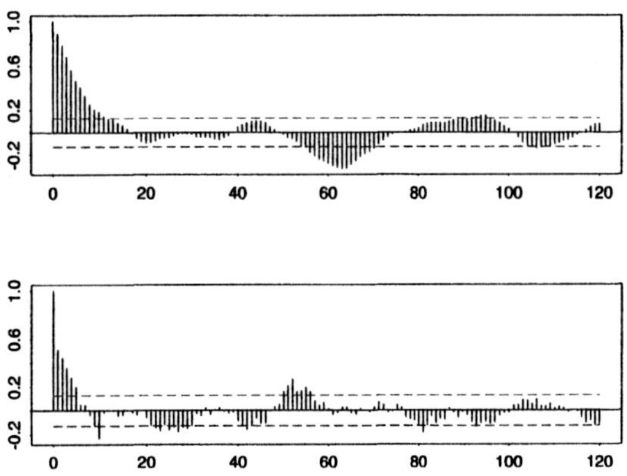

Figure 12.1: *Estimated auto-correlation functions computed from time series of length 240. The horizontal dashed lines indicate approximate critical values for testing the null hypothesis that $\rho(\tau) = 0$ for all τ at the 5% significance level.*
Top: Estimated auto-correlation function for a time series generated from an AR(1) process with $\alpha_1 = 0.9$.
Bottom: Estimated auto-correlation function for a time series generated from an MA(10) process with $\beta_1 = \cdots = \beta_{10} = 1$.

12.1.3 Example: Auto-correlation Function Estimates. Figure 12.1 shows some examples of auto-correlation function estimates computed from simulated time series of length 240. The function displayed in the upper panel was computed from time series generated from an AR(1) process with parameter $\alpha_1 = 0.9$ (see [10.3.3]); that in the lower panel was generated from an MA(10) process with parameters $\beta_1 = \cdots = \beta_{10} = 1$ (see [10.5.1]). The two standard deviation critical values estimated with (12.4) (assuming $\rho(\tau)$ is zero for all nonzero τ) are also displayed.

As we would expect from an AR(1) process, the estimated auto-correlation function in the upper panel decays more or less exponentially until about lag-15 and then varies randomly about zero at time scales that are typical of an AR(1) process with $\alpha_1 = 0.9$. As anticipated, the large lag behaviour of the estimated auto-correlation function is quite similar to that of the process itself (compare the upper panel in Figure 12.1 with the time series shown in the lower panel of Figure 10.7). Note that the estimated auto-correlation function can take large excursions from zero even when the real auto-correlation function (not shown) is effectively zero. Some of these excursions extend well beyond the approximate critical values.

The sample auto-correlation function displayed in the lower panel behaves somewhat differently. It decays to zero in about 10 lags and then varies about zero on a shorter time scale than the auto-correlation function shown in the upper panel (compare with the MA(10) time series shown in the lower panel of Figure 10.15).

While the auto-correlation function estimates are informative, it would be difficult to identify precisely the order or type of the generating process from only this display. We address the problem of process identification more fully in Section 12.2.

12.1.4 Example: Bias. An impression of the bias of auto-correlation function estimator (12.1) can be obtained from a small Monte Carlo experiment (see Section 6.3). One thousand samples of each length $T = 15$, 30, 60, and 120 were generated from AR(1) processes with parameters $\alpha_1 = 0.3$, 0.6 and 0.9. Each time series was used to estimate the auto-correlation function at lag-1 and lag-10. The results are given in the following tables.

The lag-1 correlation

α_1	$\rho(1)$	Sample length 15	30	60	120
0.3	0.3	0.16	0.23	0.27	0.28
0.6	0.6	0.36	0.47	0.54	0.57
0.9	0.9	0.54	0.72	0.81	0.86

The lag-10 correlation

α_1	$\rho(10)$	Sample length 15	30	60	120
0.3	0.0	−0.06	−0.04	−0.03	−0.01
0.6	0.01	−0.07	−0.08	−0.05	−0.02
0.9	0.35	−0.15	−0.14	0.02	0.18

We see that the auto-correlation estimates are negatively biased. The bias is small when the true correlation is small but it is large when the true correlation is large, especially when the time series is short. The bias decreases slowly with increasing sample size. Comparison with Kendall's approximation for the bias (12.3) shows that the latter breaks down when τ is large relative to T, and also that α_1 affects the goodness of the approximation.

The denominator in (12.1) is summed over more products than the numerator, but this accounts only for some of the bias. Inflating the estimated correlations by multiplying with $T/(T - |\tau|)$ to adjust for the difference in the number of products summed does not eliminate the bias. Most of the bias arises because it is necessary to remove the sample mean when estimating the auto-covariance function.

12.1.5 A Test for Serial Correlation. We introduced the Durbin–Watson statistic (8.24) in [8.3.16] as a regression diagnostic that is used to check for serial correlation in regression residuals. We mention it again here to remind readers that it can be used in contexts other than the fitting of regression models. The statistic

$$d = \frac{\sum_{t=1}^{T-1}(\mathbf{X}'_{t+1} - \mathbf{X}'_t)^2}{\sum_{t=1}^{T}(\mathbf{X}'_t)^2}$$

is essentially the sample variance of the first differences of the times series divided by the sample variance of the undifferenced time series. Subsection [8.3.16] gives references for the derivation of the distribution of d under the null hypothesis that the time series was obtained from a white noise process. Samples taken from white noise processes will have values of d near 2. Since first differencing filters out low-frequency variability and enhances high-frequency variability (cf. [11.4.4] and Figure 11.9), time series from processes more persistent than white noise will tend to have values of d less than 2. Samples from processes that have relatively more high-frequency variability than white noise will tend to have values of d greater than 2.[1]

Bloomfield [49] interprets d as an index of the 'smoothness' of the time series.

12.1.6 Estimating the Partial Auto-correlation Function. The partial auto-correlation function $\alpha_{\tau,\tau}$ (see equation (11.13) in [11.1.10]) is sometimes a useful aid for identifying the order of AR model that reasonably approximates the behaviour of a time series. In particular, if $\mathbf{X}_t$ is an AR(p) process, then $\alpha_{\tau,\tau}$ is zero for all τ greater than p.

The partial auto-correlation function can be estimated recursively by substituting the estimated auto-correlation function $r(\tau)$ (12.1) into equations (11.12, 11.13). Box and Jenkins [60] note that the recursion is sensitive to rounding errors, particularly if the parameter estimates are near the boundaries of the admissible region for weakly stationary processes. Quenouille [326] showed that if $\mathbf{X}_t$ is an AR(p) process, then

$$\mathrm{Var}(\widehat{\alpha}_{\tau,\tau}) \approx \frac{1}{T} \quad \text{for } \tau > p. \tag{12.5}$$

12.1.7 Example: Partial Auto-correlation Function Estimates. Partial auto-correlation function estimates for the examples discussed in [12.1.3] are displayed in Figure 12.2. The horizontal lines depict the two standard deviation critical values (12.5).

The estimated partial auto-correlation function displayed in the upper panel is essentially zero beyond lag-1, a characteristic that (correctly) suggests that these time series came from an AR(1) process.

In contrast, the estimated partial auto-correlation shown in the lower panel is significantly different from zero at lags 1, 2, and 11. The estimate agrees quite well with the theoretical partial auto-correlation function for the MA(10) process that generated the data, which has a sequence of damped peaks at lags $\tau = 1, 11, 21, \ldots$.

[1]An AR(1) process with negative parameter α_1 is an example of a weakly stationary process with more high-frequency variability than is expected in white noise. A time series that has been differenced to remove trend will also show excessive high-frequency variability.

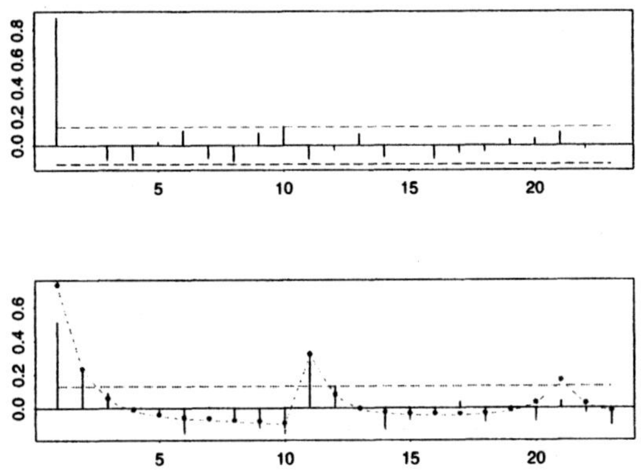

Figure 12.2: *Estimated partial auto-correlation functions computed from simulated time series of length 240. Approximate critical values for testing the null hypothesis that $\alpha_{\tau\tau}$ is zero (see equation (12.5)) at the 5% significance level are shown as dashed lines.*
Top: Partial auto-correlation function estimated from a time series generated from an AR(1) process with $\alpha_1 = 0.9$.
Bottom: Partial auto-correlation function estimated from a time series generated from an MA(10) process with $\beta_1 = \cdots = \beta_{10} = 1$. The theoretical function is shown with solid dots connected by broken lines.

12.2 Identifying and Fitting Auto-regressive Models

12.2.0 Overview. We will describe two approaches that are frequently used to identify and fit AR models to time series.

The Box–Jenkins method [60] is subjective in nature. Diagnostic aids, such as plots of the estimated auto-correlation function (cf. Section 12.1) and partial auto-correlation function (cf. [11.1.11]), and a practised eye, are used to make a first guess at the order of AR model to fit. The fitted model is then used to estimate the noise time series that forced the observed process, and the goodness-of-fit is determined by examining the estimated noise process. This process may be repeated several times, although care must be taken not to overfit the time series by choosing models with too many free parameters. An advantage of this subjective approach is that the analyst is closely involved with the data and is therefore better able to judge the goodness-of-fit of the model and the influence that idiosyncrasies in the data have on the fit.

The other approach we will discuss uses one of two objective order determining criteria (AIC, developed by Akaike [6] and BIC, developed by Swartz [360]; see [12.2.10,11]) to select the model. These criteria use penalized measures of the goodness-of-fit where the size of the penalty depends upon the number of estimated parameters in the model. The user's connection with this modelling process is not as close, and thus it is possible that an inappropriate model is fitted to a time series with some sort of pathological behaviour. On the other hand, since these methods are objective, they can be applied systematically when careful hand fitting of AR models is impractical.[2]

Both approaches require model fitting tools, a topic we will not discuss exhaustively, although [12.2.2] describes a couple common methods. Topics that we do cover include the Yule–Walker method and the method of maximum likelihood.

We assume, for now, that all processes are ergodic and weakly stationary. Beran [45], Box and Jenkins [60], Brockwell and Davis [68], and Tong [367], amongst others, describe techniques for identifying and fitting non-stationary and long memory stationary processes. Huang and North [189] and Polyak [318] are examples of authors who describe the analysis of cyclo-stationary processes in a climate research setting (cf. [10.2.6]).

However, note that the non-stationary models and methods described in the literature are often most relevant in an econometric setting. For example, Box and Jenkins [60] describe a class of models called *auto-regressive integrated moving average*, or *ARIMA*, models. ARIMA processes $\mathbf{X}_t$ are nonstationary stochastic processes that become weakly stationary ARMA processes after a differencing operator of some order has been applied. That is, they are processes that have backshift operator (cf. [10.5.5]) representation of the form

$$\phi(B)(1-B)^d\mathbf{X}_t = \theta(B)\mathbf{Z}_t \qquad (12.6)$$

where all the roots of $\phi(B)$ lie outside the unit circle. The operator $(1 - B)$ represents the first differencing operation $\mathbf{X}_t - \mathbf{X}_{t-1}$. The simplest model of this form is the random walk (cf. equation (10.4) in [10.2.8]), which has $\phi(B) = \theta(B) = 1$ and $d = 1$. As with the random walk, all ARIMA processes integrate noise without forgetting any of its effects. The ARIMA class of

[2]For example, when fitting a univariate AR model at every grid point of a time series of analysed fields, as in Trenberth [369].

models (12.6) is attractive because it provides a method that can be used to deal with many types of non-stationary behaviour (random walks, trends, explosive growth, etc.) simply by repeatedly applying the first differencing operator. Processes that can be made stationary in this way are often seen in the economic world (e.g., the accumulation of money by a financial institution) but seldom seen in the physical world except on short time scales (e.g., the accumulation of precipitation over short periods of time).

12.2.1 Making a First Guess of the Order. We will illustrate the method used to make a first guess of the order of the process with simulated time series from known processes.

First we consider the examples presented in [12.1.3] and [12.1.7].

Estimates of the full and partial auto-correlation functions computed from two time series of length 240 are shown in Figure 12.3. The samples were taken from the AR(2) processes with $(\alpha_1, \alpha_2) = (0.9, -0.8)$ and $(0.3, 0.3)$ that were discussed extensively in Chapters 10 and 11. The estimated auto-correlation functions (upper panels) are similar to their theoretical counterparts displayed in Figure 11.2b. The random perturbations observed at large lags are due to sampling variability.

Despite the similarity between the theoretical and estimated functions, the generating processes can not be unequivocally identified as AR(2) processes. However, since the estimated partial auto-correlation functions (lower panels) quickly fall to zero after lag-2, the AR(2) model would be a good first guess in both cases.

The second example we consider is a time series of length 240 generated from an MA(10) process with parameters $\beta_1 = \cdots = \beta_{10} = 1$. The full and partial auto-correlation function estimates computed from this time series are displayed in the lower panels of Figures 12.1 and 12.2.

This time series presents a greater challenge than the examples discussed above. The full auto-correlation function decays to zero more or less exponentially, suggesting that the process may be a low-order AR process. The partial auto-correlation function decays to zero after two lags, suggesting that the process is AR(2). However, there are also partial auto-correlation estimates that are significantly different from zero at lags 6, 10, 11, and 14. A skilled practitioner may suspect the process to be a pure MA process because the full auto-correlation function goes to zero quickly and the partial auto-correlation function has rather complex behaviour. On the other hand, our perception of Figures 12.1 and 12.2 is coloured by our knowledge of the model that generated the data. Making a correct first guess of the type and order of model to fit is very difficult in this case.

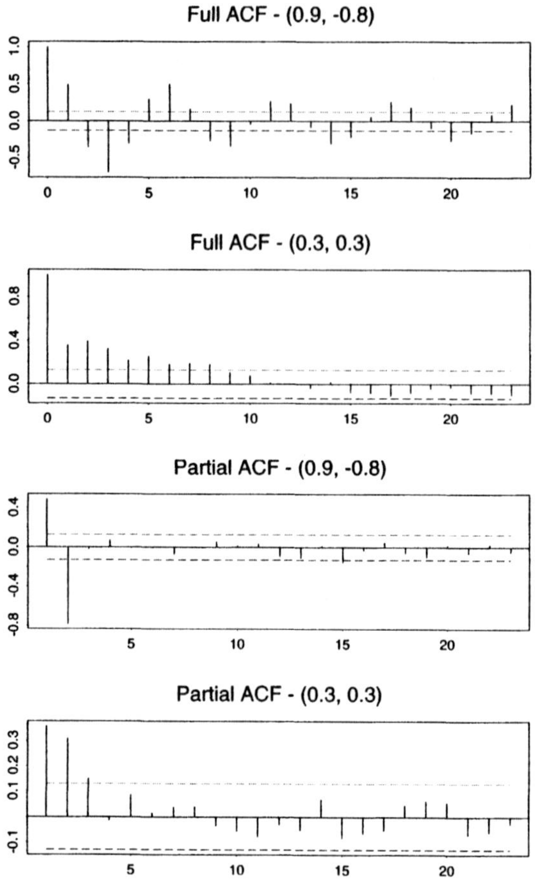

Figure 12.3: *Estimated full (upper panels) and partial (lower panels) auto-correlation functions computed from time series of length 240 generated from two AR(2) processes.*

12.2.2 Fitting AR Processes: the Yule–Walker Method. We now restrict ourselves to AR processes, both because the class of AR models is as rich as the class of ARMA models[3] and because we do not wish to consider models whose dynamics are forced by noise processes with memory (cf. [10.5.4]).

The Yule–Walker estimates of the parameters of an AR(p) process are obtained simply by plugging values of the estimated auto-covariance function $c(\tau)$ or auto-correlation function $r(\tau)$ into

[3]Note, however, that the AR approximation of a given process may not be as *parsimonious* as an ARMA approximation. See [10.5.4].

the Yule–Walker equations (11.2) (see [11.1.6]) and solving the system for the unknown process parameters. Thus the Yule–Walker estimates are given by

$$\widehat{\vec{\alpha}}_p = \mathcal{R}^{-1}\big(r(1), \dots, r(p)\big)^{\mathrm{T}}$$

where $\mathcal{R}$ is the $p \times p$ matrix with $R_{ij} = r(|i-j|)$ and $\widehat{\vec{\alpha}}_p$ is the vector of parameter estimates $(\widehat{\alpha}_1, \dots, \widehat{\alpha}_p)^{\mathrm{T}}$.

These parameter estimates can be used to estimate the variance of the forcing noise, say σ_Z^2, by computing

$$\widehat{\sigma}_Z^2 = \frac{1}{T}\sum_{t=p+1}^{T}(\mathbf{x}'_t - \widehat{\alpha}_1\mathbf{x}'_{t-1} - \cdots - \widehat{\alpha}_p\mathbf{x}'_{t-p})^2. \tag{12.7}$$

12.2.3 Example: Yule–Walker Estimates. The Yule–Walker estimates of (α_1, α_2) computed from the auto-correlation functions displayed in Figure 12.3 and the corresponding estimates of the noise variance are given in the following table.

Yule–Walker parameter estimates and noise variance estimates

	(0.9, −0.8)	(0.3, 0.3)
$\widehat{\alpha}_1$	0.868	0.358
$\widehat{\alpha}_2$	−0.784	0.308
$\widehat{\sigma}_Z^2$	1.077	1.101

Both fitted models are close to those that generated the data. The noise variance is only slightly overestimated in both cases and the errors in the parameter estimates are modest. The spectra of the fitted processes also closely match those of the generating processes. The first model has a spectral maximum at a slightly higher frequency ($\omega = 0.202$; see [11.2.6] and (11.24)) than the generating process ($\omega = 0.166$). The second has a spectral minimum at $\omega = 0.282$, which compares well with $\omega = 0.278$ for the generating process.

We conducted a small Monte Carlo experiment to obtain an impression of how the bias varies with sample size. One hundred samples of length $T = 15$, 60, and 240 were generated from AR(2) processes with parameters $(\alpha_1, \alpha_2) = (0.9, -0.8)$ and $(0.3, 0.3)$. Each sample was used to compute Yule–Walker parameter estimates.

The results, given in the table below, show that the bias is substantial when samples are very small. The bias becomes modest for samples of moderate length and, for these examples, becomes quite small for samples of length 240.

The mean of 100 Yule–Walker parameter estimates

T	(0.9, −0.8)	(0.3, 0.3)
15	(0.72, −0.63)	(0.16, 0.04)
60	(0.85, −0.75)	(0.27, 0.24)
240	(0.88, −0.78)	(0.30, 0.29)

Note that these results do not fully reflect the actual properties of the Yule–Walker estimator in practice because prior knowledge was used to choose the order of AR process to fit.

12.2.4 Fitting AR Processes: Maximum Likelihood. Most statistical and scientific subroutine packages include routines that compute maximum likelihood estimates of AR (and ARMA) parameters. We therefore give a general description of how these estimates are obtained in this subsection, and describe the estimation of their uncertainty in [12.2.6]. In most climatological research contexts, however, the Yule–Walker estimates provide close approximations to the exact maximum likelihood estimates (MLEs). Maximum likelihood estimation should be used when samples are 'small' or when the AR parameters are thought to be close to the boundaries of the admissible region. Even so, parameter estimates will be somewhat biased, as discussed at the end of this subsection.

To simplify the discussion below we assume that the observed process has zero mean. To keep our notation fairly compact, we let $\vec{\mathbf{x}}_T$ be the T-dimensional vector $(\mathbf{x}_1, \dots, \mathbf{x}_T)^{\mathrm{T}}$ that contains the observed time series. Also, we let $\vec{\mathbf{X}}_T$ be the vector that contains the corresponding segment of the stochastic process $\{\mathbf{X}_t : t \in \mathbb{Z}\}$. Vectors $\vec{\mathbf{x}}_p$ and $\vec{\mathbf{X}}_p$ are defined similarly.

Let $\{\mathbf{X}_t\colon t \in \mathbb{Z}\}$ be a stationary, normally distributed, AR(p) process that is forced by noise with variance σ_Z^2 and has parameters $\vec{\alpha}_p = (\alpha_1, \dots, \alpha_p)^{\mathrm{T}}$. Then the joint density function of $\vec{\mathbf{X}}_T$ is given by

$$f_{\vec{\mathbf{X}}_T}(\vec{\mathbf{x}}_T|\vec{\alpha}_p, \sigma_Z) = \frac{|\mathcal{M}_p|^{1/2}\, e^{-S(\vec{\alpha}_p)/(2\sigma_Z^2)}}{(2\pi\sigma_Z^2)^{T/2}} \tag{12.8}$$

where

$$\begin{aligned} S(\vec{\alpha}_p) &= \vec{\mathbf{x}}_p^{\mathrm{T}}\mathcal{M}_p\vec{\mathbf{x}}_p \\ &\quad + \sum_{t=p+1}^{T}(\mathbf{x}_t - \alpha_1\mathbf{x}_{t-1} - \cdots - \alpha_p\mathbf{x}_{t-p})^2 \\ \mathcal{M}_p &= \Sigma_p^{-1} \end{aligned}$$

and where Σ_p is the $p \times p$ matrix whose (i, j)th element is $\gamma_{xx}(|i-j|)/\sigma_Z^2$. Note that the elements of $\mathcal{M}_p$ are independent of σ_Z^2 since they describe the auto-covariance function of an AR(p) process forced with unit variance white noise.

The likelihood function (see [5.3.8]) is the probability density (12.8)

$$L(\vec{\alpha}_p, \sigma_Z|\vec{\mathbf{x}}_T) = f(\vec{\mathbf{x}}_T|\vec{\alpha}_p, \sigma_Z)$$

re-expressed as a function of the unknown parameters for a fixed realization $\vec{\mathbf{x}}_T$ of $\vec{\mathbf{X}}_T$.

With normal random variables it is easier to work with the log-likelihood function because the latter is essentially quadratic in the parameters. Here, the log-likelihood is

$$\begin{aligned} l(\vec{\alpha}_p, \sigma_Z|\vec{\mathbf{x}}_T) &= -\frac{T}{2}\left(\ln(2\pi) + \ln\sigma_Z^2\right) \\ &\quad + \frac{1}{2}\ln|\mathcal{M}_p| - \frac{S(\vec{\alpha}_p)}{2\sigma_Z^2}. \end{aligned}$$

The constant $\ln(2\pi)$ is irrelevant to the derivation of MLEs, so the log-likelihood function is usually given as

$$l(\vec{\alpha}_p, \sigma_Z|\vec{\mathbf{x}}_T) = -\frac{T\ln\sigma_Z^2}{2} + \frac{\ln|\mathcal{M}_p|}{2} - \frac{S(\vec{\alpha}_p)}{2\sigma_Z^2}. \qquad (12.9)$$

Maximum likelihood estimates are found by setting the partial derivatives of (12.9) to zero. Differentiating, we obtain

$$\frac{\partial l}{\partial \sigma_Z} = -\frac{T}{\sigma_Z} + \frac{S(\vec{\alpha}_p)}{\sigma_Z^3} \qquad (12.10)$$

$$\frac{\partial l}{\partial \alpha_k} = M_k + D_{1,k+1} - \frac{1}{\sigma_Z^2}\sum_{j=1}^{p}\alpha_j D_{j+1,k+1} \quad \text{for } k = 1, \dots, p \qquad (12.11)$$

where D_{ij} is the sum

$$D_{ij} = \mathbf{x}_i\mathbf{x}_j + \cdots + \mathbf{x}_{T+1-j}\mathbf{x}_{T+1-i}$$

and M_k is the partial derivative

$$M_k = \frac{\partial \ln|\mathcal{M}_p|}{2\partial\alpha_k}. \qquad (12.12)$$

Equations (12.10) and (12.11) are not generally used to compute maximum likelihood estimates of the AR parameters because partial derivative (12.12) is difficult to evaluate.

Instead, maximum likelihood estimates (MLEs) are obtained by using *nonlinear* numerical minimization techniques to find the minimum of $-2l(\vec{\alpha}_p, \sigma_Z|\vec{\mathbf{x}}_T)$. Ingenious methods for evaluating the log-likelihood method have been developed so that the minimization can be done efficiently (see Box and Jenkins [60, Chapter 7]; Ansley and Kohn [14]; Kohn and Ansley [228]). Also, it is difficult to constrain numerical minimization methods to the admissible parameter region for weakly stationary AR(p) processes. Thus transformations are used to map the admissible region onto the real p-dimensional vector space (see, e.g., Jones [205]). These transformations enforce stationarity in the fitted model by mapping the boundaries of the admissible region to infinity. Consequently, MLEs of AR parameters tend to be negatively biased, particularly when the time series comes from a process with parameters that are close to the edge of the admissible region. The next subsection shows, however, that the bias of ML estimates is less than that of Yule–Walker estimates.

12.2.5 Example: Maximum Likelihood Estimates. The MLEs corresponding to those displayed in [12.2.3] are

	(0.9, −0.8)	(0.3, 0.3)
$\widehat{\alpha}_1$	0.871	0.260
$\widehat{\alpha}_2$	−0.785	0.322
$\widehat{\sigma}_Z^2$	0.967	1.103

Because samples are large, these estimates appear to be only slightly different from the Yule–Walter estimates.

However, MLEs are more than worth the effort when samples are small. To illustrate, we repeated the Monte Carlo experiment described in [12.2.3], making ML estimates instead of Yule–Walker estimates.

The mean of 100
ML parameter estimates

T	(0.9, −0.8)	(0.3, 0.3)
15	(0.83, −0.73)	(0.29, 0.16)
60	(0.88, −0.78)	(0.30, 0.27)
240	(0.90, −0.80)	(0.30, 0.29)

Comparing the results in the above table with those in [12.2.3], we see that the negative bias of Yule–Walker estimates is reduced in all cases. The reduction in bias is particularly dramatic when samples are very small. In this case, the reduction of bias does not come at the cost of increased variability. The ML estimates have variance that is comparable to that of the Yule–Walker estimates. Again, be aware that these results do not fully reflect the practical properties of the ML estimator because we used prior knowledge to choose the order of AR process to fit.

12.2.6 Uncertainty of Maximum Likelihood Parameter Estimates. Software that computes MLEs also usually provides an estimate of their uncertainty. These uncertainty estimates are obtained through the use of large sample theory that approximates distributions of the AR parameter estimates (see, e.g., Box and Jenkins [60, Appendix A7.5]). The final result, an estimate of the variance-covariance matrix of the MLEs, is

$$\widehat{\Sigma}_{\hat{\vec{\alpha}}_p} = \frac{1}{T}\frac{\widehat{\sigma}_Z^2}{c(0)}\mathcal{R}^{-1} \qquad (12.13)$$

where $\widehat{\sigma}_Z^2$ is an estimate of the variance of the noise process,[4] $c(0)$ is the sample variance (12.2) of the time series, $\mathcal{R}$ is the $p \times p$ matrix that has $r(|i-j|)$ as its (i, j)th element, and $r(\tau)$ is the estimated auto-correlation function (12.1).

12.2.7 Example: Uncertainty of MLEs. We used (12.13) to estimate the standard errors and correlation of the ML parameter estimates given in [12.2.5]. We obtained

$$\widehat{\Sigma}_{\hat{\vec{\alpha}}_2} = 0.040\begin{pmatrix} 1 & -0.488 \\ -0.488 & 1 \end{pmatrix}$$

for the sample from the process with $(\alpha_1, \alpha_2) = (0.9, -0.8)$ and

$$\widehat{\Sigma}_{\hat{\vec{\alpha}}_2} = 0.061\begin{pmatrix} 1 & -0.384 \\ -0.384 & 1 \end{pmatrix}$$

for the sample from the process with $(\alpha_1, \alpha_2) = (0.3, 0.3)$. Note that the elements of $\hat{\vec{\alpha}}^2$, $\widehat{\alpha}_1$ and $\widehat{\alpha}_2$, have the same estimated variance.

Approximate 95% confidence regions for (α_1, α_2) can then be derived from these estimates as follows. We assume that

$$\widehat{\vec{\alpha}}_2 \sim \mathcal{N}(\vec{\alpha}_2, \Sigma_{\hat{\vec{\alpha}}_2}).$$

Consequently

$$\mathbf{X} = (\widehat{\vec{\alpha}}_2 - \vec{\alpha}_2)^{\mathrm{T}}\Sigma_{\hat{\vec{\alpha}}_2}^{-1}(\widehat{\vec{\alpha}}_2 - \vec{\alpha}_2) \sim \chi^2(2).$$

Thus an approximate 95% confidence region is obtained by replacing $\Sigma_{\hat{\vec{\alpha}}_2}$ with $\widehat{\Sigma}_{\hat{\vec{\alpha}}_2}$ and solving

$$(\widehat{\vec{\alpha}}_2 - \vec{\alpha}_2)^{\mathrm{T}}\widehat{\Sigma}_{\hat{\vec{\alpha}}_2}^{-1}(\widehat{\vec{\alpha}}_2 - \vec{\alpha}_2) = \mathbf{X}_{0.95}$$

where $\mathbf{X}_{0.95}$ is the 95% critical value of $\chi^2(2)$ distribution (see Appendix E).

The resulting confidence regions are shown in Figure 12.4. It is reasonable to believe that these

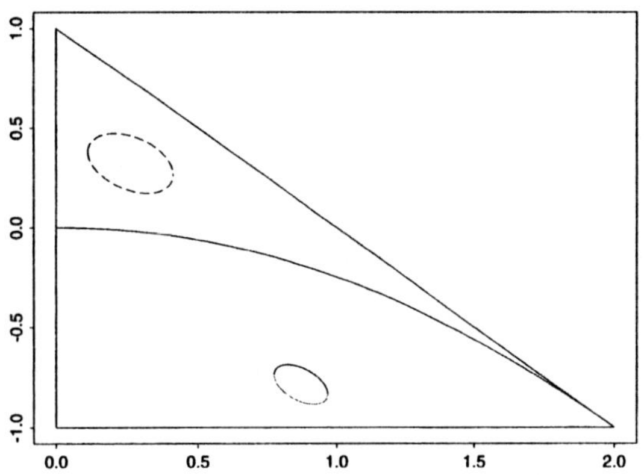

Figure 12.4: *Approximate 95% confidence regions for (α_1, α_2) computed from samples of length 240 generated from AR(2) processes with $(\alpha_1, \alpha_2) = (0.9, -0.8)$ (solid ellipse) and $(\alpha_1, \alpha_2) = (0.3, 0.3)$ (dashed ellipse). The triangle depicts the right half of the admissible region for the parameters of a stationary AR(2) process.*

confidence intervals are approximately correct in this case, since the samples are fairly large and both ellipsoids lie well within the admissible region for the parameters of an AR(2) process.

This is confirmed by extracting more information from the Monte Carlo experiment described in [12.2.5]. Each sample was used to estimate the asymptotic standard errors of $\widehat{\alpha}_1$ and $\widehat{\alpha}_2$ with (12.13). The mean estimate is compared with the actual variability of the 100 MLEs in the following table.

The observed standard deviation of 100 ML parameter estimates compared with the mean of 100 asymptotic estimates

$(\alpha_1, \alpha_2) = (0.9, -0.8)$

	Observed std. dev.		Mean
T	$\widehat{\alpha}_1$	$\widehat{\alpha}_2$	estimate
15	0.22	0.18	0.16
60	0.087	0.078	0.080
240	0.044	0.042	0.039

$(\alpha_1, \alpha_2) = (0.3, 0.3)$

	Observed std. dev.		Mean
T	$\widehat{\alpha}_1$	$\widehat{\alpha}_2$	estimate
15	0.32	0.23	0.26
60	0.12	0.10	0.13
240	0.070	0.062	0.062

Only one column is used to describe the mean asymptotic estimate of standard error since the diagonal elements of $\widehat{\Sigma}_{\hat{\vec{\alpha}}_2}$ are equal. The table

[4]This is usually provided by the ML estimation software, but is usually also closely approximated by (12.7).

shows that the large-sample theory standard error estimator performs surprisingly well even when samples are quite small.[5] Comparable performance can be expected when (12.13) is applied to Yule–Walker parameter estimates.

12.2.8 Model Diagnostics. We have now tentatively identified an AR model, estimated its parameters and perhaps also constructed an estimate of the uncertainty of the parameters with (12.13). The next step is to determine whether the model fits well. We give a very brief sketch here of a few of the ideas involved. Box and Jenkins [60] cover this topic in much more depth.

As with regression diagnostics (cf. [8.3.12–16]), it is important to plot the time series itself and to plot the estimate of the noise process

$$\widehat{\mathbf{z}}_t = \mathbf{x}'_t - \widehat{\alpha}_1\mathbf{x}'_{t-1} - \cdots - \widehat{\alpha}_p\mathbf{x}'_{t-p} \qquad (12.14)$$
$$t = p+1, \ldots, T.$$

These plots should be examined for trends, periodicities, outliers, and other evidence that the weak stationarity assumption has been violated.

It is also useful to overfit the model. If it is possible to reduce substantially the estimated error variance $\widehat{\sigma}_Z^2$ (12.7) or increase substantially the log-likelihood (12.9) by adding additional lagged terms to the AR model, then a higher-order model should be considered.

The residuals $\widehat{\mathbf{z}}_t$ (12.14) should be examined to check that they behave as white noise. They will not, of course, do so exactly because the residuals will only be asymptotically independent of one another, even when the correct model has been selected. None the less, it is useful to compute and plot the auto-correlation function of the estimated noise process. The standard errors of these auto-correlations will be approximately $1/\sqrt{T}$ at large lags.

It is also sometimes useful to compute a *portmanteau lack-of-fit statistic* such as

$$\mathbf{Q}(K) = (T-p)\sum_{\tau=1}^{K}\left(r_{\hat{\mathbf{z}}\hat{\mathbf{z}}}(\tau)\right)^2 \qquad (12.15)$$

to diagnose whether the first K lags of the auto-correlation function of the residuals jointly estimate the zero function. Note that p is the order of the fitted model and $r_{\hat{\mathbf{z}}\hat{\mathbf{z}}}(\tau)$ is the auto-correlation function of the estimated noise process. It has been shown that $\mathbf{Q}$ is distributed approximately $\chi^2(K-p)$ when the correct model has been selected, T is moderate to large, and K is of moderate size relative to T. Statistical packages such as S-Plus [78] sometimes plot $P(\mathbf{Q}(k) > \mathbf{q}(k)|H_0)$ against k for moderate values of k as a diagnostic aid. Lack-of-fit is indicated when these 'p-values' fall to near zero at some lag.

Are there hidden periodicities in the residuals $\widehat{\mathbf{z}}_t$? Truly periodic behaviour is sometimes difficult to detect in plots of the time series and the residuals, although a plot of the *normalized cumulative periodogram* as a function of frequency is often able to reveal such behaviour. The *periodogram* (cf. [11.2.0] and Section 12.3) is the squared modulus of the Fourier transform (11.16) of the residuals

$$I(\omega_j) = \frac{2}{T-p}\left(\left(\sum_{t=p+1}^{T}\widehat{\mathbf{z}}_t\cos(2\pi\omega_j t)\right)^2 + \left(\sum_{t=p+1}^{T}\widehat{\mathbf{z}}_t\sin(2\pi\omega_j t)\right)^2\right) \qquad (12.16)$$

computed at frequencies $\omega_j = j/(T-p)$, $j = 1, \ldots, (T-p-1)/2$.[6]

The normalized cumulative periodogram is computed from $I(\omega_j)$ as

$$Q(\omega_j) = \frac{1}{(T-p)\widehat{\sigma}_Z^2}\sum_{i=1}^{j} I(\omega_i),$$

where $\widehat{\sigma}_Z^2$ is the estimated variance (12.7) of the forcing noise. When the correct model has been chosen we expect $Q(\omega_j)$ to increase linearly from 0 to 1 with increasing ω_j.[7] Departures from linearity indicate either the presence of discrete periodic behaviour in $\widehat{\mathbf{z}}_t$ (and hence $\mathbf{Z}_t$) that can not be captured by an AR model, or the presence of quasi-periodic behaviour that cannot be captured by the chosen model. In the latter case, a higher-order model may be indicated.

When $\widehat{\mathbf{z}}_t$ is exactly white noise,[8] then

$$K = \max_j\left[\max[|Q(\omega_j)-2\omega_j|, |Q(\omega_j)-2\omega_{j-1}|]\right] \qquad (12.17)$$

has the same distribution as the Kolmogorov-Smirnov goodness-of-fit statistic (see [5.2.3]) for the case in which the distribution is fully specified

[5] Note, however, that the Monte Carlo experiment is conducted under ideal conditions: The process is normal and its order is known. The performance in practice will not be quite as good.

[6] We have assumed, for convenience, that $T-p$ is odd.

[7] We will see in [12.3.6] that (12.16) is an estimate of the autospectrum of $\widehat{\mathbf{z}}_t$. When $\widehat{\mathbf{z}}_t$ is white, the expected value of $I(\omega_j)$ is σ_Z^2 for all j.

[8] Even when the correct model has been chosen, $\widehat{\mathbf{z}}_t$ behaves as white noise only in the limit as $T \to \infty$.

by the null hypothesis.[9] Thus the critical value for testing the null hypothesis that the spectrum of $\widehat{\mathbf{z}}_t$ is white at the $p \times 100\%$ level is $K_p/\sqrt{\lceil \frac{T-1}{2}}$, where K_p is given in the table below and where the notation $\lceil x$ refers to the largest integer contained in x.

Significance level	K_p
0.01	1.63
0.05	1.36
0.10	1.22
0.25	1.02

12.2.9 Example: Diagnostics. The first example we consider is the time series of length 240 generated from an AR(2) process with parameters $(\alpha_1, \alpha_2) = (0.9, -0.8)$ (see also [12.2.1,3,5,7]. The full and partial auto-correlation function estimates computed from this time series are shown in Figure 12.3. Both functions show behaviour characteristic of an AR(2) model, so $p = 2$ is a good tentative choice.

Diagnostic plots of the residuals, the estimated auto-correlation function of the residuals, and the p-values of the portmanteau goodness-of-fit statistic (12.15) are shown in Figure 12.5. These plots confirm our tentative choice of model. The auto-correlation function of the residuals (middle panel) is essentially zero for nonzero lags, and all p-values of the portmanteau statistic (lower panel) are greater than the 5% critical value, which is shown as a dashed line. The upper panel hints at behaviour that might bear investigation if we had fitted this model to real data; the variability of the residuals for $t = 2$ to $t \approx 50$ seems to be somewhat less than that of subsequent residuals. The cumulative periodogram of the residuals (not shown) supports the hypothesis that the correct model has been selected.

In our second example we deliberately fit an AR(1) model to the AR(2) time series to produce an extreme example of a set of diagnostic plots (Figure 12.6) that show lack-of-fit. The plot of the residuals reveals quasi-periodic behaviour that has not been captured by the fitted model. This is also revealed in the auto-correlation function of the residuals. The p-values of the portmanteau statistics (not shown) are uniformly less than 0.05. In addition, the cumulative periodogram (Figure 12.7) shows that there is quasi-periodic variation at frequencies roughly in the interval $(0.1, 0.2)$.

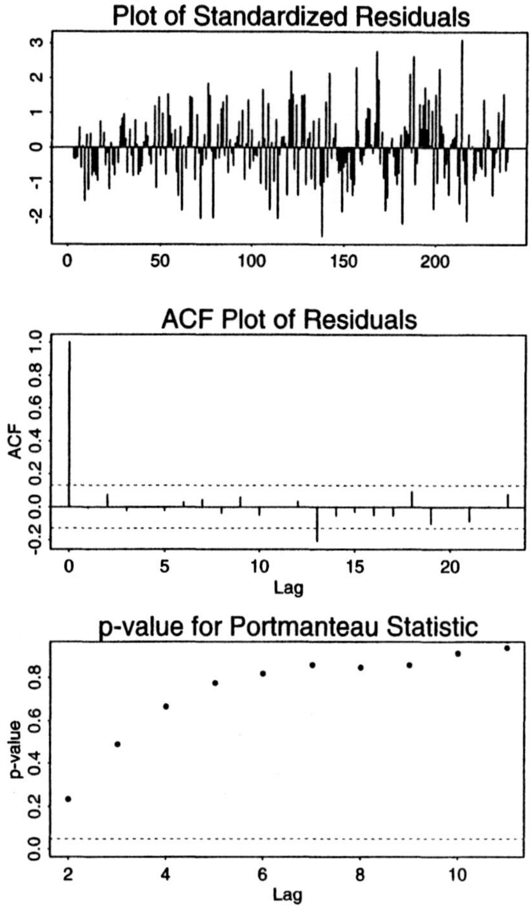

Figure 12.5: *Plots diagnosing the goodness of a maximum likelihood fit of an AR(2) model to a time series of length 240 generated from and AR(2) process with* $(\alpha_1, \alpha_2) = (0.9, -0.8)$.
Top: The residuals $\widehat{\mathbf{z}}_t$.
Middle: The auto-correlation function $r_{\widehat{\mathbf{z}}\widehat{\mathbf{z}}}(\tau)$ *of the residuals.*
Bottom: p-values of the portmanteau statistic $\mathbf{q}(k)$.

12.2.10 Objective Order Determination: AIC. The Box–Jenkins method of model identification and fitting is labour intensive: the investigator must be actively (and skilfully) involved. This is a very strong advantage, because such close interaction with the data will help to identify problems with lack-of-fit, but also a disadvantage because the method can not be practically applied to the large fields of time series often encountered in climate research. Objective order determining criteria are

[9] We will see in [12.3.5–7] that Q is the cumulative sum of independent identically distributed random variables when $\widehat{\mathbf{z}}_t$ is white. The Kolmogorov-Smirnov statistics is written in the same way as (12.17) except that Q is replaced with the empirical distribution function. The latter is also a cumulative sum of independent identically distributed random variables.

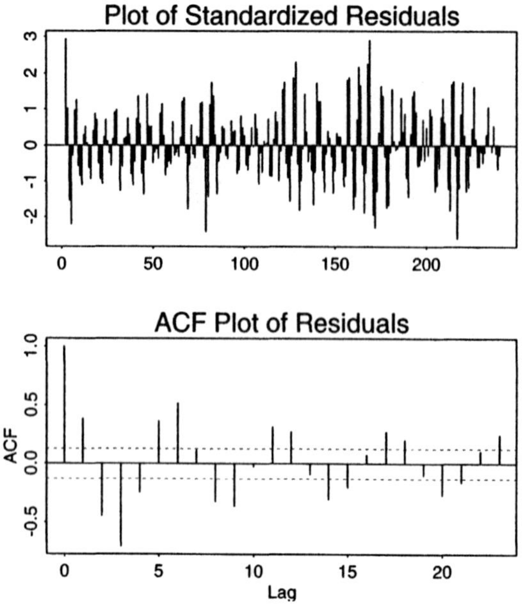

Figure 12.6: *As Figure 12.5 except these plots diagnose the lack-of-fit of an AR(1) model to the AR(2) time series.*

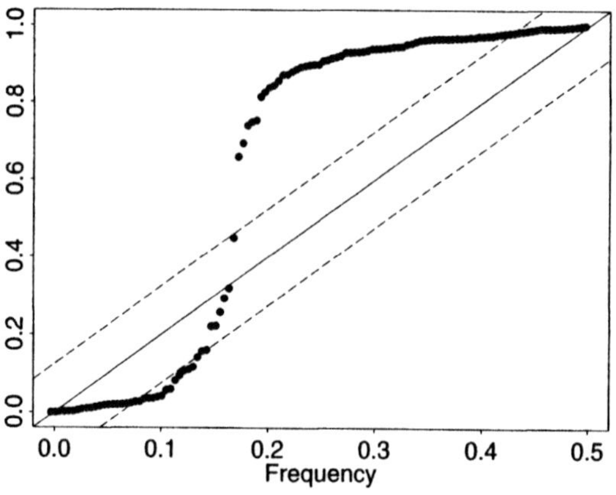

Figure 12.7: *The cumulative periodogram of the residuals obtained by fitting an AR(1) model to a time series of length 240 generated from an AR(2) process with* $(\alpha_1, \alpha_2) = (0.9, -0.8)$. *The dashed lines indicate 5% critical values for testing that the residuals are white.*

used to circumvent this problem (see, e.g., Katz [217]; Chu and Katz [85, 86]; Zwiers and von Storch [453]; Zheng, Basher, and Thompson [437]).

Two order determining criteria are commonly used. The *Akaike information criterion* (AIC; see [6][7]) determines the order by minimizing

$$\begin{aligned} AIC_p &= -2l(\widehat{\vec{\alpha}}_p, \widehat{\sigma}_Z | \vec{\mathbf{x}}_T) + 2(p+1) \\ &\approx T \ln(\widehat{\sigma}_Z^2) + 2(p+1) \end{aligned}$$

where $\widehat{\sigma}_Z^2$ is the estimated noise variance (12.7). In effect, the maximum log-likelihood obtained by fitting a model of order p is penalized by subtracting the number of parameters that were fitted. The order is chosen to be that which minimizes AIC_p.

A heuristic way to understand how AIC works is as follows. Suppose we have fitted a model of order $p+q$ and want to test the null hypothesis H_0: $\alpha_{p+1} = \cdots = \alpha_{p+q} = 0$, that the last q AR parameters are zero. H_0 can be tested with the *likelihood ratio statistic*

$$2\delta l = 2l(\widehat{\vec{\alpha}}_{p+q}, \widehat{\sigma}_Z | \vec{\mathbf{x}}_T) - 2l(\widehat{\vec{\alpha}}_p, \widehat{\sigma}_Z | \vec{\mathbf{x}}_T),$$

which is asymptotically distributed $\chi^2(q)$ under the null hypothesis. Thus $\mathcal{E}(2\delta l) \approx 2q$ when H_0 is true. That is, if the true order of the AR process is no greater than p, then the expected change between the log-likelihood of an AR(p) model and an AR($p+q$) model will be about q. The penalty compensates for this apparent increase in the log-likelihood.

However, the argument above also reveals a difficulty with the AIC that has been pointed out in the literature (see, e.g., Jones [204]; Katz [216]; Hurvich and Tsai [191]); the AIC determined order is an inconsistent estimate of the order of the process. Note that the variance of δl, and hence of the AIC, does not decrease with increasing sample size. Consequently, the sampling variability of the AIC determined order will not decrease with increasing sample size. In fact, AIC tends to overestimate the order of the process somewhat. However, these problems are not serious in practice.

The following table gives AICs for AR models of order 0–5 fitted with the Yule–Walker method to our time series of length 240 generated from the AR(2) process with $(\alpha_1, \alpha_2) = (0.9, -0.8)$.

p versus AIC_p

p	0	1	2	3	4	5
AIC_p	313	250	23.9	25.8	26.5	28.5

The minimum AIC is indeed achieved by a model of the correct order. The AIC is large for models of order less than 2 and increases slowly with p for models of order greater than 2.

We repeated this exercise 1000 times with time series of each length 60, 120, and 240 generated from AR(2) processes with $(\alpha_1, \alpha_2) = (0.9, -0.8)$ and $(0.3, 0.3)$. The results are summarized in the following table.

The frequency of AIC selected order of 1000 AR(2) time series of length T

$(\alpha_1, \alpha_2) = (0.9, -0.8)$

	Order					
T	0	1	2	3	4	≥ 5
60	0	0	736	143	46	75
120	0	0	744	120	57	79
240	0	0	742	116	62	80

$(\alpha_1, \alpha_2) = (0.3, 0.3)$

	Order					
T	0	1	2	3	4	≥ 5
60	60	220	505	101	55	59
120	1	41	715	105	58	80
240	0	1	717	111	61	111

AIC seldom underestimates the order of the process for the larger sample sizes, at least for the time series considered here. The tendency to overestimate the order appears to strengthen slightly with increasing sample size. As anticipated, the variability of the estimated order does not decrease with increasing sample size.

12.2.11 Objective Order Determination: BIC. The other order determining criterion that is often used is the *Bayesian information criterion* [360]. It is also developed around a test statistic, but in a Bayesian rather than frequentist setting. The statistic used in the development of the BIC is the *Bayes factor* $B_{(p+q),p}$ that compares the evidence for the model of order $p+q$ with that for the model of order p. The Bayes factor is similar to a likelihood ratio except that numerator and denominator are average likelihoods integrated relative to a prior distribution on the parameters of the process. When the sample is large, the prior distribution plays a relatively minor role in the Bayes factor, and it can then be shown that

$$2 \ln B_{(p+q),p} \approx 2\,\delta l - q \ln(n).$$

The BIC is consequently defined as

$$BIC_p = -2l(\widehat{\vec{\alpha}}_p, \widehat{\sigma_Z}|\vec{\mathbf{x}}_T) + (p+1)\ln(n).$$

From our perspective, of course, the main difference between the AIC and BIC is that the penalty for using an extra parameter is much greater with the latter. This penalty reflects a fundamental difference between the ways in which frequentists and Bayesians weigh evidence. These differences are beyond the scope of this book. Some readers may be interested in the highly readable discussion of this subject by Raftery [328] and discussants. Hannan [158] shows that BIC is a consistent order determining criterion.[10]

The frequency of BIC selected order of 1000 AR(2) time series of length T

$(\alpha_1, \alpha_2) = (0.9, -0.8)$

	Order					
T	0	1	2	3	4	≥ 5
60	0	0	950	40	8	2
120	0	0	969	28	3	0
240	0	0	975	22	3	0

$(\alpha_1, \alpha_2) = (0.3, 0.3)$

	Order					
T	0	1	2	3	4	≥ 5
60	196	371	403	26	6	0
120	22	198	754	18	4	4
240	0	15	967	14	4	0

The Monte Carlo experiment described in [12.2.10] was repeated using BIC. The results, which are given in the table above, illustrate that BIC tends to select more parsimonious models (i.e., models with fewer parameters) than AIC. We see that it generally identifies the correct order of the process more accurately than AIC, and that its skill improves with increasing sample size. Overall, the BIC order estimates have much lower bias and variability than their AIC counterparts. We therefore recommend the use of BIC over AIC.

12.3 Estimating the Spectrum

12.3.0 Overview. We give a brief introduction to the estimation of power spectra in this section. As in many other parts of this book, our purpose is not to be exhaustive but rather to give a flavour of the reasoning and issues involved. Jenkins and Watts [195] and Koopmans [229] give a much more detailed and competent exposition than we do here. Bloomfield [49] provides a very accessible introduction to the subject.

The fundamental tool that we will use is the periodogram.[11] The connection between the periodogram and the auto-covariance function, the statistical properties of the periodogram, and consequently the reasons for not using it as

[10] That is, $\mathcal{E}\left((\widehat{p}_{BIC} - p)^2\right) \to 0$ as $T \to \infty$ where $\widehat{p}_{BIC}$ is the order selected with BIC.

[11] We previously touched on the periodogram in [11.2.0] and [12.2.8].

an estimator of the autospectrum are discussed first. However, we will see that, despite the periodogram's poor properties as a raw spectral estimator, spectral estimators with much more acceptable characteristics can be constructed from the periodogram.

A pitfall that many have encountered is to confuse *harmonic analysis*, the detection of regular periodic signals, with *spectral* analysis, the description of how variance is distributed as a function of time scale in processes that do not vary periodically. The potential for confusion is clearly apparent from Shaw's 1936 study [347] that we cited in Chapter 10. The periodogram does not have better properties when applied to harmonic analysis than when applied to spectral analysis, but it is truly useful as a tool for harmonic analysis when the source of the periodicities is clearly understood as in the analysis of tidal variations (e.g., Zwiers and Hamilton [447]) or the analysis of emissions from a rotating star (e.g., Bloomfield's analysis of observations of a variable star [49]).

An important historical note is that Slutsky [351] was apparently suspicious of the way in which some economic data were being analysed. He showed that variance can be confined to a narrow frequency band by passing white noise through a series of summing filters

$$\mathbf{Y}_t = \mathbf{Z}_t + \mathbf{Z}_{t-1}$$

and differencing filters

$$\mathbf{Y}_t = \mathbf{Z}_t - \mathbf{Z}_{t-1}.$$

In fact, if white noise is passed through m summing filters and n differencing filters then the output process can be shown to have spectral density function

$$\Gamma_{yy}(\omega) = 2^{m+n+1}(\cos\pi\omega)^{2m}(\sin\pi\omega)^{2n}\sigma_Z^2,$$

which has a peak at

$$\omega_0 = \cos^{-1}\left(\frac{m-n}{m+n}\right).$$

In the limit, if m and n are allowed to increase infinitely in such a way that $(m-n)/(m+n)$ tends to a constant, all the energy in the spectrum is concentrated at a single frequency. Hence the limiting process is a single sinusoid. We now call this the *Slutsky effect*. Slutsky confirmed the effect by means of a simulation. Koopmans [229] points out that Slutsky's result was seminal in the development of ARMA models because it illustrated a previously unknown mechanism for generating quasi-periodic behaviour. The only time series models known before this time were combinations of simple, almost periodic functions (such as (11.25)) and white noise residuals.

The plan for the remainder of this section is as follows. We will explore the properties of the periodogram in subsections [12.3.1–7]. Data tapers, which are used to counteract problems that arise when a process has a periodic component or a spectrum with sharp peaks, are described in [12.3.8]. Spectral estimators constructed from the periodogram are covered in subsections [12.3.9–19]. The 'chunk' estimator (which is also sometimes referred to as the Bartlett estimator [12.3.9,10]), is discussed first because it is easily adapted to climate problems in which, for example, a daily time series of length 90 days is observed at the same time every year. We then go on to develop some ideas that will help readers understand how spectral estimators are constructed and interpreted. This is done in subsections [12.3.11–18] by describing smoothed periodogram estimators that are commonly used to analyse time series that are contiguous in time (as opposed to time series composed of a number of disjoint chunks). Subsection [12.3.19] contains a summary of spectral estimators constructed from the periodogram. An example intercomparing spectral estimators is presented in [12.3.20], and an alternative approach to spectral estimation is briefly discussed in [12.3.21]. The effects of *aliasing* are discussed in [12.3.22].

12.3.1 The Periodogram. Let $\{\mathbf{x}_1, \ldots, \mathbf{x}_T\}$ be a time series. Equation (C.1), which expands $\{\mathbf{x}_1, \ldots, \mathbf{x}_T\}$ in terms of complex exponentials, can be re-expressed in sine and cosine terms, as in equation (11.16), as

$$\mathbf{x}_t = a_0 + \sum_{j=1}^{q}\big(a_j\cos(2\pi\omega_j t) + b_j\sin(2\pi\omega_j t)\big), \tag{12.18}$$

where $q = \lceil\frac{T}{2}$, $\omega_j = j/T$, $j = 1, \ldots, q$, and the notation $\lceil x$ indicates the largest integer contained in x. The coefficients, given by equation (C.2), are

$$a_0 = \frac{1}{T}\sum_{t=1}^{T}\mathbf{x}_t \tag{12.19}$$

and

$$a_j = \frac{2}{T}\sum_{t=1}^{T}\mathbf{x}_t\cos(2\pi\omega_j t) \tag{12.20}$$

$$b_j = \frac{2}{T}\sum_{t=1}^{T}\mathbf{x}_t\sin(2\pi\omega_j t), \tag{12.21}$$

for $j = 1, \ldots, q$. Note that, for even T,

$$a_q = \frac{1}{T}\sum_{t=1}^{T}(-1)^q \mathbf{x}_t, \qquad (12.22)$$

$$b_q = 0. \qquad (12.23)$$

Thus the number of non-trivial coefficients a_j and b_j is always T. This is as it should be since the Fourier transform is simply a coordinate transformation in which information is neither lost nor gained.[12]

The time series can be recovered by substituting equations (12.19–12.22) into (12.18) and making use of the following orthogonality properties of the discretized sine and cosine functions:

a) $\sum_{t=1}^{T}\cos(2\pi\omega_k t)\cos(2\pi\omega_l t) = \frac{T}{2}\delta_{kl}$

b) $\sum_{t=1}^{T}\sin(2\pi\omega_k t)\sin(2\pi\omega_l t) = \frac{T}{2}\delta_{kl}$

c) $\sum_{t=1}^{T}\cos(2\pi\omega_k t)\sin(2\pi\omega_l t) = 0,$

where $\delta_{kl} = 1$ if $k = l$ and 0 otherwise.

The *periodogram* is defined in terms of the coefficients a_j and b_j as

$$I_{Tj} = \frac{T}{4}(a_j^2 + b_j^2) \qquad (12.24)$$

for $j = -\lceil\frac{T-1}{2}\rceil, \ldots, \lceil\frac{T}{2}\rceil$. For negative j, $a_j = a_{-j}$ and $b_j = -b_{-j}$. We explicitly include the subscript T to indicate that I is computed from a time series of length T. The periodogram ordinates I_{Tj} correspond to the Fourier frequencies ω_j and are sometimes referred to as *intensities*.

Note that the periodogram is symmetric in the Fourier frequencies ω_j (except for ω_q with even T) just as the spectral density function $\Gamma(\omega)$ is symmetric. In fact, we show in [12.3.6] that the periodogram is an asymptotically unbiased estimator of the spectral density. However, we first examine some other properties of the periodogram.

12.3.2 The Periodogram Distributes the Sample Variance. The intensities are interesting since they partition the *sample variance* into q components.

The argument goes as follows. Assume, for simplicity, that T is odd. Also assume that the time series $\mathbf{x}_1, \ldots, \mathbf{x}_T$ was obtained by observing an ergodic weakly stationary process. Then a natural, but slightly biased, estimator of the variance of $\mathbf{X}_T$ is

$$\widehat{\text{Var}}(\mathbf{X}_t) = \frac{1}{T}\sum_{t=1}^{T}(\mathbf{x}_t - \overline{\mathbf{x}})^2$$

where $\overline{\mathbf{x}}$ is the sample mean. Now for notational convenience, let $c_{jt} = \cos(2\pi\omega_j t)$ and define s_{jt} similarly. Then, by applying the orthogonality properties of the discretized sine and cosine, we obtain

$$\begin{aligned}
T\widehat{\text{Var}}(\mathbf{X}_t) &= \textstyle\sum_t(\mathbf{x}_t - a_0)^2\\
&= \textstyle\sum_t\big(\sum_j a_j c_{jt} + b_j s_{jt}\big)^2\\
&= \textstyle\sum_t\big(\sum_j a_j^2(c_{jt})^2 + b_j^2(s_{jt})^2\\
&\quad + \textstyle\sum_{i\neq j} a_i b_j c_{it} s_{jt}\big)\\
&= \textstyle\sum_j a_j^2\sum_t(c_{jt})^2 + \sum_j b_j^2\sum_t(s_{jt})^2 + 0\\
&= \textstyle\frac{T}{2}\sum_j(a_j^2 + b_j^2)\\
&= 2\textstyle\sum_j I_{Tj}.
\end{aligned}$$

Summations with respect to i and j are taken only over $i, j = 1, \ldots, q$ where $q = \lceil\frac{T}{2}$.

Thus when T is odd, the periodogram partitions the sample variance into q components as

$$\widehat{\text{Var}}(\mathbf{X}_t) = \frac{2}{T}\sum_{j=1}^{q} I_{Tj}.$$

When T is even, the decomposition is

$$\widehat{\text{Var}}(\mathbf{X}_t) = \frac{2}{T}\sum_{j=1}^{q-1} I_{Tj} + \frac{1}{T}I_{Tq}.$$

12.3.3 The Periodogram Carries the Same Information as the Sample Auto-covariance Function. The periodogram is the Fourier transform of the estimated auto-covariance function evaluated at the Fourier frequencies ω_j.

To show this, it is convenient to replace the sine and cosine transforms used above with the complex exponential representation of the Fourier transform:

$$I_{Tj} = \frac{T}{4}|\mathbf{z}_{Tj}|^2$$

where

$$\begin{aligned}
\mathbf{z}_{Tj} &= \frac{2}{T}\sum_{t=1}^{T}\mathbf{x}_t\, e^{-2\pi i\,\omega_j t}\\
&= a_j - i\,b_j. \qquad (12.25)
\end{aligned}$$

With this representation, it is easily shown that the periodogram (12.24) is the Fourier transform of the estimated auto-covariance function (12.2). First replace $\mathbf{x}_t$ in equations (12.25) with $\mathbf{x}_t - \overline{\mathbf{x}}$ and then substitute (12.25) into equation (12.24)

[12] Note that equations (12.19)–(12.21) describe the Fourier transform of the infinite time series $\{\mathbf{z}_t : t \in \mathbb{Z}\}$ defined by $\mathbf{z}_t = \mathbf{x}_t$ for $t = 1, \ldots, T$ and $\mathbf{z}_t = 0$ otherwise.

to obtain

$$
\begin{aligned}
I_{Tj} &= \frac{T}{4} |\mathbf{z}_{Tj}|^2 \\
&= \frac{1}{T}\Big(\sum_t (\mathbf{x}_t - \overline{\mathbf{x}})\, e^{-2\pi i \omega_j t}\Big) \\
&\qquad \times \Big(\sum_s (\mathbf{x}_s - \overline{\mathbf{x}}) e^{+2\pi i \omega_j s}\Big) \\
&= \sum_{\tau=-(T-1)}^{T-1} \frac{1}{T}\Big(\sum_{t-s=\tau} (\mathbf{x}_t - \overline{\mathbf{x}})(\mathbf{x}_s - \overline{\mathbf{x}})\Big) \\
&\qquad \times e^{-2\pi i \omega_j \tau} \\
&= \sum_{\tau=-\infty}^{\infty} c(\tau)\, e^{-2\pi\, i \omega_j \tau}. \qquad (12.26)
\end{aligned}
$$

for nonzero j and where $c(\tau)$ is zero for $|\tau|$ greater than $T-1$.

12.3.4 The Covariance Structure of the Fourier Coefficients $\mathbf{Z}_{Tj}$. We next derive the covariance structure of the Fourier coefficients $\mathbf{Z}_{Tj}$. The main result, (12.28), is used in [12.3.5,6] to show that the periodogram ordinates are asymptotically unbiased estimators of the spectral density and also to show that they are asymptotically uncorrelated.

To simplify our derivation we will assume that $\mathbf{x}_1, \ldots, \mathbf{x}_T$ come from a zero mean, ergodic, weakly stationary process so that the auto-covariance function can be estimated as

$$
c(\tau) = \frac{1}{T} \sum_{t=\tau+1}^{T} \mathbf{x}_t \mathbf{x}_{t-\tau}. \qquad (12.27)
$$

The ergodicity assumption is particularly important because it assures us that estimators such as (12.27) are consistent.

Let $\mathcal{E}_{jk} = \mathcal{E}\big(\mathbf{Z}_{Tj}\mathbf{Z}^*_{Tk}\big)$. The first step towards understanding the structure of covariance $\mathcal{E}_{jk}$ is to expand the random coefficients $\mathbf{Z}_{Tj}$ and then exchange expectation and summation operators:

$$
\begin{aligned}
\mathcal{E}_{jk} &= \frac{4}{T^2}\mathcal{E}\left(\left[\textstyle\sum_t \mathbf{X}_t e^{-2\pi i \omega_j t}\right]\right. \\
&\qquad \times \left.\left[\textstyle\sum_s \mathbf{X}_s e^{2\pi i \omega_k s}\right]\right) \\
&= \frac{4}{T^2} \textstyle\sum_t \sum_s \mathcal{E}(\mathbf{X}_t\mathbf{X}_s) e^{-2\pi i (\omega_j t - \omega_k s)} \\
&= \frac{4}{T^2} \textstyle\sum_t \sum_s \gamma(t-s) e^{-2\pi i (\omega_j t - \omega_k s)}.
\end{aligned}
$$

The next step is to replace $\gamma(\tau)$ with its Fourier transform:

$$
\begin{aligned}
\mathcal{E}_{jk} &= \frac{4}{T^2} \sum_t \sum_s \int_{-\frac{1}{2}}^{\frac{1}{2}} \Gamma(\omega) e^{2\pi i \omega (t-s)} \\
&\qquad \times e^{-2\pi i (\omega_j t - \omega_k s)}\, d\omega.
\end{aligned}
$$

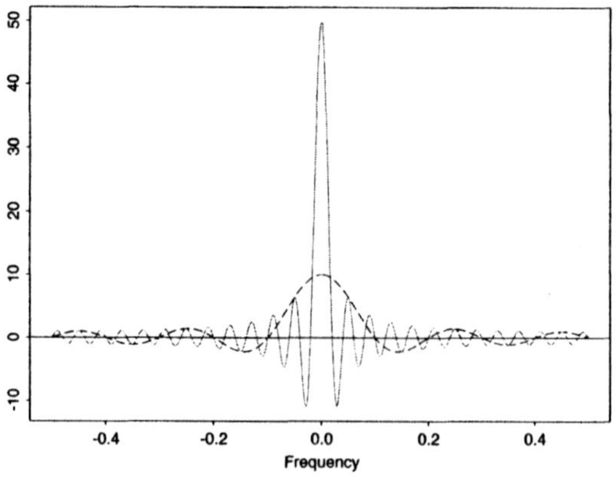

Figure 12.8: *Window function H_T (12.29) displayed for $T = 10$ (dashed curve) and $T = 50$ (solid curve).*

Finally the summation and integration operations are interchanged, and the summation is performed to obtain

$$
\begin{aligned}
\mathcal{E}_{jk} &= \frac{4}{T^2} \int_{-\frac{1}{2}}^{\frac{1}{2}} \Gamma(\omega) H_T(\omega - \omega_j) e^{(T+1)\pi i (\omega - \omega_j)} \\
&\qquad \times H_T(\omega - \omega_k)\, e^{-(T+1)\pi i (\omega - \omega_k)}\, d\omega
\end{aligned}
\qquad (12.28)
$$

where

$$
\begin{aligned}
H_T(\omega) &= \sin(T\pi\omega)/\sin(\pi\omega) \\
&= e^{-(T+1)\pi i \omega/2} \sum_{t=1}^{T} e^{2\pi i \omega t}.
\end{aligned}
\qquad (12.29)
$$

Equation (12.28) links the covariance structure of the Fourier coefficients $\mathbf{Z}_{Tj}$ to the spectral density function $\Gamma(\omega)$ of the process through the *window function* $H_T(\omega)$ given by equation (12.29). Figure 12.8 shows H_T for $T = 10$ and $T = 50$. Note that, as T increases, H_T develops into a function with a narrow central spike of height T and width $1/T$ and with side lobes that are separated by zeros at $\pm 1/T, \pm 2/T, \ldots$.

12.3.5 The Periodogram Ordinates are Asymptotically Uncorrelated. For fixed j and k and increasing T, the windows $H_T(\omega - \omega_j)$ and $H_T(\omega - \omega_k)$ tend to narrow into adjacent spikes. Therefore, since $\Gamma(\omega)$ is continuous, we can approximate

(12.28) for moderate to large T, as

$$\mathcal{E}_{jk} = \frac{4}{T^2}\Gamma\left(\frac{\omega_j+\omega_k}{2}\right)e^{(T+1)\pi i(\omega_k-\omega_j)} \times \int_{-\frac{1}{2}}^{\frac{1}{2}} H_T(\omega-\omega_j)H_T(\omega-\omega_k)\,d\omega. \quad (12.30)$$

Consequently,

$$\mathcal{E}_{jk} = \mathcal{E}\left(\mathbf{Z}_{Tj}\mathbf{Z}^*_{Tk}\right) \approx 0 \text{ for } j \neq k.$$

That is, the Fourier coefficients, and therefore the periodogram ordinates I_{Tj}, are approximately uncorrelated.

12.3.6 What does the Periodogram Estimate?

Continuing on from (12.30), we see that

$$\mathcal{E}\left(|\mathbf{Z}_{Tj}|^2\right) \approx \frac{4}{T^2}\Gamma(\omega_j)\int_{-\frac{1}{2}}^{\frac{1}{2}} H_T(\omega-\omega_j)^2\,d\omega = \frac{4}{T}\Gamma(\omega_j).$$

Consequently

$$\mathcal{E}\left(I_{Tj}\right) \approx \Gamma(\omega_j).$$

That is, the jth periodogram ordinate I_{Tj} is an asymptotically unbiased estimator of the spectral density at frequency ω_j.[13]

12.3.7 The Distribution of the Periodogram.

We need to know the distribution of an estimator to understand its properties and to use it for making inferences about the true spectrum by constructing confidence intervals and developing testing procedures.

Initially, the periodogram would appear to be a reasonable estimator of the spectrum, since it is nearly unbiased and estimates at adjacent frequencies are nearly uncorrelated. However, as we have seen before, unbiasedness is only one attribute of a good estimator. Efficiency and consistency (i.e., low variance that decreases with increasing sample size) are also very desirable attributes. Unfortunately, the periodogram lacks both of these properties.

When $\{\mathbf{X}_t : t \in \mathbb{Z}\}$ is an ergodic, weakly stationary process, the periodogram ordinates are asymptotically proportional to independent $\chi^2(2)$ random variables. In particular, it can be shown (see, e.g., Brockwell and Davis [68, p. 347]) that

$$I_{Tj} \sim \begin{cases} \Gamma(0)\chi^2(1) & j=0 \\ \frac{\Gamma(\omega_j)}{2}\chi^2(2) & 1 \leq j \leq \lceil\frac{T-1}{2}\rceil \\ \Gamma(1/2)\chi^2(1) & j=\frac{T}{2} \text{ if } T \text{ is even.} \end{cases} \quad (12.31)$$

Equation (12.31) clearly illustrates why the periodogram is such a poor spectral estimator. Although it is asymptotically unbiased, it is not consistent: its variance does not decrease with increasing sample length.

This is illustrated in the upper two panels of Figure 12.9, which shows two periodograms computed from time series of length $T = 120$ and $T = 240$ generated from a unit variance white noise process. Both periodograms vary randomly about the true spectrum. Both are equally rough and have peaks scattered randomly amongst the Fourier frequencies. Doubling the sample length has not produced a smoother estimate of the spectrum; rather, it has produced almost independent spectral estimates at twice as many frequencies. It is this property of the periodogram, its ability to extract increasing amounts of roughly independent information about the spectrum with increasing sample length, that is exploited by the spectral estimators described in the following subsections.

The third panel in Figure 12.9 shows the periodogram, computed from our now familiar AR(2) time series, on the *decibel scale*.[14] The amplitude of the variations in the periodogram reflects the magnitude of the underlying spectrum, but the periodogram itself is at best a poor estimator of the spectrum.

One other comment about equation (12.31) is in order. The statement for $j = 0$ applies in the present circumstances because we assumed that the process has mean zero and therefore did not bother to remove the sample mean from the data. In fact, the $j = 0$ statement means that, with the assumptions we have made,

$$\overline{\mathbf{X}} \sim \mathcal{N}(0, \tfrac{1}{T}\Gamma(0)).$$

In general, the mean of a time series taken from an ergodic weakly stationary process will asymptotically be a normal random variable with mean μ_X and variance $\frac{1}{T}\Gamma(0)$. However, the periodogram can not be used to estimate $\Gamma(0)$. Ordinarily $I_{T,0} = 0$, since the sample mean

[13] When $\mathbf{X}_t$ is a white noise process, it is easily shown that the periodogram ordinate I_{Tj} is an unbiased estimator of $\Gamma(\omega_j) = \mathrm{Var}(\mathbf{X}_t)$, regardless of sample size.

[14] That is, we plot ω_j versus $10\log_{10}(I_{Tj})$. See [11.2.13] for a discussion of plotting formats.

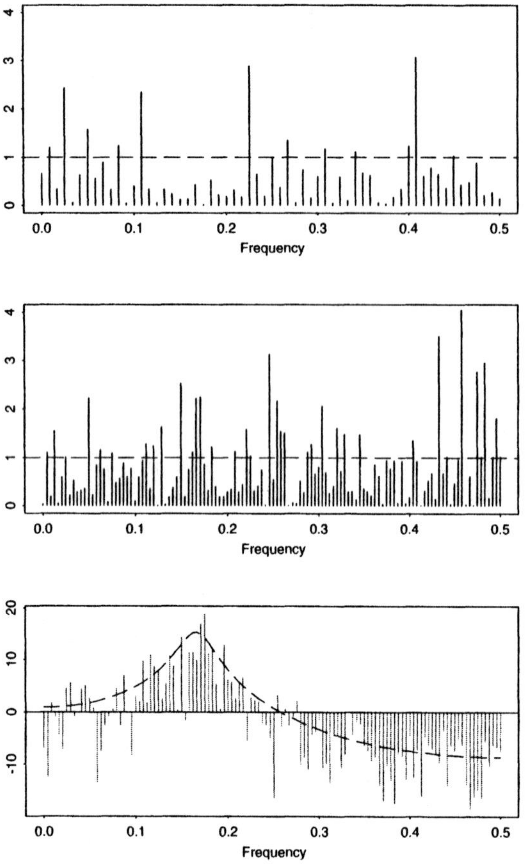

Figure 12.9: *Periodograms computed from three simulated time series. The dashed line shows the true spectral density.*
Top: The periodogram of a white noise time series of length $T = 120$.
Middle: The periodogram of a white noise time series of length $T = 240$.
Bottom: The periodogram of a time series of length $T = 240$ *generated from the AR(2) process with* $(\alpha_1, \alpha_2) = (0.9, -0.8)$. *For this panel only, the periodogram is plotted on the* decibel *scale (i.e.,* ω_j *versus* $10\log_{10}(I_{Tj})$*).*

is subtracted from the time series before the periodogram is computed. When this is not true, $I_{T,0}$ is completely confounded with the sample mean since $I_{T,0} = T\overline{X}^2$. The zero frequency periodogram ordinate is therefore useless as an estimator of the variance of the sample mean. Many people have considered the problem of estimating the variance of the sample mean including Madden [263], Thiébaux and Zwiers [363], Zwiers and von Storch [454], and Wilks [423]. The problem is discussed in some detail in Section 6.6.

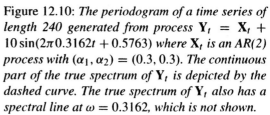

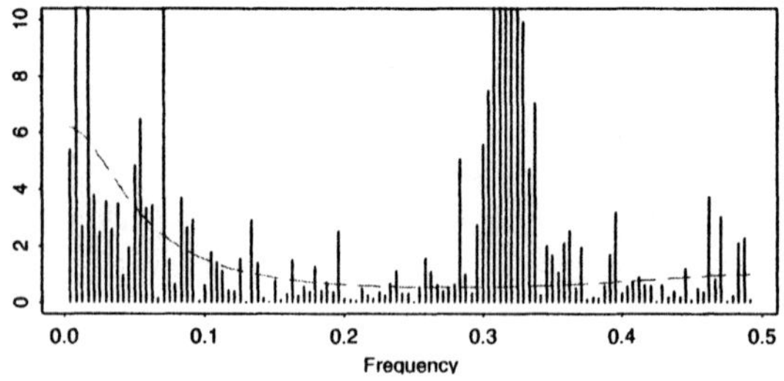

Figure 12.10: *The periodogram of a time series of length 240 generated from process* $\mathbf{Y}_t = \mathbf{X}_t + 10\sin(2\pi 0.3162t + 0.5763)$ *where* $\mathbf{X}_t$ *is an AR(2) process with* $(\alpha_1, \alpha_2) = (0.3, 0.3)$. *The continuous part of the true spectrum of* $\mathbf{Y}_t$ *is depicted by the dashed curve. The true spectrum of* $\mathbf{Y}_t$ *also has a spectral line at* $\omega = 0.3162$, *which is not shown.*

12.3.8 Tapering the Data. While the periodogram is asymptotically an unbiased estimate of the spectral density, it can have poor bias properties for finite samples if the spectrum is not very smooth or if periodic components cause lines in the spectrum (see [11.2.8]).

Equation (12.30) suggests how problems can arise. It gives the expectation of I_{Tj} as

$$\mathcal{E}(I_{Tj}) = \frac{4}{T^2}\int_{-\frac{1}{2}}^{\frac{1}{2}} H_T(\omega - \omega_j)^2 \Gamma(\omega)\, d\omega. \tag{12.32}$$

When $\Gamma(\omega)$ is not smooth or when the spectral density has a line, there can be substantial *variance leakage* through the side lobes of spectral window H_T^2 (see Figure 12.8).

The problem is illustrated in Figure 12.10. It shows the periodogram of a time series of length 240 generated from process

$$\mathbf{Y}_t = \mathbf{X}_t + 10\,\sin(2\pi 0.3162t + 0.5763)$$

where $\mathbf{X}_t$ is an AR(2) process with $(\alpha_1, \alpha_2) = (0.3, 0.3)$. The spectral density function of process $\mathbf{Y}_t$, which is depicted by the dashed curve, has a spectral line at frequency $\omega = 0.3162$. Instead of being nearly unbiased as in Figure 12.9, the periodogram now has substantial bias in a wide band centred on $\omega = 0.3162$ which is caused by variance leakage through the side lobes of the spectral window H_T^2.

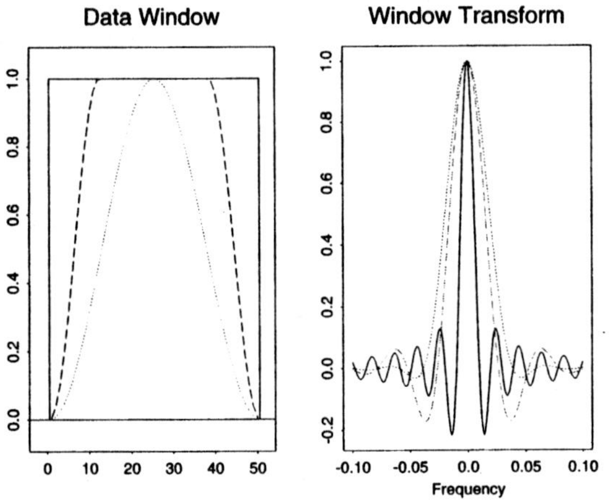

Figure 12.11: *Characteristics of some popular data tapers.*
Left: The box car (solid), Hanning (dots) and split cosine bell (dashed) data windows for a time series of length $T = 50$. The split cosine bell uses $m = T/4$ so that 25% of the data are tapered at each end of the time series.
Right: Corresponding window functions.

The problem is a side effect of the finite Fourier transform, which essentially operates on an infinite series that is abruptly 'turned on' at $t = 1$ and abruptly 'turned off' again beyond $t = T$. That is, the observed time series can be thought of as the product of the infinite series $\{\mathbf{x}_t : t \in \mathbb{Z}\}$ and a *data window*

$$h_t = \begin{cases} 1 & \text{if } 1 \le t \le T \\ 0 & \text{otherwise.} \end{cases}$$

This data window is sometimes called the *box car* taper. The result is that the periodogram is an unbiased estimator of the convolution of the true spectrum $\Gamma(\omega)$ with the square of the Fourier transform H_T of the data window (12.32). The data window h_t and corresponding spectral window function H_T^2 are shown as the solid curves in Figure 12.11.

The large side lobes of H_T^2 can be reduced by using a data window or data taper that turns on and off more smoothly. Frequently used tapers are the *Hanning* or *cosine bell* taper that has nonzero weights

$$h_t = \frac{1}{2}\Big(1 - \cos\Big(\frac{(2t-1)\pi}{T}\Big)\Big),\ 1 \le t \le T,$$

and the split cosine bell that has weights

$$h_t = \begin{cases} \frac{1}{2}\Big(1 - \cos\Big(\frac{(2t-1)\pi}{2m}\Big)\Big) & \text{if } 1 \le t \le m \\ 1 & \text{if } m+1 \le t \le T-m \\ \frac{1}{2}\Big(1 - \cos\Big(\frac{(2T-2t+1)\pi}{m}\Big)\Big) & \\ & \text{if } T-m+1 \le t \le T. \end{cases}$$

The number of non-unit weights $2m$ is typically chosen so that 10%–20% of the data are tapered. The window function corresponding to a data taper $\{h_t : t = 1, \ldots, T\}$ is[15]

$$H(\omega) = e^{-(T+1)2\pi i/2} \sum_{t=1}^{T} h_t e^{2\pi i t}.$$

The Hanning taper has very strongly reduced side lobes (see Figure 12.11). The split cosine bell taper has side lobes that are intermediate between those of the box car and Hanning tapers.

Equation (12.32) shows us that tapering induces bias in the periodogram if the weights are not suitably normalized. Dividing the periodogram by

$$U_2 = \frac{1}{T} \sum_{t=1}^{T} h_t^2 \qquad (12.33)$$

ensures that the result is an approximately unbiased estimator of the spectrum.

Split cosine bell tapers with $m = T/4$ and $T/2$ were applied to the simulated data used to produce Figure 12.10. The effectiveness of tapering in reducing the effects of variance leakage can be seen in Figure 12.12 where we show the periodogram (scaled by $1/U_2$; see (12.33)) of the tapered data. We see that spectral line appears as a narrow peak with increasing m as the amount of leakage decreases.

There are some costs to pay for reducing variance leakage by means of tapering. Smooth tapers have squared window functions with wider central peaks than the box car taper (see Figure 12.11). Thus, while contamination of the periodogram from remote frequencies is reduced, information from adjacent frequencies tends to be 'smeared' together making it more difficult to discriminate between adjacent spectral peaks and lines in the sample spectrum. Also, while the asymptotic properties of the periodogram described above still hold, larger samples are needed to achieve distributional approximations of the same quality when the data are tapered.

[15]It is easily shown that the window function for the Hanning taper is given by $H_T^H(\omega) = (H_T(\omega - \pi/T) + 2H_T(\omega) + H_T(\omega+\pi/T))/4$. The Hanning taper is thus constructed so that side lobes are destroyed by destructive interference. Bloomfield [49] gives details.

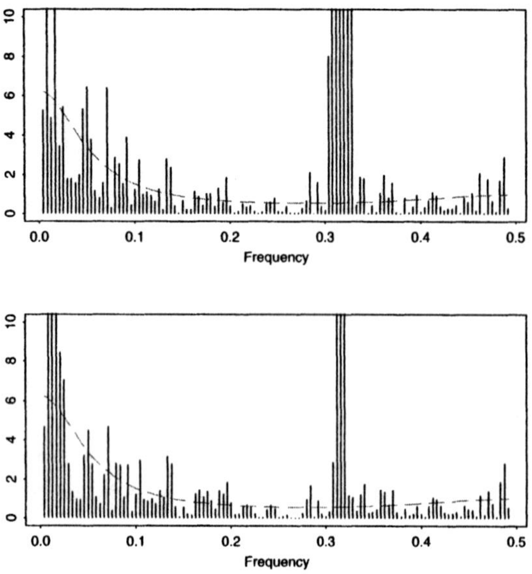

Figure 12.12: *As Figure 12.10, except the periodogram has been computed after tapering with a split cosine bell.*
Top: $m = T/4$.
Bottom: $m = T/2$.

12.3.9 The 'Chunk' Spectral Estimator: General. A spectral estimator frequently used in climatology is the 'chunk' estimator, first described by Bartlett [33] in 1948.[16] The idea is to divide the time series into a number of chunks of equal length, compute the periodogram of each chunk, and then estimate the spectrum by averaging the periodograms.

This estimator is frequently used in climatology because of the cyclo-stationary nature of the processes that are analysed. Typically, the annual cycle is removed from daily observations and it is then assumed that the remaining deviations are roughly stationary within a given season (e.g., DJF).[17] This yields one natural, disjoint chunk per year, of about 90 days in length.

A pleasing property of the chunk estimator is that its variance goes to zero as $1/m$ where m is the number of years in the data set.

A difficulty with the chunk estimator is that its bias is determined by the chunk length. In fact, the expectation of the chunk estimator is given by equation (12.32) when T is set to the chunk length. Since we generally have little control over the chunk length in climatological applications, little can be done to reduce bias. Fortunately, in most applications the true spectrum is smooth and bias is therefore not a big issue.

Variance leakage from spectral lines is a potential problem in high-frequency data sets that resolve, for example, the diurnal cycle or semi-diurnal tidal signals. In this case each chunk can be tapered (cf. [12.3.8]) separately to control variance leakage, or, if the frequency and shape of the signal are known, it can be removed before performing the spectral analysis.

12.3.10 The 'Chunk' Spectral Estimator: Details. We assume, for consistency with spectral estimators described later in this section, that we have a single, contiguous time series $\mathbf{x}_1, \dots, \mathbf{x}_T$ of length T. The chunk estimator is then computed as follows.

1 Divide the time series into m chunks of length $M = \lceil \frac{T}{m}$.

2 Compute a periodogram

$$I_{Tj}^{(\ell)}, \; j = 0, \dots, q, \; q = \lceil \tfrac{M}{2}$$

from each chunk $\ell = 1, \dots, m$.

3 Estimate the spectrum by averaging the periodograms:

$$\widehat{\Gamma}(\omega_j) = \frac{1}{m} \sum_{\ell=1}^{m} I_{Tj}^{(\ell)}. \tag{12.34}$$

The result is an estimator with approximately $2m$ degrees of freedom at each frequency ω_j (except 0 and $1/2$). The estimate at each frequency is representative of a spectral *bandwidth* of approximately $1/M$. Using (12.31), it is easily shown that

$$\widehat{\Gamma}(\omega_j) \sim \begin{cases} \frac{\Gamma(0)}{m}\chi^2(m) & j = 0 \\ \frac{\Gamma(\omega_j)}{2m}\chi^2(2m) & 1 \le j \le \lceil \frac{M-1}{2} \\ \frac{\Gamma(1/2)}{m}\chi^2(m) & j = \frac{M}{2} \; (M \text{ even}). \end{cases} \tag{12.35}$$

This estimator can be made consistent and asymptotically unbiased when the time series is contiguous by ensuring that both the number of chunks m and the chunk length M increase with increasing sample length.

We can construct an asymptotic $\tilde{p} \times 100\%$ confidence interval (see Section 5.4) for $\Gamma(\omega_j)$

[16] We use the expression 'chunk' estimator to avoid confusion with another estimator (described in [12.3.16]) that statisticians and statistical packages frequently refer to as the Bartlett estimator.

[17] While this assumption is never strictly correct, it is often accurate enough to allow use of the chunk estimator.

from the chunk estimator as follows. Equation (12.35) says that asymptotically

$$\frac{2m\widehat{\Gamma}(\omega_j)}{\Gamma(\omega_j)} \sim \chi^2(2m).$$

Therefore

$$\tilde{p} \approx P\Big(a \leq \frac{2m\widehat{\Gamma}(\omega_j)}{\Gamma(\omega_j)} \leq b\Big) = P\Big(\frac{2m\widehat{\Gamma}(\omega_j)}{b} \leq \Gamma(\omega_j) \leq \frac{2m\widehat{\Gamma}(\omega_j)}{a}\Big) \quad (12.36)$$

where a and b are the $(1 - \tilde{p})/2$ and $(1 + \tilde{p})/2$ critical values of the $\chi^2(2m)$ distribution (see Appendix E). The width of this interval can be made independent of the spectral estimate by taking logs. Re-expressed in this way, the approximate $\tilde{p} \times 100\%$ confidence interval is

$$\log\Big(\frac{2m}{b}\Big) + \log\big(\widehat{\Gamma}(\omega_j)\big) \leq \log\big(\Gamma(\omega_j)\big) \leq \log\Big(\frac{2m}{a}\Big) + \log\big(\widehat{\Gamma}(\omega_j)\big). \quad (12.37)$$

Remember that it is the *end points* of this interval that are random. For every 100 independent interval estimates that are made, the interval is expected to *cover* the true parameter $\tilde{p} \times 100$ times, on average.

12.3.11 The Daniell Spectral Estimator. We develop a number of *smoothed spectral estimators* in the following subsections and show how the user can determine their properties by controlling either a *spectral window* or a *lag window*. The estimators are typically applied to contiguous time series, but can also be applied to individual chunks and then averaged, as with the chunk estimator. A summary is available in [12.3.19].

The results of [12.3.7] suggest a natural way to reduce the variance of the periodogram, namely to smooth it, an idea that was first proposed by Daniell [99]. The simplest of all smoothed periodogram spectral estimators, which carries Daniell's name, is just a moving average of the periodogram ordinates I_{Tj}. Given an odd integer n such that $1 \leq n \leq q$, the Daniell estimator is[18]

$$\widehat{\Gamma}(\omega_j) = \frac{1}{n} \sum_{k=j-(n-1)/2}^{j+(n-1)/2} I_{Tk}. \quad (12.38)$$

The asymptotic properties of the periodogram can be extended to the Daniell estimator if n is small relative to T and if the spectral density function is smooth enough so that it is roughly constant in every frequency interval of length n/T. Under these conditions it can be shown that the Daniell estimator has the following properties for frequencies $(n+1)/2T \leq \omega_j \leq (2q - n)/2T$:[19]

1 The Daniell estimator is asymptotically unbiased. That is,

$$\mathcal{E}\big(\widehat{\Gamma}(\omega_j)\big) \approx \Gamma(\omega_j).$$

2 $\mathrm{Var}\big(\widehat{\Gamma}(\omega_j)\big) \approx (1/2n)\big(\Gamma(\omega_j)\big)^2$.

Therefore, the Daniell estimator can be made consistent by letting n tend to infinity as T tends to infinity in such a way that $n/T \to 0$.

3 $\mathrm{Cov}\big(\widehat{\Gamma}(\omega_j), \widehat{\Gamma}(\omega_k)\big) \approx$

$$\begin{cases} \frac{n-|j-k|}{n^2}\Gamma(\omega_j)\Gamma(\omega_k) & |j-k| \leq n \\ 0 & \text{otherwise.} \end{cases}$$

That is, $\widehat{\Gamma}(\omega_j)$ and $\widehat{\Gamma}(\omega_k)$ are approximately uncorrelated if frequencies ω_j and ω_k are separated by a *bandwidth* n/T or more.

4 $\widehat{\Gamma}(\omega_j) \sim \frac{\Gamma(\omega_j)}{2n}\chi^2(2n)$.

This last property allows us to construct asymptotic confidence intervals for the spectral density. Proceeding in the same way as we did with the chunk estimator, the approximate $\tilde{p} \times 100\%$ confidence interval for $\Gamma(\omega_j)$ is given by

$$\log\Big(\frac{2n}{b}\Big) + \log(\widehat{\Gamma}(\omega_j)) \leq \log(\Gamma(\omega_j)) \leq \log\Big(\frac{2n}{a}\Big) + \log(\widehat{\Gamma}(\omega_j))$$

where a and b are the $(1 - \tilde{p})/2$ and $(1 + \tilde{p}/2)$ critical values of the $\chi^2(2n)$ distribution (see Appendix E).

12.3.12 Bias Versus Variance. Although the Daniell estimator has nice asymptotic properties (cf. [12.3.11]), tradeoffs must be made between bias and variance (see [5.3.7] and Figure 5.3) when samples are finite.

[18] The Daniell estimator is defined here as the average of an odd number of periodogram ordinates. It can also be defined as the average of an even number of periodogram ordinates, in which case the estimates should be thought of as being representative of the frequencies midway between adjacent Fourier frequencies.

[19] Similar results can be obtained for frequencies j/T, $j = 1, \ldots, (n-1)/2$ and $j = ((2q - n + 1)/2T), \ldots, q/T$, where $q = \lceil T/2$.

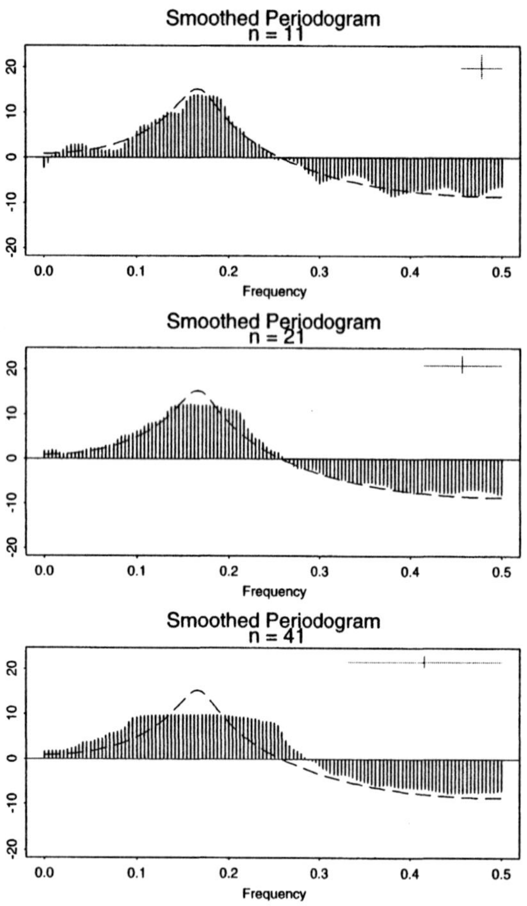

Figure 12.13: *Daniell estimates computed from the periodogram in the lower panel of Figure 12.9 and plotted on the decibel scale. The cross in the upper right corner indicates the width of the 95% confidence interval (vertical bar) and the bandwidth (horizontal bar).*

Figure 12.13 shows spectral estimates computed from the periodogram displayed in the lower panel of Figure 12.9 using the Daniell estimator with $n = 11$, 21, and 41. The dashed curve shows the spectral density that these estimators are trying to approximate. The Daniell estimator with $n = 11$, which has a bandwidth of about 0.046, is quite smooth in comparison with the periodogram (Figure 12.9) and yet is nearly unbiased. The true spectral density generally lies within the approximate 95% confidence interval. The estimators with $n = 21$ (bandwidth 0.088) and $n = 41$ (bandwidth 0.17) do not capture the spectral peak well because they smooth the periodogram excessively.

There is not a correct choice of bandwidth. In this example a bandwidth of 0.17 induces substantial bias at frequencies near the spectral peak by spreading and flattening the peak. However, when the true spectrum has no large peaks, a bandwidth this large may induce very little bias. Since we generally know little about the features of the true spectrum, balancing bias and variance in spectral estimation is a matter of subjective judgement.

12.3.13 An Alternative Representation of the Daniell Spectral Estimator. The Daniell estimator (cf. [12.3.9] and (12.38)) can be re-expressed as the convolution between the periodogram and a box car shaped *spectral window*

$$\widehat{\Gamma}(\omega_j) = \sum_{k=-q}^{q} W_D(\omega_k - \omega_j; n, T) I_{Tk} \qquad (12.39)$$

where the spectral window is given by

$$W_D(\omega; n, T) = \begin{cases} \frac{1}{n} & \text{if } |\omega| \leq (n/2T) \\ 0 & \text{otherwise.} \end{cases} \qquad (12.40)$$

We will see in the following subsections that other smoothed periodogram spectral estimators can be represented similarly.

The Daniell estimator can also be expressed as the Fourier transform of the product of the estimated auto-covariance function $c(\tau)$ and a *lag window*, $w_D(\tau; n, T)$:

$$\widehat{\Gamma}(\omega_j) = \sum_{\tau=-(T-1)}^{T-1} w_D(\tau; n, T) c(\tau) e^{-2\pi i \omega_j \tau}. \quad (12.41)$$

The lag window is derived as follows. Recall from (12.26) that the periodogram is the Fourier transform of the auto-covariance function. Therefore, expanding (12.38) we obtain

$$\begin{aligned} \widehat{\Gamma}(\omega_j) &= \frac{1}{n} \sum_{k=j-(n-1)/2}^{j+(n-1)/2} I_{Tk} \\ &= \frac{1}{n} \sum_{k=j-(n-1)/2}^{j+(n-1)/2} \sum_{\tau=-(T-1)}^{T-1} c(\tau) e^{-2\pi i \omega_k \tau}. \end{aligned}$$

Then, rearranging the order of summation, we find

$$\begin{aligned} \widehat{\Gamma}(\omega_j) &= \sum_{\tau=-(T-1)}^{T-1} c(\tau) e^{-2\pi i \omega_j \tau} \\ &\quad \times \frac{1}{n} \sum_{k=j-(n-1)/2}^{j+(n-1)/2} e^{-2\pi i (\omega_k - \omega_j)\tau} \\ &= \sum_{\tau=-(T-1)}^{T-1} c(\tau) e^{-2\pi i \omega_j \tau} w_D(\tau; n, T) \end{aligned}$$

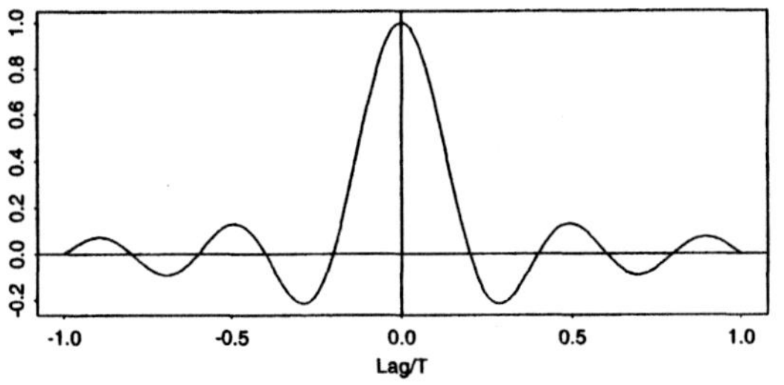

Figure 12.14: *The lag window for the Daniell spectral estimator with $n = 5$.*

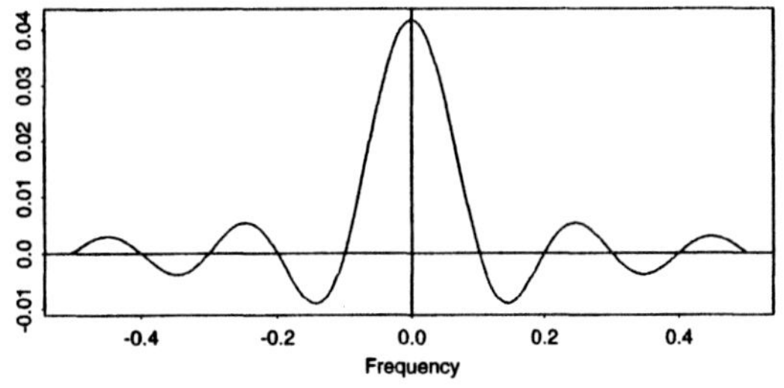

Figure 12.15: *The spectral window $W_R(\omega; M, T)$ that corresponds to the rectangular lag window (12.42) with cutoff $M = 5$ for time series of length $T = 240$.*

where

$$\begin{aligned} w_D(\tau; n, T) &= \frac{1}{n} \sum_{k=j-(n-1)/2}^{j+(n-1)/2} e^{-2\pi i(\omega_k - \omega_j)\tau} \\ &\approx \frac{T}{n} \int_{-n/2T}^{n/2T} e^{-2\pi i \omega \tau}\, d\omega \\ &= \frac{\sin(\pi n \tau / T)}{\pi n \tau / T}. \end{aligned}$$

Thus the Daniell estimator has equivalent spectral window (12.39) and lag window (12.41) representations. Since the periodogram and estimated auto-covariance function are a Fourier transform pair, smoothing in the frequency domain is equivalent to smoothing in the time domain. We will see that the same is true for other smoothed periodogram estimators as well.

The lag window representation (12.41) gives us a somewhat different and useful perspective on why the Daniell estimator has lower variance than the periodogram. The lag window, shown in Figure 12.14 for $n = 5$, decays to zero with increasing lag so that contributions to Fourier transform (12.41) from the large lag part of the estimated auto-covariance are damped. Since we expect the true auto-covariance function to decay to zero at some lag, the window can be adjusted, either in the spectral or time domains, to exclude lags for which the true auto-covariance function is expected to be zero. We can therefore avoid the noise that is contributed by these lags.

12.3.14 The Rectangular Spectral Estimator. This discussion motivates another simple, but poor, spectral estimator. One simple way to exclude the large lag part of the estimated auto-covariance function is to use a rectangular lag window

$$w_R(\tau; M, T) = \begin{cases} 1 & \text{if } |\tau| \le M \\ 0 & \text{otherwise} \end{cases} \qquad (12.42)$$

that explicitly leaves out all estimated auto-covariances beyond some predetermined lag M. The corresponding spectral window, which is shown in Figure 12.15 for $M = 5$, is

$$W_R(\omega; M, T) \approx \frac{2M}{T} \frac{\sin(2\pi\omega M)}{2\pi\omega M}.$$

The resulting spectral estimator has equivalent representations

$$\widehat{\Gamma}(\omega_j) = \sum_{\tau=-(T-1)}^{T-1} w_R(\tau; n, T) c(\tau) e^{-2\pi i \omega_j \tau} \qquad (12.43)$$

and

$$\widehat{\Gamma}(\omega_j) = \sum_{k=-q}^{q} W_R(\omega_k - \omega_j) I_{Tk}. \qquad (12.44)$$

Unfortunately, this particular estimator has some undesirable properties.

- First, the spectral window (see Figure 12.15) has large side lobes that permit variance leakage from frequencies far from ω_j. Note that this source of variance leakage is different from that discussed in [12.3.8]; it will occur whether or not the data have been tapered. This problem exists to some extent with all spectral estimators that are designed with a truncated lag window.

- Second, the spectral window has negative values at some frequencies. Consequently, equation (12.43) or (12.44) can produce negative spectral density estimates with some realizations of the periodogram.

Thus, the rectangular spectral estimator is best avoided.

12.3.15 The Bartlett Spectral Estimator. The chunk estimator is 'almost' a smoothed periodogram estimator or, equivalently, a weighted covariance estimator. When the time series is contiguous, the chunk estimator can be modified slightly to improve its properties and also permit a smoothed periodogram or weighted covariance representation. The resulting estimator is commonly known as the *Bartlett spectral estimator*.

Let $c_\ell(\tau)$ be the auto-covariance function estimate that is computed from the ℓth chunk. Then, using equation (12.26), we can write estimator (12.34) as the average of the Fourier transforms of the estimated auto-covariance functions,

$$\widehat{\Gamma}(\omega_j) = \frac{1}{m}\sum_{\ell=1}^{m}\sum_{\tau=-(M-1)}^{M-1} c_\ell(\tau)e^{-2\pi i\omega_j\tau}.$$

By rearranging the order of summation, we find that estimator (12.34) is the Fourier transform of the average estimated auto-covariance function:

$$\widehat{\Gamma}(\omega_j) = \sum_{\tau=-(M-1)}^{M-1} \bar{c}(\tau)e^{-2\pi i\omega_j\tau}. \quad (12.45)$$

Note that

$$\begin{aligned}\bar{c}(\tau) &= \frac{1}{m}\sum_{\ell=1}^{m} c_\ell(\tau)\\ &= \frac{M-|\tau|}{M}\,\frac{1}{m(M-|\tau|)}\sum_{\ell=1}^{m} Mc_\ell(\tau).\end{aligned} \quad (12.46)$$

Now assume, for convenience, that $\mathbf{X}_t$ is a zero mean process so that the auto-covariance function estimate $c_\ell(\tau)$ can be computed without subtracting the chunk mean. Then it is easily shown that

$$\frac{1}{m(M-|\tau|)}\sum_{\ell=1}^{m} Mc_\ell(\tau) \quad (12.47)$$

is an unbiased estimate of $\gamma(\tau)$. But, since estimator (12.47) does not include all possible products $X_tX_{t-|\tau|}$ that can be computed from the full time series, it is not the most efficient unbiased estimator of $\gamma(\tau)$. It therefore makes sense to replace estimator (12.47) with

$$\frac{T}{T-|\tau|}c(\tau)$$

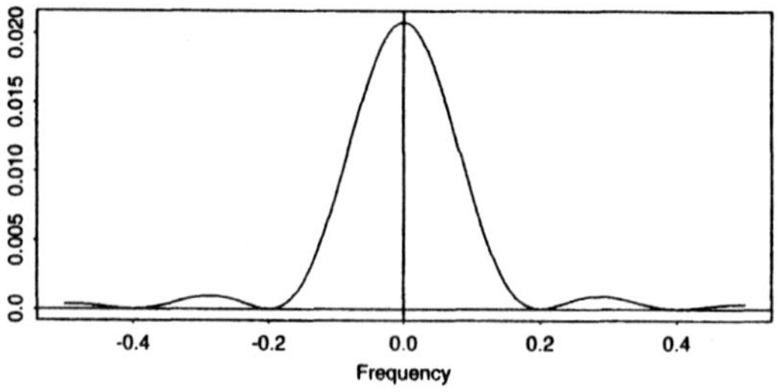

Figure 12.16: *The spectral window $W_B(\omega; M, T)$ that corresponds to the Bartlett lag window (12.48) with cutoff $M = 5$ for time series of length $T = 240$.*

where $c(\tau)$ is the auto-covariance function estimate computed from the full time series. When we do this, and then substitute back into equations (12.46) and (12.45) we see that the 'chunk' estimator can be closely approximated as

$$\widehat{\Gamma}(\omega_j) = \sum_{\tau=-(M-1)}^{M-1} \frac{1-\frac{|\tau|}{M}}{1-\frac{|\tau|}{T}} c(\tau)e^{-2\pi i\omega_j\tau}.$$

This weighted covariance spectral estimator, which has lag window

$$w(\tau; M, T) = \begin{cases} \dfrac{1-\frac{|\tau|}{M}}{1-\frac{|\tau|}{T}} & \text{if } |\tau| < M\\ 0 & \text{otherwise,}\end{cases}$$

is generally called the Bartlett estimator.

The Bartlett estimator, as it is usually computed however (see, e.g., Jenkins and Watts [195]), uses the slightly modified lag window

$$w_B(\tau; M, T) = \begin{cases} 1-\frac{|\tau|}{M} & \text{if } |\tau| < M\\ 0 & \text{otherwise}\end{cases} \quad (12.48)$$

since it is then possible to derive a closed form representation for the corresponding spectral window:

$$W_B(\omega; M, T) \approx \frac{M}{T}\Big(\frac{\sin(\pi\omega M)}{\pi\omega M}\Big)^2. \quad (12.49)$$

This window is shown in Figure 12.16 for $M = 5$. Note that this spectral window is wider than that of the rectangular spectral estimator (Figure 12.15) for the same lag-window cutoff M, but that the side lobes are substantially reduced. The degrees of freedom and bandwidth of this estimator are given in [12.3.19].[20]

[20] Degrees of freedom and bandwidth are discussed in [12.3.17].

It should be noted that the Bartlett estimator computed with (12.48) or (12.49) is *not* the 'chunk' estimator described in [12.3.9,10]. The present estimator has lower variance and greater bandwidth (see [12.3.19]).

The main problem with the Bartlett estimator is that the side lobes of its spectral window (12.49) are quite substantial when compared with those of an estimator such as the Parzen estimator (discussed in [12.3.16]). The Bartlett spectral window (see Figure 12.16) has peaks at frequencies $\pm 3/2M$ that are about 4% of the height of the central peak. Therefore, since the unsmoothed periodogram can vary randomly across a couple of orders of magnitude, the Bartlett estimator has the potential for significant unwanted variance leakage.

12.3.16 The Parzen Spectral Estimator. Another popular smoothed periodogram spectral estimator is the Parzen [305] spectral estimator. It has lag window

$$w_P(\tau; M, T) = \begin{cases} 1 - 6(\frac{|\tau|}{M})^2 + 6(\frac{|\tau|}{M})^3 & \text{if } |\tau| < \frac{M}{2} \\ 2(1 - \frac{|\tau|}{M})^3 & \text{if } \frac{M}{2} \le |\tau| \le M \\ 0 & \text{otherwise.} \end{cases} \qquad (12.50)$$

and corresponding spectral window

$$W_P(\omega; M, T) \approx \frac{3M}{4T}\Big(\frac{\sin(\pi\omega M/2)}{\pi\omega M/2}\Big)^4.$$

The primary advantage of this estimator over the Bartlett estimator is that its spectral window has virtually no side lobes (see Figure 12.17). The Parzen estimator also has somewhat lower variance than the Bartlett estimator for the same lag cutoff, since its spectral window has a wider central peak and thus more bandwidth. However, for this same reason, its estimates also have somewhat more bias when the spectrum varies quickly relative to the bandwidth. This estimator has a wider spectral peak than the Bartlett estimator because the lag window places relatively more weight on low lag covariances and less on lags near the cutoff lag M.

12.3.17 Equivalent Degrees of Freedom and Bandwidth of Smoothed Periodogram Spectral Estimators. The *bandwidth* and degrees of freedom of the Daniell estimator [12.3.11,13]

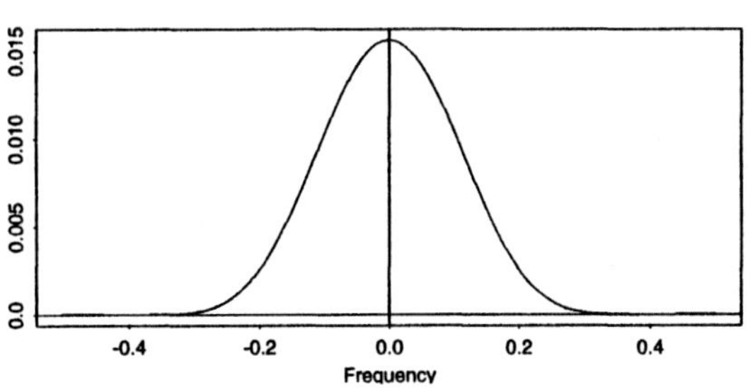

Figure 12.17: *The spectral window $W_P(\omega; M, T)$ that corresponds to the Parzen lag window (12.50) with cutoff $M = 5$ for time series of length $T =$* 240.

were easily identified because this estimator places equal weight on a fixed number of periodogram ordinates that are asymptotically independent and identically distributed as $\chi^2(2)$ random variables. However, other smoothed periodogram estimators, such as the Bartlett estimator [12.3.15] and the Parzen estimator [12.3.16] do not weight the periodogram ordinates equally. Thus it is not quite so easy to determine their bandwidth and degrees of freedom.

Inferences about spectra estimated with a general smoothed periodogram estimator are made with the help of approximating χ^2 distributions. That is, the *equivalent degrees of freedom r* is found by matching the asymptotic mean and variance of the spectral estimator with the mean and variance of a $\chi^2(r)$ random variable.[21] Standard texts, such as Koopmans [229] or Priestley [323], give the equivalent degrees of freedom of many smoothed periodogram spectral estimators. The equivalent degrees of freedom for the estimators we have described are given in [12.3.19].

Once the equivalent degrees of freedom have been determined, confidence intervals can be computed using the method outlined in [12.3.10,11].

The moment matching exercise described above essentially identifies the Daniell spectral estimator that is 'equivalent' to the smoothed periodogram estimator. Thus, in addition to identifying equivalent degrees of freedom, this exercise also identifies an *equivalent bandwidth*, namely that of the 'equivalent' Daniell estimator. Therefore, when the smoothed periodogram estimator has r equivalent degrees of freedom, its equivalent bandwidth is $r/2T$.

[21] This method of finding an approximating distribution is also used in [9.4.9].

12.3.18 Bias, Variance and Variance Leakage. Again, we emphasize the point made in [12.3.12] about the tradeoff that the practitioner must make between bias and variance when estimating spectra. It is always good to be aware of the dichotomy

Low variance	Low bias
⇕	⇕
High bandwidth	Low bandwidth
⇕	⇕
High bias	High variance

between bias and variance in spectral estimation. However, its is also important to remember that estimators with the same equivalent bandwidths and degrees of freedom (see [12.3.19]) are not created equal. These are asymptotic concepts that hold in the limit as the sample length becomes large and the equivalent bandwidth becomes small enough so that the spectral density function is approximately constant within any bandwidth. Variance leakage, that is, the contamination of the spectral estimate by contributions from periodogram ordinates at frequencies far removed from the frequency of interest, is also an important consideration in selecting a good estimator when samples are finite.

12.3.19 Summary. For easy reference, we now briefly summarize the periodogram derived spectral estimators described above.

The periodogram (see [12.3.1–7]) of a time series of length T is defined as

$$I_{Tj} = \frac{T}{4}|Z_{Tj}|^2,$$

where Z_{Tj} is the Fourier transform

$$Z_{Tj} = \frac{2}{T}\sum_{t=1}^{T}\mathbf{x}_t e^{-2\pi i\omega_j t}$$

of the time series.

The 'chunk' spectral estimator is constructed by dividing the time series into m chunks of length M (see [12.3.9,10]), separately computing the periodogram of each chunk, and then averaging the periodograms. Because climate processes are cyclo-stationary, many climate problems present the practitioner with disjoint chunks of length one season at yearly intervals such that it is possible to assume that the process is roughly stationary within chunks.

Smoothed periodogram spectral estimators (see [12.3.11–18]) are computed from contiguous time series. They can be represented as a discrete convolution

$$\widehat{\Gamma}(\omega_j) = \sum_{k=j-(n-1)/2}^{j+(n-1)/2} W(\omega_k - \omega_j)I_{Tk}$$

of the periodogram with a spectral window $W(\omega)$, and as the Fourier transform of the estimated auto-covariance function weighted by a lag window $w(\tau)$,

$$\widehat{\Gamma}(\omega_j) = \sum_{\tau=-(T-1)}^{T-1} w(\tau)c(\tau)e^{-2\pi i\omega_j\tau}.$$

The spectral and lag windows form a Fourier transform pair.

Note that the chunk estimator can be similarly represented as either a discrete convolution or as the Fourier transform of a windowed auto-covariance function estimate. The convolution form of the chunk spectral estimator is

$$\widehat{\Gamma}(\omega_j) = \sum_k W_C(\omega_k - \omega_j)\overline{I}_{Mk}$$

where

$$W_C(\omega) = \begin{cases} \frac{1}{m} & \omega = 0 \\ 0 & \text{otherwise} \end{cases}$$

and where $\overline{I}_{Mk}$ is the mean of the periodograms $I_{Mk}^{(\ell)}$ computed from the individual chunks. The Fourier transform form of the chunk estimator is

$$\widehat{\Gamma}(\omega_j) = \sum_{\tau=-(M-1)}^{M-1} w_C(\tau)\overline{c}(\tau)e^{-2\pi i\omega_j\tau}$$

where

$$w_C(\tau) = 1 \text{ for } |\tau| \le M - 1$$

and $\overline{c}(\tau)$ is the mean of the auto-covariance function estimates $c^{(\ell)}(\tau)$ computed from the individual chunks.

Asymptotic $\tilde{p} \times 100\%$ confidence intervals (see [12.3.10,11]) for the spectral density function have the form

$$\log\left(\frac{r}{b_r}\right) + \log\left(\widehat{\Gamma}(\omega_j)\right) \le \log\left(\Gamma(\omega_j)\right) \le \log\left(\frac{r}{a_r}\right) + \log\left(\widehat{\Gamma}(\omega_j)\right)$$

where a_r and b_r are the $(1 - \tilde{p})/2$ and $(1 + \tilde{p})/2$ critical values of the $\chi^2(r)$ distribution (see Appendix E) and r is the equivalent degrees of freedom (see [12.3.17]) of the periodogram derived spectral estimator.

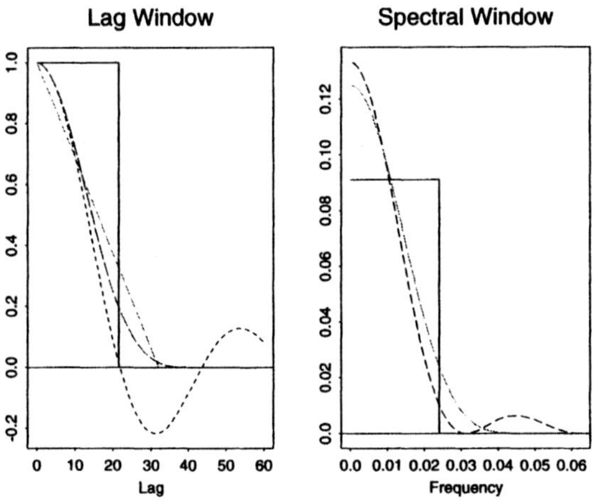

Figure 12.18: *Lag windows (left) and spectral windows (right) for four periodogram derived spectral estimators.*
Solid: *Chunk estimator with* $m = 11$.
Short-dashed curve: *Daniell estimator with* $n = 11$. *The chunk and Daniell spectral windows coincide.*
Medium-dashed curve: *Bartlett estimator with* $M = 32$.
Long-dashed curve: *Parzen estimator with* $M = 40$.
All estimators have approximately 22 equivalent degrees of freedom and an approximate bandwidth of 0.047 when the times series is of length $T = 240$.

Periodogram derived spectral estimators also have an equivalent bandwidth (see [12.3.17]) that indicates, roughly, the width of the frequency band of which an estimate $\widehat{\Gamma}(\omega_j)$ is representative. Estimates at frequencies separated by more than an equivalent bandwidth are asymptotically independent.

The spectral estimators we have discussed are summarized below. In the following, m is the number of chunks used by the chunk estimator, M is either the length of a chunk or the cutoff point of the Bartlett or Parzen lag windows, T is the length of the time series and n is the number of periodogram ordinates that are averaged to produce the Daniell estimator.

Lag Windows $w(\tau)$

- Chunk

$$\begin{cases} 1 & |\tau| \le M-1 \\ 0 & \text{otherwise} \end{cases}$$

This lag window is applied to the average of the auto-covariance function estimates computed from the individual chunks.

- Daniell (n odd)

$$\frac{\sin(\pi n\tau/T)}{\pi n\tau/T}$$

- Bartlett

$$\begin{cases} 1 - \frac{|\tau|}{M} & |\tau| < M \\ 0 & \text{otherwise.} \end{cases}$$

- Parzen

$$\begin{cases} 1 - 6\left(\frac{|\tau|}{M}\right)^2 + 6\left(\frac{|\tau|}{M}\right)^3 & |\tau| < \frac{M}{2} \\ 2\left(1 - \frac{|\tau|}{M}\right)^3 & \frac{M}{2} \le |\tau| \le M \\ 0 & \text{otherwise.} \end{cases}$$

Examples are shown in the left hand panel of Figure 12.18.

Spectral Windows $W(\omega)$

- Chunk

$$\begin{cases} \frac{1}{m} & \omega = 0 \\ 0 & \text{otherwise.} \end{cases}$$

- Daniell (n odd)

$$\begin{cases} \frac{1}{n} & |\omega| \le \frac{n}{2T} \\ 0 & \text{otherwise.} \end{cases}$$

- Bartlett

$$\frac{M}{T}\left(\frac{\sin(\pi\omega M)}{\pi\omega M}\right)^2.$$

- Parzen

$$\frac{3M}{4T}\left(\frac{\sin(\pi\omega M/2)}{\pi\omega M/2}\right)^4.$$

Examples are shown in the right hand panel of Figure 12.18.

Equivalent Degrees of Freedom (EDF) and Equivalent Bandwidth (EBW)

- The following table lists the EDF and EBW for the various spectral estimators.

Estimator	EDF	EBW
Chunk	$2m$	$1/M$
Daniell	$2n$	n/T
Bartlett	$3T/M$	$1.5/M$
Parzen	$3.71T/M$	$1.86/M$

The chunk estimator is generally suitable for problems in which there is a natural chunk length. However, if a contiguous time series is available, the use of a smoothed periodogram estimator is preferred because it better uses the information contained in the time series. We have a slight preference for the Daniell and Parzen estimators over the Bartlett estimator, for which variance leakage through side lobes is more of an issue. The rectangular spectral estimator [12.3.14] is *not recommended* because of the large negative side lobes in its spectral window; this estimator was described for pedagogical reasons.

If the spectrum is suspected to contain sharp peaks, the data should also be tapered with a data taper [12.3.8] to prevent contamination of the smooth part of the spectrum by variance leakage from the spectral peak.

Periodogram-based estimators have a number of advantages that often make them superior to other types of spectral estimators (see, e.g., [12.3.21], where we discuss maximum entropy spectral estimators).

- They are non-parametric. The only assumptions required are that the process be ergodic and weakly stationary. In addition, these estimators often make sense when the assumptions are violated, such as when the process has a periodic component.
- A well-developed asymptotic theory supports these estimators, and practical experience shows that the asymptotic results generally hold even for time series of moderate length.
- Properties of the spectral estimator, such as the bandwidth and the spectral or lag window, are easily tuned to the practitioner's own needs.
- There are many useful extensions of this methodology that we have not been able to discuss here in this short discourse.

12.3.20 Example. Spectral estimates, computed from the periodogram shown in the lower panel of Figure 12.9 with the four smoothed periodogram estimators described above, are shown in Figure 12.19. The parameters of each of the estimators has been chosen so that they have bandwidth and degrees of freedom equivalent to that of the Daniell estimator with $n = 11$. The specifics of the estimators are:

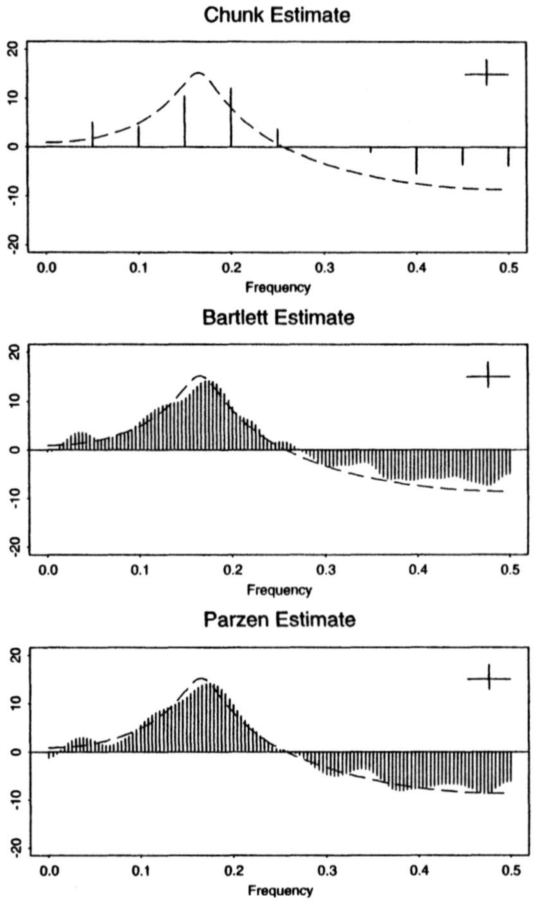

Figure 12.19: *Spectral estimates (on the decibel scale) computed from the AR(2) time series whose periodogram is shown in lower panel of Figure 12.9. The horizontal bar in the upper left hand corner indicates the bandwidth. The vertical bar indicates the width of the asymptotic confidence interval. The dashed curve displays the theoretical spectrum.*

Estimator	n or M	EBW	EDF
Chunk	21	0.0476	22
Daniell	11	0.0458	22
Bartlett	32	0.0469	22
Parzen	40	0.0465	22

The Daniell estimate is shown in the upper panel of Figure 12.13 together with two other Daniell estimates that have greater bandwidth. The

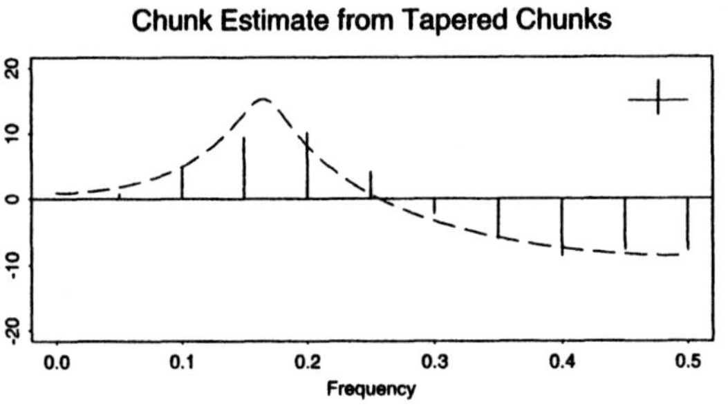

Figure 12.20: *As the top panel in Figure 12.19 except that the chunks were tapered with the cosine bell data taper.*

Chunk, Bartlett, and Daniell estimates are shown in Figure 12.19.

Note that there is little difference between the Daniell (upper panel of Figure 12.13), Bartlett, and Parzen estimators. The effects induced by the differences in window properties in this example are much less severe than the effect of oversmoothing the periodogram (Figure 12.13). The appearance of the chunk estimate is different from that of the other estimators because it only has one value for every equivalent bandwidth. The Daniell, Bartlett, and Parzen estimators are defined at every Fourier frequency. However, only points separated by at least one bandwidth can be considered roughly independent. The chunk estimator seems to have some difficulty with variance leakage in this example; the peak appears to be spreading slightly and the spectrum is overestimated at high frequencies. This behaviour is to be expected since the chunks are very short ($M = 21$).

In this case one of the smoothed periodogram estimators would definitely be preferred over the chunk estimator. However, if for some physical reason our sample consisted of disjoint junks of length 21, we would have no choice but to use the chunk estimator. In such circumstances its properties can be improved somewhat by using the cosine bell data taper [12.3.8]. The difference between Figure 12.20 and the top panel of Figure 12.19 gives an indication of the type of improvement that can be obtained in this way.

12.3.21 Auto-Regressive Spectral Estimation and Maximum Entropy. Two closely related spectral estimation methods that are also occasionally used in climatology are *maximum entropy spectral estimation* and *auto-regressive spectral estimation*.

Auto-regressive spectral estimation (see, e.g., Parzen [306] or Akaike [4, 5]) is performed by:

- assuming that the process is ergodic and weakly stationary,
- fitting an AR model of some order p. The order is chosen either objectively by means of a criterion such as AIC [12.2.10] or BIC [12.2.11], or subjectively using a procedure such as the Box–Jenkins method [12.2.1,9], and
- estimating the spectrum with the spectral density

$$\widehat{\Gamma}(\omega) = \frac{\widehat{\sigma}_Z^2}{|1 - \sum_{\ell=1}^{p} \widehat{\alpha}_\ell e^{-2\ell\pi i\omega}|^2} \qquad (12.51)$$

of the fitted AR process where, $\widehat{\alpha}_\ell$, $\ell = 1, \ldots, p$, are the estimated AR parameters and $\widehat{\sigma}_Z^2$ is the estimated noise variance.

The theoretical justification for AR spectral estimation is that any ergodic weakly stationary process can be approximated arbitrarily closely by an AR process.

This approach to spectral estimation is attractive because it describes the distribution of variance with time scale using a model of the time series that has a dynamical interpretation (cf [10.3.1]). It also produces spectral estimates that are generally smoother than those made by smoothing the periodogram. Periodic features of the process can be identified if the practitioner is willing to use AR models of high enough order.[22]

On the other hand, interpretation of the estimated spectrum is more difficult. Spectral estimates at well separated frequencies may not be approximately independent, as they are when made with a smoothed periodogram estimator, and confidence intervals are difficult to construct.

Maximum entropy spectral estimation (see Burg [73, 74], Lacoss [239], Priestley [323]) is a particular form of AR-spectral estimation. Suppose we have available estimated auto-covariances $c(0), \ldots, c(M)$. Then the maximum entropy spectral estimator $\widehat{\Gamma}(\omega)$ is the non-negative function that maximizes the *entropy*

$$\int_{-\frac{1}{2}}^{\frac{1}{2}} \ln \widehat{\Gamma}(\omega)\, d\omega \qquad (12.52)$$

[22] Tillman et al. [366], for example, use models of successively higher order to estimate the spectrum of a Martian surface pressure time series.

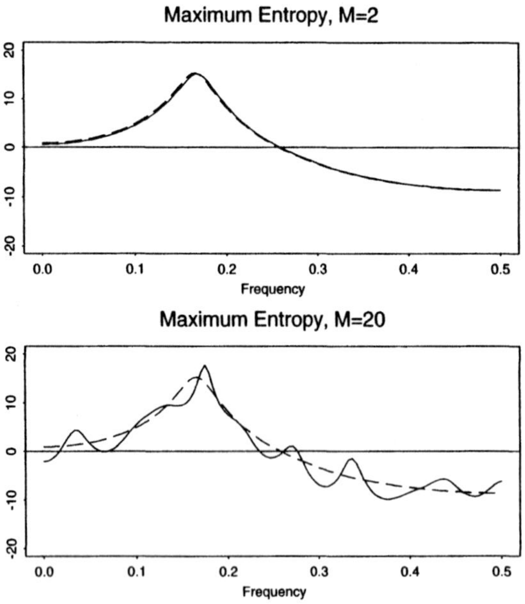

Figure 12.21: *Maximum entropy estimates of an AR(2) time series of length 240 plotted on the decibel scale. The true spectrum is dashed.*
Top: *Using an AR(2) model.*
Bottom: *Using an AR(20) model.*

subject to the constraint that

$$\int_{-\frac{1}{2}}^{\frac{1}{2}} \widehat{\Gamma}(\omega)e^{2\pi i\omega\tau}\,d\omega = c(\tau) \qquad (12.53)$$

for $\tau = 0, \dots, M$. Lacoss [239] shows that (12.52) and (12.53) have a unique solution that is given by an AR-spectral estimator (12.51) in which an AR-model of order $p = M$ is fitted using the Burg procedure.[23]

Maximum entropy spectral estimates for our familiar AR(2) time series are shown in Figure 12.21. The spectral estimate constructed with $M = 2$ very closely approximates the true density, but that constructed with $M = 20$ is considerably noisier. Note that the spectral estimate can never have more than $M/2$ peaks when the spectrum is estimated in this way. The exact number of peaks will depend upon the mix of AR(1) and AR(2) components in the AR model that is fitted. The spectral estimate shown in the lower panel of Figure 12.21 contains eight local peaks.

[23]The Burg procedure chooses the AR coefficients that minimize the sum of the forward and backward forecast squared errors (the 'back' forecast is described in [11.1.12]). Priestley [323, pp. 604–606] shows that for contiguous time series, the Burg estimates are precisely the Yule–Walker estimates (cf. [12.2.2]).

12.3.22 Aliasing. The time series objects that we have considered have a discrete time index, but they presumably represent processes that take place in continuous time. Do we need to worry about how the sampling interval is chosen?

Some years ago, a 500 hPa height time series was analysed at a tropical location. The time series was obtained from a 20-year climate simulation performed with an atmospheric General Circulation Model. The model had been sampled at 18-hour intervals because it was felt that this would produce better long-term statistics than a 12- or 24-hour sampling interval. It was argued that monthly and seasonal means would be more representative of the diurnal cycle since the 18-hour sampling strategy views the globe with the sun in four different positions. When the spectrum was analysed, a spectral line was discovered at the highest resolved frequency, one cycle per two observing times (36 hours). The source of this line was not a physical process taking place at the 36-hour time scale, but rather, one with a characteristic period of 12 hours, namely the solar–thermal tide.[24] This oscillation has a 12-hour period because the atmosphere is not deep enough to propagate the fundamental diurnal wave effectively.

The phenomenon that leads to the spectral line at the half sampling interval frequency is called *aliasing*; Figure 12.22 shows a schematic example.[25] The upper panel shows a wave with period $4\Delta/3$ (solid curve) that is sampled every Δ time intervals ($\Delta = 1\frac{1}{2}$). The resulting time series appears to contain a wave with period 4Δ (dashed curve). We say that the variation taking place at frequency $3/(4\Delta)$ has been *aliased* onto the $1/(4\Delta)$ frequency.

The highest frequency that can be resolved with an observing interval of Δ time units is $1/(2\Delta)$. The middle panel in Figure 12.22 shows that frequencies greater than $1/(2\Delta)$ are, in effect, folded back into the low frequency part of the spectrum across the $1/(2\Delta)$ line. Frequency $1/(2\Delta)$ is called the *Nyquist folding frequency*. The solid curve in the bottom panel of Figure 12.22 shows the sum of the unaliased and aliased parts of the spectrum. This distorted version of the real spectrum is the function that a good spectral

[24]A subsequent GCM experiment with a one-hour save interval showed that the model simulated the solar–thermal tide well, but that it did so for the wrong physical reasons (Zwiers and Hamilton [447]).

[25]Another example of aliasing is described by Bohle-Carbonell [54] who demonstrates how a 14-day period appears in daily Cuxhaven salinity measurements that are affected by the M2 tide (12.5 hours).

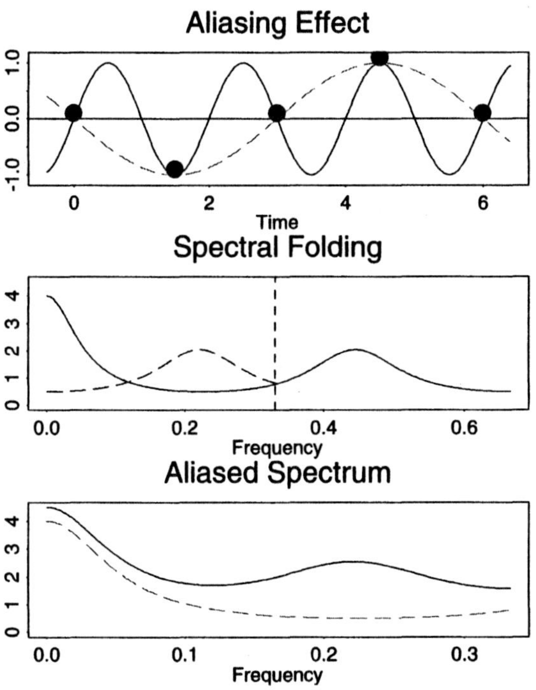

Figure 12.22: Top: *This illustrates a wave of period 2 that is sampled once every 1.5 time intervals. The resulting collection of observations is periodic with period 6. This sampling scheme has* aliased *variation at frequency $\omega = 1/2$ onto frequency $\omega = 1/6$.*
Middle: *This illustrates 'folding' of the spectral density across the Nyquist folding frequency ($\omega = 1/3$ in this example) onto frequencies less than the Nyquist folding frequency. The solid curve is the original spectrum and part that is folded back is indicated by the dashed curve.*
Bottom: *The resulting aliased spectrum (solid curve). The dashed curve indicates the real spectrum.*

estimator would be able to estimate from a time series sampled every Δ time intervals. A poor choice of sampling interval can obviously lead to a badly distorted spectrum and misleading physical interpretation.

In the thermal tide example, we sample every 18 hours, resulting in a Nyquist folding frequency of $\omega = 1/36$ hours. Variation at shorter time scales is folded accordion style onto the interval between $\omega = 0$ and $\omega = 1/36$ hours. Thus, variation at frequencies between $\omega = 1/36$ hours and $\omega = 1/18$ hours is folded onto frequencies between $\omega = 1/36$ hours and $\omega = 0$ (i.e., 18-hour variations are aliased to the mean). Variations between $\omega = 1/18$ hours and $\omega = 1/12$ hours are folded onto frequencies between $\omega = 0$ and $\omega = 1/36$ hours, and so on. The line in the unaliased spectrum at $\omega = 1/12$ hours therefore appears at the Nyquist folding frequency.

12.4 Estimating the Cross-correlation Function

12.4.1 Estimating the Cross-covariance and Cross-correlation Functions. Suppose that a sample $(\mathbf{x}_t, \mathbf{y}_t)$, $t = 1, \ldots, T$, is obtained from an ergodic weakly stationary bivariate process $\{(\mathbf{X}_t, \mathbf{Y}_t) : t \in \mathbb{Z}\}$. Then an estimator of the cross-covariance function $\gamma_{xy}(\tau)$ is

$$\begin{aligned} c_{xy}(\tau) &= \frac{1}{T}\sum_{t=1}^{T-\tau}(\mathbf{x}_t - \overline{\mathbf{x}})(\mathbf{y}_{t+\tau} - \overline{\mathbf{y}}) \text{ for } \tau \geq 0 \\ &= \frac{1}{T}\sum_{t=\tau+1}^{T}(\mathbf{x}_t - \overline{\mathbf{x}})(\mathbf{y}_{t+\tau} - \overline{\mathbf{y}}) \text{ for } \tau < 0 \\ &= 0 \quad \text{for } |\tau| \geq T, \end{aligned} \qquad (12.54)$$

and $\gamma_{yx}(\tau)$ is estimated by $c_{xy}(-\tau)$.

As with the auto-covariance function (see [12.1.1] and [12.1.2]), these estimates are sometimes inflated with a factor $T/(T - |\tau|)$. This makes the estimator unbiased if the process has zero mean and if the sample means are not subtracted in (12.54). However, this practice also inflates the variability of the estimator, particularly at large lags where the true cross-covariance function is close to zero anyway, and it affects the properties of weighted covariance spectral estimators (cf. Section 12.3) by subtly changing the lag-window.

The cross-correlation function is estimated as

$$r_{xy}(\tau) = \frac{c_{xy}(\tau)}{\big(c_{xx}(0)c_{yy}(0)\big)^{1/2}}.$$

12.4.2 Properties of the Estimated Cross-correlation Function. The types of problems that occur when estimating the auto-correlation function also occur when estimating the cross-correlation function (see the discussion in [12.1.2]). Bias is a difficulty, particularly at large lags and when the magnitude of the true cross-correlation is near 1. As with the auto-correlation function, Bartlett [34] also derived approximations for the covariance between cross-correlation function estimates at different lags (see Box and Jenkins [60, p. 376]). Using this approximation, it can be shown that, if $\rho_{xy}(\tau)$ is zero for all τ outside some range of lags

$\tau_1 \leq \tau \leq \tau_2$, then

$$\text{Var}\big(r_{xy}(\tau)\big) \approx \frac{1}{T - |\tau|} \sum_{\ell=-\infty}^{\infty} \rho_{xx}(\ell)\rho_{yy}(\ell) \tag{12.55}$$

for all τ outside the range. This result, and others similar to it that can be derived from Bartlett's approximation, can sometimes be used to determine whether an estimated cross-correlation $c_{xy}(\tau)$ is consistent with the null hypothesis that $\gamma_{xy}(\tau)$ is zero. This is done by performing an appropriate test at the 5% significance level by declaring inconsistency if $|r_{xy}(\tau)| > 2s$ where s^2 is the estimated variance of r_{xy} obtained by substituting the estimated auto-correlation functions of $\mathbf{X}_t$ and $\mathbf{Y}_t$ into equation (12.55).[26]

12.5 Estimating the Cross-spectrum

12.5.0 Introduction. Our purpose in this section is to give a brief introduction to cross-spectral estimation. We will see that many of the intuitive ideas developed in Section 12.3 naturally carry over to the multivariate setting.

The procedure used for univariate spectral analysis is also used here. We first describe the multivariate extension of the periodogram, then briefly describe the extension of the periodogram-based spectral estimators. The basic approach is to use periodogram averaging (i.e., chunks) or smoothing to construct good estimates of the cross-spectral density function or, equivalently, the co- and quad-spectra. These estimates are then used in the obvious way to estimate derived quantities, such as the coherency and phase spectra. Approximate confidence intervals are presented for both.

The tradeoff between bias and variance (see again Figure 5.3) is delicate in cross-spectral analysis. As with univariate spectral analysis, large equivalent bandwidth is associated with low variance and (potentially) large bias. The bias induced by excessive periodogram smoothing can be quite misleading since large bandwidth spectral estimators have the potential to shift the location of peaks in the coherency spectrum. Also, as with univariate spectral analysis, small equivalent bandwidth is associated with high variance. But, in contrast with univariate spectral estimators, insufficiently smoothed periodograms tend to overestimate the coherency between time series. Thus, in cross-spectral analysis one must be careful to balance the smoothing, or reduction of variance, against the biases that are associated with too much smoothing. These problems cannot be avoided by using the chunk estimator: the use of many chunks that are excessively short is equivalent to oversmoothing the periodogram of a contiguous time series, and the use of only a few chunks, each of moderate or greater length, is equivalent to insufficiently smoothing the periodogram.

[26]Note that this procedure may reject the null hypothesis $\gamma_{xy}(\tau) = 0$ when it is true more or less frequently than the nominal 5% significance level. The procedure uses an approximation in which the true cross-correlation function has been replaced with an estimate. Also, in most applications, the estimated cross-correlation function is screened at many lags τ. Thus the effect of *multiplicity* (conducting many related tests at a given significance level, cf. Section 6.8) must be accounted for when interpreting the test results.

12.5.1 Notation and Assumptions. Most of the ideas discussed in this section apply equally in bivariate and multivariate settings. However, to keep concepts as concrete as possible, vector quantities, such as the random vector $\vec{\mathbf{X}}$, will generally only be two-dimensional and, unless stated otherwise, matrices will be 2×2.

The '*' operator will denote the conjugate transpose when applied to a matrix or vector quantity.

We will use the notation $\vec{\mathbf{X}}_t$ or, more precisely, $\{\vec{\mathbf{X}}_t : t \in \mathbb{Z}\}$ to represent a bivariate stochastic process, and we will identify the components of $\vec{\mathbf{X}}_t$ as $\mathbf{X}_t$ and $\mathbf{Y}_t$.

The same assumptions made about $\vec{\mathbf{X}}_t$ in Section 12.4 also apply here.

12.5.2 The Bivariate Periodogram. Let $\vec{\mathbf{x}}_1, \ldots, \vec{\mathbf{x}}_T$ be a time series of length T that is observed from process $\{\vec{\mathbf{X}}_t : t \in \mathbb{Z}\}$.

The bivariate periodogram $\mathcal{I}_{Tj}$ is given by

$$\mathcal{I}_{Tj} = (T/4)\vec{\mathbf{Z}}_{Tj}\vec{\mathbf{Z}}_{Tj}^{*} \tag{12.56}$$

for $j = -q, \ldots, q$, $q = \lceil \frac{T}{2}$ where

$$\vec{\mathbf{Z}}_{Tj} = \frac{2}{T}\sum_{t=1}^{T} \vec{\mathbf{X}}_t e^{-2\pi i \omega_j t}. \tag{12.57}$$

We will use notation such as $\mathbf{Z}_{XTj}$ and $\mathbf{Z}_{YTj}$ to identify the elements of $\vec{\mathbf{Z}}_{Tj}$, and use I_{xxT_j}, I_{xyT_j} and I_{yyT_j} to identify the elements of the

2×2 bivariate periodogram matrix I_{T_j}. As in the univariate case, it is easily demonstrated that:

1 the bivariate periodogram distributes the total lag zero sample covariance matrix:

$$\widehat{\Sigma}_{\vec{X},\vec{X}} = \begin{pmatrix} c_{xx}(0) & c_{xy}(0) \\ c_{yx}(0) & c_{yy}(0) \end{pmatrix} = \frac{2}{T}\sum_{j=1}^{q} \mathcal{I}_{Tj};$$

2 the estimated bivariate auto-covariance function

$$\widehat{\Sigma}(\tau) = \begin{pmatrix} c_{xx}(\tau) & c_{xy}(\tau) \\ c_{yx}(-\tau) & c_{yy}(\tau) \end{pmatrix}$$

and the bivariate periodogram are a Fourier transform pair.

12.5.3 Properties of the Bi-variate Periodogram. The following are some of the properties of the bivariate periodogram (12.56).

1 The bivariate periodogram is Hermitian, that is,

$$\mathcal{I}_{Tj} = \mathcal{I}^*_{Tj}.$$

2 The bivariate periodogram ordinates are asymptotically uncorrelated. This is proven using an argument exactly analogous to that in [12.3.5].

3 The bivariate periodogram ordinates $\mathcal{I}_{Tj}$ are asymptotically unbiased estimators of the bivariate spectral density function

$$\begin{pmatrix} \Gamma_{xx}(\omega_j) & \Gamma_{xy}(\omega_j) \\ \Gamma^*_{xy}(\omega_j) & \Gamma_{yy}(\omega_j) \end{pmatrix}$$

evaluated at the Fourier frequencies ω_j. The argument is also analogous to that in [12.3.6]. In particular, the *cross-periodogram*

$$I_{xyTj} = \frac{T}{4}\mathbf{Z}_{xTj}\mathbf{Z}^*_{yTj}$$

is an unbiased estimator of the cross-spectral density $\Gamma_{xy}(\omega_j)$.

We will see below, and in [12.5.4], that I_{xyTj} is not a very good estimator of $\Gamma_{xy}(\omega_j)$.

4 The variance of the real and imaginary parts of the cross-periodogram can be approximated by (see Bloomfield [49, Section 9.4])

$$\mathrm{Var}\big(\mathrm{Re}(I_{xyTj})\big) \approx \tfrac{1}{2}\big(\Gamma_{xx}(\omega_j)\Gamma_{yy}(\omega_j) + \Lambda_{xy}(\omega_j)^2 - \Psi_{xy}(\omega_j)^2\big),$$

$$\mathrm{Var}\big(\mathrm{Im}(I_{xyTj})\big) \approx \tfrac{1}{2}\big(\Gamma_{xx}(\omega_j)\Gamma_{yy}(\omega_j) - \Lambda_{xy}(\omega_j)^2 + \Psi_{xy}(\omega_j)^2\big).$$

5 The co-variance between the real and imaginary parts of the cross-periodogram can be approximated by

$$\mathrm{Cov}\big(\mathrm{Re}(I_{xyTj}), \mathrm{Im}(I_{xyTj})\big) \approx \Lambda_{xy}(\omega_j)\Psi_{xy}(\omega_j).$$

6 The real and imaginary parts of the cross-periodogram are correlated with the periodograms of the **X** and **Y** components of $\vec{\mathbf{X}}$ (see Bloomfield [49] for details).

7 The periodograms of the **X** and **Y** components of $\vec{\mathbf{X}}$ are also correlated (see Bloomfield [49] for details).

8 The bivariate periodogram ordinates have a complex Wishart distribution that has properties analogous to those of the χ^2 distribution. The theory was developed by Goodman [144]. See Brillinger [66] or Brockwell and Davis [68] for details.

One reason the bivariate periodogram is not a good spectral estimator is that, just as with its univariate counterpart, its variability can not be reduced by taking larger and larger samples. Instead we end up with increasing numbers of periodogram ordinates, all with approximately the same information content. This is demonstrated by items 4–8 above.

Another difficulty with the bivariate periodogram is that it produces degenerate coherency estimates. To see this, let us represent the **X** and **Y** components of Fourier transform (12.57) as

$$\begin{aligned} \mathbf{Z}_{\mathbf{x}Tj} &= \mathbf{A}_{\mathbf{x}j} + i\mathbf{B}_{\mathbf{x}j} \\ \mathbf{Z}_{\mathbf{y}Tj} &= \mathbf{A}_{\mathbf{y}j} + i\mathbf{B}_{\mathbf{y}j}. \end{aligned}$$

Then

$$\begin{aligned} I_{\mathbf{xx}Tj} &= \frac{T}{4}(\mathbf{A}^2_{\mathbf{x}j} + \mathbf{B}^2_{\mathbf{x}j}) \\ I_{\mathbf{yy}Tj} &= \frac{T}{4}(\mathbf{A}^2_{\mathbf{y}j} + \mathbf{B}^2_{\mathbf{y}j}) \\ I_{\mathbf{xy}Tj} &= \frac{T}{4}\big(\mathbf{A}_{\mathbf{x}j}\mathbf{A}_{\mathbf{y}j} + \mathbf{B}_{\mathbf{x}j}\mathbf{B}_{\mathbf{y}j} \\ &\qquad + i\,(\mathbf{B}_{\mathbf{x}j}\mathbf{A}_{\mathbf{y}j} - \mathbf{A}_{\mathbf{x}j}\mathbf{B}_{\mathbf{y}j})\big). \end{aligned}$$

The resulting coherency estimate

$$\widehat{\kappa}_{xy}(\omega_j) = \frac{|I_{\mathbf{xy}Tj}|^2}{I_{\mathbf{xx}Tj}I_{\mathbf{yy}Tj}}$$

is easily shown to be unity at all Fourier frequencies ω_j.

Thus the bivariate periodogram *can not* be used to estimate the coherency spectrum directly, even though the bivariate periodogram is itself an asymptotically unbiased estimator of the cross-spectral density function.[27] This tells us that if we want to estimate the coherency well, we must construct cross-spectral estimators that average across a number of nearly independent realizations of the bivariate periodogram. The chunk and smoothed periodogram estimators discussed in Section 12.3 do exactly this. Also, it is intuitive that a relatively large number of bivariate periodogram ordinates need to be averaged to overcome the bias induced by the degeneracy of the individual bivariate periodogram ordinates. Thus the art of cross-spectral estimation involves trade offs between variance and at least two types of bias.

12.5.4 Smoothed Periodogram Estimators. Since the bivariate periodogram ordinates are asymptotically independent, cross-periodogram-based cross-spectral estimators are constructed using chunk, or smoothing, techniques in the same way that univariate spectral estimators are constructed. Equivalent bandwidths and degrees of freedom are also computed in exactly the same way, and similar considerations are made for the choice of spectral or lag-window. Goodman [144] derived the asymptotic distribution of periodogram-based bivariate spectral estimators.[28] Goodman's approximation is used to derive confidence intervals for cross-spectral parameters such as the coherency and phase (see, e.g., Koopmans [229], Hannan [157] or Brillinger [66] and also [12.5.5,6]).

12.5.5 A Confidence Interval for the Coherency Spectrum. The smoothed coherency spectrum can be thought of as a squared correlation coefficient that depends upon frequency.

This is most easily appreciated by considering the Daniell estimator, but the analogy applies equally to the other spectral estimators summarized in [12.3.19].

The Daniell cross-spectral estimator [12.3.11] is given by

$$\widehat{\Gamma}_{xy}(\omega_j) = \frac{1}{n}\sum_{k=j-(n-1)/2}^{j+(n-1)/2} I_{xyTk}.$$

We can view $\widehat{\Gamma}_{xy}(\omega_j)$ as an estimate of the (complex) covariance between processes $\mathbf{X}$ and $\mathbf{Y}$ at time scales between $\omega^{-1}_{j+(n-1)/2}$ and $\omega^{-1}_{j-(n-1)/2}$. To appreciate this, we substitute equation (12.56) for the cross-periodogram to obtain

$$\widehat{\Gamma}_{xy}(\omega_j) = \frac{T}{4n}\sum_{k=j-(n-1)/2}^{j+(n-1)/2} \mathbf{Z}_{xTk}\mathbf{Z}^*_{yTk}.$$

Except for the factor T, this expression looks just like an estimate of the (complex) covariance between a pair of zero mean random variables $\mathbf{Z}_{xT}$ and $\mathbf{Z}_{yT}$ that is computed from a sample $\{(\mathbf{Z}_{xTk}, \mathbf{Z}_{yTk})\colon k = j-(n-1)/2, \ldots, j+(n-1)/2\}$. This interpretation becomes even stronger when we assume that the cross-spectral density function is constant in the interval $(\omega_{j-(n-1)/2}, \omega_{j+(n-1)/2})$ because then the random pairs $(\mathbf{Z}_{xTk}, \mathbf{Z}_{yTk})$ are approximately independent and identically distributed.

We can estimate the correlation between the $\mathbf{X}$ and $\mathbf{Y}$ processes in the frequency range $(\omega_{j-(n-1)/2}, \omega_{j+(n-1)/2})$ by normalizing $\widehat{\Gamma}_{xy}(\omega_j)$ with estimates of the standard deviations of the $\mathbf{X}$ and $\mathbf{Y}$ in this frequency range. The latter are just the square roots of the estimated auto-spectra of $\mathbf{X}$ and $\mathbf{Y}$. Thus we have

$$\widehat{\rho}_{xy}(\omega_j) = \frac{\widehat{\Gamma}_{xy}(\omega_j)}{\big(\widehat{\Gamma}_{xx}(\omega_j)\widehat{\Gamma}_{yy}(\omega_j)\big)^{1/2}}.$$

Consequently the estimated coherency

$$\widehat{\kappa}_{xy}(\omega_j) = |\widehat{\rho}_{xy}(\omega_j)|^2$$

can be viewed as a measure of the squared correlation, or proportion of common variance that is shared by $\mathbf{X}$ and $\mathbf{Y}$ in the $\omega^{-1}_{j+(n-1)/2}$ to $\omega^{-1}_{j-(n-1)/2}$ time scale range.

This interpretation of the coherency carries over to other periodogram-based spectral estimators as well and can be used to construct confidence intervals.

[27] This is not logically inconsistent. For example, suppose that $\mathbf{Z}_1$ and $\mathbf{Z}_2$ are independent and identically distributed complex random variables such that $\mathbf{Z}_i \sim e^{i2\pi\mathbf{U}_i}$ where $\mathbf{U}_i$ is distributed uniformly on the interval [0, 1). Then $\mathbf{Z}_1\mathbf{Z}_2^*$ is an unbiased estimator of the centre of the unit circle (i.e., $\mathcal{E}(\mathbf{Z}_1\mathbf{Z}_2^*) = 0$) even though $|\mathbf{Z}_1\mathbf{Z}_2^*| = 1$. If we averaged across a large sample, say $\{(\mathbf{z}_{1,i}, \mathbf{z}_{2,i})\colon i = 1, \ldots, n\}$ we would find $|\frac{1}{n}\sum_{i=1}^n \mathbf{z}_{1,i}\mathbf{z}^*_{2,i}| \approx 0$, even though $|\mathbf{z}_{1,i}\mathbf{z}^*_{2,i}| = 1$ for all i.

[28] This distribution, known as the complex Wishart distribution, describes the behaviour of random 2 × 2 Hermitian matrices (see Brillinger [66] or Hannan [157]). It has a property similar to that of the χ^2 distribution: the sum of two independent complex Wishart random matrices again has a complex Wishart distribution.

Fisher's z-transform was used in [8.2.3] to construct confidence intervals for ordinary correlation coefficients. The same method can be used here for nonzero $\kappa_{xy}(\omega_j)$. Fisher's z-transform (8.5) of the square root of the coherency,

$$\frac{1}{2}\ln\left(\frac{1+\widehat{\kappa}_{xy}(\omega_j)^{1/2}}{1-\widehat{\kappa}_{xy}(\omega_j)^{1/2}}\right) = \tanh^{-1}(\widehat{\kappa}_{xy}(\omega_j)^{\frac{1}{2}}),$$

is approximately normally distributed with mean $\tanh^{-1}(\kappa_{xy}(\omega_j)^{1/2})$ and variance $1/r$, where r is the equivalent degrees of freedom of the spectral estimator. Therefore approximate $\tilde{p} \times 100\%$ confidence limits for the squared coherency are

$$\left(\tanh\left(\tanh^{-1}\left(\widehat{\kappa}_{xy}(\omega_j)^{1/2}\right) \pm \frac{\mathbf{Z}_{(1+\tilde{p})/2}}{\sqrt{r}}\right)\right)^2, \tag{12.58}$$

where $\mathbf{Z}_{(1+\tilde{p})/2}$ is the $(1+\tilde{p})/2$ critical value of the standard normal distribution (Appendix D).[29]

The approximation that leads to interval (12.58) breaks down when $\kappa_{xy}(\omega_j)$ is zero. Then

$$\frac{(r/2-1)\widehat{\kappa}_{xy}(\omega_j)}{1-\widehat{\kappa}_{xy}(\omega_j)}$$

is approximately distributed as an $F(2, r-2)$ random variable. Thus

$$\mathrm{H}_0\colon\ \kappa_{xy}(\omega_j) = 0 \ \text{ versus } \ \mathrm{H}_a\colon\ \kappa_{xy}(\omega_j) > 0$$

can be tested at the $(1-\tilde{p}) \times 100\%$ significance level by comparing $\widehat{\kappa}_{xy}(\omega_j)$ with

$$\frac{2F_{\tilde{p}}}{r-2+2F_{\tilde{p}}}, \tag{12.59}$$

where $F_{\tilde{p}}$ is the $\tilde{p}$ critical value of the $F(2, r-2)$ distribution (see Appendix G). Confidence intervals should only be computed when the null hypothesis that $\kappa_{xy}(\omega_j)$ is zero is rejected.

12.5.6 A Confidence Interval for the Phase Spectrum. Hannan [157, p. 257] shows that approximate $\tilde{p} \times 100\%$ confidence limits for the phase spectrum Φ_{xy} are given by

$$\widehat{\Phi}_{xy}(\omega_j) \pm \sin^{-1}\left(\frac{t_{(1+\tilde{p})/2}}{r-2}\left((\widehat{\kappa}_{xy}(\omega_j))^{-1} - 1\right)\right)$$

where $\widehat{\Phi}_{xy}(\omega_j)$ is the phase estimate obtained by substituting a periodogram-based estimator $\widehat{\Gamma}_{xy}(\omega_j)$ of the cross-spectral density into equations (11.64)–(11.66), r is the equivalent degrees of freedom of the spectral estimator, and $t_{(1+\tilde{p})/2}$ is the $(1+\tilde{p})/2$ critical value of the $t(r-2)$ distribution (see Appendix F).

12.5.7 Bias in the Coherency and Phase Spectra: An Example. We return to the problem of bias in the estimated coherency spectrum because of the conflicting demands that good coherency estimates place on the spectral estimator. In univariate spectral estimation, small numbers of equivalent degrees of freedom are associated with high variability and low bias. In cross-spectral estimation, small numbers of degrees of freedom are also associated with large positive bias in coherency estimates, which arises from the degeneracy of the coherence of the periodogram. In addition, we will see that large equivalent bandwidth leads to bias not only in the magnitude of the coherency but also in the location of coherency peaks.

We will use a time series generated from a bivariate AR(1) process with a rotational parameter matrix (11.51) to illustrate these problems (cf. [11.3.8] and [11.4.8,9]). We generate a sample of length $T = 384$ from the process with $r = 0.9$ and $u = 0.95$. It has an e-folding time of approximately 9.5 time units, a rotation frequency $\eta = 0.050$ (approximately 20 time units) and there is a peak in the coherency spectrum at $\omega_0 = 0.053$ (approximately 18 time units).

Three Daniell cross-spectral estimates with different amounts of smoothing are shown in Figure 12.23. The left hand column, with $n = 2$, uses almost no smoothing. This is the cross-spectral estimator that is obtained when adjacent bivariate periodogram ordinates are averaged. The upper panel shows the estimate of the spectrum of the $\mathbf{X}$ component of the process on the decibel scale. The true spectrum is indicated by the long-dashed curve. The spectral estimate is noisy, but otherwise satisfactory. Despite the noise, the estimate conveys useful information and gives us an indication of the shape of the spectrum and the location of the spectral peak. The middle panel shows the derived coherency estimate (solid curve) and the true coherency (long-dashed curve). Note that this estimate is very noisy with many large spikes that grossly overestimate the true coherency. It does not give any useful information about the true coherency spectrum, except to suggest that it is probably

[29] Koopmans [229, p. 283] gives a slightly refined version of this interval. He also points out that the quality of the approximation depends upon the equivalent degrees of freedom r and $\kappa_{xy}(\omega_j)$, and that it is best when $r > 40$ and $0.4 < \kappa_{xy}(\omega_j) < 0.95$. However, in our experience, interval (12.58) gives useful, although perhaps not precise, information when there are substantially fewer equivalent degrees of freedom.

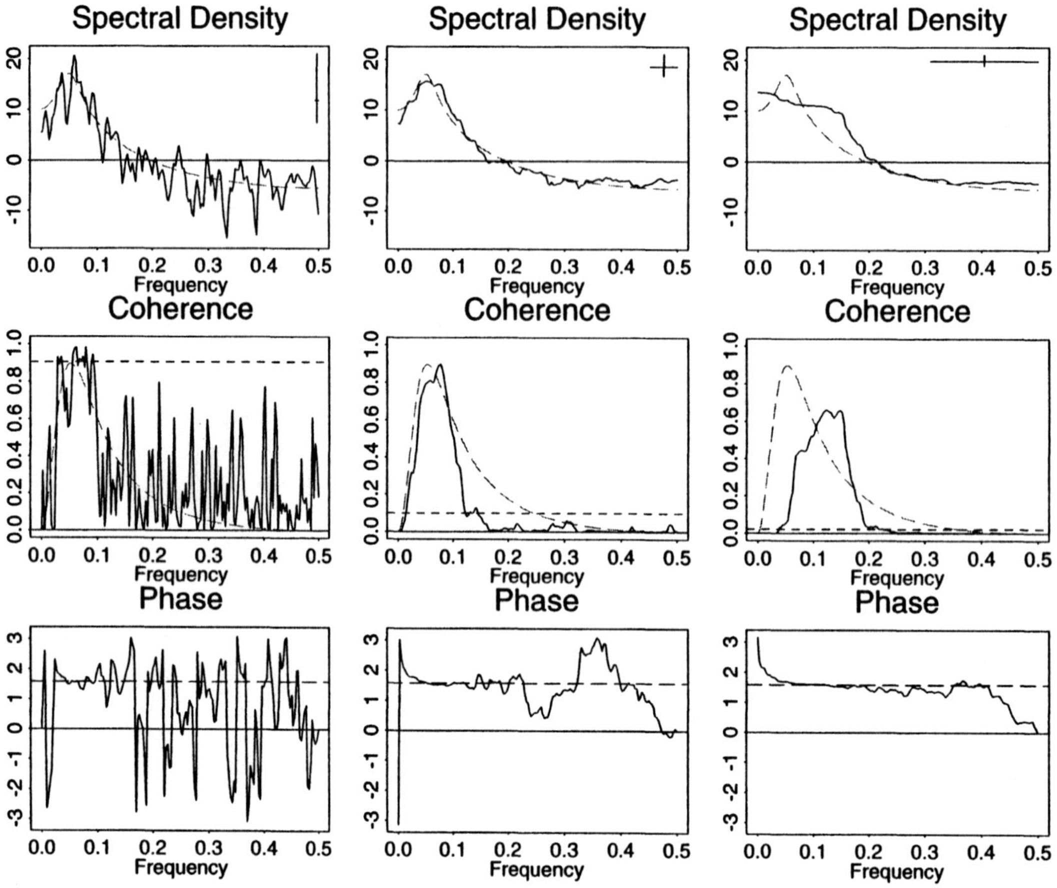

Figure 12.23: *Cross-spectral estimates computed from a time series of length $T = 384$ generated from a bivariate AR(1) process with a rotational parameter matrix (11.51) (cf. [11.3.8] and [11.4.8,9]). The columns contain Daniell estimates for $n = 2$, 16, and 64, from the left.*
Top row: *The estimated spectrum of the* **X** *component of the process, in decibels. The dashed curve indicates the true spectrum. The cross indicates the bandwidth (horizontal) and width of the 95% confidence interval (vertical).*
Middle row: *The estimated coherency. The long-dashed curve indicates the true coherency spectrum. The short-dashes indicate the critical value for the 5% significance level test of zero coherency.*
Bottom row: *The estimated phase. The dashed line indicates the true phase.*

nonzero for frequencies in the interval (0.02, 0.1). The horizontal short-dashed line in this diagram depicts the critical value from (12.59) for the 5% significance level test of the null hypothesis that $\kappa_{xy}(\omega) = 0$. Despite the noise and the many large peaks, only a few coherency estimates rise above the critical value. The bottom panel displays the corresponding phase estimates (solid) and true phase (horizontal long-dashed line). We see that the phase is reasonably well estimated in the same interval (0.02, 0.1) in which we have some indication that the true coherency is nonzero. Elsewhere, the phase estimates are of no value.

The centre column in Figure 12.23 shows the Daniell cross-spectral estimate that is obtained with a moderate amount of smoothing ($n = 16$, EDF $= 32$, EBW $= 0.042$). The spectral density is well estimated. We now have a reasonable indication of the shape of the coherence spectrum although it is severely underestimated at frequencies $\omega > 0.1$. The peak in the estimated coherence spectrum is located at a slightly higher frequency than that in the true spectrum. The phase is well estimated in the interval (0.02, 0.2).

The right hand column of Figure 12.23 illustrates the effect of over-smoothing the periodogram. We used the Daniell estimator with

$n = 64$ (EDF $= 128$, EBW $= 0.17$). The estimate of the spectral density at low frequencies is now affected by the 'peak spreading' effect of the smoothing. The coherence estimates are now strongly affected by bias as well. The peak has been shifted to the right and its magnitude has been diminished. This estimate gives quite a distorted view of the rotational properties of the sampled process. However, the phase is surprisingly well estimated over a wide frequency band.

12.5.8 Yet Another Source of Bias. Another potential source of bias occurs when one time series is *delayed* relative to another. This may happen in very simple ways, for example, by shifting the time origin of one series relative to another. For example, one could conceive of proxy data derived from tree rings, varves, ice cores, etc. in which time is measured relative to an uncertain time origin. However, it may also happen in much more complex ways, with delay occurring on some time scales but not others.

Unrecognized delay can lead to severe underestimation of the cross-spectral density function on the time scales at which delay occurs. One might even be led to the false conclusion that two strongly related time series are unrelated. This is intuitively easy to understand if we think of a weighted covariance spectral estimator with a lag window that is zero beyond lag M. Imagine a pair of strongly related processes in which the delay is ζ. Cross-covariances at lags near ζ will be large while those at other lags will be small. If the delay ζ is greater than M lags, the weighted covariance estimator will entirely miss the contributions to the cross-spectrum that are made by the large cross-covariances near lag ζ. When the delay is the same at all time scales it may be possible to correct this problem by *aligning* the components of the observed time series (see the examples in Jenkins and Watts [195, Sections 9.3.2 and 9.3.3], and also Bloomfield [49, Section 9.6]). A simple, but not very efficient, way to do this is to shift the time origin of the delayed time series by ζ time units.[30]

To examine the effect of delay on the estimated spectrum in some more detail, we consider a bivariate process $\vec{\mathbf{X}}_t$ with components that vary similarly at time scales ω_b^{-1} to ω_a^{-1}. Suppose that $\mathbf{Y}_t$ leads $\mathbf{X}_t$ by ζ time intervals on these time scales. Then

$$\Gamma_{xy}(\omega) \approx e^{2\pi i\zeta\omega}\Gamma_{xx}(\omega) \text{ for } \omega \in (\omega_a, \omega_b). \tag{12.60}$$

A very simple process of this type has

$$\mathbf{Y}_t = \mathbf{X}_{t+\zeta}$$

in which case approximation (12.60) holds for all time scales (see equation (11.72) in [11.4.4]). Generally, however, one process might lag or lead the other over only some subset of time scales.

Suppose, for simplicity, that Γ_{xy} is estimated from a time series of length T using the Daniell estimator with bandwidth n/T. The estimator of the cross-spectrum is

$$\widehat{\Gamma}_{xy}(\omega_j) = \frac{1}{n}\sum_{k=j-(n-1)/2}^{j+(n-1)/2} (I_{xy})_{Tk}.$$

The bivariate periodogram is an asymptotically unbiased estimator of the cross-spectrum. Thus

$$\mathcal{E}\big(I_{xyTk}\big) \approx e^{2\pi i\zeta\omega_k}\Gamma_{xx}(\omega_k)$$

for $\omega_k \in (\omega_a, \omega_b)$. Assuming that $\Gamma_{xx}(\omega_k)$ is approximately constant for the ω_ks lying in a bandwidth centred on ω_j, we then obtain

$$\mathcal{E}\big(\widehat{\Gamma}_{xy}(\omega_j)\big) \approx \Gamma_{xx}(\omega_j)\frac{\sum_{k=j-(n-1)/2}^{j+(n-1)/2} e^{2\pi i\zeta\omega_k}}{n}. \tag{12.61}$$

If the *delay* ζ is large (greater than about $1/\text{EBW} = T/n$), the elements of the sum in (12.61) describe points all around the unit circle and consequently we have large bias with

$$|\mathcal{E}(\widehat{\Gamma}_{xy}(\omega_j))| \ll |\Gamma_{xy}(\omega_j)|$$

at frequencies in (ω_a, ω_b).

This type of bias affects both the estimated coherency and phase spectra. The coherency spectrum will be underestimated and we might incorrectly conclude that the processes **X** and **Y** are uncorrelated at time scales between ω_b^{-1} and ω_a^{-1}. The phase spectrum will be estimated incorrectly and we may entirely miss the linear component of the variation of phase with frequency (see approximation (12.60)) that is induced by the delay.

[30] When the delay is independent of frequency, alignment is performed efficiently by estimating the delay ζ from the cross-correlation function and then multiplying the cross-periodogram I_{xyT_j} by $e^{-2\pi i\zeta\omega_j}$ before using it in a periodogram-based spectral estimator. In general, when the delay varies with time scale, simple alignment cannot be used. Hannan and Thompson [159, 160] describe a method for estimating frequency-dependent delay. See also Bloomfield [49, pp. 228–231].

Part V

Eigen Techniques

Overview

A characteristic difficulty in climate research is the size of the phase space. It is practically infinite in the case of the real system, and much smaller, though still very large, in the case of quasi-realistic models, such as atmospheric or oceanic General Circulation Models. Thus observations or simulated data sets, *per se*, are not always useful to the researcher who wants to know the dynamics controlling the developments and relationships in the system. Statistical analysis becomes an indispensable tool for helping the researcher to discriminate between a few dynamically significant components and the majority of components that are irrelevant or, in terms of frequently used slang, of 'second (or higher) order' for the problem at hand. The task of sorting out the first-order processes from myriads of second-order processes makes statistical analysis in climate research different from both conventional (mathematical) statistics and statistical mechanics. In mathematical statistics, problems are usually of low dimension, and in statistical mechanics the phase space, though infinite, is isotropic or of some simple structure. In climate problems, however, one has to expect different characteristics for each different direction in phase space. The problem is to find the relevant directions.

In this part of the book, a number of linear techniques are introduced that attempt to identify 'relevant' components in phase space. We assume that these components take the form of characteristic vectors, which can usually be represented by patterns (i.e., spatial distributions). These techniques are often based on an eigenproblem, which arises naturally when maximizing some interesting squared properties, for instance variance or correlation, under certain constraints. (The differentiation of the squared property leads to the linear problem, with the eigenvalue originating from the addition of a Lagrange multiplier.)

In Chapter 13 we begin with the problem of one random vector and its decomposition into its *Empirical Orthogonal Functions* (EOFs). The EOFs are orthogonal spatial patterns that can be thought of as empirically derived basis functions. The low-order EOFs can sometimes be interpreted as natural modes of variation of the observed system. The time coefficients obtained by projecting the observed field onto the EOFs are uncorrelated and represent the variability of the field efficiently. We also introduce two related topics, *Singular Systems Analysis* and *Rotated EOF Analysis*.

In Chapter 14 we consider a pair of random vectors and we search for pairs of directions, or patterns, that represent the strongest *joint* patterns of variations. Techniques designed for this purpose include *Canonical Correlation Analysis*, *Maximum Covariance Analysis*, and *Redundancy Analysis*.

Principal Oscillation Patterns (POPs; Chapter 15) are obtained by imposing a specific model for the time evolution of a field. This technique is useful if the system under consideration has quasi-oscillatory modes. POP analysis can be generalized to cyclo-stationary time series. The general concept of *state space models* is briefly outlined in the last section of Chapter 15.

Another approach for analysing the time evolution patterns of variability is to *complexify* the observed field (Chapter 16). This can be done by assigning the observed field to the real part of the complexified process and assigning the Hilbert transform of the observed field to the imaginary part. This approach has been used widely in conjunction with EOF analysis, but can also be used with the techniques introduced in Chapters 13–15.

13 Empirical Orthogonal Functions

13.0.0 Overview. In this chapter we present a multivariate analysis technique that is to derive the dominant patterns of variability from a statistical field (a random vector, usually indexed by location in space). *Principal Component Analysis*, or *Empirical Orthogonal Function* (EOF) Analysis as it is called in the Earth Sciences, was described by Pearson [309] in 1902 and by Hotelling [186] in 1935. EOF analysis was introduced into meteorology by Lorenz [259] in 1956.

Concepts in linear algebra that are needed to read this chapter (linear bases, matrix properties, eigenanalysis and singular value decomposition) are offered in Appendix B. Empirical Orthogonal Functions are formally defined in Section 13.1. Techniques for estimating EOFs, eigenvalues, and EOF coefficients are explained in Section 13.2. We discuss the quality of estimates in Section 13.3. Several EOF analyses of climate-related problems are given as examples in Section 13.4. *Rotated EOFs*[1] are dealt with in Section 13.5. Finally, a time series analysis technique called *Singular Systems Analysis*, which uses the same mathematics as EOF analysis, is introduced in Section 13.6.

An alternative introduction to EOFs is given by von Storch [387].

13.0.1 Introductory Example:[2] Daily Profile of Geopotential Height at Berlin. To motivate the concept of Empirical Orthogonal Functions we consider a time series of daily geopotential height profiles as obtained by radiosonde at Berlin (Germany) (Fraedrich and Dümmel [125]). A total of 1080 observations are available in each winter (NDJF) season: 120 days times 9 vertical levels between 950 hPa and 300 hPa. Thus, in a 20-year data set we have 21 600 observations at our disposal to describe the statistics of the geopotential height at Berlin in winter.

The mean state can be estimated by computing the mean value at each level. But how should we describe the variability? One way would be to compute the standard deviation at each level and to plot it in the vertical. However such a profile does not tell us how the variations are correlated in the vertical. For example, are we likely to observe a positive anomaly (i.e., a positive deviation from the mean profile) at 300 hPa and at 950 hPa at the same time?

EOF analysis is a technique that is used to identify patterns of simultaneous variation. To demonstrate the concept we let $\vec{\mathbf{x}}_t$ represent the $m = 9$ level geopotential height profile observed at time t. The mean profile is denoted by $\widehat{\vec{\mu}}$ and to describe the variability we form the *anomalies*

$$\vec{\mathbf{x}}'_t = \vec{\mathbf{x}}_t - \widehat{\vec{\mu}}.$$

These anomalies are then *expanded* into a finite series

$$\vec{\mathbf{x}}'_t = \sum_{i=1}^{k} \widehat{\alpha}_{i,t} \widehat{\vec{e}}^{\,i} \tag{13.1}$$

with time coefficients $\widehat{\alpha}_{i,t}$ and fixed *patterns* $\widehat{\vec{e}}^{\,i}$. Equality is usually only possible when $k = m$, but the variance of the time coefficients $\widehat{\alpha}_{i,t}$ usually decreases quickly with increasing index i, so that good approximations are usually possible for k much less than m. The patterns are chosen to be *orthogonal* so that optimal coefficients $\widehat{\alpha}_{i,t}$ are obtained by simply projecting the anomalies $\vec{\mathbf{x}}'_t$ onto the patterns $\widehat{\vec{e}}^{\,i}$. Moreover, the patterns can be specified such that the *error*

$$\sum_t \Big(\vec{\mathbf{x}}'_t - \sum_{i=1}^{k} \widehat{\alpha}_{i,t} \widehat{\vec{e}}^{\,i}\Big)^2$$

is minimal. The lag-0 sample cross-correlations of the optimal time coefficients are all zero,

$$\sum_t \widehat{\alpha}_{i,t} \widehat{\alpha}_{j,t} = 0$$

for $i \neq j$. The patterns $\widehat{\vec{e}}^{\,j}$ are estimated *Empirical Orthogonal Functions*.[3] The coefficients $\widehat{\alpha}_i$ are the *EOF coefficients*.[4]

[1] A misnomer.

[2] The mathematics in this subsection are explained in more detail in Sections 13.1 and 13.2.

[3] Note that the 'functions' $\widehat{\vec{e}}^{\,k}$ are really *vectors* and not *functions*.

[4] Statisticians refer to the EOF coefficients as *principal components*.

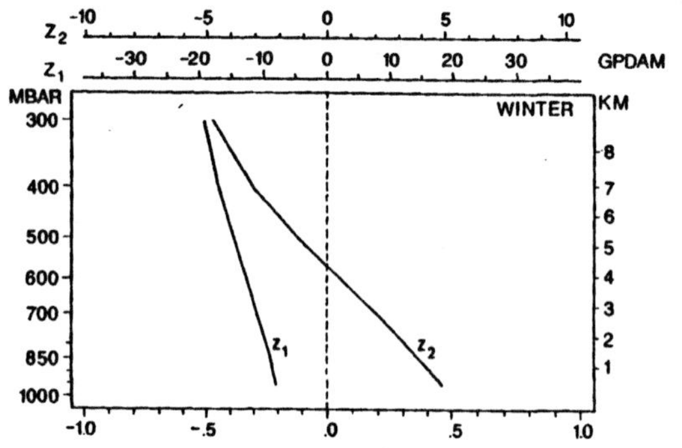

Figure 13.1: *The first two EOFs, labelled z_1 and z_2, of the daily geopotential height over Berlin in winter. From Fraedrich and Dümmel [125].*

The analysis of daily Berlin radiosonde data showed that only two patterns are required to describe most of the variability in the observed geopotential height profiles in winter (NDJF) as well as in summer (MJJA). In winter, the first EOF represents 91.2% of the variance (92.6% in summer), and the second EOF represents an additional 8.2% of the variance (7% in summer).[5] The remaining seven EOFs, which together with the first two EOFs span the full nine-dimensional space, represent only 0.6% of the variance of the height profiles (0.4% in summer). Thus, only two coefficient time series are required to represent the essential information in the time series of geopotential height at the nine levels. Instead of dealing with 1080 numbers per season, only $2 \times 120 = 240$ are needed. This demonstrates one of the advantages of EOFs, namely the ability to often identify a small subspace that contains most of the dynamics of the observed system.[6]

Another advantage is that the patterns can sometimes be seen as *modes of variability*. In the present example the two patterns $\vec{e}^{\,1}$ and $\vec{e}^{\,2}$ may be identified with the equivalent barotropic mode and the first baroclinic mode of the tropospheric circulation: The first patterns in winter (Figure 13.1) as well as in summer have the same sign throughout the troposphere, that is, they exhibit an equivalent barotropic structure. The second EOF, however, changes sign in the middle of the troposphere: it represents the first baroclinic mode.[7]

13.0.2 'Complex' EOFs. EOFs may be derived from real- or complex-valued random vectors. The latter results in complex-valued EOFs. The 'Complex EOF Analysis' (CEOF) described in the climate literature (see [181]) is a special case of the EOF analysis of complex random vectors. The time order of the observations is important for these 'CEOFs,' or 'Frequency Domain EOFs,' since they are the EOFs of a complexified time series. In contrast, the time order of observations is irrelevant in ordinary complex EOF analysis. The original real-valued time series is made complex by adding its Hilbert transform (see Section 16.2) as the imaginary component. The Hilbert transform can be thought of as the time derivative of the original process so that the EOF analysis of the complexified process reveals properties of the variability of the state and its change at the same time. To avoid confusion with the ordinary complex EOF analysis we refer to these EOFs as *Hilbert EOFs* (see Section 16.3).

13.1 Definition of Empirical Orthogonal Functions

13.1.1 Overview. EOFs are introduced formally in this section as parameters of the distribution of an m-dimensional random vector $\vec{\mathbf{X}}$.[8] For the sake of brevity we assume $\vec{\mu} = 0$. We first construct the first EOF, which is the most powerful single pattern in representing the variance of $\vec{\mathbf{X}}$. The idea is easily generalized to several patterns and in [13.1.3] the calculations are condensed into a theorem.

13.1.2 The First EOF. The first step is to find one 'pattern' $\vec{e}^{\,1}$, with $\|\vec{e}^{\,1}\| = 1$, such that

$$\epsilon_1 = \mathcal{E}\left(\|\vec{\mathbf{X}} - \langle\vec{\mathbf{X}}, \vec{e}^{\,1}\rangle\vec{e}^{\,1}\|^2\right) \tag{13.2}$$

is minimized.[9] Equation (13.2) describes the projection of the random vector $\vec{\mathbf{X}}$ onto a one-dimensional subspace spanned by the fixed vector

[5] When we say that an expansion Y 'represents' $p\%$ of the variance of X, we mean that the variance of $Y - X$ is $(100-p)\%$ of the variance of X. The word 'explains' is often used instead of the word 'represents' in the literature. This is misleading since nothing is explained causally; only part of the variability of X has been described by Y.

[6] The assumption that the subspace with maximum variance coincides with the dynamically active subspace is arbitrary. In general, it will not be valid and counter examples can easily be constructed. However, in climate research, it is often reasonable to make this assumption. An example demonstrating the dynamical dominance of EOFs is given by Selten [343].

[7] A similar result for the vertical structure of the shelf ocean has been reported by Kundu, Allen, and Smith [234].

[8] Mainly based on [392].

[9] $\|\cdot\|$ denotes the vector norm, and $\langle\cdot,\cdot\rangle$ denotes the inner product. Note that $\|\vec{\mathbf{X}}\| = \langle\vec{\mathbf{X}}, \vec{\mathbf{X}}\rangle$. See Appendix B.

$\vec{e}^{\,1}$. Minimizing ϵ_1 is equivalent to the maximizing of the variance of $\vec{\mathbf{X}}$ that is contained in this subspace:

$$\begin{aligned}\epsilon_1 &= \mathcal{E}\Big(\|\vec{\mathbf{X}}\|^2 - 2\langle\vec{\mathbf{X}},\vec{e}^{\,1}\rangle^*\vec{\mathbf{X}}^\dagger\vec{e}^{\,1} \\ &\qquad + \langle\vec{\mathbf{X}},\vec{e}^{\,1}\rangle^*\langle\vec{\mathbf{X}},\vec{e}^{\,1}\rangle\Big) \\ &= \mathcal{E}\big(\|\vec{\mathbf{X}}\|^2 - \langle\vec{\mathbf{X}},\vec{e}^{\,1}\rangle^*\langle\vec{\mathbf{X}},\vec{e}^{\,1}\rangle\big) \\ &= \mathrm{Var}(\vec{\mathbf{X}}) - \mathrm{Var}\big(\langle\vec{\mathbf{X}},\vec{e}^{\,1}\rangle\big),\end{aligned}$$

where the variance of the random vector $\vec{\mathbf{X}}$ is defined to be the sum of variances of the elements of $\vec{\mathbf{X}}$.[10] Note that

$$\mathrm{Var}\Big(\langle\vec{\mathbf{X}},\vec{e}^{\,1}\rangle\Big) = \vec{e}^{\,1\dagger}\Sigma\vec{e}^{\,1},$$

where Σ is the covariance matrix of $\vec{\mathbf{X}}$. Then minimization of equation (13.2), under the constraint $\|\vec{e}^{\,1}\| = 1$, leads to

$$\begin{aligned}&\frac{d}{d\vec{e}^{\,1}}\left[-\vec{e}^{\,1\dagger}\Sigma\vec{e}^{\,1} + \lambda\left(\vec{e}^{\,1\dagger}\vec{e}^{\,1} - 1\right)\right] \\ &\quad= 2\Sigma\vec{e}^{\,1} + 2\lambda\vec{e}^{\,1} = 0\end{aligned}$$

where λ is the Lagrange multiplier associated with the constraint $\|\vec{e}^{\,1}\| = 1$.[11] Thus, $\vec{e}^{\,1}$ is an eigenvector with a corresponding eigenvalue λ, of the covariance matrix Σ. But Σ has m eigenvectors. Therefore, to minimize ϵ_1, we select the eigenvector that maximizes

$$\begin{aligned}\mathrm{Var}\big(\langle\vec{\mathbf{X}},\vec{e}^{\,1}\rangle\big) &= \vec{e}^{\,1\dagger}\Sigma\vec{e}^{\,1} \\ &= \vec{e}^{\,1\dagger}\lambda\vec{e}^{\,1} = \lambda.\end{aligned}$$

Thus ϵ_1 is minimized when $\vec{e}^{\,1}$ is an eigenvector of Σ associated with its largest eigenvalue λ.[12] This 'pattern' is the first EOF.[13]

13.1.3 More EOFs. Having found the first EOF, we now repeat the exercise by finding the 'pattern' $\vec{e}^{\,2}$ that minimizes

$$\epsilon_2 = \mathcal{E}\big(\|(\vec{\mathbf{X}} - \langle\vec{\mathbf{X}},\vec{e}^{\,1}\rangle\vec{e}^{\,1}) - \langle\vec{\mathbf{X}},\vec{e}^{\,2}\rangle\vec{e}^{\,2}\|^2\big)$$

subject to the constraint that $\|\vec{e}^{\,2}\| = 1$. The result is that $\vec{e}^{\,2}$ is the eigenvector of Σ that corresponds to its second largest eigenvalue λ_2.[14] This second pattern is orthogonal to the first because the eigenvectors of a Hermitian matrix are orthogonal to one another.

13.1.4 Theorem. The following theorem results from the analysis presented so far.

Let $\vec{\mathbf{X}}$ be an m-dimensional random vector with mean $\vec{\mu}$ and covariance matrix Σ. Let $\lambda_1 \geq \lambda_2 \geq \cdots \geq \lambda_m$ be the eigenvalues of Σ and let $\vec{e}^{\,1}, \ldots, \vec{e}^{\,m}$ be the corresponding eigenvectors of unit length. Since Σ is Hermitian, the eigenvalues are non-negative and the eigenvectors are orthogonal.

(i) *The k eigenvectors that correspond to $\lambda_1, \ldots, \lambda_k$ minimize*

$$\epsilon_k = \mathcal{E}\big(\|(\vec{\mathbf{X}} - \vec{\mu}) - \textstyle\sum_{i=1}^{k}\langle\vec{\mathbf{X}} - \vec{\mu},\vec{e}^{\,i}\rangle\vec{e}^{\,i}\|^2\big). \tag{13.3}$$

(ii) $$\epsilon_k = \mathrm{Var}(\vec{\mathbf{X}}) - \sum_{i=1}^{k}\lambda_i. \tag{13.4}$$

(iii) $$\mathrm{Var}(\vec{\mathbf{X}}) = \sum_{i=1}^{m}\lambda_i. \tag{13.5}$$

The total variance of $\vec{\mathbf{X}}$ is broken up into m components. Each of these components is obtained by projecting $\vec{\mathbf{X}}$ onto one of the EOFs $\vec{e}^{\,i}$. The variance contribution of the kth component to the total variance $\sum_j \lambda_j$ is just λ_k. In relative terms, the proportion of the total variance represented by EOF k is $\lambda_k/\sum_j \lambda_j$. This proportion may be given as a percentage.

If the components are ordered by the size of the eigenvalues then the first component is the most important in representing variance, the second is the second most important and so forth.

Equation (13.3) gives the mean squared error ϵ_k that is incurred when approximating the full m-dimensional random vector $\vec{\mathbf{X}}$ in a k-dimensional subspace spanned by the first k EOFs. The construction of the EOFs ensures that the approximation is optimal; the use of any other k-dimensional subspace will lead to mean squared errors at least as large as ϵ_k.

13.1.5 Properties of the EOF Coefficients. The EOF coefficients, or *principal components*,

$$\alpha_i = \langle\vec{\mathbf{X}},\vec{e}^{\,i}\rangle = \vec{\mathbf{X}}^{\mathrm{T}}\vec{e}^{\,i*} = \vec{e}^{\,i\dagger}\vec{\mathbf{X}} \tag{13.6}$$

[10]That is, if $\vec{\mathbf{X}}$ has covariance matrix Σ, then we define $\mathrm{Var}\big(\vec{\mathbf{X}}\big) = \mathrm{tr}(\Sigma)$.

[11]Graybill [148, Section 10.8], describes the differentiation of quadratic forms.

[12]Recall (see Appendix B) that all eigenvalues of the Hermitian matrix $\Sigma = \mathcal{E}(\vec{\mathbf{X}}\vec{\mathbf{X}}^\dagger)$ are real and non-negative.

[13]The pattern is unique up to sign if Σ has only one eigenvector that corresponds to eigenvalue λ. Otherwise, the pattern can be any vector with unit norm that is spanned by the eigenvectors corresponding to λ. In this case, the EOF is said to be *degenerate*. See Appendix B.

[14]Note that $\lambda_1 = \lambda_2$ if $\vec{e}^{\,1}$ is degenerate. In fact, if λ_1 has k linearly independent eigenvectors, then k of the m eigenvalues of Σ will be equal to λ.

are uncorrelated, and hence independent when $\vec{\mathbf{X}}$ is multivariate normal. In fact, for $i \neq j$,

$$\begin{aligned}\mathrm{Cov}(\alpha_i, \alpha_j) &= \mathcal{E}\big(\langle(\vec{\mathbf{X}} - \vec{\mu}), \vec{e}^{\,i}\rangle\langle(\vec{\mathbf{X}} - \vec{\mu}), \vec{e}^{\,j}\rangle^*\big) \\ &= \vec{e}^{\,i\dagger}\mathcal{E}\big((\vec{\mathbf{X}} - \vec{\mu})(\vec{\mathbf{X}} - \vec{\mu})^\dagger\big)\vec{e}^{\,j} \\ &= \vec{e}^{\,i\dagger}\Sigma\vec{e}^{\,j} \\ &= \lambda_j\vec{e}^{\,i\dagger}\vec{e}^{\,j} = 0\end{aligned}$$

Therefore, the variance of $\mathbf{X}_k$, the kth component of $\vec{\mathbf{X}}$, can also be decomposed into contribution from the individual EOFs as

$$\mathrm{Var}(\mathbf{X}_k) = \sum_{i=1}^{m} \lambda_i |e_k^i|^2. \tag{13.7}$$

If the elements of $\vec{\mathbf{X}}$ represent locations in space, the spatial distribution of variance can be visualized by plotting $\mathrm{Var}(\mathbf{X}_k)$ as a function of location. Similarly, the variance contribution from the ith EOF can be visualized by plotting $\lambda_i|e_k^i|^2$ or $\lambda_i|e_k^i|^2/\mathrm{Var}(\mathbf{X}_k)$ as a function of location.

13.1.6 Interpretation. The bulk of the variance of $\vec{\mathbf{X}}$ can often be represented by the first few EOFs. If the original variable has m components the approximation of $\vec{\mathbf{X}}$ by $\vec{\alpha} = (\alpha_1, \ldots, \alpha_k)$, with $k \ll m$, leads to a significant reduction of the amount of data while retaining most of the variance. It was shown in the introductory example of Berlin geopotential height [13.0.2] that just two EOFs represent almost all of the information in the data set.

The physical interpretation of EOFs is limited by a fundamental constraint. While it is often possible to clearly associate the first EOF with a known physical process, this is much more difficult with the second (and higher-order) EOF because it is constrained to be orthogonal to the first EOF. However, real-world processes do not need to have orthogonal patterns or uncorrelated indices. In fact, the patterns that most efficiently represent variance do not necessarily have anything to do with the underlying dynamical structure.

13.1.7 Vector Notation. The random vector $\vec{\mathbf{X}}$ may conveniently be written in vector notation by

$$\vec{\mathbf{X}} = \mathcal{P}\vec{\alpha} \tag{13.8}$$

where $\mathcal{P}$ is the $m \times m$ matrix

$$\mathcal{P} = \big(\vec{e}^{\,1}|\vec{e}^{\,2}|\cdots|\vec{e}^{\,m}\big) \tag{13.9}$$

that has EOFs in its columns, and $\vec{\alpha}$ is the m-dimensional (column) vector of EOF coefficients $\alpha_1, \ldots, \alpha_m$. Because the EOFs are orthonormal, the expression (13.8) may be inverted to obtain

$$\vec{\alpha} = \mathcal{P}^\dagger\vec{\mathbf{X}}, \tag{13.10}$$

where $\mathcal{P}^\dagger$ is the conjugate transpose of $\mathcal{P}$. Another consequence of the orthonormality of the EOFs is that

$$\begin{aligned}\Sigma &= \mathrm{Cov}(\vec{\mathbf{X}}, \vec{\mathbf{X}}) \\ &= \mathcal{P}\mathrm{Cov}(\vec{\alpha}, \vec{\alpha})\mathcal{P}^\dagger \\ &= \mathcal{P}\Lambda\mathcal{P}^\dagger\end{aligned}$$

where Λ is the diagonal $m \times m$ matrix composed of the eigenvalues of Σ,

$$\Lambda = \mathrm{diag}(\lambda_1, \ldots, \lambda_m).$$

It therefore follows that

$$\begin{aligned}\mathrm{Var}(\vec{\mathbf{X}}) &= \sum_{k=1}^{m} \mathrm{Var}(\mathbf{X}_k) \\ &= \mathrm{tr}(\Sigma) \\ &= \mathrm{tr}(\mathcal{P}\Lambda\mathcal{P}^\dagger) \\ &= \mathrm{tr}(\Lambda) = \sum_{k=1}^{m} \lambda_k.\end{aligned}$$

It also follows that the eigenvalues are the m roots of the mth degree *characteristic polynomial* $p_\Sigma(\lambda) = \det(\Sigma - \lambda\mathcal{I})$, where $\mathcal{I}$ is the $m \times m$ identity matrix. In fact

$$\begin{aligned}p_\Sigma(\lambda) &= \det(\mathcal{P}\Lambda\mathcal{P}^\dagger - \lambda\mathcal{P}\mathcal{P}^\dagger) \\ &= \det\big(\mathcal{P}(\Lambda - \lambda\mathcal{I})\mathcal{P}^\dagger\big) \\ &= \det(\Lambda - \lambda\mathcal{I}) = \prod_{i=1}^{m}(\lambda_i - \lambda).\end{aligned} \tag{13.11}$$

13.1.8 Degeneracy. As noted above, EOFs are not uniquely determined. If $\lambda_\circ$ is a root of multiplicity 1 of $p_\Sigma(\lambda)$ and $\vec{e}$ is a corresponding (normalized) eigenvector, then $\vec{e}$ is unique up to sign, and either $\vec{e}$ or $-\vec{e}$ is chosen as the EOF that corresponds to $\lambda_\circ$. On the other hand, if $\lambda_\circ$ is a root of multiplicity k, the solution space of

$$\Sigma\vec{e} = \lambda_\circ\vec{e}$$

is of dimension k. The solution space is uniquely determined in the sense that it is orthogonal to the space spanned by the $m - k$ eigenvectors of Σ that correspond to eigenvalues $\lambda_i \neq \lambda_\circ$. But any orthonormal basis $\vec{e}^{\,1}, \ldots, \vec{e}^{\,k}$ for the solution space can be used as EOFs. In this case the EOFs are said to be *degenerate*. (An example is discussed in [13.1.9].)

Degeneracy can either be bad or good news. It is bad news if the EOFs are estimated from a sample of iid realizations of $\vec{\mathbf{X}}$. Then degeneracy is mostly a nuisance, because the patterns, which may represent independent processes in the underlying dynamics, can not be disentangled.

However, degeneracy may be good news if the EOFs are estimates from a realization of a stochastic process $\vec{\mathbf{X}}_t$. Suppose, for example, that $p_\Sigma(\lambda)$ has a root of multiplicity 2. By construction, the cross-correlation of the two corresponding EOF coefficient time series will be zero at lag-0. But this does not imply that the lagged cross-correlations will be zero, and, in fact, they are often nonzero. This means that a pair of EOFs and their coefficient series could represent a signal that is propagating in space.

The representation of such a spatially propagating signal requires two patterns whose coefficients vary coherently and are 90° out-of-phase. The two patterns representing a propagating signal are not uniquely determined; indeed if any two patterns represent the signal, then any linear combination of the two do so as well. Therefore, degeneracy is a necessary condition for the description of such signals.

13.1.9 Examples. To demonstrate the mathematics of EOF analysis and the phenomenon of degeneracy we now consider the case of a random vector

$$\vec{\mathbf{X}} = \sum_{k=1}^{m} \alpha_k \vec{p}^{\,k} \tag{13.12}$$

where coefficients α_k are uncorrelated real univariate random variables and $\vec{p}^{\,1}, \ldots, \vec{p}^{\,m}$ are fixed orthonormal vectors. For simplicity we assume that the αs have mean zero. Then the covariance matrix of $\vec{\mathbf{X}}$ is

$$\begin{aligned} \Sigma &= \mathcal{E}\Big(\big(\textstyle\sum_k \alpha_k \vec{p}^{\,k}\big)\big(\sum_l \alpha_l \vec{p}^{\,l}\big)^{\mathrm{T}}\Big) \\ &= \sum_j \mathrm{Var}(\alpha_j)\vec{p}^{\,j}\vec{p}^{\,j\mathrm{T}}. \end{aligned} \tag{13.13}$$

It is easily verified that $\mathrm{Var}(\alpha_k)$ is an eigenvalue of this covariance matrix with eigenvector $\vec{p}^{\,k}$:

$$\Big(\sum_j \mathrm{Var}(\alpha_j)\vec{p}^{\,j}\vec{p}^{\,j\mathrm{T}}\Big)\vec{p}^{\,k} = \mathrm{Var}(\alpha_k)\vec{p}^{\,k}.$$

Thus, the chosen orthonormal vectors are the EOFs of the random vector (13.12). The ordering is determined by the variance of the uncorrelated univariate random variables α_k.

The example has two merits. First it may be used as a recipe for constructing random vectors with a given EOF structure. To do so one has to select a set of orthonormal vectors and associate them with a set of uncorrelated random variables.

The example may also be used to demonstrate the phenomenon of degeneracy. To do so, we assume that all αs have variance 1. Then, the EOFs are degenerate and may be replaced by any other set of orthonormal vectors. One such set of orthonormal vectors are the unit vectors $\vec{u}^{\,k}$ with a 1 in the kth row and zeros elsewhere. Then, the representation (13.8), with $\mathcal{P} = (\vec{p}^{\,1}|\cdots|\vec{p}^{\,m})$, is transformed as

$$\vec{\mathbf{X}} = \mathcal{P}\vec{\alpha} = \left(\mathcal{P}\mathcal{P}^{\mathrm{T}}\right)(\mathcal{P}\vec{\alpha}) = \mathcal{U}\vec{\beta}, \tag{13.14}$$

where the new EOFs are the columns of

$$\mathcal{U} = \mathcal{P}\mathcal{P}^{\mathrm{T}} = \mathcal{I} = (\vec{u}^{\,1}|\cdots|\vec{u}^{\,m})$$

and the EOF coefficients are given by

$$\vec{\beta} = \mathcal{P}\vec{\alpha}.$$

These coefficients are uncorrelated as well because of $\mathrm{Var}(\alpha_k) = 1$ for all k:

$$\begin{aligned} \mathrm{Cov}(\vec{\beta}, \vec{\beta}) &= \mathrm{Cov}(\mathcal{P}\vec{\alpha}, \mathcal{P}\vec{\alpha}) \\ &= \mathcal{P}\,\mathrm{Cov}(\alpha, \alpha)\mathcal{P}^{\mathrm{T}} \\ &= \mathcal{P}\mathcal{I}\mathcal{P}^{\mathrm{T}} = \mathcal{I}. \end{aligned}$$

Obviously, the only meaningful information the EOF analysis offers in this case is that there is no preferred direction in the phase space. The only property that matters is the uniformity of the variance in all directions.

13.1.10 Coordinate Transformations. Let us consider two m-variate random vectors $\vec{\mathbf{X}}$ and $\vec{\mathbf{Z}}$ that are related to each other by

$$\vec{\mathbf{Z}} = \mathcal{L}\vec{\mathbf{X}} \tag{13.15}$$

where $\mathcal{L}$ is an invertible matrix so that $\vec{\mathbf{X}} = \mathcal{L}^{-1}\vec{\mathbf{Z}}$. Both vectors represent the same information but the data are given in different coordinates. The covariance matrix of $\vec{\mathbf{Z}}$ is

$$\Sigma_{ZZ} = \mathcal{L}\Sigma_{XX}\mathcal{L}^{\dagger}, \tag{13.16}$$

for $\mathcal{L}^{\dagger}$ the conjugate transpose of $\mathcal{L}$. Suppose the *transformation* is *orthogonal* (i.e., $\mathcal{L}^{-1} = \mathcal{L}^{\dagger}$), and also let λ be an eigenvalue of Σ_{XX} and let $\vec{e}^{\,X}$ be the corresponding eigenvector. Then, since $\Sigma_{XX}\vec{e}^{\,X} = \lambda\vec{e}^{\,X}$,

$$\begin{aligned} \Sigma_{ZZ}\mathcal{L}\vec{e}^{\,X} &= \mathcal{L}\Sigma_{XX}\mathcal{L}^{\dagger}\mathcal{L}\vec{e}^{\,X} \\ &= \mathcal{L}\Sigma_{XX}\vec{e}^{\,X} \\ &= \lambda\mathcal{L}\vec{e}^{\,X}. \end{aligned}$$

Thus λ is also an eigenvalue of Σ_{ZZ} and the EOFs of $\vec{\mathbf{Z}}$ are related to those of $\vec{\mathbf{X}}$ through

$$\vec{e}^{\,Z} = \mathcal{L}\vec{e}^{\,X}. \tag{13.17}$$

Thus eigenvectors are transformed (13.17) just as a random vector is transformed (13.15).

Another consequence of using an orthogonal transformation is that the EOF coefficients are invariant. To see this, let $\mathcal{P}_X$ be the matrix composed of the $\vec{\mathbf{X}}$-EOFs $\vec{e}^{\,X}$ and let $\mathcal{P}_Z$ the the corresponding matrix of $\vec{\mathbf{Z}}$-EOFs. We see from equation (13.17) that

$$\mathcal{P}_Z = \mathcal{L}\mathcal{P}_X. \tag{13.18}$$

Using equation (13.10) and transformation (13.15), the vector of $\vec{\mathbf{X}}$-EOF coefficients

$$\begin{aligned}\vec{\alpha}_X &= \mathcal{P}_X^\dagger\vec{\mathbf{X}} = \mathcal{P}_X^\dagger\mathcal{L}^\dagger\vec{\mathbf{Z}} = (\mathcal{L}\mathcal{P})^\dagger\vec{\mathbf{Z}} = \mathcal{P}_Y^\dagger\vec{\mathbf{Z}} \\ &= \vec{\alpha}_Z\end{aligned}$$

is seen to be equal to the vector of $\vec{\mathbf{Z}}$-EOF coefficients. Thus the EOF coefficients are invariant under *orthogonal transformations*. They are generally not invariant under non-orthogonal transformations. This becomes important when different variables such as precipitation and temperature are combined in a data vector. The EOFs of such a random vector depend on the units in which the variables are expressed.

A special case of transformation (13.15) occurs when $\vec{\mathbf{X}}$ has already been transformed into EOF coordinates using (13.10),

$$\vec{\mathbf{Z}} = \vec{\alpha} = \mathcal{P}^\dagger\vec{\mathbf{X}}.$$

That is, $\mathcal{L} = \mathcal{P}^\dagger$. Using transformation (13.18), we see that the $\vec{\alpha}$-EOFs are

$$\mathcal{P}_\alpha = \mathcal{L}\mathcal{P} = \mathcal{P}^\dagger\mathcal{P} = \mathcal{I}.$$

Thus, in the new coordinates the EOFs are unit vectors. This fact may be used to test EOF programs.

13.1.11 Further Aspects. Some other aspects of EOFs are worth mentioning.

- Empirical Orthogonal Functions may be generalized to continuous functions, in which case they are known as *Karhunen-Loève* functions. The standard inner product $\langle\cdot,\cdot\rangle$ is replaced by an integral, and the eigenvalue problem is no longer a matrix problem but an operator problem. (See, e.g., North et al. [296].)
- The EOFs of some random vectors or random functions are given by sets of analytic orthogonal functions. For instance, if the covariance structure of a spatial process is independent of the location, then the EOFs on a continuous or regularly discretized sphere (circle) are the spherical harmonics (trigonometric functions). See North et al. [296].
- The analysed vector $\vec{\mathbf{X}}$ may be a combination of small vectors that are expressed on different scales, such as temperature and precipitation or geopotential height at 700 and 200 hPa. Then the technique is sometimes called *Combined Principal Component Analysis* (see, e.g., Bretherton, Smith, and Wallace [64]). Vector $\vec{\mathbf{X}}$ might also consist of smaller vectors representing a single field observed at different times, in which case the technique is called *Extended EOF Analysis* (EEOF; see Weare and Nasstrom [417]) or *Multichannel Singular Spectrum Analysis* (MSSA; see Section 13.6).
- Any m-dimensional vector $\vec{\mathbf{y}}$ can be projected onto an EOF $\vec{e}$ of $\vec{\mathbf{X}}$ by computing the inner product $\langle\vec{\mathbf{y}},\vec{e}\,\rangle$. Vector $\vec{\mathbf{y}}$ can then by approximated by $\vec{\mathbf{y}} \approx \sum_{i=1}^k \langle\vec{\mathbf{y}},\vec{e}^{\,i}\rangle\vec{e}^{\,i}$.
- Where are the units? When we expand the random vector $\vec{\mathbf{X}}$ into EOFs

$$\vec{\mathbf{X}} \approx \sum_{i=1}^{k} \alpha_i\vec{e}^{\,i} \tag{13.19}$$

with $\alpha_i = \langle\vec{\mathbf{X}},\vec{e}^{\,i}\rangle$, where do we place the units of $\vec{\mathbf{X}}$ on the right side of approximation (13.19)? Formally the answer is that the coefficients carry the units while the patterns are dimensionless. However, in practice approximation (13.19) is often replaced by

$$\vec{\mathbf{X}} \approx \sum_{i=1}^{k} \alpha_i^+\vec{e}^{\,i+} \tag{13.20}$$

with re-normalized coefficients

$$\alpha_i^+ = \frac{1}{\sqrt{\lambda_i}}\alpha_i \tag{13.21}$$

and patterns

$$\vec{e}^{\,i+} = \sqrt{\lambda_i}\,\vec{e}^{\,i} \tag{13.22}$$

so that $\text{Var}(\alpha_i^+) = 1$. The re-normalized pattern then carries the units of $\vec{\mathbf{X}}$, and represents a 'typical' anomaly pattern if we regard $\alpha_i^+ = \pm 1$ as a 'typical event'.

The decomposition of the local variance, as given by equation (13.7), takes a particularly simple form with this normalization, namely

$$\text{Var}(\mathbf{X}_k) = \sum_{i=1}^{m} |e_k^{i+}|^2. \tag{13.23}$$

Note that the coefficient α_i^+ can be expressed as

$$\alpha_i^+ = \frac{1}{\lambda_i}\langle \vec{\mathbf{X}}, \vec{e}^{\,i+} \rangle. \tag{13.24}$$

13.2 Estimation of Empirical Orthogonal Functions

13.2.1 Outline. After having defined the eigenvalues and EOFs of random vector $\vec{\mathbf{X}}$ as parameters that characterize its covariance matrix, the question naturally arises as to how to *estimate* these parameters from sample $\{\vec{\mathbf{x}}_1, \ldots, \vec{\mathbf{x}}_n\}$ of realizations of $\vec{\mathbf{X}}$. It turns out that useful estimators may be defined by replacing the covariance matrix Σ with the sample covariance matrix $\widehat{\Sigma}$ and by replacing the expectation operator $\mathcal{E}(\cdot)$ with averaging over the sample. An important little trick for reducing the amount of calculation when the sample size n is less than the dimension of $\vec{\mathbf{X}}$ (as is often true) is presented in [13.2.5]. A computational alternative to solving the eigenproblem is to perform a singular value decomposition [13.2.8].

13.2.2 Strategies for Estimating EOFs. The eigenvalues and EOFs are parameters that characterize the covariance matrix of a random vector $\vec{\mathbf{X}}$. In practice, the distribution of $\vec{\mathbf{X}}$, and thus the covariance matrix Σ and its eigenvalues and eigenvectors, is unknown. They must therefore be estimated from a finite sample $\{\vec{\mathbf{x}}_1, \ldots, \vec{\mathbf{x}}_n\}$.

There are two reasonable approaches for estimation.

- Since the eigenvalues and EOFs characterize the covariance matrix Σ of $\vec{\mathbf{X}}$, one reasonable approach is to estimate the covariance matrix and then estimate the eigenvalues λ_i and eigenvectors $\vec{e}^{\,i}$ with the eigenvalues λ_j and the eigenvectors $\vec{e}^{\,j}$ of the estimated covariance matrix $\widehat{\Sigma}$.
- After [13.1.3] the EOFs form an orthonormal set of vectors that is most efficient in representing the variance of $\vec{\mathbf{X}}$ (13.3). Thus another reasonable approach is to use a set of orthonormal vectors that represent as much of the sample variance of the finite sample as possible.

The two approaches are equivalent and lead to the following.

13.2.3 Theorem. *Let* $\widehat{\Sigma} = \frac{1}{n}\sum_{j=1}^n (\vec{\mathbf{x}}_j - \widehat{\mu})(\vec{\mathbf{x}}_j - \widehat{\mu})^\dagger$, *where* $\dagger$ *indicates the conjugate transpose and* $\widehat{\mu} = \frac{1}{n}\sum_{j=1}^n \vec{\mathbf{x}}_j$, *derived from a sample* $\{\vec{\mathbf{x}}_1, \ldots, \vec{\mathbf{x}}_n\}$ *be the estimated covariance matrix of n realizations of* $\vec{\mathbf{X}}$. *Let* $\widehat{\lambda}_1 \ge \widehat{\lambda}_2 \ge \cdots \ge \widehat{\lambda}_m$ *be the eigenvalues of* $\widehat{\Sigma}$ *and let* $\widehat{\vec{e}}^{\,1}, \ldots, \widehat{\vec{e}}^{\,m}$ *be corresponding eigenvectors of unit length. Since* $\widehat{\Sigma}$ *is Hermitian, the eigenvalues are non-negative and the eigenvectors are orthogonal.*

(i) *The k eigenvectors that correspond to* $\widehat{\lambda}_1, \ldots, \widehat{\lambda}_k$ *minimize*

$$\widehat{\epsilon}_k = \sum_{j=1}^{n} \Big| \vec{\mathbf{x}}_j - \sum_{i=1}^{k} \langle \vec{\mathbf{x}}_j, \widehat{\vec{e}}^{\,i} \rangle \widehat{\vec{e}}^{\,i} \Big|^2. \tag{13.25}$$

(ii) $$\widehat{\epsilon}_k = \widehat{\text{Var}}(\vec{\mathbf{X}}) - \sum_{j=1}^{k} \widehat{\lambda}_j. \tag{13.26}$$

(iii) $$\widehat{\text{Var}}(\vec{\mathbf{X}}) = \sum_{j=1}^{m} \widehat{\lambda}_j, \tag{13.27}$$

where $\widehat{\text{Var}}(\vec{\mathbf{X}}) = \text{tr}(\widehat{\Sigma})$.

13.2.4 The Estimated Covariance Matrix $\widehat{\Sigma}$. The covariance between the jth and kth elements of $\vec{\mathbf{X}}$ is estimated by

$$\widehat{\sigma}_{jk} = \frac{1}{n}\sum_{i=1}^{n} (\mathbf{x}_{ji} - \overline{\mathbf{x}}_j)(\mathbf{x}_{ki} - \overline{\mathbf{x}}_k),$$

where $\mathbf{x}_{ji}$ and $\mathbf{x}_{ki}$ are the jth and kth elements of $\vec{\mathbf{x}}_i$. This sum of products can be expressed as a *quadratic form:*[15]

$$\widehat{\Sigma} = \frac{1}{n}\mathcal{X}(\mathcal{I} - \frac{1}{n}\mathcal{J})(\mathcal{I} - \frac{1}{n}\mathcal{J})\mathcal{X}^\dagger \tag{13.28}$$

where $\mathcal{X}$ is the *data matrix*[16]

$$\mathcal{X} = \begin{pmatrix} \mathbf{x}_{11} & \mathbf{x}_{12} & \cdots & \mathbf{x}_{1n} \\ \mathbf{x}_{21} & \mathbf{x}_{22} & \cdots & \mathbf{x}_{2n} \\ \vdots & \vdots & \ddots & \vdots \\ \mathbf{x}_{m1} & \mathbf{x}_{m2} & \cdots & \mathbf{x}_{mn} \end{pmatrix}, \tag{13.29}$$

[15] A quadratic form is a matrix product of the form $\mathcal{A}\mathcal{A}^\dagger$ or $\mathcal{A}^\dagger\mathcal{A}$.

[16] Sometimes also called the design matrix.

and $\mathcal{X}^\dagger$ is the conjugate transpose of $\mathcal{X}$, $\mathcal{I}$ is the $n \times n$ identity matrix, and $\mathcal{J}$ is the $n \times n$ matrix composed entirely of units. The n columns of the $m \times n$ data matrix $\mathcal{X}$ are the sample vectors $\vec{\mathbf{x}}_1, \dots, \vec{\mathbf{x}}_n$; the rows mark the m coordinates in the original space. The matrix product $\mathcal{X}\mathcal{X}^\dagger$ is a square matrix even if $\mathcal{X}$ is not.

13.2.5 Theorem. The following theorem is often useful when computing eigenvalues and eigenvectors [391].

Let $\mathcal{A}$ be any $m \times n$ matrix. If λ is a nonzero eigenvalue of multiplicity s of $\mathcal{A}^\dagger\mathcal{A}$ with s linearly independent eigenvectors $\vec{e}^{\,1}, \dots, \vec{e}^{\,s}$, then λ is also an s-fold eigenvalue of $\mathcal{A}\mathcal{A}^\dagger$ with s linearly independent eigenvectors $\mathcal{A}\vec{e}^{\,1}, \dots, \mathcal{A}\vec{e}^{\,s}$.

A proof is given in Appendix M.

13.2.6 Recipe. The message of Theorem [13.2.5] is that the nonzero eigenvalues of $\mathcal{A}\mathcal{A}^\dagger$ are identical to those of $\mathcal{A}^\dagger\mathcal{A}$ and that the eigenvectors of the two matrices associated with nonzero eigenvalues are related through a simple linear relationship. Thus the following recipe may be used to estimate EOFs.

- If the sample size, n, is larger than the dimension of the problem, m, then the EOFs are calculated directly as the normalized eigenvectors of the $m \times m$ matrix $\frac{1}{n}\mathcal{X}(\mathcal{I} - \frac{1}{n}\mathcal{U})(\mathcal{I} - \frac{1}{n}\mathcal{U})\mathcal{X}^\dagger$.
- If the sample size, n, is smaller than the dimension of the problem, m, the EOFs may be obtained by first calculating the normalized eigenvectors $\vec{g}$ of the $n \times n$ matrix $\frac{1}{n}(\mathcal{I} - \frac{1}{n}\mathcal{J})\mathcal{X}^\dagger\mathcal{X}(\mathcal{I} - \frac{1}{n}\mathcal{J})$ and then computing the EOFs as

$$\vec{e} = \frac{\mathcal{X}(\mathcal{I} - \frac{1}{n}\mathcal{J})\vec{g}}{\|\mathcal{X}(\mathcal{I} - \frac{1}{n}\mathcal{J})\vec{g}\|}.$$

13.2.7 Properties of the Coefficients of the Estimated EOFs. There are several properties worth noting.

- As with the true EOFs, the estimated EOFs span the full m-dimensional vector space. Random vector $\vec{\mathbf{X}}$ can therefore be expanded in terms of the estimated EOFs as $\vec{\mathbf{X}} = \sum_{j=1}^{m} \widehat{\alpha}_j \widehat{\vec{e}}^{\,j}$, where

$$\widehat{\alpha}_j = \langle \vec{\mathbf{X}}, \widehat{\vec{e}}^{\,j} \rangle. \tag{13.30}$$

- When $\vec{\mathbf{X}}$ is multivariate normal, the distribution of $\widehat{\vec{\alpha}}$, where $\widehat{\vec{\alpha}}$ is the m-dimensional vector of EOF coefficients $\widehat{\alpha}_j$, conditional upon the samples used to estimate the EOFs is multivariate normal with mean $\mathcal{E}(\widehat{\vec{\alpha}}|\vec{\mathbf{x}}_1, \dots, \vec{\mathbf{x}}_m) = \widehat{\mathcal{P}}^\dagger \vec{\mu}$ and covariance matrix $\mathrm{Cov}(\widehat{\vec{\alpha}}, \widehat{\vec{\alpha}}|\vec{\mathbf{x}}_1, \dots, \vec{\mathbf{x}}_m) = \widehat{\mathcal{P}}^\dagger \Sigma \widehat{\mathcal{P}}$. Matrix $\widehat{\mathcal{P}}$, which has $\widehat{\vec{e}}^{\,j}$ in column j, is a complicated function of $\vec{\mathbf{x}}_1, \dots, \vec{\mathbf{x}}_m$.
- $\widehat{\lambda}_j$ is the variance of the EOF coefficients computed from the sample used to estimate the EOFs. That is, if $\widehat{\alpha}_{ji} = \langle \vec{\mathbf{X}}_i, \widehat{\vec{e}}^{\,j} \rangle$, then $\frac{1}{n}\sum_{i=1}^{n} |\widehat{\alpha}_{ji} - \overline{\widehat{\alpha}}_j|^2 = \widehat{\lambda}_j$.

  Note that $\widehat{\lambda}_j$ has at least two interpretations as a variance estimate. We could regard $\widehat{\lambda}_j$ as an estimate of the variance of the true EOF coefficient $\alpha_j = \langle \vec{\mathbf{X}}, \vec{e}^{\,j} \rangle$ (see [13.3.3]). Alternatively, we could view the estimated EOFs $\widehat{\vec{e}}^{\,j}$ as fixed, not quite optimal, proxies for $\vec{e}^{\,j}$. Then $\widehat{\lambda}_j$ could be viewed as an estimator of the variance of $\widehat{\alpha}_i = \langle \vec{\mathbf{X}}, \widehat{\vec{e}}^{\,j} \rangle$ when $\widehat{\vec{e}}^{\,j}$ is fixed (see [13.3.2]). These two variances are not equal, although they become asymptotically equivalent as $n \to \infty$. Thus, at least one of the interpretations makes $\widehat{\lambda}_j$ a biased estimator. In fact, they are both poor estimators when the sample is small. In the former case there is uncertainty because the EOFs must be estimated. In the latter case the EOFs are regarded as fixed, but there is a bias because independent data are not used to estimate $\mathrm{Var}(\widehat{\alpha}_i)$. See also [13.3.2,3].
- The sample covariance of a pair of EOF coefficients computed from the sample used to estimate the EOFs is zero. That is, $\frac{1}{n}\sum_{i=1}^{n}(\widehat{\alpha}_{ji} - \overline{\widehat{\alpha}}_j)(\widehat{\alpha}_{ki} - \overline{\widehat{\alpha}}_k)^* = 0$ if $j \neq k$.

  As with $\widehat{\lambda}_j$, the covariance has two interpretations. It correctly estimates the covariance of the true EOF coefficients $\alpha_j = \langle \vec{\mathbf{X}}, \vec{e}^{\,j} \rangle$ and $\alpha_k = \langle \vec{\mathbf{X}}, \vec{e}^{\,k} \rangle$. Alternatively, if we view the estimated EOFs $\widehat{\vec{e}}^{\,j}$ as being fixed, then it incorrectly estimates $\mathrm{Cov}(\widehat{\alpha}_j, \widehat{\alpha}_k)$. The latter, the (j, k) element of $\widehat{\mathcal{P}}^\dagger \Sigma \widehat{\mathcal{P}}$, can be substantially different from zero if $\widehat{\vec{e}}^{\,j}$ and $\widehat{\vec{e}}^{\,k}$ are computed from a small sample.

13.2.8 Gappy Data. Data are often incomplete, that is, there are irregularly distributed gaps in the data vectors caused by missing observations. Estimated EOFs and EOF coefficients can be derived in this case, but the procedure is slightly different. Each element of $\boldsymbol{\Sigma}$ is estimated by forming sums of all available products

$$\widehat{\sigma}_{ij} = \frac{1}{|K_i \cap K_j|} \sum_{k \in K_i \cap K_j} (\mathbf{x}_{ki} - \widehat{\mu}_i)(\mathbf{x}_{kj} - \widehat{\mu}_j)^* \tag{13.31}$$

where $K_i = \{k$: component i of $\vec{\mathbf{x}}_k$ is not missing$\}$, and where $\widehat{\mu}_i = \frac{1}{|K_i|} \sum_{k \in K_i} \mathbf{x}_{ki}$. The estimated EOFs are then the eigenvectors $\widehat{\vec{e}}^{\,i}$ of this covariance matrix estimate. The set $K_i \cap K_j$ is the set of all indices such that $\mathbf{x}_{ki}$ and $\mathbf{x}_{kj}$ are not missing. The $|\cdot|$ notation is used to indicate the size of the enclosed set.

The EOF coefficient $\widehat{\alpha}_i$ of a gappy data vector $\vec{\mathbf{x}}$ can not be obtained as a simple dot product of the gappy data vector $\vec{\mathbf{x}}$ and the estimated EOF $\widehat{\vec{e}}^{\,i}$, as in equation (13.30), but a least squares estimate can be obtained by choosing $\widehat{\alpha}_i$ to minimize $\|\vec{\mathbf{x}} - \widehat{\alpha}_i \widehat{\vec{e}}^{\,i}\|$. The least square estimate is given by

$$\widehat{\alpha}_i = \frac{\sum_{j \in K} \mathbf{x}_j \widehat{e}_j^{\,i*}}{\sum_{j \in K} |\widehat{e}_j^{\,i}|^2} \tag{13.32}$$

where $\mathbf{x}_j$ and $\widehat{e}_j^{\,i}$ are the jth components of $\mathbf{x}$ and $\vec{e}^{\,i}$, respectively, and where $K = \{j$: $\mathbf{x}_j$ is not missing$\}$. Note that equation (13.32) reduces to $\widehat{\alpha}_i = \langle \vec{\mathbf{x}}, \widehat{\vec{e}}^{\,i} \rangle$ when there are no gaps in $\mathbf{x}$.

13.2.9 Computing Eigenvalues and Eigenvectors. One approach to computing eigenvalues and eigenvectors is to use a 'canned' eigenanalysis routine such as those that are contained in EISPACK [352], IMSL [193], or NAG [298]. Press et al. [322, p. 454] discuss the origins of these routines and give further references.

An alternative approach uses Singular Value Decomposition (Appendix B, and see also Press et al. [322, pp. 51–63] and Kelly [218, 219]). The SVD of the conjugate transpose of the $m \times n$ *centred* data matrix

$$\mathcal{X}' = \mathcal{X}\Big(\mathcal{I} - \frac{1}{n}\mathcal{J}\Big), \tag{13.33}$$

where $\mathcal{X}$ is given by equation (13.29), $\mathcal{I}$ is the $n \times n$ identity matrix, and $\mathcal{J}$ is the $n \times n$ matrix of units, is

$$\mathcal{X}'^{\dagger} = \mathcal{U}\mathcal{S}\mathcal{V}^{\dagger}, \tag{13.34}$$

where $\mathcal{U}$ is $n \times m$, $\mathcal{S}$ and $\mathcal{V}$ are each $m \times m$, n is the sample size, and m is the dimension of $\vec{\mathbf{X}}$.[17] Since $n\widehat{\boldsymbol{\Sigma}} = \mathcal{X}'\mathcal{X}'^{\dagger}$, we infer from equation (B.6) that the right singular vectors $\vec{v}^{\,i}$ are equal to the estimated EOFs $\widehat{\vec{e}}^{\,i}$. The singular values s_i are related to the estimated eigenvalues by $\widehat{\lambda}_i = \frac{1}{n}s_i^2$. The left singular vectors $\vec{u}^{\,i}$ are given by (B.5)

$$\vec{u}^{\,i} = \frac{1}{s_i}\mathcal{X}'^{\dagger}\vec{v}^{\,i}. \tag{13.35}$$

The kth column of $\mathcal{X}'$ represents the vector of deviations $\vec{\mathbf{x}}_k - \widehat{\vec{\mu}}$ so that

$$u_k^i = \frac{1}{s_i}(\vec{\mathbf{x}}_k - \widehat{\vec{\mu}})^{\dagger}\vec{v}^{\,i}. \tag{13.36}$$

Thus, u_k^i is the ith normalized EOF coefficient (13.21) of the anomalies $\vec{\mathbf{x}}_k - \widehat{\vec{\mu}}$. Note that the sample variance of the ith normalized EOF coefficient is

$$\begin{aligned} \frac{1}{n}\sum_{k=1}^{n}\Big(\frac{1}{s_i}(\vec{\mathbf{x}}_k - \widehat{\vec{\mu}})^{\dagger}\widehat{\vec{e}}^{\,i}\Big) &= \frac{1}{s_i^2}\widehat{\mathrm{Var}}(\widehat{\alpha}_i) \\ &= \frac{1}{s_i^2}\widehat{\lambda}_i = 1. \end{aligned} \tag{13.37}$$

Note also that equations (13.35)–(13.37) are only valid for those EOFs that correspond to nonzero eigenvalues. The number of nonzero eigenvalues, which is determined by the rank[18] of the centred data matrix, is no greater than $\min(m, n-1)$.

Thus SVD extracts the same information from the sample as a conventional EOF analysis.

13.3 Inference

13.3.1 General. We consider the reliability of eigenvalues and EOF estimates in this section. This is a somewhat different question from that which users generally have in mind when they enquire about the 'significance' of an EOF. The null hypothesis that is usually implicit in the latter is that the EOF in question describes only an aspect of the covariance structure of the 'noise' in the observed system, and the alternative hypothesis is that the EOF also describes at least part of the dynamics of the observed system. Unfortunately, discrimination between 'noise' and 'signal' in

[17] We have implicitly assumed here that $m \leq n$. The problem is approached similarly when $m > n$, except we begin by obtaining the SVD of $\mathcal{X}'$. Note also that in some texts $\mathcal{U}$ and $\mathcal{V}$ are $n \times n$ and $m \times m$ orthogonal matrices respectively and $\mathcal{S}$ is $n \times m$. The singular values are placed in the diagonal part of $\mathcal{S}$ and the rest of the matrix is zero. We use the decomposition given in (13.34) because it is commonly used in SVD subroutines (see, e.g., Press et al. [322]).

[18] The *rank* of a matrix is the dimension of the sub-space spanned by the columns of that matrix.

this way is fraught with difficulty. We discuss this further in [13.3.4]. However, we first briefly consider the variance of EOF coefficients in [13.3.2] and the bias of eigenvalue estimates in [13.3.3]. We consider the sampling error of the EOFs themselves in [13.3.5,6].

13.3.2 The Variance of EOF Coefficients of a Given Set of Estimated EOFs. Assume we are given a set of eigenvalues $\widehat{\lambda}_i$ and EOFs $\widehat{\vec{e}}^{\,i}$ that are derived from a finite sample $\{\vec{\mathbf{x}}_1 \dots \vec{\mathbf{x}}_n\}$. Then any random vector $\vec{\mathbf{X}}$ can be represented in the space spanned by these estimated EOFs by using the transformation $\widehat{\vec{\alpha}} = \mathcal{P}^\dagger \vec{\mathbf{X}}$. The transformed random variables $\widehat{\alpha}_i = \langle \vec{\mathbf{X}}, \widehat{\vec{e}}^{\,i} \rangle$ have their own moments, such as variances $\mathrm{Var}(\widehat{\alpha}_i)$ and covariances $\mathrm{Cov}(\widehat{\alpha}_i, \widehat{\alpha}_j)$. In the following we view the estimated EOFs $\widehat{\vec{e}}^{\,i}$ as being 'fixed' (or frozen) rather than random.

Intuitively one would hope that the variance of $\widehat{\alpha}_i$ is equal to that of the real EOF coefficient α_i. Unfortunately, this is not the case (see [13.2.6]). Consider, for example, the first EOF $\vec{e}^{\,1}$ and the corresponding EOF coefficient $\alpha_1 = \langle \vec{\mathbf{X}}, \vec{e}^{\,1} \rangle$. The first EOF minimizes

$$\epsilon_1 = \mathcal{E}\left(\|\vec{\mathbf{X}} - \langle \vec{\mathbf{X}}, \widehat{\vec{e}}^{\,1} \rangle \vec{e}^{\,1}\|^2\right).$$

Replacing $\vec{e}^{\,1}$ with any other vector, such as $\widehat{\vec{e}}^{\,1}$, increases ϵ_1. Thus

$$\begin{aligned} &\mathrm{Var}(\vec{\mathbf{X}}) - \mathrm{Var}(\alpha_1) \\ &= \mathcal{E}\left(\|\vec{\mathbf{X}} - \langle \vec{\mathbf{X}}, \vec{e}^{\,1} \rangle \vec{e}^{\,1}\|^2\right) \\ &< \mathcal{E}\left(\|\vec{\mathbf{X}} - \langle \vec{\mathbf{X}}, \widehat{\vec{e}}^{\,1} \rangle \widehat{\vec{e}}^{\,1}\|^2\right) \\ &= \mathrm{Var}(\vec{\mathbf{X}}) - \mathrm{Var}(\widehat{\alpha}_1), \end{aligned}$$

that is, $\mathrm{Var}(\alpha_1) > \mathrm{Var}(\widehat{\alpha}_1)$. Similar arguments lead to

- $\mathrm{Var}(\widehat{\alpha}_i) < \mathrm{Var}(\alpha_i)$ for the first few EOFs (those corresponding to the largest eigenvalues).

Since the total variance $\mathrm{Var}(\vec{\mathbf{X}}) = \sum_{j=1}^m \mathrm{Var}(\vec{\mathbf{x}}_j)$ is estimated with nearly zero bias by $\widehat{\mathrm{Var}}(\vec{\mathbf{X}}) = \sum_{j=1}^m \widehat{\lambda}_j$, it follows that

- $\mathrm{Var}(\widehat{\alpha}_i) > \mathrm{Var}(\alpha_i)$ for the last few EOFs.

Examples show that these deviations may be considerable, in particular for small eigenvalues [392].

13.3.3 The Bias in Estimating Eigenvalues. It is natural to ask questions about the reliability of eigenvalues and EOF estimates, such as the extent to which the estimated patterns resemble the true patterns and how close the estimated and true eigenvalues are. These questions do not have completely satisfactory answers, but there are a number of potentially useful facts. One of these facts is the following set of asymptotic formulae that apply to eigenvalue estimates computed from samples that can be represented by n independent and identically distributed normal random vectors (Lawley [245]):

$$\mathcal{E}(\widehat{\lambda}_i) = \lambda_i \left(1 + \frac{1}{n} \sum_{\substack{j=1 \\ j \neq i}}^{m} \frac{\lambda_j}{\lambda_i - \lambda_j}\right) + \mathcal{O}(n^{-2}) \tag{13.38}$$

$$\mathrm{Var}(\widehat{\lambda}_i) = \frac{2\lambda_i^2}{n} \left(1 - \frac{1}{n} \sum_{\substack{j=1 \\ j \neq i}}^{m} \left(\frac{\lambda_j}{\lambda_i - \lambda_j}\right)^2\right) + \mathcal{O}(n^{-3}). \tag{13.39}$$

As usual, m is the dimension of the random vectors. The symbol $\mathcal{O}(n^{-s})$ represents a term that converges to zero as $n \to \infty$ at least as quickly as n^{-s} does.

By equations (13.38) and (13.39), the eigenvalue estimators are consistent:

$$\lim_{n\to\infty} \mathcal{E}\left((\widehat{\lambda}_i - \lambda_i)^2\right) = 0. \tag{13.40}$$

However, something unwanted is hidden in equation (13.38), namely that the estimators of the largest and of the smallest eigenvalues are biased. For the largest eigenvalues, almost all of the denominators in equation (13.38) are positive so that the entire sum is positive, that is, $\mathcal{E}(\widehat{\lambda}_i) > \lambda_i$ for large eigenvalues. Similarly, $\mathcal{E}(\widehat{\lambda}_i) < \lambda_i$ for the smallest eigenvalues.

Together with the results from [13.3.2] this finding shows that

- for the largest eigenvalues λ_i,

$$\mathcal{E}(\widehat{\lambda}_i) > \lambda_i = \mathrm{Var}(\alpha_i) > \mathrm{Var}(\widehat{\alpha}_i), \tag{13.41}$$

- for the smallest eigenvalues λ_i,

$$\mathcal{E}(\widehat{\lambda}_i) < \lambda_i = \mathrm{Var}(\alpha_i) < \mathrm{Var}(\widehat{\alpha}_i). \tag{13.42}$$

Relations (13.41) and (13.42) illustrate that we must be cautious when using estimated eigenvalues. First, the estimates are biased: the large eigenvalues are overestimated and the small ones are underestimated. More important, though, is the inequality $\mathcal{E}(\widehat{\lambda}_i) \neq \mathrm{Var}(\widehat{\alpha}_i)$: The sample eigenvalue $\widehat{\lambda}$ is a biased estimator of the variance of $\widehat{\alpha}_i = \langle \vec{\mathbf{X}}, \widehat{\vec{e}}^{\,i} \rangle$, for any frozen set of estimated EOFs $\widehat{\vec{e}}^{\,i}$. Similarly, $\mathrm{Cov}(\widehat{\alpha}_i, \widehat{\alpha}_j) \neq \widehat{\mathrm{Cov}}(\widehat{\alpha}_i, \widehat{\alpha}_j) = 0$.

13.3.4 'Selection Rules.' Many so-called *selection rules* have been proposed that supposedly separate the physically relevant EOFs from those that are not.[19] One popular procedure of this type is 'Rule N' [321]. The basic supposition is that the full phase space can be partitioned into one subspace that contains only noise and another that contains dynamical variations (or 'signals'). It is assumed that the signal-subspace is spanned by well-defined EOFs while those in the noise-subspace are degenerate. Thus, the idea is to attempt to identify the signal-subspace as the space spanned by the EOFs that are associated with large, well-separated eigenvalues.

The selection rules compare the eigenspectrum[20] estimated from the sample with distributions of sample eigenspectra that are obtained under the assumption that all or the smallest $\widetilde{m}$ true eigenvalues are equal. The number $\widetilde{m}$ is either specified *a priori* or determined recursively. All estimated eigenvalues that are larger than, say, the 95% percentile of the (marginal) distribution of the reference 'noise' spectra, are identified as being 'significant' at the 5% level.

One problem with this approach is that this selection rule is mistakenly understood to be a *statistical test* of the null hypothesis that EOFs $\vec{e}^{\,1}, \ldots, \vec{e}^{\,\widetilde{m}}$, for $\widetilde{m} < m$, span noise against the alternative hypothesis that they span the signal-subspace. The connection between this alternative and the determination of a 'signal-subspace' is vague. Also, the approach sketched above does not consider the reliability of the estimated patterns since the selection rules are focused only on the eigenvalues.

The other problem with the 'selection rule' approach is that there need not be any connection between the shape of the eigenspectrum on the one hand and the presence or absence of 'dynamical structure' on the other. To illustrate, suppose that a process $\vec{\mathbf{X}}_t = \vec{D}_t + \vec{\mathbf{N}}_t$, containing both dynamical and noise components (recall [10.1.1]), has eigenvalues $\lambda_1, \ldots, \lambda_m$ and EOFs $\vec{e}^{\,1}, \ldots, \vec{e}^{\,m}$. Now construct a multivariate white noise $\vec{\mathbf{Z}}_t$ from iid $\mathcal{N}(0, \mathcal{I})$ random vectors. Then the multivariate white noise process $\vec{\mathbf{Y}}_t = \mathcal{P}\Lambda^{1/2}\vec{\mathbf{Z}}_t$, which is completely devoid of 'dynamics,' has the same eigenvalues and eigenvectors as $\vec{\mathbf{X}}_t$. Thus we cannot always diagnose dynamical structure from the zero lag covariance structure of a process.[21]

Our recommendation is to avoid using selection rules. We outline a better approach, based on *North's Rule-of-Thumb*, in the next subsection.

13.3.5 North's Rule-of-Thumb. Using a scale argument, North et al. [296] obtained an approximation for the 'typical' error of the estimated EOFs

$$\Delta\widehat{\vec{e}}^{\,i} \approx \sqrt{\frac{2}{n}} \sum_{\substack{j=1 \\ j\neq i}}^{m} \frac{c}{\lambda_j - \lambda_i} \vec{e}^{\,j} \tag{13.43}$$

where c is a constant and n is the number of *independent* samples. There are three things to notice about this equation.

- The first-order error $\Delta\widehat{\vec{e}}^{\,i}$ is of the order of $\sqrt{\frac{1}{n}}$. Thus convergence to zero is slow.
- The first-order error $\Delta\widehat{\vec{e}}^{\,i}$ is orthogonal to the true EOF $\vec{e}^{\,i}$.
- The estimate of the ith EOF $\vec{e}^{\,i}$ is most strongly contaminated by the patterns of those other EOFs $\vec{e}^{\,j}$ that correspond to the eigenvalues λ_j closest to λ_i. The smaller the difference between λ_j and λ_i, the more severe the contamination.

Lawley's formulae (13.38, 13.39) yield a first-order approximation of the 'typical error' in $\widehat{\lambda}_i$:

$$\Delta\lambda_i \approx \sqrt{\frac{2}{n}}\lambda_i. \tag{13.44}$$

Combining this with a simplified version of approximation (13.44), North et al. [296] finally obtain

$$\Delta\widehat{\vec{e}}^{\,i} \approx \frac{c'\Delta\lambda_i}{\lambda_j - \lambda_i} \vec{e}^{\,j} \tag{13.45}$$

where c' is a constant and λ_j is the the closest eigenvalue to λ_i. North's 'Rule-of-Thumb' follows from approximation (13.45): *'If the sampling error*

[19] See, for example, Preisendorfer, Zwiers, and Barnett [321].

[20] An *eigenspectrum* is the distribution of variance (i.e., eigenvalues), with EOF index. The eigenspectrum is an analogue of the power spectrum (see Section 11.2) since both describe the distribution of variance across the coefficients of orthonormal basis functions.

[21] We would need to also analyse at least part of the lagged covariance structure of $\vec{\mathbf{X}}_t$ to reveal the 'dynamics' in this example.

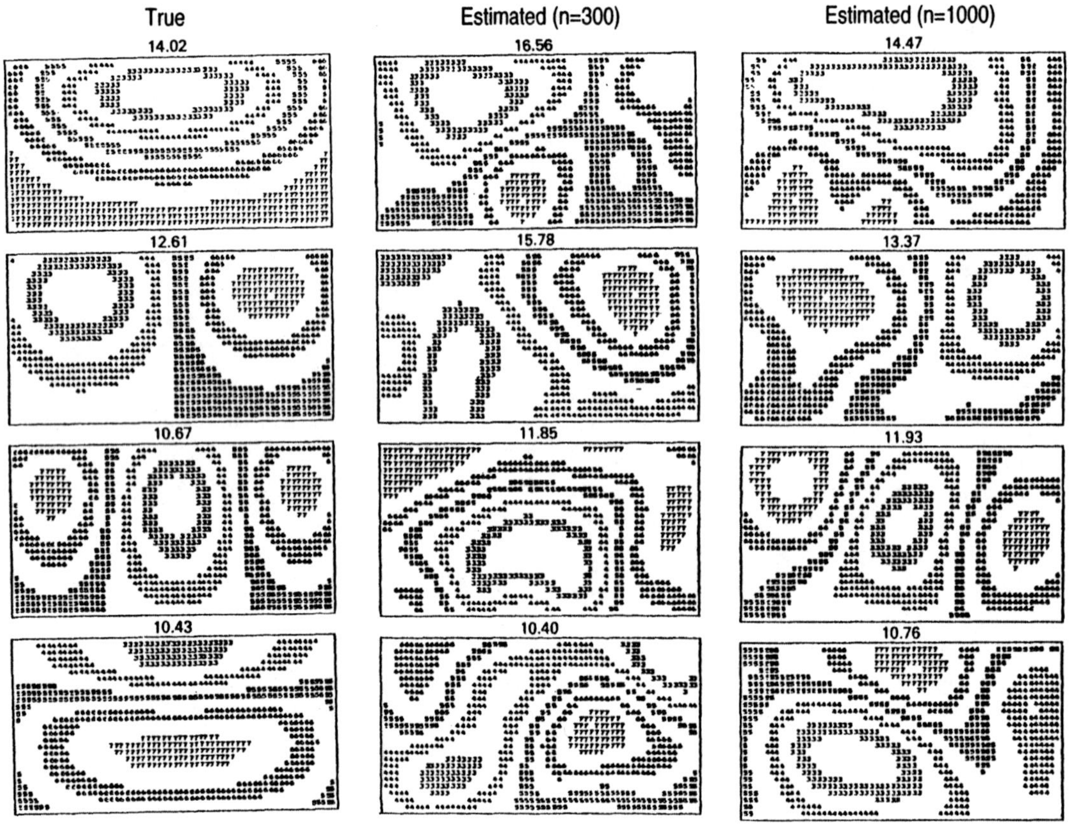

Figure 13.2: *North et al.'s [296] illustration of North's Rule-of-Thumb [13.3.5]. From [296].*
Left: The first four true eigenvalues and EOFs.
Middle: Corresponding estimates obtained from a random sample of size $n = 300$.
Right: As middle column, except $n = 1000$.

of a particular eigenvalue $\Delta\lambda$ *is comparable to or larger than the spacing between* λ *and a neighbouring eigenvalue, then the sampling error* $\Delta\vec{e}$ *of the EOF will be comparable to the size of the neighbouring EOF'.*

13.3.6 North et al.'s Example. North et al. [296] constructed a synthetic example in which the first four eigenvalues are 14.0, 12.6, 10.7 and 10.4 to illustrate North's Rule-of-Thumb [13.3.4]. The first four (true) EOFs are shown in the left hand column of Figure 13.2. According to approximation (13.44) the typical error for the first four estimated eigenvalues is $\Delta\lambda_i \approx \pm 1$ for $n = 300$ and $\Delta\lambda_i \approx \pm 0.6$ for $n = 1000$. Since $\lambda_1 - \lambda_2 = 1.4$, $\lambda_2 - \lambda_3 = 2$ and $\lambda_3 - \lambda_4 = 0.3$, one would expect the first two EOFs to be *mixed*[22] when $n = 300$ and the third and fourth EOF to be mixed for both $n = 300$ and $n = 1000$. That this is a reasonable guess is demonstrated in the middle and right hand columns of Figure 13.2, which displays EOFs estimated from random samples of size $n = 300$ and $n = 1000$, respectively.

[22] That is, we expect the first two EOFs to be a combination of the EOFs that correspond to nearby eigenvalues.

13.4 Examples

13.4.1 Overview. We will present two examples of conventional EOF analysis in this section. This first case, on the globally distributed SST, is most straight forward. The second example involves a data vector that is constructed by combining the same variable at several levels in the vertical.

13.4.2 Monthly Mean Global Sea-surface Temperature. The first two EOFs of monthly mean sea-surface temperature (SST) of the global ocean between 40°S and 60°N are shown in Figure 13.3. They represent 27.1% and 7.9% of the total variance, respectively.

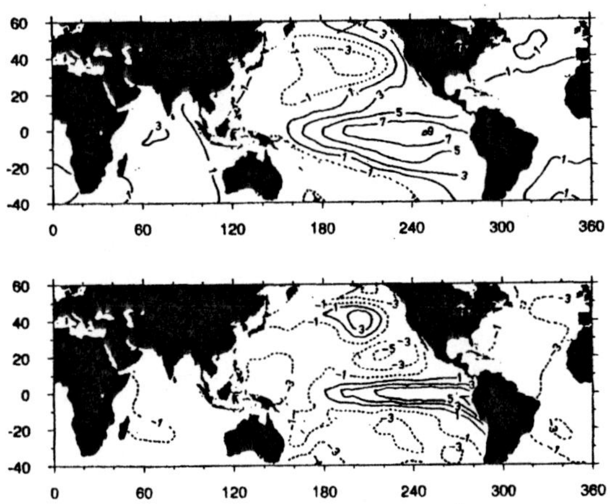

Figure 13.3: *EOFs 1 (top) and 2 (bottom) of monthly mean sea-surface temperature (SST). Units:* 10^{-2}. *Courtesy Xu.*

The first EOF, which is concentrated on the Pacific Ocean, represents ENSO. Its time coefficient, shown as curve 'D' in Figure 13.4, is highly correlated with the two Southern Oscillation indices (Darwin minus Papeete SLP, curve 'E', and SST area average, curve 'F') introduced previously. The large 'centre of action' in the North Pacific represents the oceanic response to anomalous extratropical winds which, in turn, were excited by the anomalous tropical state.

The second EOF of the SST field also involves the tropical Pacific Ocean. The most prominent feature is the narrow tongue of water in the eastern and central equatorial Pacific with temperatures that vary coherently. While the coefficient time series (Figure 13.4, curve 'A') reflects ENSO events (e.g., 1982/83) in part, the connection with the SOI is not as clear as with the first EOF. The coefficient appears to have a downward trend from about 1976 onwards, which would correspond to cooling in the eastern and central tropical Pacific or warming elsewhere. It remains to determine whether the trend is real, part of the global ocean's natural low-frequency variability, or just an artifact of the way in which these data have been collected and analysed.

13.4.3 Monthly Mean of Zonal Wind at Various Levels. The next example is on the monthly mean zonal wind in the troposphere (Xu, personal communication). A joint analysis of the wind field at the 850, 700, 500, 300 and 200 hPa levels was performed. The size of the problem was kept manageable by performing the analysis in two steps. Separate EOF analyses were first performed at each level. In each analysis, the coefficients representing 90% of the variance were retained. A combined vector, composed of EOF coefficients selected for the five levels, is used as input for the eventual EOF analysis of the three-dimensional zonal wind field.

The first two EOFs are shown in Figure 13.5, and their coefficient time series are shown as traces 'B' and 'C' in Figure 13.4. The first EOF, representing 11% of the total monthly variance, is mostly barotropic, not only in the extratropics but also in the tropics. Its coefficient time series exhibits a trend parallel to that found in the coefficient of the second SST EOF. The mean westerly winds in the Southern Hemisphere were analysed as being weaker in the 1970s than in the mid 1980s (negative sign indicates easterly wind anomalies). At the same time the mean low-level easterlies along the equatorial Pacific were weaker in the early 1970s and stronger in the mid 1980s (positive anomalies represent anomalous westerly winds). The results of the EOF analysis of the SST in [13.4.2] are consistent with this representation: The second SST EOF described an equatorial Pacific that was warmer in the early 1970s and cooler in the 1980s, a phenomenon that should be accompanied by strengthening easterly trades during this period.

Because independent analysis techniques are used to derive the SST and zonal wind fields, we can conclude that the trend found in both EOF analyses is not due to data problems. However, it is still not possible to determine whether the trend originates from a natural low-frequency variation or from some other cause.

13.5 Rotation of EOFs

13.5.1 Introduction. This section describes a class of basis vector 'rotation' procedures that is widely used in climate research. The procedures are usually applied to EOFs in the hope that the resulting 'rotated EOFs' can be more easily interpreted than the EOFs themselves. The term 'rotated EOFs' is a mild misnomer that may lead to confusion; 'rotation' transforms the Empirical Orthogonal Function into a *non-orthogonal* linear basis. Also, 'rotation' can be performed on any linear basis, not just EOFs.

We will first explain the general idea and will then describe 'varimax'-rotation in some detail. We use three examples to describe the merits

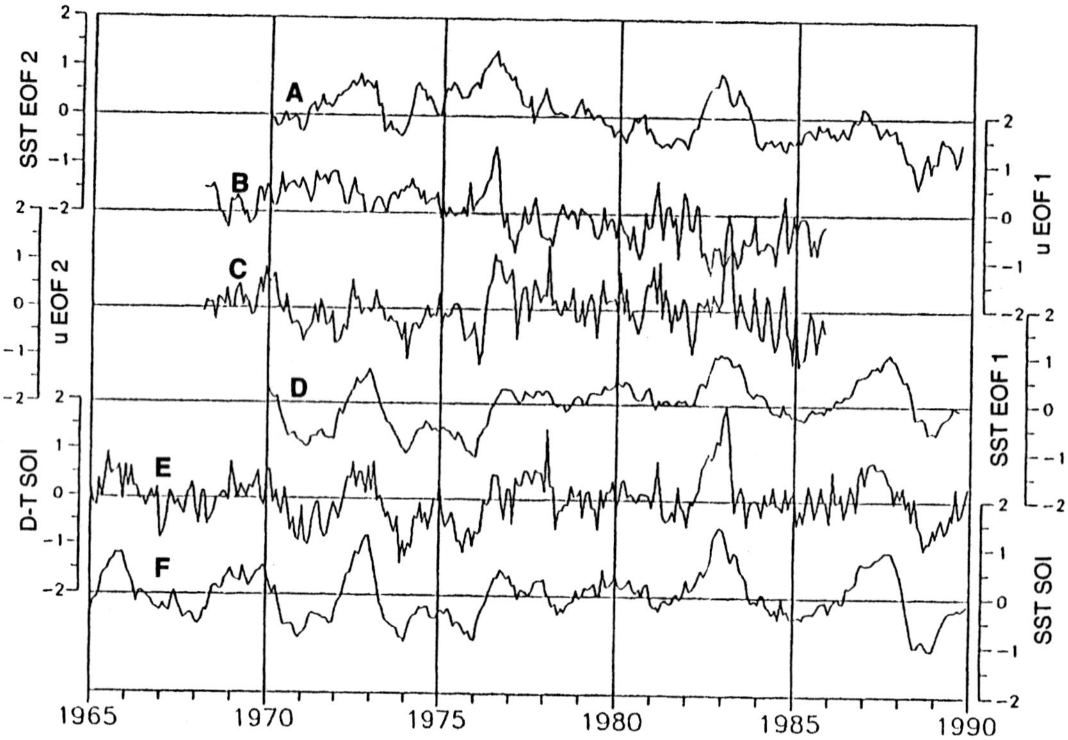

Figure 13.4: *EOF coefficients of monthly mean sea-surface temperature [13.4.2] (curves D and A), monthly mean zonal wind [13.4.3] (curves B and C) and two Southern Oscillation indices [1.2.2] (curves E and F). All data are normalized.*
A: *2nd SST EOF coefficient.* **B**: *1st zonal wind EOF coefficient.* **C**: *2nd zonal wind EOF coefficient.* **D**: *1st SST EOF coefficient.* **E**: *Darwin minus Papeete SLP index of the Southern Oscillation.* **F**: *SST index of the Southern Oscillation.*
Courtesy Xu.

of this procedure. The first of these examples is on the successful and reproducible identification of teleconnection patterns (cf. Section 17.4). The second example deals with a case in which the effect of the rotation is negligible. The third case illustrates pathological behaviour by showing that rotation sometimes splits features into different patterns even though they are part of the *same physical pattern*.

13.5.2 The Concept of 'Rotation.' Having used EOF analysis, or some other technique, to identify a low-dimensional subspace that contains a substantial fraction of the total variance, it is sometimes of interest to look for a linear basis of this subspace with specified properties, such as the following.

- Basis vectors that contain simple geometrical patterns. Simplicity could mean that the patterns are confined regionally, or that the patterns are composed of two regions, one with large positive values and another with large negative values.

- Basis vectors that have time coefficients with specific types of behaviour, such as having nonzero values only during some compact time episodes.

Richman [331] lists five vague criteria for simple structure and there are many proposals of 'simplicity' functionals. The minimization of these functionals is generally non-trivial since the functionals are nonlinear. Numerical algorithms used to obtain approximate solutions can only be applied to bases of moderate size.

The results of a rotation exercise depend on the number and length of the 'input vectors', and on the measure of simplicity. Successful application of the rotation technique requires some experience and the novice may find Richman's [331] review paper on rotation useful. Interesting examples are

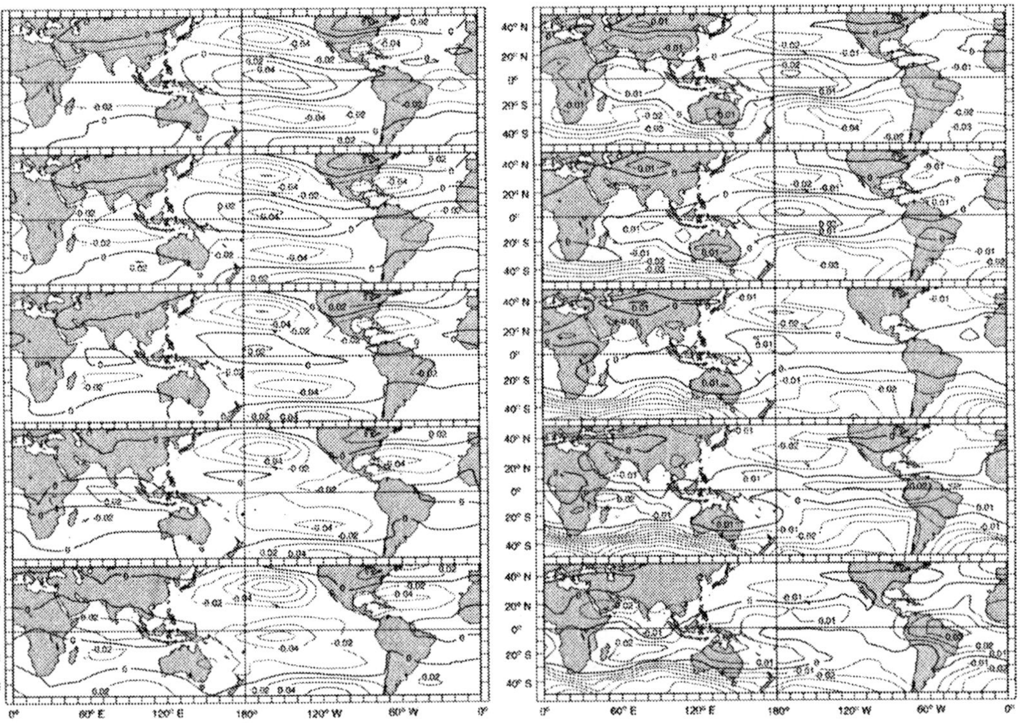

Figure 13.5: *First two EOFs of the tropospheric zonal wind between 45°S and 45°N, at 850, 700, 500, 300 and 200 hPa (from bottom to top). First EOF on the left, second on the right.*
Courtesy Xu.

offered by Barnston and Livezey [27] and Chelliah and Arkin [80], among many others.

The opinion in the community is divided on the subject of rotation. Part of the community advocates the use of rotation fervently, arguing that it is a means with which to diagnose physically meaningful, statistically stable patterns from data. Several arguments are raised in favour of the rotated EOFs.

- The technique produces compact patterns that can be used for 'regionalization,' that is, to divide an area in a limited number of homogeneous sub-areas.
- Rotated EOFs are less sensitive to the distribution of observing locations than conventional EOFs.
- Rotated EOFs are often statistically more stable then conventional EOFs (see, e.g., Cheng, Nitsche, and Wallace [82]). That is, the sampling variance of rotated EOFs is often less than that of the input vectors.

Others in the scientific community are less convinced because of the heuristic arguments that motivate the simplicity functionals, and thus the heuristic basis for the interpretation of the result. Jolliffe [198] lists four drawbacks of the routine use of rotation, namely i) the arbitrary choice of the rotation criterion, ii) the sensitivity of the result to the normalization of the EOFs (see [13.5.3]), iii) the need to redo the entire calculation if the number of EOFs is changed (see [13.5.4]), and iv) the loss of information about the dominant sources of variation in the data.

13.5.3 The Mathematics of 'Rotation.' 'Rotation' consists of the transformation of a set of 'input vectors' $\mathcal{P} = (\vec{p}^{\,1}|\cdots|\vec{p}^{\,K})$ into another set of vectors $\mathcal{Q} = (\vec{q}^{\,1}|\cdots|\vec{q}^{\,K})$ by means of an invertible $K \times K$ matrix $\mathcal{R} = (r_{ij})$:

$$\mathcal{Q} = \mathcal{P}\mathcal{R} \tag{13.46}$$

or, for each vector $\vec{q}^{\,i}$:

$$\vec{q}^{\,i} = \sum_{j=1}^{K} r_{ij}\vec{p}^{\,j}. \tag{13.47}$$

The matrix $\mathcal{R}$ is chosen from a class of matrices, such as orthonormal matrices, subject to the constraint that a functional $V(\mathcal{Q})$ is minimized. An example of such a functional is described in the next subsection.

Under some conditions, operation (13.46) can be viewed as a rotation of the 'input vectors.' Since these are often the first K EOFs, the resulting vectors $\vec{q}^{\,i}$ are called 'rotated EOFs.'

When matrix $\mathcal{R}$ is orthonormal, the operation is said to be an 'orthonormal rotation'; otherwise it is said to be 'oblique.'

Now let $\vec{\mathbf{X}}$ be a random vector that takes values in the space spanned by the input vectors. That is

$$\vec{\mathbf{X}} = \mathcal{P}\vec{\alpha} \tag{13.48}$$

where $\vec{\alpha}$ is a k-dimensional vector of random expansion coefficients. Then, because of operation (13.46)

$$\vec{\mathbf{X}} = (\mathcal{P}\mathcal{R})(\mathcal{R}^{-1}\vec{\alpha}) = \mathcal{Q}\vec{\beta} \tag{13.49}$$

where $\vec{\beta} = \mathcal{R}^{-1}\vec{\alpha}$ is the k-dimensional vector of random expansion coefficients for the rotated patterns.

Let us assume for the following that the matrix $\mathcal{R}$ is orthonormal so that $\vec{\beta} = \mathcal{R}^{\mathrm{T}}\vec{\alpha}$.[23]

- When the input vectors are orthogonal, the scalar products between all possible pairs of rotated vectors are given by the matrix

$$\mathcal{Q}^{\mathrm{T}}\mathcal{Q} = \mathcal{R}^{\mathrm{T}}\mathcal{P}^{\mathrm{T}}\mathcal{P}\mathcal{R} = \mathcal{R}^{\mathrm{T}}\mathcal{D}\mathcal{R}, \tag{13.50}$$

  where $\mathcal{D} = (\vec{p}^{\,1^{\mathrm{T}}}\vec{p}^{\,1}, \ldots, \vec{p}^{\,K^{\mathrm{T}}}\vec{p}^{\,K})$. Thus the rotated vectors are orthogonal only if $\mathcal{D} = \mathcal{I}$, or, in other words, if the input vectors are normalized to unit length.

- Similarly, if the expansion coefficients of the input vectors are pairwise uncorrelated, so that $\Sigma_{\alpha\alpha} = \mathrm{diag}(\sigma_1^2, \ldots, \sigma_K^2)$, then the coefficients of the rotated patterns are also pairwise uncorrelated only if coefficients α_j have unit variance. Then

$$\begin{aligned}\Sigma_{\beta\beta} &= \mathrm{Cov}\big(\mathcal{R}^{\mathrm{T}}\vec{\alpha}, \mathcal{R}^{\mathrm{T}}\vec{\alpha}\big) \\ &= \mathcal{R}^{\mathrm{T}}\Sigma_{\alpha\alpha}\mathcal{R}.\end{aligned} \tag{13.51}$$

Equations (13.50) and (13.51) imply that rotated patterns derived from normalized EOFs, as defined in [13.1.2,3] so that $\vec{p}^{\,j} = \vec{e}^{\,j}$, are also orthonormal, but their time coefficients are not uncorrelated. If, on the other hand, the EOFs are re-normalized as in equations (13.21) and (13.22) so that $\vec{p}^{\,j} = \vec{e}^{\,j+}$ and $\mathrm{Var}\big(\alpha_j^+\big) = 1$, then the rotated patterns are no longer orthogonal but the coefficients remain pairwise uncorrelated.

Thus two important conclusions may be drawn.

- The result of the rotation exercise depends on the lengths of the input vectors. Differently scaled but directionally identical sets of input vectors lead to sets of rotated patterns that are directionally different from one another. Jolliffe [199] demonstrates that the differences can be large.

  The rotated vectors are a function of the input vectors rather than the space spanned by the input vectors.

- After rotating EOF patterns, the new patterns and coefficients are not orthogonal and uncorrelated at the same time. When the coefficients are uncorrelated, the patterns are not orthogonal, and vice versa. Thus, the percentage of variance represented by the individual patterns is no longer additive.

13.5.4 The 'Varimax' Method. 'Varimax' is a widely used orthonormal rotation that minimizes the 'simplicity' functional

$$V(\vec{q}^{\,1}, \ldots, \vec{q}^{\,K}) = \sum_{i=1}^{K} f_V(\vec{q}^{\,i}) \tag{13.52}$$

where $\vec{q}^{\,i}$ is given by equation (13.47) and f_V is defined by

$$f_V(\vec{q}\,) = \frac{1}{m}\sum_{i=1}^{m}\left(\frac{q_i}{s_i}\right)^4 - \frac{1}{m^2}\left(\sum_{i=1}^{m}\left(\frac{q_i}{s_i}\right)^2\right). \tag{13.53}$$

The constants s_i are chosen by the user. The *raw varimax* rotation is obtained when $s_i = 1$, $i = 1, \ldots, K$, and the *normal varimax* rotation is obtained by setting $s_i = \sum_{j=1}^{K}(p_i^j)^2$. Another option is to define s_i as the standard deviation of the ith component of

$$\vec{\mathbf{X}}^{(K)} = \sum_{j=1}^{K}\alpha_j\vec{p}^{\,j},$$

which is the projection of the original full random vector $\vec{\mathbf{X}}$ onto the subspace spanned by the K vectors $\{\vec{p}^{\,1} \ldots \vec{p}^{\,K}\}$.

Note that $f_V(\vec{q}\,)$ (13.53) can be viewed as the spatial variance of the normalized squares $(q_i/s_i)^2$. That is, $f_V(\vec{q}\,)$ measures the 'weighted

[23] All matrices and vectors in this section are assumed to be real valued. Thus orthonormal matrices satisfy $\mathcal{R}\mathcal{R}^{\mathrm{T}} = \mathcal{R}^{\mathrm{T}}\mathcal{R} = \mathcal{I}$.

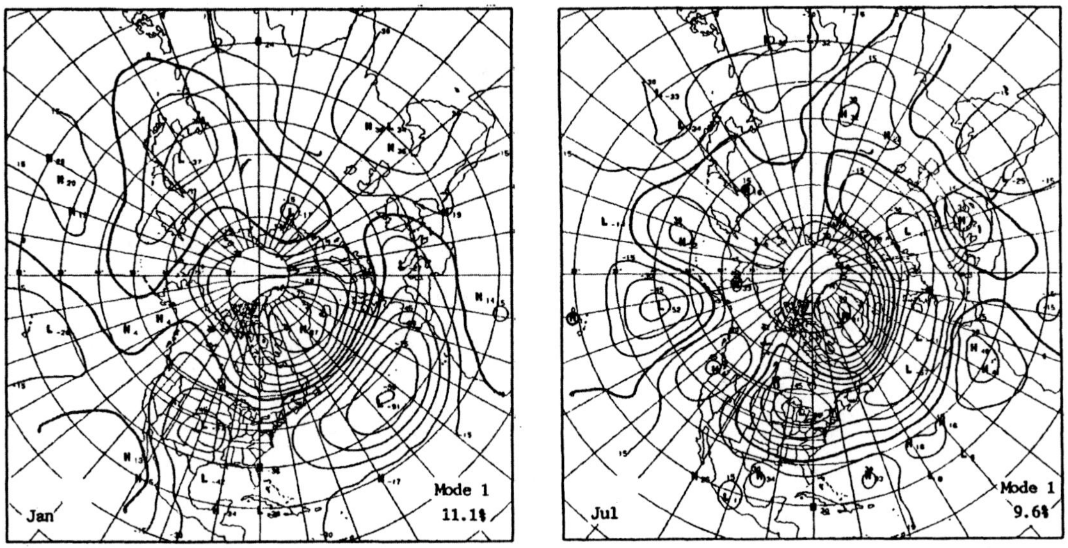

Figure 13.6: *January (left) and July (right) versions of the North Atlantic Oscillation pattern derived by Barnston and Livezey [27] by applying varimax rotation to the first 10 normalized EOFs of January and July mean 700 hPa height, respectively. Courtesy R. Livezey.*

square amplitude' variance of $\vec{q}$. Therefore, minimizing function (13.52) is equivalent to finding a matrix $\mathcal{R}$ such that the sum of the total weighted square amplitude variance of the K patterns $(\vec{q}^{\,1}|\cdots|\vec{q}^{\,k}) = \mathcal{P}\mathcal{R}$ is minimized. See Richman [331] for further details.

13.5.5 Example: Low-frequency Atmospheric Circulation Patterns. Barnston and Livezey [27] argued extensively that rotated EOF analysis is a more effective tool for the analysis of atmospheric circulation patterns than the 'teleconnection' analysis (Wallace and Gutzler [409]; see also Section 17.4). They used a varimax rotation of re-normalized EOFs (13.21, 13.22) to isolate the dominant circulation patterns in the Northern Hemisphere (NH) on the monthly time scale. The EOFs used in the study were computed separately for each month of the year from correlation matrices derived from a 35-year data set of monthly mean 700 hPa heights analysed on a 358-point grid. The data set itself was carefully screened to remove known analysis biases. Rotation was performed on the first 10 EOFs in each month. They represent about 80% of the total variance in winter and 70% in summer.

The result of the exercise is an extensive collection of NH circulation patterns. Barnston and Livezey identified 13 patterns: nine cold season patterns, two warm season patterns, and two transition season patterns. Only one pattern, the North Atlantic Oscillation (NAO, Figure 13.6) is evident in every month of the year. Barnston and Livezey estimate that it represents between 15.4% (March) and 7.4% (October) of the total variance. The NAO is the dominant circulation pattern in the solstitial seasons (DJFM and MJJAS). The NAO is characterized by a 'high' (this adjective is arbitrary since the sign of the pattern is arbitrary) that is centred, roughly, over Greenland and a low pressure band to the south. Figure 13.6 displays 'typical' configurations in winter and summer. The Greenland centre is located at about 70° N and 40–60° W in winter, and has a zero line at about 50° N. This centre retreats northward in summer and a second zero line appears at about 30–35° N.

Another pattern extracted by Barnston and Livezey that has been studied by many others is the Pacific/North American (PNA) pattern (Figure 13.7). The PNA is characterized by two centres of the same sign over the Aleutian Islands and the southeastern United States that flank a centre of opposite sign located over western North America. The PNA is evident in winter (December to April) and again in September and October. It is strongest in February when Barnston and Livezey estimate that it represents 13.2% of the total variance.

Even though the rotated EOFs appear to be less prone to 'mixing' than ordinary EOFs, a great deal of sampling variability still clouds the patterns that are produced, and a considerable

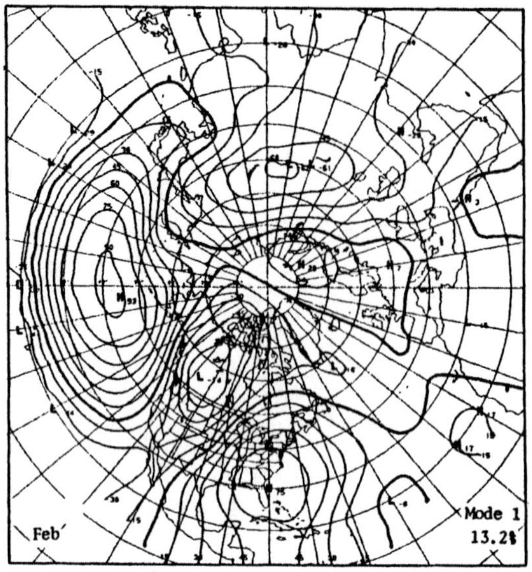

Figure 13.7: *As Figure 13.6, except the February Pacific/North American pattern is displayed. Courtesy R. Livezey.*

amount of skill and subjective judgement are needed to classify and name the patterns. This is amply illustrated by Barnston and Livezey [27], who discuss the types of latitude they permitted themselves in developing their classification. Their illustration of six renditions of the NAO obtained for different times of the year (we show two of these in Figure 13.6; see Barnston and Livezey [27, Figure 2]) demonstrates the kind of variability that the analyst must be able to penetrate when classifying estimated patterns.

13.5.6 Example: Atlantic Sea-Level Air Pressure. In this subsection we consider the EOFs and varimax-rotated EOFs of North Atlantic monthly mean SLP in DJF.[24]

The first three EOFs of SLP (Figure 13.8, left hand column) represent 41%, 26% and 9% of the total variance, respectively. The first EOF has almost uniform sign and exhibits one large feature. The second and third EOFs have dipole structures that reflect the constraint that the higher-order EOFs must be orthogonal to the first EOF.

The EOFs of North Atlantic SLP have simple structure, even without rotation. It is therefore not surprising that the application of the varimax rotation technique to these EOFs (Figure 13.8, middle and right hand columns) results in little change.

- Figure 13.8 (middle column) displays the result of the rotation using $K = 5$ normalized EOFs as input vectors. The rotated EOFs represent 38%, 24% and 10% of the total variance. Similar results are obtained when $K = 5$ non-normalized EOFs are used (not shown).

- The result of the rotation using the first $K = 10$ *non-normalized* EOFs is shown in the right hand column of Figure 13.8. The patterns represent 26%, 15%, and 13% of the total variance, respectively. These patterns deviate somewhat from those in the left hand and middle columns of the diagram. They are noisier than the other sets of patterns, including the rotated patterns derived from $K = 10$ normalized EOFs (not shown).

  Intuitively this is what we expect since the non-normalized patterns enter the minimization functional with equal weights. Thus the poorly estimated EOFs are as influential as the well-estimated EOFs in the determination of matrix $\mathcal{R}$. In contrast, normalization gives the well-estimated EOFs relatively more influence on the form of $\mathcal{R}$.

The first rotated pattern represents less variance than the first EOF, simply because the first EOF was constructed to maximize the variance. Higher-order rotated EOFs typically represent more variance than the respective EOFs (see, e.g., Table 1 of Barnston and Livezey [27]).

In this example, little is gained by processing the original EOF patterns with the varimax machinery. The rotated EOFs become noisy when too many non-normalized EOFs were used as input.

13.5.7 Example: North Atlantic Sea-surface Temperature. The first three EOFs of the monthly mean SST in DJF represent 26%, 17% and 10% of the total variance, respectively (Figure 13.9, left hand column). These EOFs do not have simple structure. The first contains three well-separated centres of location located in the West Atlantic off the North American coast, south of Greenland, and in the upwelling region off the west coast of Africa.

Varimax rotation leads to a substantially different distribution of variance between patterns (Figure 13.9, middle and right hand columns).

[24] The analysis presented here and in [13.5.7] were provided by V. Kharin (personal communication). Note that all eigenvalues, EOFs, and rotated EOFs presented here are estimates.

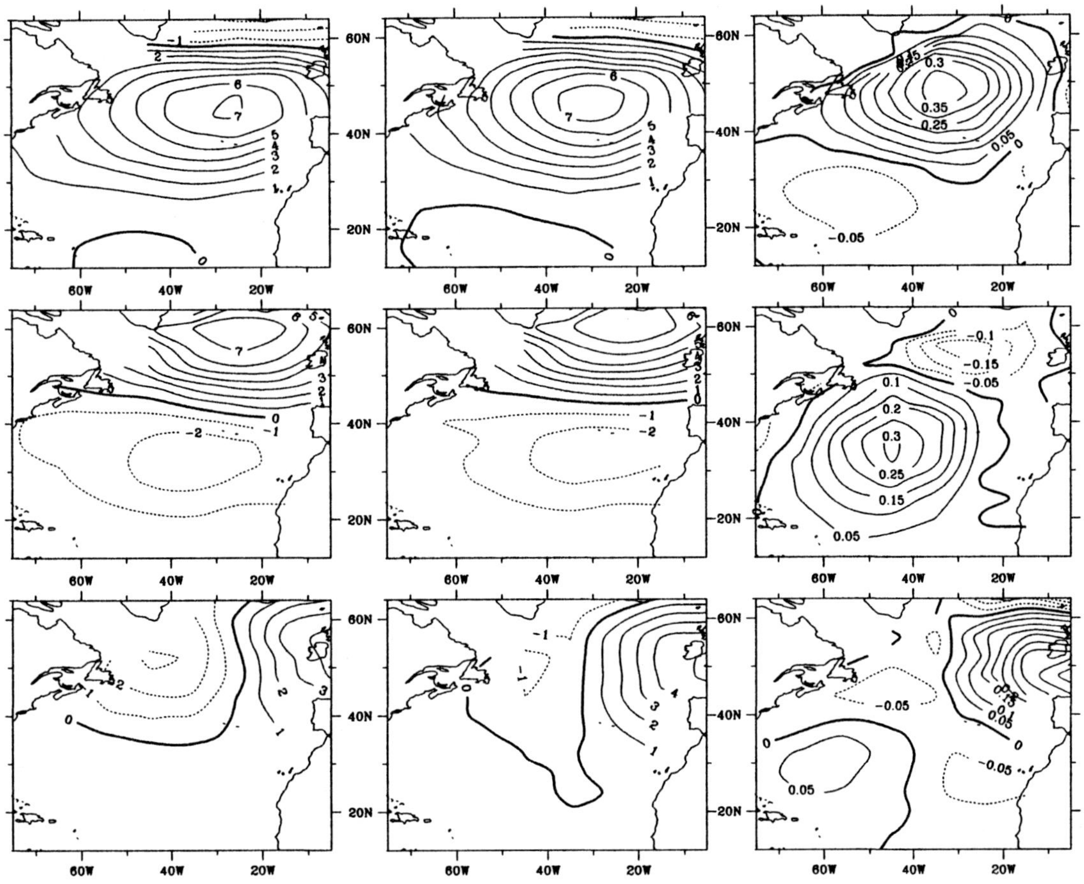

Figure 13.8: *First three rotated and unrotated EOFs of North Atlantic SLP in winter. From top to bottom $j = 1$, $j = 2$, $j = 3$. Courtesy V. Kharin.*
Left column: Normalized EOFs $\vec{e}^{\,j+}$.
Middle column: Rotated EOFs derived from $K = 5$ normalized EOFs.
Right column: Rotated EOFs derived from $K = 10$ non-normalized EOFs.

- When the input is $K = 5$ normalized EOFs (Figure 13.9, middle column) the three centres of action in the first EOF are separated and distributed to the first three rotated patterns (which represent 21%, 16% and 15% of the variance, respectively).
- When the input is $K = 5$ non-normalized EOFs (i.e., all EOFs have unit length; Figure 13.9, right hand column) the three rotated EOFs represent about the same percentage of variance, namely 15%, 15% and 13%, respectively.[25] Note that the sequence of patterns is changed from that obtained with the normalized EOFs (which have unequal lengths). When more input vectors are used, the rotated patterns become noisier and represent less variance (not shown).

We will revisit the analysis of North Atlantic SST and SLP in [14.3.1]. There we will see that the first two conventional EOFs of the North Atlantic SST reflect *two* forcing mechanisms, two characteristic variations in the large-scale atmospheric state that are encoded in the first two SLP EOFs shown in Figure 13.8. Thus, in this case, the rotation makes interpretation more difficult by masking the underlying physics (see [14.3.2]).

[25] Note that the concept of degeneracy is irrelevant for rotated EOFs, since degeneracy is immaterial for the minimization of the functional V.

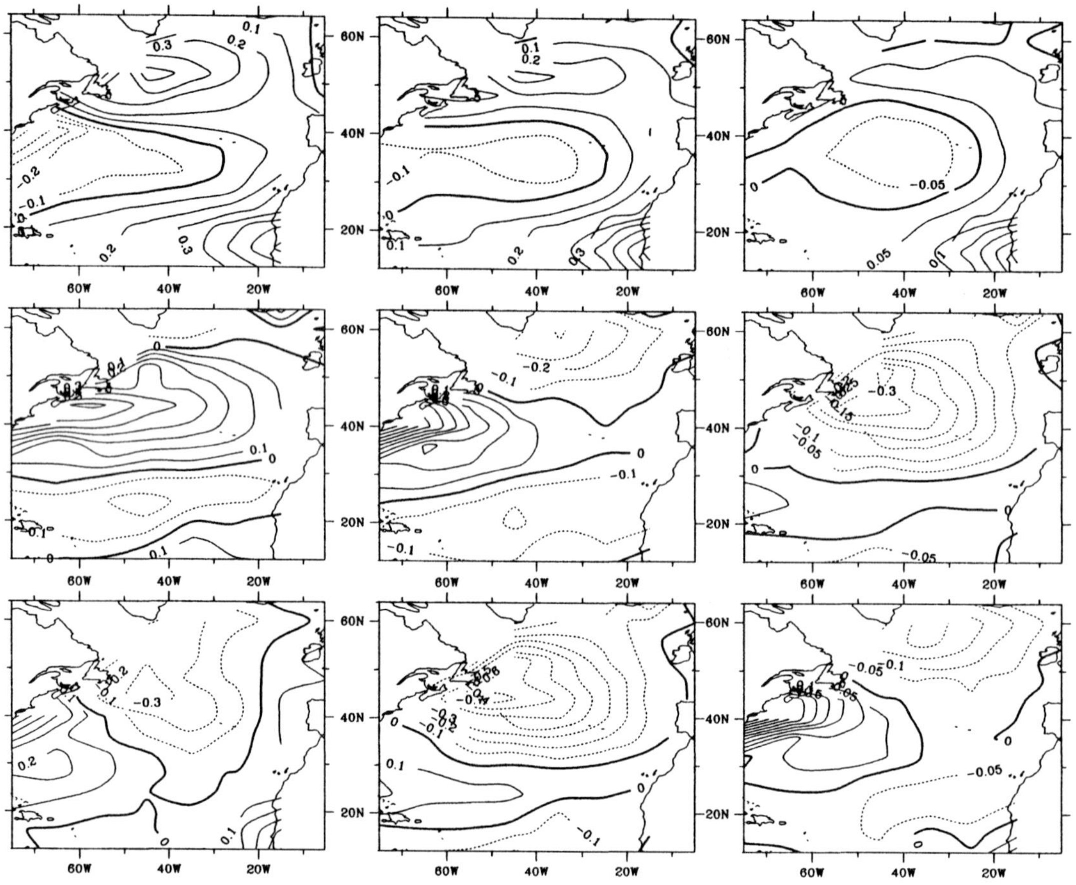

Figure 13.9: *First three unrotated (left hand column) and rotated (middle and right hand columns) EOFs of North Atlantic monthly mean SST in DJF. From top to bottom: $j = 1$, $j = 2$, $j = 3$. Courtesy V. Kharin.*
Left column: Normalized EOFs $\vec{e}^{\,j+}$.
Middle column: Rotated EOFs derived from $K = 5$ normalized EOFs.
Right column: Rotated EOFs derived from $K = 5$ non-normalized EOFs.

13.5.8 Rotation: a Postscript. EOF rotation is often useful, but it is not meant to be a default operation after every EOF analysis. Instead its use should be guided by the problem under consideration.

Jolliffe [198] points out that rotation should be used routinely for subsets of EOFs that have equal, or near-equal, eigenvalues. The corresponding EOFs are not well defined because of their degeneracy (cf. [13.1.8]), and thus the patterns contained by the degenerate EOFs may be arbitrarily rotated within the space that they span. The sensitivity of the rotation to the normalization of the EOFs becomes less relevant since all eigenvalues are similar.

13.6 Singular Systems Analysis and Multichannel SSA

13.6.1 General. The *Singular Systems Analysis* (SSA; see Vautard, Yiou, and Ghil [381] or Vautard [380]) and the *Multichannel Singular Spectrum Analysis* (MSSA, see Plaut and Vautard [317]) are time series analysis techniques used to identify recurrent patterns in univariate time series (SSA) and multivariate time series (MSSA). Mathematically, SSA and MSSA are variants of conventional EOF analysis, but the application of the mathematics is markedly different. Vautard [380] reviews recent applications of SSA and MSSA. Allen and colleagues [8, 9,

10] have investigated various aspects of these methods.

13.6.2 Singular Systems Analysis. Univariate time series $\mathbf{X}_t$ are considered in SSA. An m-dimensional vector time series $\vec{\mathbf{Y}}_t$ is derived from $\mathbf{X}_t$ by setting:

$$\vec{\mathbf{Y}}_t = (\mathbf{X}_t, \mathbf{X}_{t+1}, \dots, \mathbf{X}_{t+m-1})^{\mathrm{T}}. \tag{13.54}$$

A *Singular Systems Analysis* is an EOF analysis of $\vec{\mathbf{Y}}_t$.

The vector space occupied by $\vec{\mathbf{Y}}_t$ is called the *delay-coordinate space.*

The (zero lag) covariance matrix of $\vec{\mathbf{Y}}$, $\Sigma_{YY} = \mathrm{Cov}(\vec{\mathbf{Y}}_t, \vec{\mathbf{Y}}_t)$, is a *Töplitz matrix.*[26] Element (j,k) of Σ_{YY}, say σ_{jk}, is the covariance between the jth element of $\vec{\mathbf{Y}}_t$ $(\mathbf{X}_{t+j-1})$ and its kth element $\mathbf{X}_{t+k-1}$. Thus

$$\sigma_{jk} = \gamma_{xx}(|j-k|),$$

where $\gamma_{xx}(\cdot)$ is the auto-correlation function of $\mathbf{X}_t$. All off-diagonal elements of Σ_{YY} are identified by $|i-j| = \tau$ and have the same value $\gamma_{xx}(\tau)$. Thus, matrix Σ_{YY} is band-structured and contains all auto-covariances of $\mathbf{X}_t$ up to lag $m-1$. The covariance and correlation matrices of $\vec{\mathbf{Y}}_t$ differ by only a constant factor $(1/\sigma_X^2)$. They therefore have the same eigenvectors. The eigenvalues of the two matrices differ by the same constant factor.

The eigenvectors $\vec{e}^{\,i}$ of Σ_{YY}, sometimes called *time EOFs*, are interpreted as a sequence in time. Each eigenvector $\vec{e}^{\,i}$ is a normalized sequence of m time-ordered numbers,

$$\vec{e}^{\,i} = \left(e_0^i, \dots, e_{m-1}^i\right)^{\mathrm{T}}, \tag{13.55}$$

that may be understood as a 'typical' sequence of events. The orthogonality of the eigenvectors in the delay-coordinate space, $\vec{e}^{\,j\mathrm{T}}\vec{e}^{\,k} = \delta_{jk}$, is equivalent to the *temporal orthogonality* of any two typical sequences $(e_0^j, \dots, e_{m-1}^j)$ and $(e_0^k, \dots, e_{m-1}^k)$:

$$\sum_{i=0}^{m-1} e_i^j e_i^k = \delta_{jk}. \tag{13.56}$$

The EOF coefficients

$$\alpha_k(t) = \langle \vec{\mathbf{y}}_t, \vec{e}^{\,k}\rangle = \sum_{i=0}^{m-1} \mathbf{X}_{t+i} e_i^k \tag{13.57}$$

are empirically determined averages (recall Section 10.5) of length m. That is, $\alpha_k(t)$ is a *filtered*[27] version of the original time series $\mathbf{X}_t$, with filter weights that are given by the kth eigenvector. When $\mathbf{X}_t$ is dominated by high-frequency variations, the dominant eigenvectors will be high-pass filters, and when most of the variance of $\mathbf{X}_t$ is concentrated at low frequencies the dominant eigenvectors will act as low-pass filters. The eigenvectors will generally *not* form symmetric filters. Thus we need to be aware that operation (13.57) causes a frequency-dependent phase shift.

As with ordinary EOF analysis, SSA distributes the total variance of $\vec{\mathbf{Y}}_t$ to the m eigenvalues λ_i. The total variance of $\vec{\mathbf{Y}}_t$ is equal to m times the variance of $\mathbf{X}_t$. Thus

$$\sum_{i=1}^{m} \lambda_i = m\,\mathrm{Var}(\mathbf{X}_t). \tag{13.58}$$

The vector-matrix version of (13.57) is

$$\vec{\alpha}(t) = \mathcal{P}\vec{\mathbf{Y}}_t$$

where $\vec{\alpha}(t)$ and $\mathcal{P}$ are defined in the usual way. Thus the auto-correlation function of the multivariate coefficient process $\vec{\alpha}(t)$ is related to the auto-correlation function of $\mathbf{X}_t$ by

$$\Sigma_{\alpha\alpha}(\tau) = \mathcal{P}\Sigma_{YY}(\tau)\mathcal{P}^{\mathrm{T}}$$

where $\Sigma_{YY}(\tau)$ is the matrix whose (i,j)th entry is given by

$$[\Sigma_{YY}(\tau)]_{i,j} = \gamma_{xx}(\tau + j - i). \tag{13.59}$$

Note that

$$\Sigma_{\alpha\alpha}(0) = \Lambda = \mathrm{diag}(\lambda_1, \dots, \lambda_m).$$

13.6.3 Reconstruction in the Time Domain. Also, as with ordinary EOFs,

$$\vec{\mathbf{Y}}_t = \sum_{i=1}^{m} \alpha_i(t)\vec{e}^{\,i}. \tag{13.60}$$

Thus, using equation (13.60) to expand $\vec{\mathbf{Y}}_t, \vec{\mathbf{Y}}_{t-1}, \dots, \vec{\mathbf{Y}}_{t-m+1}$, we find that $\mathbf{X}_t$ has m equivalent time expansions in the m 'SSA-signals':

$$\begin{aligned}
\mathbf{X}_t &= \sum_{i=1}^{m} \alpha_i(t) e_1^i \\
&= \sum_{i=1}^{m} \alpha_i(t-1) e_2^i \\
&\;\;\vdots \\
&= \sum_{i=1}^{m} \alpha_i(t-m+1) e_m^i.
\end{aligned} \tag{13.61}$$

[26]The elements on each diagonal of a Töplitz matrix are equal. That is, if $\mathcal{A}$ is an $m \times m$ matrix and if there are constants $c_{(-(n-1))}, \dots, c_{(n-1)}$ such that $\mathcal{A}_{i,j} = c_{j-i}$, then $\mathcal{A}$ is Töplitz. Graybill [148] describes some of their properties (see Section 8.15).

[27]See Section 17.5.

Each of these expansions distributes the variance of the SSA-signals differently. In fact, using the orthogonality of the EOF coefficients, it is easily shown that

$$\mathrm{Var}(\mathbf{X}_t) = \sum_{i=1}^{m} \lambda_i \left(e_k^i\right)^2 \tag{13.62}$$

for *all* k.[28] If we consider the normalized representation (13.22) we find that the SSA patterns add to the same numbers:

$$\sum_i (e_k^{i+})^2 = \text{constant}, \tag{13.63}$$

for all lags k.

13.6.4 Paired Eigenvectors and Oscillatory Components. We now consider, briefly, time series that contain an oscillatory component. For simplicity, we suppose that $\mathbf{X}_t$ is pure cosine so that

$$\begin{aligned}\vec{\mathbf{Y}}_t &= (\mathbf{X}_t, \dots, \mathbf{X}_{t+m-1})^{\mathrm{T}} \qquad (13.64)\\ &= \left(\cos\left(2\pi \tfrac{t}{m}\right), \dots, \cos\left(2\pi \tfrac{t+m-1}{m}\right)\right)^{\mathrm{T}}.\end{aligned}$$

By equation (13.60), the time EOFs must be able to represent this structure. Suppose one of the time EOFs contains the cosine pattern, that is,

$$\vec{e}^{\,i} = \left(1, \cos\left(\tfrac{2\pi}{m}\right), \dots, \cos\left(\tfrac{2\pi(m-1)}{m}\right)\right)^{\mathrm{T}}.$$

Then $\vec{\mathbf{Y}}_0 = \vec{e}^{\,i}$. However, one time step later, we have

$$\begin{aligned}\vec{\mathbf{Y}}_1 &= \left(\cos\left(\tfrac{2\pi}{m}\right), \dots, \cos\left(\tfrac{2\pi(m-1)}{m}\right), 1\right)^{\mathrm{T}}\\ &= \cos\left(\tfrac{2\pi}{m}\right)\left(1, \cos\left(\tfrac{2\pi}{m}\right), \dots, \cos\left(\tfrac{2\pi(m-1)}{m}\right)\right)^{\mathrm{T}}\\ &\quad - \sin\left(\tfrac{2\pi}{m}\right)\left(0, \sin\left(\tfrac{2\pi}{m}\right), \dots, \sin\left(\tfrac{2\pi(m-1)}{m}\right)\right)^{\mathrm{T}}\\ &= \cos\left(\tfrac{2\pi}{m}\right)\vec{e}^{\,i} - \sin\left(\tfrac{2\pi}{m}\right)\vec{e}^{\,j},\end{aligned}$$

where $\vec{e}^{\,j} = \left(0, \sin\left(\frac{2\pi}{m}\right), \dots, \sin\left(\frac{2\pi(m-1)}{m}\right)\right)^{\mathrm{T}}$ is another eigenvector of $\vec{\mathbf{Y}}_t$. At time t,

$$\begin{aligned}\vec{\mathbf{Y}}_t &= \cos\left(\tfrac{2\pi t}{m}\right)\vec{e}^{\,i} - \sin\left(\tfrac{2\pi t}{m}\right)\vec{e}^{\,j}\\ &= \alpha_i(t)\vec{e}^{\,i} + \alpha_j(t)\vec{e}^{\,j}\end{aligned}$$

where $\alpha_i(t) = \cos\left(\frac{2\pi t}{m}\right)$ and $\alpha_j(t) = \sin\left(\frac{2\pi t}{m}\right)$. Note that both coefficients have the same 'variance' (i.e., $\lambda_i = \lambda_j$), and that the coefficients are 90° out-of-phase.

While the example is artificial, the properties of the eigenvectors and coefficients above characterize what happens when $\mathbf{X}_t$ contains an oscillatory signal. That is, we expect to find a pair of degenerate EOFs with coefficients that vary coherently and are 90° out-of-phase with each other.[29] The pair of patterns and their coefficients may be written as one complex pattern and one complex coefficient.

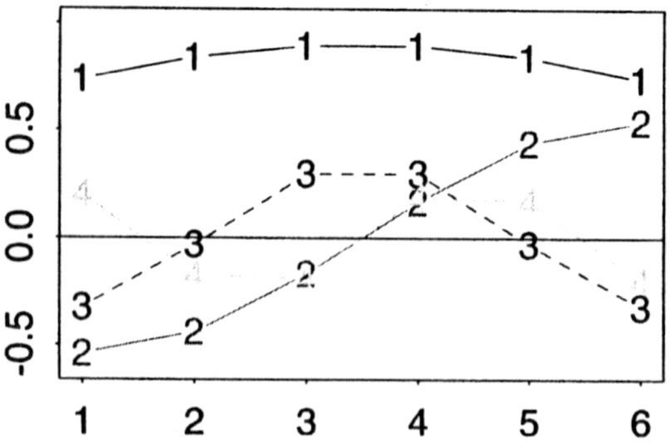

Figure 13.10: *The first four time EOFs of an AR(1) process with a = 0.8 obtained using window length m = 6. The patterns are normalized with the square root of the eigenvalue.*

13.6.5 SSA of White Noise. A white noise process $\{\mathbf{X}_t\}$ (see [10.2.3]) consists of a sequence of independent, identically distributed random variables. It has auto-covariance function $\gamma_{xx}(\tau)$ such that $\gamma_{xx}(0) = \mathrm{Var}(\mathbf{X}_t)$ and $\gamma_{xx}(\tau) = 0$ for nonzero τ. Thus

$$\mathbf{\Sigma}_{YY} = \mathrm{Var}(\mathbf{X}_t)\mathcal{I},$$

where $\vec{\mathbf{Y}}_t$ is the delay-coordinate space version of $\mathbf{X}_t$ (13.58) and $\mathcal{I}$ is the $m \times m$ identity matrix. Hence $\vec{\mathbf{Y}}_t$ has m eigenvalues $\lambda_i = \mathrm{Var}(\mathbf{X}_t)$ and m degenerate eigenvectors. One possible set of eigenvectors are the unit vectors, $\vec{e}^{\,i} = (0, \dots, 1, \dots, 0)$ with the 1 in the ith column.

13.6.6 SSA of Red Noise. Red noise processes[30] have exponentially decaying auto-covariance functions $\gamma_{xx}(\tau) = \sigma_X^2 a^{|\tau|}$, where $\sigma_X^2 = \mathrm{Var}(\mathbf{X}_t)$. Thus

$$\mathbf{\Sigma}_{YY} = \sigma_X^2 \begin{pmatrix} 1 & a & \dots & a^{m-1}\\ a & 1 & \dots & a^{m-2}\\ \vdots & \vdots & \ddots & \vdots\\ a^{m-1} & a^{m-2} & \dots & 1 \end{pmatrix}.$$

[28] Note that (13.62) is just a special case of (13.7).

[29] Compare with the discussion of complex POP coefficients in Chapter 15.

[30] AR(1), or 'red noise,' processes were introduced in [10.3.2]. They can be represented by a stochastic difference equation $\mathbf{X}_t = a\mathbf{X}_{t-1} + \mathbf{Z}_t$, where $\mathbf{Z}_t$ is white noise. The auto-covariance function was derived in [11.1.6]. We represent the lag-1 correlation coefficient by 'a' instead of 'α' to avoid confusion with our notation for the EOF coefficients.

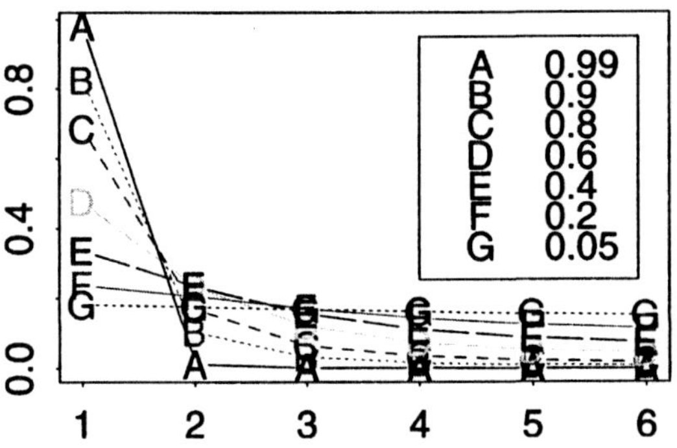

Figure 13.11: *Eigenspectra obtained with window length $m = 6$ for AR(1) processes with $a =$ 0.99, 0.90, 0.80, 0.60, 0.40, 0.20, and 0.05. The spectra are normalized by the variance of the process.*

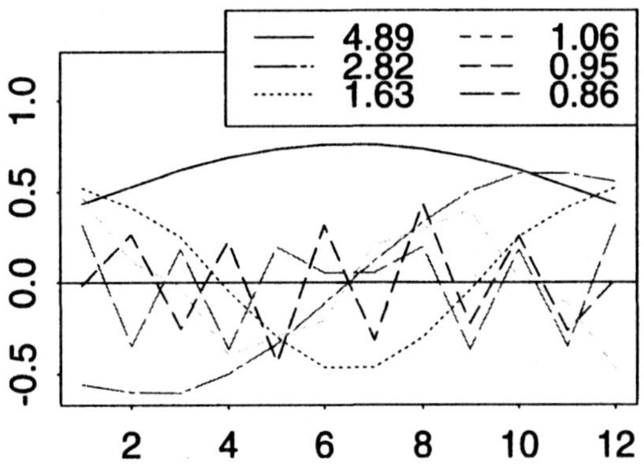

Figure 13.12: *First six time EOFs of an AR(2) process with $\vec{\alpha} = (0.3, 0.3)$ obtained using window length $m = 12$. The patterns are normalized with the square root of the eigenvalue. The eigenvalues are given at the bottom.*

When $m = 2$, Σ_{YY} has eigenvalues $(1 - a)$ and $(1 + a)$ and corresponding time EOFs $(1/\sqrt{2}, -1/\sqrt{2})^{\mathrm{T}}$ and $(1/\sqrt{2}, 1/\sqrt{2})^{\mathrm{T}}$. The order of the eigenvalues and EOFs depends upon the sign of a. Given a specific *window length m*, the same EOFs are obtained for all AR(1) processes $\mathbf{X}_t$.

The first four AR(1) time EOFs for window length $m = 6$ are shown in Figure 13.10. The patterns are multiplied by the square root of the eigenvalue, as in equation (13.22). The kth pattern crosses the zero line $k - 1$ times. Thus, the time EOFs are ordered by time scale, with most variance contributed by the variability with longest time scales.

One characteristic of the time EOFs is that no two patterns have the same number of zeros. Thus oscillatory behaviour, such as that described in [13.6.4], is not possible. This is consistent with the discussion in [11.1.2], when we also found no indication of oscillatory behaviour in AR(1) processes.

Figure 13.11 shows the eigenspectra of several AR(1) processes for the same window length m. The larger the 'memory' a, the steeper the spectrum. In the extreme case with $a = 0.99$, almost all variance is contributed by the 'almost constant' first time EOF. At the other end of the memory scale ($a = 0.05$) all time EOFs contribute about the same amount of variance.

13.6.7 SSA of an AR(2) process. An AR(2) process has the form

$$\mathbf{X}_t = a_1\mathbf{X}_{t-1} + a_2\mathbf{X}_{t-2} + \mathbf{Z}_t$$

[10.3.3]. The auto-covariance function is either the sum of the auto-covariance functions of two red noise processes (11.4), or it is a damped oscillatory function (11.9) (for details, see [11.1.9]).

The process with coefficients $a_1 = a_2 = 0.3$ was found to belong to the former 'non-oscillatory' group. The first six time EOFs obtained using window length $m = 12$ are shown in Figure 13.12. Similar to the AR(1) process, all eigenvectors have different patterns, with the kth eigenvector having $(k - 1)$ zeros. All eigenvalues are well separated. Consistent with the discussion in [11.1.7] and [11.2.6], an oscillatory mode is not identified. Note the similarity between the patterns in Figure 13.12 and the AR(1) patterns shown in Figure 13.10. (The patterns in Figure 13.12 are not sensitive to the choice of m.)

The other AR(2) process considered previously has $\alpha_1 = 0.9, \alpha_2 = -0.8$. This process has oscillatory behaviour with a 'period' of 6 time steps (see [10.3.3], [11.1.7] and [11.2.6]). The time EOFs of this process, obtained using window length $m = 12$, are shown in Figure 13.13. The first two time EOFs are sinusoidal, with a period of 6 time steps, and phase-shifted by 1 to 2 time steps (a quarter of a period, 1.5 time steps, can not be represented in time steps of 1). The two time EOFs share similar eigenvalues (4.2 and 4.1) and obviously represent an oscillatory mode as described in [13.6.4]. The higher index time EOFs are reminiscent of the time EOFs obtained for AR(1) processes.

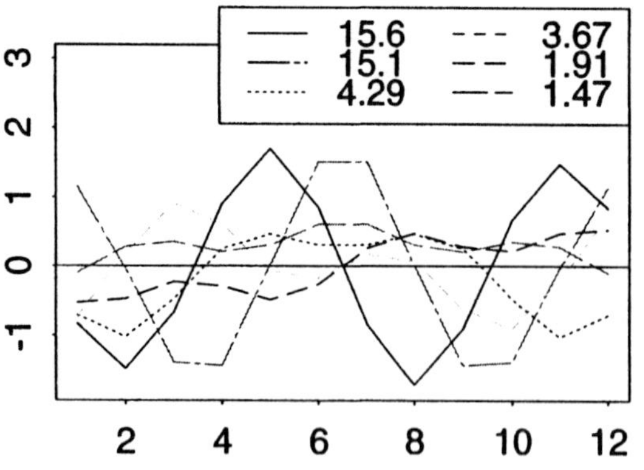

Figure 13.13: *First six time EOFs of an AR(2) process with $\vec{a} = (0.9, -0.8)$ obtained using window length $m = 12$. The patterns are normalized with the square root of the eigenvalue. The eigenvalues are given at the bottom.*

13.6.8 Multichannel Singular Spectrum Analysis. MSSA (see Vautard [380]) differs from SSA only in the dimension of the basic time series, which is now m'-dimensional rather than one-dimensional. The derived random vector $\vec{\mathbf{Y}}_t$ is therefore mm'-dimensional. Thus MSSA is *Extended EOF analysis* [13.1.8] in which m consecutively observed fields are concatenated together. The number of fields m is usually small compared with the field dimension m' in EEOF analysis. The opposite, $m > m'$, is often true in MSSA.

13.6.9 Estimation. We conclude with a brief comment on the estimation of eigenvalues and time EOFs in SSA. The same applies, by extension, to the eigenvalues and space-time EOFs in MSSA.

SSA is applied to a finite sample of observations $\{\mathbf{x}_1, \ldots, \mathbf{x}_n\}$ with $n \gg m$ by first forming $\vec{\mathbf{Y}}$-vectors,

$$\begin{aligned}
\vec{\mathbf{y}}_1 &= (\mathbf{x}_1, \ldots, \mathbf{x}_m)^{\mathrm{T}} \\
\vec{\mathbf{y}}_2 &= (\mathbf{x}_2, \ldots, \mathbf{x}_{m+1})^{\mathrm{T}} \\
&\vdots \\
\vec{\mathbf{y}}_{n-m+1} &= (\mathbf{x}_{n-m+1}, \ldots, \mathbf{x}_n)^{\mathrm{T}}.
\end{aligned}$$

Conventional EOF analysis is applied to the resulting sample of $n - m + 1$ $\vec{\mathbf{Y}}$-vectors. The estimated eigenvalues and EOFs can be computed from either the estimated covariance matrix of $\vec{\mathbf{Y}}$ (see [13.2.4]) or by means of SVD (see [13.2.8]).

Note that neither *North's Rule-of-Thumb* [13.3.5] nor Lawley's formulae (13.38, 13.39) can be used to assess the reliability of the estimate directly because consecutive realizations of $\vec{\mathbf{Y}}_t$ are auto-correlated (see (13.33)). The effects of temporal dependence must be accounted for (see Section 17.1) when using these tools.

Allen and co-workers [8, 9, 10] discuss the problem of discriminating between noisy components and truly oscillatory modes in detail.

14 Canonical Correlation Analysis

14.0.0 Overview. Just as EOF analysis (Chapter 13) is used to study the variability of a random vector $\vec{X}$, *Canonical Correlation Analysis* (CCA) is used to study the correlation structure of a pair of random vectors $\vec{X}$ and $\vec{Y}$.

CCA and EOF analyses share similar objectives and similar mathematics. One interpretation of the first EOF $\vec{e}^{\,1}$ of $\vec{X}$ is that $\vec{X}^T\vec{e}^{\,1}$ is the linear combination of elements of $\vec{X}$ with the greatest variance (subject to $\|\vec{e}^{\,1}\| = 1$). The second EOF $\vec{e}^{\,2}$ provides the linear combination $\vec{X}^T\vec{e}^{\,2}$ with greatest variance that is uncorrelated with $\vec{X}^T\vec{e}^{\,1}$, and so on. The objective of CCA is to find a pair of patterns $\vec{f}_X^{\,1}$ and $\vec{f}_Y^{\,1}$ (subject to $\|\vec{f}_X^{\,1}\| = \|\vec{f}_Y^{\,1}\| = 1$) so that the *correlation* between linear combinations $\vec{X}^T\vec{f}_X^{\,1}$ and $\vec{Y}^T\vec{f}_Y^{\,1}$ is maximized.[1] A second pair of patterns $\vec{f}_X^{\,2}$ and $\vec{f}_Y^{\,2}$ is found so that $\vec{X}^T\vec{f}_X^{\,2}$ and $\vec{Y}^T\vec{f}_Y^{\,2}$ are the most strongly correlated linear combinations of $\vec{X}$ and $\vec{Y}$ that are not correlated with $\vec{X}^T\vec{f}_X^{\,1}$ and $\vec{Y}^T\vec{f}_Y^{\,1}$, and so on.

Canonical Correlation Analysis was first described by Hotelling [187].

The 'Canonical Correlation Patterns' of a paired random vector $(\vec{X}, \vec{Y})$ are defined in Section 14.1, and their estimation is described in Section 14.2. Examples of some applications are given in Section 14.3. A closely related technique, called *Redundancy Analysis*, is described in Section 14.4.

14.0.1 Introductory Example: Large-scale Temperature and SLP over Europe and Local Weather Elements in Bern. Gyalistras et al. [152] analysed the simultaneous variations of the local climate in Bern (Switzerland) and the troposphere over the North Atlantic in DJF. The state of the local climate in a given season was represented by a 17-dimensional random vector $\vec{X}$ consisting of the number of days in the season with at least 1 mm of precipitation, and the seasonal means and daily standard deviations of the daily mean, minimum, maximum, and range of temperature, precipitation, wind speed, relative humidity, and relative sunshine duration. The large-scale state of the atmosphere was represented by a vector $\vec{Y}$ consisting of the near-surface temperature and sea-level pressure (SLP) fields over Europe and the Northeast Atlantic Ocean.

CCA was used to analyse the joint variability of $\vec{X}$ and $\vec{Y}$. As noted above, this technique finds pairs of patterns such that the correlation between two corresponding pattern coefficients is maximized.

The pair of patterns with the largest correlation is shown in Figure 14.1. The two patterns, one of which consists of two sub-patterns for the pressure and temperature (Figure 14.1, top and middle), have a meaningful physical interpretation. Below normal temperatures in Bern are associated with high pressure over the British Isles and below normal temperatures in the rest of Europe since the correlation between the local climate pattern (bottom panel, Figure 14.1) and the tropospheric pattern (top two panels, Figure 14.1) is negative. Weakened westerly flow is associated with reduced precipitation; the seasonal mean, standard deviation, and number of 'wet' days all tend to be below normal. The large-scale patterns have little effect on wind speed and relative humidity.

The link between the two patterns in Figure 14.1 is strong. The correlation between the coefficient time series (not shown) is −0.89, and the CCA pattern represents a large proportion of the variance of the local climate (Figure 14.2). More than 50% of interannual variance of the seasonal means of daily mean, minimum and maximum temperature are represented by the first CCA pair. They also represent almost 80% of the interannual variance of DJF precipitation and about 75% of the interannual variance of the number of 'wet' days.

14.1 Definition of Canonical Correlation Patterns

14.1.1 One Pair of Patterns. Let us consider an m_X-dimensional random vector $\vec{X}$ and an

[1] One could also choose $\vec{f}_X^{\,1}$ and $\vec{f}_Y^{\,1}$ to maximize the *covariance* between $\vec{X}^T\vec{f}_X^{\,1}$ and $\vec{Y}^T\vec{f}_Y^{\,1}$. Climatologists sometimes call this *SVD analysis* since the patterns are found by obtaining a singular value decomposition of the cross-covariance matrix. See [14.1.7], Bretherton, Smith, and Wallace [64] and Cherry [83].

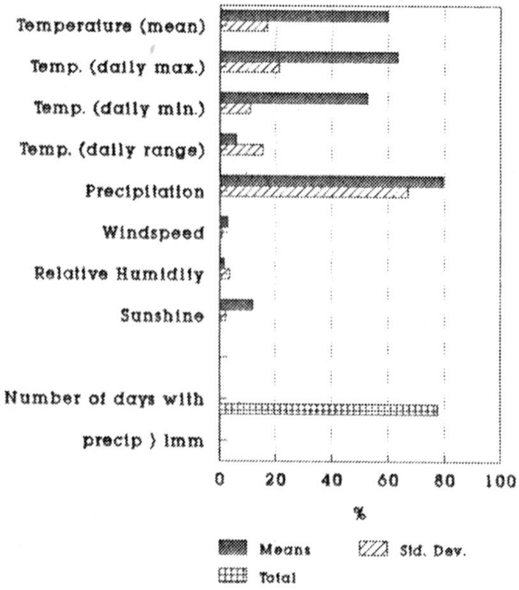

Figure 14.2: *Percentage of year-to-year variance of the local climate variables for Bern represented by the first CCA pair.*

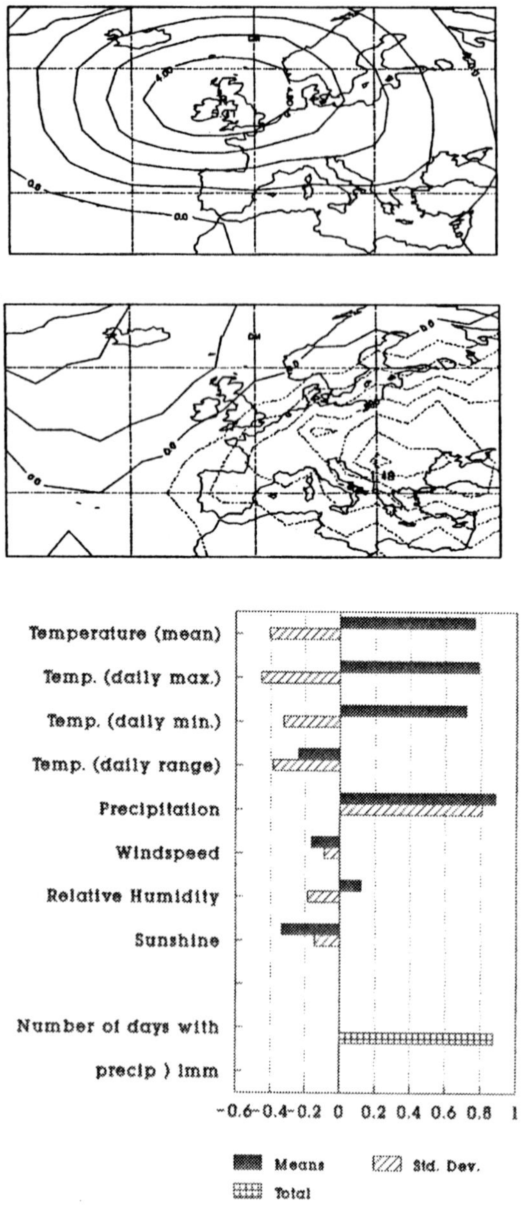

Figure 14.1: *First pair of canonical correlation patterns of $\vec{\mathbf{Y}}$ = (DJF mean SLP, DJF mean temperature) and a vector $\vec{\mathbf{X}}$ of DJF statistics of local weather elements at Bern (Switzerland).*
Top: The SLP part of the first canonical correlation pattern for $\vec{\mathbf{Y}}$.
Middle: The near-surface temperature part of the first canonical correlation pattern for $\vec{\mathbf{Y}}$.
Bottom: The canonical correlation pattern for the local variable $\vec{\mathbf{X}}$.
Note that the correlation between the corresponding pattern coefficients is negative.
From Gyalistras et al. [152].

m_Y-dimensional random vector $\vec{\mathbf{Y}}$. We require an m_X-dimensional vector $\vec{f}_X$ and an m_Y-dimensional vector $\vec{f}_Y$ such that the inner products $\beta^X = \langle \vec{\mathbf{X}}, \vec{f}_X \rangle$ and $\beta^Y = \langle \vec{\mathbf{Y}}, \vec{f}_Y \rangle$ have maximum correlation. That is, we want to maximize

$$\begin{aligned} \rho &= \frac{\operatorname{Cov}(\beta^X, \beta^Y)}{\sqrt{\operatorname{Var}(\beta^X)\operatorname{Var}(\beta^Y)}} \qquad (14.1)\\ &= \frac{\vec{f}_X^{\mathrm{T}} \operatorname{Cov}(\vec{\mathbf{X}}, \vec{\mathbf{Y}}) \vec{f}_Y}{\sqrt{\operatorname{Var}(\langle \vec{\mathbf{X}}, \vec{f}_X \rangle) \operatorname{Var}(\langle \vec{\mathbf{Y}}, \vec{f}_Y \rangle)}}. \end{aligned}$$

Note that if a pair of vectors $\vec{f}_X$ and $\vec{f}_Y$ maximizes (14.1), then all vectors $\alpha_X \vec{f}_X$ and $\alpha_Y \vec{f}_Y$ do the same for any nonzero α_X and α_Y. Thus the patterns $\vec{f}_X$ and $\vec{f}_Y$ are subject to arbitrary normalization. In particular, we can choose patterns such that

$$\operatorname{Var}(\langle \vec{\mathbf{X}}, \vec{f}_X \rangle) = \vec{f}_X^{\mathrm{T}} \Sigma_{XX} \vec{f}_X = 1 \qquad (14.2)$$

$$\operatorname{Var}(\langle \vec{\mathbf{Y}}, \vec{f}_Y \rangle) = \vec{f}_Y^{\mathrm{T}} \Sigma_{YY} \vec{f}_Y = 1, \qquad (14.3)$$

where Σ_{XX} and Σ_{YY} are the covariance matrices of $\vec{\mathbf{X}}$ and $\vec{\mathbf{Y}}$. Then equation (14.1) can be rewritten as

$$\rho = \vec{f}_X^{\mathrm{T}} \Sigma_{XY} \vec{f}_Y, \qquad (14.4)$$

where Σ_{XY} is the cross-covariance matrix

$$\Sigma_{XY} = \mathcal{E}\big((\vec{\mathbf{X}} - \vec{\mu}_X)(\vec{\mathbf{Y}} - \vec{\mu}_Y)^{\mathrm{T}} \big).$$

Vectors $\vec{f}_X$ and $\vec{f}_Y$ are found by maximizing

$$\begin{aligned}\epsilon &= \vec{f}_X^{\mathrm{T}}\Sigma_{XY}\vec{f}_Y + \zeta(\vec{f}_X^{\mathrm{T}}\Sigma_{XX}\vec{f}_X - 1)\\ &\quad + \eta(\vec{f}_Y^{\mathrm{T}}\Sigma_{YY}\vec{f}_Y - 1),\end{aligned} \tag{14.5}$$

where ζ and η are Lagrange multipliers that are used to account for constraints (14.2) and (14.3). Setting the partial derivatives of ϵ to zero, we obtain

$$\frac{\partial \epsilon}{\partial \vec{f}_X} = \Sigma_{XY}\vec{f}_Y + 2\zeta\Sigma_{XX}\vec{f}_X = 0 \tag{14.6}$$

so that

$$\Sigma_{XX}^{-1}\Sigma_{XY}\vec{f}_Y = -2\zeta\vec{f}_X, \tag{14.7}$$

and

$$\frac{\partial \epsilon}{\partial \vec{f}_Y} = \Sigma_{XY}^{\mathrm{T}}\vec{f}_X + 2\eta\Sigma_{YY}\vec{f}_Y = 0, \tag{14.8}$$

which is equivalent to

$$\Sigma_{YY}^{-1}\Sigma_{XY}^{\mathrm{T}}\vec{f}_X = -2\eta\vec{f}_Y. \tag{14.9}$$

Then (14.9) is substituted into (14.7) and vice versa to obtain a pair of eigen-equations for $\vec{f}_X$ and $\vec{f}_Y$:

$$\Sigma_{XX}^{-1}\Sigma_{XY}\Sigma_{YY}^{-1}\Sigma_{XY}^{\mathrm{T}}\vec{f}_X = 4\zeta\eta\vec{f}_X \tag{14.10}$$

$$\Sigma_{YY}^{-1}\Sigma_{XY}^{\mathrm{T}}\Sigma_{XX}^{-1}\Sigma_{XY}\vec{f}_Y = 4\zeta\eta\vec{f}_Y. \tag{14.11}$$

An argument similar to that used to establish Theorem [13.2.4] proves that the two matrices share the same non-negative eigenvalues.[2] The eigenvectors of the two matrices are related to each other through a simple equation: if $\vec{f}_X$ is a solution of equation (14.10), then $\Sigma_{YY}^{-1}\Sigma_{XY}^{\mathrm{T}}\vec{f}_X$ is a solution of equation (14.11), provided that their joint eigenvalue is nonzero. Finally, equation (14.4) is maximized by letting $\vec{f}_X$ and $\vec{f}_Y$ be the solutions of equations (14.10) and (14.11) that correspond to the largest eigenvalue $\lambda = 4\zeta\eta$.

Now that we have found the canonical random variables $\beta^X = \langle \vec{\mathbf{X}}, \vec{f}_X\rangle$ and $\beta^Y = \langle \vec{\mathbf{Y}}, \vec{f}_Y\rangle$ that are most strongly correlated, the natural next step is to find the value of ρ. Using equations (14.4), (14.6), (14.8), and (14.2), (14.3) in sequence, we find:

$$\begin{aligned}\rho^2 &= \vec{f}_X^{\mathrm{T}}\Sigma_{XY}\vec{f}_Y\vec{f}_Y^{\mathrm{T}}\Sigma_{XY}^{\mathrm{T}}\vec{f}_X\\ &= 4\eta\zeta\,\vec{f}_X^{\mathrm{T}}\Sigma_{XX}\vec{f}_X\vec{f}_Y^{\mathrm{T}}\Sigma_{YY}\vec{f}_Y\\ &= \lambda.\end{aligned}$$

Thus the correlation is the square root of the eigenvalue that corresponds to eigenvectors $\vec{f}_X$ and $\vec{f}_Y$.[3]

14.1.2 More Pairs. The derivation detailed above can now be repeated to obtain $m = \min(m_X, m_Y)$ pairs of patterns $(\vec{f}_X^{\,i}, \vec{f}_Y^{\,i})$ and m corresponding pairs of canonical variates[4]

$$\beta_i^X = \langle \vec{\mathbf{X}}, \vec{f}_X^{\,i}\rangle \tag{14.12}$$

$$\beta_i^Y = \langle \vec{\mathbf{Y}}, \vec{f}_Y^{\,i}\rangle \tag{14.13}$$

with correlation

$$\rho_i = \mathrm{Cov}(\beta_i^X, \beta_i^Y) = \sqrt{\lambda_i}.$$

The patterns and canonical variates are indexed in order of decreasing eigenvalue λ_i. Pairs of canonical variates are uncorrelated. That is, for $i \neq j$,

$$\begin{aligned}\mathrm{Cov}\big(\beta_i^X, \beta_j^X\big) &= \mathrm{Cov}\big(\beta_i^Y, \beta_j^Y\big)\\ &= \mathrm{Cov}\big(\beta_i^X, \beta_j^Y\big) = 0.\end{aligned}$$

14.1.3 The Canonical Correlation Patterns. For simplicity, we assume in this subsection that $\vec{\mathbf{X}}$ and $\vec{\mathbf{Y}}$ are of the same dimension m. Then the canonical variates $\vec{\beta}^X = (\beta_1^X, \ldots, \beta_m^X)^{\mathrm{T}}$ and $\vec{\beta}^Y = (\beta_1^Y, \ldots, \beta_m^Y)^{\mathrm{T}}$ can be viewed as the result of coordinate transforms that have been applied to $\vec{\mathbf{X}}$ and $\vec{\mathbf{Y}}$.[5] The transformations relate $\vec{\beta}^X$ and $\vec{\beta}^Y$ to $\vec{\mathbf{X}}$ and $\vec{\mathbf{Y}}$ through unknown matrices $\mathcal{F}_X$ and $\mathcal{F}_Y$:

$$\begin{aligned}\vec{\mathbf{X}} &= \mathcal{F}_X\vec{\beta}^X\\ \vec{\mathbf{Y}} &= \mathcal{F}_Y\vec{\beta}^Y.\end{aligned} \tag{14.14}$$

To find $\mathcal{F}_X$, note that

$$\begin{aligned}\vec{\beta}^X &= (\langle \vec{\mathbf{X}}, \vec{f}_X^{\,1}\rangle, \ldots, \langle \vec{\mathbf{X}}, \vec{f}_X^{\,m}\rangle)^{\mathrm{T}}\\ &= \mathbf{f}_X^{\mathrm{T}}\vec{\mathbf{X}}\end{aligned}$$

[2] Note that if $\vec{f}_X$ is a solution of equation (14.10), then $\Sigma_{XX}^{1/2}\vec{f}_X$ is an eigenvector of $(\Sigma_{XX}^{-1/2})^{\mathrm{T}}\Sigma_{XY}\Sigma_{YY}^{-1}\Sigma_{XY}^{\mathrm{T}}\Sigma_{XX}^{-1/2}$. Similarly, $\Sigma_{YY}^{1/2}\vec{f}_Y$ is an eigenvector of $(\Sigma_{YY}^{-1/2})^{\mathrm{T}}\Sigma_{XY}^{\mathrm{T}}\Sigma_{XX}^{-1}\Sigma_{XY}\Sigma_{YY}^{-1/2}$. Since these are non-negative definite matrices, their eigenvalues are real and non-negative.

[3] Note that the sign of the correlation is arbitrary since $\vec{f}_X$ and $\vec{f}_Y$ are determined uniquely only up to their signs.

[4] We assume that $(\Sigma_{XX}^{-1/2})^{\mathrm{T}}\Sigma_{XY}\Sigma_{YY}^{-1}\Sigma_{XY}^{\mathrm{T}}\Sigma_{XX}^{-1/2}$ (or, equivalently, $(\Sigma_{YY}^{-1/2})^{\mathrm{T}}\Sigma_{XY}^{\mathrm{T}}\Sigma_{XX}^{-1}\Sigma_{XY}\Sigma_{YY}^{-1/2}$) has $m = \min(m_X, m_Y)$ distinct, nonzero eigenvalues. Eigenvalues of multiplicity greater than one lead to degeneracy just as in EOF analysis. Uncorrelated canonical variates can still be constructed, but their interpretation is clouded by their non-unique determination. Tools comparable to North's Rule-of-Thumb [13.3.5] are not yet developed for CCA. Note that a pair of degenerate eigenvalues may be an indication of a propagating pattern. See Chapter 15.

[5] The discussion in this subsection is easily generalized to the case in which $\vec{\mathbf{X}}$ and $\vec{\mathbf{Y}}$ are not of the same dimension.

where $\mathbf{f}_X$ is the $m \times m$ matrix with eigenvector $\vec{f}_X^i$ in its ith column. Thus

$$\begin{aligned}\mathrm{Cov}(\vec{\mathbf{X}}, \vec{\beta}^X) &= \mathrm{Cov}(\vec{\mathbf{X}}, \mathbf{f}_X^{\mathrm{T}}\vec{\mathbf{X}}) \\ &= \mathrm{Cov}(\vec{\mathbf{X}}, \vec{\mathbf{X}})\mathbf{f}_X = \mathbf{\Sigma}_{XX}\mathbf{f}_X.\end{aligned}$$

However, substituting equation (14.14) for $\vec{\mathbf{X}}$, we also have

$$\begin{aligned}\mathrm{Cov}(\vec{\mathbf{X}}, \vec{\beta}^X) &= \mathrm{Cov}(\mathcal{F}_X\vec{\beta}^X, \vec{\beta}^X) \\ &= \mathcal{F}_X \mathrm{Cov}(\vec{\beta}^X, \vec{\beta}^X) = \mathcal{F}_X\end{aligned}$$

since $\mathrm{Cov}(\vec{\beta}^X, \vec{\beta}^X) = \mathcal{I}$. Thus

$$\mathcal{F}_X = \mathbf{\Sigma}_{XX}\mathbf{f}_X \tag{14.15}$$

and similarly

$$\mathcal{F}_Y = \mathbf{\Sigma}_{YY}\mathbf{f}_Y. \tag{14.16}$$

The columns of $\mathcal{F}_X$ and $\mathcal{F}_Y$, $\vec{F}_X^i$ and $\vec{F}_Y^i$, are called the *canonical correlation patterns*.[6] The canonical variates β_i^X and β_i^Y are also often called *canonical correlation coordinates*. Since the canonical correlation coordinates are normalized to unit variance, the canonical correlation patterns are expressed in the units of the field they represent, and they indicate the 'typical' strength of the mode of covariation described by the patterns.

While the matrix-vector representations of $\vec{\mathbf{X}}$ and $\vec{\mathbf{Y}}$ in (14.14) are convenient for the derivation of $\mathcal{F}_X$ and $\mathcal{F}_Y$, they are not very evocative. Therefore, note that (14.14) can also be written as

$$\begin{aligned}\vec{\mathbf{X}} &= \textstyle\sum_i \beta_i^X \vec{F}_X^i \\ \vec{\mathbf{Y}} &= \textstyle\sum_i \beta_i^Y \vec{F}_Y^i.\end{aligned} \tag{14.17}$$

This allows us to see more clearly that (14.14) describes an expansion of $\vec{\mathbf{X}}$ and $\vec{\mathbf{Y}}$ with respect to their corresponding canonical correlation patterns. It also suggests that it may be possible to approximate $\vec{\mathbf{X}}$ and $\vec{\mathbf{Y}}$ by truncating the summation in (14.17).

14.1.4 Computational Aspects. Once we know one set of vectors, say $\vec{f}_X^i$, all other vectors are easily obtained through simple matrix operations. Let us assume that we have the vectors $\vec{f}_X^i$. Then (14.15) yields $\vec{F}_X^i$. In [14.1.1] we noted that $\mathbf{\Sigma}_{YY}^{-1}\mathbf{\Sigma}_{XY}^{\mathrm{T}}\vec{f}_X^i$ is equal to $\vec{f}_Y^i$ after suitable normalization. Application of (14.16) gives the last set of vectors, the $\vec{\mathbf{Y}}$-canonical correlation patterns $\vec{F}_Y^i$. It is therefore necessary to solve only the smaller of the two eigenproblems (14.10) and (14.11).

14.1.5 Coordinate Transformations. What happens to the canonical correlation patterns and correlations when coordinates are transformed by an invertible matrix $\mathcal{L}$ through $\mathcal{L}\vec{\mathbf{X}} = \vec{\mathbf{Z}}$? For simplicity we assume random vector $\vec{\mathbf{Y}}$ is unchanged.

To get the same maximum correlation (14.1), we have to transform the patterns $\vec{f}_X^i$ with $\mathcal{L}^{-1}$,

$$\vec{f}_Z^i = (\mathcal{L}^{-1})^{\mathrm{T}}\vec{f}_X^i. \tag{14.18}$$

Thus the canonical correlation coordinates $\beta_i^X = \langle \vec{f}_Z^i, \vec{\mathbf{Z}}\rangle = \langle \vec{f}_X^i, \vec{\mathbf{X}}\rangle$ are unaffected by the transformation. Note that relation (14.18) can also be obtained by verifying that $\vec{f}_Z^i$ and $\vec{f}_X^i$ are eigenvectors of the CCA matrices $\mathbf{\Sigma}_{ZZ^{-1}}\mathbf{\Sigma}_{ZY}\mathbf{\Sigma}_{YY^{-1}}\mathbf{\Sigma}_{ZY}^{\mathrm{T}}$ and $\mathbf{\Sigma}_{XX^{-1}}\mathbf{\Sigma}_{XY}\mathbf{\Sigma}_{YY^{-1}}\mathbf{\Sigma}_{XY}^{\mathrm{T}}$ with the same eigenvalues.

The canonical correlation patterns $\vec{F}_X^i$ are determined by the covariance matrix of $\vec{\mathbf{X}}$ and the $\vec{f}_X^i$-pattern (14.15). Therefore,

$$\begin{aligned}\vec{F}_Z^i &= \mathbf{\Sigma}_{ZZ}\vec{f}_Z^i \\ &= \mathcal{L}\mathbf{\Sigma}_{XX}\mathcal{L}^{\mathrm{T}}(\mathcal{L}^{-1})^{\mathrm{T}}\vec{f}_X^i \\ &= \mathcal{L}\mathbf{\Sigma}_{XX}\vec{f}_X^i \\ &= \mathcal{L}\vec{F}_X^i.\end{aligned} \tag{14.19}$$

Thus the canonical correlation patterns are transformed in the same way as the random vector $\vec{\mathbf{X}}$. We may conclude that *the CCA is invariant under coordinate transformations*.

14.1.6 CCA after a Transformation to EOF Coordinates. The CCA algebra becomes considerably simpler if the data are transformed into EOF space before the analysis (Barnett and Preisendorfer [21]). Suppose that only the first k_X and k_Y EOFs are retained, so that

$$\begin{aligned}\vec{\mathbf{X}} &\approx \textstyle\sum_i^{k_X} \alpha_i^{X+}\vec{e}_X^{\,i+} \\ \vec{\mathbf{Y}} &\approx \textstyle\sum_i^{k_Y} \alpha_i^{Y+}\vec{e}_Y^{\,i+},\end{aligned} \tag{14.20}$$

where we have used the renormalized versions (13.20, 13.21) of the EOFs and their coefficients $\alpha_i^+ = (\lambda_i)^{-1/2}\alpha_i$ and $\vec{e}^{\,i+} = (\lambda_i)^{1/2}\vec{e}^{\,i}$. The CCA is then applied to the random vectors $\vec{\mathbf{X}}' = (\alpha_1^{X+}, \ldots, \alpha_{k_X}^{X+})^{\mathrm{T}}$ and $\vec{\mathbf{Y}}' = (\alpha_1^{Y+}, \ldots, \alpha_{k_Y}^{Y+})^{\mathrm{T}}$.

An advantage of this approach is that it is often possible to use only the first few EOFs.

[6]Note that neither the eigenvectors $\vec{f}_X^i$ and $\vec{f}_Y^i$ nor the canonical correlation patterns $\vec{F}_X^i$ and $\vec{F}_Y^i$ are generally orthogonal. However, the columns of $\mathbf{\Sigma}_{XX}^{1/2}\mathbf{f}_X = \mathbf{\Sigma}_{XX}^{-1/2}\mathcal{F}_X$ and $\mathbf{\Sigma}_{YY}^{1/2}\mathbf{f}_Y = \mathbf{\Sigma}_{YY}^{-1/2}\mathcal{F}_Y$ are orthonormal.

Discarding the high-index EOFs can reduce the amount of noise in the problem by eliminating poorly organized, small-scale features of the fields involved.

Another advantage is that the algebra of the problem is simplified since $\Sigma_{X'X'}$ and $\Sigma_{Y'Y'}$ are both identity matrices. Thus, according to equations (14.10, 14.11), $\vec{f}^{\,i}_{X'}$ and $\vec{f}^{\,i}_{Y'}$ are eigenvectors of $\Sigma_{X'Y'}\Sigma^{\mathrm{T}}_{X'Y'}$ and $\Sigma^{\mathrm{T}}_{X'Y'}\Sigma_{X'Y'}$ respectively. Since these are non-negative definite symmetric matrices, the eigenvectors are orthogonal. Moreover, the canonical correlation patterns $\vec{F}^{i}_{X'} = \vec{f}^{\,i}_{X'}$ and $\vec{F}^{i}_{Y'} = \vec{f}^{\,i}_{Y'}$.

A minor disadvantage is that the patterns are given in the coordinates of the re-normalized EOF space (14.20). To express the pattern in the original coordinate space it is necessary to reverse transformation (14.20) with (14.18) and (14.19):

$$\begin{aligned}
\vec{f}^{\,i}_X &= \textstyle\sum_{j=1}^{k_X}(\lambda^X_j)^{1/2}(\vec{f}^{\,i}_{X'})_j\vec{e}^{\,j}_X \\
\vec{f}^{\,i}_Y &= \textstyle\sum_{j=1}^{k_Y}(\lambda^Y_j)^{1/2}(\vec{f}^{\,i}_{Y'})_j\vec{e}^{\,j}_Y \\
\vec{F}^{i}_X &= \textstyle\sum_{j=1}^{k_X}(\lambda^X_j)^{-1/2}(\vec{f}^{\,i}_{X'})_j\vec{e}^{\,j}_X \\
\vec{F}^{i}_Y &= \textstyle\sum_{j=1}^{k_Y}(\lambda^Y_j)^{-1/2}(\vec{f}^{\,i}_{Y'})_j\vec{e}^{\,j}_Y
\end{aligned} \tag{14.21}$$

where $(\cdot)_j$ denotes the jth element of the vector contained within the brackets. The canonical correlation patterns are no longer orthogonal after this backtransformation, and vectors $\vec{f}^{\,i}$ and $\vec{F}^{i}$ are no longer identical.[7]

14.1.7 Maximizing Covariance—the 'SVD Approach.'

Another way to identify pairs of coupled patterns $\vec{p}^{\,i}_X$ and $\vec{p}^{\,i}_Y$ in random fields $\vec{\mathbf{X}}$ and $\vec{\mathbf{Y}}$ is to search for orthonormal sets of vectors such that the covariance between the expansion coefficients $\alpha^X_i = \langle \vec{\mathbf{X}}, \vec{p}^{\,i}_X\rangle$ and $\alpha^Y_i = \langle \vec{\mathbf{Y}}, \vec{p}^{\,i}_Y\rangle$,

$$\mathrm{Cov}\big(\alpha^X_i, \alpha^Y_i\big) = (\vec{p}^{\,i}_X)^{\mathrm{T}}\Sigma_{XY}\vec{p}^{\,i}_Y, \tag{14.22}$$

is maximized. Note that we explicitly require orthonormal vectors so that $\vec{\mathbf{X}}$ and $\vec{\mathbf{Y}}$ can be expanded as $\vec{\mathbf{X}} = \sum \alpha^X_i \vec{p}^{\,i}_X$ and $\vec{\mathbf{Y}} = \sum \alpha^Y_i \vec{p}^{\,i}_Y$. The solution of (14.22) is obtained as in [14.1.1] by using Lagrange multipliers to enforce the constraints $(\vec{p}^{\,i}_X)^{\mathrm{T}}\vec{p}^{\,i}_X = 1$ and $(\vec{p}^{\,i}_Y)^{\mathrm{T}}\vec{p}^{\,i}_Y = 1$. The result is a system of equations,

$$\begin{aligned}
\Sigma_{XY}\vec{p}^{\,i}_Y &= s_X\vec{p}^{\,i}_X \\
\Sigma^{\mathrm{T}}_{XY}\vec{p}^{\,i}_X &= s_Y\vec{p}^{\,i}_Y,
\end{aligned} \tag{14.23}$$

that can be solved by a singular value decomposition (Appendix B). The same solution is obtained by substituting the two equations into each other to obtain

$$\begin{aligned}
\Sigma_{XY}\Sigma^{\mathrm{T}}_{XY}\vec{p}^{\,i}_X &= \lambda_i\vec{p}^{\,i}_X \\
\Sigma^{\mathrm{T}}_{XY}\Sigma_{XY}\vec{p}^{\,i}_Y &= \lambda_i\vec{p}^{\,i}_Y,
\end{aligned}$$

where $\lambda_i = s_X s_Y$. These equations share the same eigenvalues $\lambda_i > 0$, and their normalized eigenvectors are related by

$$\begin{aligned}
\vec{p}^{\,i}_Y &= \frac{\Sigma^{\mathrm{T}}_{XY}\vec{p}^{\,i}_X}{\|\Sigma^{\mathrm{T}}_{XY}\vec{p}^{\,i}_X\|} \\
\vec{p}^{\,i}_X &= \frac{\Sigma^{\mathrm{T}}_{XY}\vec{p}^{\,i}_Y}{\|\Sigma^{\mathrm{T}}_{XY}\vec{p}^{\,i}_Y\|}.
\end{aligned}$$

It is easily shown that $\mathrm{Cov}\big(\alpha^X_i, \alpha^Y_i\big) = \lambda_i^{1/2}$. Thus the pair of patterns associated with the largest eigenvalue maximizes the covariance. The pair of patterns associated with the second largest eigenvalue and orthogonal to the first pair maximize the covariability that remains in $\vec{\mathbf{X}} - \alpha^X_1\vec{p}^{\,1}_X$ and $\vec{\mathbf{Y}} - \alpha^Y_1\vec{p}^{\,1}_Y$, and so on.

This method is often called 'SVD' analysis. This wording is misleading because it mixes the definition of a statistical parameter with the algorithm used to calculate the parameter. These patterns *can* be calculated by SVD but there are other ways, such as conventional eigen-analysis, to get the same information. Patterns $\vec{p}^{\,i}_X$ and $\vec{p}^{\,i}_Y$ are often called *left* and *right singular vectors*. The nomenclature is again misleading because the relevant property of these vectors is that they maximize covariance. We therefore call this method *Maximum Covariance Analysis* (MCA) and call the vectors *Maximum Covariance Patterns*.

Two properties of MCA are worth mentioning.

- MCA is invariant under coordinate transformation only if the transformation is orthogonal. The eigenvalues, and thus the degree of covariability, change when the transformation is non-orthonormal.
- MCA coefficients α^X_i and α^X_j, $i \neq j$, are generally correlated. They are uncorrelated when $\Sigma_{XX} = \sigma^2_X\mathcal{I}$. This also applies to $\vec{\mathbf{Y}}$-coefficients.

See Wallace, Smith, and Bretherton [411] for examples.

14.1.8 Principal Prediction Patterns.

Suppose $\{\vec{\mathbf{Z}}_t\}$ is a multivariate time series and define

$$\vec{\mathbf{X}}_t = \vec{\mathbf{Z}}_t \text{ and } \vec{\mathbf{Y}}_t = \vec{\mathbf{Z}}_{t+\tau} \tag{14.24}$$

[7] Note the similarity between this discussion and that in [14.1.4].

for some positive lag τ. Application of the CCA algorithm, with prior EOF truncation if the dimension of $\vec{\mathbf{Z}}_t$ is large, identifies patterns $\vec{F}_0^i = \vec{F}_X^i$ and $\vec{F}_\tau^i = \vec{F}_Y^i$ that tend to appear together, that is, patterns with a fixed time lag in the same variable. Thus the presence of $\vec{F}_0^i$ at a given time indicates that it is likely that pattern $\vec{F}_\tau^i$ will emerge τ time units later. Because of the properties of CCA, patterns $\vec{F}_0^i$ and $\vec{F}_\tau^i$ depict the present and future parts of $\vec{\mathbf{Z}}_t$ that are most strongly related. In other words, they are the best linearly auto-predictable components in $\vec{\mathbf{Z}}_t$.[8] An example is given in [14.3.7].

14.2 Estimating Canonical Correlation Patterns

14.2.1 Estimation. Estimates of canonical correlation patterns and coefficients are obtained in the obvious way by replacing $\mathbf{\Sigma}_{XX}$, $\mathbf{\Sigma}_{YY}$, and $\mathbf{\Sigma}_{XY}$ with corresponding estimates. We recommend that the problem be kept small by approximating the data with truncated EOF expansions (see [14.1.5] and also Bretherton et al. [64]). This has the added benefit of eliminating small-scale spatial noise.

14.2.2 Making Inferences. As noted previously, very little is known about the sampling variability of the eigenvectors or canonical correlation patterns. However, there are some useful asymptotic results for making inferences about the canonical correlations themselves.

Bartlett [32] proposed a test of the null hypothesis H$_0$: $\rho_{l+1} = \cdots = \rho_m = 0$ that the last $m - l$ canonical correlations are zero when it is *known* that the first l are nonzero. Here $m = \min(m_X, m_Y)$. Bartlett's test can be used when the canonical correlations have been estimated from a sample $\{(\vec{\mathbf{x}}_1, \vec{\mathbf{y}}_1), \ldots, (\vec{\mathbf{x}}_n, \vec{\mathbf{y}}_n)\}$ of independent realizations of random vectors $\vec{\mathbf{X}}$ and $\vec{\mathbf{Y}}$ that are jointly multivariate normal. The test statistic (Bartlett [32])

$$\chi^2 = -\Big(n - 1 - l - \frac{1}{2}(m_X + m_Y + 1)\Big) + \sum_{i=1}^{l} \widehat{\rho}_i^{-2} \ln\Big(\prod_{i=l+1}^{m} (1 - \widehat{\rho}_i^2)\Big), \quad (14.25)$$

where $\widehat{\rho}_i = \sqrt{\widehat{\lambda}_i}$, is approximately distributed as $\chi^2((m_X - l)(m_Y - l))$ under H$_0$. The test is performed at the $(1 - \tilde{p}) \times 100\%$ significance level by comparing χ^2 (14.25) against the $\tilde{p}$-quantile of the approximating χ^2 distribution (see Appendix E).

Glynn and Muirhead [142] give a bias correction for $\widehat{\rho}_i$ and also give an expression for the asymptotic variance of the corrected estimator that is useful for constructing confidence intervals. Using the Fisher z-transform (recall [8.2.3]), Glynn and Muirhead show that if

$$\theta_i - \frac{1}{2}\ln\Big(\frac{1+\rho_i}{1-\rho_i}\Big) \quad \text{and} \quad z_i = \frac{1}{2}\ln\Big(\frac{1+\widehat{\rho}_i}{1-\widehat{\rho}_i}\Big),$$

then the bias of

$$\widehat{\theta}_i = z_i - \frac{1}{2n\widehat{\rho}_i}\Big(m_X + m_Y - 2 + \widehat{\rho}_i^2 + 2(1-\widehat{\rho}_i^2)\sum_{\substack{j=1;\\ j\neq i}}^{m} \frac{\widehat{\rho}_j^2}{\widehat{\rho}_i^2 - \widehat{\rho}_j^2}\Big)$$

is approximately $\mathcal{O}(n^{-2})$ and

$$\mathrm{Var}(\widehat{\theta}_i) = \frac{1}{n} + \mathcal{O}(n^{-2}).$$

Thus the bounds for an approximate $\tilde{p} \times 100\%$ confidence interval for ρ_i are given by

$$\tanh(\widehat{\theta}_i \pm z_{(1+\tilde{p})/2}/\sqrt{n}), \quad (14.26)$$

where $z_{(1+\tilde{p})/2}$ is the $(1+\tilde{p})/2$-quantile of the standard normal distribution (Appendix D). Muirhead and Waternaux [282] show that asymptotic statistics like equations (14.25, 14.26) are not particularly robust against departures from the multivariate normal assumption. Use of the bootstrap (see Section 5.5) is probably the best practical alternative when this is a concern.

One question rarely mentioned in the context of CCA is the size of sample needed to make good estimates and inferences. Thorndike [365, pp. 183–184] suggests that $n > 10(m_X + m_Y) + 50$ is a reasonable rule of thumb, and argues that $n > (m_X + m_Y)^2 + 50$ may be needed for some purposes. Our experience, however, is that much smaller samples can provide meaningful information about the first few patterns and correlations. However, be aware that the asymptotic results discussed above are not likely to hold under these circumstances. The Monte Carlo experiments discussed in the next subsection give some further insight into what can be accomplished with small samples.

[8] It seems that the idea was first suggested by Hasselmann in an unpublished paper in 1983 but it was not pursued until 1996 [103].

mode			$k = 20$					$n = 250$		
i	ρ^i_{xy}	$n =$	50	100	500	1000	$k =$	10	30	50
1	0.69		0.96	0.83	0.70	0.69		0.68	0.71	0.74
2	0.60		0.92	0.76	0.59	0.58		0.58	0.61	0.65
3	0.37		0.79	0.51	0.33	0.31		0.30	0.36	0.43
4	0.11		0.54	0.28	0.10	0.09		0.06	0.16	0.27
5	0.07		0.46	0.23	0.08	0.06		0.03	0.13	0.25

Table 14.1: *The means of 100 canonical correlation estimates computed from simulated samples of n pairs of 251-dimensional random fields (see text). For brevity, only five of the 10 canonical correlations are listed. The true correlations ρ^i_{xy} are given in the second column; the results obtained for variable time series lengths n, with an EOF truncation of $k = 20$, are given in columns three to six. The effect of including different numbers of EOFs k, using a fixed time series length of $n = 250$, is listed in columns seven to nine. From Borgert [55].*

14.2.3 Monte Carlo Experiments. Borgert [55] conducted a Monte Carlo study of the performance of CCA on EOF truncated [14.1.6] data. He simulated a pair of 251-point random fields $\vec{\mathbf{X}}$ and $\vec{\mathbf{Y}}$ that consisted of a random linear combination of 10 pairs of patterns. Each pair of patterns was multiplied by a pair of random coefficients that were independent of all other pairs of coefficients. Thus the random coefficients are the true canonical variables. Each pair of random coefficients was generated from a different bivariate auto-regressive process. In this way the cross-correlations between the pairs of canonical variates, the true canonical correlations, were known. Thus Borgert was able to simulate a pair of random fields with known canonical correlations and patterns.

Borgert used this tool to generate 100 independent samples for a number of combinations of sample size n and EOF truncation point $k = k_X = k_Y$. A canonical correlation analysis was performed on each sample, and statistics assessing the average quality of the CCA were gathered for each combination of n and k. He found that the CCA was really able to identify the correct pairs of patterns: the estimated patterns were close to the prescribed patterns. However, as exemplified in Table 14.1, there were considerable biases in the estimated correlations if too many EOFs were retained or if the time series were too short.

Bretherton et al. [64] reviewed a number of techniques for diagnosing coupled patterns and intercomparing them in a series of small Monte Carlo experiments. They found that CCA with *a priori* EOF truncation and Maximum Covariance Analysis were more robust than the other techniques considered.

14.2.4 Irregularly Distributed Gaps in the Data. One way to cope with missing data is to fill the gaps by spatial or temporal interpolation. However, this is unsatisfactory if more than just a small amount of data is missing because we end up trying to diagnose connections between real data on the one hand and imputed data with much lower information content on the other. A better procedure is to use only the data that are actually available. This can be achieved by the procedure already outlined in [13.2.7]. The various matrices, such as Σ_{XX}, are estimated by forming sums over only the available pairs of observations (13.31):

$$\widehat{\sigma}_{ij} = \frac{1}{|K_i \cap K_j|} \sum_{k \in K_i \cap K_j} (\mathbf{x}_{ki} - \widehat{\mu}_i)(\mathbf{x}_{kj} - \widehat{\mu}_j)^*$$

where $K_i = \{k\colon \text{component } i \text{ of } \vec{\mathbf{x}}_k \text{ is not missing}\}$, the notation $|\cdot|$ indicates the number of elements in a set, and $\widehat{\mu}_i = \frac{1}{|K_i|}\sum_{k \in K_i} \mathbf{x}_{ki}$. As with EOFs, the calculation of the time coefficients can no longer be done by means of the dot products (14.12) and (14.13). Instead coefficients are determined by least squares, as in equation (13.32).

14.3 Examples

14.3.0 Overview. We will present three examples in this section. The joint variability of a pair of large-scale fields is examined for evidence of a cause-and-effect relationship between the occurrence of large-scale sea-level air pressure and sea-surface temperatures anomalies in the North

Atlantic [14.3.1,2]. In the second example, one of the vector times series is again North Atlantic sea-level pressure but the second 'partner' in the CCA is a regional scale variable, namely, precipitation on the Iberian Peninsula [14.3.3,4]. This example is used to demonstrate *statistical downscaling* of GCM output. The last example [14.3.5] illustrates the *Principal Prediction Patterns* introduced in [14.1.7].

The literature also contains many other examples of applications of CCA. Bretherton et al. [64] cite several studies, including classic papers by Barnett and Preisendorfer [21], Nicholls [293] and Barnston and colleagues [26, 28, 30, 346].

14.3.1 North Atlantic SLP and SST: Data and Results. CCA is used to analyse the relationship between $\vec{\mathbf{X}}$ = *monthly mean sea-level pressure* (SLP) and $\vec{\mathbf{Y}}$ = *sea-surface temperature* (SST) over the North Atlantic in northern winter (DJF) (see Zorita et al. [438] for details). The data are time series of monthly means of SLP and SST on a grid over the North Atlantic north of about 20° N, for DJF of 1950 to 1986. Anomalies were obtained at each grid point by subtracting the long-term monthly mean from the original values.

The coefficients of the first five EOFs of both fields were retained for the subsequent CCA. They represent 87% and 62% of the total variance or SLP and SST respectively. To check the sensitivity of the results to EOF truncation, the same calculations were performed using five SLP EOFs and either 10 or 15 SST EOFs (77% and 84%, respectively) and essentially the same results were obtained.

The CCA yields two pairs of patterns that describe the coherent variations of the SST and SLP fields. The two patterns are dominant in describing SLP and SST variance.

The first pair of patterns, $\vec{F}^1_{SLP}$ and $\vec{F}^1_{SST}$, which corresponds to a canonical correlation of 0.56, represents 21% of the variance of monthly mean SLP and 19% of the variance of monthly mean SST (Figure 1.13).[9] The two patterns are consistent with the hypothesis first suggested by Bjerknes that atmospheric anomalies cause SST anomalies. The main features of the SLP pattern are a decrease of the westerly wind at about 50° N, and an anomalous cyclonic circulation centred at 40° W and 30° N.[10] North of the cyclone, the

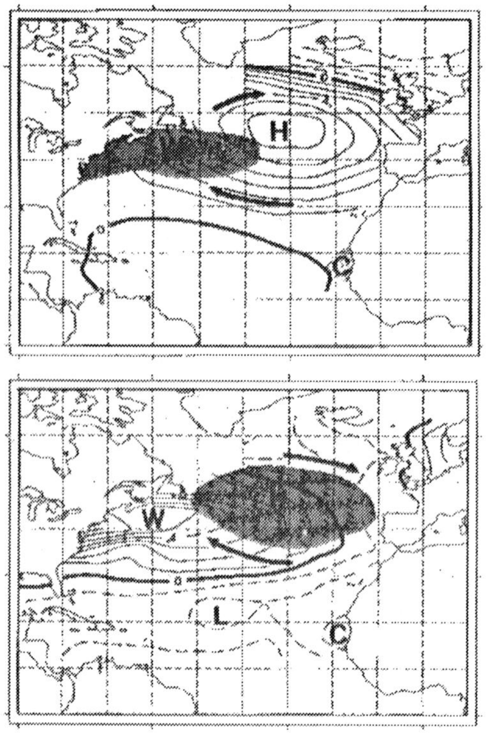

Figure 14.3: *The second pair of canonical patterns for monthly mean SLP and SST over the North Atlantic in DJF. The dark shading on each pattern identifies the main positive feature of the opposing pattern.*
Top: SLP, contour interval: 1 hPa,
Bottom: SST, contour interval: 0.1 K.
From Zorita et al. [438].

ocean surface is warmer than normal when the westerly wind is reduced. West of the cyclone, just downstream from the cold American continent, the ocean is substantially cooled. The SST anomalies off the African coast are a local response to anomalous winds; coastal upwelling is reduced when there are weaker than normal northerly winds. In contrast, when the circulation produces enhanced westerlies and anomalous anticyclonic flow in the southern part of the area, opposite SST anomalies are expected. The canonical correlation coefficient time series also support the Bjerknes hypothesis: the one month lag correlation is 0.65 when SLP leads SST but it is only 0.09 if SLP lags.

The coefficients of the second pair of patterns, $\vec{F}^2_{SLP}$ and $\vec{F}^2_{SST}$, have correlation 0.47 (Figure 14.3). The SLP pattern represents 31% of the total variance and is similar to the first SLP EOF (Figure 13.8), which is related to the North Atlantic Oscillation (see also [13.5.5] and

[9] We have dropped the '$\hat{\ }$' notation for now, but be aware that the patterns are parameter estimates. The same applies to canonical coordinate time series when they are discussed.

[10] We use the geostrophic wind relationship for the derivation of approximate wind anomalies from pressure anomalies.

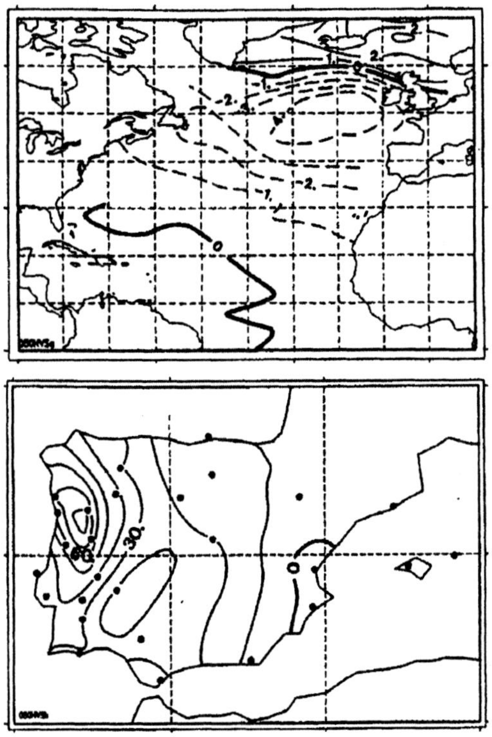

Figure 14.4: *First pair of canonical correlation patterns of the North Atlantic winter mean sea-level pressure* $\vec{\mathbf{Y}}$ *and a vector* $\vec{\mathbf{X}}$ *of seasonal means of precipitation at a number of Iberian locations [403].*

Figure 13.6). The structure of this pair of patterns is also consistent with the Bjerknes hypothesis. The one month lag correlation is 0.48 when SLP leads and 0.03 when SLP lags.

14.3.2 North Atlantic SLP and SST: Discussion. We described conventional and rotated EOF analysis of the same data in [13.5.6,7]. The CCA of SLP and SST suggests why rotation had a marked effect on the SST EOFs but not on the SLP EOFs. The coherent variations in the atmosphere (SLP) are caused by large-scale internal atmospheric processes so that the EOFs have a simple large-scale structure. In case of the ocean (SST), however, the coherent variations (EOFs) are the oceanic response to the large-scale atmospheric variations. This response really does not have simple structure (recall our description in [14.3.3] of the ocean's response to NAO variations).

14.3.3 North Atlantic SLP and Iberian Rainfall: Analysis and Historic Reconstruction. In this example, winter (DJF) mean precipitation from a number of rain gauges on the Iberian Peninsula is related to the air-pressure field over the North Atlantic (see [403] for details). CCA was used to obtain a pair of canonical correlation pattern estimates $\vec{F}^1_{SLP}$ and $\vec{F}^1_{pre}$ (Figure 14.4), and corresponding time series $\beta_1^{SLP}(t)$ and $\beta_1^{pre}(t)$ of canonical variate estimates. These strongly correlated modes of variation (the estimated canonical correlation is 0.75) represent about 65% and 40% of the total variability of seasonal mean SLP and Iberian Peninsula precipitation respectively. The two patterns represent a simple physical mechanism: when $\vec{F}^1_{SLP}$ has a strong positive coefficient, enhanced cyclonic circulation advects more maritime air onto the Iberian Peninsula so that precipitation in the mountainous northwest region ($\vec{F}^1_{pre}$) is increased.

Since the canonical correlation is large, the results of the CCA can be used to forecast or specify winter mean precipitation on the Iberian peninsula from North Atlantic SLP. The first step is to connect $\beta_1^{pre}(t)$ and $\beta_1^{SLP}(t)$ with a simple linear model $\beta_1^{pre}(t) = a\beta_1^{SLP}(t) + \epsilon$. Since $\beta_1^{pre}(t)$ and $\beta_1^{SLP}(t)$ are normalized to unit variance, the least squares estimate of coefficient a is the canonical correlation ρ_1. Given a realization of $\beta_1^{SLP}(t)$, the canonical variate for precipitation can be forecast as $\widehat{\beta}_1^{pre}(t) = \rho_1\beta_1^{SLP}(t)$, and thus the precipitation field is forecast as

$$\widehat{\vec{\mathbf{R}}} = \widehat{\beta}_1^{pre}(t)\vec{F}^1_{pre} = \rho_1\beta_1^{SLP}(t)\vec{F}^1_{pre}. \qquad (14.27)$$

Similarly, if several useful canonical correlation patterns had been found, Iberian winter mean precipitation could be forecast or specified as

$$\widehat{\vec{\mathbf{R}}} = \sum_{i=1}^{k} \rho_i\beta_i^{SLP}(t)\vec{F}^i_{pre}.$$

The analysis described above was performed with the 1950–80 segment of a data set that extends back to 1901. Since the 1901–49 segment is independent of that used to 'train' the model (14.27), it can be used to validate the model. Figure 14.5 shows both the specified and observed winter mean rainfall averaged over all Iberian stations for this period. The overall upward trend and the low-frequency variations in observed precipitation are well reproduced by the indirect method indicating the usefulness of the technique (14.27) as well as the reality of both the trend and the variations in the Iberian winter precipitation.

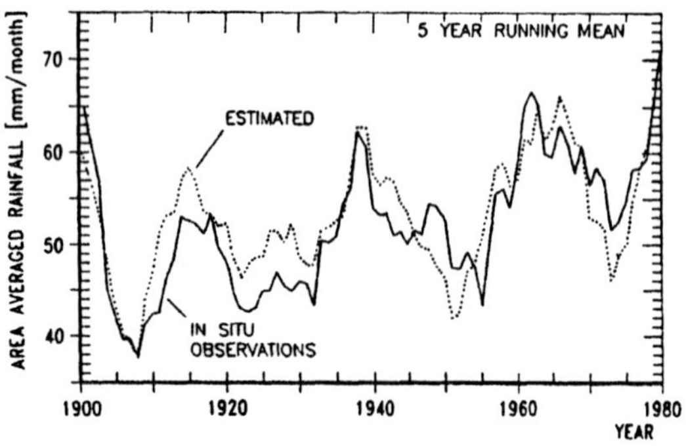

Figure 14.5: *Five-year running mean of winter mean rainfall averaged across Iberian rain gauges. The solid curve is obtained from station data, and the dotted curve is imputed from North Atlantic SLP variations [403].*

14.3.4 North Atlantic SLP and Iberian Rainfall: Downscaling of GCM output. The regression approach described above has an interesting application in climate change studies. GCMs are widely used to assess the impact that increasing concentrations of greenhouse gases might have on the climate system. But, because of their resolution, GCMs do not represent the details of regional climate change well. The *minimum scale* that a GCM is able to resolve is the distance between two neighbouring grid points whereas the *skilful scale* is generally accepted to be four or more grid lengths. The minimum scale in most climate models in the mid 1990s is of the order of 250–500 km so that the skilful scale is at least 1000–2000 km.

Thus the scales at which GCMs produce useful information does not match the scale at which many users, such as hydrologists, require information. *Statistical downscaling* [403] is a possible solution to this dilemma. The idea is to build a statistical model from historical observations that relates large-scale information that can be well simulated by GCMs to the desired regional scale information that can not be simulated. These models are then applied to the large-scale model output.

The following steps must be taken.

1 Identify a regional climate variable $\vec{\mathbf{R}}$ of interest.

2 Find a climate variable $\vec{\mathbf{L}}$ that

- controls $\vec{\mathbf{R}}$ in the sense that there is a statistical relationship between $\vec{\mathbf{R}}$ and $\vec{\mathbf{L}}$ of the form

$$\vec{\mathbf{R}} = \mathcal{G}(\vec{\mathbf{L}}, \vec{\alpha}) + \epsilon \qquad (14.28)$$

in which $\mathcal{G}(\vec{\mathbf{L}}, \vec{\alpha})$ represents a substantial fraction of the total variance of $\vec{\mathbf{R}}$. Vector $\vec{\alpha}$ contains parameters that can be used to adjust the fit of (14.28).

- is reliably simulated in a climate model.

3 Use historical realizations $(\vec{\mathbf{r}}_t, \vec{\mathbf{l}}_t)$ of $(\vec{\mathbf{R}}, \vec{\mathbf{L}})$ to estimate $\vec{\alpha}$.

4 Validate the fitted model on independent historical data or by cross-validation (see [18.5.2]).

5 Apply the validated model to GCM simulated realizations of $\vec{\mathbf{L}}$.

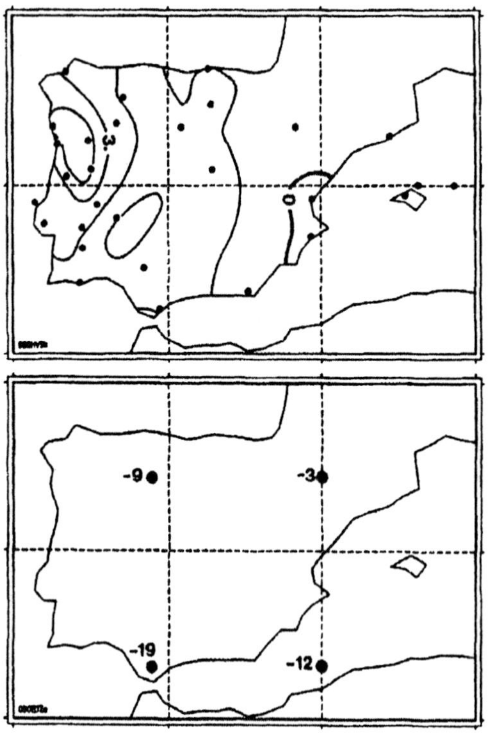

Figure 14.6: *Downscaled and grid point response of Iberian precipitation in a '2×CO_2 experiment' [403].*

This is exactly the process that was followed in the previous subsection. A model (14.27) was constructed that related Iberian rainfall $\vec{\mathbf{R}}$ to North Atlantic SLP $\vec{\mathbf{L}}$ through a simple linear functional. The adjustable parameters $\vec{\alpha}$ consisted of the canonical correlation patterns $\vec{F}^1_{pre}$ and

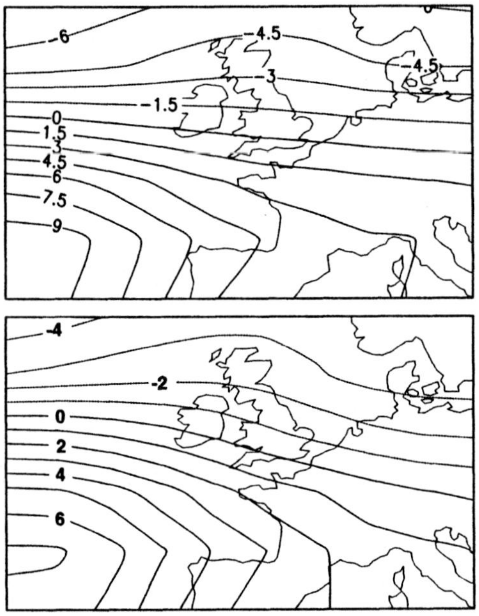

Figure 14.7: *Principal Prediction Patterns* $\vec{F}_0^1$ *(top) and* $\vec{F}_\Delta^1$ *with* $\Delta = 3$ *days (bottom) for the North Atlantic / European Sector daily winter SLP. From Dorn and von Storch [103].*

$\vec{F}_{SLP}^1$ and the canonical correlation ρ_1. These parameters were estimated from 1950 to 1980 data. Observations before 1950 have been used to validate the model.

Downscaling model (14.27) was applied to the output of a '2×CO_2' experiment performed with a GCM. Figure 14.6 compares the 'downscaled' response to doubled CO_2 with the model's grid point response. The latter suggests that there will be a marked decrease in precipitation over most of the Peninsula whereas the downscaled response is weakly positive. The downscaled response is physically more reasonable than the direct response of the model.

14.3.5 Principal Prediction Pattern of North Atlantic / European SLP. Dorn and von Storch [103] used the Principal Prediction Pattern (PPP) analysis technique to study the synoptic predictability of sea-level pressure (SLP) over the eastern North Atlantic and Western Europe. This particular field was used because a rich data set, consisting of daily analysis since approximately 1900, was available for determining the skill of the PPP model.

The PPP analysis was performed with daily winter SLP maps for 1958–88. The dimensionality of the problem was reduced by projecting the maps onto the first eight EOFs of daily winter SLP. Analyses were performed for lags $\tau = 1, \dots, 5$ days, but we discuss only the $\tau = 3$ days results below.

The first pair of PPPs is shown in Figure 14.7. The patterns are normalized such that the variance of the coefficient of $\vec{F}_0^1$ is 1, and that of the coefficient of $\vec{F}_3^1$ is $1/\sqrt{\rho_1}$. With this normalization, the coefficient for the regression of the $\vec{F}_0^1$-coefficient on the $\vec{F}_3^1$-coefficient is the identity. Also, the patterns are scaled so that if the initial state is a multiple of $\vec{F}_0^1$, then the best predictor is the same multiple of $\vec{F}_3^1$. Patterns $\vec{F}_0^1$ and $\vec{F}_3^1$ are rather similar indicating that the analysis has selected the regional SLP mode that is most persistent on synoptic time scales. The reduction of the magnitude by about $1/3$ indicates that this persistence goes with some damping. Thus, the forecast incorporated in this pair of patterns implies constancy in the pattern, but a reduction of the intensity, i.e., 'damped persistence'.

This statement also holds for the other patterns and is further supported by comparing the forecast skill, as given by the anomaly correlation coefficient[11] between the true SLP field and the field predicted by either PPP or persistence (Figure 14.8). The skill of the two forecast schemes is practically identical and exhibits the characteristic decay with increasing lag. Thus, the PPP forecast is no more skilful than the simpler 'competitor' persistence.

However, the PPP forecast scheme should not be dismissed out of hand. By conditioning on the proportion of spatial variance represented by $\vec{F}_0^1$, the PPP forecast was found to be more skilful when the proportion is large (Figure 14.8, bottom). Thus the PPP scheme also gives a forecast of forecast skill.

The utility of the PPP technique needs further exploration and the user is advised to examine all results obtained with this technique critically. In particular, surprisingly good results may be generated by using short time series or by failing to adequately reduce the degrees of freedom of the problem.

14.4 Redundancy Analysis

14.4.1 Introduction. So far, we have identified pairs of patterns by maximizing the correlation

[11]This measure of skill is explained in detail in [18.2.9]. Roughly speaking, it is the mean spatial correlation between the forecast and the verifying field.

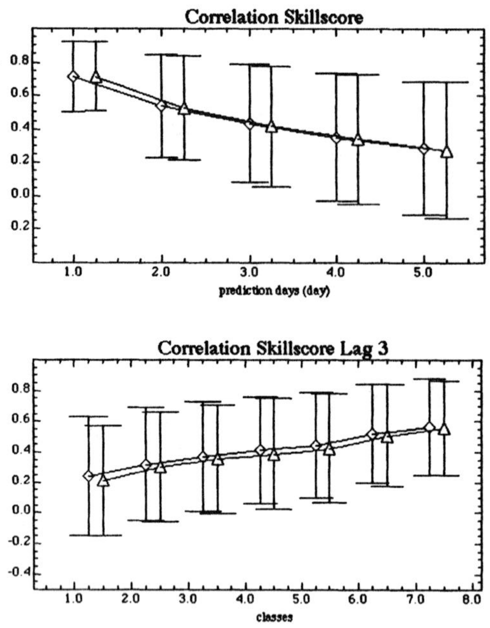

I

II

Figure 14.8: *Anomaly correlation coefficient of the PPP forecast (diamonds) and of persistence (triangles).The vertical bars indicate* $\pm\sigma$ *bands, as estimated from all forecast prepared for the winter days from 1900 until 1990. For better readability, the numbers for the two forecast schemes, persistence and PPP, are shifted horizontally.*
Top: For lags $\tau = 1, \ldots, 5$ *days.*
Bottom: For lag $\tau = 3$ *days. The anomaly correlation coefficients were classified according to the proportion of variance of the initial SLP field described by the PPP (bottom). Class 1 contains cases with proportions in the range* $[0.0, 0.4]$, *class 2 contains cases with proportions in* $(0.4, 0.5]$, *and so on up to class 7, which contains cases with proportions in* $(0.9, 1]$.

between the corresponding pattern coefficients. We then demonstrated how regression techniques can be used to specify or forecast the value of the pattern coefficients of one of the fields from those of the other field. This regression problem is generically non-symmetric because the objective is to maximize the variance of the predictand that can be represented. Properties of the predictor patterns, such as the amount of variance they represent, are irrelevant to the regression problem. Hence, there is a mismatch between CCA, which treats variables equally, and regression analysis, which focuses primarily on the predictand.

The 'redundancy analysis' technique directly addresses this problem by identifying patterns that are strongly linked through a regression model. Patterns are selected by maximizing predictand variance. This technique was developed in the late 1970s but apparently has not been introduced in climate research literature.

Here we present the *redundancy analysis* as suggested by Tyler [376]. *Note that very little experience has been collected with this technique in the field of climate research. Therefore, the technique should be applied with great care, and results should be appraised critically.*

14.4.2 Redundancy Index. Let us consider a pair of random vectors $(\vec{\mathbf{X}}, \vec{\mathbf{Y}})$ with dimensions m_X and m_Y. Let us assume further that there is a linear operator represented by a $m_X \times k$ matrix $\mathcal{Q}_k$. How much variance in $\vec{\mathbf{Y}}$ can be accounted for by a regression of $\mathcal{Q}_k^{\mathrm{T}}\vec{\mathbf{X}}$ on $\vec{\mathbf{Y}}$?[12] We assume, without loss of generality, that the expected value of both $\vec{\mathbf{X}}$ and $\vec{\mathbf{Y}}$ is zero.

The regression model that relates $\mathcal{Q}_k^{\mathrm{T}}\vec{\mathbf{X}}$ is given by

$$\vec{\mathbf{Y}} = \mathcal{R}(\mathcal{Q}_k^{\mathrm{T}}\vec{\mathbf{X}}) + \vec{\epsilon}, \tag{14.29}$$

where $\mathcal{R}$ is an $m_Y \times k$ matrix of regression coefficients. The variance represented by $(\mathcal{Q}_k^{\mathrm{T}}\vec{\mathbf{X}})$ is maximized when

$$\mathcal{R} = \Sigma_{Y,\mathcal{Q}X}\left(\Sigma_{\mathcal{Q}X,\mathcal{Q}X}\right)^{-1}, \tag{14.30}$$

where

$$\Sigma_{Y,\mathcal{Q}X} = \mathrm{Cov}\big(\vec{\mathbf{Y}}, \mathcal{Q}_k^{\mathrm{T}}\vec{\mathbf{X}}\big) = \Sigma_{YX}\mathcal{Q}_k \tag{14.31}$$

$$\Sigma_{\mathcal{Q}X,\mathcal{Q}X} = \mathcal{Q}_k^{\mathrm{T}}\Sigma_{XX}\mathcal{Q}_k. \tag{14.32}$$

Tyler [376] called the proportion of variance represented by the regression (14.29) the *redundancy index* and labelled it

$$R^2(\vec{\mathbf{Y}} : \mathcal{Q}_k^{\mathrm{T}}\vec{\mathbf{X}}) = \frac{\mathrm{tr}\big(\mathrm{Cov}(\vec{\mathbf{Y}}, \vec{\mathbf{Y}}) - \mathrm{Cov}(\vec{\mathbf{Y}} - \hat{\vec{\mathbf{Y}}}, \vec{\mathbf{Y}} - \hat{\vec{\mathbf{Y}}})\big)}{\mathrm{tr}\big(\mathrm{Cov}(\vec{\mathbf{Y}}, \vec{\mathbf{Y}})\big)} \tag{14.33}$$

where $\hat{\vec{\mathbf{Y}}} = \mathcal{R}(\mathcal{Q}_k^{\mathrm{T}}\vec{\mathbf{X}})$ is the estimated value of $\vec{\mathbf{Y}}$. The motivation of this wording is that it is a measure of how *redundant* the information in $\vec{\mathbf{Y}}$ is if one already has the information provided by $\vec{\mathbf{X}}$.

[12] The number of columns (patterns) in $\mathcal{Q}_k$ is smaller than the dimension of $\vec{\mathbf{X}}$ in most practical situations, so that $k < m_X$ or even $k \ll m_X$. Thus, the operation $\vec{\mathbf{X}} \to \mathcal{Q}_k^{\mathrm{T}}\vec{\mathbf{X}}$ represents a reduction of the phase space of $\vec{\mathbf{X}}$, as in all the other cases we have discussed in this and the previous chapter.

The numerator is the trace (sum of main diagonal elements) of the matrix

$$\begin{aligned}&\Sigma_{YY} - (\Sigma_{YY} + \Sigma_{\hat{Y}\hat{Y}} - 2\Sigma_{Y\hat{Y}})\\ &= -\mathcal{R}\Sigma_{QX,QX}\mathcal{R}^{\mathrm{T}} + 2\Sigma_{Y,QX}\mathcal{R}^{\mathrm{T}}.\end{aligned}$$

Using (14.31)–(14.33), and simplifying, we find that

$$R^2(\vec{\mathbf{Y}} : \mathcal{Q}_k^{\mathrm{T}}\vec{\mathbf{X}}) = \frac{\mathrm{tr}\big(\Sigma_{YX}\mathcal{Q}_k\left(\mathcal{Q}_k^{\mathrm{T}}\Sigma_{XX}\mathcal{Q}_k\right)^{-1}\mathcal{Q}_k^{\mathrm{T}}\Sigma_{XY}\big)}{\mathrm{tr}\big(\Sigma_{YY}\big)}. \tag{14.34}$$

14.4.3 Invariance of the Redundancy Index to Linear Transformations. The redundancy index has a number of interesting properties. One of these is its invariance to orthonormal transformations of $\vec{\mathbf{Y}}$: if $\mathcal{A}$ is orthonormal, then

$$R^2(\mathcal{A}\vec{\mathbf{Y}} : \mathcal{Q}_k^{\mathrm{T}}\vec{\mathbf{X}}) = R^2(\vec{\mathbf{Y}} : \mathcal{Q}_k^{\mathrm{T}}\vec{\mathbf{X}}). \tag{14.35}$$

The significance of this property comes from the fact that we may identify any orthonormal transformation with a linear transformation that conserves variance. Relationship (14.35) does not hold for general non-singular matrices, in particular not for transformations that change the variance since the proportion of captured variance changes when the variance of $\vec{\mathbf{Y}}$ is changed.

On the other hand, any square non-singular matrix $\mathcal{Q}_{m_X}$ used to transform the specifying variable $\vec{\mathbf{X}}$ has also no effect on the redundancy index. In that case, $(\mathcal{Q}_{m_X})^{-1}$ exists and $(\mathcal{Q}_{m_X}^{\mathrm{T}}\Sigma_{XX}\mathcal{Q}_{m_X})^{-1} = \mathcal{Q}_{m_X}^{-1}\Sigma_{XX}^{-1}(\mathcal{Q}_{m_X}^{\mathrm{T}})^{-1}$ in the numerator of (14.34), so that

$$R^2(\vec{\mathbf{Y}} : \mathcal{Q}_{m_X}^{\mathrm{T}}\vec{\mathbf{X}}) = R^2(\vec{\mathbf{Y}} : \vec{\mathbf{X}}). \tag{14.36}$$

The implication of (14.36) is that the coordinate system in which the random vector $\vec{\mathbf{X}}$ is given does not matter, so long as it describes the same linear space. This is a favourable property since the information contained in $\vec{\mathbf{X}}$ about $\vec{\mathbf{Y}}$ should not depend on the specifics of the presentation of $\vec{\mathbf{X}}$, such as the metric used to measure the components of $\vec{\mathbf{X}}$, or the order of its components.

However, if the linear transformation $\mathcal{Q}_k$ maps the m_X-dimensional variable $\vec{\mathbf{X}}$ onto a k-dimensional variable $\vec{\mathbf{X}}_k' = \mathcal{Q}_k^{\mathrm{T}}\vec{\mathbf{X}}$, the new variable contains less information about $\vec{\mathbf{Y}}$, so that

$$\begin{aligned}R^2(\vec{\mathbf{Y}} : \vec{\mathbf{X}}_k') &\le R^2(\vec{\mathbf{Y}} : \vec{\mathbf{X}}_{k+1}') \\ &\le R^2(\vec{\mathbf{Y}} : \vec{\mathbf{X}}_{m_X}') = R^2(\vec{\mathbf{Y}} : \vec{\mathbf{X}})\end{aligned} \tag{14.37}$$

provided that 'column spaces'[13] of $\mathcal{Q}_k$, $\mathcal{Q}_{k+1}$, and $\mathcal{Q}_{m_X}$ are nested and $\mathcal{Q}_{m_X}$ is invertible. If, for all k, $\mathcal{Q}_{k+1}$ is constructed by adding a column to $\mathcal{Q}_k$, then inequality (14.37) simply reflects the fact that the regression on $\vec{\mathbf{Y}}$ has k predictors in the case of $\vec{\mathbf{X}}_k'$, and the same k predictors plus one more in the case of $\vec{\mathbf{X}}_{k+1}'$.

For a given transformation $\mathcal{Q}_k$, again only the subspace spanned by the columns of $\mathcal{Q}_k$ matters. That is, for any invertible $k \times k$ matrix $\mathcal{L}$, we find

$$R^2(\vec{\mathbf{Y}} : \mathcal{L}^{\mathrm{T}}(\mathcal{Q}_k^{\mathrm{T}}\vec{\mathbf{X}})) = R^2(\vec{\mathbf{Y}} : \mathcal{Q}_k^{\mathrm{T}}\vec{\mathbf{X}}). \tag{14.38}$$

Thus, the redundancy index for two variables is a function of the subspace the variable $\vec{\mathbf{X}}$ is projected upon, and the way in which $\vec{\mathbf{Y}}$ is scaled.

Since R^2 does not depend on the specific coordinates of the variable $\vec{\mathbf{X}}$ and $\vec{\mathbf{X}}_k'$, we may assume that the columns of $\mathcal{Q}_k$ are chosen to be orthogonal with respect to $\vec{\mathbf{X}}$; that is,

$$\vec{q}^{\,k\mathrm{T}}\Sigma_{XX}\vec{q}^{\,j} = 0 \tag{14.39}$$

for any $k \neq j$. Then

$$R^2(\vec{\mathbf{Y}} : \mathcal{Q}_k^{\mathrm{T}}\vec{\mathbf{X}}) = \sum_{j=1}^{k} R^2(\vec{\mathbf{Y}} : \vec{q}^{\,j\mathrm{T}}\vec{\mathbf{X}}), \tag{14.40}$$

which may be seen as a special version of (14.37). Note that (14.39) is fulfilled if the vectors $\vec{q}^{\,j}$ are the EOFs of $\vec{\mathbf{X}}$.

14.4.4 Redundancy Analysis. The theory behind *redundancy analysis*, as put forward by Tyler [376], confirms the existence of a non-singular transformation $\mathcal{B} = (\vec{b}^1|\vec{b}^2|\cdots|\vec{b}^{m_X})$ so that the index of redundancy (i.e., the amount of $\vec{\mathbf{Y}}$-variance explained through the regression of $\mathcal{B}_k^{\mathrm{T}}\vec{\mathbf{X}}$ on $\vec{\mathbf{Y}}$) is maximized for any $k = 1, \ldots, \min(m_X, m_Y)$. Matrix $\mathcal{B}_k$ contains the first k columns of $\mathcal{B}$.

Thus redundancy analysis determines the k-dimensional subspace that allows for the most efficient regression on $\vec{\mathbf{Y}}$. Since we are free to choose the coordinates of this subspace, we may use a linear basis with k orthogonal patterns that satisfies (14.39), so that the redundancy index may be expressed specifically as (14.40).

The following theorem identifies a second set of patterns, $\mathcal{A} = (\vec{a}^1|\vec{a}^2|\cdots|\vec{a}^k)$, that represent an orthogonal partitioning of the variance of $\vec{\mathbf{Y}}$ that is accounted for by the regression of $\vec{\mathbf{X}}$ on $\vec{\mathbf{Y}}$. More specifically, the regression maps the subspace

[13] The *column space* of a matrix $\mathcal{Q}$ is the vector space spanned by the columns of $\mathcal{Q}$.

represented by $\vec{\mathbf{X}}'_k$ onto the space spanned by the first k columns of $\mathcal{A}$.

The following subsections describe the mathematics required for the determination of matrices $\mathcal{A}$ and $\mathcal{B}$. The theorems are taken from Tyler's paper [376].

14.4.5 The Redundancy Analysis Transformations. *For any random vectors $\vec{\mathbf{Y}}$ of dimension m_Y and $\vec{\mathbf{X}}$ of dimension m_X, there exists an orthonormal transformation $\mathcal{A}$ and a non-singular transformation $\mathcal{B}$ such that*

$$\mathrm{Cov}(\mathcal{B}^{\mathrm{T}}\vec{\mathbf{X}}, \mathcal{B}^{\mathrm{T}}\vec{\mathbf{X}}) = \mathcal{I} \tag{14.41}$$

$$\mathrm{Cov}(\mathcal{A}^{\mathrm{T}}\vec{\mathbf{Y}}, \mathcal{B}^{\mathrm{T}}\vec{\mathbf{X}}) = \mathcal{D} \tag{14.42}$$

where $\mathcal{D}$ is an $m_Y \times m_X$ matrix with elements $d_{ij} = 0$ for $i \neq j$ and diagonal elements $d_{jj} = \sqrt{\lambda_j}$ for $j \leq \min(m_X, m_Y)$.

The proof, which is detailed in Appendix M, revolves around two eigen-equations:

$$\Sigma_{YX}\Sigma_{XX}^{-1}\Sigma_{XY}\vec{a}^{\,j} = \lambda_j\vec{a}^{\,j} \tag{14.43}$$

$$\Sigma_{XX}^{-1}\Sigma_{XY}\Sigma_{YX}\vec{b}^{\,j} = \lambda_j\vec{b}^{\,j}. \tag{14.44}$$

Both equations have the same positive eigenvalues λ_j, and the eigenvectors $\vec{a}^{\,j}$ and $\vec{b}^{\,j}$ belonging to the same nonzero eigenvalue λ_j are related through

$$\vec{b}^{\,j} = \frac{1}{\sqrt{\lambda_j}}\Sigma_{XX}^{-1}\Sigma_{XY}\vec{a}^{\,j}. \tag{14.45}$$

The matrices $\mathcal{A}$ and $\mathcal{B}$, which are composed of eigenvectors $\vec{a}^{\,j}$ and $\vec{b}^{\,j}$, respectively, are the only matrices that satisfy the requirements of the theorem.

From the computational point of view, it is advisable to solve the eigenproblem with the Hermitian matrix (14.43), then use the identity (14.45). Since (14.43) is a Hermitian problem, all eigenvectors $\vec{a}^{\,j}$ are real valued, and since (14.45) involves only real matrices, the 'patterns' $\vec{b}^{\,j}$ are also real valued.

14.4.6 Theorem: Optimality of the Redundancy Transformation. The significance of the redundancy transformation originates from the following theorem given by Tyler [376]:

The redundancy index $R^2(\vec{\mathbf{Y}} : \mathcal{Q}_k^{\mathrm{T}}\vec{\mathbf{X}})$ is maximized by setting $\mathcal{Q}_k = \mathcal{B}_k$, where $\mathcal{B}_k$ is the $m_x \times k$ matrix that contains the k eigenvectors satisfying (14.42) that correspond to the k largest eigenvalues.

Note that the statement holds for all $k \leq m_X$. Thus, among all possible single patterns $\vec{q}$, the eigenvector $\vec{b}^{\,1}$ belonging to the largest eigenvalue of the matrix $\Sigma_{XX}^{-1}\Sigma_{XY}\Sigma_{YX}$ provides the maximum information, in a linear sense, about the variance of $\vec{\mathbf{Y}}$:

$$R^2(\vec{\mathbf{Y}} : \vec{q}^{\,\mathrm{T}}\vec{\mathbf{X}}) \leq R^2(\vec{\mathbf{Y}} : \vec{b}^{\,1^{\mathrm{T}}}\vec{\mathbf{X}}) \tag{14.46}$$

for any m_X-dimensional vector $\vec{q}$. Moreover, by equations (14.41) and (14.40), the index of redundancy takes a particularly simple form,

$$R^2(\vec{\mathbf{Y}} : \mathcal{B}_k^{\mathrm{T}}) = \sum_{j=1}^{k} R^2(\vec{\mathbf{Y}} : \vec{b}^{\,j^{\mathrm{T}}}\vec{\mathbf{X}}). \tag{14.47}$$

Also, note that inequality (14.46) may be generalized to

$$\sum_{j=1}^{k} R^2(\vec{\mathbf{Y}} : \vec{q}^{\,j^{\mathrm{T}}}\vec{\mathbf{X}}) \leq \sum_{j=1}^{k} R^2(\vec{\mathbf{Y}} : \vec{b}^{\,j^{\mathrm{T}}}\vec{\mathbf{X}}) \tag{14.48}$$

for any set of vectors $\vec{q}^{\,1}, \ldots, \vec{q}^{\,k}$.

14.4.7 The Role of Matrix $\mathcal{A}$. Since $\mathcal{B} = (\vec{b}^{\,1}|\cdots|\vec{b}^{\,m_X})$ is non-singular, random vector $\vec{\mathbf{X}}$ can be expanded in the usual manner as

$$\vec{\mathbf{X}} = \sum_{j=1}^{m_X}(\vec{\mathbf{X}}^{\mathrm{T}}\vec{b}^{\,j})\vec{p}^{\,j}, \tag{14.49}$$

where the *adjoint patterns* $\mathcal{P} = (\vec{p}^{\,1}|\cdots|\vec{p}^{\,m_X})$ are given by $\mathcal{P}^{\mathrm{T}} = \mathcal{B}^{-1}$. When re-expressed in matrix-vector form, equation (14.49) simply reads as

$$\vec{\mathbf{X}} = \mathcal{P}\mathcal{B}^{\mathrm{T}}\vec{\mathbf{X}}.$$

Similarly, since $\mathcal{A}$ is orthonormal, the part of $\vec{\mathbf{Y}}$ that can be represented by $\vec{\mathbf{X}}$, that is, $\hat{\vec{\mathbf{Y}}}$, can be expanded as

$$\hat{\vec{\mathbf{Y}}} = \mathcal{A}\mathcal{A}^{\mathrm{T}}\hat{\vec{\mathbf{Y}}} = \sum_j(\hat{\vec{\mathbf{Y}}}^{\mathrm{T}}\vec{a}^{\,j})\vec{a}^{\,j}. \tag{14.50}$$

When we regress $\vec{\mathbf{Y}}$ on $\vec{\mathbf{X}}$, we find that $\hat{\vec{\mathbf{Y}}} = \Sigma_{YX}\Sigma_{XX}^{-1}\vec{\mathbf{X}}$. Thus the expansion coefficients in (14.50), the elements of $\mathcal{A}^{\mathrm{T}}\hat{\vec{\mathbf{Y}}}$, are given by

$$\mathcal{A}^{\mathrm{T}}\hat{\vec{\mathbf{Y}}} = \mathcal{A}^{\mathrm{T}}\Sigma_{YX}\Sigma_{XX}^{-1}\vec{\mathbf{X}}.$$

Now, from equations (14.41) and (14.42) we have that $\Sigma_{XX}^{-1} = \mathcal{B}\mathcal{B}^{\mathrm{T}}$ and $\mathcal{A}^{\mathrm{T}}\Sigma_{YX}\mathcal{B} = \mathcal{D}$. Thus

$$\mathcal{A}^{\mathrm{T}}\hat{\vec{\mathbf{Y}}} = \mathcal{A}^{\mathrm{T}}\Sigma_{YX}\mathcal{B}\mathcal{B}^{\mathrm{T}}\vec{\mathbf{X}} = \mathcal{D}\mathcal{B}^{\mathrm{T}}\vec{\mathbf{X}}.$$

Hence the expansion coefficients in (14.50) are given by

$$\widehat{\vec{\mathbf{Y}}}^{\mathrm{T}}\vec{a}^{\,j} = \sqrt{\lambda_j}\,\vec{\mathbf{X}}^{\mathrm{T}}\vec{b}^{\,j}. \tag{14.51}$$

Considering both (14.49) and (14.50), we see that the regression maps variations in the amplitude of $\vec{\mathbf{X}}$ patterns $\vec{p}^{\,j}$ onto variations in the amplitude of $\widehat{\vec{\mathbf{Y}}}$ patterns $\vec{a}^{\,j}$. On average, $\widehat{\vec{\mathbf{Y}}}_j = \sqrt{\lambda_j}\,\vec{a}^{\,j}$ when $\vec{\mathbf{X}}_j = \vec{p}^{\,j}$ (cf. (14.49) and (14.51)). It is easily shown that the patterns themselves are related by[14]

$$\mathcal{A}\mathcal{D} = \mathbf{\Sigma}_{YX}\mathbf{\Sigma}_{XX}^{-1}\mathcal{P}.$$

That is, the $\vec{\mathbf{X}}$-patterns are transformed into scaled versions of the $\widehat{\vec{\mathbf{Y}}}$ patterns by the regression operator.

Thus, redundancy analysis offers a number of useful insights. First, it helps us to identify an efficient way of specifying a maximum of variance in one random vector from the information provided by another vector. It also guides us in finding those components of the specifying variable that contain the most information about the variable to be specified. Finally, it offers pairs of patterns that are mapped onto each other. If we observe the pattern $\vec{p}^{\,j}$ in the specifying field, then the likelihood of observing pattern $\vec{a}^{\,j}$ in the field to be specified is increased.

If we consider the full $\vec{\mathbf{X}}$-space, we find that

$$\mathbf{\Sigma}_{\hat{Y}\hat{Y}} = \mathbf{\Sigma}_{YX}\mathbf{\Sigma}_{XX}^{-1}\mathbf{\Sigma}_{XY}. \tag{14.52}$$

When comparing this expression with the eigenproblem (14.43), it becomes obvious that the $\vec{a}$-vectors are the EOFs of $\widehat{\vec{\mathbf{Y}}}$. Thus the $\vec{a}^{\,1}$ coefficient accounts for the largest amount of $\widehat{\vec{\mathbf{Y}}}$ variance (i.e., λ_1), $\vec{a}^{\,2}$ accounts for the second largest amount of variance λ_2, and so on. The total variance of the regressed vector $\widehat{\vec{\mathbf{Y}}}$ is $\sum_j \lambda_j$. Since $\mathbf{\Sigma}_{Y\hat{Y}} = \mathbf{\Sigma}_{\hat{Y}\hat{Y}}$, we have

$$\begin{aligned} R^2(\vec{\mathbf{Y}} : \widehat{\vec{\mathbf{Y}}}) &= R^2(\vec{\mathbf{Y}} : \vec{\mathbf{X}}) \\ &= \frac{\mathrm{tr}(\mathbf{\Sigma}_{\hat{Y}\hat{Y}})}{\mathrm{tr}(\mathbf{\Sigma}_{YY})} = \frac{\sum_j \lambda_j}{\mathrm{tr}(\mathbf{\Sigma}_{YY})}. \end{aligned} \tag{14.53}$$

When we truncate (14.49) to the k components of $\vec{\mathbf{X}}$ that carry the most information about $\vec{\mathbf{Y}}$, we find that

$$R^2(\vec{\mathbf{Y}} : \widehat{\vec{\mathbf{Y}}}) = R^2(\vec{\mathbf{Y}} : \mathcal{B}_k^{\mathrm{T}}\vec{\mathbf{X}}) = \frac{\sum_{j=1}^{k}\lambda_j}{\mathrm{tr}(\mathbf{\Sigma}_{YY})}.$$

[14]The proof is straightforward: $\mathcal{R}\mathcal{P} = \mathbf{\Sigma}_{YX}\mathbf{\Sigma}_{XX}^{-1}(\mathcal{B}^{\mathrm{T}})^{-1} = \mathbf{\Sigma}_{YX}\mathbf{\Sigma}_{XX}^{-1}\mathbf{\Sigma}_{XX}\mathcal{B} = \mathbf{\Sigma}_{YX}\mathcal{B} = \mathcal{A}\mathcal{D}$.

14.4.8 Comparison with CCA. Let us now consider the special case in which $\mathbf{\Sigma}_{XX}$ and $\mathbf{\Sigma}_{YY}$ are both identity matrices. Then $\mathcal{B}$ and $\mathcal{P}$ are also identity matrices, and the regressed patterns $\vec{a}$, the EOFs $\widehat{\vec{\mathbf{Y}}}$, are the eigenvectors of $\mathbf{\Sigma}_{YX}\mathbf{\Sigma}_{XY}$. That is, $\vec{\mathbf{X}}$ provides the most information about the component of $\vec{\mathbf{Y}}$ that lies in the $\vec{a}^{\,1}$ direction, where $\vec{a}^{\,1}$ is the first eigenvector of $\mathbf{\Sigma}_{YX}\mathbf{\Sigma}_{XY}$. The best predictor of this component is $\vec{\mathbf{X}}^{\mathrm{T}}\mathbf{\Sigma}_{XY}\vec{a}^{\,1}$.

When we perform CCA on the same system we must solve the paired eigenvalue problem

$$\begin{aligned} \mathbf{\Sigma}_{XY}\mathbf{\Sigma}_{YX}\vec{f}_X &= \lambda\vec{f}_X \\ \mathbf{\Sigma}_{YX}\mathbf{\Sigma}_{XY}\vec{f}_Y &= \lambda\vec{f}_Y. \end{aligned}$$

The first pair of eigenvectors of this system is given by $\vec{f}_X = \mathbf{\Sigma}_{XY}\vec{a}^{\,1}$, and $\vec{f}_Y = \vec{a}^{\,1}$, indicating that $\vec{\mathbf{X}}^{\mathrm{T}}\mathbf{\Sigma}_{XY}\vec{a}^{\,1}$ is the $\vec{\mathbf{X}}$-component most strongly correlated with $\vec{\mathbf{Y}}^{\mathrm{T}}\vec{a}^{\,1}$.

Thus redundancy analysis and CCA are equivalent in this special case: both identify the same $\vec{\mathbf{X}}$ and $\vec{\mathbf{Y}}$ directions.

In general, however, the methods are not equivalent. Redundancy analysis finds the best predicted (or specified) components of $\vec{\mathbf{Y}}$ by finding the eigenvectors $\vec{a}$ of

$$\mathbf{\Sigma}_{YX}\mathbf{\Sigma}_{XX}^{-1}\mathbf{\Sigma}_{XY}$$

and then finding the patterns $\vec{p}$ of $\vec{\mathbf{X}}$-variations that carry this information. CCA, on the other hand, finds the most strongly correlated components of $\vec{\mathbf{Y}}$ by finding the eigenvectors $\mathbf{\Sigma}_{YY}^{-1/2}\vec{f}_Y$ of

$$(\mathbf{\Sigma}_{YY}^{-1/2})^{\mathrm{T}}\mathbf{\Sigma}_{YX}\mathbf{\Sigma}_{XX}^{-1}\mathbf{\Sigma}_{XY}\mathbf{\Sigma}_{YY}^{-1/2}.$$

That is, CCA does redundancy analysis on $\vec{\mathbf{Y}}' = (\mathbf{\Sigma}_{YY}^{-1/2})^{\mathrm{T}}\vec{\mathbf{Y}}$, the random vector that is obtained by projecting $\vec{\mathbf{Y}}$ onto its EOFs and scaling each component by its standard deviation. We can therefore anticipate that the two techniques will produce similar results if $\vec{\mathbf{Y}}$ is projected onto a small number of EOFs with similar eigenvalues.

14.4.9 Example: Interdecadal Variability of Intramonthly Percentiles of Significant 'Brent' Wave Height. We now describe an application in which we use redundancy analysis to specify monthly wave height statistics at the Brent oil field, located northeast of Scotland in the North Atlantic at (61° N, 1.5° E). Wave height (sea state) data are available from visual assessments made on ships of opportunity, at light houses, from wave rider buoys, and shipborne instruments at ocean weather stations. Also, wave height maps

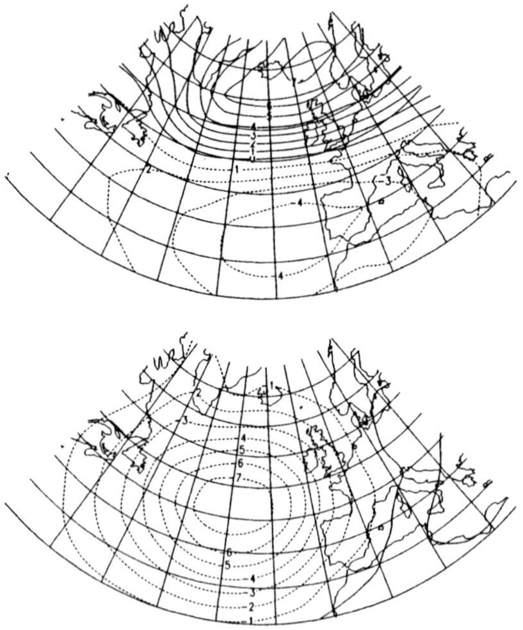

Figure 14.9: *First two monthly mean air-pressure anomaly distributions $\widehat{\vec{p}}^{\,k}$ identified in a redundancy analysis as being most strongly linked to simultaneous variations of the intramonthly percentiles of significant wave height in the Brent oil field (61° N, 1.5° E; northeast of Scotland).*

have been constructed from wind analyses for the purpose of ship routing (Bouws et al. [57]). These data are sparse and suffer from various inhomogeneities. Also, the records are generally too short to allow an assessment of changes during the past century.

Thus, observational data alone do not contain sufficient information about the interdecadal variability of wave statistics. One solution is a combined statistical/dynamical reconstruction of the past that uses a dynamical wave model. The model is forced with recent wind data that are believed to be fairly reliable and not strongly affected by improving analysis techniques.[15] The wave heights derived from the hindcast simulation are treated as observations and are used to build a statistical model linking the wave heights to surface air pressure. Finally, the resulting

	Wave height percentile		
	50%	80%	90%
$\widehat{\vec{a}}^{\,1}$	−81	−107	−114
$\widehat{\vec{a}}^{\,2}$	32	2	−25

Table 14.2: *The vectors $\widehat{\vec{a}}^{\,k}$ of anomalous intramonthly percentiles of significant wave height are given as rows in the following table. Units: cm.*

statistical model is fed with the observed air pressure from the beginning of the century onward, thereby producing a plausible estimate of wave height statistics for the entire century. The statistical model is presented below.

In this case we bring together 'apples' and 'oranges', that is, two vector quantities that are not directly linked. One vector time series, $\vec{\mathbf{X}}_t$, represents the winter (DJF) monthly mean surface air-pressure distributions in the North Atlantic. The other vector time series, $\vec{\mathbf{Y}}_t$, is a three-dimensional random vector consisting of the 50th, 80th, and 90th percentiles of the *intramonthly distributions* of significant wave height[16] in the *Brent* oil field at (61° N, 1.5° E). Both vector time series are assumed to be centred, so that the air-pressure values and percentiles are deviations from their respective long-term means.

The monthly mean of North Atlantic SLP is indirectly linked to the intramonthly percentiles, since storms affect both the monthly mean air-pressure distribution and the distribution of wave heights within a month at a specific location. Of course, the storm activity may also be seen as being conditioned by the monthly mean state.

The daily wave height data are taken from a 40-year 'hindcast' simulation (Günther et al. [153]). The following analysis assumes that the hindcasts and windfield analyses both represent the real world well enough for statistical relationships between the wave and wind fields on the monthly time scale to be reliably diagnosed.

A redundancy analysis of the two vector time series is performed to detect the dominant coupled anomaly patterns in the mean air pressure and in the intramonthly wave height percentiles. The SLP patterns $\widehat{\vec{p}}^{\,1}$ and $\widehat{\vec{p}}^{\,2}$ are shown in Figure 14.9 and the corresponding intramonthly percentiles $\widehat{\vec{a}}^{\,1}$ and $\widehat{\vec{a}}^{\,2}$ are listed in Table 14.2. The time coefficients are normalized to unit variance so that

[15]Note that the homogeneity of weather maps and their surface winds is difficult to assess. Analysis system improvements can introduce artificial signals, such as increasing frequencies of extreme events, into the hindcast. Improved analyses procedures, be it more or better observations or more intelligently designed dynamical and statistical analysis tools, lead to the emergence of more details in weather maps and, therefore, larger extremes.

[16]Significant wave height is a physical parameter that describes the wave field on the sea surface. The word 'significant' does not imply a significance test in this context. See [3.2.4].

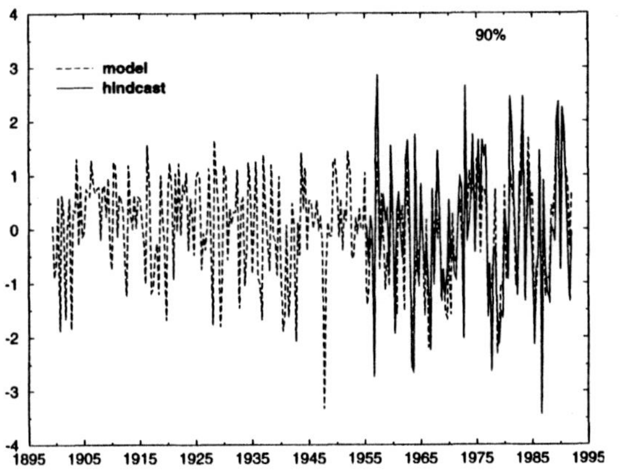

Figure 14.10: *Reconstructed (dashed line) and hindcasted (continuous line; 1955–94) anomalies of the 90th percentile of significant wave height at in the Brent oil field. Units: m.*

the three components of $\widehat{\vec{a}}^k$ may be interpreted as typical anomalies that occur when the pressure field anomalies are given by $\frac{1}{\sqrt{\lambda_k}}\widehat{\vec{p}}^k$.

The pattern $\vec{a}^1$ accounts for 94% of the variance of $\vec{\mathbf{Y}}$, and $\vec{a}^2$ for 5%. The correlation between the coefficient time series of the first pair of vectors is 0.84 while that between the second pair is only 0.08. Thus, the first pair establishes a regression representing $R^2(\vec{\mathbf{Y}} : \mathcal{B}_1^T\vec{\mathbf{X}}) = 94\% \times 0.84^2 = 66\%$ of the variance of $\vec{\mathbf{Y}}$, whereas the second pair represents only $5\% \times 0.08^2 < 0.1\%$ of variance. Thus the redundancy index for $k = 1$ (0.66) can not be usefully increased by adding a second vecor.

The first air-pressure pattern is closely related to the *North Atlantic Oscillation* (see [13.5.5] and Figure 13.6). A weakening of the NAO is associated with a decrease in all three intramonthly percentiles of significant wave height. In effect, this pattern describes a shift of the intramonthly distribution towards smaller waves.

The second pattern describes a mean south-easterly flow across the northern North Sea. The 50th percentile of the significant wave heights is increased by 32 cm, while the 90th percentile is reduced by 25 cm. Thus there is a tendency for the wave height distribution to be widened when pressure anomaly pattern $\widehat{\vec{p}}^2$ prevails. The reversed pattern goes with a narrowed intramonthly distribution of wave heights. This pair of patterns accounts for only 5% of the predictable wave height variance.

The regression model incorporated in the redundancy analysis was used to estimate the time series of the percentiles of significant wave height in the Brent oil field from the observed monthly mean air pressure anomaly fields between 1899 and 1994. The last 40 years may be compared with the hindcast data, whereas the first 50 years represent our best guess and can not be verified at this time. The 90th percentiles of the reconstructed wave height time series for 1899–94 and the corresponding hindcasted time series for 1955–94 are shown in Figure 14.10. The link appears to be strong, as is demonstrated by the correlations and the proportion of described variance, during the overlapping period:

	Wave height percentile		
	50%	80%	90%
Correlation	0.83	0.82	0.77
Described variance	0.70	0.66	0.60

The amount of percentile variance represented by the SLP patterns is consistent with the redundancy index (14.53), which has value 0.66. As with all regression models, the variance of the estimator is smaller than the variance of the original variable. This makes sense, since the details of the wave action in a month are not completely determined by the monthly mean air-pressure field. It is also affected by variations in surface wind that occur on shorter time scales.

15 POP Analysis

15.0.1 Summary. The *Principal Oscillation Pattern* (POP) analysis is a linear multivariate technique used to empirically infer the characteristics of the space-time variations of a complex system in a high-dimensional space [167, 389]. The basic approach is to identify and fit a linear low-order system with a few free parameters. The space-time characteristics of the fitted system are then assured to be representative of the full system.

This chapter is organized as follows. POPs are introduced as normal modes of a discretized linear system in Section 15.1. Three POP analyses are given in Section 15.2. Since a POP analysis includes the fitting of a time series model to data, the POP approach has predictive potential (Section 15.3). Cyclo-stationary POP analysis is explained in Section 15.4. Another generalization, the Hilbert or 'complex' POPs, is introduced briefly in [16.3.15].

POP models may also be viewed as simplified state space models. Such models, and in particular the Principal Interaction Pattern (PIP) ansatz[1] (Hasselmann [167]), are a fairly general approach which allow for a large variety of complex scenarios. The merits and limitations of this ansatz are discussed in Section 15.5.

15.0.2 Applications of POP Analysis. POP analysis is a tool [136] that is now routinely used to diagnose the space-time variability of the climate system. Processes that have been analysed with POPs include the Madden-and-Julian Oscillation (MJO; also called the 30–60 day oscillation) [388, 389, 399, 401], oceanic variability [275, 421], the stratospheric Quasi-Biennial Oscillation (QBO) [431], the El Niño/Southern Oscillation (ENSO) phenomenon [20, 50, 75, 242, 243, 429, 430, 432], and others, tropospheric baroclinic waves [341], and low-frequency variability in the coupled atmosphere–ocean system [431].

The POP method is not useful in all applications. If the analysed vector time series exhibits strongly nonlinear behaviour, as in, for example, the day-to-day weather variability in the extratropical atmospheric flow, a POP analysis will not be useful because a low-dimensional linear subsystem does not control a significant portion of the variability. The POP method will be useful if there are *a priori* indications that the processes under consideration are linear to first approximation.

15.1 Principal Oscillation Patterns

15.1.1 Normal Modes. The normal modes of a discretized real linear system

$$\vec{\mathbf{X}}_{t+1} = \mathcal{A}\vec{\mathbf{X}}_t \tag{15.1}$$

are the eigenvectors $\vec{p}$ of the matrix $\mathcal{A}$. In general, $\mathcal{A}$ is not symmetric and some or all of its eigenvalues λ and eigenvectors $\vec{p}$ are complex. However, since $\mathcal{A}$ is a real matrix, the complex conjugates λ^* and $\vec{p}^*$ are also eigenvalues and eigenvectors of $\mathcal{A}$.

The eigenvectors of $\mathcal{A}$ form a linear basis when all of its eigenvalues are nonzero. Thus any state $\vec{\mathbf{X}}$ may be uniquely expressed in terms of the eigenvectors as

$$\vec{\mathbf{X}} = \sum_j z_j \vec{p}^{\,j} \tag{15.2}$$

where the pattern coefficients z_j are given by the inner product of $\vec{\mathbf{X}}$ with the normalized eigenvectors $\vec{p}_a^{\,j}$ of $\mathcal{A}^{\mathrm{T}}$.[2]

[1]The word 'ansatz' is causing some confusion in the scientific community. In contrast with meteorologists and statisticians, theoretical physicists and non-statistical applied mathematicians are generally acquainted with this word. It is of German origin and means an 'educated guess' that may or may not lead to a successful line of analysis.

[2]The eigenvectors of $\mathcal{A}$ are linearly independent if all of the eigenvalues of $\mathcal{A}$ are distinct. Making this assumption, it is then easily shown that $\mathcal{A}^{\mathrm{T}}$ has the same eigenvalues as $\mathcal{A}$, and that the eigenvectors $\vec{p}_a^{\,j}$ of $\mathcal{A}^{\mathrm{T}}$ are columns of a matrix $(\mathcal{P}^{-1})^{\mathrm{T}}$, where the columns of $\mathcal{P}$ are the eigenvectors of $\mathcal{A}$. Then $\vec{\mathbf{X}}$ can be expanded as $\vec{\mathbf{X}} = \sum_j z_j \vec{p}^{\,j}$, where $z_j = \vec{p}_a^{\,j\mathrm{T}}\vec{\mathbf{X}}$, because

$$\sum_j z_j \vec{p}^{\,j} = \sum_j \vec{p}_a^{\,j\mathrm{T}}\vec{\mathbf{X}}\vec{p}^{\,j} = \Big(\sum_j \vec{p}^{\,j}\vec{p}_a^{\,j\mathrm{T}}\Big)\vec{\mathbf{X}} = \mathcal{P}\big((\mathcal{P}^{-1})^{\mathrm{T}}\big)^{T}\vec{\mathbf{X}} = \vec{\mathbf{X}}.$$

The eigenvectors $\vec{p}_a^{\,j}$ are called *adjoint patterns*.

Inserting (15.2) into (15.1), we find that the coupled system (15.1) becomes uncoupled, yielding m single equations,[3]

$$z_{t+1}\vec{p} = \lambda z_t \vec{p} \tag{15.3}$$

where m is the dimension of the process $\vec{\mathbf{X}}_t$. Thus, if $z_0 = 1$,

$$z_t \vec{p} = \lambda^t \vec{p}\,. \tag{15.4}$$

Now let $\vec{\mathbf{P}}_t$ be the vector

$$\vec{\mathbf{P}}_t = z_t\vec{p} + z_t^*\vec{p}^{\,*}. \tag{15.5}$$

Then

$$\vec{\mathbf{P}}_t = z_t^r\vec{p}^{\,r} + z_t^i\vec{p}^{\,i} \tag{15.6}$$

where $z_t^r = 2\mathrm{Re}\{z_t\}$, $z_t^i = -2\mathrm{Im}\{z_t\}$, $\vec{p}^{\,r} = \mathrm{Re}\{\vec{p}\,\}$, and $\vec{p}^{\,i} = \mathrm{Im}\{\vec{p}\,\}$. When $z_0 = 1$, we find, by substituting (15.4) into (15.6), that

$$\begin{aligned}\vec{\mathbf{p}}_0 &= \vec{p}^{\,r}\\ \vec{\mathbf{p}}_t &= \xi^t\big(\cos(\eta t)\vec{p}^{\,r} - \sin(\eta t)\vec{p}^{\,i}\big)\end{aligned} \tag{15.7}$$

where ξ and η satisfy $\lambda = \xi e^{-i\eta}$.

The geometrical and physical interpretation of (15.6) and (15.7) is as follows. When λ is complex, the corresponding eigenvector $\vec{p}$ is also complex, and $(\lambda^*, \vec{p}^{\,*})$ is also an eigenvalue/eigenvector pair. Thus the sum (15.5) describes two of the terms in expansion (15.2). Equation (15.6) shows that this sum is real and that it describes variations in a two-dimensional subspace of the full m-dimensional space that is spanned by the real and imaginary parts of $\vec{p}$. Equation (15.7) shows how system (15.1) evolves if its initial state is $\vec{p}^{\,r}$. If $\xi = 1$, then pattern $\vec{p}^{\,r}$ evolves into pattern $-\vec{p}^{\,i}$ in $\pi/(2\eta)$ time steps, then evolves to pattern $-\vec{p}^{\,r}$ at time π/η, and eventually returns to pattern $\vec{p}^{\,r}$ in period $T = 2\pi/\eta$. Schematically,

$$\cdots \to \vec{p}^{\,r} \to -\vec{p}^{\,i} \to -\vec{p}^{\,r} \to \vec{p}^{\,i} \to \vec{p}^{\,r} \to \cdots \tag{15.8}$$

In the real world, $\xi < 1$ (otherwise (15.1) would describe explosive behaviour; see below). Thus the amplitude of the sequence of patterns decays exponentially in time with an *e-folding time* $\tau = -1/\ln(\xi)$ so that $\vec{\mathbf{p}}_t$ of (15.7) evolves as the spiral displayed in Figure 15.1.

Note that any eigenvector $\vec{p}$ is determined up to a complex scalar α. To make things unique up to sign, one can choose α in such a way that $\vec{p}^{\,r}$ and $\vec{p}^{\,i}$ are orthogonal and $\|\vec{p}^{\,r}\| \geq \|\vec{p}^{\,i}\|$.

The modes may be represented either by the two patterns $\vec{p}^{\,r}$ and $\vec{p}^{\,i}$, or by plots of the local *wave amplitude* pattern $\vec{A}^2$ defined by $A_j^2 = (p_j^r)^2 + (p_j^i)^2$, $j = 1, \ldots, m$, and the local *relative phase* pattern $\vec{\psi}$ defined by $\psi_j = \tan^{-1}(p_j^i/p_j^r)$, $j = 1, \ldots, m$ (Figure 15.2a). The evolution depicted by (15.8) can describe a travelling wave form. For example, if we re-express $\vec{p}^{\,r}$ and $\vec{p}^{\,i}$ as functions of a location vector $\vec{r}$, it may turn out that $\vec{p}^{\,i}$ is just a translated version of $\vec{p}^{\,r}$ (i.e., that $p^i(\vec{r}) = p^r(\vec{r} - \vec{r}_\circ)$ for some displacement $\vec{r}_\circ$). If so, evolution (15.8) describes a wave train that propagates in the $\vec{r}_\circ$ direction and has wavelength $4\|\vec{r}_\circ\|^2$. Amphidromal (rotating) wave forms (Figure 15.2b) can also be represented by (15.8).

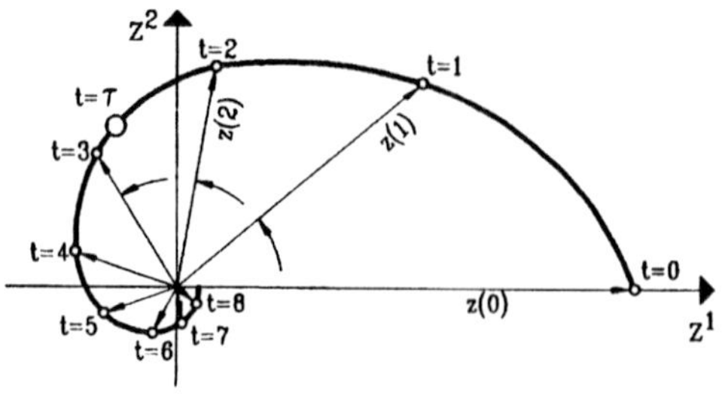

Figure 15.1: *Schematic diagram of the time evolution of POP coefficients z_t with an initial value $z_0 = (z^r, z^i) = (0, 1)$. The rotation time is slightly more than eight time steps. The e-folding time τ is indicated by the large open circle [400].*

15.1.2 POPs. The only information used so far is the existence of linear equation (15.1) and the assumption that coefficient matrix $\mathcal{A}$ has no repeated eigenvalues. No assumption was made about the origins of this matrix. In dynamical theory, equations such as (15.1) arise from linearized and discretized differential equations. In POP analysis, the state vector $\vec{\mathbf{X}}$ is assumed to satisfy a stochastic difference equation of the form

$$\vec{\mathbf{X}}_{t+1} = \mathcal{A}\vec{\mathbf{X}}_t + \text{noise} \tag{15.9}$$

Multiplying (15.9) on the right hand side by $\vec{\mathbf{X}}_t^{\mathrm{T}}$ and taking expectations leads to

$$\mathcal{A} = \mathcal{E}\big(\vec{\mathbf{X}}_{t+1}\vec{\mathbf{X}}_t^{\mathrm{T}}\big)\big(\mathcal{E}(\vec{\mathbf{X}}_t\vec{\mathbf{X}}_t^{\mathrm{T}})\big)^{-1}. \tag{15.10}$$

The normalized eigenvectors of (15.10) are called *Principal Oscillation Patterns*, and the coefficients z are called *POP coefficients*. Their time evolution is given by (15.3), except that it is forced by noise:

$$z_{t+1} = \lambda z_t + \text{noise}. \tag{15.11}$$

[3]Indices are dropped in the following for convenience.

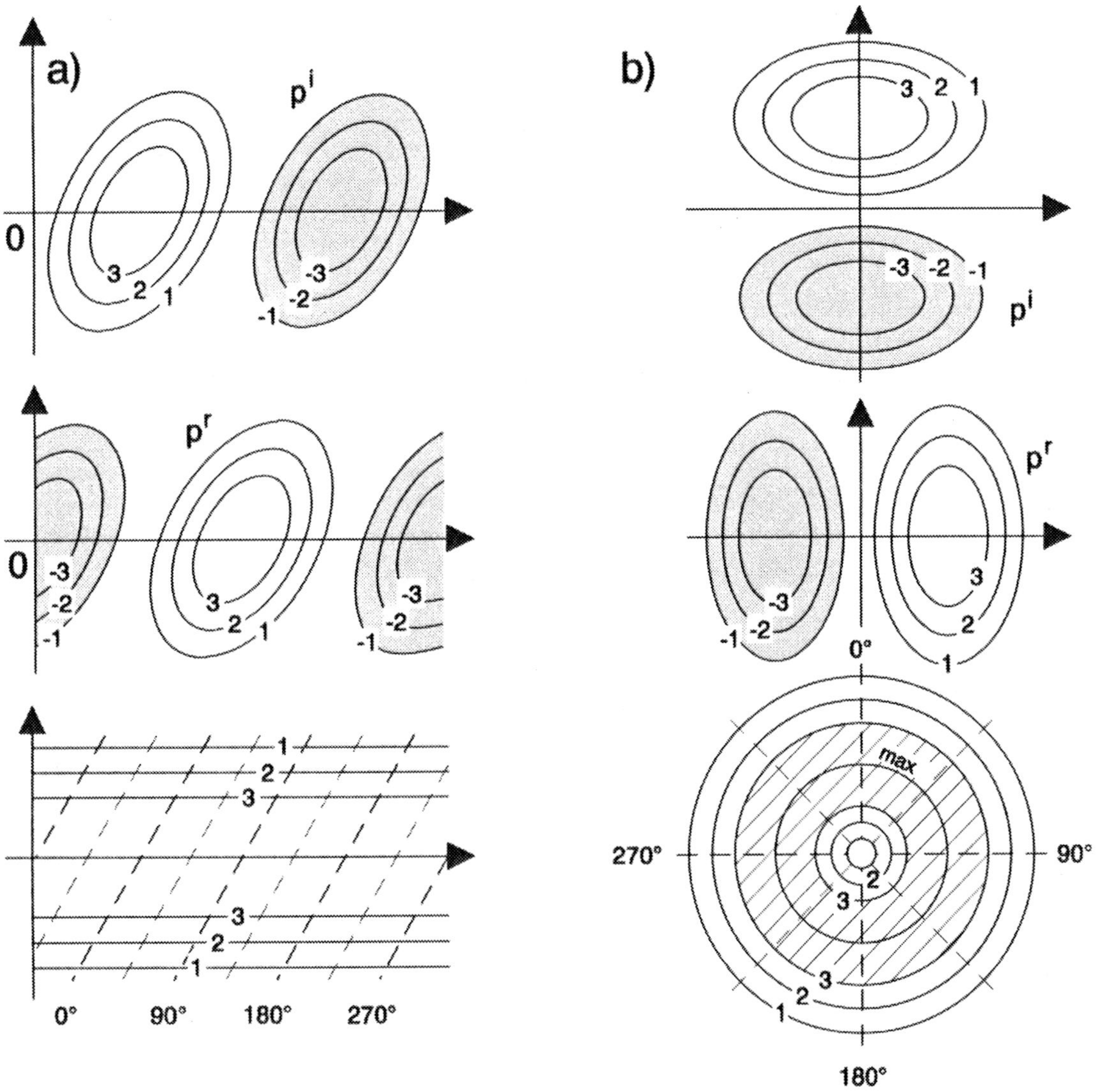

Figure 15.2: *Schematic examples representing a complex-valued POP $\vec{p} = \vec{p}^{\,r} + i\vec{p}^{\,i}$ with their imaginary and real parts parts $\vec{p}^{\,i}$ (top) and $\vec{p}^{\,r}$ (middle). The corresponding phase (ψ) and amplitude A are shown in the bottom panel.*
a) A linearly propagating wave is shown. If the initial state of the system is $\vec{\mathbf{P}} = \vec{p}^{\,i}$ (top), then its state a quarter of a period later will be $\vec{\mathbf{P}} = \vec{p}^{\,r}$ (middle). The wave propagates to the right with a constant phase speed (bottom), and the amplitude is constant along horizontal lines with maximum values in the centre.
b) A clockwise rotating wave is displayed. The evolution of the top pattern to the middle pattern takes one-quarter of a period. The amplitude (bottom) is zero in the centre, and the lines of constant amplitude form concentric circles around the centre. From [389].

The stationarity of (15.11) requires $|\lambda| < 1$ (see (10.12)).

15.1.3 Transformation of Coordinates. Suppose the original time series $\vec{\mathbf{X}}_t$ is transformed into another time series $\vec{\mathbf{Y}}_t$ by means of $\vec{\mathbf{Y}}_t = \mathcal{L}\vec{\mathbf{X}}_t$ with an invertible matrix $\mathcal{L}$. The eigenvalues are unchanged by the transform. The eigenvectors transform as $\vec{\mathbf{X}}$, and the adjoints are transformed by $(\mathcal{L}^{-1})^{\mathrm{T}}$, as

$$\begin{aligned} \vec{p}_Y &= \mathcal{L}\vec{p}_X \\ \vec{p}_{aY} &= (\mathcal{L}^{-1})^{\mathrm{T}}\vec{p}_{aX}. \end{aligned} \tag{15.12}$$

The POP coefficients are unaffected by the transformation because

$$(\vec{p}_{aY})^{\mathrm{T}}\vec{\mathbf{Y}} = (\vec{p}_{aX})^{\mathrm{T}}\mathcal{L}^{-1}\mathcal{L}\vec{\mathbf{X}} = (\vec{p}_{aX})^{\mathrm{T}}\vec{\mathbf{X}}.$$

15.1.4 Estimating POPs. In practice, when only a finite time series $\vec{\mathbf{x}}_1, \dots, \vec{\mathbf{x}}_n$ is available, $\mathcal{A}$ is estimated by first computing the sample lag-1 covariance matrix

$$\widehat{\Sigma}_1 = \frac{1}{n}\sum_t (\vec{\mathbf{x}}_{t+1} - \bar{\vec{\mathbf{x}}})(\vec{\mathbf{x}}_t - \bar{\vec{\mathbf{x}}})^{\mathrm{T}}$$

and the sample covariance matrix

$$\widehat{\Sigma}_0 = \frac{1}{n}\sum_t (\vec{\mathbf{x}}_t - \bar{\vec{\mathbf{x}}})(\vec{\mathbf{x}}_t - \bar{\vec{\mathbf{x}}})^{\mathrm{T}}$$

and then forming

$$\widehat{\mathcal{A}} = \widehat{\Sigma}_1\widehat{\Sigma}_0^{-1}. \tag{15.13}$$

The eigenvalues of this matrix always satisfy $|\lambda| < 1$ (i.e., $\xi < 1$).

In many applications the data are first subjected to a truncated EOF expansion to reduce the number of spatial degrees of freedom. POP analysis is then applied to the vector of the first EOF coefficients.[4] A positive byproduct of this procedure is that noisy components can be excluded from the analysis. Also, the sample covariance matrix $\widehat{\Sigma}_0$ is made diagonal.

It is often best to time-filter the data prior to the POP analysis if there is prior information that the expected signal is located in a certain frequency band. A somewhat milder way to focus on selected time scales is to derive the EOFs from time-filtered data, but to project the unfiltered data onto these EOFs. Note that the resulting sample covariance matrix is no longer diagonal.

Criteria for distinguishing between POPs that contain useful information and those that reflect primarily sample effects are given in [389]. The most important rule-of-thumb is related to the cross-spectrum of the POP coefficients z^r and z^i: The coefficient time series should vary coherently and be 90° out-of-phase in the neighbourhood of the POP frequency η.

15.1.5 Estimating POP Coefficients. Two approaches can be used to estimate the POP coefficients z_t. The straightforward approach is to

- compute the eigenvectors $\widehat{\vec{p}}$ of $\widehat{\mathcal{A}}$, (15.1),
- form $\widehat{\mathcal{P}} = (\widehat{\vec{p}}^{\,1}|\cdots|\widehat{\vec{p}}^{\,m})$, and
- compute the matrix of adjoint patterns $\widehat{\mathcal{P}}_a = (\widehat{\mathcal{P}}^{-1})^{\mathrm{T}}$.

However, because $\widehat{\mathcal{A}}$ is subject to some sampling variability, this approach will produce some POPs (those with near-zero eigenvalues and poorly organized spatial structure) that reflect mostly noise. These noisy POPs affect all of the adjoint patterns through the computation of $\widehat{\mathcal{P}}^{-1}$.

The solution to this problem is to add subjective judgement to the POP ansatz by using experience and physical knowledge to identify the POPs that are related to the dynamics of the system. The coefficients and adjoint patterns of these *useful* POPs can be estimated by least squares, essentially by assuming that the eigenvalues of the other POPs are zero.

Suppose, for simplicity, that there is only one useful POP. Then the POP coefficient can be estimated by minimizing

$$\|\vec{\mathbf{x}}_t - \widehat{z}_t^r\,\widehat{\vec{p}}^{\,r} - \widehat{z}_t^i\,\widehat{\vec{p}}^{\,i}\|^2 \tag{15.14}$$

if $\vec{p}$ is complex, or

$$\|\vec{\mathbf{x}}_t - z_t\widehat{\vec{p}}\,\|^2 \tag{15.15}$$

if $\vec{p}$ is real. The solution of (15.14) is

$$\begin{pmatrix}\widehat{z}_t^r\\ \widehat{z}_t^i\end{pmatrix} = \begin{pmatrix}\widehat{\vec{p}}^{\,r\mathrm{T}}\widehat{\vec{p}}^{\,r} & \widehat{\vec{p}}^{\,r\mathrm{T}}\widehat{\vec{p}}^{\,i}\\ \widehat{\vec{p}}^{\,r\mathrm{T}}\widehat{\vec{p}}^{\,i} & \widehat{\vec{p}}^{\,i\mathrm{T}}\widehat{\vec{p}}^{\,i}\end{pmatrix}^{-1}\begin{pmatrix}\widehat{\vec{p}}^{\,r\mathrm{T}}\\ \widehat{\vec{p}}^{\,i\mathrm{T}}\end{pmatrix}\vec{\mathbf{x}}_t \tag{15.16}$$

and that of (15.15) is

$$\widehat{z}_t = \frac{\widehat{\vec{p}}^{\,\mathrm{T}}\vec{\mathbf{x}}_t}{\widehat{\vec{p}}^{\,\mathrm{T}}\widehat{\vec{p}}}. \tag{15.17}$$

Note that (15.16) can be written in terms of estimated adjoint patterns as

$$\widehat{z}_t = \widehat{\vec{p}}_a^{\,\mathrm{T}}\vec{\mathbf{x}}_t$$

where $\widehat{\vec{p}}_a = \widehat{\vec{p}}_a^{\,r} + i\widehat{\vec{p}}_a^{\,i}$,

$$\begin{pmatrix}\widehat{\vec{p}}_a^{\,r\mathrm{T}}\\ \widehat{\vec{p}}_a^{\,i\mathrm{T}}\end{pmatrix} = \kappa\begin{pmatrix}\widehat{\vec{p}}^{\,i\mathrm{T}}\widehat{\vec{p}}^{\,i} & -\widehat{\vec{p}}^{\,r\mathrm{T}}\widehat{\vec{p}}^{\,i}\\ -\widehat{\vec{p}}^{\,r\mathrm{T}}\widehat{\vec{p}}^{\,i} & \widehat{\vec{p}}^{\,r\mathrm{T}}\widehat{\vec{p}}^{\,r}\end{pmatrix}\begin{pmatrix}\widehat{\vec{p}}^{\,r\mathrm{T}}\\ \widehat{\vec{p}}^{\,i\mathrm{T}}\end{pmatrix}$$

and $\kappa = \big((\widehat{\vec{p}}^{\,r\mathrm{T}}\widehat{\vec{p}}^{\,r})(\widehat{\vec{p}}^{\,i\mathrm{T}}\widehat{\vec{p}}^{\,i}) - (\widehat{\vec{p}}^{\,r\mathrm{T}}\widehat{\vec{p}}^{\,i})^2\big)^{-1}$. Equation (15.17) can also be interpreted as a projection onto an estimated adjoint pattern. When there are two or more useful POPs, the coefficients are estimated *simultaneously* by minimizing

$$\|\vec{\mathbf{x}}_t - \textstyle\sum_j \widehat{z}_{jt}\vec{p}_j\|^2$$

where the sum is taken over the useful POPs.

[4] Note, however, that there is a small cost. The results of such a POP analysis will generally change if the data are transformed to another coordinate system since EOFs are invariant only under orthonormal transformations.

15.1.6 Associated Correlation Patterns. The POP coefficients can often be regarded as an *index* of some process, such as the MJO or ENSO. It is then often desirable to be able to relate the index to other fields. This can be achieved by means of the *associated correlation patterns* [389] discussed in [17.3.4]. The MJO is presented as an example in [17.3.5].

15.1.7 POPs and Hilbert EOFs. The POP method is an approach for identifying modal structures in a vector time series that has been demonstrated to work well in real applications. There are certainly other techniques that can be used successfully for similar purposes. An alternative is *Hilbert Empirical Orthogonal Function* analysis [19, 408].[5] The Hilbert EOFs of a field $\vec{\mathbf{X}}_t$ are EOFs of the complex vector field that has $\vec{\mathbf{X}}_t$ as its real part and the Hilbert transform of $\vec{\mathbf{X}}_t$ as its imaginary part.[6]

The main differences between Hilbert EOFs and POPs are that Hilbert EOFs are orthogonal and they maximize explained variance. The proportion of variance represented by the POP is not optimal, and it must be diagnosed from the POP coefficients after the POP analysis has been completed. Another difference is that the period and e-folding time (i.e., damping rate) are not an immediate result of the Hilbert EOF analysis; they must be derived empirically from the Hilbert EOF coefficient time series. The POPs, on the other hand, are constructed to satisfy a dynamical equation, and the characteristic times are an output of the analysis. A third difference is that the POP coefficients z_t are not pairwise orthogonal. This makes the mathematics less elegant, but it is not a physical drawback because there is usually no reason to assume that different geophysical processes are stochastically independent of each other.

15.1.8 POPs as Multivariate Spectral Analysis. The power spectrum of the POP coefficients, $\Gamma_{zz}(\omega)$, is determined by the eigenvalue λ and the power spectrum $\Gamma_{nn}(\omega)$ of the noise:

$$\Gamma_{zz}(\omega) = \frac{\Gamma_{nn}(\omega)}{|e^{i\omega} - \lambda|^2}. \tag{15.18}$$

Assume that $\lambda = \xi e^{-i\eta}$ and that the noise is approximately white, that is, $\Gamma_{nn}(\omega) \approx$ constant. Then the power spectrum of z_t has a single maximum at frequency $\omega = \eta$ that is different from zero when λ is complex. The width of the spectral peak is determined by η. As ξ becomes smaller, the spectrum becomes broader (in the limit as $\xi \to 0$, the spectrum becomes white).

Thus, the POP analysis yields a multivariate AR spectral analysis of a vector time series [167]. A first attempt to simultaneously derive several signals with different spectra from a high-dimensional data set was made by Xu [431]. For a more complete discussion of the POP technique as a type of multivariate spectral analysis, refer to J. von Storch [405].

[5] *Hilbert EOFs* are frequently referred to as *Complex* EOFs in the climate literature. However, the term 'complex EOFs' is a misnomer (see [16.1.1]).

[6] For details, see Section 16.2.

15.2 Examples

15.2.1 Overview. Three examples of POP analysis are presented in this section. The purpose of the first example, which is of the tropospheric baroclinic waves [341], is to demonstrate the normal mode interpretation of the POPs. The best defined POP coincides, to good approximation, with the most unstable modes obtained from a conventional stability analysis of the linearized dynamical equations. The other two examples show that the POP analysis can detect signals in different situations. A joint POP analysis of tropospheric and stratospheric data [430] identifies two independent modes with similar time scales, the Southern Oscillation (SO) and the Quasi-Biennial Oscillation (QBO). A POP analysis of the Madden-and-Julian Oscillation (MJO; [401]), shows that its signal has a well-defined signature all along the equator. We will see that this is a very robust signal. It is possible to detect the signal in data that are restricted to 90° subsectors on the equator, and in two-year sub-samples of the full five-year data set.

15.2.2 Tropospheric Rossby Waves, from POP and Stability Analyses. POPs can be seen as empirical estimates of the normal modes of a linear approximation to a dynamical system. The estimated normal modes are the eigenvectors of a matrix $\widehat{\mathcal{A}}$ (15.13). An alternative to estimating $\mathcal{A}$ is to derive it by linearizing the dynamical equation that governs the system. The eigenmodes of the linearized system can then be computed directly.

Schnur et al. [341] compared these two approaches in the context of the tropospheric baroclinic waves that are responsible for much of the high-frequency atmospheric variability

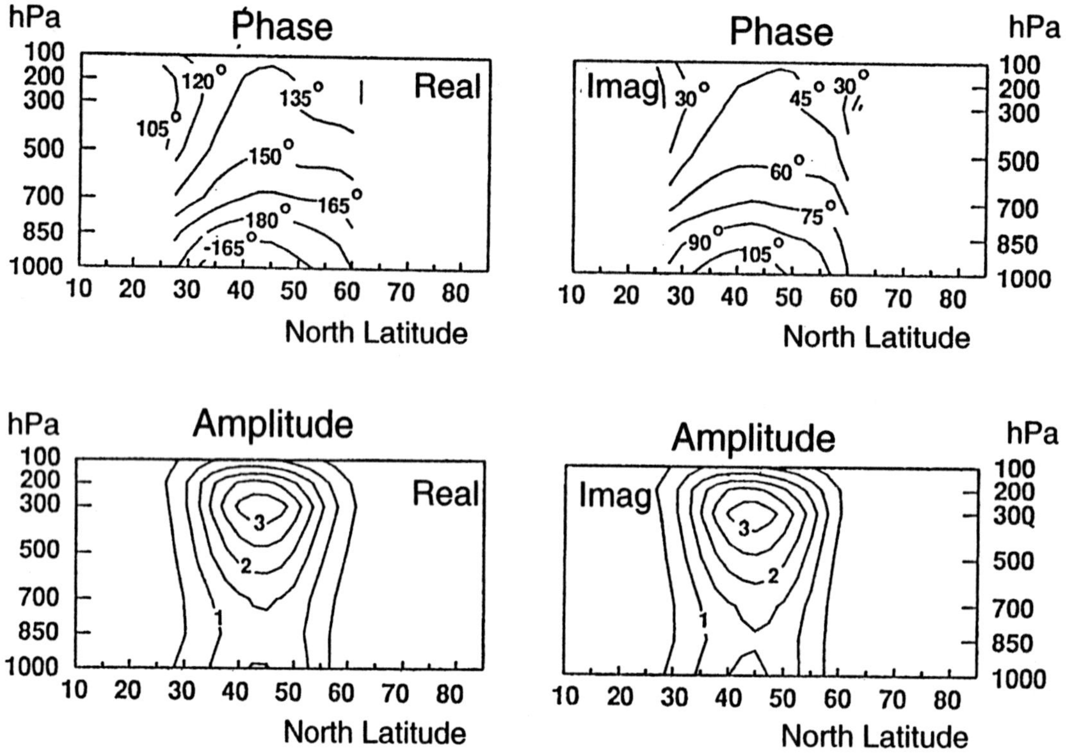

Figure 15.3: *Baroclinic waves: The Northern Hemisphere zonal wavenumber 8 POP. This mode represents 54% of the total zonal wavenumber 8 variance in the 3–25 day time scale. The oscillation period is 4 days and the e-folding time is 8.6 days. The amplitude A^r and A^i (bottom) and phase $\vec{\Xi}^r$ and $\vec{\Xi}^i$ (top) of the real and imaginary parts of the POP are shown (see text). From Schnur et al. [341].*

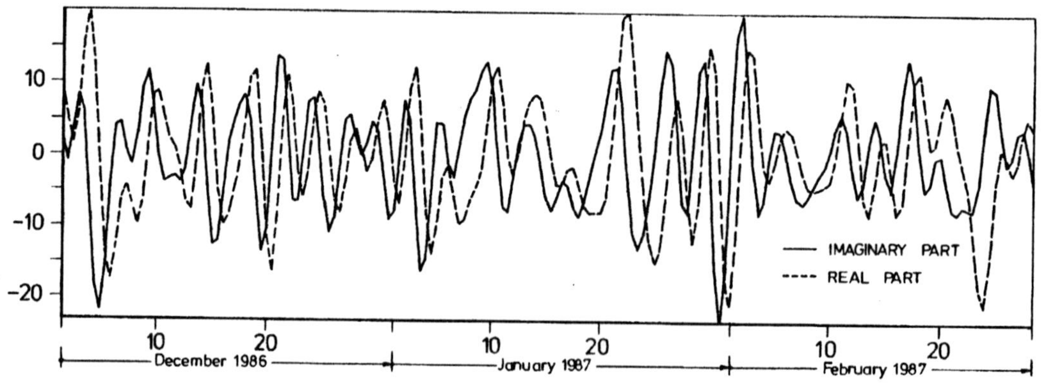

Figure 15.4: *Baroclinic waves: The coefficient time series $\hat{z}^r$ (dashed) and $\hat{z}^i$ (solid) of the POP shown in Figure 15.3. From Schnur et al. [341].*

in midlatitudes. A POP analysis of twice-daily geopotential heights at various tropospheric levels and a conventional linear stability analysis of the quasi-geostrophic vorticity equation were compared. Both analyses are expected to detect signals that propagate more or less zonally on a symmetric mean state. Such waves can be represented in a semi-spectral form as

$$\Psi(\theta,\phi,z,t) = \Psi_1(\phi,z,t)\cos(k\theta) + \Psi_2(\phi,z,t)\sin(k\theta) \quad (15.19)$$

where θ is longitude, ϕ is latitude, z represents height, and k is the zonal wavenumber. Note that (15.19) can be re-expressed as

$$\begin{aligned}\Psi(\theta,\phi,z,t) &= \mathrm{Re}\big((\Psi_1(\phi,z,t)+i\Psi_2(\phi,z,t)) \\ &\qquad \times e^{-ik\theta}\big) \\ &= \mathrm{Re}\big(A(\phi,z,t)e^{-i(k\theta-\Xi(\phi,z,t))}\big)\end{aligned} \quad (15.20)$$

where $A(\phi,z,t)$ and $\Xi(\phi,z,t)$ are the amplitude and phase of zonal wavenumber k. This representation will be used in the diagrams.

A separate POP analysis was performed for each wavenumber k on the random vector composed of trigonometric coefficients Ψ_1 and Ψ_2 of geopotential height at all latitudes and heights. The system matrix $\mathcal{A}$ of (15.1) was estimated from winter (DJF) observations for 1984/85 through 1986/87.

The data were band-pass filtered to remove variability on time scales shorter than 3 days and longer than 25 days. Also, the dimensionality of the problem was reduced by using a truncated EOF expansion. The first 18 EOFs, which represent more than 95% of the total variance for each wavenumber, were retained.

Here we discuss only the POP obtained for Northern Hemisphere wavenumber 8. The POP represents 54% of the wavenumber 8 variance, has a period of 4.0 days, and an e-folding time of time 8.1 days. Note that the decay time is sensitive to the type of time-filter.

Since the state vector $\vec{\mathbf{X}}_t$ consists of the sine and cosine coefficients of zonal wavenumber 8, both the real and the imaginary part of the complex POP, $\vec{p} = \vec{p}^{\,r} + i\,\vec{p}^{\,i}$, must also be interpreted as vectors of sine and cosine coefficients. These, in turn, can be represented as amplitude patterns A^r and A^i composed of amplitudes $A^r(\phi,z)$ and $A^i(\phi,z)$, respectively, and corresponding phase patterns $\vec{\Xi}^r$ and $\vec{\Xi}^i$. These patterns are shown in Figure 15.3 as height-latitudinal distributions. The amplitude fields $\vec{A}^r$ and $\vec{A}^i$ are almost identical, and the phase distribution $\vec{\Xi}^r$ is shifted 90° eastward relative to $\vec{\Xi}^i$ at those latitudes where the amplitudes are large. We therefore conclude that the diagnosed POP describes an eastward travelling pattern.

The estimated coefficient time series $\widehat{z}_t^r$ and $\widehat{z}_t^i$ vary coherently, with $\widehat{z}_t^r$ lagging $\widehat{z}_t^i$ by one or two days (Figure 15.4). This visual interpretation is substantiated by the cross-spectral analysis[7] of the two coefficient time series (Figure 15.5).

[7] Spectral and cross-spectral estimation techniques are described in Sections 12.3 and 12.5.

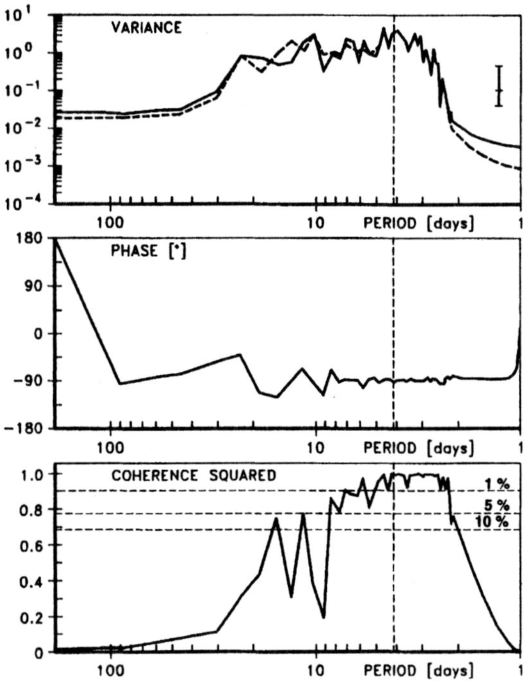

Figure 15.5: *Baroclinic waves: Cross-spectral analysis of the POP coefficient time series shown in Figure 15.4. The vertical dashed line marks the POP period T. The horizontal dashed lines in the coherence plot (bottom) depict critical values for tests of no coherence null hypothesis at the 10%, 5%, and 1% significance levels. From Schnur et al. [341].*

The maximum variance is found in the three- to five-day time scale, the phase difference is uniformly 90°, as it should be, and the coherence is high in the neighbourhood of the POP period of four days.

The system matrix $\mathcal{A}$ in (15.1) can also be obtained from theoretical considerations. Schnur et al. [341] did this by using a standard perturbation analysis to linearize the quasi-geostrophic vorticity equation on a sphere around the observed zonally averaged mean winter state. The linearized system was then discretized. The resulting system equation for the streamfunction Ψ was expressed in the form of (15.1) by using representation (15.19) for the streamfunction Ψ for each wavenumber k and forming the (unknown) state vector $\vec{\mathbf{X}}$ from Ψ_1 and Ψ_2 as above.

The resulting system matrix $\mathcal{A}$ has complex eigenvectors $\vec{q} = \vec{q}^{\,r} + i\,\vec{q}^{\,i}$. The complex eigenvalue that is connected with the pattern $\vec{q}$ can be written as $\lambda = \xi e^{-i\eta}$, where $T = 2\pi/\eta$ is the period of a cyclical sequence like (15.8) involving the real and imaginary parts of $\vec{q}$, and where the value of ξ determines whether the system amplifies or damps these oscillations. Thus, as with POP analysis, the normal modes represent propagating waves. The phase direction depends on the eigenvalue.

However, there are also important differences between the POP and perturbation analysis techniques. We mentioned that POP analysis of stationary data yields eigenvalues $|\lambda| < 1$. POP analysis based on the estimated matrix $\mathcal{A}$ preferentially 'sees' oscillations in their mature state (i.e., when noise is comparatively small and when there is damping by nonlinear and other processes). In contrast, the system matrix $\mathcal{A}$ obtained from perturbation analysis describes the early evolution of small deviations from a specified basic state. This system will amplify many of these initial perturbations, and these are in fact the solutions that are of interest. Thus it is the modes with eigenvalues $|\lambda| \geq 1$ that describe the growing oscillations that the POP analysis eventually detects.

Just as with POPs, both $\vec{q}^{\,r}$ and $\vec{q}^{\,i}$ can be represented by amplitude and phase patterns. However, since the system matrix depends only on a zonally averaged basic state, the solutions must be invariant with respect to zonal rotation (unlike the POPs). It can therefore be shown that $\vec{q}^{\,r}$ and $\vec{q}^{\,i}$ have equal amplitude and that the phase of $\vec{q}^{\,i}$ is just that of $\vec{q}^{\,r}$ shifted by $-90°$. That is, $\vec{q}^{\,i}$ is redundant.

The most unstable normal mode (i.e., with the greatest eigenvalue $|\lambda| \geq 1$) obtained for Northern Hemisphere wavenumber 8 has a period of 3.9 days. This is an eastward propagating growing mode that increases amplitude e-fold in 2.2 days.

The amplitude pattern $\vec{A}^r$ (Figure 15.6) of this normal mode is almost identical to the amplitude patterns of the POP shown in Figure 15.3. The normal mode has a large maximum near the surface at 40° N because the perturbation analysis did not account for friction. The phase pattern $\vec{\Xi}^r$ differs from the POP phases $\vec{\Xi}^i = \vec{\Xi}^r - \pi/2$ by only a constant angle.

In summary, POP analysis, which estimates the system matrix from observations, finds modes similar to those found by conventional perturbation analysis, which obtains the matrix from first-principle dynamical reasoning.

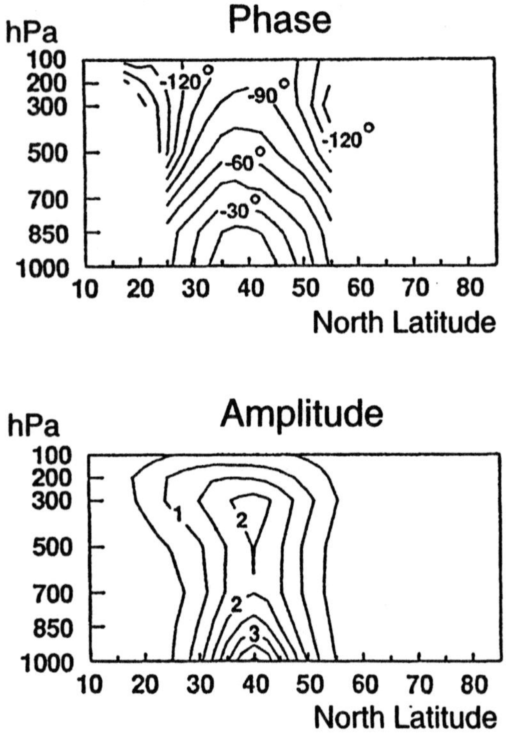

Figure 15.6: *Baroclinic waves: The amplitude and phase of the fastest growing Northern Hemisphere zonal wavenumber 8 normal mode. The mode was obtained from a perturbation analysis of the discretized quasi-geostrophic vorticity equation linearized about the observed zonal mean state in Northern winter. The amplitude grows e-fold in 2.2 days, and the period is 3.9 days. From Schnur et al. [341].*

15.2.3 The Southern Oscillation and the Quasi-Biennial Oscillation.

In this subsection we describe how POP analysis was used by Xu [430] to examine two oscillations in the tropical atmosphere with similar oscillation period: the stratospheric *Quasi-Biennial Oscillation* (QBO) and the tropospheric Southern Oscillation (SO).

The QBO can be observed in the stratospheric equatorial zonal wind with time series available at six stratospheric levels. POP analysis was performed on deviations from the long-term mean. No time-filtering was done for this data set.

Monthly mean anomalies of the 10 m zonal wind along the equator between 50° E and 80° W and of the equatorial sea-surface temperature (SST) anomalies are used to describe the SO signal. These data were low-pass filtered to remove variability on time scales shorter than 15 months.

The three equatorial data sets, stratospheric wind, zonal surface wind, and SST, were subjected to a joint POP analysis. The three components were normalized so that they contributed equal amounts of variance to the combined data set.

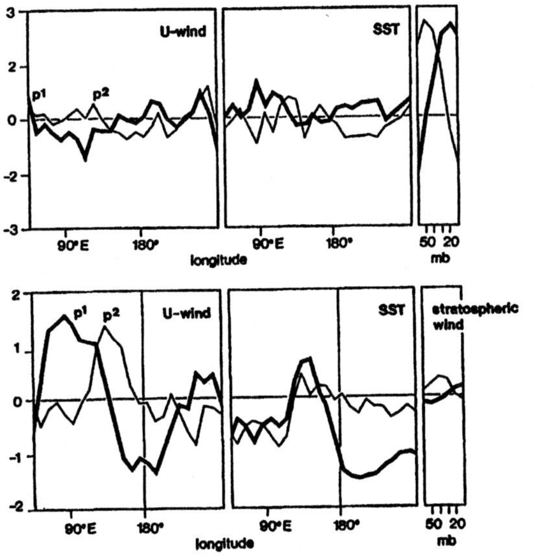

Figure 15.7: *QBO and SO: Two POPs obtained from a joint POP analysis of zonal 10 m wind, SST, and stratospheric zonal wind. The real part of each POP (light curve) is labelled* p^2 *and the imaginary part (heavy curve),* p^1*. From Xu [430].*
Top: The 28-month mode representing the Quasi-Biennial Oscillation (QBO),
Bottom: The 30-month mode representing the Southern Oscillation (SO).

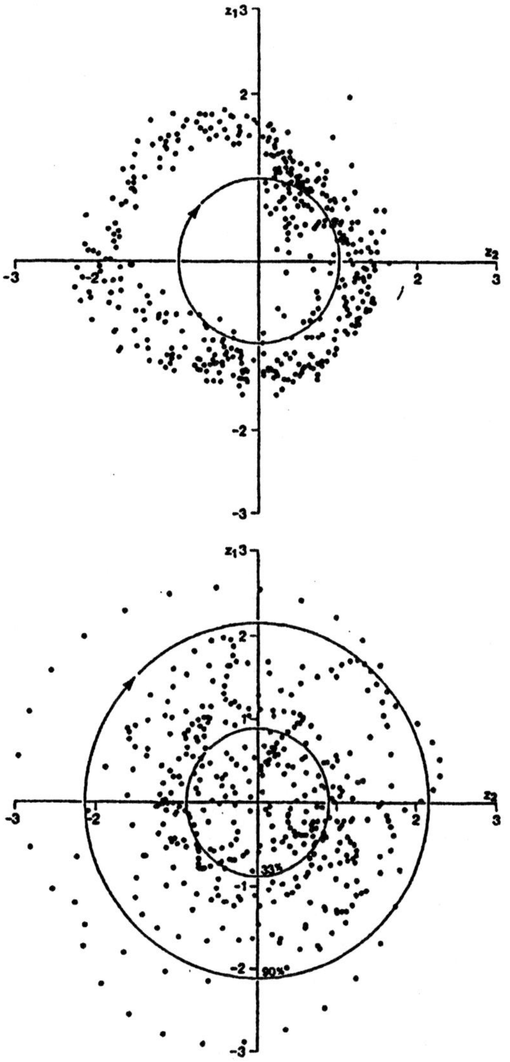

Figure 15.8: *QBO and SO: scatter plots of the complex POP coefficients associated with the patterns shown in Figure 15.7. From Xu [430].*
Top: The coefficients of the QBO mode.
Bottom: The coefficients of the SO mode.

Two significant POP pairs were found, one with an oscillation period of 28 months, and the other with a period of 45 months. Cross-spectral analysis of the POP coefficients (not shown) indicates that the 28-month period is reliably estimated, but that the period of the '45-month' POP is overestimated. A more realistic estimate of its oscillation period is approximately 30 months. The two modes are shown in Figure 15.7.

The first mode (Figure 15.7, upper panel) carries useful information only in the stratosphere where it represents the downward propagation of a signal from the upper stratosphere to the lower stratosphere over a 14-month period. The POP coefficient time series oscillates regularly (not shown), and occupies a torus-shaped region in phase space (Figure 15.8, top).

The second mode, on the other hand, only carries useful information at the surface in the SST and 10 m zonal wind. It describes a 10 m wind signal that propagates eastward from the Indian Ocean into the Pacific, and an almost stationary feature of SST variability. The POP coefficient time series sometimes oscillate regularly, and the occurrence of El Niño and La Niña events coincide with the oscillatory intervals. When the Southern Oscillation is quiet, the POP coefficients are small and noisy. The POP coefficients have a unimodal distribution in phase space (Figure 15.8, bottom).

These modes represent the QBO and the SO, respectively. They are essentially uncorrelated.

15.2.4 The Madden-and-Julian Oscillation: Sensitivity to Analysis Time-interval and Analysis Area. The Madden-and-Julian Oscillation (MJO), also known as the tropical 30–60 day oscillation, is particularly well represented in equatorial tropospheric velocity potential. This subsection describes a POP analysis of five years NMC[8]-analysed 200 hPa velocity potential from which the annual cycle was removed. The data cover the period May 1984 to April 1989.

Six POP analyses were performed in total on various subsets of the data (see [401]). Two analyses, 'A' and 'B', use data along the entire equator. 'A' uses a two-year subset and 'B' uses the whole five-year data set. Four additional analyses, labelled 'C' to 'F', use spatial subsets of the data that extend over the full five years. 'C' uses data between 0° and 90° W, 'D' uses data from 90° W to the date line, and so on.

One physically important POP was identified in each of the six analyses. The POPs from analyses 'B' to 'F' were rotated so that their $\vec{p}^{\,r}$ patterns match that obtained from analysis 'A' as closely as possible.[9]

The POP obtained in the 'A'-analysis has a period of 44 days, and an e-folding time of 13 days (about 30% of the period). The squared coherency of the POP coefficients is larger than 68% on time scales between 20 and 50 days with a maximum value of 96% at 50 days. The real and imaginary parts of the POP are shown as solid lines in Figure 15.9a. They are zonal wavenumber 1 type patterns with one minimum and one maximum. The two patterns are about 90° out-of-phase, indicating eastward propagation of the signal. The trough and the crest do not move at a constant rate.

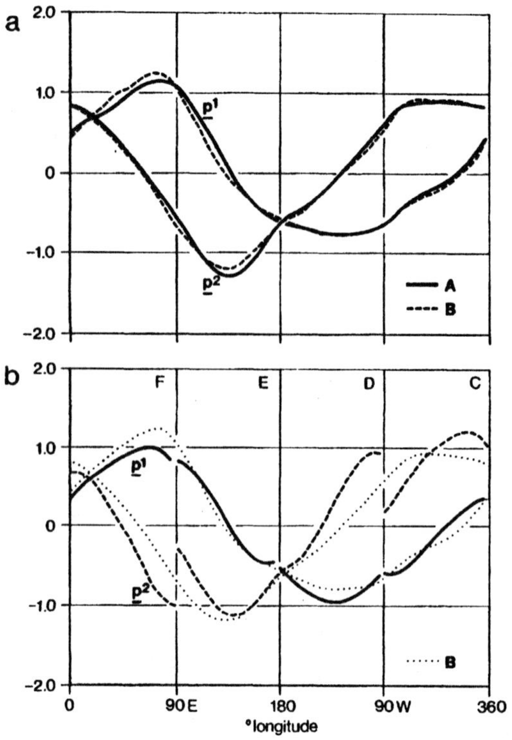

Figure 15.9: *MJO: The real (labelled p^1) and imaginary (labelled p^2) POPs of equatorial 200 mb velocity potential. From von Storch and Xu [401].*

a) Analysis of equatorial data from a two-year subset (analysis 'A'; solid line) and from the complete five-year data set (analysis 'B'; dashed line).

b) The 'C' to 'F' analyses for 90°-sectors along the equator. The real patterns are plotted with a solid line, and the imaginary patterns with a dashed line. Patterns from analysis 'A' are shown in dots for comparison.

The pattern in Figure 15.9a is very robust: the extra three years of data in the 'B' analysis (dashed curve) resulted in very little change.

Data in adjacent 90° sectors were considered in analyses 'C' to 'F'. The 90°-sector patterns resemble the full 360° patterns (Figure 15.9b) closely. The $\widehat{\vec{p}}^{\,r}$ patterns appear to match their 'A' counterpart somewhat better than the $\widehat{\vec{p}}^{\,i}$ patterns because the rotation was optimized on the former.

The e-folding times in the 90° sectors are considerably smaller than in 'A' and 'B'. This difference is reasonable, since the POPs describe a global, travelling feature. Thus the memory in the system is retained for a longer time in the full 360° circle than in the 90° sectors. Interestingly, the damping time in the eastern hemisphere (7 days) is about double that in the western hemisphere (4 days). This finding is consistent with the observation that the 30- to 60-day oscillation is markedly stronger in the eastern hemisphere.

The differences in the periods in the four 90° sectors are consistent with the variable longitudinal phase speed of the MJO. The 30–60 day waves travel most slowly in the 90° E to 180° sector: the period in this sector was found to be 62 days. The waves travel most quickly, and the period is shortest (33 days), in the 180° to 90° W sector. The average period for analyses 'C' to 'F' is 45 days, which is nearly identical to the value obtained in analyses 'A' and 'B'. Thus 'C' to 'F' further emphasize the robustness of the MJO signal extracted using the POP method.

[8]National Meteorological Center.

[9]That is, $\widehat{\vec{p}} = \widehat{\vec{p}}^{\,r} + i\,\widehat{\vec{p}}^{\,i}$ was multiplied by $e^{i\theta}$ for a suitably chosen θ. This is acceptable since eigenvector $\widehat{\vec{p}}$ of $\widehat{\mathcal{A}}$ can be uniquely determined only up to a factor $e^{i\theta}$.

15.3 POPs as a Predictive Tool

15.3.1 The POP forecast technique. Forecasting is a natural part of the POP ansatz (see, e.g., [432, 429, 401]), because the POP coefficient time series evolve similarly to AR(1) processes (i.e., as in (15.11)). Assuming that the forcing noise in (15.11) is white, the optimal lag-τ forecast of $z_{t+\tau}$ from z_t is given by

$$\widehat{z}^F_{t+\tau} = \xi^\tau e^{-i\frac{2\pi\tau}{T}} z_t \tag{15.21}$$

where $T = 2\pi/\eta$ is the period of the POP and $\xi = |\lambda|$. Equation (15.21) describes a damped persistence forecast in the complex plane (Figure 15.1) which corresponds to a damped propagating mode in physical space. Forecasts are made by identifying the current state of the POP coefficient process and then applying (15.21). Depending upon whether the practitioner thinks a mature or growing oscillatory mode has been detected, the forecast will either be a damped persistence forecast (i.e., $\xi = |\lambda| < 1$ in (15.21)) or a persistence forecast in terms of amplitude (i.e., $\xi = 1$ in (15.21)). These forecasts will have some skill at short leads, but at longer lead times the built-in linearity of the POP analysis, as well as the unpredictable noise, will result in a deterioration of forecast skill.

A basic limitation of POP forecasts is that, although they can predict the regularly changing phase of the oscillation, they cannot predict an intensification of amplitude. However, a phase forecast is valuable even if the amplitude is not well predicted.

Forecasting is complicated by the substantial amount of noise in the analysed field, resulting in estimates of the POP coefficient that may not be very reliable on a given day. Thus some sort of 'initialization' is necessary. 'Time filtering' initialization [432] uses a one-sided digital filter to suppress variance on short time scales before estimating the POP coefficient in the usual way. 'Time averaging' initialization begins with direct estimates of the POP coefficients realized at the last few time steps, say $\widehat{z}_t, \widehat{z}_{t-1}, \ldots, \widehat{z}_{t-\tau}$. Then (15.21) is used to produce a one-lag ahead forecast $\widehat{z}_t^{F_1}$ of $\widehat{z}_t$ from $\widehat{z}_{t-1}$, a two-lag forecast $\widehat{z}_t^{F_2}$ of $\widehat{z}_t$ from $\widehat{z}_{t-2}$, and so on. Finally, an improved estimate of z_t is obtained by computing a weighted average of $\widehat{z}_t, \widehat{z}_t^{F_1}, \ldots, \widehat{z}_t^{F_\tau}$. More weight is given to the recent information than the older information.

Small POP coefficients that move irregularly in the two-dimensional phase space indicate that the process represented by the POP is not active, in which case it is reasonable not to rely on the formal POP forecast. An appropriate POP forecast in this case is that the system will stay in its 'quiet phase.'

15.3.2 Measures of Skill. The quality of the POP forecasts can be determined with the correlation skill score ρ_τ (18.3) and the root mean square error $\mathcal{S}_\tau$ (18.1)[10]

$$\rho_\tau = \frac{\mathrm{Cov}(\widehat{z}_t^{F_\tau}, \widehat{z}_t)}{\sqrt{\mathrm{Var}(\widehat{z}_t^{F_\tau})\mathrm{Var}(\widehat{z}_t)}} \tag{15.22}$$

$$\mathcal{S}_\tau = \sqrt{\mathcal{E}(|\widehat{z}_t^{F_\tau} - \widehat{z}_t|)}, \tag{15.23}$$

where, as above, $\widehat{z}_t^{F_\tau}$ is the (complex) forecast of z_t made at time $t - \tau$, and $\widehat{z}_t$ is the estimated state at time t that is used to verify the forecast. Note that the diagnosed forecast skill depends upon the skill of both $\widehat{z}^{F_\tau}$ and $\widehat{z}_t$ as estimators of z_t.

The correlation skill score ρ_τ is an indicator only of phase errors since it is insensitive to amplitude errors. This makes ρ_τ a suitable skill score for POP forecasts since we anticipate that most of their utility lies in the phase component. The mean squared error $\mathcal{S}_\tau$, which is sensitive to both phase and amplitude errors, tends to be less flattering of POP forecasts.

The skill of the POP forecast is put in perspective by comparing with the skill of the *persistence forecast* $\widehat{z}_t^{P_\tau} = \widehat{z}_{t-\tau}$, which freezes patterns in time and space. As shown by (15.21), persistence and POP forecasts are close neighbours in the hierarchy of forecast schemes. Thus comparison of their skills is well justified.

15.3.3 Example: The Madden-and-Julian Oscillation. The skill of the POP forecasts of the MJO (see [15.2.4]) was examined in [388, 401]. The forecasts were initialized with the 'time averaging' technique using information from days 0 through −4 (i.e., $l = 4$; see [15.3.1]). The POP amplitude $|z_t|$ was predicted by persistence (i.e., $\xi = 1$ in (15.21)).

Individual forecasts are presented as *harmonic dials* that display the evolution of the POP coefficients before and after the forecast date, and the forecast itself. Two cases are considered: 30 January 1985 and 1 December 1988. Dynamical forecasts, produced with the NCAR CCM, were also made for a number of cases.[11]

[10] See [18.2.3] for details about these measures of forecast skill.

[11] The dynamical model was used to forecast 15 cases. According to the correlation skill score, the POP forecasts outperformed the dynamical forecasts in these cases (see Figure 18.8 and [18.4.4]).

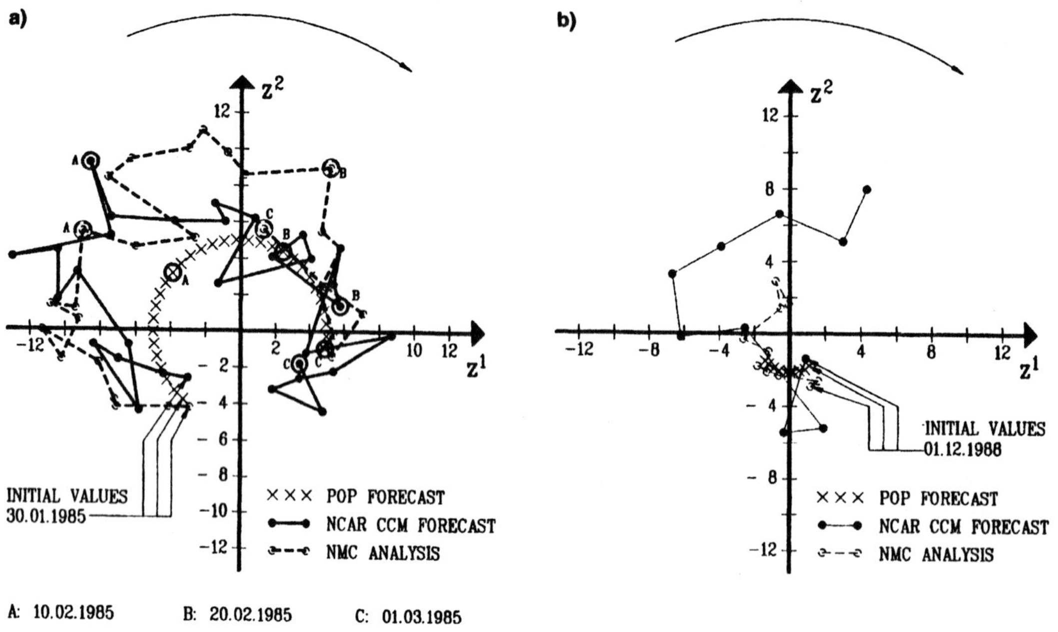

Figure 15.10: *MJO: Forecasts of the POP coefficient z_t. The forecasts are presented in the two-dimensional POP-coefficient plane with the x-axis representing the z^r-coefficient, and the y-axis the z^i-coefficient. The POP forecast model (15.21) implies a trajectory that rotates clockwise.*
The dashed line that connects the open circles represents the observed evolution, the solid line that connects the solid circles represents a dynamical forecast, and the POP forecast is given by the crosses. From von Storch and Baumhefner [388].
a) Initialized 30 January 1985.
b) Initialized 1 December 1988.

Figure 15.10a shows the predicted and analysed evolution for 30 days beginning on 30 January 1985. The MJO evolved smoothly, with a clockwise rotation in the POP coefficient plane, until about 25 February. It reversed direction after that day. Both the POP forecast and the NCAR CCM forecast are skilful in predicting the regular evolution in the first 25 days, but they fail to predict the phase reversal on 25 February.

Figure 15.10b shows the less successful forecast of 1 December 1988. The MJO POP coefficient was small at the time of initialization and remained so. The velocity potential field did not contain a well-defined wavenumber 1 pattern, and thus the failure of both forecasts is not unexpected.

The correlation skill score, ρ_τ, and the root mean square error, $\mathcal{S}_\tau$, derived from a large ($n \approx 1500$) ensemble of forecast experiments are shown in Figure 15.11 for the POP scheme and for persistence. Persistence is more skilful than the POP forecast during the first 2 days, but rapidly loses skill at longer leads. Persistence has a minimum in ρ_τ at about 20 days, consistent with the 30–60 day period of the MJO. The mean squared error, $\mathcal{S}_\tau$, reaches its saturation level at about the same time. The skill of the POP forecast decreases more slowly with time, reaching a value of 0.5 at a lead of 9 days. Also note that the mean squared error of the POP forecast has not yet reached saturation at a 24-day lead.

15.4 Cyclo-stationary POP Analysis

The POP analysis described in Section 15.1 assumes temporal stationarity while observed processes are often *cyclo-stationary*, that is, the first and second moments depend on an external cycle, such as the annual cycle. In this section we present a generalization of the conventional POP analysis that explicitly accounts for this non-stationarity.[12]

[12]Cyclo-stationary POP analysis was first suggested by Klaus Hasselmann in an unpublished manuscript in 1985. Two groups, namely Maria Ortiz and her colleagues at the University of Alcala in Spain and Benno Blumenthal from the Lamont Doherty Geological Observatory in Palisades, New York, showed how to implement the cyclo-stationary POP analysis independently in 1989/1990. Only Blumenthal published his results [50].

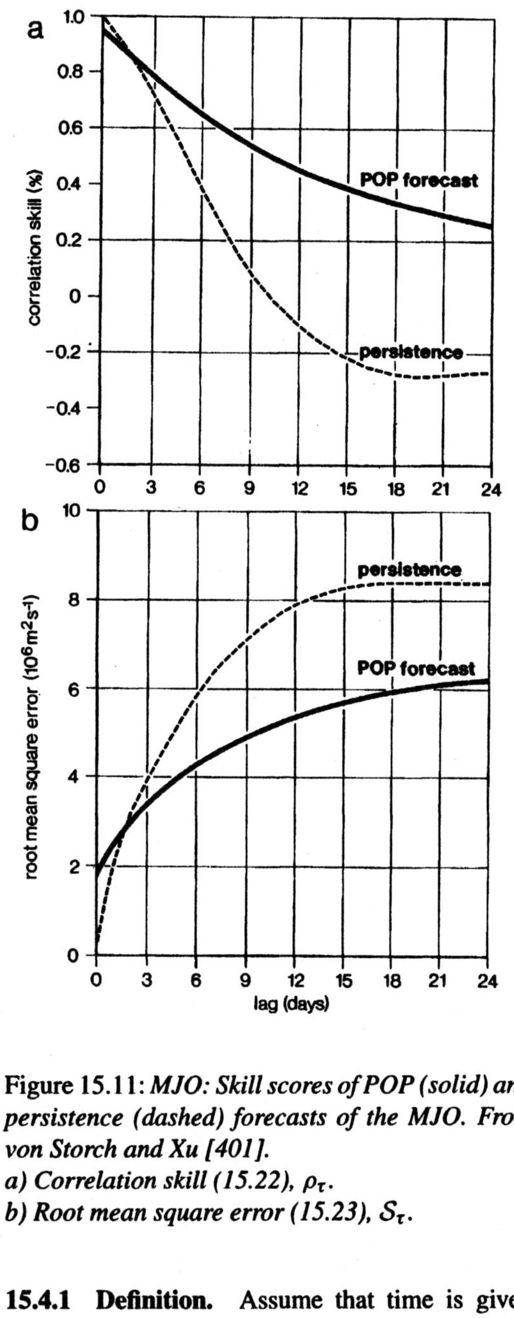

Figure 15.11: *MJO: Skill scores of POP (solid) and persistence (dashed) forecasts of the MJO. From von Storch and Xu [401].*
a) Correlation skill (15.22), ρ_τ.
b) Root mean square error (15.23), S_τ.

15.4.1 Definition. Assume that time is given by a pair of integers (t, τ), where t counts the cycles (e.g., annual cycle), and τ indicates the 'seasonal date' (e.g., months), or time steps within a cycle. Assume that a cycle has n time steps so that $\tau = 1, \ldots, n$. Note that $(t, n+1) = (t+1, 1)$ or, generally, $(t, \tau+n) = (t+1, \tau)$. As with ordinary POP analysis, we then assume that the cyclo-stationary process can be approximated by

$$\vec{\mathbf{X}}_{t,\tau+1} = \mathcal{A}_\tau \vec{\mathbf{X}}_{t,\tau} + \text{noise} \tag{15.24}$$

where $\vec{\mathbf{X}}_{t,\tau+n} = \vec{\mathbf{X}}_{t+1,\tau}$ and $\mathcal{A}_{\tau+n} = \mathcal{A}_\tau$. Substituting (15.24) into itself n consecutive times, we find

$$\vec{\mathbf{X}}_{t+1,\tau} = \mathcal{B}_\tau \vec{\mathbf{X}}_{t,\tau} + \text{noise} \tag{15.25}$$

where

$$\mathcal{B}_\tau = \prod_{s=1}^{n} \mathcal{A}_{\tau+s-1} \tag{15.26}$$

and where the noise in (15.25) is a moving average of n consecutive noise terms from (15.24) (recall [10.5.5]). We assume that this *integrated* noise is white on the inter-cycle time scale.

A conventional POP analysis can be applied to each of the n models described by (15.25). This results in n collections of eigenvectors $\vec{p}^{\,\tau}$ and eigenvalues λ_τ that are obtained from the n eigenproblems

$$\mathcal{B}_\tau \vec{p}^{\,\tau} = \lambda_\tau \vec{p}^{\,\tau}. \tag{15.27}$$

As usual, all eigenvectors are normalized to unit length. Note that the eigenvalues λ_τ are independent of τ, because

$$\begin{aligned} \mathcal{B}_\tau \vec{p}^{\,\tau} &= \lambda_\tau \vec{p}^{\,\tau} \\ \Leftrightarrow \mathcal{A}_{\tau+n} \mathcal{B}_\tau \vec{p}^{\,\tau} &= \lambda_\tau \mathcal{A}_{\tau+n} \vec{p}^{\,\tau} \\ \Leftrightarrow \mathcal{B}_{\tau+1}\big(\mathcal{A}_\tau \vec{p}^{\,\tau}\big) &= \lambda_\tau \big(\mathcal{A}_\tau \vec{p}^{\,\tau}\big). \end{aligned}$$

The last step is a consequence of (15.26) and the periodicity of $\mathcal{A}_\tau$.

Thus we now have a recursive relationship that can generate eigenvectors for all n eigenproblems by solving only the first problem. That is, $\mathcal{A}_\tau \vec{p}^{\,\tau}$ is an eigenvector of $\mathcal{B}_{\tau+1}$ when $\vec{p}^{\,\tau}$ is an eigenvector of $\mathcal{B}_\tau$. These eigenvectors are unique up to multiplication by a complex constant.[13] If we now normalize $\vec{p}^{\,1}$ to unit length and set

$$\vec{p}^{\,\tau+1} = \big(r_\tau^{-1} e^{i\phi}\big) \mathcal{A}_\tau \vec{p}^{\,\tau} \tag{15.28}$$

where $r_\tau = \|\mathcal{A}_\tau \vec{p}^{\,\tau}\|$, $\phi = \eta/n$, and η satisfies $\lambda_\tau = \lambda = \xi e^{-1\eta}$, then the resulting eigenvectors will be unique up to multiplication by a factor $e^{i\theta}$, and will be periodic (i.e., $\vec{p}^{\,\tau+n} = \vec{p}^{\,\tau}$). Thus the cyclo-stationary POP is damped by the factor ξ and rotated by an angle $-\eta$ in one cycle.

The cyclo-stationary POP coefficients evolve in time as a cyclo-stationary auto-regression that is similar to the auto-regression (15.10) that applies to ordinary POP coefficients. Specifically,

$$z_{t,\tau+1} = r_\tau e^{-i\phi} z_{t,\tau} + \text{noise}. \tag{15.29}$$

[13] We assume throughout that $\mathcal{B}_\tau$, $\tau = 1, \ldots, n$, (and hence $\mathcal{A}_\tau$) are non-singular and that all eigenvalues of $\mathcal{B}_\tau$ are distinct.

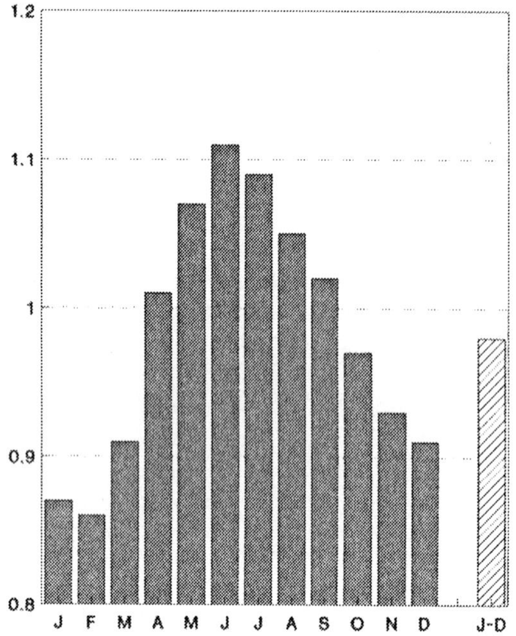

Figure 15.12: *ENSO: Amplitudes obtained in the conventional and cyclo-stationary POP analyses of equatorial 10 m wind and sea-surface temperature. Bars labelled 'J', 'F', etc., indicate the amplitudes obtained from the cyclo-stationary analysis in January, February, etc. The bar labelled 'J–D' is the amplitude obtained from the conventional analysis.*

Substituting (15.29) into itself n times, we obtain a conventional auto-regression

$$\begin{aligned} z_{t+1,\tau} &= \Big(\prod_{s=1}^{n} r_{\tau+s+1}\Big) e^{-i\eta} z_{t,\tau} + \text{noise} \\ &= \lambda z_{t,\tau} + \text{noise} \end{aligned}$$

for POP coefficients at one cycle increments that is consistent with model (15.25).

The time coefficients at a given time t may be obtained by projecting the full field $\vec{\mathbf{X}}_{t,\tau}$ onto the respective adjoint $\vec{p}_a^{\,\tau}$ or by using a least square approximation similar to (15.14) and (15.15). The adjoint patterns $\vec{p}_a^{\,\tau}$ and $\vec{p}_a^{\,\tau+1}$ are related to each other through a simple formula similar to (15.28):

$$\vec{p}_a^{\,\tau} = r_\tau^{-1} e^{i\phi} \mathcal{A}_\tau^{\mathrm{T}} \vec{p}_a^{\,\tau+1}. \tag{15.30}$$

The cyclo-stationary system matrices $\mathcal{A}_\tau$ can be estimated with (15.6) for each $\tau = 1, \ldots, n$ as

$$\widehat{\mathcal{A}}_\tau = \widehat{\Sigma}_{\tau,1} \widehat{\Sigma}_{\tau,0}^{-1},$$

where $\widehat{\Sigma}_{\tau,1}$ is the estimated lag-1 cross-covariance matrix between $\vec{\mathbf{X}}_{t,\tau}$ and $\vec{\mathbf{X}}_{t,\tau+1}$, and $\widehat{\Sigma}_{\tau,0}$ is the estimated covariance matrix of $\vec{\mathbf{X}}_{t,\tau}$.

15.4.2 Example: The Southern Oscillation. Time series of surface wind and SST along the equator between 50° E and 80° W (described in [15.2.2]) are good candidates for a cyclo-stationary POP analysis because the Southern Oscillation is known to be phase-locked to the annual cycle [330]. Monthly anomalies are analysed so that $n = 12$. The data are time-filtered to suppress the month-to-month variability. A conventional POP analysis was performed for comparison.

Both analyses identified a single dominant POP with comparable periods (31 months for the cyclo-stationary analysis, 34 months for the conventional analysis). The mode identified in the conventional analysis is similar to the ENSO mode described in [15.2.3] (see Figure 15.7, bottom, and Figure 15.8b).

The amplitude, r_τ, exhibits a marked annual cycle (Figure 15.12) which is strongly non-sinusoidal. Amplification takes place from April to September, with a maximum in June. The process is damped from October to March, with a minimum in February. Note that the amplitude increases from minimum to maximum in only four months, but then it takes eight months to return to minimum. The annually averaged amplitude is almost identical to the amplitude obtained in the conventional analysis.

The *zonal wind* patterns (Figure 15.13, left column) show eastward progression of the main centre of action with the annual cycle. The imaginary component is strongest during the first half of the year whereas the real component is strongest during the second half.

The imaginary part of the SST patterns (Figure 15.13, top right) has substantial amplitude in the Indian Ocean and East Pacific in northern winter, but not at other times of year. In contrast, the real component (Figure 15.13, bottom right) has large amplitude (at least 0.2) throughout the year in the East Pacific that coincides with large amplitudes of opposite sign in the West Pacific. The signal in the real pattern is strongest in the East Pacific in northern fall.

The average of these cyclo-stationary modes is similar to the pattern obtained from the conventional POP analysis described in [15.2.3] (Figure 15.8, bottom).

Note that the wind data were normalized into unit variance before the POP analysis. To transform the patterns to meaningful physical units, the wind patterns (Figure 15.13, left) must be multiplied by 0.45 times the standard deviations of the POP coefficients (Figure 15.14). Similarly, typical SST amplitudes are obtained

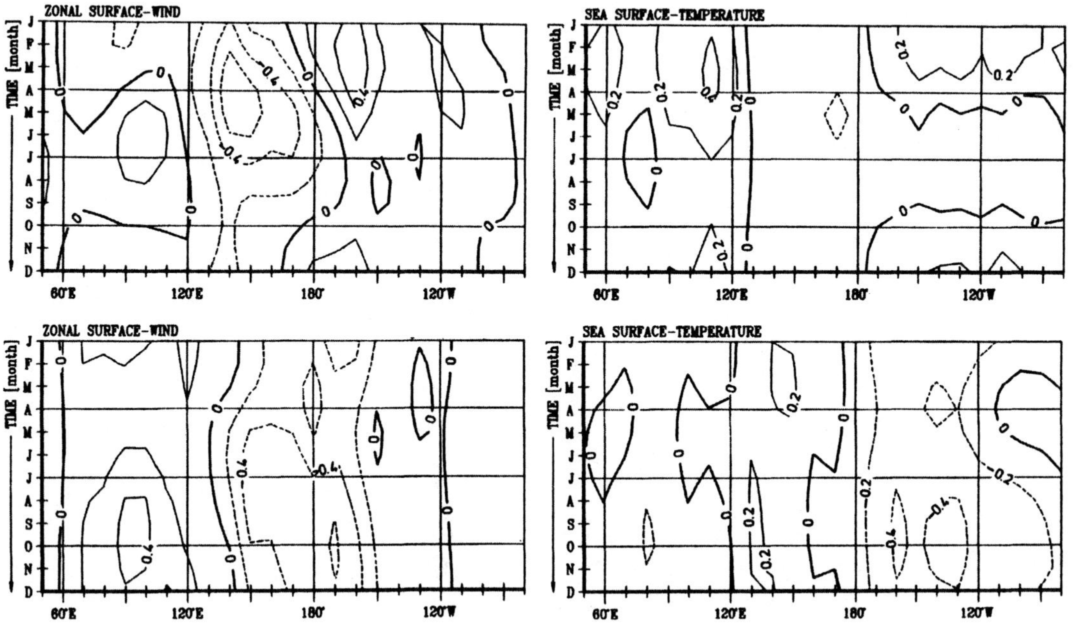

Figure 15.13: *ENSO: Cyclo-stationary POPs analysed from a combined normalized zonal wind/SST data set. The horizontal axis represents the longitude along the equator, and the vertical axis the annual cycle of the patterns.*
Top row: Imaginary part. Bottom row: Real part.

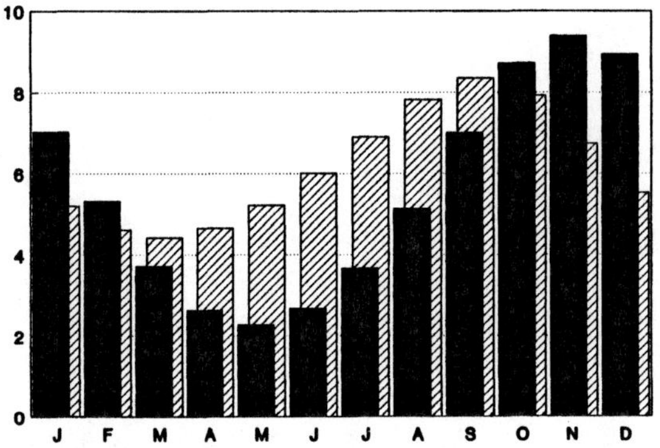

Figure 15.14: *ENSO: Annual cycle of the variance of the cyclo-stationary POP-coefficients. (Solid: imaginary component; hatched: real component.)*

by multiplying the patterns (Figure 15.13, right) by 0.60 times the standard deviation of the POP coefficients.

Note that the variance of the POP coefficients has a marked annual cycle (Figure 15.14). The annual average is about five. Both components have maximum variability in northern autumn, but they are not phased identically. Also, note that the variance extremes are delayed relative to those of the amplitudes.

As with conventional POPs, it is possible to build scenarios that describe the 'typical' evolution of the field from a given initial state. Suppose that a field $\vec{\mathbf{X}}$ is well represented by a cyclo-stationary POP and that the initial state at time τ in the cycle is

$$\vec{\mathbf{x}}_{0,\tau} \approx 2\,\mathrm{Re}(z_{0,\tau}\vec{p}^{\,\tau}).$$

Then its future state δ time units later is given by

$$\vec{\mathbf{x}}_{0,\tau+\delta} \approx 2\,\mathrm{Re}\Big(z_{0,\tau}\vec{p}^{\,\tau+\delta}\prod_{s=1}^{\delta} r_{\tau+s}e^{-i\eta/n}\Big) + \text{noise}.$$

This yields a typical evolution in time from $\vec{\mathbf{x}}_{0,\tau}$ when the noise is set to zero.

Figure 15.15 shows the typical evolution of equatorial zonal 10 m wind (left panel) and SST (right panel) when the initial state is given by the imaginary part of the cyclo-stationary POP in January. The diagram illustrates that, depending upon the sign, a La Niña or El Niño typically evolves from the January state depicted in Figure 15.13 (top row).

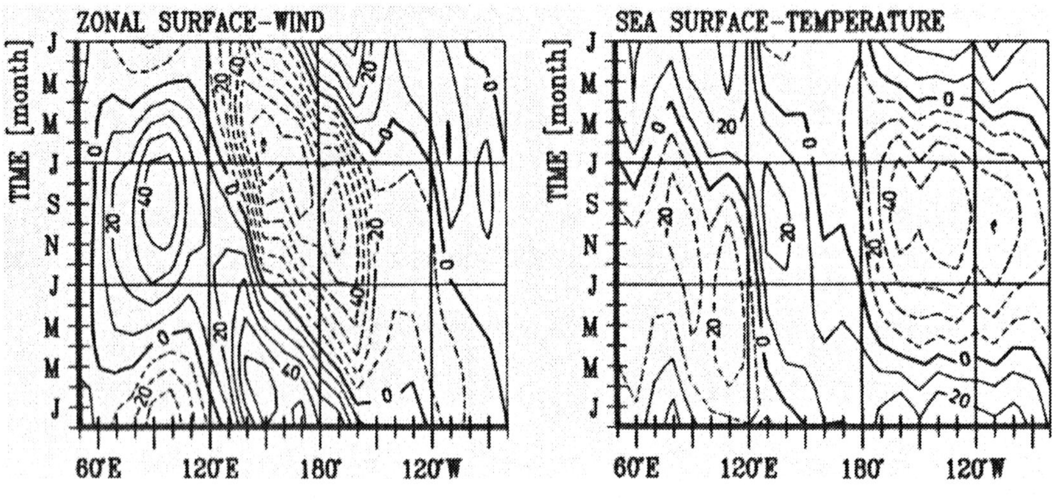

Figure 15.15: *ENSO: 'Typical' evolution of SST and zonal wind from a prescribed initial state. The horizontal axes represent the longitudinal position, and the vertical axis represents time over 24 months (time increasing downwards). The imaginary component of the POP (see Figure 15.13, top row) in January is the initial state.*

15.5 State Space Models

15.5.1 Overview. We have described POPs as eigenmodes of an empirically determined system matrix. However, POPs can be placed in a much more general setting as members of the class of *state space models*. We will explain this concept in the next subsection, briefly describe its merits, and introduce the *Principal Interaction Patterns* (PIPs).

While the general idea is ubiquitous in climate research, specific attempts to explicitly and objectively determine reduced phase spaces have been made only recently. So far, these attempts have dealt with simplified systems and have mostly addressed the complicated methodical and conceptual aspects of the problem; there is still a way to go until these techniques will be applied routinely by researchers trying to understand the dynamics of the real ocean and the real atmosphere. This field is certainly a frontier of climate research, and we may expect new developments in the future.

15.5.2 State Space Models. A complex dynamical system with an m-dimensional state vector $\vec{\mathbf{X}}_t$ can often be approximated as being driven by a simpler dynamical system with a state vector $\vec{\mathbf{Z}}_t$ of dimension $k < m$. Mathematically, such processes can be approximated by a state space model. These models consist of a discrete (or continuous) *system equation* for the k dynamical variables $\vec{\mathbf{Z}} = (\mathbf{Z}_1, \ldots, \mathbf{Z}_k)^{\mathrm{T}}$,

$$\vec{\mathbf{Z}}_{t+1} = \mathcal{F}_D(\vec{\mathbf{Z}}_t, \vec{\alpha}, t) + \text{noise} \qquad (15.31)$$

and an *observation equation* for the observed variables $\vec{\mathbf{X}} = (\mathbf{X}_1, \ldots, \mathbf{X}_m)^{\mathrm{T}}$,

$$\begin{aligned} \vec{\mathbf{X}}_t &= \mathcal{P}^{\mathrm{T}}\vec{\mathbf{Z}}_t + \text{noise} \\ &= \textstyle\sum_{j=1}^{k} \mathbf{Z}_{t,j}\vec{p}_j + \text{noise}. \end{aligned} \qquad (15.32)$$

Operator $\mathcal{F}_D$ represents a class of models that may be nonlinear in the dynamical variables $\vec{\mathbf{Z}}_t$ and depends on a set of free parameters $\vec{\alpha} = (\alpha_1, \alpha_2, \ldots)$.

Matrix $\mathcal{P}$ generally has many more columns (m) than rows (k). The system equations (15.31) therefore describe a dynamical system in a smaller phase space than the space that contains $\vec{\mathbf{X}}_t$. Ideally in applications, a reduced system governed by the same dynamics as the full system can be identified.

The advantage of such low-order systems over the original high-dimensional system is, at least in theory, that the low-order system is easier to 'understand.' Experience, however, suggests that the system state vector must have very low dimension if the dynamics are to be analytically tractable.

15.5.3 State Space Models as Conceptual Tools and as Numerical Approximations. One application of the state space models is the conceptualization of hypotheses without determining the unknown parameters $\vec{\alpha}$ and $\mathcal{P}$. Indeed, almost all dynamical reasoning can be expressed as a state space model. For example, the barotropic vorticity equation may be seen as a state space model in which the system state vector evolves in a space that excludes a large class of waves. Time series models, such as the Box–Jenkins ARMA models described in [10.5.5,6] can also be expressed in state space model form.[14]

In other applications, attempts are made to actually determine the underlying dynamical variables $\vec{\mathbf{Z}}_t$ and the unknown parameters $\vec{\alpha}$ for a given class of dynamical operators $\mathcal{F}$. The *Principal Interaction Pattern* ansatz proposed by Hasselmann [167] is probably the most general formalization of this type (see [15.5.4] below).

The noise term in (15.31) is often disregarded in nonlinear dynamical analyses. However, disregarding the noise in low-order systems ($k < 10$) usually changes the dynamics of the system significantly since the low-order system is a closed system without noise. However, components of the climate system, such as the tropical troposphere or the thermohaline circulation in the ocean, are never closed; they continuously respond to 'noise' from other parts of the climate system, hence the noise term in (15.31). It is doubtful if the fundamental assumption, namely that the low-order system is governed by the same dynamics as the full system, is satisfied when the noise is turned off.

15.5.4 Principal Interaction Patterns. Since $k \leq m$, the time coefficients $\mathbf{Z}_{t,j}$ of a pattern $\vec{p}_j$ at a time t are not uniquely determined by $\vec{\mathbf{X}}_t$. Thus the time coefficients are determined by least squares as

$$\vec{\mathbf{Z}}_t = (\mathcal{P}^{\mathrm{T}}\mathcal{P})^{-1}\mathcal{P}^{\mathrm{T}}\vec{\mathbf{X}}_t. \qquad (15.33)$$

When fitting the state space model from equations (15.31) and (15.32) to a time series, the following must be specified: the class of models $\mathcal{F}$, the patterns $\mathcal{P}$, the free parameters α and the dimension of the reduced system k. The class of models $\mathcal{F}$, must be selected *a priori* on the basis of physical reasoning. The number k might also be specified *a priori*. The parameters α and the patterns $\mathcal{P}$ are fitted simultaneously to a time series by minimizing the mean square error $\epsilon[\mathcal{P};\vec{\alpha}]$ of the approximation of the (discretized) time derivative of the observations $\vec{\mathbf{X}}$ by the state space model:

$$\epsilon[\mathcal{P};\vec{\alpha}] = \mathcal{E}\Big(\| \vec{\mathbf{X}}_{t+1} - \vec{\mathbf{X}}_t - \mathcal{P}(\mathcal{F}[\vec{\mathbf{Z}}_t,\vec{\alpha},t] - \vec{\mathbf{Z}}_t) \|^2 \Big). \qquad (15.34)$$

The patterns $\mathcal{P}$ that minimize (15.34) are called *Principal Interaction Patterns* (PIPs) [167]. If only a finite time series of observations $\vec{\mathbf{X}}$ is available, the expectation $\mathcal{E}(\cdot)$ is replaced by a summation over time.

In general, minimization of (15.34) does not result in a unique solution. In particular, if $\mathcal{L}$ is any non-singular matrix, and if $\mathcal{P}$ minimizes (15.34), then the set of patterns $\mathcal{P}' = \mathcal{PL}$ will also minimize (15.34) as long as the corresponding model $\mathcal{F}' = \mathcal{L}^{-1}\mathcal{F}$ belongs to the *a priori* specified class of models. This problem may be solved by imposing a constraint. For example, one might require that the linear term in the Taylor expansion of $\mathcal{F}$ is a diagonal matrix.

Successful applications of the PIP idea to dynamical systems with different degrees of complexity have been presented by Achatz and colleagues [1, 2], Kwasniok [236, 237], and Selten [345, 344].

15.5.5 POPs as Simplified PIPs. The Principal Oscillation Patterns can be understood as a kind of simplified Principal Interaction Patterns. For that assume $m = n$. Then, the patterns $\mathcal{P}$ span the full $\vec{\mathbf{X}}$-space, and their choice does not affect $\epsilon[\mathcal{P};\vec{\alpha}]$. Also, let $\mathcal{F}$ be a linear model $\mathcal{F}[\vec{\mathbf{Z}}_t,\vec{\alpha}] = \mathcal{A}\vec{\mathbf{Z}}_t$, where the parameters $\vec{\alpha}$ are the entries of $\mathcal{A}$. Then the dynamical equation (15.31) is identical to (15.10). The constraint mentioned above results in PIPs (of the admittedly simplified state space model) that are given by the eigenvectors of $\mathcal{A}$.

---

[14] See, for example, Priestley [323, Section 10.4.4].

16 Complex Eigentechniques

16.1 Introduction

16.1.1 Modelling the State and the 'Momentum.' The purpose of EOF analysis (Chapter 13) is simply to identify patterns that efficiently characterize variations in the current 'state' or 'location' of a vector field $\vec{\mathbf{X}}$. Consequently, the technique completely ignores the time evolution of the analysed field.

POP analysis (Chapter 15) accounts for patterns that evolve in time by representing the observed field as a vector AR(1) process, so that information about the present state is transferred to the next state. Such a system can describe oscillatory behaviour since any m-dimensional system of first-order difference equations is equivalent to one mth order difference equation.

A generalization of this approach is to model not only the 'state' $\vec{\mathbf{X}}_t$ but also an indicator of its tendency $\delta\vec{\mathbf{X}}_t$ (Wallace and Dickinson [408]). Such an approach is related to the Hamiltonian principle in mechanics that the future of a system is described by a set of first-order differential equations for the location (state) and the momentum.

The Hilbert transform $\vec{\mathbf{X}}_t^{\mathrm{H}}$ (see Section 16.2) is a reasonable measure of 'momentum' $\delta\vec{\mathbf{X}}_t$ when variations in $\vec{\mathbf{X}}_t$ are confined to a relatively narrow time scale. Then the conventional eigentechniques, such as EOFs and POPs, are applied to the *complexified time series* $\vec{\mathbf{X}}_t + i\vec{\mathbf{X}}_t^{\mathrm{H}}$.

16.1.2 Confusing Names. There is some confusion in the literature about what to call the EOFs or POPs of the complexified process.

The EOFs of the complexified process are sometimes called 'frequency domain EOFs' or 'FDEOFs', since they may be understood as eigenvectors of the cross-spectral matrix averaged over some frequency interval (see below). When applied to narrowly band-pass filtered data this name makes sense, but the technique may also be used for broad-band features.

The term 'complex EOFs' or 'CEOFs' is also sometimes used to refer to the EOFs of $\vec{\mathbf{X}}_t + i\,\vec{\mathbf{X}}_t^{\mathrm{H}}$, but this usage is ambiguous since it also applies to the eigenvectors of any general complex vector process. Similar ambiguity occurs when the POPs of the complexified process are called 'Complex POPs' or 'CPOPs' (see, e.g., Bürger [75]).

Therefore, for conceptual clarity, we revive a suggestion first made by Rasmusson et al. [329] in 1981; we refer to the EOFs of the complexified process as *Hilbert* EOFs, and to the corresponding POPs as *Hilbert* POPs.[1]

16.1.3 Outlook. The *Hilbert transform* is introduced in Section 16.2 and we define the *Hilbert EOFs* in Section 16.3, where we also deal briefly with *Hilbert POPs*.

Canonical Correlation Analysis, rotated EOFs, redundancy analysis, and other pattern analysis techniques can all be extended to complexified processes. Attempts in this respect are currently underway, but no applications seem to have been published in the geophysical literature so far.[2]

16.2 Hilbert Transform

16.2.1 Motivation and Heuristic Introduction. If X_t is a real time series with Fourier decomposition

$$X_t = \sum_{\omega} \zeta(\omega)e^{-2\pi i\,\omega t} \tag{16.1}$$

then its *Hilbert transform* is

$$X_t^{\mathrm{H}} = \sum_{\omega} \zeta^{\mathrm{H}}(\omega)e^{-2\pi i\,\omega t} \tag{16.2}$$

where $\zeta^{\mathrm{H}}(\omega)$ is defined to be

$$\zeta^{\mathrm{H}}(\omega) = \begin{cases} i\,\zeta(\omega) & \text{for } \omega \le 0 \\ -i\,\zeta(\omega) & \text{for } \omega > 0. \end{cases} \tag{16.3}$$

The Hilbert transform X^{H} is identical to original time series X_t except for a $\pi/2$ phase-shift of ξ

[1] Rasmusson et al. [329] used the expression 'Hilbert Singular Decomposition' (HSD).

[2] Brillinger [66] deals with the CCA of complexified processes, and Horel [181] discusses rotated Hilbert EOFs.

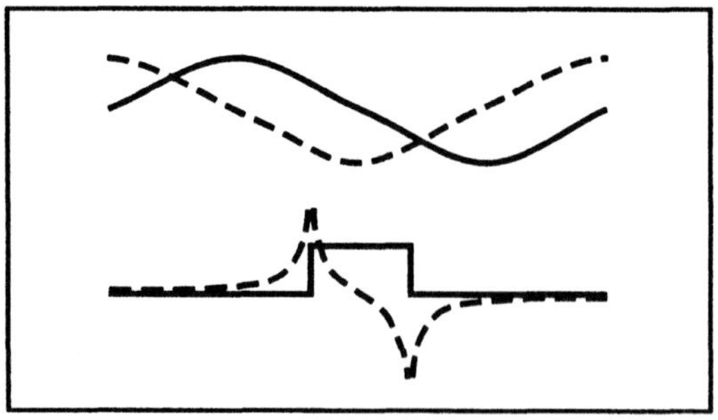

Figure 16.1: *A schematic illustration of the effect of the Hilbert transform. The solid curves depict the input time series, and the dashed curves depict the corresponding Hilbert transforms. After Horel [181, Fig. 1, p. 1662].*

that is performed separately at each frequency ω. For instance, if

$$X_t = 2\cos(2\pi\omega_0 t) \qquad (16.4)$$

for some fixed ω_0, then $\zeta(\pm\omega_0) = 1$, $\zeta^{\mathrm{H}}(\pm\omega_0) = \mp i$, and

$$X_t^{\mathrm{H}} = -2\sin(2\pi\omega_0 t). \qquad (16.5)$$

That is, the Hilbert transform shifts X_t a quarter of a period to the right. Another interpretation, in this example, is that X_t^{H} provides information about the rate of change of X_t at time t.

To illustrate, Figure 16.1 depicts two idealized input time series and their Hilbert transforms. If the input is monochromatic, the transform produces the same output, only advanced by a quarter period. When the input is not monochromatic, there is a quarter period advance at every frequency, with the result that the Hilbert transform can appear to be quite different from the input. For example, if X_t is the rectangular phase function, the Hilbert transform will have spikes at the beginning and end of the pulse. This is because the decomposition of the pulse function into trigonometric components requires contributions from many components, and each of the components is shifted by its own quarter of a period. This example indicates that the Hilbert transform can only be interpreted as a 'time rate of change' when most of the variability of X_t is confined to a relatively narrow frequency band.

The Hilbert transform is used to augment the information contained in a vector time series by adding information about its future behaviour. This is accomplished by combining the original vector time series $\vec{\mathbf{X}}_t$ and its Hilbert transform $\vec{\mathbf{X}}_t^{\mathrm{H}}$ into a new complex vector time series

$$\vec{\mathbf{Y}}_t = \vec{\mathbf{X}}_t + i\vec{\mathbf{X}}_t^{\mathrm{H}}. \qquad (16.6)$$

Conventional techniques, such as EOFs or POPs (see Sections 16.3 and 16.2) are then applied to these 'complexified' time series (16.6).

We complete this section by introducing the Hilbert transform in mathematically rigorous terms and describing its estimation. The 'Hilbert EOFs' and 'Hilbert POPs' will be introduced in Section 16.3 and the former will be discussed in terms of examples.

16.2.2 Derivation of the Hilbert Transform.

The motivation behind the Hilbert EOF and POP analysis is the creation of a process $\mathbf{X}_t^{\mathrm{H}}$ that is something like 'momentum'. Physical arguments tell us that the 'momentum' process $\mathbf{X}_t^{\mathrm{H}}$ should be related to the original process through a linear filter operator, that is,

$$\mathbf{X}_t^{\mathrm{H}} = \sum_{\delta=-\infty}^{\infty} h_\delta \mathbf{X}_{t+\delta}. \qquad (16.7)$$

Also it should be out-of-phase by $\pi/2$ for all frequencies ω with the 'change' $\mathbf{X}_t^{\mathrm{H}}$ leading the 'state' $\mathbf{X}_t$, that is,

$$\Phi_{x^{\mathrm{H}}x}(\omega) = \pi/2 \ \text{ for } \omega > 0. \qquad (16.8)$$

To construct the filter (16.7) we note that the cross-spectrum (11.74) between $\mathbf{X}_t^{\mathrm{H}}$ and $\mathbf{X}_t$ satisfies

$$\Gamma_{x^{\mathrm{H}}x}(\omega) = H(\omega)\Gamma_{xx}(\omega). \qquad (16.9)$$

Since the autospectrum Γ_{xx} is real, the phase spectrum satisfies (16.8) if and only if $H(\omega)$ is imaginary and anti-symmetric, with a negative imaginary component for positive frequencies, as in

$$H(\omega) = \begin{cases} -i & \text{for } \omega > 0 \\ i & \text{for } \omega < 0. \end{cases} \qquad (16.10)$$

Thus

$$\begin{aligned} |H(\omega)| &= 1 \\ \Gamma_{x^{\mathrm{H}}x^{\mathrm{H}}}(\omega) &= \Gamma_{xx}(\omega) \qquad (16.11) \\ \mathrm{Var}(\mathbf{X}^{\mathrm{H}}) &= \mathrm{Var}(\mathbf{X}) \qquad (16.12) \end{aligned}$$

and

$$\begin{aligned} \Psi_{x^{\mathrm{H}}x}(\omega) &= \begin{cases} -\Gamma_{xx}(\omega) & \text{for } \omega \geq 0 \\ \Gamma_{xx}(\omega) & \text{for } \omega < 0 \end{cases} \\ \Lambda_{x^{\mathrm{H}}x}(\omega) &= 0. \qquad (16.13) \end{aligned}$$

Note also that

$$\kappa_{x\text{H}_x}(\omega) = 1 \quad \text{for all } \omega \neq 0.$$

That is, there is perfect coherence between $\mathbf{X}_t$ and its Hilbert transform at all nonzero frequencies. This is as it should be since $\mathbf{X}_t^{\text{H}}$ is just a phase-shifted version of $\mathbf{X}_t$ at each frequency.

So far we have defined the Hilbert transform in the frequency domain. To obtain the filter in the time domain we use the following theorem from Brillinger [66, pp. 31,395]:

If $\mathbf{X}_t$ is a stationary multivariate process with absolutely summable auto-covariance function γ_{xx}, then the process

$$\mathbf{Y}_t = \lim_{T\to\infty} \mathbf{Y}_t^T \tag{16.14}$$

where

$$\mathbf{Y}_t^T = \sum_{\delta=-T}^{T} h_\delta \mathbf{X}_{t-\delta} \tag{16.15}$$

and

$$h_\delta = \int_{-\frac{1}{2}}^{\frac{1}{2}} H(\omega) e^{2\pi i \delta\omega}\, d\omega \tag{16.16}$$

exists and has finite variance.

The application of (16.16) to (16.10) yields (cf. Rasmusson et al. [329])

$$h_\delta = \begin{cases} \frac{2}{\delta\pi} & \text{if } \delta \text{ is odd} \\ 0 & \text{if } \delta \text{ is even.} \end{cases} \tag{16.17}$$

Note that $h_\delta \leq 0$ for negative δ and $\sum_\delta h_\delta = 0$ so that the time mean of $\mathbf{X}_t^{\text{H}}$ is zero.

Thus, the Hilbert transform $\mathbf{X}_t^{\text{H}}$ *in the time domain* of a stationary process $\mathbf{X}_t$ is

$$\mathbf{X}_t^{\text{H}} = \sum_{\delta=0}^{\infty} \frac{2}{(2\delta+1)\pi} \big(\mathbf{X}_{t+2\delta+1} - \mathbf{X}_{t-(2\delta+1)}\big). \tag{16.18}$$

Note that the series in (16.18) does *not* converge for sine time functions and other non-stationary time series because their auto-covariance functions are not absolutely summable.

16.2.3 Examples: The Hilbert Transform of AR Processes. We now apply the Hilbert transform to the AR(1) and AR(2) processes discussed in Chapter 11.

Figure 16.2 displays realizations of AR(1) processes with $\alpha_1 = 0.9$ and $\alpha_1 = 0.3$ and their Hilbert transforms (using (16.15) and $T = 20$). Since AR(1) processes have a red spectrum and no preferred frequency, the connection between the input and its Hilbert transform is rather loose. The Hilbert transforms lead the input series. Visually, the lead seems to be longer when $\alpha_1 = 0.9$ than when $\alpha_1 = 0.3$. This impression is substantiated by the cross-covariance between the input time series and its Hilbert transform (Figure 16.3). Maximum cross-correlations for the short memory process are obtained for lag-1, while the long memory process exhibits almost uniform lag correlations for a wide range of lags.

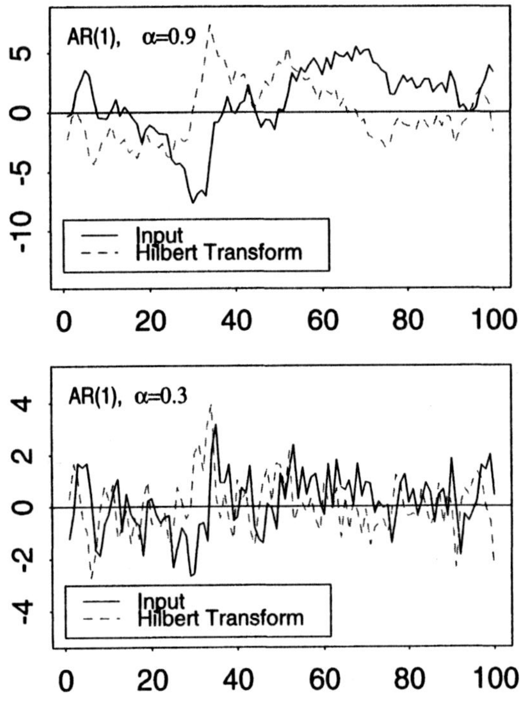

Figure 16.2: *Realizations of AR(1) processes (solid) with $\alpha_1 = 0.9$ (top) and $\alpha_2 = 0.3$ (bottom) and their Hilbert transforms $\mathbf{X}_t^{\text{H}}$ (dashed) computed with (16.15) and $T = 20$.*

There is a more rigid link between the input and its Hilbert transform when the input is the AR(2) process with $\alpha_1 = 0.9$ and $\alpha_2 = -0.8$, which is shown in Figure 16.4. This process is quasi-oscillatory with a period of about 6 time steps (cf. [10.3.4–6], [11.1.7] and [11.2.6]). Since this process has a preferred frequency, the phase shift between the Hilbert transform and the input series is about 1.5 time steps. Large Hilbert transform values regularly precede large changes of the input series, confirming the interpretation of the Hilbert transform as the 'momentum' of the input process. This impression is further substantiated by the lagged cross-covariance function (Figure 16.3),

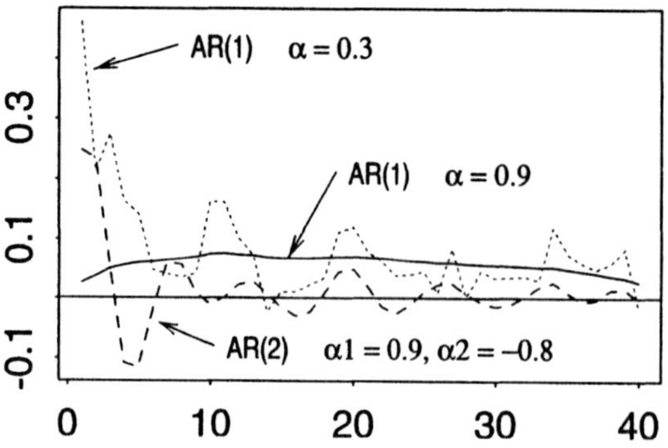

Figure 16.3: *Cross-correlation functions between the input series and corresponding Hilbert transform shown in Figures 16.2 and 16.4. The cross-covariance functions have been estimated from finite time series.*

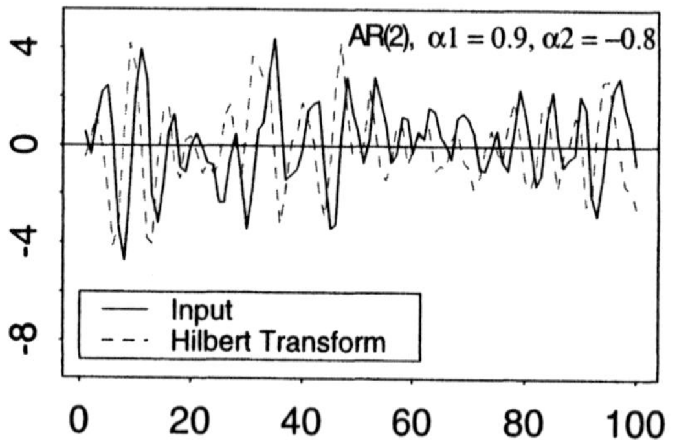

Figure 16.4: *A realization of an AR(2) process (solid) with $\vec{\alpha} = (0.9, -0.8)$ and its Hilbert transform $\mathbf{X}_t^{\mathrm{H}}$ (dashed) computed with (16.15) and $T = 20$.*

which has a maximum at lags-1 and 2, a zero at lag-3, a negative minimum at lag-5, and so forth.

The results are virtually unchanged if a longer filter window with $T > 20$ is used.

16.2.4 Estimating the Hilbert Transform from a Finite Time Series. Two different approaches may be used to estimate the Hilbert transform of a finite time series (cf. Barnett [19]).

In the time domain we can use the approximate filter (16.15) with some finite T. Obviously the first and last T values of the Hilbert transform are not as well estimated since the filter length must either be reduced, or filter (16.15) must be used in an asymmetric manner.

The Hilbert transforms displayed in Figures 16.2 and 16.4 were derived in this way, but end-effects can not be seen because the middle of a longer time series is shown.

The filter length T is determined by iteratively increasing T until there is little change in the estimated transform.

An alternative approach is to re-express the finite time series $\{\mathbf{x}_1, \ldots, \mathbf{x}_n\}$ in its trigonometric expansion (see [12.3.1] and (C.1))

$$\mathbf{x}_t = \sum_k a_k \cos\left(\frac{2\pi kt}{n}\right) + b_k \sin\left(\frac{2\pi kt}{n}\right) \tag{16.19}$$

and then estimating $\mathbf{X}_t^{\mathrm{H}}$ with

$$\begin{aligned} \widehat{\mathbf{x}}_t^{\mathrm{H}} &= \sum_k a_k \cos\left(\frac{2\pi kt}{n} + \frac{\pi}{2}\right) \\ &\qquad + b_k \sin\left(\frac{2\pi kt}{n} + \frac{\pi}{2}\right) \\ &= \sum_k b_k \cos\left(\frac{2\pi kt}{n}\right) - a_k \sin\left(\frac{2\pi kt}{n}\right). \end{aligned} \tag{16.20}$$

This estimate matches equations (16.4) and (16.5).

The frequency domain approach has two advantages over the time domain approach. First, it is not necessary to choose the filter length T. Second, it appears that data near the endpoints need not be treated specially. Thus the frequency domain approach seems to be more robust than the time domain approach. However, this is not really the case. The trigonometric expansion (16.19) implicitly assumes that the discrete finite time series represents one chunk of a periodic process with period $n + 1$. This is generally not the case. The numbers $\{\mathbf{x}_1, \mathbf{x}_2, \ldots\}$ are *not* a smooth continuation of $\{\ldots, \mathbf{x}_{n-1}, \mathbf{x}_n\}$, and the shift (16.20) of the entire non-periodic time series transports the 'discontinuity' into the middle of the transformed time series. The problem will be more severe for shorter time series and longer time scales. As in spectral analysis, the problem can be reduced by using a data taper (cf. [12.3.8]).

Again, we advise making plots of the input time series together with the estimated Hilbert transform to ensure that there are no unpleasant surprises.

16.2.5 Properties of the Hilbert Transformed Process. The cross-covariance function between a process and its Hilbert transform is anti-symmetric since their co-spectrum vanishes (cf. (16.13, 11.68));

$$\gamma_{x^{\mathrm{H}}x}(\delta) = -\gamma_{xx^{\mathrm{H}}}(\delta), \tag{16.21}$$

and in particular

$$\gamma_{x^{\mathrm{H}}x}(0) = 0. \tag{16.22}$$

Thus, the process and its Hilbert transform are uncorrelated at lag zero.

When the Hilbert transform is applied twice, then the original time series appears with reversed sign:

$$(\mathbf{X}^{\mathrm{H}})^{\mathrm{H}}_t = -\mathbf{X}_t. \tag{16.23}$$

Also, the Hilbert transform is a linear operation. Thus

$$(\mathbf{X} + \beta\mathbf{Y})^{\mathrm{H}}_t = \mathbf{X}^{\mathrm{H}}_t + \beta\mathbf{Y}^{\mathrm{H}}_t. \tag{16.24}$$

The relationship between a process and its Hilbert transformed process, as represented by the covariance matrix or the spectrum, is described in [16.2.7]. This relationship will be used in Section 16.3. We briefly introduce the spectral matrix next.

16.2.6 The Spectral Matrix of a Random Vector. In Section 11.4 we defined the cross-spectrum of two processes $\mathbf{X}_{1t}$ and $\mathbf{X}_{2t}$ as the Fourier transform of their cross-covariance function. We now generalize these definitions to vector random variables.

The lag covariance matrix of an m-dimensional random vector $\vec{\mathbf{X}}_t = (\mathbf{X}_{1t}, \ldots, \mathbf{X}_{mt})^{\mathrm{T}}$ is the $m \times m$ matrix

$$\Sigma_{xx}(\tau) = \mathcal{E}\big((\vec{\mathbf{X}}_t - \mathcal{E}(\vec{\mathbf{X}}_t))(\vec{\mathbf{X}}_{t+\tau} - \mathcal{E}(\vec{\mathbf{X}}_{t+\tau}))^{\dagger}\big).$$

The spectrum of the vector process is defined as the Fourier transform of the lag covariance matrix

$$\boldsymbol{\Gamma}_{xx}(\omega) = \sum_{\tau=-\infty}^{\infty} \Sigma_{xx}(\tau) e^{-2\pi i \omega\tau} \tag{16.25}$$

or, in short,

$$\boldsymbol{\Gamma}_{xx} = \mathcal{F}\{\Sigma_{xx}\}. \tag{16.26}$$

The complex $m \times m$ matrix $\boldsymbol{\Gamma}_{xx}$ is called the *spectral matrix*. The element in the jth row and lth column is the cross-spectrum $\Gamma_{x_j x_l}$ between the jth and the lth components of $\vec{\mathbf{X}}$. Thus the matrix is Hermitian, that is, $\boldsymbol{\Gamma}^{\dagger}_{xx}(\omega) = \boldsymbol{\Gamma}_{xx}(\omega)$, and its main diagonal contains the autospectra $\Gamma_{x_k x_k}(\omega)$.

The lag covariance matrix can be recovered from the spectral matrix by inverting the Fourier transform. Thus

$$\Sigma_{xx}(\tau) = \int_{-\frac{1}{2}}^{\frac{1}{2}} \boldsymbol{\Gamma}_{xx}(\omega) e^{2\pi i \tau\omega}\, d\omega \tag{16.27}$$

and in particular, the covariance matrix is given by

$$\begin{aligned} \Sigma_{xx}(0) &= \Sigma_{xx} = \int_{-\frac{1}{2}}^{\frac{1}{2}} \boldsymbol{\Gamma}_{xx}(\omega)\, d\omega \qquad (16.28)\\ &= 2\int_0^{\frac{1}{2}} \Lambda_{xx}(\omega)\, d\omega \end{aligned}$$

where the co-spectrum matrix $\Lambda_{xx}(\omega)$ is the real part of the spectral matrix. Similarly, the quadrature spectrum matrix $\Psi_{\vec{x}}$ is the imaginary part of the spectral matrix (see [11.4.1]).

It follows from (16.28) that the conventional EOFs are the eigenvectors of the co-spectrum matrix of the process $\vec{\mathbf{X}}$.

When two different random vectors $\vec{\mathbf{X}}$ and $\vec{\mathbf{Y}}$ with dimensions m_x and m_y are considered, then the rectangular $m_x \times m_y$ cross-covariance matrix $\Sigma_{xy} = \big(\mathrm{Cov}(\mathbf{X}_j, \mathbf{Y}_k)\big)_{jk}$ describes the covariability of the two vectors. The $m_x \times m_y$ matrix of Fourier transforms of the entries in the cross-covariance matrix is known as the cross-spectral matrix and denoted by $\boldsymbol{\Gamma}_{xy}$.

16.2.7 Hilbert Transform and the Spectral Matrix. The covariance matrix of the Hilbert transform is equal to the covariance matrix of the original process. This follows directly from (16.11) and (16.28).

We saw in [16.2.2] that the Hilbert transform may be viewed as a linear filter h. It therefore follows from (11.74) that the cross-spectral matrix between $\vec{\mathbf{X}}_t$ and $\vec{\mathbf{Y}}_t$ is given by

$$\begin{aligned} \boldsymbol{\Gamma}_{x^{\mathrm{H}}x}(\omega) &= \mathcal{F}\{h\}(\omega)\boldsymbol{\Gamma}_{xx}(\omega) \qquad (16.29)\\ &= H(\omega)[\boldsymbol{\Lambda}_{xx} + i\boldsymbol{\Psi}_{xx}](\omega)\\ &= \begin{cases} (\boldsymbol{\Psi}_{xx} - i\,\boldsymbol{\Lambda}_{xx})(\omega) & \text{if } \omega > 0\\ (\boldsymbol{\Psi}_{xx} + i\,\boldsymbol{\Lambda}_{xx})(\omega) & \text{if } \omega < 0. \end{cases} \end{aligned}$$

Therefore

$$\Sigma_{x^{\mathrm{H}}x} = -\Sigma_{xx^{\mathrm{H}}} = 2\int_0^{\frac{1}{2}} \boldsymbol{\Psi}_{xx}(\omega)\, d\omega. \tag{16.30}$$

16.3 Complex and Hilbert EOFs

16.3.1 Outline. EOFs were defined in Chapter 13 not only for real vectors $\vec{\mathbf{X}}$ but also for complex random vectors $\vec{\mathbf{Y}}$ (although we showed examples only for real vectors).[3] In this section we introduce the Hilbert EOFs that are a special case of complex EOFs, that is, EOFs derived

[3] Here we use $\vec{\mathbf{Y}}$ to denote complex vectors and reserve $\vec{\mathbf{X}}$ for real vectors.

from complex random vectors. We first review the concept of complex EOFs in [16.3.2–4].

The straightforward way to define Hilbert EOFs is to complexify a random vector by adding its Hilbert transform as an artificial imaginary component. Then the Hilbert EOFs are simply the complex EOFs of this complexified random vector. This is discussed in [16.3.6]. The direct approach is useful when most of the variability is confined to a relatively narrow frequency. When this is not the case, the approach described in [16.3.7] may be useful. It involves computing eigenvectors from the spectral matrix after it has been averaged over a frequency band. Some computational aspects of complex EOF analysis are explored in [16.3.8,9] and examples are presented in [16.3.10,11]. Their interpretation and estimation is briefly considered in [16.3.12,13] and a further example is presented in [16.3.15].

16.3.2 Reminder: Complex EOFs. We know from the conventional EOF analysis, (Chapter 13) that the eigenvectors $\vec{e}^{\,k}$ of the covariance matrix Σ_{yy} of a complex random vector $\vec{\mathbf{Y}}$ form a basis such that $\vec{\mathbf{Y}}_t$ can be expanded as

$$\vec{\mathbf{Y}}_t = \sum_k \alpha_k(t)\vec{e}^{\,k} \tag{16.31}$$

with the 'principal components'

$$\alpha_k(t) = \langle \vec{\mathbf{Y}}_t, \vec{e}^{\,k}\rangle = \mathbf{Y}_t^{\dagger}\vec{e}^{\,k}. \tag{16.32}$$

The basis is 'optimal' in the sense that, for every $K = 1, \ldots, m$, the expected error

$$\begin{aligned}\epsilon_K &= \Big\|\vec{\mathbf{Y}}_t - \sum_{k=1}^{K}\alpha_k(t)\vec{e}^{\,k}\Big\|^2 \\ &= \mathrm{Var}(\vec{\mathbf{Y}}) - \sum_{k=1}^{K}\lambda_k \end{aligned}\tag{16.33}$$

is smaller for the EOFs than for any other basis.

The complex EOFs may be displayed as a pair of patterns, representing the real and imaginary components $\vec{e}^{\,k}_R$ and $\vec{e}^{\,k}_I$. An alternative representation uses polar coordinates:

$$e^k_j = A^k_j \exp\big(i\,\phi^k_j\big) \tag{16.34}$$

for each component $j = 1, \ldots, m$ of the m-dimensional vector $\vec{e}^{\,k}$. Thus, the kth complex EOF may also be plotted as a pattern of two-dimensional vectors, with vector of A^k_j and angle ϕ^k_j plotted at each point in much the same way that we plot the vector wind. Note that complex eigenvectors are unique only up to a constant $\vec{e}^{\,i\xi}$ where ξ is an arbitrary angle. Thus the angles ϕ^k_j may be expressed relative to any *a priori* specified angle.

This ambiguity with respect to the angle of complex EOFs may be used to rotate the EOFs in the complex domain so that either the imaginary and the real parts of each EOF are orthogonal (i.e., $\vec{e}^{\,k\,\mathrm{T}}_R\vec{e}^{\,k}_I = 0$), *or* the real and imaginary components of the EOF coefficients are uncorrelated (i.e., $\mathrm{Cov}(\mathrm{Re}(\alpha_k), \mathrm{Im}(\alpha_k)) = 0$).

The complex EOF coefficient may be written in polar coordinates as

$$\alpha_k(t) = a_k(t)\exp\big(i\psi_k(t)\big) \tag{16.35}$$

The part of the field or *signal* that is represented by the kth EOF at time t is given by

$$\alpha_k(t)\vec{e}^{\,k} = a_k(t)\vec{A}^k \exp\big((i\psi_k(t) + \vec{\phi}^k)\big)$$

where $\vec{A}^k$ is the vector of amplitudes $(A^k_1, \ldots, A^k_m)^{\mathrm{T}}$ and $\vec{\phi}^k$ is the corresponding vector of angles $(\phi^k_1, \ldots, \phi^k_m)^{\mathrm{T}}$. Thus the spatial distribution of a signal $\alpha_k(t)\vec{e}^{\,k}$ at a given time t is obtained by rotating the elements of vector $\vec{e}^{\,k}$ through a common angle $\psi_k(t)$ and scaling the elements with a common factor $a_k(t)$.

The eigenvalues obtained in an EOF analysis indicate the variance of the input vector that is carried by the corresponding principal component (EOF coefficient). This statement is also valid for complex input vectors $\vec{\mathbf{Y}}$. However, no general statement can be made about the amount of variance that is represented by just the real or imaginary part of the principal component.[4]

We present an example of a complex EOF analysis in the next subsection.

16.3.3 An Example of a Complex EOF Analysis: An Analysis of Velocities and Wind Stress Currents at a Coastal Mooring. Several moored sensors were used in an observational campaign to measure surface variables such as wind stress and sub-surface variables in the Santa Barbara channel of the coast of California. The observational campaign extended over 60 days, during which velocities were recorded every 7.5 min at five depths and wind stress was recorded hourly at two neighbouring locations (see Brink and Muench [67]). Figure 16.5 shows the mooring location, the mean wind stress vectors, and the mean current vectors. The wind stress is directed

[4] For example, it is easy to construct a complex random vector that has a first complex EOF $\vec{e}^{\,1}$ such that $\mathrm{Var}(\mathrm{Re}(\alpha_1)) = 0$.

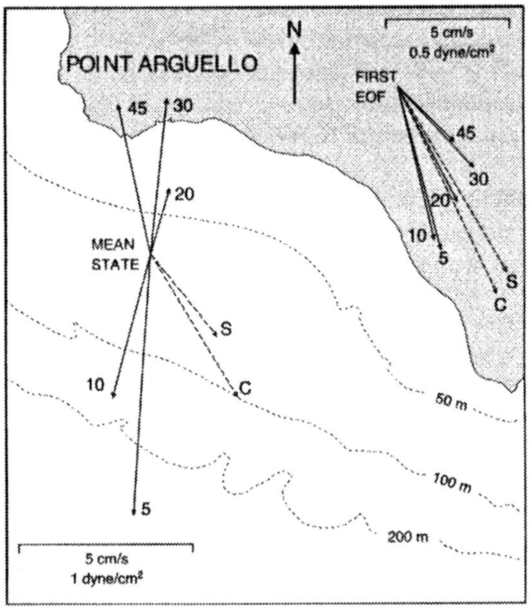

Figure 16.5: *Mean and first complex EOF of currents (solid arrows; depth in metres given by numbers) at a mooring in the Santa Barbara Channel and wind stress at neighbouring buoys (labelled S and C). The mean state is the time average of the currents and the wind stress, and the first complex EOF was calculated separately for the wind stress and for the currents.*
The mooring is located at the origin of the 'mean state' vector bundle. The EOF vector bundle is drawn at another point for convenience.
Adapted from Brink and Muench [67].

towards the southeast on average. Consistent with Ekman theory, the near-surface currents are southerly (i.e., to the right of the mean wind stress) and deeper currents, between 20 m and the bottom at about 60 m, are northerly.

Brink and Muench [67] performed separate complex EOF analyses for the horizontal velocities at five depths and for the wind stress at two locations. If $\mathbf{U}_j$ and $\mathbf{V}_j$ are the zonal and meridional velocities at depth j, then the complex random vector is $\vec{\mathbf{X}} = \mathbf{U}_j + i\mathbf{V}_j$, where $\vec{\mathbf{U}} = (\mathbf{U}_1, \ldots, \mathbf{U}_5)^{\mathrm{T}}$ and $\vec{\mathbf{V}} = (\mathbf{V}_1, \ldots, \mathbf{V}_5)^{\mathrm{T}}$. The two-dimensional complex wind stress vector is constructed similarly from the zonal and meridional components of the wind stress at the two locations.

The first complex EOFs of the velocity vectors and of the two wind stress vectors are also shown in Figure 16.5 as a vector bundle in the upper right hand corner. As mentioned above, complex EOFs have arbitrary base angles. Thus the orientation of the velocity and wind stress EOFs was chosen to maximize the correlation (0.62) between the corresponding EOF coefficients.

The first velocity CEOF consists of a rather uniform set of anomalies even though the mean state varies considerably with depth in terms of speed and direction. The most important pattern of current variability is characterised by a maximum current speed anomaly at the surface and counterclockwise veering with increasing depth. Thus positive current anomalies near the surface tend to be associated with weaker anomalies at depths related to the left of the near-surface anomaly.

The first CEOF of the wind stress indicates that it varies very similarly at the two locations. Current anomalies near the surface tend to lie to the right of the wind stress anomalies, and those at greater depths tend to lie to the left.

16.3.4 Complex EOF Analysis and Propagating Waves. Horel [181] points out that under special circumstances, such as waves associated with out-of-phase zonal and meridional currents, propagating oscillating signals may be identified through a complex EOF analysis by attributing the zonal current to the real part of a complex vector field, and the meridional current to the imaginary part. Studies pursuing this idea are listed by Horel [181]. In general, though, such an approach is unable to detect propagating signals.

16.3.5 EOFs of the Complexified Process. We now consider the complexified process

$$\vec{\mathbf{Y}} = \vec{\mathbf{X}} + i\vec{\mathbf{X}}^{\mathrm{H}} \tag{16.36}$$

where $\vec{\mathbf{X}}^{\mathrm{H}}$ is the Hilbert transform (16.2) of $\vec{\mathbf{X}}$. Without loss of generality, we can assume that the process has zero mean, and we find that (cf. (16.30))

$$\begin{aligned}\Sigma_{yy} &= \mathcal{E}\Big((\vec{\mathbf{X}}_t + i\vec{\mathbf{X}}_t^{\mathrm{H}})(\vec{\mathbf{X}}_t + i\vec{\mathbf{X}}_t^{\mathrm{H}})^{\dagger}\Big) \\ &= 2\Sigma_{xx} + i(\Sigma_{x^{\mathrm{H}}x} - \Sigma_{xx^{\mathrm{H}}}) \\ &= 2\big(\Sigma_{xx} + i\Sigma_{x^{\mathrm{H}}x}\big). \end{aligned} \tag{16.37}$$

This is a Hermitian matrix and therefore has a set of orthogonal complex eigenvectors $\vec{e}^{\,k}$ with real non-negative eigenvalues λ_k. These eigenvectors are said to be the *Hilbert EOFs* of the process $\vec{\mathbf{X}}$.

The principal components (or EOF coefficients, cf. (16.32)) of the Hilbert EOFs have special properties. If we write the Hilbert EOFs as

$\vec{e}^{\,k} = \vec{e}_R^{\,k} + i\vec{e}_I^{\,k}$, then the EOF coefficient may be expanded as

$$\begin{aligned}\alpha_k(t) &= (\vec{\mathbf{X}}_t)^{\mathrm{T}}\vec{e}_R^{\,k} + (\vec{\mathbf{X}}_t^{\mathrm{H}})^{\mathrm{T}}\vec{e}_I^{\,k} \\ &\quad + i\big((\vec{\mathbf{X}}_t)^{\mathrm{T}}\vec{e}_I^{\,k} - (\vec{\mathbf{X}}_t^{\mathrm{H}})^{\mathrm{T}}\vec{e}_R^{\,k}\big).\end{aligned} \tag{16.38}$$

Then, if we take the Hilbert transform of the EOF coefficients themselves, we see that

$$(\alpha_k(t))^{\mathrm{H}} = -i\alpha_k(t) \tag{16.39}$$

(a proof is given in Appendix M). Thus, the Hilbert transform of the EOF coefficient is just the untransformed coefficient rotated 90° in the complex domain. It therefore follows that the real and imaginary parts of the complex EOF coefficients are related through their Hilbert transforms by

$$\mathrm{Re}(\alpha_k(t)) = -\big(\mathrm{Im}(\alpha_k(t))\big)^{\mathrm{H}} \tag{16.40}$$

$$\mathrm{Im}(\alpha_k(t)) = \big(\mathrm{Re}(\alpha_k(t))\big)^{\mathrm{H}}, \tag{16.41}$$

and that their variances are equal:

$$\mathrm{Var}(\mathrm{Im}(\alpha_k)) = \mathrm{Var}(\mathrm{Re}(\alpha_k)). \tag{16.42}$$

The EOF expansion (16.31) of the complexified process (16.36) also has special properties. Expanding $\vec{\mathbf{Y}}_t$ as

$$\vec{\mathbf{Y}}_t = \sum_k \vec{\mathbf{Y}}_t^k,$$

where $\vec{\mathbf{Y}}_t^k = \alpha_k(t)\vec{e}^{\,k}$, and equating with the real and imaginary parts of (16.36), we find that

$$\vec{\mathbf{X}}_t = \sum_k \mathrm{Re}(\vec{\mathbf{Y}}_t^k) \tag{16.43}$$

$$\vec{\mathbf{X}}_t^{\mathrm{H}} = \sum_k \mathrm{Im}(\vec{\mathbf{Y}}_t^k) = \sum_k \big(\mathrm{Re}(\vec{\mathbf{Y}}_t^k)\big)^{\mathrm{H}}.$$

Thus the real and imaginary parts of the complexified process (16.36) have the same Hilbert EOF expansion. This is easily confirmed with (16.40) and (16.41) by noting that

$$\begin{aligned}\mathrm{Re}(\vec{\mathbf{Y}}_t^k) &= \mathrm{Re}\big(\alpha_k(t)\big)\vec{e}_R^{\,k} - \mathrm{Im}\big(\alpha_k(t)\big)\vec{e}_I^{\,k} \\ \mathrm{Im}(\vec{\mathbf{Y}}_t^k) &= \mathrm{Re}\big(\alpha_k(t)\big)\vec{e}_I^{\,k} + \mathrm{Im}\big(\alpha_k(t)\big)\vec{e}_R^{\,k} \\ &= \big(\mathrm{Re}\big(\alpha_k(t)\big)\vec{e}_R^{\,k} - \mathrm{Im}\big(\alpha_k(t)\big)\vec{e}_I^{\,k}\big)^{\mathrm{H}} \\ &= \big(\mathrm{Re}(\vec{\mathbf{Y}}_t^k)\big)^{\mathrm{H}}\end{aligned} \tag{16.44}$$

It follows, therefore, that the Hilbert EOF represents equal amounts of variance in the input time series and its Hilbert transform.

16.3.6 The Spectral Matrix of the Complexified Process. Equation (16.37), together with (16.28) and (16.30), tell us that the covariance matrix of the complexified process equals the integral of the spectral matrix of $\vec{\mathbf{X}}$ over all *positive* frequencies:

$$\mathbf{\Sigma}_{yy} = 4\int_0^{\frac{1}{2}} \mathbf{\Gamma}_{xx}(\omega)\,d\omega \tag{16.45}$$

where $\vec{\mathbf{Y}}$ is the complexified process (16.36). Thus, the Hilbert EOFs are not only the eigenvectors of the covariance matrix of the complexified process, but also the eigenvectors of the frequency integrated spectral matrix of process $\vec{\mathbf{X}}$.[5]

16.3.7 Frequency Domain EOFs. The Hilbert EOFs can be interpreted as characteristic patterns of the spectral matrix of $\vec{\mathbf{X}}$ when the variability of the process is confined to a narrow frequency band $\omega_0 \pm \delta\omega$. In that case

$$\mathbf{\Sigma}_{yy} \propto \mathbf{\Gamma}_{xx}(\omega_0) \tag{16.46}$$

and the Hilbert EOFs are the eigenvectors of the spectral matrix at frequency ω_0. It is therefore natural to extend the Hilbert EOF analysis to the frequency domain by applying it to the spectral matrix $\mathbf{\Gamma}_{xx}(\omega)$ so that the characteristic modes of variation can be identified for arbitrary time scales ω^{-1} where $\omega \in [0, 1/2]$.

16.3.8 Equivalence of Complex and Real Eigenproblems. The real and imaginary parts of the Hilbert EOFs are related to the cross-covariances between the components of the input vector and its Hilbert transform. This relationship is easier to see when the eigenproblem is expressed in real terms. We therefore describe the corresponding real eigenproblem here, and then return to the role of the cross-covariances in [16.3.9].

It is easily shown that $\vec{e}^{\,k} = \vec{e}_R^{\,k} + i\,\vec{e}_I^{\,k}$ is an eigenvector of the complex Hermitian matrix $\mathbf{\Sigma}_{yy} = \mathbf{\Sigma}_R + i\,\mathbf{\Sigma}_I$ with eigenvalue λ_k if and only if $\vec{e}^{\,k}$ satisfies the real eigen-equation

$$\begin{pmatrix} \mathbf{\Sigma}_R & -\mathbf{\Sigma}_I \\ \mathbf{\Sigma}_I & \mathbf{\Sigma}_R \end{pmatrix}\begin{pmatrix} \vec{e}_R^{\,k} \\ \vec{e}_I^{\,k} \end{pmatrix} = \lambda_k \begin{pmatrix} \vec{e}_R^{\,k} \\ \vec{e}_I^{\,k} \end{pmatrix}. \tag{16.47}$$

[5] Note that $\mathbf{\Gamma}_{xx}(\omega)$ is only integrated over *positive* frequencies. When $\mathbf{\Gamma}_{xx}(\omega)$ is integrated over both positive and negative frequencies, the contribution from the anti-symmetric quadrature spectrum is cancelled and we arrive at the real covariance matrix and the conventional EOFs since

$$\mathbf{\Sigma}_{xx} = \int_{-\frac{1}{2}}^{\frac{1}{2}} \mathbf{\Gamma}_{xx}(\omega)\,d\omega = 2\int_0^{\frac{1}{2}} \mathbf{\Lambda}_{xx}(\omega)\,d\omega.$$

The eigenvectors of $\mathbf{\Sigma}_{yy}$ are orthogonal and are ordinarily chosen with unit length so that

$$\langle \vec{e}^{\,k}, \vec{e}^{\,j} \rangle = (\vec{e}^{\,k})^{\dagger} \vec{e}^{\,j} = \delta_{kl}. \tag{16.48}$$

In real terms, equation (16.48) reads

$$\begin{aligned} (\vec{e}^{\,k}_R)^{\mathrm{T}} \vec{e}^{\,j}_R + (\vec{e}^{\,k}_I)^{\mathrm{T}} \vec{e}^{\,j}_I &= \delta_{kl} \\ (\vec{e}^{\,k}_R)^{\mathrm{T}} \vec{e}^{\,j}_I - (\vec{e}^{\,k}_I)^{\mathrm{T}} \vec{e}^{\,j}_R &= 0. \end{aligned} \tag{16.49}$$

The complex eigenproblem $\mathbf{\Sigma}_{yy} \vec{e}^{\,k} = \lambda_k \vec{e}^{\,k}$ has m real eigenvalues and m complex eigenvectors $\vec{e}^{\,k}$. The real eigenproblem (16.47) has $2m$ real eigenvalues $\lambda_1, \lambda_1, \ldots, \lambda_m, \lambda_m$, where $\lambda_1, \ldots, \lambda_m$ are the eigenvalues of the complex eigenproblem. The corresponding set of $2m$ eigenvectors is given by

$$\left\{ \begin{pmatrix} \vec{e}^{\,k}_R \\ \vec{e}^{\,k}_I \end{pmatrix}, \begin{pmatrix} -\vec{e}^{\,k}_I \\ \vec{e}^{\,k}_R \end{pmatrix} : k = 1, \ldots, m \right\}.$$

Equation (16.49) can be used to verify that these vectors are orthonormal.

16.3.9 Real Eigenproblems for the Determination of Hilbert EOFs. We may use the result of the preceding subsection to characterize the frequency domain EOFs as eigenvectors of a real, frequency-integrated matrix. Let $\int \mathbf{\Gamma}_{xx}$ denote the integral of the spectral matrix over a frequency band $\omega_0 \pm \delta\omega$:

$$\int \mathbf{\Gamma}_{xx} = \int_{\omega_0 - \delta\omega}^{\omega_0 + \delta\omega} \mathbf{\Gamma}_{xx}(\omega)\, d\omega.$$

The complex $m \times m$ matrix $\int \mathbf{\Gamma}_{xx}$ corresponds to the $2m \times 2m$ real matrix

$$\begin{pmatrix} \int \mathbf{\Lambda}_{xx} & -\int \mathbf{\Psi}_{xx} \\ \int \mathbf{\Psi}_{xx} & \int \mathbf{\Lambda}_{xx} \end{pmatrix}. \tag{16.50}$$

where $\int \mathbf{\Lambda}_{xx}$ and $\int \mathbf{\Psi}_{xx}$ are the corresponding integrated $m \times m$ co-spectrum and quadrature spectrum matrices. The frequency band $\omega_0 \pm \delta\omega$ could encompass all or part of $[0, 1/2]$.

Equation (16.37) shows that there is also a real $2m \times 2m$ real counterpart to the $m \times m$ complex covariance matrix $\mathbf{\Sigma}_{yy}$ of the complexified process:

$$\begin{pmatrix} \mathbf{\Sigma}_{xx} & -\mathbf{\Sigma}_{x\mathrm{H}_x} \\ \mathbf{\Sigma}_{x\mathrm{H}_x} & \mathbf{\Sigma}_{xx} \end{pmatrix}. \tag{16.51}$$

Thus we see that both the Hilbert and frequency domain EOFs depend upon the cross-covariances of the input series and its Hilbert transform.

16.3.10 Example: Several Uncorrelated Processes. What are the Hilbert EOFs of a stationary process $\vec{\mathbf{X}}_t = (\mathbf{X}_{1t}, \ldots, \mathbf{X}_{mt})^{\mathrm{T}}$ such that the cross-covariance function (and thus quadrature spectra) between any two components is zero? Under these circumstances the complex covariance matrix $\mathbf{\Sigma}_{yy}$ equals twice the real covariance matrix $\mathbf{\Sigma}_{xx}$. Thus the conventional EOFs are also the Hilbert EOFs.[6]

What are the coefficients of the Hilbert EOFs in this case? If we assume that the Hilbert EOFs $\vec{e}^{\,c}$ have been normalized so that they equal the conventional EOFs $\vec{e}^{\,r}$, then the coefficient α^c of the Hilbert EOF $\vec{e}^{\,c}$ is the dot product of the complexified process $\vec{\mathbf{X}} + i\vec{\mathbf{X}}^{\mathrm{H}}$ with the conventional real EOF $\vec{e}^{\,r}$:

$$\begin{aligned} \alpha^c &= \langle \vec{\mathbf{X}}, \vec{e}^{\,r} \rangle - i \langle \vec{\mathbf{X}}^{\mathrm{H}}, \vec{e}^{\,r} \rangle \\ &= \alpha^r - i{\alpha^r}^{\mathrm{H}} \end{aligned}$$

where $\alpha^r = \langle \vec{\mathbf{X}}, \vec{e}^{\,r} \rangle$ is the conventional EOF coefficient.

In summary, Hilbert EOF analysis has no advantages over the conventional EOF analysis when $\vec{\mathbf{X}}_t$ consists of uncorrelated processes. Note also that neither of these approaches can provide information that is useful for characterizing the temporal correlation of the time series that comprise $\vec{\mathbf{X}}_t$.

16.3.11 Example: The POP Case. Another situation occurs when the two processes are linked through a lag relationship. A prototype of this situation is the bivariate POP case discussed in [11.3.8] and [11.4.10]. We will consider a bivariate AR(1) process (cf. (11.45)) of the form

$$\vec{\mathbf{X}}_t = r \begin{pmatrix} u & -v \\ v & u \end{pmatrix} \vec{\mathbf{X}}_{t-1} + \vec{\mathbf{Z}}_t$$

where $|r| < 1$, $u^2 + v^2 = 1$, and $\vec{\mathbf{Z}}_t$ is a bivariate white noise process with covariance matrix

$$\mathbf{\Sigma}_{zz} = \sigma^2 \begin{pmatrix} 1 & 0 \\ 0 & 1 \end{pmatrix}.$$

The system generates oscillatory behaviour with $\mathbf{X}_{1t}$ leading $\mathbf{X}_{2t}$ when v is positive. Note also that processes $\mathbf{X}_{1t}$ and $\mathbf{X}_{2t}$ are uncorrelated at lag zero. In fact,

$$\mathbf{\Sigma}_{xx}(0) = \frac{\sigma^2}{1 - r^2} \begin{pmatrix} 1 & 0 \\ 0 & 1 \end{pmatrix}.$$

[6] But note that Hilbert EOFs may be multiplied by any complex number whereas ordinary EOFs may only be multiplied by real numbers.

Therefore, the conventional EOFs $\vec{e}^{\,k}$ of $\vec{\mathbf{X}}_t$ are degenerate; specific choices of $\vec{e}^{\,k}$ are the two unit vectors $(0,1)^{\mathrm{T}}$ and $(1,0)^{\mathrm{T}}$ (cf. [13.1.9]).[7]

Recalling equation (11.81), we find that the spectral matrix of $\vec{\mathbf{X}}_t$ is

$$\begin{aligned}\mathbf{\Gamma}_{xx}(\omega) &= \begin{pmatrix}\Gamma_{11}(\omega) & \Gamma_{12}(\omega)\\ \Gamma_{21}(\omega) & \Gamma_{22}(\omega)\end{pmatrix}\\ &= \begin{pmatrix}\Gamma_{11}(\omega) & i\Psi_{12}(\omega)\\ -i\,\Psi_{12}(\omega) & \Gamma_{11}(\omega)\end{pmatrix}.\end{aligned}$$

The Hilbert EOFs are the solutions of the eigenproblem

$$\left(\int_0^{\frac{1}{2}}\mathbf{\Gamma}_{xx}(\omega)\,d\omega\right)\vec{e}^{\,k} = \lambda_k\vec{e}^{\,k}. \tag{16.52}$$

The eigenvalues are

$$\begin{aligned}\lambda_1 &= \textstyle\int_0^{\frac{1}{2}}\big(\Gamma_{11}(\omega)-\Psi_{12}(\omega)\big)\,d\omega\\ \lambda_2 &= \textstyle\int_0^{\frac{1}{2}}\big(\Gamma_{11}(\omega)+\Psi_{12}(\omega)\big)\,d\omega,\end{aligned} \tag{16.53}$$

and the corresponding eigenvectors are

$$\vec{e}^{\,1} = \begin{pmatrix}1\\ i\end{pmatrix} \text{ and } \vec{e}^{\,2} = \begin{pmatrix}1\\ -i\end{pmatrix}. \tag{16.54}$$

The larger of the two eigenvalues is λ_1 since Ψ_{12} is negative for positive v.

Note that the Hilbert EOFs are markedly different from the conventional EOFs.

The time coefficients α_k of the Hilbert EOFs are

$$\begin{aligned}\alpha_1 &= (\vec{\mathbf{X}}_t + i\,\vec{\mathbf{X}}_t^{\mathrm{H}})^{\dagger}\vec{e}^{\,1}\\ &= \mathbf{X}_{1t} + \mathbf{X}_{2t}^{\mathrm{H}} + i\,(\mathbf{X}_{2t} - \mathbf{X}_{1t}^{\mathrm{H}})\\ &= \mathbf{X}_{1t} + \mathbf{X}_{2t}^{\mathrm{H}} - i\,(\mathbf{X}_{2t} + \mathbf{X}_{1t}^{\mathrm{H}})^{\mathrm{H}}\end{aligned} \tag{16.55}$$

which is consistent with equations (16.40) and (16.41). Similarly

$$\alpha_2 = \mathbf{X}_{1t} - \mathbf{X}_{2t}^{\mathrm{H}} + i\,(\mathbf{X}_{2t} - \mathbf{X}_{1t}^{\mathrm{H}})^{\mathrm{H}}.$$

The 'signal' represented by the first Hilbert EOF is

$$\alpha_1\vec{e}^{\,1} = \mathrm{Re}(\alpha_1\vec{e}^{\,1}) + i\,\mathrm{Im}(\alpha_1\vec{e}^{\,1})$$

where

$$\mathrm{Re}(\alpha_1\vec{e}^{\,1}) = \begin{pmatrix}\mathbf{X}_{1t} + \mathbf{X}_{2t}^{\mathrm{H}}\\ (\mathbf{X}_{1t} + \mathbf{X}_{2t}^{\mathrm{H}})^{\mathrm{H}}\end{pmatrix} \tag{16.56}$$

and

$$\mathrm{Im}(\alpha_1\vec{e}^{\,1}) = \begin{pmatrix}-(\mathbf{X}_{1t} + \mathbf{X}_{2t}^{\mathrm{H}})^{\mathrm{H}}\\ \mathbf{X}_{1t} + \mathbf{X}_{2t}^{\mathrm{H}}\end{pmatrix}$$

[7] This example is easily generalized to the case in which the noise components are correlated: then $\mathbf{\Sigma}_{xx}(0) = \frac{1}{1-r^2}\mathbf{\Sigma}_{zz}$, and the conventional EOFs of $\vec{\mathbf{X}}_t$ coincide with those of $\vec{\mathbf{Z}}_t$.

so that, consistent with (16.41),

$$\big(\mathrm{Im}(\alpha_1\vec{e}^{\,1})\big)^{\mathrm{H}} = \mathrm{Re}(\alpha_1\vec{e}^{\,1}). \tag{16.57}$$

For the second EOF we find

$$\mathrm{Re}(\alpha_2\vec{e}^{\,2}) = \begin{pmatrix}\mathbf{X}_{1t} - \mathbf{X}_{2t}^{\mathrm{H}}\\ (\mathbf{X}_{1t} - \mathbf{X}_{2t}^{\mathrm{H}})^{\mathrm{H}}\end{pmatrix} \tag{16.58}$$

and

$$\mathrm{Im}(\alpha_2\vec{e}^{\,2}) = \begin{pmatrix}(\mathbf{X}_{1t} - \mathbf{X}_{2t}^{\mathrm{H}})^{\mathrm{H}}\\ -(\mathbf{X}_{1t} - \mathbf{X}_{2t}^{\mathrm{H}})\end{pmatrix}.$$

Thus, for both 'signal' time series, the second element is the Hilbert transform of the first.

We showed in [11.4.11] that the specific system considered here tends to form 'typical' $\vec{\mathbf{x}}_t$ sequences of the type (11.83),

$$\begin{aligned}\cdots\to\begin{pmatrix}1\\0\end{pmatrix}\to\begin{pmatrix}0\\1\end{pmatrix}\to\begin{pmatrix}-1\\0\end{pmatrix}\\ \to\begin{pmatrix}0\\-1\end{pmatrix}\to\begin{pmatrix}1\\0\end{pmatrix}\to\begin{pmatrix}0\\1\end{pmatrix}\to\cdots\end{aligned} \tag{16.59}$$

when v is positive (cf. Figure 15.1). Therefore, since these are oscillatory processes, it is reasonable to interpret the Hilbert transform as a rate of change. Our system tends to generate $\vec{\mathbf{x}}_t^{\mathrm{H}}$-sequences identical to (16.59) but shifted in time by a quarter of a period so that $\vec{\mathbf{x}}_t = (1,0)^{\mathrm{T}}$ and $\vec{\mathbf{x}}_t^{\mathrm{H}} = (0,1)^{\mathrm{T}}$ appear together. (The 'change' $\vec{\mathbf{X}}_t^{\mathrm{H}}$ leads the 'state' $\vec{\mathbf{X}}_t$.) The 'state' of the second component equals the 'change' of the first component, and the 'state' of the first component is the reversed 'change' of the second.

The two 'signals' represented by the two Hilbert EOFs, (16.56) and (16.58), may then be characterized by sequences of the type (16.59) as well. The sequence (16.59) implies

$$\mathbf{X}_2^{\mathrm{H}} \approx \mathbf{X}_1 \tag{16.60}$$

so that

$$\mathrm{Re}(\alpha_1\vec{e}^{\,1}) \approx 2\begin{pmatrix}\mathbf{X}_1\\ \mathbf{X}_1^{\mathrm{H}}\end{pmatrix} \tag{16.61}$$

and

$$\mathrm{Re}(\alpha_2\vec{e}^{\,2}) \approx \begin{pmatrix}0\\0\end{pmatrix}. \tag{16.62}$$

Thus the first Hilbert EOF describes the dominant rotational behaviour of the system whereas the second Hilbert EOF represents just the 'residual' which is small.

Note that this interpretation is independent of the frequency interval used to form the integrals in (16.52) and (16.53).

16.3.12 Interpretation of Hilbert EOFs. The heuristic argument of the previous subsection is generally used to interpret the outcome of a Hilbert EOF analysis. Its validity depends crucially on the validity of (16.60), which is by no means a trivial assumption as demonstrated by the AR examples studied in [16.2.3].

There are no generally applicable techniques for deciding whether an estimated Hilbert EOF describes a real oscillatory signal. Prudence is clearly advisable. If possible, the data should be divided into 'learning' and 'validation' subsets so that phase relationships identified in the learning data set can be verified independently in the validation data set.

16.3.13 Estimating Hilbert EOFs. The two different definitions of Hilbert EOFs, either directly by means of the covariance matrix of the complexified process or by means of the frequency integrated spectral matrix, provide two different approaches for estimating the Hilbert EOFs.

The estimation can be done in the time domain, in which case the Hilbert transform is first estimated. This can be done either with the truncated time domain filter (16.18) or by a Fourier decomposition and phase-shifted reconstruction (see [16.2.4]). Then the complex covariance matrix is computed in the usual manner by computing

$$\frac{1}{n}\sum_{t=1}^{n}\left(\vec{\mathbf{x}}'_j + i\,\widehat{(\vec{\mathbf{x}}'_t)^{\mathrm{H}}}\right)\left(\vec{\mathbf{x}}_j + i\,\widehat{(\vec{\mathbf{x}}'_t)^{\mathrm{H}}}\right)^{\dagger} \qquad (16.63)$$

where $\vec{\mathbf{x}}_1, \ldots, \vec{\mathbf{x}}_n$ form a sample of size n and $\vec{\mathbf{x}}'_1, \ldots, \vec{\mathbf{x}}'_n$ are deviations from the sample mean. Finally, the eigenvectors of this matrix are determined.

Estimation can also be done in the frequency domain. First the width $2\delta\omega$ and the centre ω_0 of the frequency band of interest are selected. Next, an estimate of the spectral matrix with equivalent bandwidth $2\delta\omega$ is constructed (see Section 12.3 for a description of spectral estimation). This estimator is evaluated at frequency ω_0, and eigenvectors are found.

For the estimation in the frequency domain the spectral matrix is estimated for all frequencies $\omega \geq 0$ in the frequency band of interest, and then the spectral matrices are summed.

16.3.14 Applications of Hilbert EOF Analysis. Hilbert EOF analysis has been pursued extensively in climate research, for instance by Barnett [19] who pioneered this technique, Wallace and Dickinson [408], Brillinger [66], Rasmusson et al. [329], Horel [181], Wang and Mooers [413], Johnson and McPhaden [196], Trenberth and Shin [373], just to mention a few.

16.3.15 Example: Tropical Pacific Sea-surface Temperatures. We now describe a Hilbert EOF analysis of monthly mean SST anomalies in the tropical Pacific Ocean between 20° S and 20° N. The data used in this example were obtained from COADS (Woodruff et al. [425]) and cover the period 1951–90.

The annual cycle was removed by subtracting the 40-year mean for each month of the year and variations on time scales shorter than a year were removed by low-pass filtering the anomalies from the annual cycle. The filtered time series was then Hilbert transformed (16.19, 16.20). Finally, the covariance matrix of the complexified process was estimated with equation (16.63). The eigenvectors of this matrix are the estimated Hilbert EOFs.

The dominant Hilbert EOF, which represents 40% of the variance of both the filtered SST anomalies and the filtered complexified process, is shown in Figure 16.6. The real part, shown in the upper panel, depicts the mature phase of El Niño when the corresponding EOF coefficient, $\widehat{\alpha}_1(t)$, is real and positive. It also approximates the mature phase of La Niña when $\widehat{\alpha}_1(t)$ is real and negative. The imaginary part of the first Hilbert EOF, shown in the lower panel, depicts a transition phase between the warm El Niño and the cool La Niña.

The time series of complex time coefficients of the first Hilbert EOF is shown in Figure 16.7. The imaginary part, given by the dashed curve, is the Hilbert transform of the real part (recall (16.42)). By focusing on the 1982/83 El Niño event, we can see that the Hilbert transform can indeed be interpreted as a crude derivative: the imaginary part is positive until the warm event peaks in late 1982/early 1983 and then becomes negative as the event fades.

Figure 16.8 shows the same time series in polar coordinate form. The upper panel displays the amplitude of the EOF coefficient as a function of time, and the lower panel displays the phase in radians. We see that the complex coefficient tends to rotate in a clockwise direction, but not at uniform speed. The amplitude varies irregularly in time. Each 'sawtooth' in the lower panel of Figure 16.8 depicts one ENSO-like cycle. It begins with the cold version of Figure 16.6a, then rotates to the warm version in Figure 16.6b with weak warm anomalies over most of the tropical Pacific one-quarter of a period later. This

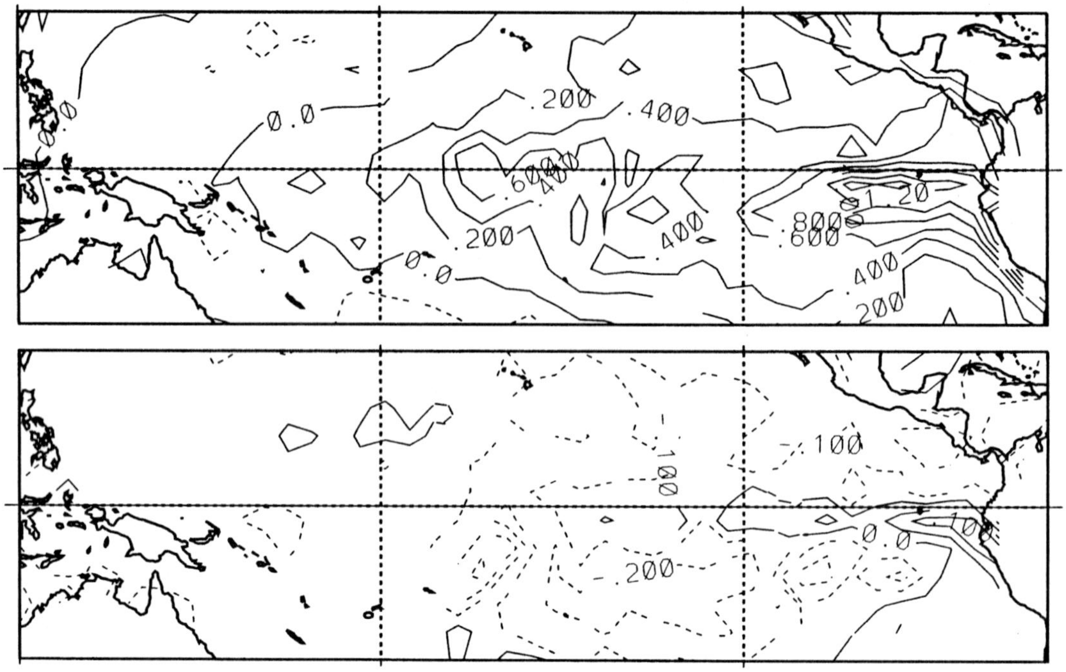

Figure 16.6: *The first Hilbert EOF of low-pass filtered tropical Pacific sea-surface temperatures. Courtesy E. Zorita. a) Real part (top). b) Imaginary part (bottom).*

is followed by the mature warm phase (positive version of Figure 16.6a) halfway through the cycle and weak negative SST anomalies (Figure 16.6b) three-quarters of the way through the cycle. The cycle is completed with the mature cold phase (Figure 16.6a multiplied by -1).

It is clear from Figure 16.8b that there is significant variability in the length of an ENSO cycle. The vertical lines in Figure 16.8 give the approximate time of warm events (short dashes) and cold events (long dashes) as identified by Kiladis and Diaz [222]. Warm events tend to occur within one radian of zero phase while cold events tend to occur 180° later. The amplitude is often, but not always, large when a warm or cold event is identified, perhaps because there is large variability from event to event in the precise spatial structure of the SST anomalies.

In summary, the Hilbert EOF analysis of the filtered SST anomalies captures the essential features of ENSO. We have found a pair of patterns that depict a substantial fraction of the ENSO cycle. The length of the cycle varies from 2 to 11 years, with a mean of about 4.5 years. Within cycles, the progression between warm and cold phases is irregular.

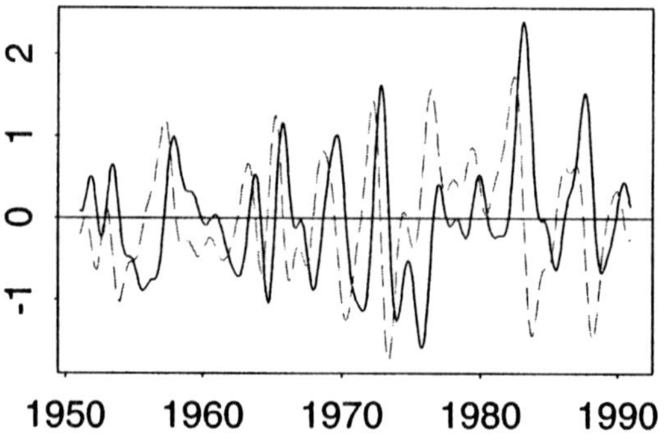

Figure 16.7: *The time series of complex time coefficients of the first Hilbert EOF of low-pass filtered tropical Pacific SST anomalies. Units: °C.*

16.3.16 Hilbert POPs. When the POP analysis (see Chapter 15) is applied to the complexified process, complex patterns are derived. As with the Hilbert EOF, these may be interpreted as specifying the 'state' and the 'rate of change' of the process. Bürger [75] has pioneered this technique and offers as an example the analysis of El Niño/Southern Oscillation in terms of monthly SST along the equatorial Pacific.

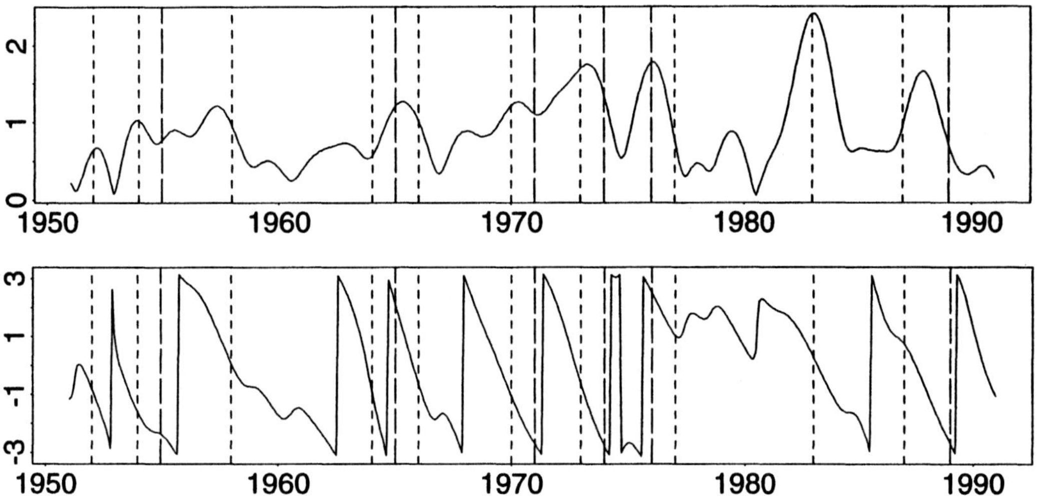

Figure 16.8: *As Figure 16.7, except in polar-coordinate form.*
a) Amplitude (top). Units: °C. b) Phase (bottom). Units: radians.

Part VI

Other Topics

Overview

The last part of the book features aspects of applications of statistical concepts that are specific to climate research and, with the exception of a section on time filters, are usually not found in other fields of statistical applications.

Chapter 17 contains those aspects that could not be logically included in the more systematically designed earlier parts of the book. In fact, many concepts in Chapter 17 overlap with material presented earlier. The so-called decorrelation time is related to the distribution of mean values calculated from serially correlated data; the potential predictability may be considered a special variant of ANOVA; teleconnections are a special representation of spatial correlations; associated correlation patterns are an offspring of regression analysis. We tried, however, to write this chapter such that the material may be understood without in-depth study of the previous chapters.

Most of the techniques in Chapter 17 were developed by climatologists while struggling with specific problems; as such, many of them are based on ad-hoc heuristic ideas with interpretations that may or may not hold in real world situations. We have presented two cases of such heuristically motived approaches, namely the frequency–wavenumber analysis and the quadrature EOFs. A typical case to this end is 'potential predictability.' We try to clarify the methodical basis of the various techniques, so that the reader may use them in a more objective manner without using tacitly inadequate intuitive interpretations (such as the misconception that the decorrelation time is the time between two independent observations).

Chapter 18 describes a classical problem in meteorology, namely a variety of techniques designed to measure the relative advantages and disadvantages of (weather) forecasts.

17 Specific Statistical Concepts in Climate Research

17.0.0 Overview. In this chapter we review several additional topics that are important in atmospheric, oceanic or other geo-environmental sciences. These topics are as follows.

In Section 17.1 we discuss the 'decorrelation time', a concept that is often misunderstood, because of its confusing name. The term suggests that it is a *physical* time scale that represents the interval between consecutive, uncorrelated observations. In fact, it is a statistical measure that compares the information content of correlated observations with that of uncorrelated observations. If a sample of n' uncorrelated observations gives a particular amount of information *about the population mean* then $n = n' \times$ 'decorrelation time' is the number of correlated observations required to obtain the same amount of information about the population mean. Similarly, other 'decorrelation times' can be derived for other parameters such as the population variance or the lag-1 correlation (cf. Trenberth [368]) by comparing the information contained about the parameters in samples of independent and dependent observations. Not only is the nomenclature confusing, but its meaning is highly dependent upon the parameter of interest.

We describe a concept called potential predictability in Section 17.2. Measures of potential predictability determine whether the variation in seasonal mean climate variables is caused by anything other than daily weather variations. If seasonal means have more variance than can be accounted for by weather noise, then part of the seasonal mean variance may be predictable from slowly varying external sources.

Processes, such as El Niño/Southern Oscillation or the Madden-and-Julian Oscillation, are often described by an *index*. It is therefore often of interest to describe how field variables, such as the sea-surface temperature distribution or the oceanic 'meridional overturning stream function',[1] evolve with the indexed process (Section 17.3). Two techniques are frequently used: 'regression' or 'associated correlation pattern' analysis and 'composite pattern' or 'epoch' analysis.

A popular and simple method for identifying dynamical links between well-separated areas is 'teleconnection' analysis (Section 17.4), which is essentially the mapping of fields of correlations.

Digital filters (Section 17.5) are tools that can be used to remove variation on time scales unrelated to the phenomenon under study. This is useful because the climate varies on many time scales, from day-to-day weather variability to the 'slow' variations connected with the coming and going of the Ice Ages. Depending upon the researcher's goals, much of the variability in the observed record may be regarded as 'noise' that obscures the 'signal' of interest. Filters can remove much of this noise.

17.1 The Decorrelation Time

17.1.1 Motivation and Definition. We defined the characteristic time τ_M in [10.2.1] as the time that is required for a system to forget its current state. This has meaning for some processes, such as MA(q) processes (cf. [10.5.2]) for which $\tau_M = q$, but not for others, such as AR(p) processes, for which $\tau_M = \infty$.

In this section we introduce another 'characteristic time,' labelled τ_D. The basic idea originates from the observation that the mean of n iid random variables $\mathbf{X}_1, \dots, \mathbf{X}_n$ has variance

$$\mathrm{Var}(\overline{\mathbf{X}}) = \frac{\sigma_X^2}{n}, \tag{17.1}$$

while the mean of n identically distributed but correlated random variables has variance

$$\mathrm{Var}(\overline{\mathbf{X}}) = \frac{\sigma_X^2}{n'} \tag{17.2}$$

where $n' \neq n$ depends upon the correlations between $\mathbf{X}_1, \dots, \mathbf{X}_n$. We call n' the *equivalent*

[1] A measure of the strength of the deep ocean circulation.

sample size.[2] The *decorrelation time* is then defined as

$$\tau_D = \lim_{n\to\infty} \frac{n}{n'}. \qquad (17.3)$$

We will show in [17.1.2] that

$$n' = \frac{n}{1+2\sum_{k=1}^{n-1}\left(1-\frac{k}{n}\right)\rho(k)} \qquad (17.4)$$

$$\tau_D = 1+2\sum_{k=1}^{\infty}\rho(k), \qquad (17.5)$$

where $\rho(\cdot)$ is the auto-correlation function of $\mathbf{X}_t$.

The decorrelation time defined in (17.5) is dimensionless since the time increment is implicitly assumed to be dimensionless. A proper dimensional definition would be

$$\tilde{\tau}_D = (\Delta t)\tau_D \qquad (17.6)$$

where Δt is the time increment.

Equation (17.5) is the appropriate definition of decorrelation time when we use the sample mean to make inferences about the population mean. However, its arbitrariness in defining a *characteristic time scale* becomes obvious when we reformulate our problem by replacing the *mean* in (17.2) with, for instance, the *variance* or the *correlation* of two processes $\mathbf{X}_t$ and $\mathbf{Y}_t$. When this is done, the appropriate characteristic times are given by (Trenberth [368])

$$\tau = 1+2\sum_{k=1}^{\infty}\rho^2(k) \qquad (17.7)$$

and

$$\tau = 1+2\sum_{k=1}^{\infty}\rho_X(k)\rho_Y(k) \qquad (17.8)$$

respectively. Thus the definition of the characteristic time is strongly dependent on the *statistical* problem under consideration. *In general, these numbers do not correspond directly to the important physical time scales of the process under study.*

17.1.2 Calculation of the τ_D. We now prove (17.4) by deriving $\mathrm{Var}(\overline{\mathbf{X}})$. Without loss of generality we assume that $\mathcal{E}(\mathbf{X}_t) = 0$ so that $\mathrm{Var}(\mathbf{X}_t) = \mathcal{E}(\mathbf{X}_t^2)$. Then, for an arbitrary time t,

$$\mathcal{E}(\overline{\mathbf{X}}^2) = \frac{1}{n^2}\sum_{i,j=0}^{n-1}\mathcal{E}(\mathbf{X}_{t+i}\mathbf{X}_{t+j})$$

$$= \frac{1}{n^2}\sum_{i,j=0}^{n-1}\gamma(i-j) \qquad (17.9)$$

$$= \frac{1}{n}\sum_{k=-n+1}^{n-1}\left(1-\frac{|k|}{n}\right)\gamma(k).$$

The last expression is obtained by gathering terms in (17.9) with identical differences $i-j$. Equation (17.5) follows by taking the limit as $n\to\infty$.[3]

17.1.3 Estimation of the Decorrelation Time. A straightforward way to estimate n' is to substitute the estimated auto-correlation function directly into (17.4). Another approach is to fit an AR(p) model to the data and then use the derived auto-correlation function correlation function in (17.4). A third approach is based on the observation that

$$\mathrm{Var}(\overline{\mathbf{X}}) = \frac{\Gamma_{xx}(0)}{n}$$

in the limit as $n\to\infty$, where $\Gamma_{xx}(\omega)$ is the spectral density function of $\mathbf{X}_t$. Therefore

$$n' \approx \frac{\sigma_X^2}{\Gamma_{xx}(0)}n$$

and

$$\tau_D \approx \frac{\Gamma_{xx}(0)}{\sigma_X^2}.$$

Thus τ_D can also be estimated by estimating the spectral density at frequency zero.[4]

Thiébaux and Zwiers [363] examined various approaches and found that the true n' values are difficult to estimate accurately. In particular, the first approach performed very poorly. The second approach (see also Zwiers and von Storch [454]) is the best of the three methods when samples are large, and the spectral approach produces better estimates when the samples are moderate to small.

Prior knowledge can sometimes be used to improve the estimate of n'. For example, we know that $n' < n$ when the observed process is 'red'

[2]See also [6.6.8], where this number comes up in the context of testing hypotheses about the mean.

[3]To be precise we must assume that the auto-correlation function is absolutely summable. This is frequently called a 'mixing condition' in the time series literature (see [10.3.0] and also texts such as [323] or [66]).

[4]The usual approach is to use a good spectral estimator (cf. Section 10.3) to estimate the spectral density at a frequency near zero and then to *extrapolate* this estimate to $\omega = 0$. Madden [263] calls this the 'low-frequency white noise' extension of the estimated spectral density. This approach works because the spectral density functions of many weakly stationary ergodic processes are continuous and symmetric about the origin, and therefore have zero slope at the origin. These processes are approximately white at long time scales.

because then $\Gamma_{xx}(0) > \sigma_X^2$, as in the top panel of Figure 17.1. Thus it would be reasonable to truncate any estimate $\widehat{n}' > n$ to n. Similarly, we know that $n' > n$ when the observed process is 'blue' as in the lower panel of Figure 17.1.[5] Knowledge that the process tends to oscillate near a given frequency in the interior of the frequency interval (0, 1/2) is less useful for isolating the possible range of values for n' because we then can not be sure about whether $\Gamma_{xx}(0)/\sigma_X^2 < 1$.

17.1.4 The Decorrelation Time of AR(p) Processes. The decorrelation times for AR(p) processes with $p = 0$, 1 and 2 are easily computed.

- For $p = 0$, the 'white noise' process without any memory, the auto-correlation function $\rho(k)$ is zero for nonzero k so that

$$\tau_D = 1. \tag{17.10}$$

- For an AR(1) process the decorrelation time is

$$\tau_D = 1 + 2\sum_{k=1}^{\infty} \alpha_1^k = \frac{1+\alpha_1}{1-\alpha_1}. \tag{17.11}$$

  The decorrelation time for the two 'red' noise examples discussed in Chapters 10 and 11 are $\tau_D = 1.9$ for $\alpha_1 = 0.3$ and $\tau_d = 19$ for $\alpha_1 = 0.9$. Note that $\lim_{\alpha_1 \to +1} \tau_D = +\infty$. That is, the decorrelation time becomes infinite when the process becomes non-stationary.

- Recall from [11.1.9] that the auto-correlation function of an AR(2) process can take one of two different forms, depending upon whether the characteristic polynomial (10.11) has real or complex roots.

  When (10.11) has real roots y_1 and y_2, the auto-correlation function is given by

$$\rho(k) = a_1 y_1^{-|k|} + a_2 y_2^{-|k|},$$

  where a_1 and a_2 are given by (11.7). Then, since the decorrelation time is linear in ρ, we have

$$\tau_D = a_1 \frac{y_1 + 1}{y_1 - 1} + a_2 \frac{y_2 + 1}{y_2 - 1}.$$

  The AR(2) process with $\alpha_1 = \alpha_2 = 0.3$ is of this type. It has $y_1 = 1.39$, $y_2 = -2.39$, $a_1 = 0.74$, and $a_2 = 0.26$. Thus $\tau_D = 4.53 + 0.11 = 4.64$.

  When (10.11) has complex roots, the auto-correlation function is of the form

$$\rho(k) = \frac{\sqrt{1 + 4\,\mathrm{Im}(a)^2}}{r^k} \cos(k\phi + \psi)$$

  where constants a, r, ϕ and ψ are determined as in [11.1.9]. The corresponding decorrelation time is

$$\tau_D = 1 + 2\sqrt{1 + 4\,\mathrm{Im}(a)^2} \sum_{k=1}^{\infty} \frac{\cos(k\phi + \psi)}{r^k}.$$

  The AR(2) process with $(\alpha_1, \alpha_2) = (0.9, -0.8)$ has $a = 0.5 - i\,0.032$, $r = 1.12$, $\phi \approx \pi/3$, and $\psi \approx -\pi/50$ so that $\tau_D = 0.33$.

[5] 'Blue' noise processes tend to oscillate about the mean more frequently than white noise processes, so they produce observations that 'bracket' the mean much more quickly than 'white' or 'red' noise processes. Intuitively, then, it makes sense that $n' > n$.

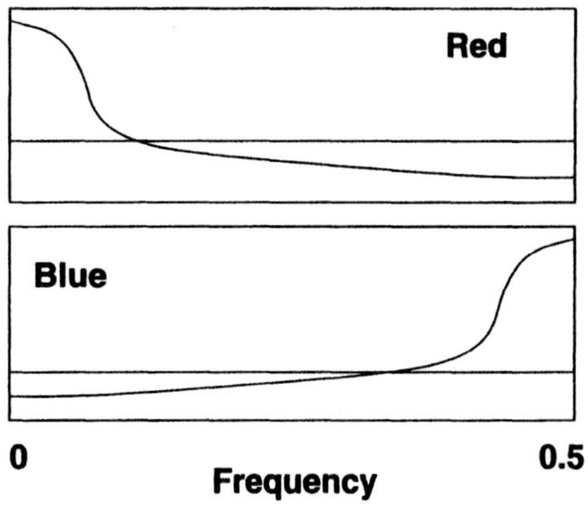

Figure 17.1: *A schematic illustration of the spectra of 'red' and 'blue' noise processes. The horizontal line indicates the variance of the process. The 'red' process has $n' < n$ and $\tau_D > 1$. The 'blue' process has $n' > n$ and $\tau_D < 1$.*

17.1.5 The 'Decorrelation Time': a 'Characteristic Time Scale?'

We now briefly discuss the extent to which τ_D can be interpreted as a physical time scale in AR(p) processes.

- The decorrelation time (17.10) for an AR(0) process makes physical sense since these processes are devoid of temporal continuity.
- Decorrelation time (17.11) has a reasonable physical interpretation as an indicator of the

'memory' or persistence of AR(1) processes with positive α_1 (i.e., $\tau_D > 1$). Trenberth [369] calls it a 'persistence time scale' and maps it for the Southern Hemisphere geopotential height field. Processes with negative α_1 are the ultimate weakly stationary 'oscillatory' processes because they tend to change sign at every time step. Thus τ_D is less than 1 even though the auto-correlation 'envelope' $|\rho(k)|$ decays at the same rate as that of an AR(1) process with coefficient $|\alpha_1|$. Thus a reasonable indicator of memory or persistence that applies to all AR(1) processes is

$$\begin{aligned}\tau'_D &= \frac{1+|\alpha_1|}{1-|\alpha_1|}\\ &= \begin{cases}\tau_D & \text{if } \alpha_1 > 0\\ \tau_D^{-1} & \text{if } \alpha_1 < 0.\end{cases}\end{aligned}$$

- Similar difficulties occur with oscillatory AR(2) processes since the decorrelation time tends to be smaller than that indicated by the decay of the correlation envelope $\sqrt{1+4\,\mathrm{Im}(a)^2}r^{-k}$. In this case, a better indicator of physical memory is

$$\begin{aligned}\tau'_D &= 1+2\sqrt{1+4\,\mathrm{Im}(a)^2}\sum_{k=1}^{\infty} r^{-k}\\ &= \frac{2r\sqrt{1+4\,\mathrm{Im}(a)^2}+r-1}{r-1}\end{aligned}$$

  where a and r are defined as before. The AR(2) process with $(\alpha_1, \alpha_2) = (0.9, -0.8)$ has $\tau_D = 0.33$ and $\tau'_D \approx 20$.

- We showed in [11.1.9] that the auto-correlation function of an AR(p) process can be decomposed into a sum of decaying persistent and oscillatory terms. As above, a meaningful indicator of physical memory can be obtained by summing the envelope that contains all of these terms.

In summary, τ_D must be interpreted carefully. It represents a physical time scale only when the auto-correlation function coincides with the 'auto-correlation envelope', as it does in white noise processes.[6]

17.1.6 The Dependence of the Decorrelation Time on the Time Increment. If

$$\mathbf{X}_t = \alpha_1 \mathbf{X}_{t-1} + \mathbf{Z}_t \tag{17.12}$$

[6]But see the caveat discussed in the next subsection.

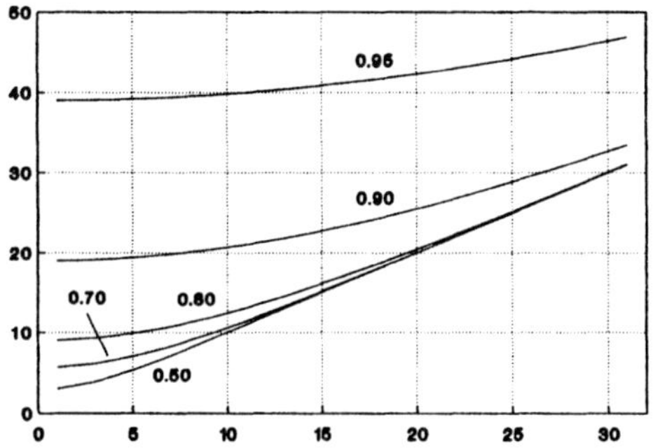

Figure 17.2: *The dependency of the dimensional decorrelation time $\tilde{\tau}_{D,k}$ on the time increment k and on the coefficient α.*

is an AR(1) process with a unit time increment, then we can construct other AR(1) processes with k unit time increments by noting that

$$\mathbf{X}_t = \alpha_1^k \mathbf{X}_{t-k} + \mathbf{Z}'_t \tag{17.13}$$

where $\mathbf{Z}'_t = \sum_{l=0}^{k-1} \alpha_1^l \mathbf{Z}_{t-l}$. The corresponding dimensional decorrelation times (17.6) are

$$\begin{aligned}\tilde{\tau}_{D,1} &= \frac{1+\alpha_1}{1-\alpha_1}\\ \tilde{\tau}_{D,k} &= k\frac{1+\alpha_1^k}{1-\alpha_1^k}.\end{aligned}$$

Thus $\tilde{\tau}_{D,k} \geq k$ for all k when $\alpha_1 \geq 0$. That means that the decorrelation time is at least as long as the time increment. In the case of white noise, with $\alpha_1 = 0$, the decorrelation time is always equal to the time increment. Some dimensional decorrelation times are plotted in Figure 17.2. The longer the time increment, the larger the decorrelation time. Note that $\tilde{\tau}_{D,k} = k$ for sufficiently large time increments. For small α_1-values, such as $\alpha_1 = 0.5$, $\tilde{\tau}_{D,k} = k$ for $k \geq 5$. If $\alpha_1 = 0.8$ then $\tilde{\tau}_{D,1} = 9$, $\tilde{\tau}_{D,11} = 13.1$ and $\tilde{\tau}_{D,21} = 21.4$. Thus the decorrelation time of an $\alpha_1 = 0.8$ process is 9 days or 21 days depending on whether we sample the process once a day or once every 21 days.

We conclude that the specific value of the decorrelation time may not be very informative. However, comparison between time series with the same sampling interval helps us identify which processed have larger memory.

17.2 Potential Predictability

17.2.1 Concept of 'Potential Predictability.' It is generally accepted that the skill of the short-term 'climate' forecasts[7] derives primarily from the persistence of the atmosphere's lower boundary conditions. Thus the *potential* for short-term climate predictability, or *potential predictability*, is often estimated using a time-domain analysis of variance technique. This technique assumes that variations in, say, seasonal mean sea-level pressure arise from two sources: one source represents the effect of the daily weather variations and the other reflects the effect of presumably unrelated processes, such as tropical sea-surface temperature or the presence of volcanic aerosols in the atmosphere. Variation from the first source is known to be unpredictable for lead times of more than, say, 10 days, but the second source is thought to be predictable, at least in principle. Madden [263] first described a time domain ANOVA technique for diagnosing potential predictability in 1976. His technique tries to infer from time series the strength of the predictable contribution without identifying its dynamical source. The statistical aspects were further elaborated by Zwiers [440]. See also Zwiers et al. [449] and [9.4.7–11].

17.2.2 Formal Definition of Potential Predictability.[8] The following statistical model is used in the analysis of potential predictability. The variable, say the temperature, $\mathbf{T}_t$, is assumed to be the sum of two independent processes $\mathbf{T}^S$ and $\mathbf{T}^F$:

$$\mathbf{T}_t = \mathbf{T}_t^S + \mathbf{T}_t^F. \qquad (17.14)$$

Furthermore, $\mathbf{T}^S$ is assumed to vary slowly, and $\mathbf{T}^F$ quickly. The latter is sometimes assumed to be a red noise process (e.g., see [9.4.10] and [449]). We also assume that the averaging time τ is short relative to the characteristic time for the slow process so that

$$\overline{\mathbf{T}}_t^{\tau} = \mathbf{T}_t^S + \overline{\mathbf{T}}_t^{F\,\tau}, \qquad (17.15)$$

where $\overline{\mathbf{X}}^{\tau}$ indicates averaging over an interval of length τ. Thus the τ-mean of $\mathbf{T}_t$ is controlled by two mechanisms: the slow process $\mathbf{T}^S$ and the integrated fast process $\mathbf{T}^F$. In terms of the temperature example, we would typically set τ to 90 days, interpret $\mathbf{T}^F$ as the weather noise, and assume that $\mathbf{T}^S$ is the slow variability from the atmosphere's lower boundary and other sources not related to the weather variability.

Since $\mathbf{T}^S$ and $\mathbf{T}^F$ are assumed to be independent, the variance of the τ-mean of $\mathbf{T}_t$ may be separated into a part reflecting the integrated weather noise and another part stemming from the low-frequency process(es):

$$\mathrm{Var}\big(\overline{\mathbf{T}_t}^{\tau}\big) = \mathrm{Var}\big(\mathbf{T}_t^S\big) + \mathrm{Var}\big(\overline{\mathbf{T}_t^F}^{\tau}\big). \qquad (17.16)$$

Since the weather fluctuations are 'unpredictable' on time scales of the order of τ, only that part of the τ-mean of $\mathbf{T}_t$ accounted for by the slow process is potentially predictable. Thus a reasonable measure of the relative importance of the potentially predictable component in (17.15) is the variance ratio

$$S_\tau = \frac{\mathrm{Var}\big(\overline{\mathbf{T}_t}^{\tau}\big)}{\mathrm{Var}\big(\overline{\mathbf{T}_t^F}^{\tau}\big)}. \qquad (17.17)$$

A variance ratio $S_\tau = 1$ indicates that all low-frequency variability originates from weather noise whereas $S_\tau > 1$ indicates that $\overline{\mathbf{T}_t}^{\tau}$ contains more variability than can be explained by weather noise alone. Hence there is the potential to forecast some of the variance of $\overline{\mathbf{T}_t}^{\tau}$.

17.2.3 Estimating the Variance Ratio S_τ. The numerator and denominator of variance ratio S_τ (17.17) are estimated separately. We assume that we have a sample $\{\mathbf{t}_{jk} : j = 1, \ldots, n; k = 1, \ldots, \tau\}$ that consists of n chunks of τ consecutive observations. In typical applications, each chunk is a daily time series observed over a season, say DJF, and different chunks represent different years. Ordinarily, the annual cycle is removed so that, to first order, the chunks can be assumed to be independent realizations of a weakly stationary time series of length τ.

The 'inter-chunk' variability is used to estimate the variance of the τ-means of T, $\mathrm{Var}\Big(\overline{\mathbf{T}_t}^{\tau}\Big)$. A 'chunk-mean'

$$\overline{\mathbf{t}_j}^{\tau} = \frac{1}{\tau}\sum_{k=1}^{\tau} \mathbf{t}_{jk} \qquad (17.18)$$

is first computed from each chunk, and these, in turn, are used to estimate the 'inter-chunk' variance

$$\widehat{\mathrm{Var}}\big(\overline{\mathbf{T}_t}^{\tau}\big) = \frac{1}{n-1}\sum_{j=1}^{n}\big(\overline{\mathbf{t}_j}^{\tau} - \overline{\mathbf{t}_\circ}^{\tau}\big)^2$$

[7]That is, forecasts of the monthly or seasonal mean conditions made at leads of up to about a year.

[8]The methods described here are appropriate for quantitative values that vary continuously in time, such as near-surface temperature or sea-level pressure. Wang and Zwiers [414] describe methods suitable for use with quantities that vary episodically, such as precipitation.

where

$$\overline{\mathbf{t}_{\circ}}^{\tau} = \frac{1}{n}\sum_{j=1}^{n}\overline{\mathbf{t}_j}^{\tau}.$$

The next step is to understand the properties of this estimator. Our assumptions about model (17.15) can be used to show that $\widehat{\mathrm{Var}}\left(\overline{\mathbf{T}_t}^{\tau}\right)$ is an unbiased estimate of the sum of the variances of the slow process $\mathbf{T}_t^S$ and the integrated fast process $\overline{\mathbf{T}_t^F}^{\tau}$. That is,

$$\mathcal{E}\left(\widehat{\mathrm{Var}}(\overline{\mathbf{T}_t}^{\tau})\right) = \mathrm{Var}(\mathbf{T}_t^S) + \mathrm{Var}(\overline{\mathbf{T}_t^F}^{\tau}).$$

If we also make the distributional assumptions that $\mathbf{T}_t^S$ and $\overline{\mathbf{T}_t^F}^{\tau}$ are normal,[9] then we can also show that

$$(n-1)\widehat{\mathrm{Var}}(\overline{\mathbf{T}_t}^{\tau}) \sim \left(\mathrm{Var}(\mathbf{T}_t^S) + \mathrm{Var}(\overline{\mathbf{T}_t^F}^{\tau})\right)\chi^2(n-1).$$

To test the null hypothesis that potential predictability is absent (i.e., to test H_0: $S_\tau = 1$ or, equivalently, H_0: $\mathrm{Var}(\mathbf{T}_t^S) = 0$), it is necessary to obtain a statistically independent estimator of $\widehat{\mathrm{Var}}(\overline{\mathbf{T}_t}^{\tau})$. This is done by using the 'intra-chunk' variations $\mathbf{t}'_{jk} = \mathbf{t}_{jk} - \overline{\mathbf{t}_j}^{\tau}$ to infer the *inter*-chunk variance of $\overline{\mathbf{T}_t^F}^{\tau}$.

Several methods can be used to infer $\mathrm{Var}(\overline{\mathbf{T}_t^F}^{\tau})$ from the intra-chunk variations. One approach is based on the observation that

$$\mathrm{Var}(\overline{\mathbf{T}_t^F}^{\tau}) = \frac{1}{\tau'}\mathrm{Var}(\mathbf{T}_t^F)$$

where τ' is the 'equivalent chunk length'. Thus $\mathrm{Var}\left(\overline{\mathbf{T}_t^F}^{\tau}\right)$ can be estimated as

$$\widehat{\mathrm{Var}}\left(\overline{\mathbf{T}_t^F}^{\tau}\right) = \frac{1}{\widehat{\tau'}}\widehat{\mathrm{Var}}\left(\mathbf{T}_t^F\right)$$

where

$$\widehat{\mathrm{Var}}\left(\mathbf{T}_t^F\right) = \frac{1}{n(\tau-1)}\sum_{j=1}^{n}\sum_{k=1}^{\tau}(\mathbf{t}'_{jk})^2$$

and τ' is estimated by using one of the methods discussed in [17.1.3]. A suitable method is to

1 compute a 'pooled' estimate of the intra-chunk auto-correlation function

$$\widehat{\rho}(l) = \frac{\sum_{j=1}^{n}\sum_{k=1}^{\tau-l}\mathbf{t}'_{jk}\mathbf{t}'_{j(k+l)}}{\sum_{j=1}^{n}\sum_{k=1}^{\tau}(\mathbf{t}'_{jk})^2},$$

2 fit a low-order AR model using the Yule–Walker method [12.2.2]. The order can be determined either from physical considerations or by means of an objective criterion such as the AIC [12.2.10] or BIC [12.2.11], and

3 substitute the auto-correlation function derived from the fitted AR model into (17.4) to obtain $\widehat{\tau}'$.

The problem with this approach is that the distributional properties of

$$\widehat{\mathrm{Var}}(\overline{\mathbf{T}_t^F}^{\tau}) = \frac{1}{\widehat{\tau}'}\widehat{\mathrm{Var}}(\mathbf{T}_t^F) \qquad (17.19)$$

are not well known.

An alternative approach is based on the observation that

$$\mathrm{Var}(\overline{\mathbf{T}_t^F}^{\tau}) \approx \frac{1}{\tau}\Gamma_{FF}(0),$$

where $\Gamma_{FF}(\omega)$ is the spectral density function of the weather noise. As discussed in [17.1.3], $\Gamma_{FF}(0)$ can be estimated by assuming that the spectrum is white near the origin. Thus a reasonable estimator of $\mathrm{Var}(\overline{\mathbf{T}_t^F}^{\tau})$ is

$$\widehat{\mathrm{Var}}(\overline{\mathbf{T}_t^F}^{\tau}) = \frac{1}{\tau}\widehat{\Gamma}_{FF}\left(\frac{1}{\tau}\right)$$

where $\widehat{\Gamma}_{FF}(1/\tau)$ is the chunk estimator of $\Gamma_{FF}(1/\tau)$ (see [12.3.9,10]). An advantage of this approach is that the asymptotic distributional properties of $\widehat{\mathrm{Var}}(\overline{\mathbf{T}_t^F}^{\tau})$ are well known. In fact, asymptotically

$$2n\,\widehat{\mathrm{Var}}(\overline{\mathbf{T}_t^F}^{\tau}) \sim \mathrm{Var}(\overline{\mathbf{T}_t^F}^{\tau})\chi^2(2n).$$

See Madden [263] and Zwiers [440] for more discussion.

17.2.4 Testing the Null Hypothesis H_0: $S_\tau = 1$. Now that we have estimates of $\mathrm{Var}(\overline{\mathbf{T}_t}^{\tau})$ and $\mathrm{Var}(\overline{\mathbf{T}_t^F}^{\tau})$, we can estimate S_τ with

$$\widehat{S}_\tau = \frac{\widehat{\mathrm{Var}}(\overline{\mathbf{T}_t}^{\tau})}{\widehat{\mathrm{Var}}(\overline{\mathbf{T}_t^F}^{\tau})} \qquad (17.20)$$

and use $\widehat{S}_\tau$ to test H_0: $S_\tau = 1$. The test is performed at the $(1-\tilde{p})$ significance level by rejecting H_0 when $\widehat{S}_\tau$ is greater than the appropriate critical value $S_{\tau,\tilde{p}}$.

The method used to estimate the variance of the integrated weather noise affects the choice of critical value. When the spectral approach is used, the numerator and denominator are asymptotically

[9] The Central Limit Theorem [2.7.5] often ensures that $\overline{\mathbf{T}_t^F}^{\tau}$ is close to normal.

proportional to independent χ^2 random variables under H_0 with $\tau - 1$ and $2n$ degrees of freedom respectively. Thus $\widehat{S}_\tau \sim F(\tau - 1, 2n)$ under H_0 so that $S_{\tau,\tilde{p}}$ is the $\tilde{p}$-quantile of the F distribution with $\tau - 1$ and $2n$ degrees of freedom (Appendix G). This test will be nearly unbiased (i.e., it will operate at the specified significance level) when the weather noise spectrum has a moderate peak or trough at zero frequency. It will tend to be liberal (i.e., reject H_0 more frequently than specified) when $\Gamma_{FF}(\omega)$ has a strong peak at zero frequency because the extrapolation of spectral estimate at frequency $1/\tau$ to frequency zero will negatively bias the estimate of $\mathrm{Var}(\overline{\mathbf{T}_t^F}^\tau)$. This results in $\widehat{S}_\tau$-values that tend to be slightly larger than 1 under H_0. The opposite happens when $\Gamma_{FF}(\omega)$ has a strong trough at the zero frequency.

It is more difficult to determine an appropriate critical value for $\widehat{S}_\tau$ when the 'equivalent chunk length' approach is used. One solution is to argue that both estimates in (17.19) have little sampling variability since they are obtained from a large number (τn) of deviations $\mathbf{t}'_{jk}$. This reasoning allows us to ignore uncertainty in the denominator of (17.20), with the result that the test can be conducted by comparing $(\tau - 1)\widehat{S}_\tau$ with $\chi^2(\tau - 1)$ critical values (Appendix E). The resulting test will tend to be liberal because the variability in $\widehat{\mathrm{Var}}(\overline{\mathbf{T}_t^F}^\tau)$ has been ignored. It may also be adversely affected by bias in this estimator. A better approach is to estimate the distribution of $\widehat{S}_\tau$ under H_0 by applying the moving blocks bootstrapping procedure (cf. [5.5.3]) to the deviations $\mathbf{t}'_{jk}$. An example can be found in [414].

17.2.5 Example. Zwiers [440] analysed the potential predictability of the climate simulated by a GCM in a 20-year run. Sea-surface temperature and sea ice were specified from the same climatological annual cycle in each of the 20 years. Land surface conditions (snow cover, and soil temperature, moisture content and albedo) were computed interactively. Except for these land surface processes, the only other source of interannual variability in the GCM simulation is 'internal' variability.

Daily surface air pressure was gathered into 20 chunks, one for each DJF season. Thus $\overline{\mathbf{t}_j}^\tau$ is the DJF-mean of surfaced pressure during the jth DJF season at each grid point. The interannual variance of these seasonal means is the estimated 'inter-chunk' variance that constitutes the numerator of $\widehat{S}_\tau$ (17.20). The spectral approach was used to obtain the 'intra-chunk' variance. The resulting

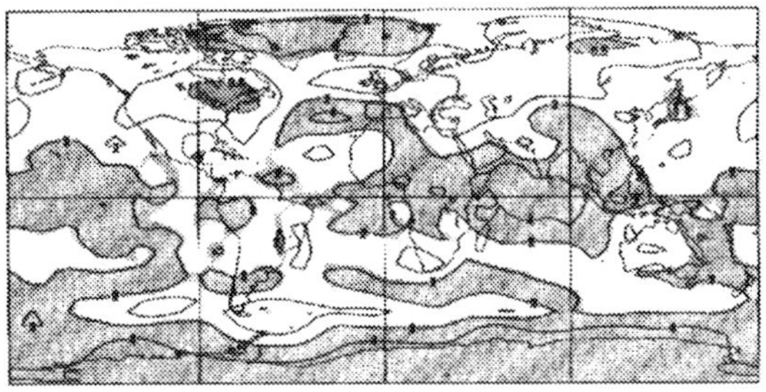

Figure 17.3: *Horizontal distribution of the estimated $\widehat{S}_{DJF}$ potential predictability ratios obtained from a 20-year run with an atmospheric GCM. Stippled areas mark grid points where the local null hypothesis of no potential predictability is rejected at the 5% significance level. Hatched areas represent regions where the $\widehat{S}_{DJF}$-ratios have values in the lower 5%-tail of the respective F distribution. From [440].*

F test was performed independently at the 5% significance level at each of the 2080 grid points. As a guide to interpretation, a rejection rate of approximately 10% would be field significant at the 5% significance level in a field with 100 spatial degrees of freedom (cf. [6.8.3] and Figure 6.12).

The results are shown in Figure 17.3. Several things can be noticed.

- The variance ratio $\widehat{S}_{DJF}$ is of the order 2 over large areas, particularly in the tropics and in the Southern Hemisphere, suggesting that only half of the interannual variability may be the integrated effect of weather variability. The local null hypothesis of no potential predictability is rejected at about 40% of all grid points making it unlikely that all rejections are due to chance.

- There are other areas in which the variance ratio $\widehat{S}_{DJF}$ is less than 1. Although puzzling at first glance, this is consistent with our model (17.14) since it is a natural consequence of the sampling variability of $\widehat{S}_{DJF}$. The number of grid points with $\widehat{S}_{DJF}$ ratios in the lower 5% tail of the F distribution is about 5%, indicating that our basic assumptions, which lead us to the F distribution, are approximately correct.

Further analysis indicated that the large $\widehat{S}_{DJF}$-values were not related to the surface hydrology, soil moisture, and snow cover terms in the GCM. The potential predictability in this simulated climate seems to arise from the occurrence of

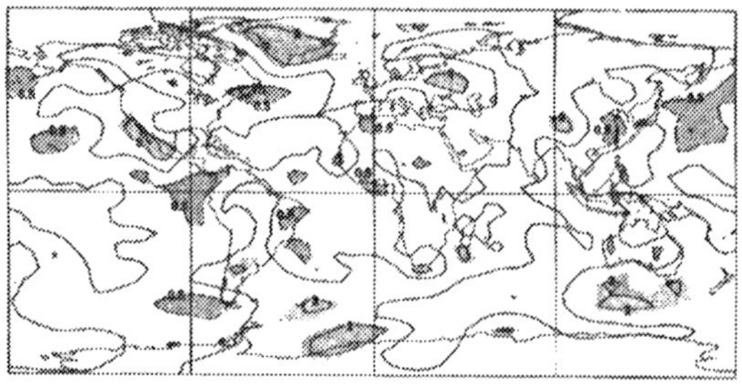

Figure 17.4: *Same as Figure 17.3 but for the SON season. From [440].*

a single large anomaly extending over a period of about a season, during which atmospheric mass is systematically shifted from the tropics to the high latitudes of the Southern Hemisphere. Such large extended anomalies have also been observed in other climate simulations and in the real atmosphere.

The same potential predictability analysis was conducted in other seasons. The SON (September-October-November) map is shown in Figure 17.4. In this season the areas with $\widehat{S}_{SON}$ in the upper 5% tail of the F distribution are small and the 'significant areas' cover roughly 5% of the globe. Thus the data do not contradict the null hypothesis of no potential predictability.

17.3 Composites and Associated Correlation Patterns

17.3.1 Introduction. An important part of climate research deals with the identification, description and understanding of *processes*, such as the El Niño/Southern Oscillation (ENSO) or the Madden-and-Julian Oscillation (MJO). Univariate and bivariate indices are frequently used to identify and characterize such signals.

For example, many aspects of the temporal behaviour of ENSO are captured by the conventional Southern Oscillation Index, the surface air-pressure difference between Darwin (Australia) and Papeete (Tahiti) (see Figure 1.2). Wright [427] found that many, roughly equivalent, ENSO indices can be defined (see, for example, Figure 1.4, which displays the SOI and a related tropical Pacific SST index).

Another example is the MJO. In this case a bivariate index is required to capture information about the propagating feature of this process. One such bivariate index, derived through a POP analysis, is shown in Figure 10.3.

Further insight into the way a signal is expressed in other variables can often be obtained with *composite analysis*[10] (discussed in [17.3.2,3]) and *associated correlation* or *regression* patterns [17.3.4,5].

In the following we assume that we have either a univariate index $\mathbf{z}_t$ or a bivariate index $\vec{\mathbf{z}}_t = (\mathbf{z}_{1t}, \mathbf{z}_{2t})^T$. The variable in which we want to identify the signal represented by the index is labelled $\vec{\mathbf{V}}_t$.

17.3.2 Composites. The general idea is to form sets Θ of the index $\vec{\mathbf{z}}$ and to estimate the expected value of $\vec{\mathbf{V}}$ conditional on $\vec{\mathbf{z}} \in \Theta$. Formally, the composite $\vec{\mathbf{V}}_\Theta$ is given by

$$\vec{\mathbf{V}}_\Theta = \mathcal{E}(\vec{\mathbf{V}}_t | \vec{\mathbf{z}}_t \in \Theta). \tag{17.21}$$

In practice, the expectation operator in (17.21) is replaced by a sum to obtain an *estimate* of the composite

$$\widehat{\vec{\mathbf{V}}}_\Theta = \frac{1}{k}\sum_{j=1}^{k} \vec{\mathbf{v}}_{t_j} \tag{17.22}$$

where the sum is taken over the observing times $t_1, \ldots, t_k$ for which $\vec{\mathbf{z}}_{t_j} \in \Theta$.

There are several things to note about this approach.

- It does not make any specific assumptions about the link between $\vec{\mathbf{Z}}$ and $\vec{\mathbf{V}}$. This link may be linear or nonlinear.
- The basic idea with composites is to construct 'typical' states of $\vec{\mathbf{V}}$ conditional on the value of the external index. It achieves this goal in the sense that we obtain estimates of the mean state. However, there may be considerable variability around each composite, and thus the composite may not be representative of the typical state of $\vec{\mathbf{V}}$ when $\vec{\mathbf{Z}} \in \Theta$ (recall the discussion in [1.2.1]).
- One way to determine whether aspects of the signal captured by $\vec{\mathbf{Z}}$ are expressed in $\vec{\mathbf{V}}$ is to test null hypotheses of the form

$$H_0\colon \vec{\mathbf{V}}_{\Theta_1} = \vec{\mathbf{V}}_{\Theta_2} \tag{17.23}$$

for appropriately chosen disjoint subsets Θ_1 and Θ_2. This is often done with one of the difference of means tests discussed in Section 6.6. An example is given in the next subsection.

[10]Composite analysis is also sometimes called *epoch analysis*.

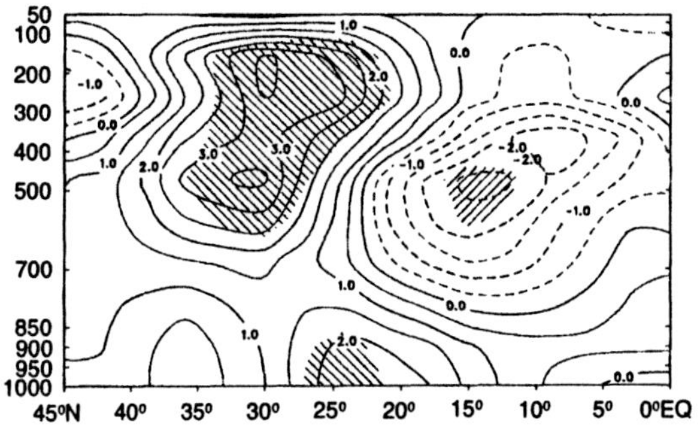

Figure 17.5: *Composite analysis of the latitude/height distribution of the zonal wind in JJA averaged between 60° E and 90° E longitude as simulated in a 20-year integration with an AGCM. The difference between the six cases with strongest Southeast Asian monsoon precipitation signal and the six cases with weakest precipitation is shown. Local t tests were performed to test the stability of the difference. Rejection of the local zero difference null hypothesis at the 5% significance level is indicated by cross-hatching. From [443].*

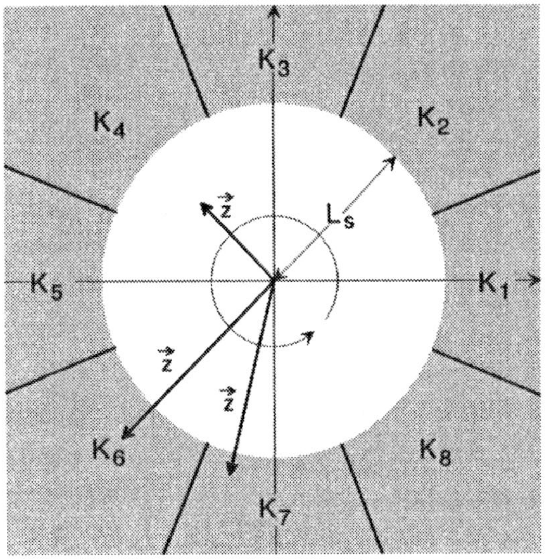

Figure 17.6: *Division of the two-dimensional plane containing the bivariate MJO index $\vec{\mathbf{Z}}$ into nine sectors. The composite sets Θ_j are labelled K_j. Realizations $\vec{\mathbf{z}}$ that fall into the inner white region, such as the short heavy vector, are not classified. Realizations that fall into the outer shaded sectors are classified as belonging to classes K_1 to K_8.*

17.3.3 Examples of Composite Analyses. We will discuss two examples in this subsection. The first example, from [443], deals with a univariate index and demonstrates a test of the null hypothesis (17.23). The second example, from [389], is on the oscillatory MJO and features eight different composites which supposedly represent canonical sequences of events.

Zwiers [443] analysed the variability of the Asian summer monsoon simulated in a 20-year GCM experiment. The study included a Canonical Correlation Analysis (CCA, see Chapter 14) of the surface heat flux on the Tibetan Plateau and rainfall over Southeast Asia.[11] The CCA exercise produced a univariate index $\mathbf{z}_t$ that represented a significant part of the simulated interannual variability of the monsoon rainfall. Since 20 years were simulated, a sample of 20 indices $\mathbf{z}_1, \dots, \mathbf{z}_{20}$ were available.

A composite analysis was performed to determine whether large-scale circulation changes that are associated with the monsoon in the real atmosphere also occur in the model.[12] Thus $\vec{\mathbf{V}}$ was set to the latitude/height distribution of the zonal wind averaged between 60° E and 90° W longitude. Two sets Θ_i and Θ_r, representing 'intensified' and 'reduced' Southeast Asian rainfall, were formed: Θ_r consisted of the six smallest values of $\mathbf{z}_t$, and Θ_i the six largest. The estimated composite difference $\widehat{\vec{\mathbf{V}}}_{\Theta_i} - \widehat{\vec{\mathbf{V}}}_{\Theta_r}$ is shown in Figure 17.5.

The circulation differences seen in Figure 17.5 are similar to differences between weak and strong monsoon years seen in the real atmosphere. The t test indicates that the upper tropospheric anticyclonic flow resulting from heating of the Tibetan Plateau is significantly stronger in strong monsoon years than in weak monsoon years. At the same time we also see some evidence of a significantly enhanced Somali jet near the near the surface between 20° N and 25° N. Thus all three centres with statistically significant wind changes are part of the same physical signal.

von Storch et al. [389] derived composites from an extended GCM simulation that are supposedly representative of different parts of the lifecycle of the MJO. The two-dimensional plane that contains the bivariate MJO index $\vec{\mathbf{z}}$ (see Figure 10.3 and [17.2.4]) was divided into nine regions (see Figure 10.5 and the sketch in Figure 17.6). The 'inner circle' set, which covers all indices with small amplitudes (i.e., $|\mathbf{z}|^2 = (\mathbf{z}_1)^2 + (\mathbf{z}_2)^2 < L_s$) is disregarded in the analysis. The remaining eight

[11] Heating of the Tibetan Plateau is thought to influence the strength of the Asian summer monsoon.

[12] The westerly zonal jet shifts northward and an easterly jet develops to the south at the onset of the Asian summer monsoon in response to the heating of the Tibetan Plateau.

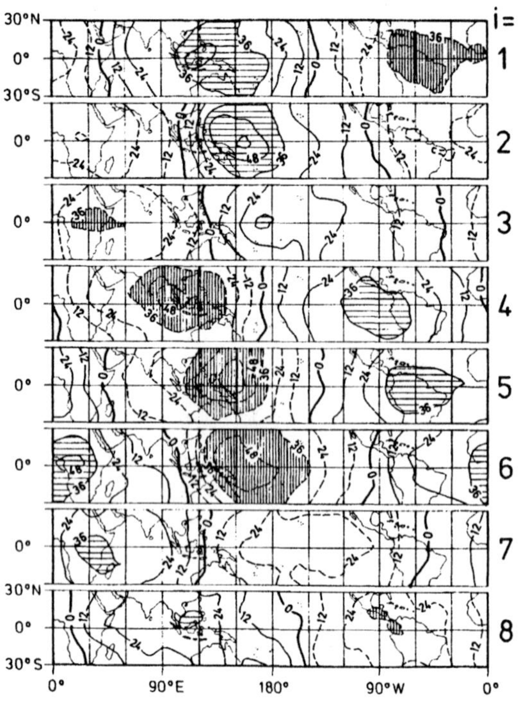

Figure 17.7: *The composite mean tropical velocity potential anomalies $\widehat{\vec{\mathbf{V}}}_{kj}$ derived from MJO indices $\vec{\mathbf{z}} \in K_j$. From [389].*

sectors, which each represent a 45° segment, are labelled $K_1, \dots, K_8$.

Composite means of tropical velocity potential anomalies at 200 hPa were computed for each sector (Figure 17.7). All eight composites exhibit a zonal wavenumber 1 pattern with maximum values on the equator.

The POP model from which the index was derived shows that the index tends to rotate counterclockwise in the two-dimensional plane (as indicated by the circular arrow in Figure 17.6). We may therefore interpret the composites as a sequence of patterns that appear consecutively in time. The main features propagate eastward, and intensification occurs whenever a maximum or minimum enters a region with active convection. Note that composite $\widehat{\vec{\mathbf{V}}}_{K_2}$ is almost a mirror image of $\widehat{\vec{\mathbf{V}}}_{K_6}$.

17.3.4 Associated Correlation Patterns. Composite analysis is 'non-parametric' in the sense that no assumption is made about the structure of the connection between the index $\vec{\mathbf{Z}}_t$ and the analysed vector variable $\vec{\mathbf{V}}_t$. The 'associated correlation pattern' approach, on the other hand, is based on a linear statistical model which relates the index or indices $\mathbf{z}_{i,t}$ with the vector variable $\vec{\mathbf{V}}_t$:

$$\vec{\mathbf{V}}_t = \sum_i \tilde{\mathbf{z}}_{i,t} \vec{q}^{\,i} + \text{noise}, \tag{17.24}$$

where $\vec{\mathbf{V}}_t$ usually represents anomalies (i.e., $\mathcal{E}(\vec{\mathbf{V}}_t) = 0$), and $\tilde{\mathbf{z}}_i$ is usually a *normalized* index given by $\tilde{\mathbf{z}} = (\mathbf{z} - \mu_z)/\sigma_z$. Patterns $\vec{q}^{\,i}$ usually carry the same units as $\vec{\mathbf{V}}_t$ because of the normalization of the index time series $\tilde{\mathbf{z}}_{i,t}$.

We now briefly discuss the one and two index versions of (17.24).

Only one pattern $\vec{q} = \vec{q}^{\,1}$ is obtained when a single index, $\mathbf{z}_t = \mathbf{z}'_t$, is used. This pattern is often called the 'regression pattern' for obvious reasons. Then

$$\vec{\mathbf{V}}_t = \tilde{\mathbf{z}}_t \vec{q} + \text{noise}. \tag{17.25}$$

The interpretation of $\vec{q}$ is that we observe pattern $\vec{q}$, on average, when $\tilde{\mathbf{z}} = 1$ and $-\vec{q}$ when $\tilde{\mathbf{z}} = -1$. More precisely, (17.25) says that

$$\mathcal{E}\big(\vec{\mathbf{V}}_t | \tilde{\mathbf{z}} = \alpha\big) = \alpha \vec{q} \tag{17.26}$$

for any number α, provided that the noise has mean zero.

Pattern $\vec{q}$ must be estimated from data. This can be done by minimizing the expected mean squared error,

$$\epsilon = \mathcal{E}\big(\|\vec{\mathbf{V}}_t - \tilde{\mathbf{z}}\vec{q}\,\|^2\big), \tag{17.27}$$

that is, by finding the vector $\vec{q}$ such that

$$\frac{\partial \epsilon}{\partial \vec{q}} = 0. \tag{17.28}$$

By differentiating, we find that $\vec{q}$ satisfies

$$-2\mathcal{E}\big(\tilde{\mathbf{z}}_t \vec{\mathbf{V}}_t\big) + 2\mathcal{E}\big(\tilde{\mathbf{z}}_t^2\big)\vec{q} = 0.$$

Since $\mathcal{E}\big(\tilde{\mathbf{z}}_t^2\big) = \text{Var}(\tilde{\mathbf{z}}) = 1$, the solution of (17.28) is given by

$$\vec{q} = \text{Cov}\big(\tilde{\mathbf{z}}_t, \vec{\mathbf{V}}_t\big). \tag{17.29}$$

Thus the associated correlation pattern $\vec{q}$ consists of the regressions between the index time series and the components of the analysed vector time series.

Two patterns are obtained when $\vec{\mathbf{Z}}_t$ is a bivariate index. Such indices arise naturally in many circumstances, including POP analysis (see [17.1.8] and Figure 10.3) and complex EOF analysis (see Chapter 16). When $\vec{\mathbf{Z}}_t = (\tilde{\mathbf{z}}_{1t}, \tilde{\mathbf{z}}_{2t})^{\mathrm{T}}$, equation (17.24) takes the form

$$\vec{\mathbf{V}}_t = \tilde{\mathbf{z}}_{1t} \vec{q}^{\,1} + \tilde{\mathbf{z}}_{2t} \vec{q}^{\,2} + \text{noise}. \tag{17.30}$$

This model states that the conditional mean of $\vec{\mathbf{V}}_t$ is given by

$$\mathcal{E}\big(\vec{\mathbf{V}}_t|\tilde{\mathbf{z}}_1 = \alpha, \tilde{\mathbf{z}}_2 = \beta\big) = \alpha\vec{q}^{\,1} + \beta\vec{q}^{\,2}. \quad (17.31)$$

The interpretation is that $\vec{\mathbf{V}}_t = \vec{q}^{\,1}$, on average, when $\tilde{\mathbf{z}}_1 = 1$ and $\tilde{\mathbf{z}}_2 = 0$, and that $\vec{\mathbf{V}}_t = \vec{q}^{\,2}$, on average, when $\tilde{\mathbf{z}}_1 = 0$ and $\tilde{\mathbf{z}}_2 = 1$. However, individual realizations $\vec{\mathbf{v}}_t$ may differ substantially from these long-term mean states.

The *associated correlation patterns* $\vec{q}^{\,1}$ and $\vec{q}^{\,2}$ are derived by minimizing the expected mean squared error,

$$\epsilon = \mathcal{E}\big(\|\vec{\mathbf{V}}_t - \tilde{\mathbf{z}}_{1t}\vec{q}^{\,1} - \tilde{\mathbf{z}}_{2t}\vec{q}^{\,2}\|^2\big), \quad (17.32)$$

that is, $\vec{q}^{\,1}$ and $\vec{q}^{\,2}$ are the solutions of

$$\frac{\partial\epsilon}{\partial\vec{q}^{\,1}} = \frac{\partial\epsilon}{\partial\vec{q}^{\,2}} = 0.$$

By differentiating and taking expectations we find

$$\begin{pmatrix}\sigma_1^2 & \sigma_{12}\\ \sigma_{12} & \sigma_2^2\end{pmatrix}\begin{pmatrix}(\vec{q}^{\,1})^{\mathrm{T}}\\ (\vec{q}^{\,2})^{\mathrm{T}}\end{pmatrix} = \begin{pmatrix}(\vec{\sigma}_{1v})^{\mathrm{T}}\\ (\vec{\sigma}_{2v})^{\mathrm{T}}\end{pmatrix} \quad (17.33)$$

where $\sigma_1^2 = \mathrm{Var}(\tilde{\mathbf{z}}_1)$, $\sigma_2^2 = \mathrm{Var}(\tilde{\mathbf{z}}_2)$, $\sigma_{12} = \mathrm{Cov}(\tilde{\mathbf{z}}_1, \tilde{\mathbf{z}}_2)$, $\vec{\sigma}_{1v} = \mathrm{Cov}(\tilde{\mathbf{z}}_1, \vec{\mathbf{V}}_t)$, and $\vec{\sigma}_{2v} = \mathrm{Cov}(\tilde{\mathbf{z}}_2, \vec{\mathbf{V}}_t)$. Equation (17.33) has solution

$$\vec{q}^{\,1} = \frac{\sigma_2^2\vec{\sigma}_{1v} - \sigma_{12}\vec{\sigma}_{2v}}{\sigma_1^2\sigma_2^2 - \sigma_{12}^2} \quad (17.34)$$

$$\vec{q}^{\,2} = \frac{\sigma_1^2\vec{\sigma}_{2v} - \sigma_{12}\vec{\sigma}_{1v}}{\sigma_1^2\sigma_2^2 - \sigma_{12}^2}. \quad (17.35)$$

The relative importance of associated correlation patterns can be measured by the 'proportion of variance' they represent, either locally or in total. The proportion of the total variance represented by the patterns is given by

$$r_{v|z}^2 = \frac{\mathcal{E}(\vec{\mathbf{V}}_t^{\mathrm{T}}\vec{\mathbf{V}}_t) - \epsilon}{\mathcal{E}(\vec{\mathbf{V}}_t^{\mathrm{T}}\vec{\mathbf{V}}_t)}$$

where ϵ is given by (17.27) or (17.32). Locally, the proportion is given by

$$r_{v|z,j}^2 = \frac{\mathcal{E}(\mathbf{V}_{jt}^2) - \epsilon_j}{\mathcal{E}(\mathbf{V}_{jt}^2)},$$

where $\mathbf{V}_{jt}$ is the jth element of $\vec{\mathbf{V}}_t$ and ϵ_j is the local version of ϵ.[13] The last representation is useful because it can be displayed as a function of location.

[13]Note that both (17.27) and (17.32) can be easily re-expressed as $\epsilon = \sum_j \epsilon_j$.

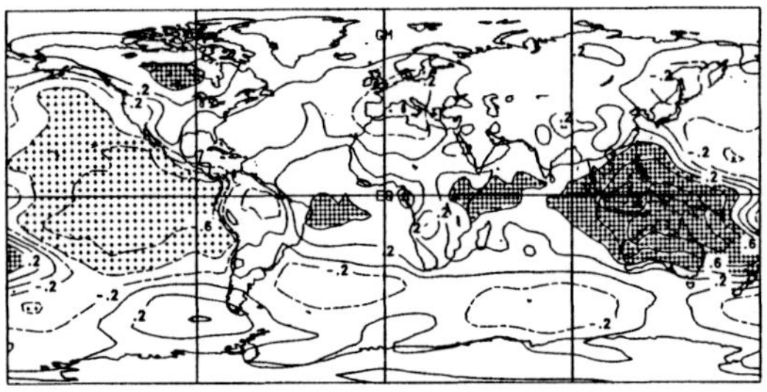

Figure 17.8: *Map of correlations between annual mean sea-level pressure at a point ('Darwin') over north Australia with annual mean SLP everywhere else on the globe. The data are taken from a 47-year GCM experiment with prescribed observed sea-surface temperatures.*

17.3.5 Examples. We now consider two examples: one with a univariate Southern Oscillation index and the other with a bivariate MJO index.

In the first example, $\mathbf{Z}_t$ is the annual mean sea-level pressure near Darwin, Australia, in a 47-year GCM simulation in which SST and sea-ice observations are prescribed from observations. The Darwin pressure index is a widely used ENSO index that carries information similar to the standard SOI [426]. The field $\vec{\mathbf{V}}_t$ in this example is the corresponding annual mean sea-level pressure. The associated correlation pattern that is obtained is shown in Figure 17.8. As expected, sea-level pressure variations occur coherently over broad regions, and variations in the eastern tropical Pacific are opposite in sign to those occurring over the western tropical Pacific. The present diagram compares favourably with similar diagrams computed from observations. See, for example, Peixoto and Oort [311, p. 492], or Trenberth and Shea [372].[14]

Our second example uses the same bivariate MJO index (Figure 10.3) employed in the composite analysis of [17.3.3] (Figure 17.7). The estimated associated correlation patterns are shown in Figure 17.9 [389]. As explained in Section 15.1, the two POP coefficients $(\mathbf{z}_t^1, \mathbf{z}_t^2)$ tend to have quasi-oscillatory variations of the type

$$\cdots \to (0,1) \to (1,0) \to (0,-1) \to (-1,0) \to (0,1) \to \cdots.$$

At the same time, the $\vec{\mathbf{V}}$ field tends to evolve as

$$\cdots \to \vec{q}^{\,2} \to \vec{q}^{\,1} \to -\vec{q}^{\,2} \to -\vec{q}^{\,1} \to \vec{q}^{\,2} \to \cdots$$

[14]See also Berlage [46], who published a similar diagram in 1957.

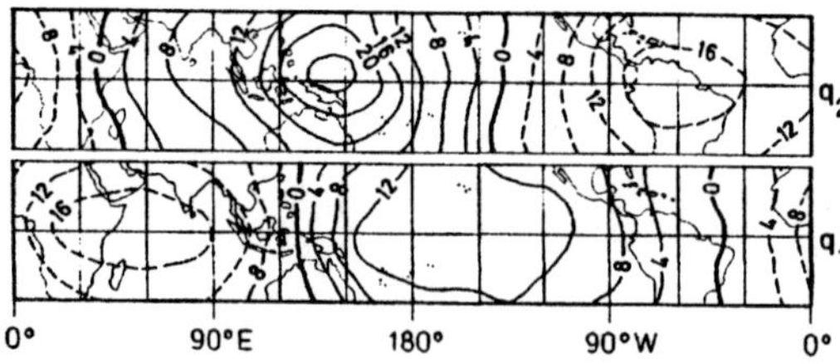

Figure 17.9: *The associated correlation patterns* $\widehat{\vec{\mathbf{V}}}_{K_j}$ *of tropical velocity potential anomalies derived from MJO indices* $\vec{\mathbf{z}}$. *Compare with the composites shown in Figure 17.7. From [389].*

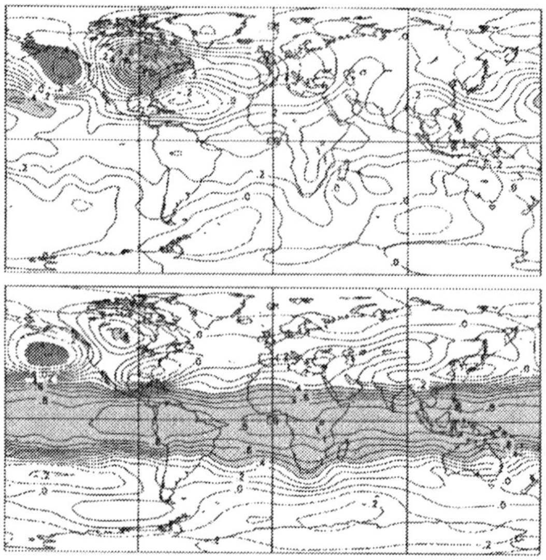

Figure 17.10: *Top: The correlation between DJF monthly mean 500 hPa geopotential height simulated by a GCM at (50° N, 90° W) and all other points in the model's grid.*
Bottom: As top, except the reference point is located at (2° N, 90° W).

Thus the patterns in Figure 17.9 provide the same information as the composites in Figure 17.7. The signal has a zonal wavenumber 1 structure that propagates eastward around the world. The oscillation is most energetic when the 'wave crest' (or 'valley') is positioned over the Maritime Continent.

A significant conclusion from the discussions here and in [17.3.3] is that both techniques provide useful information. The associated correlation pattern technique is superior in the MJO case since fewer parameters must be estimated from the available data (specifically, two patterns instead of eight patterns).

17.4 Teleconnections

17.4.1 Example: 500 hPa Geopotential Height. A classical method for exploring the spatial structure of climate variability is to compute cross-correlations between a variable at a fixed location and the same or another variable elsewhere. The resulting map of cross-correlation coefficients is called a *teleconnection* pattern. When the same variable is considered at two nearby locations, the correlation will tend to be large and positive (compare with the argument in [1.2.2]). Sometimes variables at two well-separated locations are also highly—often negatively—correlated.[15]

We demonstrate with DJF monthly mean 500 hPa geopotential height from an ensemble of six 10-year GCM simulations. The SST and sea-ice extent were specified from 1979–88 observations so the simulated atmosphere experienced realistic lower boundary variations (see [444]).

The upper panel in Figure 17.10 shows teleconnections for a fixed point located over Lake Superior. The main feature is an arched wave train that extends from the eastern Pacific, across North America, and into the western Atlantic. The decorrelation length scale, which is of the order of 3000 km, compared well with that of the observations (see, for example, Thiébaux [361], Fraedrich et al. [127], and Figure 2.8). This length scale is typical of that of teleconnection patterns that can be computed for other locations in the midlatitudes of both hemispheres.

In contrast, the lower panel of Figure 17.10 displays the teleconnection map that is obtained for a reference point off the coast of Peru at approximately 2° N, 90° W. Here we see that the entire simulated tropical 500 hPa geopotential height field varies more or less in unison on the monthly time scale. Much the same pattern can be obtained for virtually any reference point near the equator.[16] Also note the model's relatively weak rendition of the Pacific/North American pattern (cf. [13.5.5] and Figure 13.7).

In the following we will present an approach that is used to screen large data sets for such teleconnections systematically. It was pioneered

[15] The most prominent example of such a *teleconnection* is the *Southern Oscillation* discussed in [1.2.2].

[16] Tropical geopotential height variations are small, and primarily reflect variations in the temperature of the lower tropical troposphere.

by Wallace and Gutzler [409] and Horel and Wallace [182]. See also the review by Navarra [290], the example discussed in [2.8.8], and Figure 2.8.

17.4.2 The Wallace and Gutzler Approach. Let $\vec{\mathbf{X}}$ represent a gridded data variable, such as SLP or 500 hPa geopotential height, and let $\mathcal{R}_{xx}$ be the corresponding matrix of estimated cross-correlations. The jth column of $\mathcal{R}_{xx}$, $\widehat{\vec{\rho}}_j$, contains the estimated cross-correlations between $\mathbf{X}_j$ and $\vec{\mathbf{X}}$. Thus, if $\vec{\mathbf{X}}$ is m-dimensional, $\mathcal{R}_{xx}$ can be shown as m maps. All maps have unit value at the base point j, and for most variables they will have relatively large positive values in a neighbourhood of the base point. Points that are outside the 'region of influence' of the base point are not considered interesting. Such correlation maps are called *teleconnection patterns*.

In the next subsection we will present some results from Wallace and Gutzler's [409] original analysis, and then define a measure of the strength of the teleconnections in [17.4.4].

17.4.3 The PNA, WA, WP, EA, and EU-Family of Teleconnections. We briefly discussed Wallace and Gutzler's [409] identification of characteristic patterns of the month-to-month variability of winter 500 hPa height in [3.1.6]. Five 'significant'[17] correlation maps with sequences of large positive and negative *centres of action* were found. These patterns, called the Eastern Atlantic (EA), Pacific/North American (PNA), Eurasian (EU), West Pacific (WP), and West Atlantic (WA) patterns, are shown in Figure 3.9.

The locations of maxima and minima of a teleconnection pattern (i.e., the centres of action) can be used to define time-dependent teleconnection indices. Wallace and Gutzler define such an index as a weighted sum of the heights at the centres of action. In the PNA case, the centres of action are located at (20° N, 160° W), (55° N, 115° W), (45° N, 165° W) and (30° N, 85° W). The first two points are associated with maxima and the last two with minima. Thus the contribution from the last two centres of action in the *PNA index* is negatively weighted:

$$\mathrm{PNA} = \frac{1}{4}\big(z_{20^\circ N,160^\circ W} - z_{45^\circ N,165^\circ W} + z_{55^\circ N,115^\circ W} - z_{30^\circ N,85^\circ W}\big) \tag{17.36}$$

where z is 500 hPa geopotential height. Similar indices are obtained for the other four teleconnection patterns.

Such indices may be used to derive composites [17.3.2] or associated correlation patterns [17.3.4], to monitor the strength of a teleconnection, or they can be fed into predictive schemes. Also the correlation *between* teleconnections can be quantified. Wallace and Gutzler found moderate correlations between patterns with spatial overlap and small correlations between patterns with little or no spatial overlap.

The teleconnections patterns depend somewhat on the choice of the base point. The base point for the PNA pattern shown in the upper panel of Figure 3.9 was (45° N, 165° W). Similar patterns are obtained if another centre of action, (20° N, 160° W), (55° N, 115° W), or (30° N, 85° W), is used as the base point.

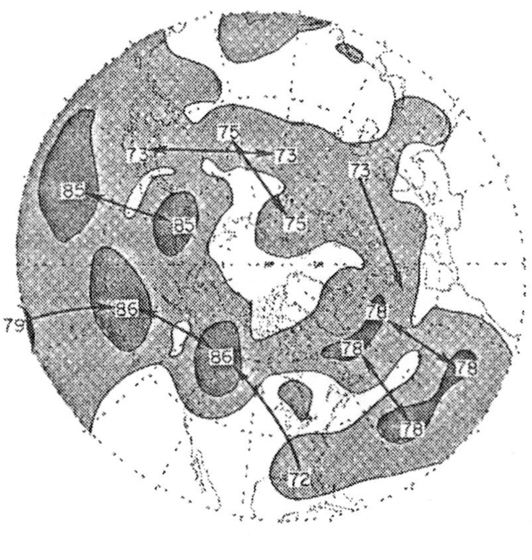

Figure 17.11: *Teleconnectivity map of 500 hPa height in northern winter. From [409].*

17.4.4 The 'Teleconnectivity.' A typical feature of teleconnection maps is the presence of large *negative* correlations. It therefore makes sense to define the 'teleconnectivity' T_j of a base point j as the maximum of all *negative correlations*:

$$T_j = -\min_j \widehat{\rho}_{ij}. \tag{17.37}$$

[17] Wallace and Gutzler use the word 'significant' in a somewhat pragmatic sense. They split the data set into two subsets, and used one subset to establish the teleconnection patterns and the second subset to assess (successfully) the stability of the patterns. In this way they determined a rule-of-thumb that correlations $|\widehat{\rho}| > 0.75$ should be reproducible in samples of size 15. Reproducibility is a stronger criterion that statistical significance since $|\widehat{\rho}| > 0.5$ is sufficient to reject $H_0: \rho \neq 0$ at approximately the 5% significance level (see [8.2.3] and David [100]). Wallace and Gutzler also found consistent patterns with an EOF analysis (see Chapter 13).

T_j can then be plotted as a spatial distribution; Wallace and Gutzler's example is shown as Figure 17.11. We occasionally find that the maxima in the teleconnectivity map are connected by a common point, that is, that T_j and T_k obtain their values from $\widehat{\rho}_{lj}$ and $\widehat{\rho}_{lk}$ for a common point l. These conditions, which hint at physical relationships, can be displayed with arrows as in Figure 17.11.

17.4.5 Generalizations. The basic idea of mapping correlations between one variable at a *base point* and another variable at many other geographically distributed points can be applied to any two climate variables. Indeed the square *correlation matrix* $\mathcal{R}$ can be replaced with a rectangular *cross-correlation matrix* $\mathcal{R}_{zx}$. Also, the variables may be lagged relative to each other so that the cross-correlation matrix is really a *lagged* cross-correlation matrix (see, for example, Horel and Wallace [182], who correlated the Southern Oscillation Index with 500 hPa height throughout the Northern Hemisphere).

Thus the basic idea is very general and can be applied in many different settings. A key limitation, however, is that the method can only be used to diagnose linear relationships. The term 'teleconnection' is usually reserved for cases in which a correlation (rather than a cross-correlation) matrix is analysed.

Teleconnection patterns are closely related to associated correlation patterns derived for a single index ([17.3.4]). If we normalize the base point time series $\tilde{\mathbf{z}}_t$, then the teleconnection is a point i given by $\mathrm{Cov}(\tilde{\mathbf{z}}_t, \mathbf{X}_{it})/\sigma_{X_{it}}$ (cf. equation (17.29)).

17.4.6 Assessing 'Significance.' As with EOFs and other patterns, there is a tendency to confuse physical and statistical significance of teleconnection patterns. In general, the patterns are worthy of physical interpretation when the basic structure is not strongly affected by sampling variability (i.e., when there is reproducibility).

The best way to assess reproducibility is to ensure that the pattern reappears in independent data sets and with other analysis techniques. Wallace and Gutzler used this approach by keeping part of the data to assess the stability of their patterns in a second step. Barnston and Livezey [27] subsequently reproduced the results of Wallace and Gutzler using rotated EOFs (Section 13.5). There is little doubt of the reality of the teleconnections discussed so far.

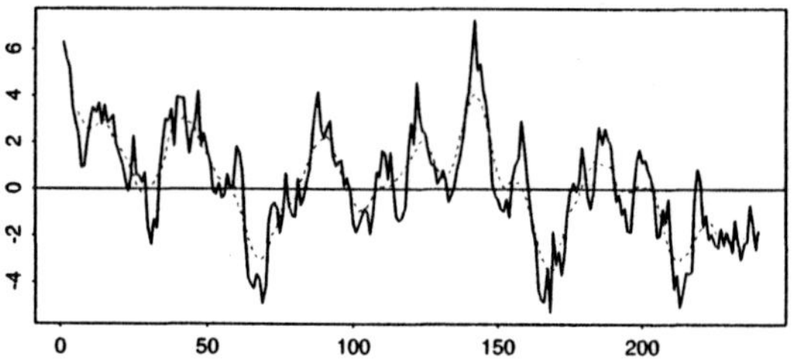

Figure 17.12: *A realization of an AR(1) time series with lag-1 correlation coefficient* $\alpha_1 = 0.9$ *(solid curve), and its 11-term running mean (dashed curve).*

Another way to assess reproducibility is to test $H_0: \rho = 0$ at every point in a teleconnection map (methods are described in [8.2.3]). The 'field significance' of the resulting map of reject decisions can then be assessed using the techniques described in Section 6.8 (see also the related discussion in [6.5.2]). Note that the local rejection rate will tend to be greater than the nominal level (often 5%) because correlations near the base point will be large. Care must be exercised to account for this phenomenon when determining whether H_0 can be rejected globally. Note also that this problem is amplified in teleconnection analysis because many maps are screened. Despite these difficulties, local significance tests are useful because they identify important features in the teleconnection maps.

17.5 Time Filters

17.5.0 General. We have often used the concept that the variability of a time series may be caused by different processes that are characterised by their 'time scales'. It is therefore useful to split a time series into certain components, such as

$$\mathbf{T}_t = \mathbf{T}_t^F + \mathbf{T}_t^S \qquad (17.38)$$

where $\mathbf{T}^F$ and $\mathbf{T}^S$ represent stationary components with 'fast' and 'slow' variability, respectively. The *time filters* described in this section are designed for this purpose.

17.5.1 Time Filters—Concepts. One of the simplest filtering operations is to smooth a time series by computing its *running mean.* For example, Figure 17.12 shows smoothed and unsmoothed versions of an AR(1) time series with

$\alpha_1 = 0.9$. The smoothed version (dashed curve) is given by

$$\mathbf{y}_t = \frac{1}{11}\sum_{k=-5}^{5} \mathbf{x}_{t+k}. \tag{17.39}$$

The large, slow variations remain but the small, fast variations have almost been eliminated.

Running mean (17.39) is an example of a *digital filter* given by

$$\mathbf{Y}_t = \sum_{k=-K}^{K} a_k \mathbf{X}_{t+k}, \tag{17.40}$$

where $\{a_{-K}, \ldots, a_K\}$ is a set of $2K + 1$ real *weights*. The weights can be tailored so that the filter retains variation on long, short, or intermediate time scales. Filters with these characteristics are known as *low-*, *high-*, and *band-pass* filters.

Suppose now that we have a digital filter of the form (17.40). It is then easily shown that the spectral density function of the *output* $\vec{\mathbf{Y}}_t$ is related to that of the *input* $\vec{\mathbf{X}}_t$ by

$$\Gamma_{yy}(\omega) = |c(\omega)|^2 \Gamma_{xx}(\omega) \tag{17.41}$$

where $c(\omega)$ is the *frequency response function*

$$c(\omega) = \sum_{k=-K}^{K} a_k e^{2\pi i k\omega} \tag{17.42}$$

of the filter. This can be proved as follows. Beginning with (11.8), we express the spectral density function of $\mathbf{Y}_t$ in terms of its auto-covariance function:

$$\Gamma_{yy}(\omega) = \sum_{\tau=-\infty}^{\infty} \gamma_{yy}(\tau) e^{2\pi i\tau\omega}. \tag{17.43}$$

But the auto-covariance function of the output is related to that of the input by

$$\begin{aligned}\gamma_{yy}(\tau) &= \mathrm{Cov}\Big(\sum_{k=-K}^{K} a_k \mathbf{X}_{t+k}, \sum_{l=-K}^{K} a_l \mathbf{X}_{t+\tau-l}\Big)\\ &= \sum_{k=-K}^{K}\sum_{l=-K}^{K} a_k a_l \mathrm{Cov}(\mathbf{X}_{t+k}, \mathbf{X}_{t+\tau-l})\\ &= \sum_{k=-K}^{K}\sum_{l=-K}^{K} a_k a_l \gamma_{xx}(\tau - l + k).\end{aligned} \tag{17.44}$$

Substituting (17.44) into (17.43), and changing the order of summation, we find

$$\begin{aligned}\Gamma_{yy}(\omega) &= \sum_{k=-K}^{K}\sum_{l=-K}^{K} a_k a_l \sum_{\tau=-\infty}^{\infty} e^{-2\pi i\tau\omega} \times \gamma_{xx}(\tau - l + k)\\ &= \sum_{k=-K}^{K} a_k e^{2\pi i k\omega} \sum_{l=-K}^{K} a_l e^{-2\pi i l\omega}\\ &\quad \sum_{\tau=-\infty}^{\infty} e^{-2\pi i(\tau-l+k)} \gamma_{xx}(\tau - l + k)\\ &= |c(\omega)|^2 \Gamma_{xx}(\omega),\end{aligned}$$

thus proving (17.41) and (17.42).

Now suppose the input contains a monochromatic signal, say $\cos(2\pi\omega t)$, and that the purpose of the filtering is to isolate this signal. Certainly we do *not* want the filter to shift the signal's phase. That is, if $\cos(2\pi\omega t)$ is input to the filter, we require that the output be of the form $r\cos(2\pi\omega t)$. Substituting

$$\cos(2\pi\omega t) = \frac{1}{2}\big(e^{2\pi i\omega t} + e^{-2\pi i\omega t}\big)$$

into (17.40) we find

$$\begin{aligned}&\sum_{k=-K}^{K} a_k \cos\big(2\pi\omega(t+k)\big)\\ &= \frac{1}{2}\sum_{k=-K}^{K} a_k\big(e^{2\pi i\omega(t+k)} + e^{-2\pi i\omega(t+k)}\big)\\ &= \frac{1}{2}\big(e^{2\pi i\omega t} c(\omega) + e^{-2\pi i\omega t} c^*(\omega)\big).\end{aligned}$$

The latter is again a zero phase cosine $c(\omega)\cos(2\pi\omega t)$ only when $c(\omega)$ is real. Thus the weights a_k must be symmetric in k, that is, $a_k = a_{-k}$. Therefore

$$c(\omega) = a_0 + 2\sum_{k=1}^{K} a_k \cos(2\pi k\omega). \tag{17.45}$$

A final detail that is important in some applications is that it may be necessary to preserve the time average of the input, in which case the weights should also be constrained so that $c(0) = a_0 + 2\sum_{k=1}^{K} a_k = 1$. On the other hand, the time average (i.e., the zero frequency component) of the input can be removed by selecting weights such that $c(0) = 0$.

The remainder of this section is laid out as follows. We explore the so-called '(1-2-1)-filter' and further examine the running mean filter in [17.5.2]. Then, in [17.5.3], we consider the effect of adding filters, and applying them in sequence. The latter is a technique that is frequently

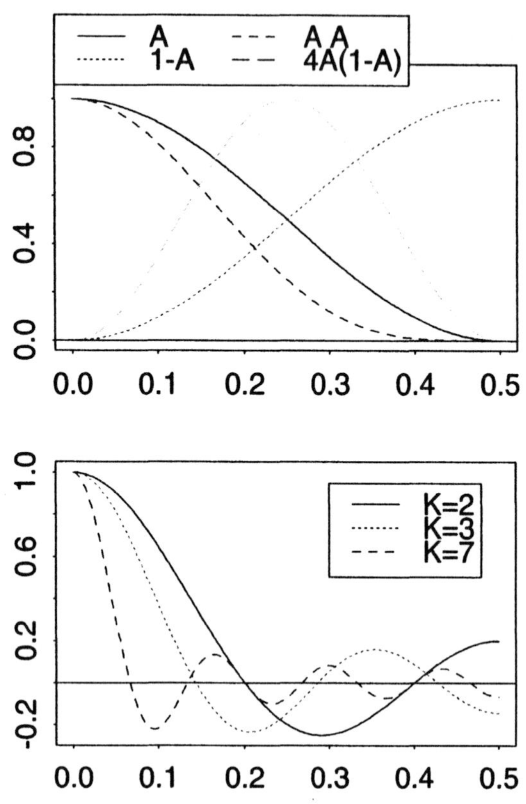

Figure 17.13: *Response functions of a number of simple filters. The abscissa is the frequency $\omega \in [0, 1/2]$.*
a) The plain low-pass (1-2-1)-filter $\mathcal{A}$, the squared (1-2-1)-filter $\mathcal{A} \cdot \mathcal{A}$, the high-pass '1-(1-2-1)'-filter $1 - \mathcal{A}$ and the band-pass filter $4 \times \mathcal{A} \cdot (1 - \mathcal{A})$.
b) Low-pass running mean filters with $K = 2$, 3, and 7.

used to construct a complex filter with desirable properties from simple building blocks such as the (1-2-1)-filter. This approach is discussed in [17.5.4]. Specific filters that are frequently used in atmospheric science are discussed in [17.5.5], some examples are mentioned in [17.5.6], and we wind up the section by describing a technique that can be used to custom design filters. For further reading, see standard texts such as Brockwell and Davis [68], Jenkins and Watts [195], or Koopmans [229].

17.5.2 The (1-2-1)-Filter and the 'Running Mean' Filter. These two symmetric filters are simple tools for suppressing high-frequency variability. They preserve the mean since their weights add to 1.

The filter with weights

$$a_k = \begin{cases} \frac{1}{2} & \text{for } k = 0 \\ \frac{1}{4} & \text{for } k = 1 \\ 0 & \text{for } k \geq 2 \end{cases} \tag{17.46}$$

is named the '(1-2-1)-filter' since two units are given to the central weight whereas only one unit is given to each of the two outer weights. This filter may be seen as an 'integrator' since the integral $\int_0^2 f(t)\,dt$ can be approximated as

$$\begin{aligned} \int_0^2 f(t)\,dt &\approx \frac{1}{2}\Big(\frac{f(0)+f(1)}{2} + \frac{f(1)+f(2)}{2}\Big) \\ &= a_1 f(0) + a_0 f(1) + a_1 f(2). \end{aligned}$$

The response function, shown in Figure 17.13a, decreases smoothly from 1 at $\omega = 0$ to zero at $\omega = 1/2$. The 'half-power' point at which the spectral density of the output is half of that of the input (i.e., $|c(\omega)|^2 = 1/2$) occurs at $\omega \approx 0.18$.

The running mean filters, such as (17.39), have weights

$$a_k = \begin{cases} \frac{1}{2K+1} & \text{for } k \leq K \\ 0 & \text{for } k > K. \end{cases} \tag{17.47}$$

The response functions for the three running mean filters with $K = 2$, 3, and 7 are shown in Figure 17.13b. Note that the frequency response functions have strong side lobes and zeros at frequencies $\frac{j}{2K+1}$, $j = 1, \ldots, K$. The running mean filter suppresses all oscillatory components with wavelengths such that the 'filter length' $2K + 1$ is an integer multiple of the wavelength. Residual amounts of all other waves remain after averaging because the running mean does not 'sample' the positive and negative halves of these waves symmetrically.

From (17.41) we see that the side lobes in Figure 17.13b have a significant effect on the spectral density of the output of the running mean filter. For example, if weakly persistent red noise were input into the five-term running mean filter, the output might appear to have a broad spectral peak near $\omega = 0.3$.

17.5.3 Combining Filters. Let $\mathcal{A}$ and $\mathcal{B}$ be two filters with weights a_k and b_k, respectively. Then, if $\mathbf{X}_t$ is an input series, we denote the output by

$$\begin{aligned} [\mathcal{A}(\mathbf{X})]_t &= \sum_k a_k \mathbf{X}_{t+k} \\ [\mathcal{B}(\mathbf{X})]_t &= \sum_k b_k \mathbf{X}_{t+k}. \end{aligned}$$

The corresponding response functions are denoted by $c_{\mathcal{A}}$ and $c_{\mathcal{B}}$. The weighted sum $\alpha\mathcal{A}+\beta\mathcal{B}$ is again a filter with weights $\alpha a_k + \beta b_k$ and

$$[(\alpha\mathcal{A}+\beta\mathcal{B})(\mathbf{X})]_t = \alpha\,[\mathcal{A}(\mathbf{X})]_t + \beta\,[\mathcal{B}(\mathbf{X})]_t\,.$$

Since filtering is a linear operation, linearly combining output from two filters is equivalent to passing the input through the combined filter. Similarly, the response function of the combined filter is the linear combination of response functions:

$$c_{\alpha\mathcal{A}+\beta\mathcal{B}}(\omega) = \alpha\, c_{\mathcal{A}}(\omega) + \beta\, c_{\mathcal{B}}(\omega).$$

In particular, suppose that A is a low-pass filter, that is, a filter designed to remove high-frequency variations. Also, suppose that B is the 'do nothing' filter that leaves the input unchanged. Filter B has weights $b_0 = 1$ and $b_k = 0$ for $k \neq 0$, and is denoted $\mathcal{B} = 1$. A *high-pass* filter C can be constructed from A and B by setting $\alpha = -1$ and $\beta = 1$ to obtain $\mathcal{C} = 1 - \mathcal{A}$ with weights $c_0 = 1 - a_0$ and $c_k = 1 - a_k$ for $k \neq 0$. Note that if $c_{\mathcal{A}}(0) = 1$, then $c_{1-\mathcal{A}}(0) = 0$, in which case the output of the high-pass filter has time mean zero.

The convolution filter $\mathcal{A}\cdot\mathcal{B}$ is constructed by applying filters A and B in sequence:

$$[\{\mathcal{A}\cdot\mathcal{B}\}(\mathbf{X})]_t = [\mathcal{A}(\mathcal{B}(\mathbf{X}))]_t.$$

The filter weights p_k of the product $\mathcal{P} = \mathcal{A}\cdot\mathcal{B}$ are given by the convolution

$$p_k = \sum_l a_l b_{k-l} \tag{17.48}$$

and the response function is the product of the response functions of $\mathcal{A}$ and $\mathcal{B}$:

$$c_{\mathcal{AB}}(\omega) = c_{\mathcal{A}}(\omega)c_{\mathcal{B}}(\omega). \tag{17.49}$$

Thus convolution in the time domain is equal to multiplication in the frequency domain (and vice versa).

Since a filter is uniquely determined by its response function (17.49) proves that the sequence of the application of the two filters is irrelevant. That is,

$$\mathcal{A}\cdot\mathcal{B} = \mathcal{B}\cdot\mathcal{A}. \tag{17.50}$$

17.5.4 Further Simple Filters. The results of the preceding subsection may be used to construct other simple filters from the (1-2-1)-filter ([17.5.2]).

The plain (1-2-1)-filter may be applied repeatedly to suppress the high-frequency variability more efficiently. For example, the response function of the (1-2-1)·(1-2-1)-filter is shown in Figure 17.13a. Using (17.48), we see that the filter weights p_k of $\mathcal{A}\cdot\mathcal{A}$ are:

$$p_k = \begin{cases} a_0^2 + 2a_1^2 & = \frac{3}{8} \quad \text{for } k = 0 \\ a_{-1}a_0 + a_0a_1 & = \frac{1}{4} \quad \text{for } |k| = 1 \\ a_{-1}a_1 & = \frac{1}{16} \quad \text{for } |k| = 2 \\ 0 & \quad\quad \text{otherwise.} \end{cases}$$

A high-pass filter can be derived from the (1-2-1)-filter by forming the filter 1 − (1-2-1). Its response function is a mirror image of the response function of the low-pass filter (Figure 17.13a) and its weights are

$$b_k = \begin{cases} \frac{1}{2} & \text{for } k = 0 \\ -\frac{1}{4} & \text{for } |k| = 1 \\ 0 & \text{otherwise.} \end{cases}$$

Finally, a band-pass filter $\mathcal{B}$ may be obtained by combining the (1-2-1)-filter $\mathcal{A}$ with the 1 − (1-2-1)-filter $1-\mathcal{A}$ and setting $\mathcal{B} = 4\cdot\mathcal{A}(1-\mathcal{A})$. The response function $c_{\mathcal{A}}(\omega)c_{1-\mathcal{A}}(\omega)$ has zeros at both ends of the frequency interval [0, 1/2] and a maximum at $\omega = 0.25$. The factor 4 was chosen to ensure that the filter does not attenuate variability at its point of peak response (Figure 17.13a). This filter has weights

$$b_k = \begin{cases} \frac{1}{2} & \text{for } k = 0 \\ 0 & \text{for } |k| = 1 \\ -\frac{1}{4} & \text{for } |k| = 2 \\ 0 & \text{otherwise.} \end{cases}$$

17.5.5 Some Filters That Discriminate Between Time Scales. The simple filters discussed up to this point do not have particularly desirable properties. The (1-2-1)-filter and its relatives 'cut off' slowly by gradually changing the attenuation of variance with frequency. The running-mean filter also cuts off slowly, but it also has large sidelobes that allow variance leakage from high frequencies.

In contrast, the ideal low-pass filter has a box-car shaped frequency response function that cuts off sharply at a prescribed cut-off frequency (see Figure 17.14). Unfortunately, the ideal digital filter can not actually be used because it has infinitely many nonzero weights. The ideal low-pass filter has frequency response function $c(\omega) = 1$ for $|\omega| \leq \omega_0$ and $c(\omega) = 0$ elsewhere. It has weights $a_0 = 2\omega_0$ and $a_k = \frac{1}{\pi|k|}\sin(2\pi|k|\omega_0)$ for $|k| > 0$. Simply truncating the weights at

Δ	Blackmon's filters low-pass	Blackmon's filters band-pass	Blackmon's filters high-pass	Filters for daily data low-pass	Filters for daily data band-pass	Wallace et al.'s filters ≤ 10 days	Wallace et al.'s filters ≤ 5 days
0	0.09747	0.27769	0.47626	0.21196	0.45221	0.82119	0.66850
1	0.09547	0.14335	−0.31860	0.19744	−0.07287	−0.16871	−0.27390
2	0.08963	−0.10201	0.01975	0.15769	−0.28851	-0.14062	−0.13432
3	0.08049	−0.19477	0.10098	0.10288	0.09733	−0.10059	0.00000
4	0.06883	−0.09233	−0.01860	0.04625	0.03951	−0.05682	0.06204
5	0.05564	0.02830	−0.05468	0.0	0.02833	−0.01752	0.04669
6	0.04196	0.04193	0.01678	−0.02820	0.03316	0.01118	0.00000
7	0.02882	0.00335	0.03331	−0.03684	−0.07089	0.02646	−0.02810
8	0.01707	0.00411	−0.01445	−0.03003	−0.00227	0.02906	−0.02194
9	0.00734	0.03281	−0.02073	−0.01518	0.00302	0.02250	0.00000
10	0.0	0.03043	0.01179	−0.0	0.00708	0.01163	0.02193
11	−0.00488	−0.00200	0.01257	—	—	0.00101	0.00965
12	−0.00748	−0.01917	−0.00900	—	—	−0.00624	0.00000
13	−0.00818	−0.00967	−0.00715	—	—	−0.00903	−0.00461
14	−0.00749	−0.00013	0.00627	—	—	−0.00801	−0.00274
15	−0.00596	-0.00304	0.00362	—	—	−0.00491	0.00000

Table 17.1: *Digital filters designed to extract variability on specific time scales. The filters are symmetric (i.e., $a_k = a_{-k}$).*
Blackmon's [47] filters for separating low-frequency, baroclinic, and high-frequency time scales from 12-hourly data are listed in columns 2 to 4. The response functions are shown in Figure 17.15.
Two additional filters are listed in columns 5 and 6.
Wallace et al.'s [410] high-pass filters are listed in columns 7 and 8. The response functions are shown in Figure 17.17.

a fixed lag K does not yield a particularly ideal filter. As illustrated in Figure 17.14, this results in a filter with frequency response function that has large Gibbsian overshoots and side lobes.[18] The solution to this problem is to strive for a frequency response function shape that does not cut off as abruptly as the 'ideal' filter. In this way, excellent digital filters can be constructed by carefully selecting a finite number of weights.

A set of three such filters designed by Blackmon [47] for the twice-daily data that are often used in atmospheric science. The low-pass filter in this set suppresses variability related to day-to-day weather events and keeps variability on time scales of weeks (such as 'blocking' events). The band-pass filter extracts variability in the baroclinic time scale (approximately 2.25 to 5 days), and the high-pass filter retains only variability on the one- to two-day time scale (see below for the exact definition).[19]

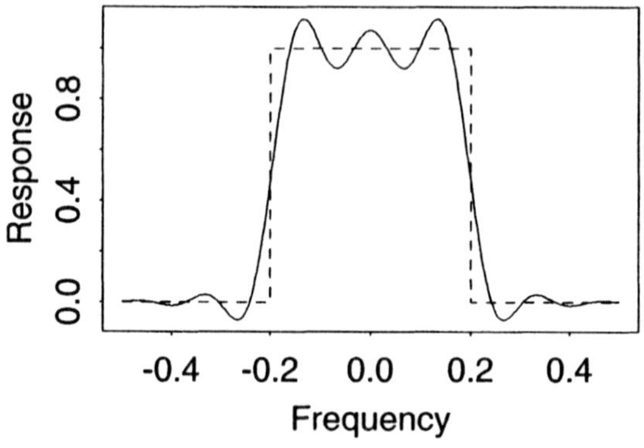

Figure 17.14: *The frequency response function of the ideal low-pass filter (dashed curve) and the filter obtained by truncating the weights of the ideal filter at $K = 7$.*

Blackmon's filters have $2K + 1 = 31$ weights, which are listed in Table 17.1. The response functions are plotted in Figure 17.15.

Another pair of filters designed to retain low-frequency and baroclinic variability in daily data

[18]This problem is similar to the one that motivates the use of data tapers. (cf. [12.3.8]).

[19]These filters are applied in [3.1.6].

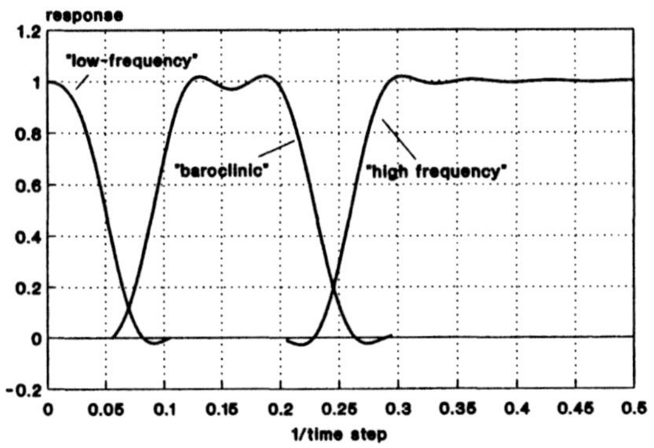

Figure 17.15: *The response function of Blackmon's filters (Table 17.1, columns 2 to 4).*

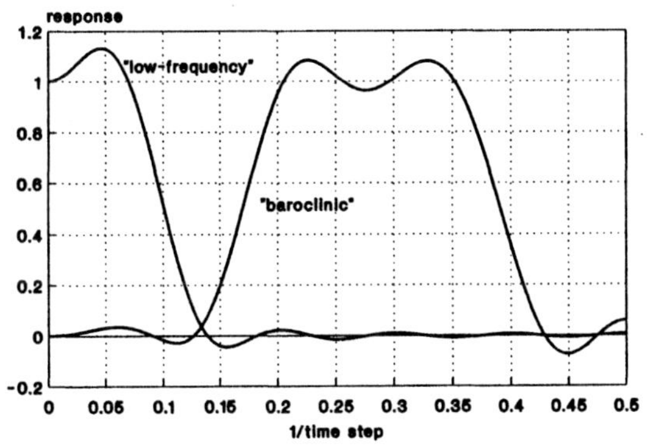

Figure 17.16: *Two filters for isolating low-frequency and baroclinic variations for daily data (Table 17.1, columns 5 and 6).*

are also given in Table 17.1. The response functions are shown in Figure 17.16.

A further pair of high-pass filters, from Wallace et al. [410], are listed in Table 17.1 and shown in Figure 17.17. When applied to daily observations the cut-offs are approximately 5 and 10 days.

17.5.6 Examples. Several examples of applications of filters similar to those described above are discussed in Section 3.1. The day-to-day variability of DJF 500 hPa height is separated into three *time windows* by means of Blackmon's filters (Figure 17.15) in [3.1.5], and the Northern Hemispheric distributions of the variances attributed to the three windows are shown in Figure 3.8. The skewness of the low-pass filtered 500 hPa height and its relationship to the location of the storm-tracks is discussed in [3.1.8] (see Figure 3.11).

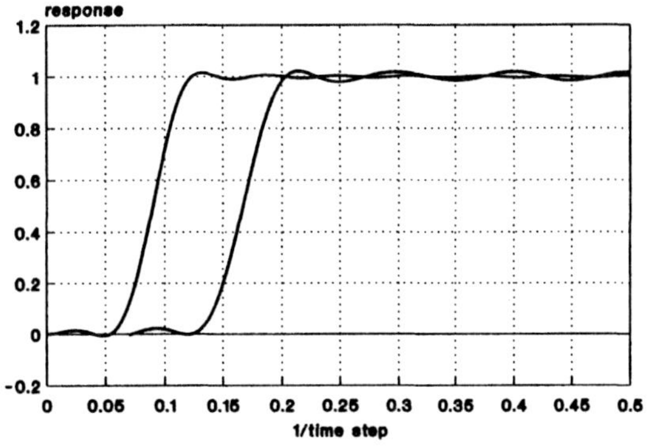

Figure 17.17: *Wallace et al.'s high-pass filters (Table 17.1, columns 7 and 8).*

17.5.7 Construction of Filters. Representation (17.42) of the response function may be used to choose filter weights so that the resulting response function $c(\omega)$ closely approximates a specified form $\tilde{c}(\omega)$. The weights for a specified filter length $2K+1$ are obtained by minimizing

$$\begin{aligned}\epsilon &= \int_{-\frac{1}{2}}^{\frac{1}{2}} \big(c(\omega) - \tilde{c}(\omega)\big)^2\, d\omega \\ &= \int_{-\frac{1}{2}}^{\frac{1}{2}} \Big(a_0 + 2\sum_{k=1}^{K} a_k \cos(2\pi k\omega) - \tilde{c}(\omega)\Big)^2\, d\omega.\end{aligned}$$

Taking derivatives with respect to a_0 and a_k, and setting the derivatives to zero, we find

$$a_k = \int_{-\frac{1}{2}}^{\frac{1}{2}} \tilde{c}(\omega)\cos(2\pi k\omega)\, d\omega$$

for all k. Thus the 'optimal' filter with $2K+1$ weights is formed simply by truncating the Fourier transform of $\tilde{c}(\omega)$. However, as we saw above, the 'optimal' $2K+1$ weight filter may have undesirable properties, such as large side lobes, if $\tilde{c}(\omega)$ has discontinuous low-order derivatives. The 'best' $2K+1$ weight filter will be found by striving for a response function $\tilde{c}(\omega)$ that varies smoothly with ω for all $\omega \in [-1/2, 1/2]$.

A strategy that results in good filters is to taper the weights of the optimal filter that is obtained by truncating the Fourier transform of $\tilde{c}(\omega)$ with a Hanning taper (see [12.3.8]). For example, consider a low-pass filter with cut-off frequency ω_0. The 'optimal' $2K+1$ weight filter has weights $a_0 = 2\omega_0$ and $a_k = \frac{1}{\pi|k|}\sin(2\pi|k|\omega_0)$ for $0 < |k| \le K$. The variance leakage problems associated with this filter are largely eliminated when these weights are tapered with factors $h_k = \frac{1}{2}(1 + \cos(\pi|k|/(K+1))$ and then renormalized.

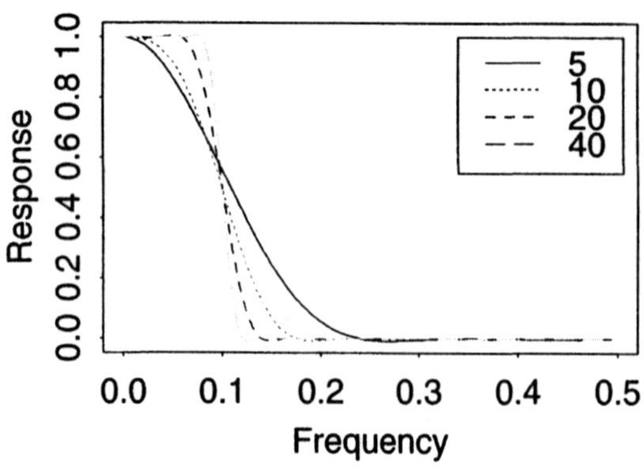

Figure 17.18: *Low-pass filters with cut-off frequency* $\omega_0 = 0.1$ *with 11, 21, 41 and 81 weights (i.e., K* = 5, 10, 20, *and* 40*) constructed by tapering the weights of the ideal low-pass filter.*

The resulting low-pass filter shuts off smoothly with increasing frequency. The amplitude of the variations at frequency ω_0 are attentuated by 50% with the result that only 25% of the variance at this frequency is passed by the filter. The sharpness of the cut-off is determined by the number of weights. Figure 17.18 displays the response function for filters with cut-off frequency $\omega_0 = 0.1$ and 11, 21, 41 and 81 weights (i.e., $K = 5, 10, 20$, and 40).

18 Forecast Quality Evaluation

18.0.1 Summary. Here we continue a discussion that we began in [1.2.4], by extending our treatment of some aspects of the art of forecast evaluation.[1] We describe statistics that can be used to assess the *skill* of categorical and quantitative forecasts in Sections 18.1 and 18.2.[2] The utility of the correlation skill score is discussed by illustrating that it can be interpreted as a summary statistic that describes properties of the probability distribution of future states conditional upon the forecast. The Murphy–Epstein decomposition is used to explain the relationships between commonly used skill scores (Section 18.3). Some of the common pitfalls in forecast evaluation problems are discussed in Section 18.4.

18.0.2 The Ingredients of a Forecast. In this section, we consider the problem of quantifying the *skill* of a forecast such as that of monthly mean temperature at a certain location. We use the symbols $\mathbf{F}_\tau(t)$ to denote the forecast for the time t with a lead time of τ (e.g., in units of months) and $\mathbf{P}(t)$ to denote the verifying observations, or the *predictand* at time t. We generally omit the suffix (t) and the index τ in our notation unless they are needed for clarity.

Note that in some applications there may be substantial differences between $\mathbf{P}$ and the true observations. These differences might arise from biases induced by analysis systems or from errors induced by observing systems. An example of the latter are the various biases that are inherent in the many different rain gauge designs used throughout the world [247]. We ignore these biases in this chapter and assume that $\mathbf{P}$ represents the true verifying observations.

The method used to produce the forecast is *not* important in the context of this chapter. The forecast may have been produced using a sophisticated dynamical model, but it may also have been based on a coin tossing procedure. The information used to produce a forecast, called the *predictor*, is also not relevant here.

A forecast must be precise in time and space. That is, the time t for which the forecast $\mathbf{F}_\tau(t)$ is issued must be clearly stated and it must correspond precisely with the time t of the verifying analysis $\mathbf{P}(t)$. Thus, statements of the form 'there will be a thunderstorm at the end of August' do not qualify as a forecast.

18.0.3 Forecasts and Random Variables. In this chapter, both the predictand $\mathbf{P}$ and the forecast $\mathbf{F}$ are treated as random variables. An actual predictand, or actual observation, is denoted $\mathbf{p}$, that is, as a realization of the random variable $\mathbf{P}$. Accordingly, a single forecast is denoted $\mathbf{f}$.

Skill parameters that measure the ensemble quality of forecasting system are parameters that characterize some aspect of the distribution of the bivariate random variable $(\mathbf{F}, \mathbf{P})$. In practice, where a skill parameter is derived from a finite number of forecasts, the skill is an estimate of the true unknown parameter. Therefore, forecast skill evaluation can be thought of as a form of parameter estimation (see Chapter 5), even though the problem of forecast skill evaluation is generally not considered as such by the practitioners of the parameter estimation art.

18.0.4 Categorical and Quantitative Forecasts. We restrict ourselves to examples in which the forecast $\mathbf{F}$ and the predictand $\mathbf{P}$ are either both quantitative (i.e., a number such as '13 °C') or categorical statements (such as 'warmer than normal'). If the forecast is categorical, we require that the category boundaries ('normal') are unequivocally defined. We will not discuss *probabilistic*

[1] This chapter is based in part on Livezey [255], Murphy and Epstein [285] and Stanski, Wilson, and Burrows [355]. For further reading, we also recommend Murphy and Daan [284], Murphy and Winkler [286], Murphy, Brown, and Chen [283], Barnston [25] and Livezey [256].

[2] We call these statistics *skill scores* for convenience. We use the phrase 'skill score' somewhat less formally than is dictated by statistical convention, where this expression is limited to statistics that have a specific functional form, such as the Heidke skill score (18.1) and the Brier skill score (18.5). These formal scores compare the actual rate of success with the success rate of a reference forecast.

Verifying analysis	Forecast *above*	*below normal*	
above normal	p_{aa}	p_{ba}	p_a^P
below normal	p_{ab}	p_{bb}	p_b^P
	p_a^F	p_b^F	1

Table 18.1: *An illustration of a* 2×2 *contingency table used to summarize the performance of a categorical forecasting system.*

forecasts such as 'The chance of precipitation tomorrow is 70%.' We begin by discussing categorical forecasts in Section 18.1. Quantitative forecasts are discussed in Section 18.2.

18.1 The Skill of Categorical Forecasts

18.1.1 Categorical Forecasts. Categorical forecasts are often made in two or three (or more) classes, such as *above normal*, *near normal* and *below normal*, that are clearly defined in terms of *a priori* specified threshold values. For example, two-class forecasts often specify either *above normal* or *below normal*, where the threshold *normal* is the long-term mean of the predicted parameter. The outcome of a two-class categorical forecasting scheme can be summarized in a 2×2 *contingency table* (see Table 18.1).

The entries in the table are defined as follows. The probability that the forecast **F** and the predictand **P** jointly fall in the *above normal* category is p_{aa}. Similarly, p_{ab} is the probability that the forecast falls into the *above normal* normal category and the predictand falls into the *below normal* category. Probabilities p_{ba} and p_{bb} are defined analogously.

The marginal probability distribution (cf. [2.3.12]) of the forecast **F** is given by

$$p_a^F = P(\mathbf{F} = above\ normal) = p_{aa} + p_{ab}$$
$$p_b^F = P(\mathbf{F} = below\ normal) = p_{bb} + p_{ba}.$$

The marginal probability distribution for the predictand **P** are defined similarly.

18.1.2 The Heidke Skill Score. A useful measure of the skill of a two-class categorical forecasting scheme is the *Heidke skill score* (Heidke [173]), which is given by

$$S = \frac{p_C - p_E}{1 - p_E} \tag{18.1}$$

where p_C is the probability of a correct forecast, given by

$$p_C = p_{aa} + p_{bb},$$

and p_E is the probability of a correct forecast when the forecast carries no information about the subsequent observation (a 'random forecast'). We obtain a random forecast when **F** and **P** are independent, and therefore find that

$$p_E = p_a^P p_a^F + p_b^P p_b^F.$$

If *above normal* and *below normal* classes are equally likely for both **F** and **P**, then $p_E = 0.5$ because $p_a^F = p_b^F = p_a^P = p_b^P = 0.5$. On the other hand, the two classes may not be equally likely. For example, we might have $p_a^F = p_a^P = 0.6$ and $p_b^F = p_b^P = 0.4$. Then $p_E = 0.4^2 + 0.6^2 = 0.52$.

It is easily demonstrated that the skill S of a random forecast is zero and that the skill of a perfect forecast (i.e., $p_C = 1$) is 1. If there is perfect *reverse* reliability, that is, every forecast is wrong, then $p_C = 0$ and $S = -p_E/(1 - p_E)$. In this case we obtain $S = -1$ if both classes are equally likely for **F** and **P**.

When sample sizes are finite, the Heidke skill score (18.1) is often written as

$$S = \frac{n\hat{p}_C - n\hat{p}_E}{n - n\hat{p}_E} \tag{18.2}$$

where n is the number of $(\mathbf{F}, \mathbf{P})$ realizations in the sample, and the hat notation, as usual, indicates that the probability is estimated. In this expression, $n\hat{p}_C$ is the number of correct forecasts and $n\hat{p}_E$ is an estimate of the expected number of correct random forecasts.

The Heidke skill score may be extended to categorical forecasts with more than two categories. Many other useful skill scores for categorical forecasts may also be defined (see, for example, Stanski et al. [355]). Also, the term p_E in (18.1), which represents the probability of correct random forecasts, may be replaced by the probability of a correct forecast produced by any other reference forecasting system (such as persistence, in which the class the predictand will occupy at the next verifying time is forecast to be the class currently occupied by the predictand).

18.1.3 Example: The Old Farmer's Almanac. The following example is taken from Walsh and Allen [412] who evaluated five years of regular monthly mean temperature forecasts for the USA issued by *The Old Farmer's Almanac* [364]. The success rate for temperature was 50.7%. The corresponding rate for precipitation was 51.9%. *The Old Farmer's Almanac*'s forecasts have some skill, with $S = 7/500 = 0.014$ for temperature and $S = 0.038$ for precipitation, if we assume that the monthly means have symmetric distributions. However, the distributions are actually somewhat skewed: there are fewer (but larger) above-normal temperature extremes than below-normal extremes. If we assume that $p_a^F = p_a^P = 0.45$ for temperature, then $p_E = 0.45^2 + 0.55^2 = 0.505$, so that the actual skill of the Almanac is

$$\begin{aligned} S &= (p_C - p_E)/(1 - p_E) \\ &= (0.507 - 0.505)/(1 - 0.505) = 4 \times 10^{-3}. \end{aligned}$$

Similarly, if $p_a^F = p_a^P = 0.60$ for precipitation, then $p_E = 0.52$ and

$$S = \frac{0.519 - 0.520}{1 - 0.520} = -2 \times 10^{-3}.$$

Apparently the skill of the *Farmer's Almanac* is no greater than that of a forecast constructed by drawing random numbers from slightly skewed distributions.

18.1.4 Mixing Forecasts of Unequal Skill. Let us now consider a hypothetical forecasting scheme that operates throughout the year. During winter, the scheme produces random forecasts so that the number of correct forecasts in winter is $p_{C_w} = p_{E_w}$ and S_w is zero. In summer, however, the scheme is better than chance and produces forecasts for which $p_{C_s} = 1.5 p_{E_s}$. Then $S_s = (p_{C_s} - p_{E_s})/(1 - p_{E_s}) = 0.5$. For simplicity, we assume that $p_{E_w} = p_{E_s}$ and that the number of winter and summer forecasts are equal. Then, over summer and winter, the Heidke skill score S_{w+s} is larger than the winter score and smaller than the summer score:

$$S_{w+s} = \frac{\frac{1}{2}(p_{C_w} + p_{C_s}) - p_{E_w}}{1 - p_{E_2}} = 0.25.$$

Thus, if we add random forecasts to a set of skilful forecasts, the overall skill score will be lowered. If we avoid making forecasts when the forecast scheme is unable to use the information contained in the predictor, the skill score will be enhanced.

18.1.5 The Skill Score is Subject to Sampling Variation. The Heidke skill score (18.1) is a one number summary of the forecasting scheme performance relative to a competing reference scheme. As noted in [18.0.3], the forecast and predictand should be viewed as a (hopefully) correlated pair of random variables $(\mathbf{F}, \mathbf{P})$ and the skill score S_{FP} should properly be viewed as an estimator of some characteristic of the joint distribution of **F** and **P**. One might therefore ask how accurate this estimate is. One might also ask what the likelihood is of obtaining a positive realization of the skill score from a finite sample of random forecasts. There are no general answers to these questions. Radok [327] however, has suggested an estimate of the sampling error of the Heidke score. *Monte Carlo* techniques might also be helpful for making inferences about the skill parameter (see Section 6.3).

18.1.6 Example: Prediction of Snowfall in Ontario. Burrows [76] designed a forecast scheme to predict 'lake-effect' snowfall for a number of stations leeward of Lake Huron in Ontario, Canada. The predictors were designed to be useful when the synoptic situation is favourable for the occurrence of lake-effect snow and the only cases considered were those in which the weather map forecast a synoptic situation conducive to lake-effect snow. Categorical forecasts were prepared at 28 stations. Five categories were used, with **F** and **P** defined as follows:

Category (**F** and **P**)	Snow amount (cm)
1	[0,trace]
2	(trace, 5]
3	(5, 12.5]
4	(12.5, 22.5]
5	> 22.5

Figure 18.1 shows a typical field of predictand **P** (snow amount category actually observed) and the corresponding field of forecasts. An asterisk in the forecast field indicates that a forecast was not made at that location. The overall performance of the forecasting scheme is summarized in Table 18.2.

Burrows [76] rated a forecast that was one category different from the predictand (i.e., $|\mathbf{p} - \mathbf{f}| = 1$) a better forecast than a forecast which was two categories different (i.e., $|\mathbf{p} - \mathbf{f}| = 2$) and so on. Entries on the diagonals in Table 18.2 were therefore weighted depending upon the value of $|\mathbf{p} - \mathbf{f}|$. Counts in the table for which $|\mathbf{P} - \mathbf{F}| = k$ were multiplied by $\gamma_k = 1 - k/4$. The

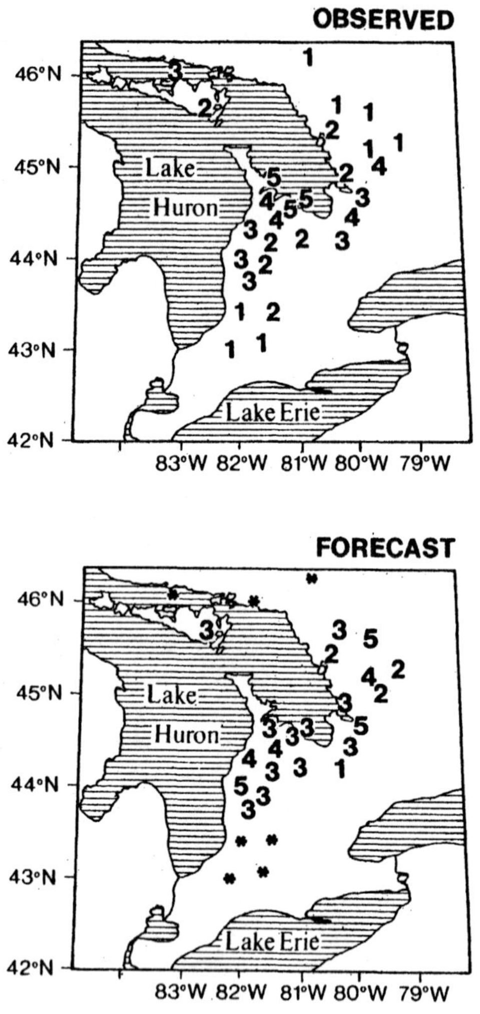

Figure 18.1: *An example of a categorical forecast of snow amount at 28 stations in southern Ontario leeward of Lake Huron. The observed snow category (see text) is shown in the top panel. The corresponding forecasts are shown in the lower panel. From Burrows [76].*

	Forecast					
Observed	1	2	3	4	5	
1	14	13	1	1	0	29
2	12	26	14	2	0	54
3	2	12	14	5	5	38
4	0	2	4	2	1	9
5	0	0	0	0	0	0
	28	53	33	10	6	130

Table 18.2: *A 5 × 5 contingency table summarizing the performance of a lake-effect snow forecasting scheme. From Burrows [76].*

weighted counts were then totalled for the entire table and used as a measure, say s_b, of the number of 'correct' forecasts. A similar measure of the number of 'correct' random forecasts was computed by estimating the entries of Table 18.2 under the assumption that **F** and **P** are independent. The estimated distribution of random counts was obtained by multiplying the row total by the column total and dividing by the table total (130). Burrows then weighted and summed the entries in this new table as before to produce a corresponding measure, say s_b^{random}, of the number of 'correct' forecasts expected by chance. Finally, a skill score analogous to the Heidke score was computed as

$$S_B = \frac{s_b - s_b^{random}}{n - s_b^{random}},$$

where n is the total number of forecasts made. Note that if we set γ_0 to 1 and γ_k to zero for nonzero k, then S_B reduces to the Heidke skill score (18.1). Like the Heidke score, S_B is zero for random forecasts and 1 for perfect forecasts.

The S_B value for Table 18.2 is 33%. The finding that the forecasts are skilful is also supported by two other skill scores computed by Burrows.

The *critical success index* is defined for each category $k = 1, \ldots, 5$ as the ratio of number of occasions C_k on which $\mathbf{f} = \mathbf{p} = k$ and the sum of number of occasions on which either $\mathbf{p} = k$ or $\mathbf{f} = k$ minus C_k. This index is 33% for $k = 0$, 32% for $k = 1$, 25% for $k = 2$, 12% for $k = 3$, and 0% for $k = 4$. The critical success ratio for category k is simply an estimate of the probability of forecast conditional upon either forecasting or observing category k. The critical skill index can be compared with that expected under no skill by recomputing the contingency table under the assumption of independence. The corresponding critical success indices expected for random forecasts are 12% for $k = 0$, 26% for $k = 1$, 16% for $k = 2$, 4%, for $k = 3$, and 0% and for $k = 4$.

The *probability of detection (POD)* is defined as the ratio of $\sum_k C_k$ divided by the number of all forecasts T. In this case, the POD is $56/130 = 43\%$. The probability of detection is simply the probability of making a correct forecast. The estimated POD for a random forecast is 30% in this case.

18.1.7 Comments. The skill score S_B introduced in [18.1.6] is a modified Heidke score. Barnston [25] points out that the original Heidke score has two undesirable properties. First, the Heidke score increases as the number of categories decreases. For example, for a broad range of moderately skilful forecast sets, the two-class Heidke skill score will be about double the five-class score. Second, if the reference forecast is the random forecast, and if classes are not observed (or forecast) with equal frequency, then the Heidke skill score is not *equitable*. That is, the Heidke score will favour a biased forecast unfairly. An example of this property is given in [18.4.2]. Barnston [25], and also Ward and Folland [415] designed modified Heidke skill scores that are independent of the number of classes and equitable.

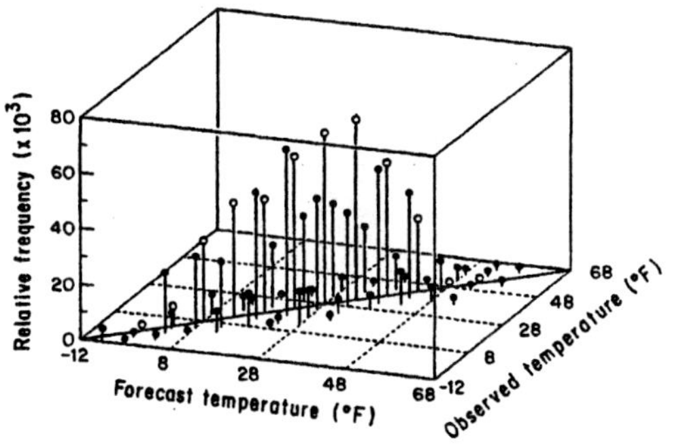

Figure 18.2: *Estimated joint distributions of forecasts and observations* (**F**,**P**)*. All data are collected into bins of* 5 °F × 5 °F*. Values for* **f** = **p** *are indicated by open circles to facilitate identification. From Murphy et al. [283].*

18.2 The Skill of Quantitative Forecasts

18.2.1 Forecast and Predictand as Bivariate Random Variable. As we noted in [18.0.3], the forecast/predictand pair $(\mathbf{F}, \mathbf{P})$ form a bivariate random variable with a joint density function f_{FP}. The conditional density functions $f_{F|P=p}$ and $f_{P|F=f}$ tell us something about the performance of the forecast. (For a detailed discussion see Murphy and Winkler [286] and Murphy et al. [283].)

First, one would hope that $\mathcal{E}(\mathbf{F}|\mathbf{P} = \mathbf{p}) = \mathbf{p}$ and that $\mathcal{E}(\mathbf{P}|\mathbf{F} = \mathbf{f}) = \mathbf{f}$. That is, the mean of all forecasts **F**, given a predictand $\mathbf{P} = \mathbf{p}$, is **p**, and, the mean of all predictands **P** is **f** when averaged over all occasions when **F**= **f**. If the former condition is satisfied, the forecast is called *conditionally unbiased*.

The conditional variances $\mathrm{Var}(\mathbf{F}|\mathbf{P} = \mathbf{p})$ and $\mathrm{Var}(\mathbf{P}|\mathbf{F} = \mathbf{f})$ are ideally small.

Note that the forecast **F** and the predictand **P** can be statistically associated. Let us choose a and b so that

$$\mathcal{E}\big((\mathbf{F} - (a + b\mathbf{P}))^2\big)$$

is minimized. The line $\alpha + \beta\mathbf{P}$ is the regression of **F** on **P**. Two necessary conditions for the forecast to be unbiased are that a is zero and b is 1 such that the regression line is the 45° diagonal in the two-dimensional $(\mathbf{F}, \mathbf{P})$-plane.

18.2.2 Joint distributions. The joint $(\mathbf{F}, \mathbf{P})$-density may be crudely estimated by plotting a *scatter diagram*, in which each realization $(\mathbf{f}, \mathbf{p})$ is marked by a dot. Alternatively one could group all realizations $(\mathbf{f}, \mathbf{p})$ into small boxes and display the number of entries per box. An example of such a diagram is shown in Figure 18.2 (Murphy et al. [283]). The forecast is for temperature for Minneapolis (Minnesota) at a 24-hour lead during winter. Forecasts of 'correct' or 'near correct' are marked by open circles. The maximum density estimate usually lies on the diagonal **f** = **p**, but for forecasts $\mathbf{F} \leq 28$ °F the corresponding observed temperatures tend to be systematically lower than the forecast by a few degrees. The conditional standard deviations of the forecast errors are of the order of 5 °F, and forecast errors larger than 20 °F never occur. Very little can be learned about the skill of forecasts below 8 °F and above 48 °F because of poor sampling.

An example of an estimated conditional distribution $f_{P|F=f}$ is the estimated distribution of Minneapolis temperature observations **P** given the forecast $\mathbf{F} = \mathbf{f}$, which is shown in the upper panel of Figure 18.3. The 10%, 25%, 50%, 75%, and 90% quantiles of the observations are derived and plotted for each 5 °F bin of the forecast. Ideally, the solid curve, representing the conditional 50% quantile, will lie on the diagonal. This is not so. In particular, the mean observed temperature is about 3 °F lower when temperatures below 20 °F are forecast. When temperatures below 12 °F are forecast, about 75% of observations are actually less than the forecast. The typical forecast error is generally independent of the forecast itself.

Estimates of the density function of the forecast conditional upon a fixed observed temperature **p** (i.e., $\hat{f}_{f|P=p}$) are displayed in the lower panel of Figure 18.3 for $\mathbf{p}$ = 24 °F, 25 °F, 34 °F. The two conditional F distributions for $\mathbf{p}$ = 14 °F and $\mathbf{p}$ = 25 °F are almost symmetric, but since

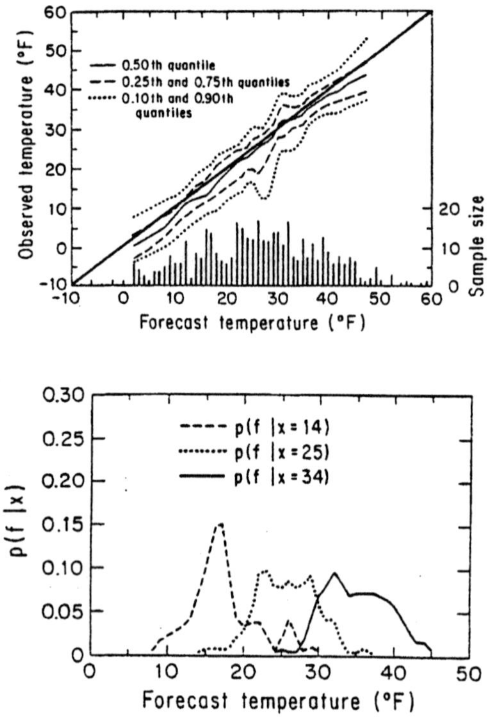

Figure 18.3: *Estimated conditional probability density functions of the Minneapolis temperature forecast. From Murphy et al. [283].*
Top: Quantiles of the distribution of the predictand **P** *conditional on the forecast* **F** = **f**. *The frequency of the forecasts is also shown so that the credibility of the conditional quantiles can be judged.*
Bottom: Distribution of the forecast **F** *conditional on the value of the predictand for* **P** $= 14°F, 25°F, 34°F$. *The '$p(f|x)$' in the diagram is Murphy's notation for the conditional probability density function $f_{F|P=p}$.*

$\mathcal{E}(\mathbf{F}|\mathbf{P} = 14°\text{F}) > 14°\text{F}$, it is evident that the forecast is biased.

18.2.3 Skill Scores. Several measures are frequently used to describe the skill of quantitative forecasts. These measures include the *correlation skill score*, the *mean squared error*, the *Brier skill score* and the *proportion of explained variance*.[3]

The correlation between the forecast **F** and the verifying observation **P** is called the *correlation skill score* and is given by

$$\rho_{FP} = \frac{\text{Cov}(\mathbf{F}, \mathbf{P})}{\sqrt{\text{Var}(\mathbf{F})\text{Var}(\mathbf{P})}}. \tag{18.3}$$

[3]A more complex measure of skill than those defined here is the 'linear error in the probability space' (LEPS) score introduced by Ward and Folland [415].

The *mean squared error* is the expected (i.e., long-term average) squared error which is defined by

$$S_{FP}^2 = \mathcal{E}\big((\mathbf{F} - \mathbf{P})^2\big). \tag{18.4}$$

The *Brier skill score* is a measure of the skill of the forecast **F** relative to a reference forecast **R** of the same predictand **P**. The comparison is made on the basis of the mean square error of the individual forecasts. The Brier score is given by

$$\begin{aligned} B_{FRP} &= 1 - \frac{S_{FP}^2}{S_{RP}^2} \\ &= \frac{S_{RP}^2 - S_{FP}^2}{S_{RP}^2}. \end{aligned} \tag{18.5}$$

The *proportion of explained variance* is the percentage of **P**-variance that is explained by **F**,

$$\begin{aligned} R_{FP}^2 &= \frac{\text{Var}(\mathbf{P}) - \text{Var}(\mathbf{F} - \mathbf{P})}{\text{Var}(\mathbf{P})} \\ &= 1 - \frac{\text{Var}(\mathbf{F} - \mathbf{P})}{\text{Var}(\mathbf{P})}. \end{aligned} \tag{18.6}$$

18.2.4 Skill Score Ranges. For a perfect forecast, that is, $\mathbf{F} = \mathbf{P}$, the correlation skill score ρ_{FP} is 1, the mean squared error S_{FP}^2 is zero and the percentage of explained variance R_{FP}^2 is 100%.

If **F** is the climatological forecast (i.e., $\mathbf{F} = \mathcal{E}(\mathbf{P})$), then ρ_{FP} and R_{FP}^2 are zero and $S_{FP}^2 = \text{Var}(\mathbf{P})$.

If **F** is a random forecast, with the same mean and variance as **P** then ρ_{FP} is zero and $S_{FP}^2 = \text{Var}(\mathbf{F} - \mathbf{P}) = \text{Var}(\mathbf{F}) + \text{Var}(\mathbf{P}) = 2\text{Var}(\mathbf{P})$. The explained variance is $R_{FP}^2 = 1 - 2\text{Var}(\mathbf{P})/\text{Var}(\mathbf{P}) = -1$.

Thus, the skill scores ρ_{FP} and R_{FP}^2 are constructed so that they have value 1 for a perfect forecast and zero or less than zero for trivial reference forecasts.

18.2.5 Skill Score Characteristics. The correlation skill score is insensitive to some types of systematic error. In particular, skill is not affected if the forecasts contain a constant bias or if the amplitude of two differ by a constant factor. That is, for two forecasts **F** and $\mathbf{G} = a\mathbf{F} + b$ for some constants a and b, then **F** and **G** have the same correlation skill score. On the other hand, mean squared error is very sensitive to such systematic errors. The results of many years of weather forecasting have shown that the mean squared error favours forecasting schemes that avoid extremes and tend not to deviate greatly from climatology (because the penalty grows as the *square* of the error [179]).

18.2.6 Correlation Skill Score and Probability Statements. Some appreciation for the interpretation of the correlation skill score can be obtained from the following thought exercise, which is based on the usual normal assumptions.[4] Suppose that we are given a pair of realizations **f** and **p** for the forecast and the verifying observation. Then the correlation skill score may be used to derive statements about the probability that $\mathbf{P} \geq \mathbf{p}$ conditional on $\mathbf{F} \geq \mathbf{f}$, for any **p** and **f**. We assume that $\mathcal{E}(\mathbf{F}) = \mathcal{E}(\mathbf{P})$ and that we are somehow able to identify, or reliably estimate, the covariance matrix

$$\Sigma = \begin{pmatrix} \sigma_F^2 & \gamma \\ \gamma & \sigma_P^2 \end{pmatrix}$$

of the bivariate random variable $(\mathbf{F}, \mathbf{P})$. For simplicity we assume that $\mathcal{E}(\mathbf{P})$ (and thus $\mathcal{E}(\mathbf{F})$) is zero.

Let us first consider $\mathbf{p} = 0$ and $\mathbf{f} = 0$. Then the probability of observing non-negative **P** given that non-negative **F** was predicted is the *conditional* probability

$$\begin{aligned} P(\mathbf{P} \geq 0 | \mathbf{F} \geq 0) &= \frac{P(\mathbf{P} \geq 0 \text{ and } \mathbf{F} \geq 0)}{P(\mathbf{F} \geq 0)} \\ &= \frac{\int_0^\infty \int_0^\infty \mathrm{f}(f,p)\, df\, dp}{\int_0^\infty \mathrm{f}(f)\, df}. \end{aligned} \tag{18.7}$$

Now let $A = 1/\sqrt{1-\gamma^2}$. Then

$$P(\mathbf{P} \geq 0 \text{ and } \mathbf{F} \geq 0) = \tag{18.8}$$
$$\frac{A}{2\pi\sigma_F\sigma_P} \int_0^\infty \int_0^\infty \exp\left\{ -\frac{A^2}{2} \times \left(\left(\frac{f}{\sigma_F}\right)^2 + \left(\frac{p}{\sigma_P}\right)^2 - \frac{2\gamma f p}{\sigma_F \sigma_P} \right) \right\} dp\, df.$$

Now, note that the exponentiated quadratic form in (18.8) may be written

$$\begin{aligned} &\frac{A^2}{2}\left(\left(\frac{f}{\sigma_F}\right)^2 + \left(\frac{p}{\sigma_P}\right)^2 - \frac{2\gamma f p}{\sigma_F\sigma_P} \right) \\ &= \frac{A^2}{2}\left((1-\gamma^2)\left(\frac{f}{\sigma_F}\right)^2 \right) \\ &\quad + \frac{A^2}{2\sigma_P^2}\left(p - \frac{\sigma_P}{\sigma_F} \frac{\gamma f}{(\sigma_P/A)} \right)^2. \end{aligned}$$

We may therefore write equation (18.8) as

$$P(\mathbf{P} \geq 0 \text{ and } \mathbf{F} \geq 0) = \tag{18.9}$$
$$\int_0^\infty \mathrm{g}(f; 0, \sigma_F^2) \times \int_0^\infty \mathrm{g}\left(p; \left(\frac{\sigma_P}{\sigma_F}\right)\gamma f, \left(\frac{\sigma_P}{A}\right)^2\right) dp\, df.$$

In this last expression $\mathrm{g}(f; 0, \sigma_F^2)$ represents the normal probability density function with mean 0 and variance σ_F^2. Similarly, $\mathrm{g}(p; (\sigma_P/\sigma_F)\gamma f, (\sigma_P/A)^2)$ represents the normal probability density function with mean $(\sigma_P/\sigma_F)\gamma f$ which depends upon the realized value of the forecast $\mathbf{F} = \mathbf{f}$ and variance σ_P^2/A^2, which is the conditional variance of **P** given $\mathbf{F} = \mathbf{f}$.[5] Finally, substituting (18.9) into (18.7), we obtain

$$P(\mathbf{P} \geq 0 | \mathbf{F} \geq 0) = \tag{18.10}$$
$$\frac{\int_0^\infty \mathrm{g}(f; 0, \sigma_F^2) \int_0^\infty \mathrm{g}\left(p; \frac{\sigma_P}{\sigma_F}\gamma f, (\frac{\sigma_P}{A})^2\right) dp\, df}{\int_0^\infty \mathrm{g}(f; 0, \sigma_F^2)\, df}.$$

The last formula becomes simpler if the forecast **F** and the predictand **P** are normalized so that $\sigma_F = \sigma_P = 1$. Then the covariance γ becomes the correlation ρ_{FP} and

$$P(\mathbf{P} \geq 0 | \mathbf{F} \geq 0) \tag{18.11}$$
$$\begin{aligned} &= \frac{\int_0^\infty \mathrm{g}(f; 0, 1) \int_0^\infty \mathrm{g}\left(p; \rho f, (1-\rho^2)\right) dp\, df}{\int_0^\infty f_{\mathcal{N}}(f; 0, 1)\, df} \\ &= 2\int_0^\infty \mathrm{g}(f; 0, 1) \int_0^\infty \mathrm{g}(p; \rho f, (1-\rho^2))\, dp\, df. \end{aligned}$$

Similar expressions for $P(\mathbf{P} > \mathbf{p} | \mathbf{F} > \mathbf{f})$ are easily obtained. In fact,

$$P(\mathbf{P} \geq \mathbf{p} | \mathbf{F} \geq \mathbf{f}) \tag{18.12}$$
$$= \frac{\int_{\mathbf{f}}^\infty \mathrm{g}(f; 0, 1) \int_{\mathbf{p}}^\infty \mathrm{g}\left(p; \rho f, (1-\rho^2)\right) dp\, df}{\int_{\mathbf{f}}^\infty f_{\mathcal{N}}(f; 0, 1)\, df}.$$

Appendix L contains these probabilities for a few values of ρ, **p** and **f** for the case $\sigma_F = \sigma_P = 1$. However, for most practical purposes, equation (18.12) must be calculated manually.

We can use equations (18.11) and (18.12) to make the following general deductions about forecast skill in terms of the correlation skill score when **F** and **P** are jointly normal with the same means and variances.

- If $\rho = 0$ then $\mathcal{N}_P(\rho f, \sqrt{1-\rho^2}) = \mathcal{N}_P(0, 1)$. Therefore the inner integral

[4] We assume that forecast **F** and observation **P** are jointly normal.

[5] This same decomposition has been encountered in [2.8.6] and in Section 8.2.

in (18.11) is 0.5 and consequently $P(\mathbf{P} \geq 0|\mathbf{F} \geq 0) = 0.5$. That is, a forecast which has a correlation skill score of zero is no more skilful than a toss of a coin.

- If ρ is positive, then for all positive F we have $\int_0^\infty g\big(p; \rho f, (1-\rho^2)\big)\,dp > 1/2$ and therefore $P(\mathbf{P} \geq 0|\mathbf{F} \geq 0) > 1/2$.
- Similarly $P(\mathbf{P} \geq 0|\mathbf{F} \geq 0) < 1/2$ if ρ is negative.

A positive correlation skill score ρ indicates that the forecast is useful, whereas a negative score indicates that the forecast with reversed sign has some skill.

Note that this exercise may be interpreted as the transformation of a quantitative forecast into a categorical forecast.

18.2.7 Conditional Moments. Using (2.36, 2.37) we see that

$$\begin{aligned}\mathcal{E}(\mathbf{P}|\mathbf{F}=\mathbf{f}) &= \mathcal{E}(\mathbf{P}) + \frac{\gamma}{\sigma_F^2}(\mathbf{f} - \mathcal{E}(\mathbf{F})) \\ &= a + b\mathbf{f} \qquad (18.13)\\ \mathrm{Var}(\mathbf{P}|\mathbf{F}=\mathbf{f}) &= \sigma_P^2 - \frac{\gamma^2}{\sigma_F^2}. \qquad (18.14)\end{aligned}$$

The conditional expectation consists of a constant term a and a term $b\mathbf{f}$ that is linear in $\mathbf{f}$. The conditional variance is independent of $\mathbf{f}$.[6]

The moments of the forecast $\mathbf{F}$ conditional on the observation $\mathbf{p}$ may be derived.

18.2.8 Improvement of a Quantitative Forecast. Any forecast $\mathbf{F}$ can be improved statistically if we have access to a large sample of previous forecasts $\mathbf{f}_t$ and corresponding predictands $\mathbf{p}_t$. This improvement can be obtained by regressing the predictand on the forecast using a simple linear regression model of the form suggested by equation (18.13). Least squares estimators of a bias correction $\widehat{a}$ and an amplitude correction $\widehat{b}$ are obtained in the familiar way. We will assume that the forecasts are already unbiased, so that $\widehat{a}$ is approximately zero, even though bias correction does not affect the correlation skill score.

The outcome of the exercise might be an amplitude correction $\widehat{b} = 1$ so that an improved forecast, say $\widetilde{\mathbf{F}}$, is given by $\widetilde{\mathbf{F}} = \mathbf{F}$. In this case nothing is gained. The tools of Chapter 8 show us that for large samples $\widehat{b} \approx \mathrm{Cov}(\mathbf{F},\mathbf{P})/\mathrm{Var}(\mathbf{F})$. Thus, asymptotically, the improved forecast scheme is given by

$$\widetilde{\mathbf{F}} = \rho_{FP}\frac{\sigma_P}{\sigma_F}\mathbf{F} \qquad (18.15)$$

and the proportion of the variance of $\mathbf{P}$ that is explained by $\widetilde{\mathbf{F}}$ is

$$\frac{\mathrm{Var}(\mathbf{P}) - \mathrm{Var}\big(\mathbf{P} - \widetilde{\mathbf{F}}\big)}{\mathrm{Var}(\mathbf{P})} = \rho_{FP}^2.$$

The proportion of variance explained by the improved forecast is given by the squared correlation skill score.

In contrast, when $\mathrm{Var}(\mathbf{P}) = \mathrm{Var}(\mathbf{F})$, the proportion of variance explained by the unimproved forecast, R_{FP}^2, is (18.6)

$$\begin{aligned}R_{FP}^2 &= \frac{\mathrm{Var}(\mathbf{P}) - \mathrm{Var}(\mathbf{F}-\mathbf{P})}{\mathrm{Var}(\mathbf{P})} \\ &= 2\rho_{FP} - 1. \qquad (18.16)\end{aligned}$$

The improved forecast is always more skilful than the unimproved forecast under these circumstances if ρ_{FP} is less than 1.

18.2.9 Comparing a Predicted Field and its Predictand. So far we have considered the prediction of a single number. Evaluation of such a forecast requires many samples in order to estimate the skill scores in [18.2.3]. When we have a vector or field of forecasts, the skill of a given forecast can be estimated using scores such as the *anomaly correlation coefficient* $\rho_{FP}^A(t)$ or the *mean squared error* $S_{FP}^2(t)$ which measure the similarity of two fields relative to a given climatology $\vec{C}$.

Suppose $\vec{\mathbf{f}}_\tau(t)$ is a forecast of a field, say Southern Hemisphere 500 mb height, for the time t prepared τ days in advance, and suppose that the predictand $\vec{\mathbf{p}}(t)$ is the analysis of that field on the day t. Let $\vec{C}$ be the observed long-term mean field. Then the anomaly fields,

$$\vec{\mathbf{f}}'_\tau(t) = \vec{\mathbf{f}}_\tau(t) - \vec{C} \text{ and } \vec{\mathbf{p}}'(t) = \vec{\mathbf{p}}(t) - \vec{C},$$

are compared using the anomaly correlation,

$$\rho_{FP}^A(t) = \frac{\langle(\vec{\mathbf{f}}'_\tau - \langle\vec{\mathbf{f}}'_\tau\rangle)(\vec{\mathbf{p}}' - \langle\vec{\mathbf{p}}'\rangle)\rangle}{\sqrt{\langle(\vec{\mathbf{f}}'_\tau - \langle\vec{\mathbf{f}}'_\tau\rangle)^2\rangle\langle(\vec{\mathbf{p}}' - \langle\vec{\mathbf{p}}'\rangle)^2\rangle}}, \qquad (18.17)$$

where the notation $\langle\cdot\rangle$ denotes an area weighted mean. The time argument (t) has been suppressed on the right hand side of (18.17) for convenience.

The mean squared error is computed similarly as

$$(S_{FP}^A)^2(t) = \frac{\langle(\vec{\mathbf{p}}'(t) - \vec{\mathbf{f}}'_\tau(t))^2\rangle}{\langle\vec{1}\rangle}. \qquad (18.18)$$

[6] See also Section 8.2.

The quantity in the denominator is the sum of the area weights.

Note that both the anomaly correlation coefficient and the mean squared error are defined for an individual forecast. Therefore, an annual cycle of these scores can be calculated, and the gradual improvement of weather forecast models can be monitored by these measures. An interesting aspect of these scores is that forecasts can be stratified by their success. Thus it may be possible to understand empirically why some forecasts are more successful than others.

18.2.10 Example: US NMC Weather Forecasts. Branstator [61] and Kalnay, Kanamitsu, and Baker [209] analysed the quality of the operational forecasts prepared by the US National Meteorological Center (US NMC). Both considered the Northern Hemisphere 500 mb height field, and monitored the forecast performance using the anomaly correlation coefficients ρ^A_{FP}.

Branstator evaluated three-day forecasts for 11 winters (defined as November to March–NDJFM) from 1974 to 1985. Time series of ρ^A_{FP} are shown in Figure 18.4 for three winters, 1974/75, 1978/79, and 1982/83. We see that the anomaly skill score of the forecasts gradually improved, from approximately 0.65 in 1974/75, to approximately 0.75 in 1978/79 and 0.80 in the winter 1982/83. However, the anomaly correlation skill score shows a remarkable variability within a winter. There are periods (e.g., mid January 1978 to mid February 1978) when the forecast scores are consistently better than during other periods (e.g., after mid February 1978). It is not clear if these variations tell us something about the numerical weather prediction model (i.e., that the model scores better with certain initial states than with others), or if they tell us something about variations in the *predictability* of the atmospheric circulation.[7]

The distribution of ρ^A_{FP}, shown in Figure 18.5 for the first six winters and the last five winters, is clearly not normal. The skill varies between 0.5 and 0.9 during first six winters, and it varies between 0.65 and 0.95 during the last five winters.

Kalnay et al. [209] calculated the DJF seasonal mean anomaly correlation coefficient for *lags* from 1 to 10 days (Figure 18.6) for the period 1981/82 to 1989/90. The curves lie above the magical 60% line (see [18.3.5]) for about 4.5 days in the early 1980s. By the end of the decade, the skill curves stayed above this skill threshold for about 7 days. Figure 18.6 is a representation of forecast skill which is typically used by operational weather forecast centres to document their progress. We return to this example in [18.4.5].

[7]Branstator [61] performed a spectral analysis of the skill score time series and found a *red* power spectrum, similar to that of the (predicted) height field. He suggested that this similarity could indicate that the swings in the skill score reflect the varying predictability of the atmosphere.

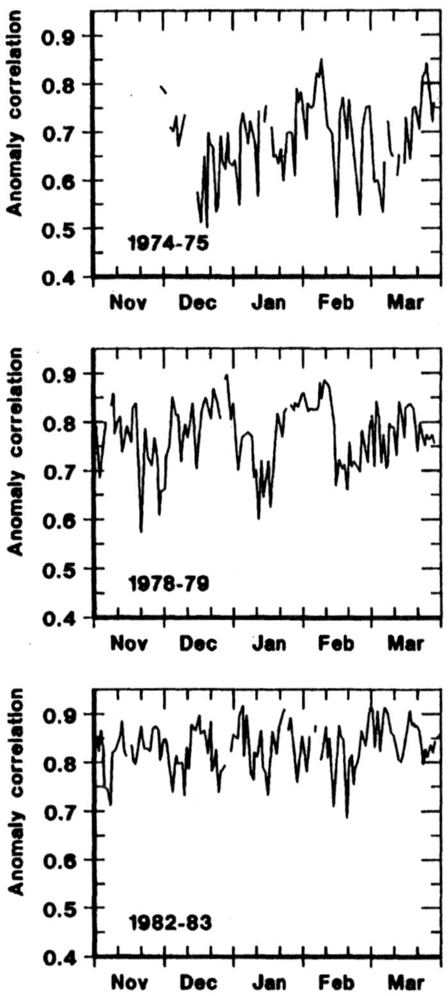

Figure 18.4: *Daily time series of the anomaly correlation coefficient ρ^A_{FP} of three-day forecasts of Northern Hemisphere 500 mb height field prepared by the US National Meteorological Center during the winters of 1974/75, 1978/79 and 1982/83. Winter is defined as the November to March cold season. From Branstator [61].*

18.3 The Murphy–Epstein Decomposition

18.3.1 Introduction. In this Section we introduce the *Murphy–Epstein* decomposition of the

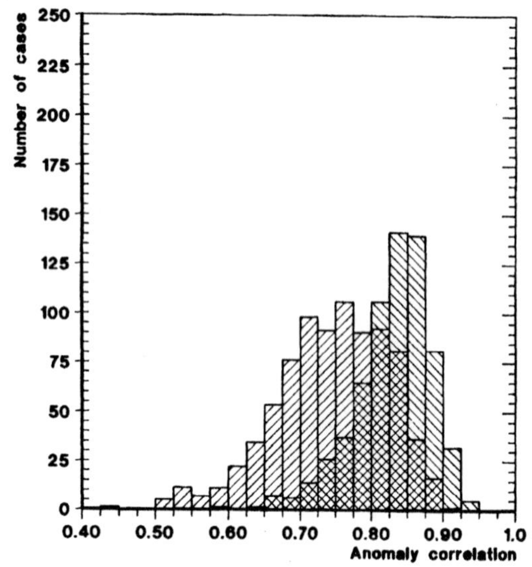

Figure 18.5: *Frequency distributions of the anomaly correlation coefficient ρ^A_{FP} of three-day forecasts of Northern Hemisphere 500 mb height field prepared by the US National Meteorological Center during the winters of 1974/75 to 1979/80 and during the winters 1980/81 to 1984/85. Winter is defined as the November to March cold season. From Branstator [61].*

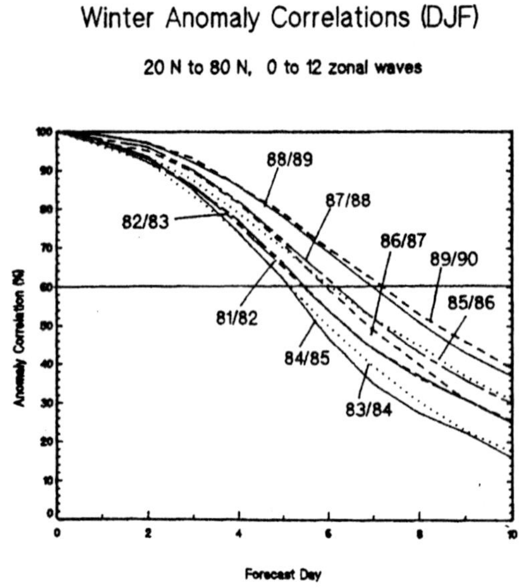

Figure 18.6: *Seasonal mean anomaly correlation coefficients ρ^A_{FP} for 1- to 10-day lead forecasts of the Northern Hemisphere 500 mb height field prepared by the US National Meteorological Center during the winters (DJF) of 1981/82 to 1989/90. Operational weather forecasters usually consider 60% as a threshold for useful forecasts. From Kalnay et al. [209].*

Brier skill score [285]. This decomposition provides useful insights into the interpretation of the Brier skill score and both the time and anomaly correlation skill scores. The Murphy–Epstein decomposition will not be used subsequently in this book so readers may feel free to skip this section.

Suppose that a set of n forecasts $\mathbf{F}_i$ and n predictands $\mathbf{P}_i$ are available for the verification of forecasts. Usually the index i refers either to time or to space. In the former case the index i refers to different times, so that $\mathbf{P}_i$ is a predictand such as temperature observed at a fixed location at time $t = i$. In the latter case the index i refers to a location x, so that $\mathbf{P}_i$ represents the predictand at a fixed time at location $x = i$. We can use the Murphy–Epstein decomposition of the Brier skill score in both cases.

18.3.2 The Correlation Decomposition. Let C be any reference climatology. If the forecasts are indexed by the time (i.e., $t = i$), this reference C is the (constant) long-term mean so that $C = \mathcal{E}(\mathbf{P})$. If the index refers to space (i.e., $x = i$), then C is the long-term mean field at the location x, that is, $C_x = \mathcal{E}(\mathbf{P}_x)$. In either case, we use $\mathbf{F}' = \mathbf{F} - C$ and $\mathbf{P}' = \mathbf{P} - C$ to represent the corresponding anomalies.

If forecasts are verified over time at a fixed location (i.e., $t = i$) we find that the mean squared error and correlation skill scores (18.3, 18.4) are given by

$$S^2_{FP} = \mathcal{E}\big((\mathbf{F}-\mathbf{P})^2\big) = \mathcal{E}\big((\mathbf{F}'-\mathbf{P}')^2\big)$$

$$\rho_{FP} = \frac{\mathcal{E}\big(\mathbf{F}'\mathbf{P}'\big)}{\sqrt{\mathrm{Var}\big(\mathbf{P}'\big)\mathrm{Var}\big(\mathbf{F}'\big)}}$$

If forecasts are verified across space at a fixed time (i.e., $x = i$) then the mean squared error and the anomaly correlation coefficient (18.17) (the time t is omitted) are

$$(S^A)^2_{FP} = \langle(\mathbf{F}'_x - \mathbf{P}'_x)^2\rangle$$

$$\rho^A_{FP} = \frac{\langle(\mathbf{F}'_x - \langle\mathbf{F}'\rangle)(\mathbf{P}'_x - \langle\mathbf{P}'\rangle)\rangle}{\sqrt{\langle(\mathbf{P}'_x - \langle\mathbf{P}'\rangle)^2\rangle\langle(\mathbf{F}_x - \langle\mathbf{F}'\rangle)^2\rangle}}$$

where $\langle\mathbf{F}'\rangle$ and $\langle\mathbf{P}'\rangle$ represent the spatial means.

The derivation of the Murphy–Epstein decomposition is formally carried out for the $t = i$ case, but it can be done in the same way for the $x = i$ case. To accomplish this the correlation skill score ρ_{FP} must be replaced by the anomaly correlation coefficient ρ^A_{FP} and the $\mathcal{E}(\cdot)$-operator has to be replaced by the spatial averaging operator $\langle\cdot\rangle$. Some of the terms in the decomposition vanish in the $t = i$ case but have been retained because they are needed for the $x = i$ case.

By simultaneously adding and subtracting $\mathcal{E}(\mathbf{P}')$ and $\mathcal{E}(\mathbf{F}')$ to the mean square error S^2_{FP} we find, after some algebraic manipulation, that

$$S^2_{PF} = \mathrm{Var}(\mathbf{P}') + \mathrm{Var}(\mathbf{F}') - 2\mathrm{Cov}(\mathbf{P}', \mathbf{F}') + \left[\mathcal{E}(\mathbf{P}') - \mathcal{E}(\mathbf{F}')\right]^2. \quad (18.19)$$

If we replace $\mathbf{F}$ with C in this formula we find

$$S^2_{PC} = \mathrm{Var}(\mathbf{P}') + \mathcal{E}(\mathbf{P}')^2 \quad (18.20)$$

simply because $C' = 0$. Finally, after some further manipulation, we see that the Brier skill score may be expressed as

$$B_{FCP} = \frac{A^2 - B^2 - D^2 + E^2}{1 + E^2} \quad (18.21)$$

where

$$A = \rho_{PF}, \quad B = \rho_{PF} - \frac{\sigma_{F'}}{\sigma_{P'}}$$

$$D = \frac{\mathcal{E}(\mathbf{P}') - \mathcal{E}(\mathbf{F}')}{\sigma_{P'}}, \quad \text{and } E = \frac{\mathcal{E}(\mathbf{P}')}{\sigma_{P'}}.$$

Decomposition (18.21) is the Murphy–Epstein decomposition of the Brier skill score.

The first term A^2 in (18.21) is the correlation skill score squared.

To understand the second term B^2 we will assume that $\mathbf{P}'$ and $\mathbf{F}'$ are jointly normal so that we can use (18.13) and write the expected value of $\mathbf{P}'$ conditional on $\mathbf{F}' = \mathbf{f}'$ as

$$\mathcal{E}(\mathbf{P}'|\mathbf{F}' = \mathbf{f}') = a + b\mathbf{f}'. \quad (18.22)$$

We see from (18.13) that $\rho_{PF} = (\sigma_{F'}/\sigma_{P'})b$ so that

$$B^2 = \left((b-1)(\sigma_{F'}/\sigma_{P'})\right)^2.$$

This term vanishes only if $b = 1$, that is, if the forecasts are not systematically biased.[8] Murphy and Epstein [285] call this term the *conditional bias* because it reflects the extent to which the mean observation $\mathcal{E}(\mathbf{P}')$ (conditional upon a forecast $\mathbf{f}'$) reflects that forecast.

The third term in (18.21) D^2 vanishes only if $\mathcal{E}(\mathbf{F}') = \mathcal{E}(\mathbf{P}')$. This term therefore represents the *unconditional bias* of the forecast. If (18.22) holds then $\mathcal{E}(\mathbf{P}') = a + b\mathcal{E}(\mathbf{F}')$, so that for an unconditionally unbiased forecast $a = (1 - b)\mathcal{E}(\mathbf{F}')$. In the special case of $a = 0$ and $\mathcal{E}(\mathbf{F}') = \mathcal{E}(\mathbf{P}') \neq 0$ we find $b = 1$ and therefore that the forecast is not only unconditionally unbiased but also conditionally unbiased.

[8] Note that $b = 1$ does not imply $\sigma_{F'} = \sigma_{P'}$ but rather that $\sigma_F < \sigma_P$ if $\rho_{PF} < 1$.

18.3.3 Forecasts of the Same Predictands at Different Times. In the $t = i$ case, when $\mathbf{F}_t$ is a series of forecasts of the same predictand $\mathbf{P}_t$ at various times t, C is taken as the climatology of that predictand and thus $C = \mathcal{E}(\mathbf{P})$. Therefore, because $\mathcal{E}(\mathbf{P}') = 0$, we have

$$B_{FCP} = \rho^2_{P'F'} - \left(\rho_{P'F'} - \frac{\sigma_{F'}}{\sigma_{P'}}\right)^2 - \left(\frac{\mathcal{E}(\mathbf{F}')}{\sigma_{P'}}\right)^2 \quad (18.23)$$

and

$$S^2_{PC} = \mathrm{Var}(\mathbf{P}). \quad (18.24)$$

If $\mathbf{F}'$ is unconditionally unbiased then $\mathcal{E}(\mathbf{F}')$ is zero and we find the Brier skill score is identical to the proportion of explained variance R^2_{FP} (18.6). Even when this happens we still have (see equation(18.16))

$$B_{FCP} < \rho^2_{P'F'}. \quad (18.25)$$

Thus we see that, as a general rule, the correlation skill score overestimates the 'true' forecast skill. This is why the correlation skill should generally be regarded as a measure of *potential* skill; it only represents the actual skill if the forecast is unbiased. In this case the Brier skill score, the squared correlation skill and the proportion of explained variance are equivalent.

18.3.4 Forecasts of Different Predictands at the Same Time. In the $x = i$ case the reference forecast is climatology C so that the unconditional bias represents the error in predicting the spatial mean. This bias might be large if the forecast region is small. Murphy and Epstein [285] computed the relative contributions of the terms in the Murphy–Epstein decomposition to the Brier skill score for a series of medium range forecasts prepared by the US National Meteorological Center (NMC) in the mid 1980s.

18.3.5 The Correlation Skill Score and the Mean Squared Error. Equation (18.19) provides a decomposition of the mean square error [286]. We will assume that the forecast $\mathbf{F}$ is unconditionally unbiased (i.e., $\mathcal{E}(\mathbf{F}') = \mathcal{E}(\mathbf{P}')$) so that the last term in (18.19) vanishes, and now

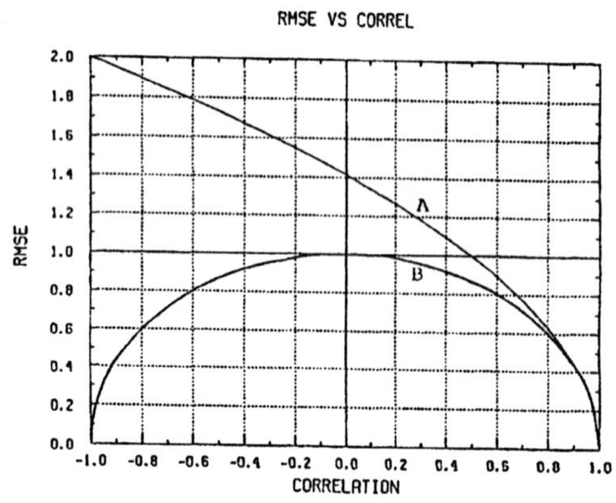

Figure 18.7: *The root of the mean squared error $\mathcal{S}_{FP}$, labelled 'RMSE,' is displayed as a function of the correlation skill score ρ_{FP} for the two cases discussed in [18.3.5]. Curve 'A' illustrates the case in which the forecast and the observations have the same expected value and the same variance. Curve 'B' holds for the improved forecast [18.2.7]. From Barnston [25].*

consider two cases (Barnston [25]). First, suppose $\text{Var}(\mathbf{F}') = \text{Var}(\mathbf{P}')$. Then

$$\begin{aligned}\mathcal{S}^2_{PF} &= 2\text{Var}(\mathbf{P}') - 2\rho_{FP}\text{Var}(\mathbf{P}')\\ &= 2\text{Var}(\mathbf{P}')(1-\rho_{FP}).\end{aligned}$$

The relationship between the correlation ρ_{FP} and the mean square error $\mathcal{S}^2_{PF}$ is illustrated in Figure 18.7 as curve 'A.' When the correlation is zero then the mean square error is $2\text{Var}(\mathbf{P}')$, which is twice the expected error of the climatology forecast. When the correlation is negative, the mean square error becomes even larger than twice that of the climatology forecast.

Second, suppose the improved forecast [18.2.7] is $\widetilde{\mathbf{F}}' = b\mathbf{F}'$, where $b = \rho_{FP}\sigma_p/\sigma_f$, and suppose also that $\mathcal{E}(\widetilde{\mathbf{F}}') = \mathcal{E}(\mathbf{P}') = 0$. Then the improved forecast $\widetilde{\mathbf{F}}'$ is unconditionally unbiased (i.e., $\mathcal{E}(\widetilde{\mathbf{F}}') = \mathcal{E}(\mathbf{P}')$) and it is also conditionally unbiased because $\frac{\sigma_{\tilde{F}}}{\sigma_P} = \rho_{\tilde{F}P}$. Thus (18.19) simplifies to

$$\mathcal{S}^2_{P\tilde{F}} = \text{Var}(\mathbf{P}')(1-\sigma^2_{FP}).$$

The resulting relationship between the correlation and the mean squared error is shown as curve 'B' in Figure 18.7. The improved forecast always has mean squared error that is less than, or in the case of zero correlation equal to, that of the climatology forecast.

18.3.6 Correlation Skill Score Thresholds at which the Brier Skill Score Becomes Positive. If we accept the notion that the Brier skill score as the best indicator of the presence or absence of skill relative to a reference forecast, we can derive a threshold for the correlation skill score (in the $t = i$ case) and for the anomaly correlation coefficient (in the $x = i$ case) at which the Brier score becomes positive [285]. To derive the threshold we assume that $\mathbf{F}$ is an unconditionally unbiased forecast of $\mathbf{P}$, that $\mathcal{E}(\mathbf{P}') = 0$, and that $\text{Var}(\mathbf{F}') = \text{Var}(\mathbf{P}')$. Then, for the $t = i$ case, (18.21) becomes

$$\begin{aligned}B_{FCP} &= A^2 - B^2 = \rho^2_{FP} - (\rho_{FP}-1)^2\\ &= 2\rho_{FP} - 1\end{aligned}$$

so that

$$B_{FCP} \geq 0 \Leftrightarrow \rho_{FP} \geq 0.5. \tag{18.26}$$

Similarly, for the $x = i$ case,

$$B_{FCP} \geq 0 \Leftrightarrow \rho^A_{FP} \geq 0.5. \tag{18.27}$$

The experience of several decades of operational weather forecasting has led weather forecasters to use a larger threshold for the anomaly correlation coefficient, namely 0.6. This choice is based on the subjective assessment that a predicted field with $\rho^A_{FP} \geq 0.6$ bears sufficient resemblance to the observed field for the forecast to be of use to at least some users of the forecast product.

18.4 Issues in the Evaluation of Forecast Skill

18.4.1 The Reference Forecast. A forecasting scheme can not be accepted as being useful if it yields skill scores that can be obtained by means of less sophisticated forecasting procedures. That is, any forecasting scheme must be compared against a *reference forecast* which is easier to prepare than the forecast under consideration.

Some standard reference forecasts are:

- the *random forecast*, $\mathbf{F}$, which is simply a random variable with the same statistical properties as the predictand $\mathbf{P}$;
- the *persistence forecast* $\mathbf{F}_\tau(t) = \mathbf{P}(t-\tau)$;
- the *damped persistence forecast* $\mathbf{F}_\tau(t) = \xi^\tau \mathbf{P}(t-\tau)$ with $0 < \xi < 1$ and $\mathcal{E}(\mathbf{P}) = 0$;
- the *climatological forecast* $\mathbf{F}_\tau(t) = \mathbf{C}$.

	Forecast	
Observation	Tornado	No Tornado
Tornado	28	23
No Tornado	72	2680

Table 18.3: *Finley's [116] success in predicting tornados.*

Another reference forecast which is suitable for quasi-cyclic processes is the *POP forecast* (see Section 15.3).

The Heidke skill score of a categorical forecast (as defined in [18.1.1]) uses the random forecast as its reference. The Heidke score can be modified to assess skill relative to another reference forecast by defining p_E in (18.1) as the success rate of this other reference (see [18.1.2]).

We illustrate the idea of a reference forecast with the following examples.

18.4.2 Example: The Old Farmer's Almanac. We again consider the forecasts of monthly mean temperature and precipitation issued by the *Old Farmer's Almanac* [364] (see [18.2.3]). Because the forecasting algorithm used by the *Old Farmer's Almanac* is unpublished the complexity of the procedure is unknown. We might therefore ask if there exists a trivial forecasting scheme which would do better than the *Old Farmer's Almanac*. The answer is yes. The constant forecast $\mathbf{F}$ = *above normal* has better skill than the *Old Farmer's Almanac* for both precipitation and temperature.

In the case of temperature, we have $p_a^F = 1$, $p_a^P = 0.55$, $p_b^F = 0$ and $p_b^P = 0.45$ so that $p_C = (1\times0.55+0\times0.45)$, $p_E = 0.505$ and consequently $S = (0.55 - 0.505)/(1 - 0.505) = 9.1 \times 10^{-2}$. In contrast, we showed in [18.1.3] that the skill of the *Old Farmer's Almanac* is only 4×10^{-3}.

For precipitation, the constant 'below normal' forecast yields $p_a^F = 0$, $p_a^P = 0.40$, $p_b^F = 1$ and $p_b^P = 0.60$ so that $C = 0.60 \times T$ and $S = (0.60 - 0.52)/0.48 = 0.16$. This is much larger than the *Almanac*'s skill of -2×10^{-3}.

This example illustrates that the Heidke skill score is *inequitable* [18.1.7]. In these examples two competing forecasts, both of which are statistically independent of the predictand, have different Heidke skill scores.

18.4.3 Example: Finley's Tornado Forecast. In the late nineteenth century, Finley [116] (see also Stanski, Wilson, and Burrows [355]) prepared three months of daily forecasts for 18 US districts east of the Rocky Mountains which predicted whether conditions would be favourable for the development of tornados. Daily weather maps served as the predictor. A total of $n = 2803$ forecasts were prepared. Tornado were observed on 51 of these occasions. The 2×2 contingency table describing the results of Finley's efforts is given in Table 18.3.

The number of correct forecasts, or *hits*, was $C = 2708$ whereas the number of expected random hits would be $(51^2 + 2680^2)/T = 2703$. Thus, the Heidke skill score (18.2) is

$$(C - E)/(T - E) = 5.6\%.$$

Is there a simple reference forecast which does better? Consider the constant 'no tornado' forecast. Then the number of hits is equal to the number of occasions with no tornado (i.e., $C = 2752$) and the Heidke skill score is $49/105 = 48\%$. Thus, the verdict of the Heidke skill score is to abandon Finley's forecast and to use the trivial competitor $\mathbf{F}$= 'no tornado' instead.

But is that a fair answer? The 'no tornado' forecast would have a *false alarm rate* of zero, but it would not have warned of any tornados. Finley, on the other hand, had a false alarm rate of $72/(72 + 28) = 72\%$, but correctly warned of a tornado on $28/(28 + 23) = 55\%$ of all tornado days.

What see then is that there are no universal rules that can be used to judge the performance of each and every forecast. Each case must be judged separately while keeping in mind the various pitfalls.

18.4.4 Example: The Madden-and-Julian Oscillation. We evaluate the outcome of two series of forecasts of an index of the Madden-and-Julian Oscillation using the correlation skill score ρ. Forecasts were prepared from 15 sets of initial conditions with the POP method[9] and with a dynamical forecast model. The correlation skill score was calculated for the two forecasting schemes for various temporal *lags* τ (Figure 18.8). In these experiments the POP forecast scores better than the sophisticated dynamical model. Therefore the substantial computational cost of the dynamical model is not rewarded with increased forecast skill in this particular case. (See also [15.3.3].)

[9]POP is an abbreviation for *Principal Oscillation Pattern*. See sec. 15.3.

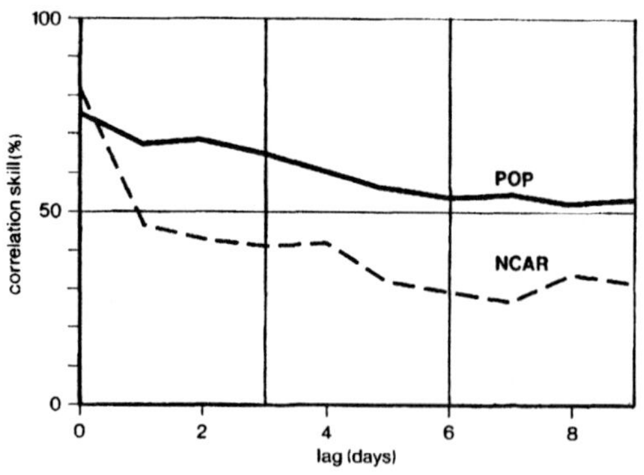

Figure 18.8: *The correlation skill scores ρ_τ of two sets of forecasts of an index of the Madden-and-Julian Oscillation. Both series are constructed from 15 trials using the same initial conditions. One series (solid) was prepared with the POP method [15.3.3], the other with a dynamical forecast model (dashed). From [388].*

18.4.5 Example: The Skill of Weather Prediction. In [18.2.8] we examined the performance of the operational weather forecasts of the US National Meteorological Center, and displayed a plot of winter mean anomaly correlation coefficients from from Kalnay et al. [209]. They compared the operational forecast against the persistence forecast. The lead time beyond which the skill of the forecasts fall below the 60% threshold (see [18.3.5]) is shown in Figure 18.9 for both the operational and the persistence forecasts. Clearly, the operational forecast outperforms persistence. Also, the diagram shows that the improvement in the operational forecasts after 1985 is not due to increased persistence of the Northern Hemisphere circulation.

18.4.6 Effect of Trends. Many meteorological time series exhibit a *trend* on decadal and longer time scales. That is, the series contains either a deterministic or a low-frequency component. These trends reflect a variety of processes, both natural and anthropogenic origins.[10]

The definition of a (trivial) reference forecast may become difficult in the presence of a trend. The skill of the persistence forecast is generally not affected much by a trend because the amplitudes of trends are generally small relative to the natural variability of the forecasted process. On the other hand, the climatological forecast might become

[10]For a short discussion of processes potentially responsible for trends, see [1.2.3].

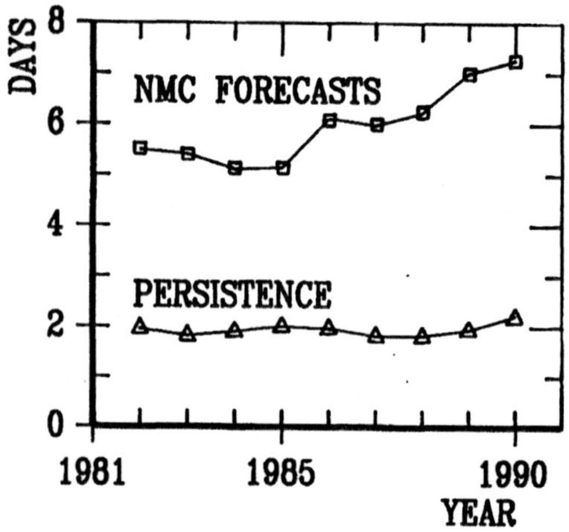

Figure 18.9: *The average lead time, in days, at which the winter (DJF) mean anomaly correlation coefficient $\rho_F^A P$ of the forecasts of Northern Hemisphere 500 mb height field fall below 60%. The boxes indicate that the average lead time of forecasts prepared by the US National Meteorological Center during the winters (DJF) of 1981/82 to 1989/90 falls below 60%. The triangles indicate the corresponding average lead time for persistence forecasts. From Kalnay et al. [209].*

useless because the climatology may no longer be the mean value of the present observations. The random forecast is undefined simply because the statistical parameters $\mathcal{E}(\mathbf{P})$ and Var($\mathbf{P}$) have become moving targets.

Livezey [255] presents an interesting and convincing example of a forecasting scheme whose reputed merits were entirely due to the systematic exploitation of the urbanization effect [1.2.3]. Several of the scores introduced in this chapter sometimes exhibit pathological behaviour if sufficient care is not exercised in designing the forecast evaluation.

18.4.7 Artificial skill. Skill scores should be constructed so that they give an unbiased view of the true utility of the forecasting scheme. This requirement is violated when statistical forecast schemes are built if the same data are used to develop the scheme and evaluate its skill. Quite often, the statistical forecast model is fitted to the data by maximizing a skill score or a quantity, such as mean squared error, that is related to a skill score. If the sample size is small, or if the number of parameters fitted to the data is large relative to the sample size, the skill score is artificially enhanced because in such circumstances the fitted model is able to *adapt* itself to the available data. The sample used to fit the model is often called the *training sample*. The estimate of skill obtained from the training sample is called the *hindcast skill*. The hindcast skill is always somewhat greater than the forecast skill, and this optimistic bias in estimated skill is called *artificial skill*.

Techniques, such as *cross-validation* (see Section 18.5) and *bootstrapping* (see Section 5.5) can sometimes be used to provide good estimates of the forecast skill.

A notoriously efficient manner in which to introduce artificial skill into (time series) forecast models is to time-filter the analysed data in order to suppress high-frequency variations, and to fit and verify the forecast model against these smoothed observations. The time filtering makes future information about the predictand available at the time of the forecast; in a real-time setup this future information would not be available.

18.4.8 Skill Scores Derived from Non-Randomly Chosen Subsets. In real applications the skill score is derived from a finite ensemble of forecasts. These ensembles can usually be thought of as random samples representative of the process being forecast. For example, an ensemble might consist of all cases during a certain time period. The ensemble is sometimes also sub-sampled using criteria available at forecast time in order to make inferences about forecast skill under prescribed conditions (an example can be found in [18.1.6]). Both of these approaches to estimating skill are perfectly legitimate.

On the other hand, it is somewhat misleading to sub-sample an ensemble using criteria that are available only at verification (rather than forecast) time. An example of such a criterion is the strength of the predictand. This type of sub-sampling criterion automatically enhances the correlation skill score ρ_{FP} and the proportion of explained variance R^2_{FP} (cf. [18.1.4]).

To demonstrate this, we consider the $x = i$ case and a forecast of the form $\mathbf{F} = \mathbf{P} + \mathbf{N}$ where $\mathbf{N}$ is random error independent of $\mathbf{P}$. Then

$$\begin{aligned}
\rho_{FP} &= \sqrt{V/(V+N)} \\
R^2_{FP} &= 1 - (N/V) \\
S^2_{FP} &= N,
\end{aligned}$$

where $V = \mathrm{Var}(\mathbf{P})$ and $N = \mathrm{Var}(\mathbf{N})$. If we calculate the skill scores for the subset of forecasts for which $|\mathbf{P}| \geq \tilde{p}$ we find with $\tilde{V} = \mathrm{Var}(\mathbf{P}|\mathbf{P} > \tilde{p}) > V$,

$$\begin{aligned}
\tilde{\rho}_{FP} &= \sqrt{\tilde{V}/(\tilde{V}+N)} > \rho_{FP} \\
\tilde{R}^2_{FP} &= 1 - (N/\tilde{V}) > R^2_{FP} \\
\tilde{S}^2_{FP} &= N = S_{FP}.
\end{aligned}$$

18.5 Cross-validation

18.5.1 General. It is generally desirable to be able to estimate the skill of a forecast or specification model before it is actually applied.[11] However, skill estimates that are obtained from the data used to identify and fit the model tend to be overly optimistic (see Davis [101]) because the fitting process, by definition, chooses parameters that 'adapt' the model to the data as closely as possible. This phenomenon, called artificial skill, is of particular concern in small samples.[12]

One simple way to avoid the artificial skill effect is to divide the data into 'learning' and 'validation' data sets; the model is fitted to the learning data and tested on the independent information contained in the validation subset. However, the data sets in which artificial predictability is particularly troublesome are not large enough to use this strategy effectively. These samples are too small to withhold a substantial fraction of the sample for validation, but if only a few observations are withheld, validation can not be performed effectively.

18.5.2 Cross-validation. Cross-validation avoids the difficulty described above, in essence, by making all of the data available for validation. The procedure is simple to apply provided that the model fitting can be automated. The first step is to withhold a small part of the sample. For example, one might withhold 1 or 2 years of data when building a model for seasonal climate forecasting from a 45-year data base. The model is fitted to the data that is retained and is used to make forecasts or specifications of the data that are withheld. These steps are performed separately, either until no new verification data sets can be selected or until there are enough forecast/verification or specification/verification pairs to estimate skill accurately. See Michaelson

[11] A forecast model is used to extrapolate into the future; specification models are used to estimate present or past unobserved values.

[12] Small is a relative term in this context. The reliability of an internal estimate of skill increases with sample size but decreases with the number of free parameters in the fitted model.

[273] (and also Barnston [25] and the references therein) for more information.

18.5.3 Some Cautions. Care must be taken to ensure that the information used to fit the model in each cross-validation step is completely independent of the information that is withheld for the validation data. Barnston and van den Dool [29] and van den Dool [379] document problems that can occur when there is dependence between the two samples.

Avoiding dependence is more difficult than it sounds. If the data are serially correlated it may be necessary to separate the learning data from the validation data in every cross-validation iteration by a buffer of observations that is long enough to ensure that the learning and validation data are statistically independent. Even when serial correlation is not a problem, there are still a variety ways in which the model fitted to the learning data can be influenced by the information in the validation data. For example, in many analyses the annual mean is first estimated and removed, and models are subsequently fitted to the anomalies that remain. If cross-validation is performed by repeatedly dividing the anomalies into learning and verification subsets, the model fitted to the learning subset will also 'learn' about the verification subset because the sum of anomalies across both subsets is constrained to total zero. The distortion in skill estimates that are caused by this kind of geometrical dependence can be large when the validation subsets are small.[13] It is therefore imperative that the *entire* process that turns data into a fitted model, including the calculation of climatologies, anomalies, and so on, be cross-validated.

[13]When the validation subset is of size 1, which is often the case, the validation anomaly is completely determined by the sum of anomalies in the learning subset.

Part VII

Appendices

A Notation

Throughout this book we use the following notation.

- Real- and complex-valued univariate random variables are given as bold-faced upper-case letters, such as $\mathbf{A}$ or $\mathbf{X}$.
- A random sample of size n from a univariate population is generally represented by a collection of independent and identically distributed (iid) random variables $\{\mathbf{X}_1, \ldots, \mathbf{X}_n\}$.
- The kth order statistic (see [2.6.9]) is denoted by $\mathbf{X}_{(k|n)}$.
- Vector random variables are given as bold-faced upper-case letters with a vector on top, for example, $\vec{\mathbf{A}}$ or $\vec{\mathbf{X}}$. The components of a vector are labelled by subscripts, for instance $\vec{\mathbf{X}} = (\mathbf{X}_1, \ldots, \mathbf{X}_m)^{\mathrm{T}}$, where m is the length of the vector.
- A random sample of size n from a multivariate population is generally represented by a collection of iid random vectors $\{\vec{\mathbf{X}}_1, \ldots, \vec{\mathbf{X}}_n\}$. The kth element of $\vec{\mathbf{X}}_j$ is identified as $\vec{\mathbf{X}}_{jk}$, $\vec{\mathbf{X}}_{j,k}$, or sometimes $(\vec{\mathbf{X}}_j)_k$.
- Univariate stochastic processes in discrete time are identified by $\{\mathbf{X}_t : t \in \mathbb{Z}\}$ or sometimes simply as $\{\mathbf{X}_t\}$. Multivariate stochastic processes are denoted analogously.
- Realizations of a random variable, for example, $\mathbf{B}$ or $\vec{\mathbf{B}}$, are denoted by bold faced lower case letters, such as $\mathbf{b}$ or $\vec{\mathbf{b}}$.
- Matrices are denoted with calligraphic letters, such as $\mathcal{A}$ or $\mathcal{X}$. An $m \times n$ matrix has m rows and n columns. The matrix element in the ith row and the jth column is denoted a_{ij}.
- Sets of numbers or vectors are denoted by upper case Greek characters, such as Θ.
- Statistical parameters are denoted by lower case Greek characters, such as θ or α, or upper case letters in italics, such as T.
- Estimated statistical parameters are denoted with a '$\widehat{\cdot}$', as in $\widehat{\theta}$ or $\widehat{T}$.
- Definitions are stated in *italics*. When new expressions are introduced, they are often written in *italics* or enclosed in quotation marks.
- Footnotes contain additional comments that are not important for the development of the arguments or concepts. They are sometimes used to explain expressions that may be unknown to some readers.

Special Conventions

- The sample space is given by $\mathcal{S}$. Subsets of $\mathcal{S}$ (i.e., events) are indicated by upper case italics, such as A or B.
- The probability of an event $A \in \mathcal{S}$ is given by $P(A)$.
- Sample sizes are usually denoted by n and the dimension of a vector by m.
- We use the notation f_X to represent the probability function or the probability density function of a continuous random variable **X**. Likewise, the distribution function is given by F_X.
- A vertical bar '|' is used to denote conditioning, as in $P(A|B)$ or $f_{X|Y}(x|Y = y)$.
- The expectation operator, applied to a random variable **X**, is indicated by $\mathcal{E}(\mathbf{X})$.
- Averaging in time or over a sample is indicated by a horizontal over-bar, as in $\overline{\mathbf{x}} = \frac{1}{n}\sum_{i=1}^{n} \mathbf{x}_i$.
- The covariance matrix of a random vector $\vec{\mathbf{X}}$ is represented by $\boldsymbol{\Sigma}$ or $\boldsymbol{\Sigma}_{xx}$. The covariance between vector elements $\mathbf{X}_i$ and $\mathbf{X}_j$ is denoted σ_{ij}.
- The cross-covariance matrix between random vectors $\vec{\mathbf{X}}$ and $\vec{\mathbf{Y}}$ is denoted $\boldsymbol{\Sigma}_{xy}$.
- Correlation is denoted by ρ. Estimated correlations are denoted by $\widehat{\rho}$ or r.
- Lags in time are denoted by τ.
- The auto-covariance function of a weakly stationary time series $\{\mathbf{X}_t : t \in \mathbb{Z}\}$ is denoted $\gamma(\tau)$ or $\gamma_{xx}(\tau)$. The corresponding estimator is denoted $\widehat{\gamma}(\tau)$, $\widehat{\gamma}_{xx}(\tau)$, or sometimes $c_{xx}(\tau)$.
- The cross-covariance function of a weakly stationary bivariate time series $\{(\mathbf{X}_t, \mathbf{Y}_t)^{\mathrm{T}} : t \in \mathbb{Z}\}$ is denoted $\gamma_{xy}(\tau)$. The corresponding estimator is denoted $\widehat{\gamma}_{xy}(\tau)$ or $c_{xy}(\tau)$.
- The spectral density function of a weakly stationary time series $\{\mathbf{X}_t : t \in \mathbb{Z}\}$ is denoted $\Gamma(\omega)$ or $\Gamma_{xx}(\tau)$. The cross-spectral density density function of a weakly stationary bi-variate time series $\{(\mathbf{X}_t, \mathbf{Y}_t)^{\mathrm{T}} : t \in \mathbb{Z}\}$ is denoted $\Gamma_{xy}(\omega)$. $\Lambda_{xy}(\omega)$ and $\Psi_{xy}(\omega)$ denote the co- and quadrature spectra; $A_{xy}(\omega)$ and $\Phi_{xy}(\omega)$ denote the amplitude and phase spectra; $\kappa_{xy}(\omega)$ denotes the (squared) coefficiency spectrum.
- The symbols μ_x and $\vec{\mu}_x$ are reserved for ensemble mean values of a random variable **X** and a random vector $\vec{\mathbf{X}}$ respectively. Subscript 'x' will often be omitted for convenience.
- The symbol σ represents a standard deviation, its square σ^2 is a variance. If required for clarity, the name of the random variable is added as a subscript, for example, σ_x denotes the standard deviation of **X**.
- We write $\mathbf{X} \sim \mathcal{N}(\mu, \sigma^2)$ if **X** is normally distributed with mean μ and variance σ^2 (see [2.7.3] and Appendix D). We write $\mathbf{X} \sim \mathcal{B}(n, p)$ and say discrete random variable **X** has a binomial distribution when **X** is the number of successes in n independent Bernoulli trials with probability p of success on any trial (see [2.2.2]). We write $\mathbf{X} \sim \chi^2(k)$ if **X** has a χ^2 distribution with k degrees of freedom (see [2.6.8] and Appendix E). Similarly, we write $\mathbf{X} \sim t(k)$ if **X** has a t distribution with k degrees of freedom (see [2.6.8] and Appendix F). We indicate that **X** has an F distribution with k and l degrees of freedom by writing $F(k, l)$ (see [2.6.10] and Appendix G).

  Because of their historical background, the normal distribution and the t distribution are often called the *Gaussian distribution* and *Student's t distribution*, respectively. To preserve simplicity and clarity in our notation, we do not use these expressions.
- Geographical latitude and longitude are denoted (λ, ϕ). The vertical coordinate is labelled z or p.
- The symbol λ is also used to identify eigenvalues.

- The size of a confidence interval is denoted as $\tilde{p} \times 100\%$, where $\tilde{p}$ is a probability between 0 and 1 (a typical value is 0.95). Significance levels are denoted as $(1 - \tilde{p}) \times 100\%$.

Mathematical Operators

- The *complex conjugate* of a complex number x is indicated with a star: x^*.
- The *transpose of a matrix* $\mathcal{A}$ or a vector $\vec{x}$ is denoted with a superscript T: $\mathcal{A}^{\mathrm{T}}$ or $\vec{x}^{\mathrm{T}}$.
- The *complex conjugate* of a complex matrix $\mathcal{C}$ or vector $\vec{c}$ is denoted $\mathcal{C}^*$ or $\vec{c}^*$. The conjugate transpose operation is indicated by $\mathcal{C}^\dagger$ or $\vec{c}^{\,\dagger}$.
- The *dot product* (also scalar or inner product) of two vectors $\vec{a}$ and $\vec{b}$ is given by: $\langle \vec{a}, \vec{b} \rangle = \vec{a}^{\mathrm{T}}\vec{b}^* = \vec{b}^\dagger \vec{a} = \sum_i a_i b_i^*$.
- The *norm of a vector* $\vec{a}$ is given by $\|\vec{a}\| = \sqrt{\langle \vec{a}, \vec{a} \rangle}$.
- The *sign operator* is given by $\operatorname{sgn}(x) = -1$ if $x < 0$ and $\operatorname{sgn}(x) = 1$ if $x \geq 0$.
- The symbol $\binom{p}{q}$ represents $\frac{p!}{q!(p-q)!}$ for integers p, q with $p \geq q$, where $0! = 1$ and $p! = 1 \times 2 \times \cdots \times p$.
- The *Fourier transform* $\mathcal{F}$ is an operator that operates on series s_t with $\sum_{t=-\infty}^{\infty} |s_t| < \infty$, such that $\mathcal{F}\{s\}(\omega) = \sum_{t=-\infty}^{\infty} s_t e^{-i2\pi t\omega}$. The result of the Fourier transform, $\mathcal{F}\{s\}$, is a complex function defined on the real interval $[-\frac{1}{2}, \frac{1}{2}]$. See also Appendix C.

A brief summary of some essentials about linear bases, eigenvalues, and eigenvectors can be found in Appendix B.

Abbreviations and Technical Expressions

Frequently used abbreviations include:

- AGCM, or simply GCM: (Atmospheric) General Circulation Model. These are detailed models that describe the atmosphere's fluid- and thermodynamics; its transport and conversion of moisture; its radiative properties; and its interaction with the land, water, and ice surfaces of the planet. Most models include at least a crude interactive land surface processes model. In addition, some AGCMs are coupled to thermodynamic models of sea ice and the mixed layer of the ocean, while others have been coupled to fully dynamic Ocean GCMs (OGCMs). AGCMs, OGCMs, and coupled GCMs are essential tools of climate research.
- AIC: Akaike information Criterion.
- ARMA: auto-regressive moving average.
- BIC: Bayesian information Criterion.
- CCA: Canonical Correlation Analysis. See Chapter 14.
- DJF, MAM, JJA and SON: December-January-February, March-April-May, etc.
- EBW: equivalent bandwidth
- EDF: equivalent degrees of freedom.
- EEOF or simply EOF: (Extended) Empirical Orthogonal Function. See Chapter 13.

- MCA: Maximum Covariance Analysis. See [14.1.7].
- MJO: Madden-and-Julian Oscillation. See footnote 10 in [1.2.3].
- MLE: Maximum Likelihood Estimator.
- MOS: model output statistics.
- MSSA: Multichannel Singular Spectrum Analysis.
- NAO: North Atlantic Oscillation.
- PIP: Principal Interaction Pattern.
- PNA: Pacific–North American pattern. See [13.5.5] and Section 17.4.
- POP: Principal Oscillation Pattern. See Chapter 15.
- QBO: Quasi-Biennial Oscillation.
- SLP: sea-level pressure.
- SO and ENSO: Southern Oscillation and El Niño/Southern Oscillation. See footnote 1.2 in [1.2.2] for a short description.
- SOI: Southern Oscillation Index, defined as the pressure difference between Darwin (Australia) and Papeete (Tahiti). An index defined by sea-surface temperature anomalies in the Central Pacific is sometimes used as an alternative SOI, and is called the 'SST index.' See Figure 1.3.
- SVD: Singular Value Decomposition. See Appendix B.
- SST: sea-surface temperature.
- UTC is time independent of time zone: 'Universal Time Co-ordinated.'

The word 'zonal' denotes the east–west direction, and the *zonal wind* is the eastward component of the wind. Similarly, 'meridional' indicates the north–south direction, and the *meridional wind* is the northward component of the wind.

B Elements of Linear Analysis

In this subsection we briefly review some basic concepts of linear algebra, particularly linear bases and eigenvalues and eigenvectors. The notation used is described in Appendix A.

Eigenvalues and Eigenvectors

Let $\mathcal{A}$ be an $m \times m$ matrix. A real or complex number λ is said to be an *eigenvalue* of $\mathcal{A}$ if there is a nonzero m-dimensional vector $\vec{e}$ such that

$$\mathcal{A}\vec{e} = \lambda\vec{e} . \tag{B.1}$$

Vector $\vec{e}$ is said to be a (right) *eigenvector* of $\mathcal{A}$.[1] Eigenvectors are not uniquely determined; since it is clear that, if $\vec{e}$ is an eigenvector of $\mathcal{A}$, then $\alpha\vec{e}$ is also for any number α. However, when an eigenvector is simple (i.e., any other eigenvector with the same eigenvalue is a scalar multiple of this eigenvector), then it uniquely determines a direction in the m-dimensional vector space.

It is possible that a real matrix $\mathcal{A}$ has a complex eigenvalue λ. Then, the eigenvector $\vec{e}$ is also complex (otherwise $\mathcal{A}\vec{e} \in \mathbb{R}^m$ but $\lambda\vec{e} \in \mathbb{C}^m$). Because $\mathcal{A} = \mathcal{A}^*$, the complex conjugate eigenvalue λ^* is an eigenvalue of the real matrix $\mathcal{A}$ as well, with eigenvector $\vec{e}^{\,*}$:

$$\mathcal{A}\vec{e}^{\,*} = \mathcal{A}^*\vec{e}^{\,*} = (\mathcal{A}\vec{e}\,)^* = (\lambda\vec{e}\,)^* = \lambda^*\vec{e}^{\,*}.$$

A square matrix $\mathcal{A}$ is said to be *Hermitian* if $\mathcal{A}^\dagger = \mathcal{A}$, where $\mathcal{A}^\dagger$ is the conjugate transpose of $\mathcal{A}$. Real Hermitian matrices are *symmetric*. Hermitian matrices have real eigenvalues only.

One eigenvalue may have several linearly independent eigenvectors $\vec{e}^{\,i}$. In that case the eigenvectors are said to be *degenerate* since their directions are no longer uniquely determined. The simplest example of a matrix with degenerate eigenvectors is the identity matrix. It has only one eigenvalue $\lambda = 1$, which has m linearly independent eigenvectors $\vec{e}^{\,i} = (0, \ldots, 0, 1, 0, \ldots, 0)^{\mathrm{T}}$ with a unit in the ith position. In general, when λ is a degenerate eigenvalue with linearly independent eigenvectors $\vec{e}^{\,i}$, $i = 1, \ldots, m_\lambda$, any linear combination $\sum_i \alpha_i\vec{e}^{\,i}$ is also an eigenvector with eigenvalue λ. Note that a given eigenvector is associated with only one eigenvalue.

Bases

A collection of vectors $\{\vec{e}^{\,1}, \ldots, \vec{e}^{\,m}\}$ is said to be a *linear basis* for an m-dimensional vector space V if for any vector $\vec{a} \in V$ there exist coefficients α_i, $i = 1, \ldots, m$, such that $\vec{a} = \sum_i \alpha_i\vec{e}^{\,i}$. An *orthogonal* basis is a linear basis consisting of vectors $\vec{e}^{\,i}$ that are mutually orthogonal, that is, $\langle\vec{e}^{\,i}, \vec{e}^{\,j}\rangle = 0$ if $i \neq j$. The set of vectors is called *orthonormal* if $\|\vec{e}^{\,i}\| = 1$ for all $i = 1, \ldots, m$.

[1] A nonzero m-dimensional vector $\vec{f}$ is said to be a left eigenvector of $\mathcal{A}$ if $\vec{f}^{\mathrm{T}}\mathcal{A} = \lambda\vec{f}^{\mathrm{T}}$ for some nonzero λ. The left eigenvectors of $\mathcal{A}$ are right eigenvectors of $\mathcal{A}^{\mathrm{T}}$, and vice versa. We use the term *eigenvector* to denote a right eigenvector.

Orthonormal Transformations

If $\{\vec{e}^{\,1}, \ldots, \vec{e}^{\,m}\}$ is an orthonormal basis and $\vec{y} = \sum_i \alpha_i \vec{e}^{\,i}$, then

$$\langle \vec{y}, \vec{e}^{\,j} \rangle = \sum_i \alpha_i \langle \vec{e}^{\,i}, \vec{e}^{\,j} \rangle = \alpha_j \tag{B.2}$$

$$\vec{y} = \sum_i \langle \vec{y}, \vec{e}^{\,i} \rangle \vec{e}^{\,i}. \tag{B.3}$$

Equation (B.3) describes a transformation from standard coordinates $(y_1, \ldots, y_m)^{\mathrm{T}}$ to a new set of coordinates $(\langle \vec{y}, \vec{e}^{\,1} \rangle, \ldots, \langle \vec{y}, \vec{e}^{\,m} \rangle)^{\mathrm{T}}$.

Continue to assume that $\{\vec{e}^{\,1}, \ldots, \vec{e}^{\,m}\}$ is an orthonormal basis. The expectation $\mathcal{E}(\vec{\mathbf{Y}})$ of a random vector $\vec{\mathbf{Y}}$ in standard coordinates transforms in the same way as the coordinates:

$$\mathcal{E}\big(\langle \vec{\mathbf{Y}}, \vec{e}^{\,j} \rangle\big) = \langle \mathcal{E}(\vec{\mathbf{Y}}), \vec{e}^{\,j} \rangle.$$

The covariance matrix of $\vec{\mathbf{Y}}$, with respect to the standard coordinates,

$$\mathbf{\Sigma} = \mathcal{E}\big((\vec{\mathbf{Y}} - \mu_y)(\vec{\mathbf{Y}} - \mu_y)^{\dagger}\big),$$

is related to the covariance matrix $\mathbf{\Sigma}'$ of the transformed vector $(\langle \vec{\mathbf{Y}}, \vec{e}^{\,1} \rangle, \ldots, \langle \vec{\mathbf{Y}}, \vec{e}^{\,m} \rangle)^{\mathrm{T}}$ through

$$\mathbf{\Sigma}' = \mathcal{P}^{\dagger} \mathbf{\Sigma} \mathcal{P}$$

where $\mathcal{P}^{\dagger}$ is the conjugate transpose of $\mathcal{P}$ and the columns of $\mathcal{P}$ are the m vectors $\vec{e}^{\,1}, \ldots, \vec{e}^{\,m}$. Note that, since the basis is orthonormal, $\mathcal{P}^{\dagger}\mathcal{P} = \mathcal{P}\mathcal{P}^{\dagger} = \mathcal{I}$. The trace of the covariance matrix (i.e., the sum of the variances of all components) is invariant under the transformation (B.2):

$$\sum_j \sigma^2_{\mathbf{Y}_j} = \mathrm{tr}(\mathcal{P}^{\dagger}\mathbf{\Sigma}\mathcal{P}) = \mathrm{tr}(\mathbf{\Sigma}) = \mathrm{tr}(\mathcal{P}\mathcal{P}^{\dagger}\mathbf{\Sigma}) = \mathrm{tr}(\mathbf{\Sigma}') = \sum_j \sigma^2_{\alpha_j}$$

where $\alpha_j = \langle \mathbf{Y}, \vec{e}^{\,j} \rangle$.

Square Root of a Positive Definite Symmetric Matrix

The square root of a positive definite symmetric matrix $\mathbf{\Sigma}$ is given by $\mathbf{\Sigma}^{1/2} = \Lambda^{1/2}\mathcal{P}^{\mathrm{T}}$, where $\mathcal{P}$ is an orthonormal matrix of eigenvectors of $\mathbf{\Sigma}$, $\Lambda = \mathrm{diag}(\lambda_1, \ldots, \lambda_m)$ is the corresponding diagonal matrix of eigenvalues, and $\Lambda^{1/2} = \mathrm{diag}(\lambda_1^{1/2}, \ldots, \lambda_m^{1/2})$. Then $\mathbf{\Sigma} = (\mathbf{\Sigma}^{1/2})^{\mathrm{T}}\mathbf{\Sigma}^{1/2}$. The inverse square root of $\mathbf{\Sigma}$ is given by $\mathbf{\Sigma}^{-1/2} = \mathcal{P}\Lambda^{-1/2}$, where $\Lambda^{-1/2} = \mathrm{diag}(\lambda_1^{-1/2}, \ldots, \lambda_m^{-1/2})$. Note that $\mathbf{\Sigma}^{1/2}\mathbf{\Sigma}^{-1/2} = \mathbf{\Sigma}^{-1/2}\mathbf{\Sigma}^{1/2} = \mathcal{I}$ and that $\mathbf{\Sigma}^{-1} = \mathbf{\Sigma}^{-1/2}(\mathbf{\Sigma}^{-1/2})^{\mathrm{T}}$. See Graybill [148] for more details.

Normal and Orthonormal Matrices

A *normal matrix* is a square matrix $\mathcal{A}$ for which $\mathcal{A}^{\dagger}\mathcal{A} = \mathcal{A}\mathcal{A}^{\dagger}$, where $\mathcal{A}^{\dagger}$ is the conjugate transpose of $\mathcal{A}$. Normal matrices are special because they have m eigenvectors that form a linear basis for the vector space. Note that Hermitian matrices are normal.

An *orthonormal matrix* is a square matrix $\mathcal{A}$ such that its conjugate transpose $\mathcal{A}^{\dagger}$ is its inverse, that is, $\mathcal{A}\mathcal{A}^{\dagger} = \mathcal{A}^{\dagger}\mathcal{A} = \mathcal{I}$.

Singular Value Decomposition[2]

Any $m \times n$ matrix $\mathcal{A}$ can be given a *Singular Value Decomposition* (SVD)

$$\mathcal{A} = \mathcal{U}\mathcal{S}\mathcal{V}^{\dagger} \tag{B.4}$$

where $\mathcal{U}$ is $m \times n$, $\mathcal{S}$ is $n \times n$, $\mathcal{V}$ is $n \times n$, and $\mathcal{V}^{\dagger}$ is the conjugate transpose of $\mathcal{V}$. The first $\min(m, n)$ columns of $\mathcal{U}$ and $\mathcal{V}$ are orthonormal vectors of dimension n and m and are called *left* and *right singular vectors*, respectively. Matrix $\mathcal{S}$ is a diagonal matrix with non-negative elements $s_{ii} = s_i$, $i = 1, \ldots, \min(m, n)$, called *singular values*. All other elements of $\mathcal{S}$ are zero.

When $m \geq n$:

- $\mathcal{U}^{\dagger}\mathcal{U} = \mathcal{I}_n$, where $\mathcal{I}_n$ is the $n \times n$ identity matrix,
- $\mathcal{V}^{\dagger}\mathcal{V} = \mathcal{V}\mathcal{V}^{\dagger} = \mathcal{I}_n$,
- $\mathcal{S} = \mathrm{diag}(s_1, \ldots, s_n)$.

Note that

$$\begin{aligned} \mathcal{A}^{\dagger}\mathcal{U} &= \mathcal{V}\mathcal{S}\mathcal{U}^{\dagger}\mathcal{U} &= \mathcal{V}\mathcal{S} \\ \mathcal{A}\mathcal{V} &= \mathcal{U}\mathcal{S}\mathcal{V}^{\dagger}\mathcal{V} &= \mathcal{U}\mathcal{S}. \end{aligned} \tag{B.5}$$

Therefore

$$\begin{aligned} \mathcal{A}\mathcal{A}^{\dagger}\mathcal{U} &= \mathcal{U}\mathcal{S}\mathcal{V}^{\dagger}\mathcal{V}\mathcal{S} &= \mathcal{U}\mathcal{S}^2 \\ \mathcal{A}^{\dagger}\mathcal{A}\mathcal{V} &= \mathcal{V}\mathcal{S}\mathcal{U}^{\dagger}\mathcal{U}\mathcal{S} &= \mathcal{V}\mathcal{S}^2. \end{aligned} \tag{B.6}$$

That is, the columns of $\mathcal{V}$ are the eigenvectors of $\mathcal{A}^{\dagger}\mathcal{A}$, the squares of the singular values s_i are the eigenvalues of $\mathcal{A}^{\dagger}\mathcal{A}$, and the columns of $\mathcal{U}$ are the eigenvectors of $\mathcal{A}\mathcal{A}^{\dagger}$ that correspond to these eigenvalues.

When $m < n$:

A similar singular value decomposition can be constructed when $m < n$. We first write

$$\mathcal{A}^{\dagger} = \mathcal{U}'\mathcal{S}'\mathcal{V}'^{\dagger}$$

where $\mathcal{U}'$ is $n \times m$, $\mathcal{S}'$ is $m \times m$, and $\mathcal{V}'$ is $m \times m$, all with properties as described above. Thus

$$\mathcal{A} = \mathcal{V}'\mathcal{S}'\mathcal{U}'^{\dagger}.$$

Now construct an $m \times n$ matrix $\mathcal{U} = (\mathcal{V}'|\vec{0} \cdots \vec{0})$ by adding $n - m$ columns of zeros to $\mathcal{V}'$, construct an $n \times n$ matrix $\mathcal{S}$ by placing $\mathcal{S}'$ in the upper left corner and padding the rest of the matrix with zeros, and construct an $n \times n$ matrix $\mathcal{V} = (\mathcal{U}'|\vec{g}^1 \cdots \vec{g}^{n-m})$, where $\vec{g}^1, \ldots, \vec{g}^{n-m}$ are chosen so that the columns of $\mathcal{V}$ form an orthonormal basis for the n-dimensional vector space. Then we again have a decomposition in the form of equation (B.4) that has properties analogous to those described for the $m \geq n$ case.

The algorithms in the *Numerical Recipes* [322] or other software libraries can be used to perform an SVD, or first solve one of the eigen-equations (B.6) and then calculate the other set of singular vectors from (B.5). Navarra [290] points out that the first approach is numerically more robust than the second.

An interesting byproduct of this subsection is that the eigenvectors and eigenvalues of a matrix of the form $\mathcal{A}\mathcal{A}^{\dagger}$ may be derived through an SVD of the matrix $\mathcal{A}$. When estimating *Empirical Orthogonal Functions* (see Section 13.3), the eigenvalues of the estimated covariance matrix must be calculated. This estimated covariance matrix can be written as $\frac{1}{n}\mathcal{X}\mathcal{X}^{\dagger}$, where $\mathcal{X}$ is an $m \times n$ matrix with m the dimension of the random vector and n the number of realizations of the vector in the sample. The columns of $\mathcal{X}$ consist of deviations from the vector of sample means.

---

[2] See also Navarra's summary [290] or Golub and van Loan's [143] detailed presentation of the topic .

C Fourier Analysis and Fourier Transform

Fourier Analysis and Fourier Transform

Fourier analysis and the Fourier transform are mathematically different and can not be applied to the same objects. The two approaches should not be confused.

Fourier analysis is a geometrical concept. It offers two equivalent (i.e., isomorphic) descriptions of a discrete or continuous *periodic* function.

- In case of discrete functions $(X_0, \ldots, X_{T-1})$ with $X_T = X_0$ and T even, the trigonometric expansion is

$$X_t = \sum_{k=-n}^{n-1} a_k e^{i2\pi kt/T} \tag{C.1}$$

for $t = 0, \ldots, T-1$, and the coefficients are given by

$$a_k = \frac{1}{T} \sum_{t=0}^{T-1} X_t e^{-i2\pi kt/T} \tag{C.2}$$

for $k = -n, \ldots, n-1$. A similar formula holds for odd T.

- Very similar formulae hold for continuous periodic functions, namely

$$X_t = \sum_{k=-\infty}^{\infty} a_k e^{i2\pi kt/T} \tag{C.3}$$

for $t \in [0, T]$, with coefficients

$$a_k = \frac{1}{T} \int_0^T X_t e^{-i2\pi kt/T} \, dt \tag{C.4}$$

for $k = 0, \pm 1, \pm 2, \ldots, \pm\infty$.

Note that Fourier *analysis* can not be applied to a summable function, such as the auto-covariance function, since such a function can *not* be *periodic*.

The *Fourier transform* is a mapping from a set of discrete, summable series to the set of real functions defined on the interval $[-\frac{1}{2}, \frac{1}{2}]$. The auto-covariance function is summable in all ordinary cases, but stationary time series are *not* summable. If s is such a summable discrete series, then its Fourier transform $\mathcal{F}\{s\}$ is a function that, for any real $\omega \in [-\frac{1}{2}, \frac{1}{2}]$, takes the value

$$\mathcal{F}\{s\}(\omega) = \sum_{j=-\infty}^{\infty} s_j e^{-i2\pi\omega j}. \tag{C.5}$$

The variable ω is usually named 'frequency'. The Fourier transform mapping is invertible,

$$s_j = \int_{-\frac{1}{2}}^{\frac{1}{2}} \mathcal{F}\{s\}(\omega) e^{i2\pi\omega j}\, d\omega, \tag{C.6}$$

so that the infinite series s and the function $\mathcal{F}\{s\}$ are isomorphic and represent the same information.

Note that a Fourier *transform* can not be obtained for a *periodic* function.

The definition of the Fourier transform is arbitrary in detail. In the present definition there no minus sign in the exponent of the 'reconstruction' equation (C.6). One could insert a minus sign in equation (C.6), but then the minus in the 'decomposition' equation (C.5) must be removed.

Some Properties of the Fourier Transform

The following computational rules are easily derived from the definition of the Fourier transform.

- The Fourier transform is linear, that is, if f and g are summable series and if α is a real number, then:

  $$\mathcal{F}\{\alpha f + g\} = \alpha \mathcal{F}\{f\} + \mathcal{F}\{g\}. \tag{C.7}$$

  If we denote the shift operator with the superscript τ so that $f_t^\tau = f_{t+\tau}$ and the reversal operator with superscript r so that $f_t^r = f_{-t}$, then

  $$\begin{aligned} \mathcal{F}\{f^\tau\}(\omega) &= e^{2\pi i \tau\omega}\mathcal{F}\{f\}(\omega) \\ \mathcal{F}\{f^r\}(\omega) &= \mathcal{F}\{f\}(-\omega) = \mathcal{F}\{f\}^*(\omega). \end{aligned} \tag{C.8}$$

- The Fourier transform of a symmetric series ($f_t = f_{-t}$) is real, and that of an anti-symmetric series is imaginary.

- Every real or complex series f_t may be decomposed into a symmetric part $f_t^s = \frac{1}{2}(f_t + f_{-t})$ and an anti-symmetric part $f_t^a = \frac{1}{2}(f_t - f_{-t})$. Then, using equation (C.7),

  $$\mathcal{F}\{f\} = \mathcal{F}\{f^s\} + \mathcal{F}\{f^a\} \tag{C.9}$$

  or, with the finding that the Fourier transform of a symmetric series is real and that of an anti-symmetric series is imaginary,

  $$\mathrm{Re}\,(\mathcal{F}\{f\}) = \mathcal{F}\{f^s\} \tag{C.10}$$

  $$i\,\mathrm{Im}\,(\mathcal{F}\{f\}) = \mathcal{F}\{f^a\}. \tag{C.11}$$

- The Fourier transform of a real symmetric series is symmetric, that is,

  $$\mathcal{F}\{f^s\}(\omega) = \mathcal{F}\{f^s\}(-\omega) \tag{C.12}$$

  and that of a real anti-symmetric series is anti-symmetric

  $$\mathcal{F}\{f^a\}(\omega) = -\mathcal{F}\{f^a\}(-\omega). \tag{C.13}$$

- If f_t is an absolutely summable series and

  $$F(f)_t = \sum_{k=-\infty}^{\infty} a_k f_{t+k} \tag{C.14}$$

where a_k is also absolutely summable, then, using equation (C.7) and operation (C.8),

$$\mathcal{F}\{F(f)\}(\omega) = \sum_{k=-\infty}^{\infty} a_k \mathcal{F}\{f^k\}(\omega) = \sum_k a_k e^{2\pi i \tau \omega} \mathcal{F}\{f\}(\omega) = \mathcal{F}\{a^r\}(\omega) \cdot \mathcal{F}\{f\}(\omega). \quad \text{(C.15)}$$

Similarly

$$\mathcal{F}\{F^r(f)\} = \mathcal{F}\{a\} \cdot \mathcal{F}\{f\} \quad \text{(C.16)}$$

$$\mathcal{F}\{F^*(f)\} = \mathcal{F}\{a\}^* \cdot \mathcal{F}\{f\}, \quad \text{(C.17)}$$

where $F^r(f)_t = \sum a_{-k} f_{t+k}$ and $F^*(f)_t = \sum a_k^* f_{t+k}$.

Operator (C.14), which transforms the series f_t into $F(f)_t$, is called a *linear filter*. Mathematically it takes the form of a *convolution*. Equations (C.15), (C.16), and (C.17)) state that a convolution 'in the time domain' becomes a multiplication 'in the frequency domain'.

D Normal Density and Cumulative Distribution Function

Values of the standard normal distribution $\mathcal{N}(\mu, \sigma^2)$ with mean $\mu = 0$ and variance $\sigma^2 = 1$. The density function $f_{\mathcal{N}}(z)$ is given by equation (2.25) and the exact cumulative distribution function $F_{\mathcal{N}}(z) = \int_{-\infty}^{z} f_{\mathcal{N}}(x)\,dx$ is given by equation (2.26). The column labelled $F^*_{\mathcal{N}}(z)$ contains the approximated cumulative distribution function given by the right hand side of equation (2.27).

z	$f_{\mathcal{N}}(z)$	$F_{\mathcal{N}}(z)$	$F^*_{\mathcal{N}}(z)$	z	$f_{\mathcal{N}}(z)$	$F_{\mathcal{N}}(z)$	$F^*_{\mathcal{N}}(z)$	z	$f_{\mathcal{N}}(z)$	$F_{\mathcal{N}}(z)$	$F^*_{\mathcal{N}}(z)$
−3.00	0.004	0.001	0.001	−2.00	0.054	0.023	0.020	−1.00	0.242	0.159	0.156
−2.95	0.005	0.002	0.001	−1.95	0.060	0.026	0.023	−0.95	0.254	0.171	0.169
−2.90	0.006	0.002	0.001	−1.90	0.066	0.029	0.026	−0.90	0.266	0.184	0.182
−2.85	0.007	0.002	0.001	−1.85	0.072	0.032	0.029	−0.85	0.278	0.198	0.196
−2.80	0.008	0.003	0.002	−1.80	0.079	0.036	0.033	−0.80	0.290	0.212	0.210
−2.75	0.009	0.003	0.002	−1.75	0.086	0.040	0.037	−0.75	0.301	0.227	0.225
−2.70	0.010	0.003	0.002	−1.70	0.094	0.045	0.041	−0.70	0.312	0.242	0.241
−2.65	0.012	0.004	0.003	−1.65	0.102	0.049	0.046	−0.65	0.323	0.258	0.257
−2.60	0.014	0.005	0.003	−1.60	0.111	0.055	0.051	−0.60	0.333	0.274	0.273
−2.55	0.016	0.005	0.004	−1.55	0.120	0.061	0.057	−0.55	0.343	0.291	0.290
−2.50	0.018	0.006	0.005	−1.50	0.130	0.067	0.063	−0.50	0.352	0.308	0.308
−2.45	0.020	0.007	0.005	−1.45	0.139	0.073	0.070	−0.45	0.361	0.326	0.326
−2.40	0.022	0.008	0.006	−1.40	0.150	0.081	0.078	−0.40	0.368	0.345	0.344
−2.35	0.025	0.009	0.007	−1.35	0.160	0.088	0.085	−0.35	0.375	0.363	0.363
−2.30	0.028	0.011	0.009	−1.30	0.171	0.097	0.094	−0.30	0.381	0.382	0.382
−2.25	0.032	0.012	0.010	−1.25	0.183	0.106	0.103	−0.25	0.387	0.401	0.401
−2.20	0.036	0.014	0.011	−1.20	0.194	0.115	0.112	−0.20	0.391	0.421	0.421
−2.15	0.040	0.016	0.013	−1.15	0.206	0.125	0.122	−0.15	0.395	0.440	0.440
−2.10	0.044	0.018	0.015	−1.10	0.218	0.136	0.133	−0.10	0.397	0.460	0.460
−2.05	0.049	0.020	0.017	−1.05	0.230	0.147	0.145	−0.05	0.398	0.480	0.480
−2.00	0.054	0.023	0.020	−1.00	0.242	0.159	0.156	0.00	0.399	0.500	0.500

z	$f_{\mathcal{N}}(z)$	$F_{\mathcal{N}}(z)$	$F^*_{\mathcal{N}}(z)$	z	$f_{\mathcal{N}}(z)$	$F_{\mathcal{N}}(z)$	$F^*_{\mathcal{N}}(z)$	z	$f_{\mathcal{N}}(z)$	$F_{\mathcal{N}}(z)$	$F^*_{\mathcal{N}}(z)$
0.00	0.399	0.500	0.500	1.00	0.242	0.841	0.844	2.00	0.054	0.977	0.980
0.05	0.398	0.520	0.520	1.05	0.230	0.853	0.855	2.05	0.049	0.980	0.983
0.10	0.397	0.540	0.540	1.10	0.218	0.864	0.867	2.10	0.044	0.982	0.985
0.15	0.395	0.560	0.560	1.15	0.206	0.875	0.878	2.15	0.040	0.984	0.987
0.20	0.391	0.579	0.579	1.20	0.194	0.885	0.888	2.20	0.036	0.986	0.989
0.25	0.387	0.599	0.599	1.25	0.183	0.894	0.897	2.25	0.032	0.988	0.990
0.30	0.381	0.618	0.618	1.30	0.171	0.903	0.906	2.30	0.028	0.989	0.991
0.35	0.375	0.637	0.637	1.35	0.160	0.912	0.915	2.35	0.025	0.991	0.993
0.40	0.368	0.655	0.656	1.40	0.150	0.919	0.922	2.40	0.022	0.992	0.994
0.45	0.361	0.674	0.674	1.45	0.139	0.927	0.930	2.45	0.020	0.993	0.995
0.50	0.352	0.692	0.692	1.50	0.130	0.933	0.937	2.50	0.018	0.994	0.995
0.55	0.343	0.709	0.710	1.55	0.120	0.939	0.943	2.55	0.016	0.995	0.996
0.60	0.333	0.726	0.727	1.60	0.111	0.945	0.949	2.60	0.014	0.995	0.997
0.65	0.323	0.742	0.743	1.65	0.102	0.951	0.954	2.65	0.012	0.996	0.997
0.70	0.312	0.758	0.759	1.70	0.094	0.955	0.959	2.70	0.010	0.997	0.998
0.75	0.301	0.773	0.775	1.75	0.086	0.960	0.963	2.75	0.009	0.997	0.998
0.80	0.290	0.788	0.790	1.80	0.079	0.964	0.967	2.80	0.008	0.997	0.998
0.85	0.278	0.802	0.804	1.85	0.072	0.968	0.971	2.85	0.007	0.998	0.999
0.90	0.266	0.816	0.818	1.90	0.066	0.971	0.974	2.90	0.006	0.998	0.999
0.95	0.254	0.829	0.831	1.95	0.060	0.974	0.977	2.95	0.005	0.998	0.999
1.00	0.242	0.841	0.844	2.00	0.054	0.977	0.980	3.00	0.004	0.999	0.999

The following table lists the upper tail critical values of the standard normal distribution commonly used in tests of hypothesis. These values are the solutions of $\tilde{p} = F_{\mathcal{N}}(z)$. Lower tail critical values are given by $Z_{\tilde{p}} = -Z_{1-\tilde{p}}$.

$\tilde{p}$	0.900	0.950	0.975	0.990	0.995	0.999
$Z_{\tilde{p}}$	1.282	1.645	1.960	2.326	2.576	3.080

E The χ^2 Distribution

The following tables list selected critical values of the χ^2 distribution with k degrees of freedom, that is, the solution x of

$$\tilde{p} = \int_{-\infty}^{x} \frac{t^{(k-2)/2} e^{-t/2}}{\Gamma(k/2) 2^{k/2}} \, dt$$

where Γ denotes the Gamma function. The χ^2 distribution is discussed in [2.7.8]. Examples of $\chi^2(k)$ distributions are plotted in Figure 2.5.

	Degrees of freedom									
$\tilde{p}$	1	2	3	4	5	6	7	8	9	10
0.005	0.0000393	0.0100	0.0717	0.207	0.412	0.676	0.989	1.34	1.73	2.16
0.010	0.000157	0.0201	0.115	0.297	0.554	0.872	1.24	1.65	2.09	2.56
0.025	0.000982	0.0506	0.216	0.484	0.831	1.24	1.69	2.18	2.70	3.25
0.050	0.00393	0.103	0.352	0.711	1.15	1.64	2.17	2.73	3.33	3.94
0.100	0.0158	0.211	0.584	1.06	1.61	2.20	2.83	3.49	4.17	4.87
0.250	0.102	0.575	1.21	1.92	2.67	3.45	4.25	5.07	5.90	6.74
0.500	0.455	1.39	2.37	3.36	4.35	5.35	6.35	7.34	8.34	9.34
0.750	1.32	2.77	4.11	5.39	6.63	7.84	9.04	10.2	11.4	12.5
0.900	2.71	4.61	6.25	7.78	9.24	10.6	12.0	13.4	14.7	16.0
0.950	3.84	5.99	7.81	9.49	11.1	12.6	14.1	15.5	16.9	18.3
0.975	5.02	7.38	9.35	11.1	12.8	14.4	16.0	17.5	19.0	20.5
0.990	6.63	9.21	11.3	13.3	15.1	16.8	18.5	20.1	21.7	23.2
0.995	7.88	10.6	12.8	14.9	16.7	18.5	20.3	22.0	23.6	25.2

$\tilde{p}$	Degrees of freedom 11	12	13	14	15	16	17	18	19	20
0.005	2.60	3.07	3.57	4.07	4.60	5.14	5.70	6.26	6.84	7.43
0.010	3.05	3.57	4.11	4.66	5.23	5.81	6.41	7.01	7.63	8.26
0.025	3.82	4.40	5.01	5.63	6.26	6.91	7.56	8.23	8.91	9.59
0.050	4.57	5.23	5.89	6.57	7.26	7.96	8.67	9.39	10.1	10.9
0.100	5.58	6.30	7.04	7.79	8.55	9.31	10.1	10.9	11.7	12.4
0.250	7.58	8.44	9.30	10.2	11.0	11.9	12.8	13.7	14.6	15.5
0.500	10.3	11.3	12.3	13.3	14.3	15.3	16.3	17.3	18.3	19.3
0.750	13.7	14.8	16.0	17.1	18.2	19.4	20.5	21.6	22.7	23.8
0.900	17.3	18.5	19.8	21.1	22.3	23.5	24.8	26.0	27.2	28.4
0.950	19.7	21.0	22.4	23.7	25.0	26.3	27.6	28.9	30.1	31.4
0.975	21.9	23.3	24.7	26.1	27.5	28.8	30.2	31.5	32.9	34.2
0.990	24.7	26.2	27.7	29.1	30.6	32.0	33.4	34.8	36.2	37.6
0.995	26.8	28.3	29.8	31.3	32.8	34.3	35.7	37.2	38.6	40.0

$\tilde{p}$	Degrees of freedom 21	22	23	24	25	26	27	28	29	30
0.005	8.03	8.64	9.26	9.89	10.5	11.2	11.8	12.5	13.1	13.8
0.010	8.90	9.54	10.2	10.9	11.5	12.2	12.9	13.6	14.3	15.0
0.025	10.3	11.0	11.7	12.4	13.1	13.8	14.6	15.3	16.0	16.8
0.050	11.6	12.3	13.1	13.8	14.6	15.4	16.2	16.9	17.7	18.5
0.100	13.2	14.0	14.8	15.7	16.5	17.3	18.1	18.9	19.8	20.6
0.250	16.3	17.2	18.1	19.0	19.9	20.8	21.7	22.7	23.6	24.5
0.500	20.3	21.3	22.3	23.3	24.3	25.3	26.3	27.3	28.3	29.3
0.750	24.9	26.0	27.1	28.2	29.3	30.4	31.5	32.6	33.7	34.8
0.900	29.6	30.8	32.0	33.2	34.4	35.6	36.7	37.9	39.1	40.3
0.950	32.7	33.9	35.2	36.4	37.7	38.9	40.1	41.3	42.6	43.8
0.975	35.5	36.8	38.1	39.4	40.6	41.9	43.2	44.5	45.7	47.0
0.990	38.9	40.3	41.6	43.0	44.3	45.6	47.0	48.3	49.6	50.9
0.995	41.4	42.8	44.2	45.6	46.9	48.3	49.6	51.0	52.3	53.7

F Student's t Distribution

The following tables list the upper tail critical values of the t distribution with k degrees of freedom, that is, the solution x of

$$\tilde{p} = \int_{-\infty}^{x} \frac{\Gamma((k+1)/2)(1+t^2/k)^{-(k+1)/2}}{\sqrt{k\pi}\,\Gamma(k/2)}\,dt.$$

The $t(k)$ distribution is discussed in [2.7.9]. Examples of $t(k)$ distributions are plotted in Figure 2.6. Lower tail critical values are given by $t_{\tilde{p}} = -t_{1-\tilde{p}}$.

	$\tilde{p}$						
df	0.750	0.900	0.950	0.975	0.990	0.995	0.999
1	1.000	3.078	6.314	12.706	31.821	63.657	318.313
2	0.816	1.886	2.920	4.303	6.965	9.925	22.327
3	0.765	1.638	2.353	3.182	4.541	5.841	10.214
4	0.741	1.533	2.132	2.776	3.747	4.604	7.173
5	0.727	1.476	2.015	2.571	3.365	4.032	5.893
6	0.718	1.440	1.943	2.447	3.143	3.707	5.208
7	0.711	1.415	1.895	2.365	2.998	3.499	4.785
8	0.706	1.397	1.860	2.306	2.896	3.355	4.501
9	0.703	1.383	1.833	2.262	2.821	3.250	4.297
10	0.700	1.372	1.812	2.228	2.764	3.169	4.144
11	0.697	1.363	1.796	2.201	2.718	3.106	4.025
12	0.695	1.356	1.782	2.179	2.681	3.055	3.930
13	0.694	1.350	1.771	2.160	2.650	3.012	3.852
14	0.692	1.345	1.761	2.145	2.624	2.977	3.787
15	0.691	1.341	1.753	2.131	2.602	2.947	3.733
16	0.690	1.337	1.746	2.120	2.583	2.921	3.686
17	0.689	1.333	1.740	2.110	2.567	2.898	3.646
18	0.688	1.330	1.734	2.101	2.552	2.878	3.610
19	0.688	1.328	1.729	2.093	2.539	2.861	3.579
20	0.687	1.325	1.725	2.086	2.528	2.845	3.552
21	0.686	1.323	1.721	2.080	2.518	2.831	3.527
22	0.686	1.321	1.717	2.074	2.508	2.819	3.505
23	0.685	1.319	1.714	2.069	2.500	2.807	3.485
24	0.685	1.318	1.711	2.064	2.492	2.797	3.467
25	0.684	1.316	1.708	2.060	2.485	2.787	3.450
26	0.684	1.315	1.706	2.056	2.479	2.779	3.435
27	0.684	1.314	1.703	2.052	2.473	2.771	3.421
28	0.683	1.313	1.701	2.048	2.467	2.763	3.408
29	0.683	1.311	1.699	2.045	2.462	2.756	3.396
30	0.683	1.310	1.697	2.042	2.457	2.750	3.385

G The F Distribution

The following tables list upper tail critical values of the $F(k, l)$ distribution for $\tilde{p} = 0.90, 0.95$, and 0.99. Lower tail critical values can be obtained by noting that if $F_{\tilde{p}}$ is the $\tilde{p}$-quantile of the $F(k, l)$ distribution, then $1/F_{\tilde{p}}$ is the $(1 - \tilde{p})$-quantile of the $F(l, k)$ distribution.

The $F(k, l)$ distribution is discussed in [2.7.10]. Examples of $F(k, l)$ distributions are plotted in Figure 2.7.

90% quantiles of $F(k, l)$

l \ k	1	2	3	4	5	6	7	8	9	10
1	39.86	49.50	53.59	55.84	57.24	58.20	58.91	59.44	59.86	60.19
2	8.53	9.00	9.16	9.24	9.29	9.33	9.35	9.37	9.38	9.39
3	5.54	5.46	5.39	5.34	5.31	5.28	5.27	5.25	5.24	5.23
4	4.54	4.32	4.19	4.11	4.05	4.01	3.98	3.95	3.94	3.92
5	4.06	3.78	3.62	3.52	3.45	3.40	3.37	3.34	3.32	3.30
6	3.78	3.46	3.29	3.18	3.11	3.05	3.01	2.98	2.96	2.94
7	3.59	3.26	3.07	2.96	2.88	2.83	2.78	2.75	2.72	2.70
8	3.46	3.11	2.92	2.81	2.73	2.67	2.62	2.59	2.56	2.54
9	3.36	3.01	2.81	2.69	2.61	2.55	2.51	2.47	2.44	2.42
10	3.29	2.92	2.73	2.61	2.52	2.46	2.41	2.38	2.35	2.32
11	3.23	2.86	2.66	2.54	2.45	2.39	2.34	2.30	2.27	2.25
12	3.18	2.81	2.61	2.48	2.39	2.33	2.28	2.24	2.21	2.19
13	3.14	2.76	2.56	2.43	2.35	2.28	2.23	2.20	2.16	2.14
14	3.10	2.73	2.52	2.39	2.31	2.24	2.19	2.15	2.12	2.10
15	3.07	2.70	2.49	2.36	2.27	2.21	2.16	2.12	2.09	2.06
16	3.05	2.67	2.46	2.33	2.24	2.18	2.13	2.09	2.06	2.03
17	3.03	2.64	2.44	2.31	2.22	2.15	2.10	2.06	2.03	2.00
18	3.01	2.62	2.42	2.29	2.20	2.13	2.08	2.04	2.01	1.98
19	2.99	2.61	2.40	2.27	2.18	2.11	2.06	2.02	1.98	1.96
20	2.97	2.59	2.38	2.25	2.16	2.09	2.04	2.00	1.96	1.94
21	2.96	2.57	2.36	2.23	2.14	2.08	2.02	1.98	1.95	1.92
22	2.95	2.56	2.35	2.22	2.13	2.06	2.01	1.97	1.93	1.90
23	2.94	2.55	2.34	2.21	2.11	2.05	1.99	1.95	1.92	1.89
24	2.93	2.54	2.33	2.19	2.10	2.04	1.98	1.94	1.91	1.88
25	2.92	2.53	2.32	2.18	2.09	2.02	1.97	1.93	1.89	1.87
26	2.91	2.52	2.31	2.17	2.08	2.01	1.96	1.92	1.88	1.86
27	2.90	2.51	2.30	2.17	2.07	2.00	1.95	1.91	1.87	1.85
28	2.89	2.50	2.29	2.16	2.06	2.00	1.94	1.90	1.87	1.84
29	2.89	2.50	2.28	2.15	2.06	1.99	1.94	1.89	1.86	1.83
30	2.88	2.49	2.28	2.14	2.05	1.98	1.93	1.88	1.85	1.82
40	2.84	2.44	2.23	2.09	2.00	1.93	1.87	1.83	1.79	1.76
60	2.79	2.39	2.18	2.04	1.95	1.87	1.82	1.78	1.74	1.71
120	2.75	2.35	2.13	1.99	1.90	1.82	1.77	1.72	1.68	1.65
∞	2.71	2.30	2.08	1.94	1.85	1.77	1.72	1.67	1.63	1.60

90% quantiles of $F(k, l)$

l \ k	12	15	20	24	30	40	50	60	120	∞
1	60.71	61.22	61.74	62.00	62.26	62.53	62.69	62.79	63.06	63.33
2	9.41	9.42	9.44	9.45	9.46	9.47	9.47	9.47	9.48	9.49
3	5.22	5.20	5.18	5.18	5.17	5.16	5.16	5.15	5.14	5.13
4	3.90	3.87	3.84	3.83	3.82	3.80	3.80	3.79	3.78	3.76
5	3.27	3.24	3.21	3.19	3.17	3.16	3.15	3.14	3.12	3.11
6	2.90	2.87	2.84	2.82	2.80	2.78	2.77	2.76	2.74	2.72
7	2.67	2.63	2.59	2.58	2.56	2.54	2.52	2.51	2.49	2.47
8	2.50	2.46	2.42	2.40	2.38	2.36	2.35	2.34	2.32	2.29
9	2.38	2.34	2.30	2.28	2.25	2.23	2.22	2.21	2.18	2.16
10	2.28	2.24	2.20	2.18	2.16	2.13	2.12	2.11	2.08	2.05
11	2.21	2.17	2.12	2.10	2.08	2.05	2.04	2.03	2.00	1.97
12	2.15	2.10	2.06	2.04	2.01	1.99	1.97	1.96	1.93	1.90
13	2.10	2.05	2.01	1.98	1.96	1.93	1.92	1.90	1.88	1.85
14	2.05	2.01	1.96	1.94	1.91	1.89	1.87	1.86	1.83	1.80
15	2.02	1.97	1.92	1.90	1.87	1.85	1.83	1.82	1.79	1.76
16	1.99	1.94	1.89	1.87	1.84	1.81	1.79	1.78	1.75	1.72
17	1.96	1.91	1.86	1.84	1.81	1.78	1.76	1.75	1.72	1.69
18	1.93	1.89	1.84	1.81	1.78	1.75	1.74	1.72	1.69	1.68
19	1.91	1.86	1.81	1.79	1.76	1.73	1.71	1.70	1.67	1.63
20	1.89	1.84	1.79	1.77	1.74	1.71	1.69	1.68	1.64	1.61
21	1.87	1.83	1.78	1.75	1.72	1.69	1.67	1.66	1.62	1.59
22	1.86	1.81	1.76	1.73	1.70	1.67	1.65	1.64	1.60	1.57
23	1.84	1.80	1.74	1.72	1.69	1.66	1.64	1.62	1.59	1.55
24	1.83	1.78	1.73	1.70	1.67	1.64	1.62	1.61	1.57	1.53
25	1.82	1.77	1.72	1.69	1.66	1.63	1.61	1.59	1.56	1.52
26	1.81	1.76	1.71	1.68	1.65	1.61	1.60	1.58	1.54	1.50
27	1.80	1.75	1.70	1.67	1.64	1.60	1.58	1.57	1.53	1.49
28	1.79	1.74	1.69	1.66	1.63	1.59	1.57	1.56	1.52	1.48
29	1.78	1.73	1.68	1.65	1.62	1.58	1.56	1.55	1.51	1.47
30	1.77	1.72	1.67	1.64	1.61	1.57	1.55	1.54	1.50	1.46
40	1.71	1.66	1.61	1.57	1.54	1.51	1.48	1.47	1.42	1.38
60	1.66	1.60	1.54	1.51	1.48	1.44	1.41	1.40	1.35	1.29
120	1.60	1.55	1.48	1.45	1.41	1.37	1.34	1.32	1.26	1.19
∞	1.55	1.49	1.42	1.38	1.34	1.30	1.28	1.24	1.17	1.00

95% quantiles of $F(k, l)$

l \ k	1	2	3	4	5	6	7	8	9	10
1	161.4	199.5	215.7	224.6	230.2	234.0	236.8	238.9	240.5	241.9
2	18.51	19.00	19.16	19.25	19.30	19.33	19.35	19.37	19.38	19.40
3	10.13	9.55	9.28	9.12	9.01	8.94	8.89	8.85	8.81	8.79
4	7.71	6.94	6.59	6.39	6.26	6.16	6.09	6.04	6.00	5.96
5	6.61	5.79	5.41	5.19	5.05	4.95	4.88	4.82	4.77	4.74
6	5.99	5.14	4.76	4.53	4.39	4.28	4.21	4.15	4.10	4.06
7	5.59	4.74	4.35	4.12	3.97	3.87	3.79	3.73	3.68	3.64
8	5.32	4.46	4.07	3.84	3.69	3.58	3.50	3.44	3.39	3.35
9	5.12	4.26	3.86	3.63	3.48	3.37	3.29	3.23	3.18	3.14
10	4.96	4.10	3.71	3.48	3.33	3.22	3.14	3.07	3.02	2.98
11	4.84	3.98	3.59	3.36	3.20	3.09	3.01	2.95	2.90	2.85
12	4.75	3.89	3.49	3.26	3.11	3.00	2.91	2.85	2.80	2.75
13	4.67	3.81	3.41	3.18	3.03	2.92	2.83	2.77	2.71	2.67
14	4.60	3.74	3.34	3.11	2.96	2.85	2.76	2.70	2.65	2.60
15	4.54	3.68	3.29	3.06	2.90	2.79	2.71	2.64	2.59	2.54
16	4.49	3.63	3.24	3.01	2.85	2.74	2.66	2.59	2.54	2.49
17	4.45	3.59	3.20	2.96	2.81	2.70	2.61	2.55	2.49	2.45
18	4.41	3.55	3.16	2.93	2.77	2.66	2.58	2.51	2.46	2.41
19	4.38	3.52	3.13	2.90	2.74	2.63	2.54	2.48	2.42	2.38
20	4.35	3.49	3.10	2.87	2.71	2.60	2.51	2.45	2.39	2.35
21	4.32	3.47	3.07	2.84	2.69	2.57	2.49	2.42	2.37	2.32
22	4.30	3.44	3.05	2.82	2.66	2.55	2.46	2.40	2.34	2.30
23	4.28	3.42	3.03	2.80	2.64	2.53	2.44	2.37	2.32	2.27
24	4.26	3.40	3.01	2.78	2.62	2.51	2.42	2.36	2.30	2.25
25	4.24	3.39	2.99	2.76	2.60	2.49	2.40	2.34	2.28	2.24
26	4.23	3.37	2.98	2.74	2.59	2.47	2.39	2.32	2.27	2.22
27	4.21	3.35	2.96	2.73	2.57	2.46	2.37	2.31	2.25	2.20
28	4.20	3.34	2.95	2.71	2.56	2.45	2.36	2.29	2.24	2.19
29	4.18	3.33	2.93	2.70	2.55	2.43	2.35	2.28	2.22	2.18
30	4.17	3.32	2.92	2.69	2.53	2.42	2.33	2.27	2.21	2.16
40	4.08	3.23	2.84	2.61	2.45	2.34	2.25	2.18	2.12	2.08
60	4.00	3.15	2.76	2.53	2.37	2.25	2.17	2.10	2.04	2.00
120	3.92	3.07	2.68	2.45	2.29	2.17	2.09	2.02	1.96	1.91
∞	3.84	3.00	2.60	2.37	2.21	2.10	2.01	1.94	1.88	1.83

95% quantiles of $F(k, l)$

l \ k	12	15	20	24	30	40	50	60	120	∞
1	243.9	245.9	248.0	249.1	250.1	251.1	251.8	252.2	253.3	254.3
2	19.41	19.43	19.45	19.45	19.46	19.47	19.48	19.48	19.49	19.50
3	8.74	8.70	8.66	8.64	8.62	8.59	8.58	8.57	8.55	8.53
4	5.91	5.86	5.80	5.77	5.75	5.72	5.70	5.69	5.66	5.63
5	4.68	4.62	4.56	4.53	4.50	4.46	4.44	4.43	4.40	4.36
6	4.00	3.94	3.87	3.84	3.81	3.77	3.75	3.74	3.70	3.67
7	3.57	3.51	3.44	3.41	3.38	3.34	3.32	3.30	3.27	3.23
8	3.28	3.22	3.15	3.12	3.08	3.04	3.02	3.01	2.97	2.93
9	3.07	3.01	2.94	2.90	2.86	2.83	2.80	2.79	2.75	2.71
10	2.91	2.85	2.77	2.74	2.70	2.66	2.64	2.62	2.58	2.54
11	2.79	2.72	2.65	2.61	2.57	2.53	2.51	2.49	2.45	2.40
12	2.69	2.62	2.54	2.51	2.47	2.43	2.40	2.38	2.34	2.30
13	2.60	2.53	2.46	2.42	2.38	2.34	2.31	2.30	2.25	2.21
14	2.53	2.46	2.39	2.35	2.31	2.27	2.24	2.22	2.18	2.13
15	2.48	2.40	2.33	2.29	2.25	2.20	2.18	2.16	2.11	2.07
16	2.42	2.35	2.28	2.24	2.19	2.15	2.12	2.11	2.06	2.01
17	2.38	2.31	2.23	2.19	2.15	2.10	2.08	2.06	2.01	1.96
18	2.34	2.27	2.19	2.15	2.11	2.06	2.04	2.02	1.97	1.92
19	2.31	2.23	2.16	2.11	2.07	2.03	2.00	1.98	1.93	1.88
20	2.28	2.20	2.12	2.08	2.04	1.99	1.97	1.95	1.90	1.84
21	2.25	2.18	2.10	2.06	2.01	1.96	1.94	1.92	1.87	1.81
22	2.23	2.15	2.07	2.03	1.98	1.94	1.91	1.89	1.84	1.78
23	2.20	2.13	2.05	2.01	1.96	1.91	1.89	1.86	1.81	1.76
24	2.18	2.11	2.03	1.98	1.94	1.89	1.86	1.84	1.79	1.73
25	2.16	2.09	2.01	1.96	1.92	1.87	1.84	1.82	1.77	1.71
26	2.15	2.07	1.99	1.95	1.90	1.85	1.82	1.80	1.75	1.69
27	2.13	2.06	1.97	1.93	1.88	1.84	1.81	1.79	1.73	1.67
28	2.12	2.04	1.96	1.91	1.87	1.82	1.79	1.77	1.71	1.65
29	2.10	2.03	1.94	1.90	1.85	1.81	1.78	1.75	1.70	1.64
30	2.09	2.01	1.93	1.89	1.84	1.79	1.76	1.74	1.68	1.62
40	2.00	1.92	1.84	1.79	1.74	1.69	1.66	1.64	1.58	1.51
60	1.92	1.84	1.75	1.70	1.65	1.59	1.56	1.53	1.47	1.39
120	1.83	1.75	1.66	1.61	1.55	1.50	1.46	1.43	1.35	1.25
∞	1.75	1.67	1.57	1.52	1.46	1.39	1.38	1.32	1.22	1.00

99% quantiles of $F(k, l)$

l \ k	1	2	3	4	5	6	7	8	9	10
1	4052.	4999.5	5403.	5625.	5764.	5859.	5928.	5982.	6022.	6056.
2	98.50	99.00	99.17	99.25	99.30	99.33	99.36	99.37	99.39	99.40
3	34.12	30.82	29.46	28.71	28.24	27.91	27.67	27.49	27.35	27.23
4	21.20	18.00	16.69	15.98	15.52	15.21	14.98	14.80	14.66	14.55
5	16.26	13.27	12.06	11.39	10.97	10.67	10.46	10.29	10.16	10.05
6	13.75	10.92	9.78	9.15	8.75	8.47	8.26	8.10	7.98	7.87
7	12.25	9.55	8.45	7.85	7.46	7.19	6.99	6.84	6.72	6.62
8	11.26	8.65	7.59	7.01	6.63	6.37	6.18	6.03	5.91	5.81
9	10.56	8.02	6.99	6.42	6.06	5.80	5.61	5.47	5.35	5.26
10	10.04	7.56	6.55	5.99	5.64	5.39	5.20	5.06	4.94	4.85
11	9.65	7.21	6.22	5.67	5.32	5.07	4.89	4.74	4.63	4.54
12	9.33	6.93	5.95	5.41	5.06	4.82	4.64	4.50	4.39	4.30
13	9.07	6.70	5.74	5.21	4.86	4.62	4.44	4.30	4.19	4.10
14	8.86	6.51	5.56	5.04	4.69	4.46	4.28	4.14	4.03	3.94
15	8.68	6.36	5.42	4.89	4.56	4.32	4.14	4.00	3.89	3.80
16	8.53	6.23	5.29	4.77	4.44	4.20	4.03	3.89	3.78	3.69
17	8.40	6.11	5.18	4.67	4.34	4.10	3.93	3.79	3.68	3.59
18	8.29	6.01	5.09	4.58	4.25	4.01	3.84	3.71	3.60	3.51
19	8.18	5.93	5.01	4.50	4.17	3.94	3.77	3.63	3.52	3.43
20	8.10	5.85	4.94	4.43	4.10	3.87	3.70	3.56	3.46	3.37
21	8.02	5.78	4.87	4.37	4.04	3.81	3.64	3.51	3.40	3.31
22	7.95	5.72	4.82	4.31	3.99	3.76	3.59	3.45	3.35	3.26
23	7.88	5.66	4.76	4.26	3.94	3.71	3.54	3.41	3.30	3.21
24	7.82	5.61	4.72	4.22	3.90	3.67	3.50	3.36	3.26	3.17
25	7.77	5.57	4.68	4.18	3.85	3.63	3.46	3.32	3.22	3.13
26	7.72	5.53	4.64	4.14	3.82	3.59	3.42	3.29	3.18	3.09
27	7.68	5.49	4.60	4.11	3.78	3.56	3.39	3.26	3.15	3.06
28	7.64	5.45	4.57	4.07	3.75	3.53	3.36	3.23	3.12	3.03
29	7.60	5.42	4.54	4.04	3.73	3.50	3.33	3.20	3.09	3.00
30	7.56	5.39	4.51	4.02	3.70	3.47	3.30	3.17	3.07	2.98
40	7.31	5.18	4.31	3.83	3.51	3.29	3.12	2.99	2.89	2.80
60	7.08	4.98	4.13	3.65	3.34	3.12	2.95	2.82	2.72	2.63
120	6.85	4.79	3.95	3.48	3.17	2.96	2.79	2.66	2.56	2.47
∞	6.63	4.61	3.78	3.32	3.02	2.80	2.64	2.51	2.41	2.32

99% quantiles of $F(k, l)$

l \ k	12	15	20	24	30	40	50	60	120	∞
1	6106.	6157.	6209.	6235.	6261.	6287.	6303.	6313.	6339.	6366.
2	99.4	99.43	99.45	99.46	99.47	99.47	99.48	99.48	99.49	99.50
3	27.1	26.87	26.69	26.60	26.50	26.41	26.35	26.32	26.22	26.13
4	14.4	14.20	14.02	13.93	13.84	13.75	13.69	13.65	13.56	13.46
5	9.89	9.72	9.55	9.47	9.38	9.29	9.24	9.20	9.11	9.02
6	7.72	7.56	7.40	7.31	7.23	7.14	7.09	7.06	6.97	6.88
7	6.47	6.31	6.16	6.07	5.99	5.91	5.86	5.82	5.74	5.65
8	5.67	5.52	5.36	5.28	5.20	5.12	5.07	5.03	4.95	4.86
9	5.11	4.96	4.81	4.73	4.65	4.57	4.52	4.48	4.40	4.31
10	4.71	4.56	4.41	4.33	4.25	4.17	4.12	4.08	4.00	3.91
11	4.40	4.25	4.10	4.02	3.94	3.86	3.81	3.78	3.69	3.60
12	4.16	4.01	3.86	3.78	3.70	3.62	3.57	3.54	3.45	3.36
13	3.96	3.82	3.66	3.59	3.51	3.43	3.38	3.34	3.25	3.17
14	3.80	3.66	3.51	3.43	3.35	3.27	3.22	3.18	3.09	3.00
15	3.67	3.52	3.37	3.29	3.21	3.13	3.08	3.05	2.96	2.87
16	3.55	3.41	3.26	3.18	3.10	3.02	2.97	2.93	2.84	2.75
17	3.46	3.31	3.16	3.08	3.00	2.92	2.87	2.83	2.75	2.65
18	3.37	3.23	3.08	3.00	2.92	2.84	2.78	2.75	2.66	2.57
19	3.30	3.15	3.00	2.92	2.84	2.76	2.71	2.67	2.58	2.49
20	3.23	3.09	2.94	2.86	2.78	2.69	2.64	2.61	2.52	2.42
21	3.17	3.03	2.88	2.80	2.72	2.64	2.59	2.55	2.46	2.36
22	3.12	2.98	2.83	2.75	2.67	2.58	2.53	2.50	2.40	2.31
23	3.07	2.93	2.78	2.70	2.62	2.54	2.49	2.45	2.35	2.26
24	3.03	2.89	2.74	2.66	2.58	2.49	2.44	2.40	2.31	2.21
25	2.99	2.85	2.70	2.62	2.54	2.45	2.40	2.36	2.27	2.17
26	2.96	2.81	2.66	2.59	2.50	2.42	2.36	2.33	2.23	2.13
27	2.93	2.78	2.63	2.55	2.47	2.38	2.33	2.29	2.20	2.10
28	2.90	2.75	2.60	2.52	2.44	2.35	2.30	2.26	2.17	2.06
29	2.87	2.73	2.57	2.49	2.41	2.33	2.27	2.23	2.14	2.03
30	2.84	2.70	2.55	2.47	2.39	2.30	2.25	2.21	2.11	2.01
40	2.66	2.52	2.37	2.29	2.20	2.11	2.06	2.02	1.92	1.80
60	2.50	2.35	2.20	2.12	2.03	1.94	1.88	1.84	1.73	1.60
120	2.34	2.19	2.03	1.95	1.86	1.76	1.70	1.66	1.53	1.38
∞	2.18	2.04	1.88	1.79	1.70	1.59	1.56	1.47	1.32	1.00

H Table-Look-Up Test

The following tables are for use with the 'Table-Look-Up test' of equality of means when observations are serially correlated (see [6.6.9]). Critical levels are listed for one- and two-sided tests at the significance level given in each heading. The number α is the *estimated* lag-1 correlation; n is the sample size. Dashes indicate that sample correlations of that particular magnitude were not observed in the simulations used to create the table. Note that the table was constructed by means of Monte Carlo simulation (see Zwiers and von Storch [454]) and thus its entries are subject to some sampling variability.

Table-Look-Up test: critical values

Significance level: 20% for a two-sided test
10% for a one-sided test

α \ n	10	15	20	25	30	45	60	75	90	120	180	240
−0.35	2.30	—	—	—	—	—	—	—	—	—	—	—
−0.30	2.34	1.87	—	—	—	—	—	—	—	—	—	—
−0.25	2.41	1.88	1.70	—	—	—	—	—	—	—	—	—
−0.20	2.46	1.92	1.71	1.62	1.54	—	—	—	—	—	—	—
−0.15	2.52	1.99	1.75	1.64	1.55	1.45	—	—	—	—	—	—
−0.10	2.61	2.04	1.80	1.67	1.59	1.47	1.44	1.42	1.39	1.38	—	—
−0.05	2.67	2.09	1.84	1.68	1.64	1.50	1.47	1.44	1.39	1.39	1.36	1.33
0.00	2.75	2.18	1.88	1.72	1.66	1.53	1.49	1.46	1.43	1.40	1.38	1.35
0.05	2.82	2.26	1.97	1.79	1.71	1.60	1.55	1.47	1.50	1.46	1.43	1.42
0.10	2.94	2.35	2.08	1.87	1.76	1.64	1.58	1.55	1.50	1.48	1.46	1.40
0.15	3.10	2.48	2.18	1.96	1.82	1.70	1.66	1.60	1.58	1.57	1.53	1.51
0.20	3.23	2.59	2.27	2.08	1.91	1.77	1.70	1.66	1.64	1.63	1.59	1.57
0.25	3.36	2.75	2.42	2.22	2.05	1.87	1.78	1.75	1.77	1.69	1.69	1.70
0.30	3.48	2.96	2.57	2.36	2.22	1.96	1.91	1.91	1.86	1.86	1.79	1.77
0.35	3.61	3.20	2.79	2.56	2.38	2.12	2.02	2.03	1.96	1.88	1.92	1.90
0.40	3.77	3.46	3.02	2.74	2.61	2.31	2.22	2.13	2.11	2.07	2.00	2.02
0.45	3.95	3.66	3.38	3.05	2.79	2.55	2.36	2.25	2.23	2.19	2.14	2.13
0.50	4.13	3.92	3.71	3.41	3.24	2.73	2.57	2.47	2.46	2.39	2.31	2.30
0.55	4.27	4.23	4.13	3.77	3.55	3.03	2.86	2.65	2.61	2.56	2.50	2.43
0.60	4.45	4.59	4.47	4.26	3.94	3.45	3.14	3.03	2.83	2.75	2.65	2.72
0.65	4.55	4.83	4.85	4.72	4.49	3.89	3.60	3.26	3.16	3.02	2.91	2.94
0.70	4.56	5.17	5.37	5.28	5.25	4.49	4.01	3.74	3.59	3.36	3.27	3.33
0.75	—	5.37	5.71	5.82	5.85	5.35	4.91	4.40	4.14	3.92	3.69	3.55
0.80	—	—	5.99	6.49	6.49	6.42	6.00	5.51	5.18	4.67	4.28	4.25
0.85	—	—	—	6.66	7.26	7.82	7.33	7.28	6.76	6.09	5.41	5.05
0.90	—	—	—	—	7.31	8.77	9.47	9.45	9.01	8.59	7.55	6.97
0.95	—	—	—	—	—	—	9.93	10.7	11.3	13.2	13.3	12.5

(α = estimated lag-1 correlation; n = sample size)

Table-Look-Up test: critical values

Significance level: 10% for a two-sided test
5% for a one-sided test

	n											
α	10	15	20	25	30	45	60	75	90	120	180	240
−0.35	3.46	—	—	—	—	—	—	—	—	—	—	—
−0.30	3.55	2.57	—	—	—	—	—	—	—	—	—	—
−0.25	3.68	2.59	2.26	—	—	—	—	—	—	—	—	—
−0.20	3.80	2.66	2.28	2.15	2.03	—	—	—	—	—	—	—
−0.15	3.85	2.76	2.35	2.17	2.04	1.89	—	—	—	—	—	—
−0.10	4.03	2.89	2.44	2.22	2.09	1.92	1.85	1.83	1.81	1.78	—	—
−0.05	4.21	2.98	2.52	2.24	2.13	1.96	1.90	1.88	1.81	1.80	1.73	1.72
0.00	4.34	3.15	2.59	2.31	2.18	2.02	1.92	1.89	1.84	1.83	1.75	1.74
0.05	4.48	3.30	2.76	2.40	2.26	2.09	1.98	1.89	1.94	1.86	1.84	1.84
0.10	4.75	3.47	2.88	2.56	2.38	2.15	2.07	2.01	1.96	1.91	1.89	1.85
0.15	5.07	3.71	3.06	2.69	2.43	2.24	2.14	2.09	2.04	2.02	1.96	1.96
0.20	5.27	3.96	3.26	2.85	2.58	2.33	2.21	2.17	2.14	2.12	2.05	2.03
0.25	5.42	4.25	3.63	3.07	2.79	2.48	2.35	2.28	2.29	2.21	2.19	2.20
0.30	5.63	4.60	3.82	3.32	3.07	2.66	2.53	2.51	2.42	2.37	2.33	2.26
0.35	5.90	5.10	4.20	3.72	3.35	2.85	2.63	2.65	2.56	2.47	2.44	2.43
0.40	6.17	5.56	4.74	4.07	3.67	3.11	2.94	2.81	2.76	2.68	2.56	2.56
0.45	6.44	5.60	5.28	4.59	4.09	3.53	3.19	2.97	2.94	2.84	2.79	2.76
0.50	6.80	6.54	5.93	5.23	4.80	3.79	3.52	3.26	3.19	3.11	3.02	2.95
0.55	7.00	6.93	6.64	5.96	5.31	4.39	3.85	3.52	3.49	3.33	3.25	3.13
0.60	7.15	7.45	7.36	6.96	6.14	5.05	4.29	4.10	3.76	3.61	3.44	3.54
0.65	7.20	7.91	8.03	7.67	7.15	5.80	5.03	4.49	4.24	4.06	3.79	3.78
0.70	7.21	8.39	8.70	8.72	8.50	6.87	5.74	5.30	4.85	4.53	4.34	4.21
0.75	—	8.57	9.13	9.44	9.22	8.43	7.19	6.43	5.80	5.23	4.89	4.67
0.80	—	—	9.65	10.3	10.5	10.4	9.11	8.28	7.64	6.55	5.79	5.50
0.85	—	—	9.67	10.4	11.4	12.8	11.8	11.4	10.3	9.02	7.40	6.72
0.90	—	—	—	—	11.4	13.8	14.4	14.9	14.9	13.4	10.9	9.58
0.95	—	—	—	—	—	—	14.9	16.4	17.3	20.3	20.7	19.8

(α = estimated lag-1 correlation; n = sample size)

Table-Look-Up test: critical values

Significance level: 5% for a two-sided test
2.5% for a one-sided test

	n											
α	10	15	20	25	30	45	60	75	90	120	180	240
−0.35	5.18	—	—	—	—	—	—	—	—	—	—	—
−0.30	5.32	3.44	—	—	—	—	—	—	—	—	—	—
−0.25	5.56	3.47	2.84	—	—	—	—	—	—	—	—	—
−0.20	5.71	3.61	2.86	2.65	2.50	—	—	—	—	—	—	—
−0.15	5.76	3.75	2.98	2.70	2.50	2.34	—	—	—	—	—	—
−0.10	6.25	3.88	3.11	2.76	2.56	2.34	2.22	2.23	2.19	2.10	—	—
−0.05	6.48	4.05	3.19	2.81	2.62	2.40	2.28	2.28	2.19	2.12	2.08	2.05
0.00	6.75	4.44	3.34	2.90	2.71	2.46	2.33	2.28	2.31	2.19	2.11	2.20
0.05	7.17	4.68	3.59	3.05	2.82	2.53	2.37	2.31	2.31	2.25	2.18	2.20
0.10	7.49	5.16	3.82	3.29	2.99	2.65	2.53	2.42	2.34	2.32	2.21	2.21
0.15	7.91	5.55	4.15	3.55	3.12	2.77	2.63	2.52	2.46	2.43	2.38	2.35
0.20	8.38	6.00	4.61	3.74	3.34	2.89	2.70	2.64	2.60	2.59	2.45	2.43
0.25	8.52	6.50	5.20	4.13	3.52	3.07	2.89	2.78	2.76	2.69	2.66	2.66
0.30	8.81	7.17	5.57	4.52	4.05	3.32	3.10	3.05	2.91	2.86	2.76	2.70
0.35	9.11	7.93	6.25	5.23	4.48	3.60	3.28	3.21	3.11	2.92	2.93	3.00
0.40	9.55	8.74	7.19	5.88	5.05	3.96	3.61	3.45	3.34	3.28	3.13	3.11
0.45	9.91	9.36	8.25	6.72	5.92	4.50	3.95	3.75	3.51	3.48	3.36	3.29
0.50	10.4	9.84	9.10	8.00	7.01	4.99	4.36	3.97	3.90	3.78	3.70	3.54
0.55	10.6	10.6	10.2	9.21	7.79	5.82	5.04	4.41	4.28	4.09	4.00	3.85
0.60	10.7	11.2	11.4	11.0	9.22	6.98	5.49	5.31	4.69	4.47	4.18	4.25
0.65	10.6	12.1	12.1	11.9	11.0	8.33	6.68	5.81	5.36	5.04	4.67	4.57
0.70	10.6	12.5	13.2	13.5	13.0	10.1	8.20	7.07	6.25	5.71	5.25	5.04
0.75	—	12.3	13.8	14.2	14.2	13.1	10.7	9.12	7.77	6.58	5.98	5.62
0.80	—	—	14.2	15.2	15.6	15.2	13.6	12.1	11.0	8.60	7.22	6.64
0.85	—	—	—	15.1	16.5	18.8	17.3	16.9	15.4	12.9	9.50	8.37
0.90	—	—	—	—	16.5	20.0	21.2	22.0	21.2	20.0	15.1	12.6
0.95	—	—	—	—	—	—	20.9	23.4	24.3	27.4	28.8	29.5

(α = estimated lag-1 correlation; n = sample size)

Table-Look-Up test: critical values

Significance level: 2% for a two-sided test
1% for a one-sided test

	n											
α	10	15	20	25	30	45	60	75	90	120	180	240
−0.35	8.76	—	—	—	—	—	—	—	—	—	—	—
−0.30	9.03	5.02	—	—	—	—	—	—	—	—	—	—
−0.25	9.26	5.12	3.80	—	—	—	—	—	—	—	—	—
−0.20	9.98	5.43	3.89	3.34	3.07	—	—	—	—	—	—	—
−0.15	10.1	5.67	4.03	3.42	3.10	2.79	—	—	—	—	—	—
−0.10	10.9	5.87	4.21	3.52	3.19	2.82	2.68	2.68	2.63	2.51	—	—
−0.05	11.1	6.49	4.38	3.67	3.30	2.97	2.77	2.69	2.61	2.51	2.52	2.45
0.00	11.9	7.35	4.71	3.91	3.44	3.04	2.85	2.76	2.69	2.61	2.50	2.41
0.05	12.4	7.97	5.30	4.07	3.60	3.08	2.89	2.82	2.76	2.69	2.61	2.59
0.10	12.8	8.92	5.72	4.49	3.95	3.28	3.03	2.88	2.78	2.71	2.66	2.64
0.15	14.0	10.1	6.51	4.78	4.26	3.45	3.24	3.08	3.01	2.92	2.83	2.86
0.20	14.6	10.5	7.18	5.28	4.40	3.60	3.39	3.19	3.06	3.09	2.94	2.96
0.25	14.7	11.1	8.59	6.16	4.81	3.81	3.55	3.43	3.27	3.21	3.19	3.23
0.30	15.2	12.8	8.94	6.88	5.68	4.26	3.97	3.72	3.57	3.42	3.31	3.21
0.35	15.4	13.7	10.5	8.26	6.58	4.70	4.05	3.99	3.82	3.51	3.49	3.56
0.40	15.7	14.9	12.8	9.54	7.85	5.26	4.60	4.37	4.07	3.85	3.84	3.79
0.45	16.4	16.3	14.9	11.4	8.92	6.32	5.14	4.73	4.33	4.28	4.07	3.92
0.50	17.1	17.4	16.3	13.8	11.4	6.99	5.81	5.18	4.79	4.58	4.39	4.35
0.55	16.9	17.9	17.3	15.6	12.6	8.73	6.91	5.65	5.34	5.05	4.85	4.57
0.60	17.2	18.6	18.7	18.9	14.8	10.6	7.53	6.78	5.94	5.63	4.94	5.10
0.65	16.9	19.8	19.2	20.6	18.9	14.2	10.1	7.76	7.05	6.24	5.62	5.37
0.70	16.9	19.3	21.3	22.0	22.2	17.6	12.4	9.77	8.35	7.32	6.51	6.12
0.75	—	18.8	21.4	22.1	23.0	21.9	18.1	14.2	11.0	8.56	7.64	6.98
0.80	—	—	20.7	22.9	24.6	24.6	21.3	20.0	16.6	12.6	9.15	8.18
0.85	—	—	—	22.7	25.1	28.6	28.4	30.3	24.6	19.1	12.9	11.1
0.90	—	—	—	—	25.0	27.9	31.0	33.7	33.1	31.1	23.8	17.8
0.95	—	—	—	—	—	—	29.7	32.2	33.4	37.4	42.3	45.8

(α = estimated lag-1 correlation; n = sample size)

Table-Look-Up test: critical values

Significance level: 1% for a two-sided test
0.5% for a one-sided test

α \ n	10	15	20	25	30	45	60	75	90	120	180	240
−0.35	13.0	—	—	—	—	—	—	—	—	—	—	—
−0.30	13.3	6.82	—	—	—	—	—	—	—	—	—	—
−0.25	14.0	7.01	4.70	—	—	—	—	—	—	—	—	—
−0.20	15.1	7.42	4.83	3.87	3.58	—	—	—	—	—	—	—
−0.15	15.5	7.70	5.12	3.98	3.58	3.20	—	—	—	—	—	—
−0.10	16.2	8.41	5.26	4.16	3.69	3.23	3.02	2.96	2.97	2.82	—	—
−0.05	16.2	9.41	5.45	4.47	3.80	3.34	3.20	2.98	2.94	2.83	2.80	2.79
0.00	18.0	10.6	5.91	4.77	4.01	3.48	3.24	3.05	2.99	2.93	2.82	2.67
0.05	19.0	11.6	7.22	5.11	4.21	3.56	3.20	3.20	3.13	3.00	2.92	2.88
0.10	19.2	13.7	8.28	5.80	4.77	3.66	3.43	3.31	3.16	2.99	2.95	2.93
0.15	20.5	14.9	9.44	6.20	5.27	3.96	3.66	3.44	3.37	3.30	3.16	3.13
0.20	20.6	15.5	11.0	7.08	5.37	4.25	3.93	3.60	3.47	3.55	3.29	3.28
0.25	20.5	16.6	12.1	8.80	6.17	4.43	4.04	3.85	3.70	3.65	3.50	3.49
0.30	21.2	18.4	12.7	10.1	7.51	5.25	4.57	4.29	3.93	3.95	3.76	3.59
0.35	21.5	19.8	15.9	12.0	8.80	5.71	4.63	4.50	4.26	3.89	3.90	3.93
0.40	21.8	20.9	19.7	14.1	11.3	6.32	5.44	4.96	4.58	4.36	4.25	4.22
0.45	22.3	23.0	23.0	17.6	13.6	7.65	6.20	5.49	4.92	4.83	4.61	4.41
0.50	23.3	23.7	24.9	19.6	16.4	9.56	7.13	6.00	5.49	5.17	4.90	4.75
0.55	23.3	25.0	25.7	22.6	19.2	12.0	8.75	6.60	6.25	5.82	5.56	5.07
0.60	22.9	25.9	26.3	26.1	22.0	14.5	9.83	8.50	7.01	6.43	5.63	5.62
0.65	22.1	26.8	26.3	28.6	27.6	20.6	14.4	9.76	8.43	7.29	6.43	5.88
0.70	22.0	26.3	27.8	30.8	30.4	26.5	17.8	12.8	10.7	8.69	7.72	7.09
0.75	—	24.3	27.3	29.9	29.7	29.8	25.0	20.6	14.1	10.6	8.97	7.96
0.80	—	—	26.8	29.8	31.3	33.5	32.0	28.5	26.8	15.7	10.7	9.92
0.85	—	—	—	29.3	30.9	36.3	38.4	41.2	35.1	26.5	15.8	13.2
0.90	—	—	—	—	30.8	33.8	39.1	42.9	42.9	41.1	35.8	21.4
0.95	—	—	—	—	—	—	37.6	39.5	41.4	46.3	54.3	61.5

(α = estimated lag-1 correlation; n = sample size)

I Critical Values for the Mann–Whitney Test

The following tables are for use with the non-parametric 'Mann–Whitney test' for the comparison of the means μ_X and μ_Y of two independent random variables **X** and **Y** (see [6.6.11]). The null hypothesis is either two-sided, H_0: $\mu_X = \mu_Y$, or one-sided, H_0: $\mu_X \geq \mu_Y$. The distributions of **X** and **Y** are assumed to be identical apart from their means. Samples $\{\mathbf{x}_1, \ldots, \mathbf{x}_{n_X}\}$ and $\{\mathbf{y}_1, \ldots, \mathbf{y}_{n_Y}\}$ are assumed to consist of n_X and n_Y independent realizations of **X** and **Y**, respectively.

The test statistic S is the sum of the ranks of $\mathbf{x}_1, \ldots, \mathbf{x}_{n_X}$ in the combined sample $\{\mathbf{x}_1, \ldots, \mathbf{x}_{n_X}, \mathbf{y}_1, \ldots, \mathbf{y}_{n_Y}\}$. The jth observation $\mathbf{x}_j$ has rank $R_j = k$ if it is the kth smallest observation in the combined sample. With this definition, $S = \sum_{j=1}^{n_X} R_j$.

The following tables give critical thresholds $\kappa_{\tilde{p}}$ for testing

- the two-sided null hypothesis at the $(1 - \tilde{p}) \times 100\%$ significance level; H_0: $\mu_X = \mu_Y$ is rejected when $S < \kappa_{(1-\tilde{p})/2}$ or $S > S_{max} - \kappa_{(1+\tilde{p})/2}$, or
- the one-sided null hypothesis at the $(1 - \tilde{p})/2 \times 100\%$ significance level; H_0: $\mu_X \geq \mu_Y$ is rejected when $S < \kappa_{1-\tilde{p}}$.

The maximum possible value of S is $S_{max} = n_X(n_X + 2n_Y + 1)/2$. The smallest possible value of S is $n_X(n_X + 1)/2$.

Critical values appropriate for one- and two-sided tests at a variety of significance levels are listed. The critical values were determined by Monte Carlo simulation, with 10 000 independent trials for each combination of n_X and n_Y. The sample values for each trial were generated from the $\mathcal{U}(0, 1)$ distribution. Asterisks indicate that the null hypothesis can never be rejected with the given combination of sample sizes and significance level.

For large sample sizes, the approximation (6.37)

$$\kappa_{\tilde{p}} \approx n_X(n_X + n_Y + 1)/2 - Z_\alpha\sqrt{n_X n_Y(n_X + n_Y + 1)/12}$$

is useful, where Z_α is the α-quantile of the normal distribution (for numerical values see Appendix D) with $\alpha = \tilde{p}$ in the case of the one-sided test and $\alpha = (1 + \tilde{p})/2$ in the case of the two-sided test.

Significance level: 20% for a two-sided test
10% for a one-sided test

n_X \ n_Y	1	2	3	4	5	6	7	8	9	10	11	12	13	14
2	*	*	4	4	5	5	5	6	6	7	7	8	8	9
3	*	7	8	8	9	10	11	12	12	13	14	15	16	16
4	*	11	12	13	15	16	17	18	19	21	22	23	24	26
5	*	17	18	20	21	23	24	26	28	29	31	33	34	36
6	*	23	25	27	29	31	33	35	37	39	41	43	45	47
7	*	30	33	35	37	40	42	45	47	50	52	54	57	60
8	*	39	42	44	47	50	53	55	59	61	64	67	70	73
9	46	48	51	55	58	61	64	67	71	74	77	81	84	88
10	56	59	62	66	69	73	77	80	84	88	92	95	99	103
11	67	70	74	78	82	86	90	94	99	103	107	111	115	119
12	79	82	87	91	96	100	105	109	114	118	123	128	132	137
13	92	96	101	105	110	115	120	125	130	135	140	146	150	155
14	106	110	116	121	126	131	137	142	147	153	159	164	169	175
15	121	126	131	137	143	148	154	160	166	172	177	183	190	195
16	137	142	148	154	160	166	173	179	185	192	198	204	210	217
17	154	160	166	173	179	186	192	199	206	212	219	226	233	240
18	172	178	185	192	199	205	213	220	227	234	242	249	256	263
19	191	198	205	212	219	227	234	242	249	257	265	272	280	288
20	212	218	226	233	241	249	256	264	273	280	289	297	305	313
21	233	239	247	255	263	272	280	288	297	305	315	323	331	339
22	255	262	270	279	287	296	305	313	322	332	340	348	358	367
23	278	285	294	303	312	321	330	339	349	358	368	377	386	395
24	302	310	319	328	338	347	356	367	376	386	396	406	416	425
25	327	335	345	354	364	374	384	394	404	414	425	436	445	456
26	353	362	372	382	392	402	412	423	433	444	456	466	477	488
27	380	389	399	410	421	431	442	454	464	475	486	498	509	521
28	408	418	428	439	450	462	473	485	496	507	519	530	542	554
29	438	447	458	469	480	492	504	516	528	540	552	564	576	587
30	468	478	489	500	513	524	536	548	562	573	587	599	611	623
31	499	509	521	533	545	558	570	582	595	609	622	634	647	660
32	531	542	554	566	579	592	605	618	631	643	658	670	685	698
33	564	574	588	601	613	627	640	654	667	681	695	709	722	737
34	598	609	622	636	649	663	676	690	705	720	733	748	762	776
35	633	645	658	672	686	700	714	728	743	757	772	787	802	817
36	669	682	695	709	723	738	753	767	783	798	813	828	843	858
37	706	719	732	748	762	777	792	807	822	839	854	869	885	900
38	744	757	772	787	801	817	832	849	864	881	896	912	928	945
39	784	797	811	827	842	858	874	890	906	923	940	956	972	990
40	823	837	853	868	884	901	917	933	950	967	985	1000	1018	1036
41	865	878	894	911	927	943	960	977	995	1012	1029	1046	1064	1081
42	907	921	937	954	971	988	1005	1022	1040	1058	1076	1094	1110	1128
43	950	964	981	998	1015	1032	1050	1068	1085	1104	1122	1141	1159	1178
44	994	1010	1026	1043	1062	1078	1096	1116	1134	1152	1171	1190	1208	1226
45	1039	1054	1072	1089	1108	1126	1144	1163	1182	1201	1220	1239	1258	1279
46	1085	1100	1119	1137	1155	1174	1193	1211	1231	1251	1270	1288	1310	1329
47	1132	1148	1166	1185	1204	1222	1242	1261	1283	1302	1321	1340	1362	1382
48	1180	1196	1215	1234	1253	1273	1293	1313	1333	1353	1373	1393	1414	1437
49	1229	1246	1265	1284	1304	1324	1344	1365	1385	1407	1426	1447	1468	1490
50	1280	1296	1315	1335	1356	1377	1397	1418	1439	1459	1483	1503	1524	1545

Significance level: 10% for a two-sided test
5% for a one-sided test

n_X \ n_Y	1	2	3	4	5	6	7	8	9	10	11	12	13	14
2	*	*	*	*	4	4	4	5	5	5	5	6	6	7
3	*	*	7	7	8	9	9	10	11	11	12	12	13	14
4	*	*	11	12	13	14	15	16	17	18	19	20	21	22
5	*	16	17	18	20	21	22	24	25	26	28	29	31	32
6	*	22	24	25	27	29	30	32	34	36	38	39	41	43
7	*	29	31	33	35	37	40	42	44	46	48	50	53	55
8	*	38	40	42	45	47	50	52	55	57	60	63	65	68
9	*	47	49	52	55	58	61	64	67	70	73	76	79	82
10	*	57	60	63	66	70	73	76	80	83	87	90	93	97
11	*	68	72	75	79	83	86	90	94	98	102	105	110	113
12	*	81	84	88	92	96	100	105	109	113	117	121	126	130
13	*	94	98	102	107	111	116	120	125	129	134	139	144	148
14	*	109	113	117	122	127	132	137	142	147	152	157	162	167
15	*	124	129	134	138	144	149	154	160	165	170	176	182	187
16	*	140	145	150	156	162	167	173	179	185	190	197	202	208
17	*	157	163	168	174	181	187	193	199	205	211	218	224	231
18	*	176	182	187	194	200	207	213	220	227	233	240	247	254
19	*	195	201	208	214	221	228	235	242	249	256	263	271	277
20	211	215	222	229	236	243	250	257	266	272	280	288	296	303
21	232	237	243	251	258	266	273	281	289	297	305	313	321	329
22	254	259	266	274	282	289	298	306	314	324	330	339	348	356
23	277	282	290	298	306	314	323	332	340	349	358	367	376	384
24	301	307	315	323	332	340	349	358	368	376	385	395	404	413
25	326	332	340	349	358	367	376	385	395	404	414	425	434	444
26	352	358	367	376	385	395	404	414	424	434	444	455	464	475
27	379	385	394	404	413	424	434	444	454	465	475	486	497	508
28	407	414	423	433	443	454	465	475	486	496	508	518	529	541
29	436	443	453	463	473	484	495	507	518	529	540	552	563	573
30	466	473	483	494	506	516	527	539	551	562	574	586	598	609
31	497	505	515	527	538	549	561	572	584	597	609	621	633	646
32	529	537	548	559	571	583	595	608	620	631	645	657	672	682
33	562	570	582	594	605	618	630	642	655	668	682	695	707	721
34	596	605	616	628	640	653	666	678	694	707	719	732	746	761
35	631	640	652	665	677	690	703	716	731	744	758	772	787	801
36	667	676	689	701	715	728	742	755	770	785	798	813	827	841
37	704	714	726	740	753	767	781	795	810	825	840	853	868	884
38	742	752	765	779	793	807	821	836	851	867	882	896	911	928
39	781	791	805	818	833	848	862	878	893	909	924	940	955	973
40	821	832	845	860	874	889	905	919	936	952	970	983	1001	1018
41	863	873	887	902	916	932	949	964	980	996	1014	1029	1046	1063
42	905	915	930	945	961	976	992	1008	1025	1043	1060	1077	1092	1110
43	948	959	973	989	1004	1020	1037	1054	1071	1089	1105	1123	1142	1158
44	992	1003	1018	1034	1051	1067	1084	1102	1118	1136	1155	1172	1189	1206
45	1037	1048	1064	1079	1097	1115	1131	1149	1167	1185	1203	1221	1240	1258
46	1083	1094	1110	1127	1144	1161	1180	1197	1215	1234	1254	1270	1291	1310
47	1130	1142	1158	1175	1193	1210	1229	1247	1267	1286	1302	1321	1341	1361
48	1178	1190	1206	1224	1242	1260	1279	1298	1318	1335	1355	1375	1393	1414
49	1227	1240	1256	1274	1292	1311	1329	1350	1369	1390	1407	1428	1448	1468
50	1277	1290	1306	1325	1343	1363	1382	1402	1421	1441	1463	1483	1503	1523

Significance level: 5% for a two-sided test
2.5% for a one-sided test

n_X \ n_Y	1	2	3	4	5	6	7	8	9	10	11	12	13	14
2	*	*	*	*	*	*	*	4	4	4	4	5	5	5
3	*	*	*	*	7	8	8	9	9	10	10	10	11	12
4	*	*	*	11	12	13	14	14	15	16	17	18	19	20
5	*	*	16	17	18	19	21	22	23	24	26	27	28	29
6	*	*	23	24	25	27	28	30	32	33	35	36	38	40
7	*	*	30	32	34	35	37	39	41	43	45	47	49	51
8	*	37	39	41	43	45	47	49	52	54	57	59	61	64
9	*	46	48	51	53	55	58	61	63	66	69	71	74	77
10	*	56	58	61	64	67	70	73	76	79	82	85	89	92
11	*	67	70	73	76	80	83	87	90	93	97	100	104	107
12	*	80	82	86	90	93	97	101	105	108	112	117	120	124
13	*	93	96	100	104	108	112	116	120	125	129	133	137	142
14	*	107	111	115	119	123	128	133	137	142	146	151	156	161
15	*	122	126	131	135	141	145	150	155	159	165	169	176	180
16	*	138	143	147	153	158	163	168	174	179	185	190	195	201
17	*	156	160	165	171	176	182	187	194	199	205	211	217	223
18	*	174	179	184	190	196	202	208	215	220	227	232	240	245
19	*	193	198	204	210	216	222	229	236	242	249	256	263	269
20	*	213	219	225	232	238	244	252	259	266	273	280	287	295
21	*	235	240	247	253	261	268	275	283	289	297	305	312	321
22	*	257	263	270	277	284	292	299	307	316	323	330	339	347
23	*	280	287	294	301	309	317	325	333	342	350	358	366	374
24	*	304	311	318	327	335	343	352	360	368	377	387	395	403
25	*	330	337	345	352	361	370	378	387	396	405	415	424	433
26	*	356	363	371	380	389	398	407	416	426	435	444	454	464
27	*	383	390	399	408	417	427	437	447	456	466	476	487	496
28	*	411	419	428	438	448	457	468	477	487	497	507	519	529
29	*	441	448	458	468	478	488	499	509	519	530	541	553	561
30	*	471	479	489	499	509	520	530	542	552	564	575	587	597
31	*	502	511	521	531	541	553	563	575	586	599	610	622	634
32	*	534	544	553	564	575	587	599	610	621	634	645	659	671
33	*	567	577	588	598	610	622	634	645	657	671	684	695	709
34	*	601	611	622	634	645	657	669	683	696	708	720	733	747
35	*	637	647	658	670	682	695	707	721	732	746	759	774	787
36	*	673	683	695	707	720	732	746	760	773	787	800	814	827
37	*	711	720	733	745	758	772	785	799	814	827	840	855	870
38	*	748	759	772	786	798	813	826	840	854	870	882	897	913
39	*	787	799	811	825	839	852	867	882	897	911	925	940	958
40	*	828	840	853	866	880	895	908	924	940	956	969	987	1003
41	862	869	880	894	908	923	939	953	967	984	1000	1014	1031	1047
42	904	911	924	938	952	966	982	996	1012	1030	1045	1063	1077	1093
43	947	955	967	980	996	1011	1027	1041	1059	1075	1091	1108	1126	1142
44	991	999	1011	1026	1042	1057	1073	1090	1105	1123	1141	1157	1173	1191
45	1036	1043	1057	1071	1088	1105	1120	1136	1154	1171	1188	1205	1224	1241
46	1082	1090	1103	1119	1134	1151	1168	1185	1203	1220	1239	1254	1273	1292
47	1129	1137	1151	1167	1183	1200	1218	1234	1252	1272	1287	1305	1324	1344
48	1177	1186	1199	1215	1232	1250	1267	1285	1304	1319	1340	1358	1376	1395
49	1226	1235	1249	1265	1282	1300	1316	1337	1354	1374	1391	1410	1431	1450
50	1276	1285	1299	1316	1334	1351	1369	1388	1406	1426	1447	1465	1485	1504

Significance level: 2% for a two-sided test
1% for a one-sided test

n_X \ n_Y	1	2	3	4	5	6	7	8	9	10	11	12	13	14
2	*	*	*	*	*	*	*	*	*	*	*	*	4	4
3	*	*	*	*	*	*	7	7	8	8	8	9	9	10
4	*	*	*	*	11	12	12	13	14	14	15	16	16	17
5	*	*	*	16	17	18	19	20	20	21	23	23	24	26
6	*	*	*	23	24	25	26	27	29	30	31	33	34	36
7	*	*	29	30	32	33	35	36	38	40	41	43	45	47
8	*	*	37	39	41	42	45	46	48	50	52	54	56	59
9	*	*	47	49	51	53	55	57	60	62	65	67	69	72
10	*	*	57	59	61	64	67	69	72	75	78	80	84	86
11	*	*	68	71	74	77	79	83	86	89	92	95	98	101
12	*	*	81	83	86	90	93	96	100	103	107	110	114	118
13	*	92	94	97	101	104	108	111	115	119	123	127	130	135
14	*	106	108	112	116	119	123	127	132	136	141	144	149	153
15	*	121	124	127	132	137	140	144	150	154	158	162	168	172
16	*	137	140	144	148	153	157	163	168	173	178	183	187	193
17	*	154	158	162	167	172	177	181	188	193	197	203	209	214
18	*	172	176	180	186	190	197	202	208	213	218	225	232	236
19	*	191	195	200	206	212	217	222	229	236	241	248	254	259
20	*	212	216	221	227	233	238	245	252	258	264	271	278	285
21	*	233	237	243	248	255	262	268	275	280	289	296	302	309
22	*	255	259	266	272	278	286	292	299	308	315	320	329	336
23	*	278	283	289	296	303	310	318	325	333	341	349	356	363
24	*	302	307	314	321	328	336	344	352	359	367	375	383	390
25	*	327	333	340	346	354	363	370	377	387	395	404	413	420
26	*	353	359	365	374	382	390	399	408	417	425	433	443	452
27	*	380	387	394	402	410	419	427	437	446	455	465	475	482
28	*	409	415	422	431	441	449	459	468	476	486	497	506	517
29	*	438	444	453	462	470	480	489	498	509	519	528	539	548
30	*	468	475	483	492	502	511	521	531	542	552	563	573	584
31	*	499	507	515	524	533	544	554	565	576	587	598	609	620
32	*	531	539	547	557	567	578	589	599	610	620	631	644	656
33	*	564	571	581	590	602	613	623	635	646	658	669	680	693
34	*	599	606	616	626	636	648	659	671	684	694	707	718	731
35	*	634	642	652	662	673	685	696	707	720	733	745	760	771
36	*	670	678	688	699	712	722	735	748	759	773	785	798	813
37	*	707	716	726	738	750	762	774	787	802	815	827	840	776
38	*	745	754	764	777	788	804	815	827	841	856	868	882	897
39	*	784	793	804	817	829	840	856	869	884	896	910	925	938
40	*	824	834	845	858	870	884	896	910	927	941	954	970	984
41	*	866	874	887	900	913	927	941	955	969	984	998	1014	1029
42	*	908	918	931	943	956	970	985	999	1015	1029	1047	1059	1074
43	*	951	961	972	986	1001	1015	1028	1044	1059	1074	1091	1109	1125
44	*	995	1005	1017	1032	1046	1061	1077	1092	1107	1123	1141	1155	1172
45	*	1040	1050	1062	1077	1093	1108	1122	1138	1156	1172	1187	1206	1221
46	*	1086	1097	1110	1124	1140	1154	1170	1189	1203	1220	1236	1254	1272
47	*	1133	1145	1158	1173	1189	1205	1221	1237	1255	1271	1286	1305	1325
48	*	1181	1193	1206	1221	1238	1254	1270	1289	1304	1322	1338	1356	1375
49	*	1231	1242	1256	1271	1287	1303	1321	1340	1356	1373	1391	1410	1430
50	*	1281	1291	1307	1322	1338	1355	1371	1390	1409	1426	1445	1465	1485

Significance level: 0.2% for a two-sided test
0.1% for a one-sided test

n_X \ n_Y	1	2	3	4	5	6	7	8	9	10	11	12	13	14
2	*	*	*	*	*	*	*	*	*	*	*	*	*	*
3	*	*	*	*	*	*	*	*	*	*	*	*	*	*
4	*	*	*	*	*	*	*	*	11	11	11	12	12	12
5	*	*	*	*	*	*	*	16	16	17	18	18	19	20
6	*	*	*	*	*	*	22	23	24	25	25	27	27	29
7	*	*	*	*	*	29	31	31	32	33	36	35	37	38
8	*	*	*	*	37	38	40	41	42	43	45	46	48	49
9	*	*	*	*	47	48	51	50	54	54	57	57	60	62
10	*	*	*	56	57	59	61	62	63	67	68	71	73	74
11	*	*	*	67	68	71	73	75	76	80	81	84	87	89
12	*	*	*	79	81	84	85	88	91	94	97	98	102	105
13	*	*	*	92	94	97	100	103	106	109	110	114	118	121
14	*	*	*	107	109	112	115	118	122	124	129	133	134	138
15	*	*	*	122	126	129	131	134	139	144	147	147	153	157
16	*	*	137	139	142	145	147	153	155	162	165	168	172	178
17	*	*	154	157	160	163	168	170	177	181	182	189	192	199
18	*	*	172	174	177	183	186	190	197	201	203	209	214	221
19	*	*	191	195	197	201	206	210	215	221	227	232	235	242
20	*	*	211	214	219	223	228	232	239	241	249	253	261	266
21	*	*	233	236	239	244	249	254	259	262	271	279	283	293
22	*	*	255	258	262	268	274	277	286	294	297	303	306	316
23	*	*	278	283	285	292	298	303	308	315	320	327	334	342
24	*	*	303	306	311	315	322	327	336	343	347	353	361	365
25	*	*	328	331	336	343	350	355	361	369	375	383	392	396
26	*	*	354	357	363	369	375	382	392	398	405	413	420	424
27	*	*	381	385	390	396	404	409	418	427	433	445	452	455
28	*	*	409	414	420	428	434	442	450	453	460	474	484	488
29	*	*	437	444	448	454	464	469	480	489	495	499	513	520
30	*	*	468	473	480	488	492	503	511	521	532	543	546	549
31	*	*	500	505	510	517	530	537	546	553	564	574	579	591
32	*	*	531	536	543	554	560	569	577	588	597	605	620	626
33	*	*	565	570	578	584	596	600	614	624	632	642	653	662
34	*	*	599	606	610	623	631	637	651	661	670	680	687	700
35	*	*	634	639	647	655	662	677	684	697	701	717	728	734
36	*	*	670	676	685	697	704	711	725	733	748	761	764	785
37	*	*	707	715	722	732	744	753	762	773	784	796	809	823
38	*	*	746	753	762	773	781	794	802	818	825	836	850	865
39	*	*	785	792	801	810	821	832	845	852	866	883	894	899
40	*	*	825	831	841	853	867	871	886	899	912	923	939	944
41	*	862	865	874	884	891	905	918	927	941	957	966	979	996
42	*	904	909	917	926	937	949	957	971	987	997	1016	1024	1039
43	*	947	953	959	969	980	991	1002	1017	1033	1040	1055	1074	1089
44	*	991	996	1003	1016	1024	1040	1050	1066	1079	1090	1105	1122	1137
45	*	1036	1040	1049	1058	1071	1083	1095	1113	1123	1137	1153	1172	1183
46	*	1082	1087	1095	1106	1116	1129	1141	1155	1173	1186	1198	1219	1232
47	*	1129	1134	1143	1156	1168	1180	1195	1207	1220	1242	1250	1270	1284
48	*	1177	1183	1191	1202	1217	1231	1247	1259	1273	1284	1304	1310	1332
49	*	1226	1231	1240	1251	1264	1277	1290	1311	1327	1334	1356	1375	1386
50	*	1276	1282	1291	1304	1316	1327	1340	1359	1373	1391	1409	1423	1445

J Quantiles of the Squared-ranks Test Statistic

Critical values for the sum of squared-ranks test for the comparison of two variances (see [6.7.4]) computed from samples of size n_X and n_Y. The test statistic is $T = \sum_{i=1}^{n_X} R_i^2$, where R_i is the rank of $|\mathbf{x}_i - \overline{\mathbf{x}}|$ amongst the combined sample of absolute deviations

$$\{|\mathbf{x}_1 - \overline{\mathbf{x}}|, \ldots, |\mathbf{x}_{n_X} - \overline{\mathbf{x}}|, |\mathbf{y}_1 - \overline{\mathbf{y}}|, \ldots, |\mathbf{y}_{n_Y} - \overline{\mathbf{y}}|\}.$$

When n_X or n_Y is greater than 10, the $(1 - \tilde{p})$-critical values can be approximated by

$$T_{(1-\tilde{p})} = \frac{n_Y(N+1)(2N+1)}{6} - Z_{\tilde{p}}\sqrt{\frac{n_X n_Y(N+1)(2N+1)(8N+1)}{180}}.$$

where $Z_{\tilde{p}}$ is the $\tilde{p}$-quantile of the standard normal distribution (Appendix D). After Conover [88, Table A9, pp. 454–455].

		n_X							
n_Y	p	3	4	5	6	7	8	9	10
3	0.005	14	14	14	14	14	14	21	21
	0.010	14	14	14	14	21	21	26	26
	0.025	14	14	21	26	29	30	35	41
	0.050	21	21	26	30	38	42	49	54
	0.100	26	29	35	42	50	59	69	77
	0.900	65	90	117	149	182	221	260	305
	0.950	70	101	129	161	197	238	285	333
	0.975	77	110	138	170	213	257	308	362
	0.990	77	110	149	194	230	285	329	394
	0.995	77	110	149	194	245	302	346	413
4	0.005	30	30	30	39	39	46	50	54
	0.010	30	30	39	46	50	51	62	66
	0.025	30	39	50	54	63	71	78	90
	0.050	39	50	57	66	78	90	102	114
	0.100	50	62	71	85	99	114	130	149
	0.900	111	142	182	222	270	321	375	435
	0.950	119	154	197	246	294	350	413	476
	0.975	126	165	206	255	311	374	439	510
	0.990	126	174	219	270	334	401	470	545
	0.995	126	174	230	281	351	414	494	567
5	0.005	55	55	66	75	79	88	99	110
	0.010	55	66	75	82	90	103	115	127
	0.025	66	79	88	100	114	130	145	162
	0.050	75	88	103	120	135	155	175	195
	0.100	87	103	121	142	163	187	212	239
	0.900	169	214	264	319	379	445	514	591
	0.950	178	228	282	342	410	479	558	639
	0.975	183	235	297	363	433	508	592	680
	0.990	190	246	310	382	459	543	631	727
	0.995	190	255	319	391	478	559	654	754
6	0.005	91	104	115	124	136	152	167	182
	0.010	91	115	124	139	155	175	191	210
	0.025	115	130	143	164	184	208	231	255
	0.050	124	139	164	187	211	239	268	299
	0.100	136	163	187	215	247	280	315	352
	0.900	243	300	364	435	511	592	679	772
	0.950	255	319	386	463	545	634	730	831
	0.975	259	331	406	486	574	670	771	880
	0.990	271	339	424	511	607	706	817	935
	0.995	271	346	431	526	624	731	847	970

		n_X							
n_Y	p	3	4	5	6	7	8	9	10
7	0.005	140	155	172	195	212	235	257	280
	0.010	155	172	191	212	236	260	287	315
	0.025	172	195	217	245	274	305	338	372
	0.050	188	212	240	274	308	344	384	425
	0.100	203	236	271	308	350	394	440	489
	0.900	335	407	487	572	665	764	871	984
	0.950	347	428	515	608	707	814	929	1051
	0.975	356	443	536	635	741	856	979	1108
	0.990	364	456	560	664	779	900	1032	1172
	0.995	371	467	571	683	803	929	1067	1212
8	0.005	204	236	260	284	311	340	368	401
	0.010	221	249	276	309	340	372	408	445
	0.025	249	276	311	345	384	425	468	513
	0.050	268	300	340	381	426	473	524	576
	0.100	285	329	374	423	476	531	590	652
	0.900	447	536	632	735	846	965	1091	1224
	0.950	464	560	664	776	896	1023	1159	1303
	0.975	476	579	689	807	935	1071	1215	1368
	0.990	485	599	716	840	980	1124	1277	1442
	0.995	492	604	731	863	1005	1156	1319	1489
9	0.005	304	325	361	393	429	466	508	549
	0.010	321	349	384	423	464	508	553	601
	0.025	342	380	423	469	517	570	624	682
	0.050	365	406	457	510	567	626	689	755
	0.100	390	444	501	561	625	694	766	843
	0.900	581	689	803	925	1056	1195	1343	1498
	0.950	601	717	840	972	1112	1261	1420	1587
	0.975	615	741	870	1009	1158	1317	1485	1662
	0.990	624	757	900	1049	1209	1377	1556	1745
	0.995	629	769	916	1073	1239	1417	1601	1798
10	0.005	406	448	486	526	573	620	672	725
	0.010	425	470	513	561	613	667	725	785
	0.025	457	505	560	616	677	741	808	879
	0.050	486	539	601	665	734	806	883	963
	0.100	514	580	649	724	801	885	972	1064
	0.900	742	866	1001	1144	1296	1457	1627	1806
	0.950	765	901	1045	1197	1360	1533	1715	1907
	0.975	778	925	1078	1241	1413	1596	1788	1991
	0.990	793	949	1113	1286	1470	1664	1869	2085
	0.995	798	961	1130	1314	1505	1708	1921	2145

K Quantiles of the Spearman Rank Correlation Coefficient

Critical values for a non-parametric test at the $(1 - \tilde{p}) \times 100\%$ significance level of the null hypothesis that two random variables are uncorrelated using the Spearman rank correlation coefficient (8.8) (see [8.2.3]). Approximate quantiles for $n > 30$ are given by $Z_{\tilde{p}}/\sqrt{n-1}$, where $Z_{\tilde{p}}$ is the corresponding quantile of the standard normal distribution (see Appendix D). After Conover [88, Table A10, p. 456].

	p					
n	0.900	0.950	0.975	0.990	0.995	0.999
4	0.8000	0.8000	—	—	—	—
5	0.7000	0.8000	0.9000	0.9000	—	—
6	0.6000	0.7714	0.8286	0.8857	0.9429	—
7	0.5357	0.6789	0.7450	0.8571	0.8929	0.9643
8	0.5000	0.6190	0.7143	0.8095	0.8571	0.8929
9	0.4667	0.5833	0.6833	0.7667	0.5167	0.9000
10	0.4424	0.5515	0.6364	0.7333	0.7818	0.8667
11	0.4182	0.5273	0.6091	0.7000	0.7455	0.8364
12	0.3986	0.4965	0.5804	0.6713	0.7273	0.8182
13	0.3791	0.4780	0.5549	0.6429	0.6978	0.7912
14	0.3626	0.4593	0.5341	0.6220	0.6747	0.7670
15	0.3500	0.4429	0.5179	0.6000	0.6536	0.7464
16	0.3382	0.4265	0.5000	0.5824	0.6324	0.7265
17	0.3260	0.4118	0.4853	0.5637	0.6152	0.7083
18	0.3148	0.3994	0.4716	0.5480	0.5975	0.6904
19	0.3070	0.3895	0.4579	0.5333	0.5825	0.6737
20	0.2977	0.3789	0.4451	0.5203	0.5684	0.6586
21	0.2909	0.3688	0.4351	0.5078	0.5545	0.6455
22	0.2829	0.3597	0.4241	0.4963	0.5426	0.6318
23	0.2767	0.3518	0.4150	0.4852	0.5306	0.6186
24	0.2704	0.3435	0.4061	0.4748	0.5200	0.6070
25	0.2646	0.3362	0.3977	0.4654	0.5100	0.5962
26	0.2588	0.3299	0.3894	0.4564	0.5002	0.5856
27	0.2540	0.3236	0.3822	0.4481	0.4915	0.5757
28	0.2490	0.3175	0.3749	0.4401	0.4828	0.5660
29	0.2443	0.3113	0.3685	0.4320	0.4744	0.5567
30	0.2400	0.3059	0.3620	0.4251	0.4665	0.5479

L Correlations and Probability Statements

In [18.2.6], we derived an expression for the probability of a future event **P** conditional on a forecast **F**. Here we present tables of these conditional probabilities $P(\mathbf{P} > \mathbf{p} \mid \mathbf{F} > \mathbf{f})$ for various correlations ρ_{FP} and for various numbers **p** and **f**. It is assumed that both the forecast **F** and the verifying analysis **P** have been normalized so that they are distributed as standard normal random variables. The headline in each table gives the value of **p**.

$P(\mathbf{P} > 0 \mid \mathbf{F} > \mathbf{f})$
$P(\mathbf{P} \geq 0) = 0.50$

	ρ														
f	0.95	0.90	0.85	0.80	0.75	0.70	0.65	0.60	0.55	0.50	0.45	0.40	0.30	0.20	0.10
0.0	0.91	0.87	0.83	0.81	0.78	0.76	0.74	0.71	0.70	0.68	0.66	0.64	0.61	0.57	0.54
0.2	0.96	0.92	0.88	0.85	0.82	0.80	0.77	0.75	0.72	0.70	0.68	0.66	0.62	0.58	0.55
0.4	0.99	0.95	0.92	0.89	0.86	0.83	0.81	0.78	0.75	0.73	0.71	0.68	0.64	0.60	0.55
0.6	1.00	0.98	0.95	0.92	0.89	0.87	0.84	0.81	0.78	0.76	0.73	0.71	0.66	0.61	0.56
0.8	1.00	0.99	0.97	0.95	0.92	0.90	0.87	0.84	0.81	0.78	0.76	0.73	0.67	0.62	0.56
1.0	1.00	1.00	0.99	0.97	0.95	0.92	0.90	0.87	0.84	0.81	0.78	0.75	0.69	0.63	0.57
1.2	1.00	1.00	0.99	0.98	0.96	0.94	0.92	0.89	0.86	0.84	0.80	0.77	0.71	0.64	0.58
1.4	1.00	1.00	1.00	0.99	0.98	0.96	0.94	0.91	0.89	0.86	0.83	0.80	0.73	0.66	0.58
1.6	1.00	1.00	1.00	0.99	0.99	0.97	0.95	0.93	0.91	0.88	0.85	0.82	0.74	0.67	0.59
1.8	1.00	1.00	1.00	1.00	0.99	0.98	0.97	0.95	0.93	0.90	0.87	0.84	0.76	0.68	0.60
2.0	1.00	1.00	1.00	1.00	1.00	0.99	0.98	0.96	0.94	0.92	0.89	0.85	0.78	0.69	0.60
2.2	1.00	1.00	1.00	1.00	1.00	0.99	0.98	0.97	0.95	0.93	0.90	0.87	0.80	0.71	0.61
2.4	1.00	1.00	1.00	1.00	1.00	1.00	0.99	0.98	0.96	0.94	0.92	0.89	0.81	0.72	0.62
2.6	1.00	1.00	1.00	1.00	1.00	1.00	0.99	0.98	0.97	0.95	0.93	0.90	0.83	0.73	0.62

$P(\mathbf{P} > 0.5 \mid \mathbf{F} > \mathbf{f})$
$P(\mathbf{P} \geq 0.5) = 0.31$

f	ρ 0.95	0.90	0.85	0.80	0.75	0.70	0.65	0.60	0.55	0.50	0.45	0.40	0.30	0.20	0.10
0.0	0.63	0.61	0.59	0.57	0.55	0.53	0.51	0.50	0.48	0.46	0.45	0.43	0.40	0.37	0.35
0.2	0.72	0.69	0.66	0.63	0.60	0.58	0.56	0.53	0.51	0.49	0.47	0.46	0.42	0.38	0.35
0.4	0.82	0.77	0.73	0.69	0.66	0.63	0.60	0.57	0.55	0.52	0.50	0.48	0.44	0.40	0.36
0.6	0.91	0.84	0.79	0.75	0.71	0.68	0.64	0.61	0.58	0.56	0.53	0.50	0.45	0.41	0.36
0.8	0.96	0.91	0.85	0.81	0.77	0.73	0.69	0.66	0.62	0.59	0.56	0.53	0.47	0.42	0.37
1.0	0.99	0.95	0.90	0.86	0.82	0.78	0.74	0.70	0.66	0.63	0.59	0.56	0.49	0.43	0.37
1.2	1.00	0.98	0.94	0.90	0.86	0.82	0.78	0.74	0.70	0.66	0.62	0.58	0.51	0.44	0.38
1.4	1.00	0.99	0.97	0.94	0.90	0.86	0.82	0.78	0.74	0.69	0.65	0.61	0.53	0.46	0.39
1.6	1.00	1.00	0.98	0.96	0.93	0.89	0.85	0.81	0.77	0.73	0.68	0.64	0.55	0.47	0.39
1.8	1.00	1.00	0.99	0.98	0.95	0.92	0.89	0.85	0.80	0.76	0.71	0.67	0.58	0.49	0.40
2.0	1.00	1.00	1.00	0.99	0.97	0.94	0.91	0.88	0.83	0.79	0.74	0.70	0.60	0.50	0.41
2.2	1.00	1.00	1.00	0.99	0.98	0.96	0.93	0.90	0.86	0.82	0.77	0.72	0.62	0.51	0.41
2.4	1.00	1.00	1.00	1.0	0.99	0.97	0.95	0.92	0.89	0.84	0.80	0.75	0.64	0.53	0.42
2.6	1.00	1.00	1.00	1.00	0.99	0.98	0.97	0.94	0.91	0.87	0.82	0.77	0.66	0.54	0.43

$P(\mathbf{P} > 1.0 \mid \mathbf{F} > \mathbf{f})$
$P(\mathbf{P} \geq 1.0) = 0.16$

f	ρ 0.95	0.90	0.85	0.80	0.75	0.70	0.65	0.60	0.55	0.50	0.45	0.40	0.30	0.20	0.10
0.0	0.33	0.33	0.32	0.32	0.31	0.30	0.29	0.28	0.27	0.26	0.25	0.24	0.22	0.20	0.18
0.2	0.39	0.39	0.38	0.36	0.35	0.34	0.33	0.31	0.30	0.29	0.27	0.26	0.24	0.21	0.19
0.4	0.47	0.46	0.44	0.42	0.40	0.38	0.36	0.35	0.33	0.31	0.30	0.28	0.25	0.22	0.19
0.6	0.58	0.55	0.51	0.48	0.46	0.43	0.41	0.38	0.36	0.34	0.32	0.30	0.26	0.23	0.20
0.8	0.70	0.64	0.60	0.56	0.52	0.48	0.45	0.42	0.40	0.37	0.35	0.32	0.28	0.24	0.20
1.0	0.82	0.74	0.68	0.63	0.58	0.54	0.50	0.47	0.44	0.40	0.38	0.35	0.30	0.25	0.20
1.2	0.92	0.83	0.76	0.70	0.65	0.60	0.56	0.52	0.48	0.44	0.41	0.37	0.31	0.26	0.21
1.4	0.97	0.90	0.83	0.77	0.71	0.66	0.61	0.56	0.52	0.48	0.44	0.40	0.33	0.27	0.21
1.6	0.99	0.95	0.89	0.83	0.77	0.72	0.66	0.61	0.56	0.52	0.47	0.43	0.35	0.28	0.22
1.8	1.00	0.98	0.93	0.88	0.83	0.77	0.71	0.66	0.61	0.55	0.51	0.46	0.37	0.29	0.22
2.0	1.00	0.99	0.96	0.92	0.87	0.82	0.76	0.71	0.65	0.59	0.54	0.49	0.39	0.31	0.23
2.2	1.00	1.00	0.98	0.95	0.91	0.86	0.81	0.75	0.69	0.63	0.58	0.52	0.41	0.32	0.23
2.4	1.00	1.00	0.99	0.97	0.94	0.90	0.85	0.79	0.73	0.67	0.61	0.55	0.44	0.33	0.24
2.6	1.00	1.00	1.00	0.98	0.96	0.92	0.88	0.83	0.77	0.71	0.64	0.58	0.46	0.34	0.25

$P(\mathbf{P} > 1.5 \mid \mathbf{F} > \mathbf{f})$
$P(\mathbf{P} \geq 1.5) = 0.0.7$

f	ρ 0.95	0.90	0.85	0.80	0.75	0.70	0.65	0.60	0.55	0.50	0.45	0.40	0.30	0.20	0.10
0.0	0.14	0.14	0.14	0.14	0.14	0.13	0.13	0.13	0.12	0.12	0.12	0.11	0.10	0.09	0.08
0.2	0.17	0.17	0.17	0.16	0.16	0.16	0.15	0.15	0.14	0.13	0.13	0.12	0.11	0.10	0.08
0.4	0.20	0.20	0.20	0.19	0.19	0.18	0.17	0.17	0.16	0.15	0.14	0.13	0.12	0.10	0.08
0.6	0.26	0.25	0.24	0.24	0.22	0.21	0.20	0.19	0.18	0.17	0.16	0.15	0.13	0.11	0.09
0.8	0.33	0.32	0.30	0.29	0.27	0.25	0.23	0.22	0.20	0.19	0.17	0.16	0.13	0.11	0.09
1.0	0.43	0.40	0.37	0.35	0.32	0.30	0.27	0.25	0.23	0.21	0.19	0.18	0.15	0.12	0.09
1.2	0.56	0.51	0.46	0.42	0.38	0.35	0.32	0.29	0.26	0.24	0.22	0.20	0.16	0.12	0.09
1.4	0.71	0.62	0.55	0.50	0.45	0.40	0.37	0.33	0.30	0.27	0.24	0.22	0.17	0.13	0.10
1.6	0.84	0.73	0.65	0.58	0.52	0.47	0.42	0.38	0.34	0.30	0.27	0.24	0.18	0.14	0.10
1.8	0.93	0.83	0.74	0.66	0.59	0.53	0.47	0.42	0.38	0.33	0.30	0.26	0.20	0.15	0.10
2.0	0.98	0.90	0.82	0.74	0.66	0.60	0.53	0.47	0.42	0.37	0.33	0.28	0.21	0.15	0.11
2.2	0.99	0.95	0.88	0.81	0.73	0.66	0.59	0.53	0.47	0.41	0.36	0.31	0.23	0.16	0.11
2.4	1.00	0.98	0.93	0.87	0.79	0.72	0.65	0.58	0.51	0.45	0.39	0.34	0.25	0.17	0.11
2.6	1.00	0.99	0.96	0.91	0.85	0.78	0.70	0.63	0.56	0.49	0.43	0.37	0.26	0.18	0.12

$P(\mathbf{P} > 2.0 \mid \mathbf{F} > \mathbf{f})$
$P(\mathbf{P} \geq 2.0) = 0.02$

f	ρ 0.95	0.90	0.85	0.80	0.75	0.70	0.65	0.60	0.55	0.50	0.45	0.40	0.30	0.20	0.10
0.0	0.05	0.05	0.05	0.05	0.05	0.05	0.05	0.05	0.05	0.04	0.04	0.04	0.04	0.03	0.03
0.2	0.06	0.06	0.06	0.06	0.06	0.06	0.05	0.05	0.05	0.05	0.05	0.05	0.04	0.03	0.03
0.4	0.07	0.07	0.07	0.07	0.07	0.07	0.06	0.06	0.06	0.06	0.05	0.05	0.04	0.04	0.03
0.6	0.09	0.09	0.09	0.09	0.08	0.08	0.08	0.07	0.07	0.07	0.06	0.06	0.05	0.04	0.03
0.8	0.11	0.11	0.11	0.11	0.10	0.10	0.09	0.09	0.08	0.08	0.07	0.06	0.05	0.04	0.03
1.0	0.15	0.15	0.15	0.14	0.13	0.12	0.11	0.10	0.10	0.09	0.08	0.07	0.06	0.04	0.03
1.2	0.21	0.20	0.19	0.18	0.16	0.15	0.14	0.13	0.11	0.10	0.09	0.08	0.06	0.05	0.03
1.4	0.29	0.27	0.25	0.23	0.21	0.19	0.17	0.15	0.13	0.12	0.11	0.09	0.07	0.05	0.04
1.6	0.41	0.37	0.33	0.29	0.26	0.23	0.20	0.18	0.16	0.14	0.12	0.10	0.08	0.06	0.04
1.8	0.57	0.48	0.42	0.36	0.32	0.28	0.24	0.21	0.19	0.16	0.14	0.12	0.09	0.06	0.04
2.0	0.73	0.61	0.52	0.45	0.39	0.34	0.29	0.25	0.22	0.19	0.16	0.13	0.09	0.06	0.04
2.2	0.86	0.73	0.62	0.53	0.46	0.40	0.34	0.29	0.25	0.21	0.18	0.15	0.10	0.07	0.04
2.4	0.94	0.83	0.72	0.62	0.54	0.46	0.40	0.34	0.29	0.24	0.20	0.17	0.11	0.07	0.04
2.6	0.98	0.90	0.81	0.71	0.62	0.53	0.46	0.39	0.33	0.28	0.23	0.19	0.12	0.08	0.04

$P(\mathbf{P} > 2.5 \mid \mathbf{F} > \mathbf{f})$
$P(\mathbf{P} \geq 2.5) = 0.01$

f	ρ 0.95	0.90	0.85	0.80	0.75	0.70	0.65	0.60	0.55	0.50	0.45	0.40	0.30	0.20	0.10
0.0	0.01	0.01	0.01	0.01	0.01	0.01	0.01	0.01	0.01	0.01	0.01	0.01	0.01	0.01	0.01
0.2	0.02	0.02	0.02	0.02	0.02	0.02	0.02	0.02	0.01	0.01	0.01	0.01	0.01	0.01	0.01
0.4	0.02	0.02	0.02	0.02	0.02	0.02	0.02	0.02	0.02	0.02	0.02	0.02	0.01	0.01	0.01
0.6	0.02	0.02	0.02	0.02	0.02	0.02	0.02	0.02	0.02	0.02	0.02	0.02	0.01	0.01	0.01
0.8	0.03	0.03	0.03	0.03	0.03	0.03	0.03	0.03	0.03	0.02	0.02	0.02	0.02	0.01	0.01
1.0	0.04	0.04	0.04	0.04	0.04	0.04	0.04	0.03	0.03	0.03	0.03	0.02	0.02	0.01	0.01
1.2	0.06	0.06	0.06	0.05	0.05	0.05	0.05	0.04	0.04	0.03	0.03	0.03	0.02	0.01	0.01
1.4	0.08	0.08	0.08	0.07	0.07	0.06	0.06	0.05	0.05	0.04	0.04	0.03	0.02	0.02	0.01
1.6	0.12	0.12	0.11	0.10	0.09	0.08	0.07	0.07	0.06	0.05	0.04	0.04	0.03	0.02	0.01
1.8	0.18	0.17	0.16	0.14	0.12	0.11	0.09	0.08	0.07	0.06	0.05	0.04	0.03	0.02	0.01
2.0	0.28	0.25	0.22	0.19	0.16	0.14	0.12	0.10	0.09	0.07	0.06	0.05	0.03	0.02	0.01
2.2	0.42	0.35	0.29	0.25	0.21	0.18	0.15	0.13	0.10	0.09	0.07	0.06	0.04	0.02	0.01
2.4	0.59	0.47	0.39	0.32	0.27	0.22	0.19	0.15	0.13	0.10	0.08	0.07	0.04	0.02	0.01
2.6	0.76	0.60	0.49	0.41	0.34	0.28	0.23	0.19	0.15	0.12	0.10	0.08	0.05	0.03	0.01

M Some Proofs of Theorems and Equations

In this Appendix we have put together the proofs of some theorems given in main part of this monograph. The proofs will be of little interest for most readers, and they are given here for the sake of completeness.

Proof of Theorem [11.1.10]

The existence and uniqueness of an AR(p) process that satisfies (11.11) are results of the invertibility of the matrix $L_p[\vec{\gamma}]$ used with the Yule–Walker equations (11.2).

For the proof of the recursive formulae (11.12, 11.13) we resort to the technique of 'complete induction.' In the first step we show that (11.13) holds for $p = 1$ and 2, and that (11.12) holds for $p = 2$. In the second step, the 'induction step,' we show that we may infer from the validity of (11.13, 11.12) for $k \leq p$ the validity of (11.13, 11.12) for $k = p + 1$.

For $p = 1$ the trivial solution of the Yule–Walker equations is $\alpha_1^1 = \rho(1)/\rho(0)$ which is (11.13) for $p = 1$ with $\rho_1(1) = \rho(1)$ after (11.11). For $p = 2$ the Yule–Walker equations are

$$\alpha_1^2 + \rho_2(1)\alpha_2^2 = \rho_1(1) \tag{M.1}$$

$$\rho_2(1)\alpha_1^2 + \alpha_2^2 = \rho_2(2). \tag{M.2}$$

If we insert into (M.1) $\alpha_1^1 = \rho(1)$ we find $\alpha_1^2 = \alpha_1^1 - \alpha_2^2\alpha_1^1$ which is just the equation (11.12) for $p = 2$. From (M.2) we infer the validity of (M.3) for $p = 2$.

In the 'induction step' we show that we solve the Yule–Walker equations $L_{p+1}(\vec{\alpha}^{p+1}) = \vec{\gamma}_{p+1}$ with the process parameters $\vec{\alpha}^{\,p+1}$ as inferred through (11.12) and (11.13) from the set of process parameters $\vec{\alpha}^{\,p}$, which satisfy the Yule–Walker equations $L_p(\vec{\alpha}^p) = \vec{\gamma}_p$. After division with $\gamma(0)$ these Yule–Walker equations are written in terms of the auto-correlation function ρ as

$$\sum_{k=1}^{p} \alpha_k^p \rho(k-j) - \rho(j) = 0, \tag{M.3}$$

for $j = 1 \dots p$. After replacement of j by $p - j$ and relabelling $p + 1 - i = k$ and the exploitation of $\rho(i) = \rho(-i)$:

$$\sum_{i=1}^{p} \alpha_{p-i}^p \rho(i-j) - \rho(p+1-j) = 0.$$

After multiplication with $\alpha_{p+1}^{p+1} \neq 0$ and subtraction of (M.3):

$$\begin{aligned} 0 &= \sum_{k=1}^{p} \alpha_k^p \rho(k-j) - \rho(j) - \alpha_{p+1}^{p+1}\left[\sum_{i=1}^{p} \alpha_{p-i}^p \rho(i-j) - \rho(p+1-j)\right] \\ &= \sum_{k=1}^{p}\left[\alpha_k^p - \alpha_{p+1}^{p+1}\alpha_{p-k}^p\right]\rho(k-j) + \alpha_{p+1}^{p+1}\rho(p+1-j) - \rho(j) \end{aligned}$$

the application of (11.12) yields the first $j \dots p$ lines of the Yule–Walker equations $L_{p+1}[\vec{\alpha}^{p+1}] = \vec{\gamma}_{p+1}$:

$$\sum_{k=1}^{p} \alpha_k^{p+1} \rho(k-j) + \alpha_{p+1}^{p+1}\rho(p+1-j) = \rho(j).$$

To get the last line of the Yule–Walker equations, with row index $p+1$, (11.13) is rewritten such that

$$\alpha_{p+1}^{p+1}\left[1-\sum_{k=1}^{p}\alpha_{p+1-k}^{p}\rho(p+1-k)\right]+\sum_{k=1}^{p}\alpha_k^p\rho(p+1-k)=\rho(p+1).$$

Then

$$\sum_{k=1}^{p}\left[\alpha_k^p-\alpha_{p+1}^{p+1}\alpha_{p+1-k}^{p}\right]\rho(p+1-k)+\alpha_{p+1}^{p+1}\rho(0)=\rho(p+1)$$

so that, by means of (11.12),

$$\sum_{k=1}^{p}\alpha_k^{p+1}\rho(k-(p+1))+\alpha_{p+1}^{p+1}\rho(p+1-(p+1))=\rho(p+1),$$

which is just the $(p+1)$th line of the Yule–Walker equations for the AR$(p+1)$ process.

Proof of [13.2.5]

Let us assume $\mathcal{X}^{\mathrm{T}*}\mathcal{X}\vec{e}=\lambda\vec{e}$. Then

$$\lambda\mathcal{X}\vec{e}=\mathcal{X}(\mathcal{X}^{\mathrm{T}*}\mathcal{X})\vec{e}=(\mathcal{X}\mathcal{X}^{\mathrm{T}*})(\mathcal{X}\vec{e})$$

that is, $\mathcal{X}\vec{e}$ is an eigenvector of $\mathcal{X}\mathcal{X}^{\mathrm{T}*}$ if $\mathcal{X}\vec{e}\neq 0$. $\mathcal{X}\vec{e}=0$ would imply $\mathcal{X}^{\mathrm{T}*}\mathcal{X}\vec{e}=0$ and thus $\lambda\vec{e}=0$ contradicting the assumption of $\lambda\neq 0$.

Let us now assume that $\vec{e}^{\,k}$ and $\vec{e}^{\,j}$ are two linearly independent eigenvectors to the same eigenvalue. Then it has to be shown that $\mathcal{X}\vec{e}^{\,k}$ and $\mathcal{X}\vec{e}^{\,j}$, $j\neq k$ are linearly independent as well. If α_j,α_k are two numbers with $\alpha_k\mathcal{X}\vec{e}^{\,k}+\alpha_j\mathcal{X}\vec{e}^{\,j}=0$, then $0=\alpha_k\mathcal{X}^{\mathrm{T}*}\mathcal{X}\vec{e}^{\,k}+\alpha_j\mathcal{X}^{\mathrm{T}*}\mathcal{X}\vec{e}^{\,j}=\lambda(\alpha_k\vec{e}^{\,k}+\alpha_j\vec{e}^{\,j})$ since both eigenvectors belong to the same eigenvalue. Since $\lambda\neq 0:\alpha_k\vec{e}^{\,j}+\alpha_j\vec{e}^{\,j}=0$. Since two eigenvectors $\vec{e}^{\,k}$ and $\vec{e}^{\,j}$ are linearly independent it follows that $\alpha_k=\alpha_j=0$ so that $\mathcal{X}\vec{e}^{\,k}$ and $\mathcal{X}\vec{e}^{\,j}$ are linearly independent.

Proof of Theorem [14.4.5]

We first restate the theorem: *For any random vectors* $\vec{\mathbf{Y}}$ *of dimension* m_Y *and* $\vec{\mathbf{X}}$ *of dimension* m_X*, there exists an orthonormal transformation* $\mathcal{A}$ *and a non-singular transformation* $\mathcal{B}$ *such that*

$$\Sigma_{BX,BX}=\mathcal{I}\qquad\text{(M.4)}$$

$$\Sigma_{AY,BX}=\mathcal{D}\qquad\text{(M.5)}$$

where $\mathcal{D}$ *is an* $m_Y\times m_X$ *matrix for which all entries are zero except for non-negative diagonal elements* $d_{jj}=\sqrt{\lambda_j}$, $j\leq\min(m_X,m_Y)$.

The theorem is proved in two steps. First, we derive two eigen-equations for the matrices $\mathcal{A}$ and $\mathcal{B}$ and a linear link between these two matrices as necessary conditions. In the second step, we show that the solutions of the eigen-equations satisfy equations (M.4)(M.5).

Let us assume that we have determined two matrices $\mathcal{A}$ and $\mathcal{B}$ that satisfy the theorem. Then equations (M.5)(M.4) may be rewritten as

$$\mathcal{A}^{\mathrm{T}}\Sigma_{YX}\mathcal{B}=\mathcal{D}\qquad\text{(M.6)}$$

$$\Sigma_{XX}^{-1}=\mathcal{B}\mathcal{B}^{\mathrm{T}}.\qquad\text{(M.7)}$$

Multiplying (M.6) with itself leads to

$$\mathcal{A}^{\mathrm{T}}\Sigma_{YX}\mathcal{B}\mathcal{B}^{\mathrm{T}}\Sigma_{XY}\mathcal{A}=\mathcal{A}^{\mathrm{T}}\Sigma_{XY}\Sigma_{XX}^{-1}\Sigma_{XY}\mathcal{A}=\mathcal{D}\mathcal{D}^{\mathrm{T}}\qquad\text{(M.8)}$$

where $\mathcal{D}\mathcal{D}^{\mathrm{T}}$ is a diagonal $m_Y \times m_Y$ matrix with positive entries $d_{jj}^2 = \lambda_j$, $j = \min(m_X, m_Y)$. Since $\mathcal{A}$ is orthonormal, we can multiply (M.8) on the left by $\mathcal{A}$ to obtain the first eigen-equation

$$\Sigma_{XY}\Sigma_{XX}^{-1}\Sigma_{XY}\mathcal{A} = \mathcal{A}\mathcal{D}\mathcal{D}^{\mathrm{T}}. \tag{M.9}$$

That is, the columns of $\mathcal{A}$ satisfy (14.43)

$$\Sigma_{YX}\Sigma_{XX}^{-1}\Sigma_{XY}\vec{a}^{\,j} = \lambda_j\vec{a}^{\,j}. \tag{M.10}$$

Equation (M.10) has $\min(m_X, m_Y)$ positive eigenvalues $\lambda_j = d_{jj}^2$. Beginning again with the transpose of (M.6), we find

$$\mathcal{B}^{\mathrm{T}}\Sigma_{XY}\mathcal{A}\mathcal{A}^{\mathrm{T}}\Sigma_{YX}\mathcal{B} = \mathcal{B}^{\mathrm{T}}\Sigma_{XY}\Sigma_{YX}\mathcal{B} = \mathcal{D}^{\mathrm{T}}\mathcal{D}. \tag{M.11}$$

Re-expressing (M.7) as $\mathcal{B}^{\mathrm{T}} = \mathcal{B}^{-1}\Sigma_{XX}^{-1}$ and substituting $\mathcal{B}^{\mathrm{T}}$ into (M.11), we obtain the second eigen-equation

$$\Sigma_{XX}^{-1}\Sigma_{XY}\Sigma_{YX}\mathcal{B} = \mathcal{D}^{\mathrm{T}}\mathcal{D}\mathcal{B}.$$

That is, the columns of $\mathcal{B}$ satisfy (14.44)

$$\Sigma_{XX}^{-1}\Sigma_{XY}\Sigma_{YX}\vec{b}^{\,j} = \lambda_j\vec{b}^{\,j}. \tag{M.12}$$

This completes the first part of the proof.

We now define matrices $\mathcal{A}$ and $\mathcal{B}$ as the matrices of eigenvectors of $\Sigma_{YX}\Sigma_{XX}^{-1}\Sigma_{XY}$ and $\Sigma_{XX}^{-1}\Sigma_{XY}\Sigma_{YX}$, respectively, and show that $\mathcal{A}$ and $\mathcal{B}$ satisfy the requirements of the theorem.

Since Σ_{XX} is positive-definite symmetric, it may be written as $\Sigma_{XX} = (\Sigma_{XX}^{1/2})^{\mathrm{T}}\Sigma_{XX}^{1/2}$ (see Appendix B). Thus, vector $\vec{b}$ solves (M.12) with eigenvalue λ if and only if

$$\vec{c} = \Sigma_{XX}^{1/2}\vec{b} \tag{M.13}$$

solves the eigen-equation

$$[\mathcal{C}^{\mathrm{T}}\mathcal{C}]\vec{c} = \lambda\vec{c} \tag{M.14}$$

where

$$\mathcal{C} = \Sigma_{YX}\Sigma_{XX}^{-1/2}. \tag{M.15}$$

Since $\mathcal{C}^{\mathrm{T}}\mathcal{C}$ is Hermitian, all of its eigenvalues are non-negative reals, and it has m_X orthonormal eigenvectors. Thus, eigenproblem (M.12) has m_X linearly independent solutions $\vec{b}^{\,j} = \Sigma_{XX}^{-1/2}\vec{c}^{\,j}$ that satisfy (M.4):

$$(\vec{b}^{\,i})^{\mathrm{T}}\Sigma_{XX}\vec{b}^{\,j} = (\vec{c}^{\,i})^{\mathrm{T}}(\Sigma_{XX}^{-1/2})^{\mathrm{T}}\Sigma_{XX}\Sigma_{XX}^{-1/2}\vec{c}^{\,j} = \delta_{ij}.$$

The eigenproblem (M.10) may be written as

$$[\mathcal{C}\mathcal{C}^{\mathrm{T}}]\vec{a} = \lambda\vec{a}, \tag{M.16}$$

which has the same eigenvalues as $\mathcal{C}^{\mathrm{T}}\mathcal{C}$ (see Theorem [13.2.5]). If $\vec{c}$ is an eigenvector of $\mathcal{C}^{\mathrm{T}}\mathcal{C}$ with eigenvalue λ, then

$$\vec{a} = \frac{1}{\sqrt{\lambda}}\mathcal{C}\vec{c} \tag{M.17}$$

is an eigenvector of $\mathcal{C}\mathcal{C}^{\mathrm{T}}$ with the same eigenvalue.

It remains to be shown that these vectors fulfil (M.5). Let r be the number of eigenvectors of $\mathcal{C}^{\mathrm{T}}\mathcal{C}$ and $\mathcal{C}\mathcal{C}^{\mathrm{T}}$ that correspond to nonzero eigenvalues. For all indices j and $i \le r$, we find that

$$\begin{aligned}
(\vec{a}^{\,i})^{\mathrm{T}}\Sigma_{YX}\vec{b}^{\,j} &= \Big(\frac{1}{\sqrt{\lambda_i}}\mathcal{C}\vec{c}^{\,r}\Big)^{\mathrm{T}}\Sigma_{YX}(\Sigma_{XX}^{-1/2}\vec{c}^{\,j} \\
&= \frac{1}{\sqrt{\lambda_i}}(\vec{c}^{\,i})^{\mathrm{T}}\mathcal{C}^{\mathrm{T}}\Sigma_{XY}\Sigma_{XX}^{-1/2}\vec{c}^{\,j} \\
&= \frac{1}{\sqrt{\lambda_i}}(\vec{c}^{\,i})^{\mathrm{T}}\mathcal{C}^{\mathrm{T}}\mathcal{C}\vec{c}^{\,j} = \sqrt{\lambda_j}\,\delta_{ij}.
\end{aligned}$$

When $i > r$

$$(\vec{a}^i)^{\mathrm{T}}\Sigma_{YX}\vec{b}^j = (\vec{a}^i)^{\mathrm{T}}\mathcal{C}\vec{c}^j = 0$$

because $(\vec{a}^i)^{\mathrm{T}}\mathcal{C} = \vec{0}$. We can show this by contradiction. Suppose $(\vec{a}^i)^{\mathrm{T}} \neq \vec{0}$. Then we would have $(\vec{a}^i)^{\mathrm{T}}\mathcal{C}\mathcal{C}^{\mathrm{T}}\vec{a}^i = \|(\vec{a}^i)^{\mathrm{T}}\mathcal{C}\| > 0$, which implies that $\mathcal{C}\mathcal{C}^{\mathrm{T}}\vec{a}^i \neq \vec{0}$. But this contradicts the fact that $\vec{a}^i$ belongs to a zero eigenvalue of $\mathcal{C}\mathcal{C}^{\mathrm{T}}$. This therefore completes the proof that matrices $\mathcal{A}$ and $\mathcal{B}$ satisfy (M.5).

Proof of Equation (16.39)

We prove here equation (16.39), which states that the Hilbert transform of the complex EOF coefficient $\alpha = (\vec{\mathbf{X}} + i\,\vec{\mathbf{X}}^{\mathrm{H}})^{\dagger}$ is equal to the coefficient itself multiplied by $-i$. First note that, if $Y = X + i\,X^{\mathrm{H}}$, then by (16.23) and (16.24)

$$Y^{\mathrm{H}} = (X + i\,X^{\mathrm{H}})^{\mathrm{H}} = x^{\mathrm{H}} - i\,X = (-i)(X + i\,X^{\mathrm{H}}) = -i\,Y. \tag{M.18}$$

By repeatedly using (M.18), we infer that

$$[Y^{\mathrm{H}}]^* = [(-i)Y]^* = i\,Y^* = i/(-i)[Y^*]^{\mathrm{H}} = -[Y^*]^{\mathrm{H}}.$$

Then, with (M.18), we have

$$\alpha^{\mathrm{H}} = \left[\vec{\mathbf{Y}}^{\dagger}\vec{p}\,\right]^{\mathrm{H}} = \left[\vec{\mathbf{Y}}^{\dagger}\right]^{\mathrm{H}}\vec{p} = -[\vec{\mathbf{Y}}^{\mathrm{H}}]^{\dagger}\vec{p} = [i\,\vec{\mathbf{Y}}]^{\dagger}\vec{p} = (-i)\vec{\mathbf{Y}}^{\dagger}\vec{p} = -i\alpha.$$

References

[1] U. Achatz and G. Schmitz. On the closure problem in the reduction of complex atmospheric models by PIPs and EOFs: A comparison for the case of a two-layer model with zonally symmetric forcing. *J. Atmos. Sci.*, 54:2452–2474, 1997.

[2] U. Achatz, G. Schmitz, and K.-M. Greisiger. Principal interaction patterns in baroclinic wave life cycles. *J. Atmos. Sci.*, 52:3201–3213, 1995.

[3] A. Adebar. *Die Welt nördlich und südlich der Elbe*. Ekelacker-Verlag, Baikal-Hüll, 1992.

[4] H. Akaike. Fitting autoregressive models for prediction. *Ann. Inst. Statist. Math.*, 21:243–247, 1969.

[5] H. Akaike. Power spectrum estimation through auto-regressive model fitting. *Ann. Inst. Statist. Math.*, 21:407–419, 1969.

[6] H. Akaike. Information Theory and an Extension of the Maximum Likelihood Principle. In B.N. Petrov and F. Csàki, editors, *Second International Symposium on Information Theory*, pages 267–281. Akademia Kiadó, Budapest, 1973.

[7] H. Akaike. A new look at the statistical model identification. *IEEE Trans. Auto. Control*, 19:716–723, 1974.

[8] M.R. Allen and A.W. Robertson. Distinguishing modulated oscillations from coloured noise in multivariate datasets. *Climate Dynamics*, 12:775–784, 1996.

[9] M.R. Allen and L.A. Smith. Investigating the origins and significance of low-frequency modes of climate variability. *Geophys. Res. Letters*, 21:883–886, 1994.

[10] M.R. Allen and L.A. Smith. Monte Carlo SSA: Detecting irregular oscillations in the presence of coloured noise. *J. Climate*, 9:3373–3404, 1996.

[11] J.L. Anderson and W.F. Stern. Evaluating the potential predictive utility of ensemble forecasts. *J. Climate*, 9:260–269, 1996.

[12] T.W. Anderson. *An Introduction to Multivariate Statistical Analysis*. Wiley & Sons, second edition, 1984. 675pp.

[13] V.L. Anderson and R.A. McLean. *Design of Experiments: A Realistic Approach*. Marcel Dekker, New York, 1974.

[14] C.F. Ansley and R. Kohn. Estimation, filtering and smoothing in state space models with incompletely specified initial conditions. *Ann. Statist.*, 13:1286–1316, 1985.

[15] G. Arfken. *Mathematical Methods for Physicists*. Academic Press, 1985. 985pp.

[16] S.A. Arrhenius. On the influence of carbonic acid in the air upon the temperature of the ground. *Philosophical Magazine and Journal of Science*, 41:237–276, 1896.

[17] H.W. Barker. A parameterization for computing grid-averaged solar fluxes for marine boundary layer clouds. Part I: Methodology and homogeneous biases. *J. Atmos. Sci.*, 53:2289–2303, 1996.

[18] H.W. Barker, B.A. Wielicki, and L. Parker. A parameterization for computing grid-averaged solar fluxes for inhomogeneous marine boundary layer clouds. Part II: Validation using satellite data. *J. Atmos. Sci.*, 53:2304–2316, 1996.

[19] T.P. Barnett. Interaction of the Monsoon and Pacific trade wind system interannual time scale. Part I. *Mon. Wea. Rev.*, 111:756–773, 1983.

[20] T.P. Barnett, M. Latif, N. Graham, M. Flügel, S. Pazan, and W. White. ENSO and ENSO-related predictability. Part I: Prediction of Equatorial Pacific sea surface temperature with a hybrid coupled ocean-atmosphere model. *J. Climate*, 8:1545–1566, 1993.

[21] T.P. Barnett and R. Preisendorfer. Origins and levels of monthly and seasonal forecast skill for United States surface air temperature determined by canonical correlation analysis. *Mon. Wea. Rev.*, 115:1825–1850, 1987.

[22] T.P. Barnett, R.W. Preisendorfer, L.M. Goldstein, and K. Hasselmann. Significance tests for regression model hierarchies. *J. Phys. Oceano.*, 11:1150–1154, 1981.

[23] T.P. Barnett and M.E. Schlesinger. Detecting changes in global climate induced by greenhouse gas. *J. Geophys. Res*, 92:14,772–14,780, 1987.

[24] T.P. Barnett, M.E. Schlesinger, and X. Jiang. On greenhouse gas signal detection strategies. In M.E. Schlesinger, editor, *Greenhouse-Gas-Induced Climate Change: A Critical Appraisal of Simulations and Observations*, pages 537–558. Elsevier Science Publishers B.V., 1991.

[25] A.G. Barnston. Correspondence among the correlation, RMSE, and Heidke forecast verification measures; Refinement of the Heidke score. *Weather and Forecasting*, 7:699–709, 1992.

[26] A.G. Barnston. Linear statistical short-term climate predictive skill in the Northern Hemisphere. *J. Climate*, 7:1513–1564, 1994.

[27] A.G. Barnston and R.E. Livezey. Classification, seasonality and persistence of low-frequency atmospheric circulation patterns. *Mon. Wea. Rev.*, 115:1825–1850, 1987.

[28] A.G. Barnston and C.F. Ropelewski. Prediction of ENSO episodes using canonical correlation analysis. *J. Climate*, 5:1316–1345, 1992.

[29] A.G. Barnston and H.M. van den Dool. A degeneracy in cross-validated skill in regression-based forcasts. *J. Climate*, 6:963–977, 1993.

[30] H.G. Barnston and T.M. Smith. Specification and prediction of global surface temperature and precipitation from global SST using CCA. *J. Climate*, 9:2660–2697, 1996.

[31] M.S. Bartlett. On the theoretical specification of sampling properties of auto-correlated time series. *J. R. Stat. Soc.*, B8:27–41, 1946.

[32] M.S. Bartlett. The general canonical correlation distribution. *Ann. Math. Stat.*, 18:1–17, 1947.

[33] M.S. Bartlett. Smoothing periodograms from time series with continuous spectra. *Nature*, 161:686–687, 1948.

[34] M.S. Bartlett. *An Introduction to Stochastic Processes with special reference to methods and applications.* Cambridge University Press, 1955.

[35] D.M. Bates and D.G. Watts. *Nonlinear regression analysis and its applications.* Wiley, New York, 1988.

[36] R.A. Becker, J.M. Chambers, and A.R. Wilks. *The New S Language.* Wadsworth & Brooks/Cole, 1988. 702pp.

[37] T.L. Bell. Optimal weighting of data to detect climate change: Applications to the carbon dioxide problem. *J. Geophys. Res.*, 87:11 161–11 170, 1982.

[38] T.L. Bell. Optimal weighting of data to detect climate change: Application to the carbon dioxide problem. *J. Geophys. Res.*, 87:11 161–11 170, 1982.

[39] T.L. Bell. Theory of optimal weighting of data to detect climate change. *J. Atmos. Sci.*, 43:1694–1710, 1986.

[40] T.L. Bell and R. Suhasini. Principal modes of variation of rain-rate probability distributions. *J. Appl. Meteor.*, 33:1067–1078, 1994.

[41] D.A. Belsley, E. Kuh, and R.E. Welsch. *Regression Diagnostics.* Wiley & Sons, New York, Chichester, Brisbane, Toronto, Singapore, 1980.

[42] L. Bengtsson, M. Botzet, and M. Esch. Hurricane-type vortices in a General Circulation Model. Part I. *Tellus*, 47A:175–196, 1995.

[43] L. Bengtsson, M. Botzet, and M. Esch. Will greenhouse gas-induced warming over the next 50 years lead to higher frequency and greater intensity of hurricanes? *Tellus*, 48A:57–73, 1996.

[44] L. Bengtsson, M. Kanamitsu, P. Kallberg, and S. Uppsala. First GARP Global Experiment 4-dimensional assimilation at ECMWF. *Bull. Am. Meteor. Soc.*, 63:29–43, 1982.

[45] J. Beran. *Statistical Methods for Long Memory Processes.* Chapman & Hall, Boca Raton, Florida, 1994, 326pp.

[46] H.P. Berlage. Schommelingen van de algemene luchtcirculatie met perioden van meer dan een jaar, hun aard en betekenis voor de weersverwachting op langtermijn (Fluctuations of the general atmospheric circulation of more than one year, their nature and prognostic value). Medelingen en Verhandelingen 69, Koninklijk Nederlands Meteorologisch Instituut, 1957. 152pp.

[47] M.L. Blackmon. A climatological spectral study of the 500 mb geopotential height of the Northern Hemisphere. *J. Atmos. Sci.*, 33:1607–1623, 1976.

[48] M.L. Blackmon, J.E. Geisler, and E.J. Pitcher. A general circulation model study of January climate patterns associated with interannual variations of Equatorial Pacific sea surface temperature. *J. Atmos. Sci.*, 40:1410–1425, 1983.

[49] P. Bloomfield. *Fourier Analysis of Time Series: An Introduction.* Wiley, 1976. 258pp.

[50] B. Blumenthal. Predictability of a coupled ocean-atmosphere model. *J. Climate*, 4:766–784, 1991.

[51] G.J. Boer. Modelling the Atmospheric Response to the 1982/83 El Niño. In J.C.J Nihoul, editor, *Coupled Ocean-Atmosphere Models*, pages 7–17. Elsevier Science Publishers B.V., Amsterdam, 1985.

[52] G.J. Boer, N.A. McFarlane, and M. Lazare. Greenhouse gas-induced climate change simulated with the CCC Second-Generation General Circulation Model. *J. Climate*, 5:1045–1077, 1992.

[53] G.J. Boer, N.A. McFarlane, R. Laprise, J.D. Henderson, and J.-P. Blanchet. The Canadian Climate Centre spectral atmospheric general circulation model. *Atmosphere-Ocean*, 22:397–429, 1984.

[54] M. Bohle-Carbonell. Pitfalls in sampling, comments on reliability and suggestions for simulation. *Cont. Shelf Res.*, 12:3–24, 1992.

[55] H. Borgert. Praktische Erprobung der Kanonischen Korrelationsanalyse mit Monte-Carlo-Experimenten und Messdaten. Diplomarbeit am Fachbereich Geowissenschaften der Universität Hamburg, 1990.

[56] H. Böttger and K. Fraedrich. Disturbances in the wavenumber-frequency domain observed along 50°N. *Beitr. Phys. Atmos.*, 53:90–105, 1980.

[57] E. Bouws, D. Jannink, and G.J. Komen. On increasing wave height in the North Atlantic ocean. *Bull. Am. Meteor. Soc.*, 77:2275–2277, 1996.

[58] K.P. Bowman, J. Sacks, and Y.-F. Chang. Design and analysis of numerical experiments. *J. Atmos. Sci.*, 50:1267–1278, 1993.

[59] G.E. Box, W.G. Hunter, and S. Hunter. *Statistics for Experimenters.* Wiley, 1978.

[60] G.E.P. Box and G.M. Jenkins. *Time series analysis, forecasting and control.* Holden-Day, San. Fran., revised edition, 1976. 575pp.

[61] G.W. Branstator. The variability in skill of 72-hour global-scale NMC forecasts. *Mon. Wea. Rev.*, 114:2628–2639, 1986.

[62] L. Breiman. *Statistics: With a View Towards Applications.* Houghton Mifflin, Boston, 1973. 399pp.

[63] L. Breiman, J.H. Friedman, R.A. Olshen, and C.J. Stone. *Classification and Regression Trees.* Wadsworth & Brooks/ Cole Advanced Books and Software, Monterey, California, 1984.

[64] C.S. Bretherton, C. Smith, and J.M. Wallace. Intercomparison of methods for finding coupled patterns in climate data. *J. Climate*, 5:541–560, 1992.

[65] K.R. Briffa. Interpreting high-resolution proxy climate data — The example of dendroclimatology. In H. von Storch and A. Navarra, editors, *Analysis of Climate Variability: Applications of Statistical Techniques*, pages 77–94. Springer Verlag, 1995.

[66] D.R. Brillinger. *Time Series Data Analysis and Theory: Expanded Edition*. Holden Day, 1981. 540pp.

[67] K.H. Brink and R.D. Muench. Circulation in the Point Conception–Santa Barbara Channel region. *J. Geophys. Res.*, 91C:877–895, 1986.

[68] P.J. Brockwell and R.A. Davis. *Time Series: Theory and Methods*. Springer-Verlag, second edition, 1991. 577pp.

[69] J. Bruce. Atlas of rainfall intensity-duration frequency data for Canada. Technical report, AES Climatological Studies No. 8, Downsview (Ontario), 1968.

[70] E. Brückner. *Klimaschankungen seit 1700 nebst Bemerkungen über die Klimaschwankungen der Diluvialzeit*. Geographische Abhandlungen herausgegeben von Prof. Dr. Albrecht Penck in Wien; Wien and Olmütz, E.D. Hölzel, 1890.

[71] N. Brunet, R. Verret, and N. Yacowar. An objective comparison of model output statistics and perfect prog systems in producing numerical weather element forecasts. *Weather and Forecasting*, 3:273–283, 1988.

[72] K. Bryan. A numerical method for the study of the circulation of the world ocean. *J. Comput. Phys.*, 4:347–376, 1969.

[73] J.P. Burg. Maximum entropy spectral analysis. In *37th annual international meeting*. Soc. of Explor. Geophysics, Oklahoma City, OK, 1967. Oct. 31.

[74] J.P. Burg. A new analysis technique for time series data. In *Advanced Study Institute on Signal Processing*. NATO, Enschede, Netherlands, 1968.

[75] G. Bürger. Complex Principal Oscillation Patterns. *J. Climate*, 6:1972–1986, 1993.

[76] W.R. Burrows. Objective guidance for 0–24 hour and 24–48 hour mesoscale forecasts on lake-effect snow using CART. *Weather and Forecasting*, 6:357–378, 1991.

[77] G. Casella. An introduction to empirical Bayesian data analysis. *Amer. Stat.*, 39:83–87, 1985.

[78] J.M. Chambers and T.J. Hastie. *Statistical Models in S*. Wadsworth & Brooks/Cole, 1992. 608pp.

[79] W.L. Chapman, W.J. Welch, W.J. Bowman, K.P. Sacks, and J.E. Walsh. Arctic sea ice variability: Model sensitivities and a multidecadal simulation. *J. Geophys. Res.*, 99C:919–935, 1994.

[80] M. Chelliah and P. Arkin. Large-scale interannual variability of monthly outgoing longwave radiation anomalies over global tropics. *J. Climate*, 5:371–389, 1992.

[81] X. Chen. Sea level changes since early 1920s from the long records of two tidal gauges in Shanghai, China. In Y. Qin and S. Zhao, editors, *Quarternary Coastline Changes in China*, pages 127–141. China Ocean Press Beijing, 1991.

[82] X. Cheng, G. Nitsche, and J.M. Wallace. Robustness of low-frequency circulation patterns derived from EOF and rotated EOF analysis. *J. Climate*, 8:1709–1713, 1995.

[83] S. Cherry. Singular value analysis and canonical correlation analysis. *J. Climate*, 9:2003–2009, 1996.

[84] R.M. Chervin and S.H. Schneider. On determining the statistical significance of climate experiments with general circulation models. *J. Atmos. Sci.*, 33:405–412, 1976.

[85] P.-S. Chu and R.W. Katz. Modeling and forecasting the Southern Oscillation: A time domain approach. *Mon. Wea. Rev.*, 113:1876–1888, 1985.

[86] P.-S. Chu and R.W. Katz. Measures of predictability with applications to the Southern Oscillation. *Mon. Wea. Rev.*, 115:1542–1549, 1987.

[87] W.G. Cochran and G.M. Cox. *Experimental Designs*. Wiley & Sons, New York, second edition, 1957.

[88] W.J. Conover. *Practical Nonparametric Statistics*. Wiley & Sons, New York, Chichester, Brisbane, Toronto, Singapore, second edition, 1980.

[89] N.J. Cook. Towards better estimation of extreme winds. *Journal of Wind Eng. and Industrial Aerodynamics*, 9:295–323, 1981.

[90] R.D. Cook and S. Weisburg. *Residuals and Influence in Regression*. Chapman and Hall, London, 1982.

[91] J.B. Copas. Regression, prediction and shrinkage (with discussion). *J. Royal Stat. Soc. Bull.*, 45:311–354, 1983.

[92] D.R. Cox and D.V. Hinkley. *Theoretical Statistics*. Chapman and Hall, 1974. 511pp.

[93] M. Cox. A primitive equation, three dimensional model of the ocean. Technical report, GFDL Ocean Tech. Report No. 1, Princeton (NJ), 1984.

[94] C.L. Crozier, P.I. Joe, J.W. Scott, H.N. Herscovitch, and T.R. Nichols. The King City operational doppler radar: Development, all-season application and forecasting. *Atmosphere-Ocean*, 29:479–516, 1991.

[95] U. Cubasch. The mean response of the ECMWF global model to the El Niño anomaly in extended range prediction experiments. *Atmosphere-Ocean*, 23:43–66, 1985.

[96] U. Cubasch, K. Hasselmann, H. Höck, E. Maier-Reimer, U. Mikolajewicz, B.D. Santer, and R. Sausen. Time dependent greenhouse warming computations with a coupled ocean-atmosphere model. *Climate Dynamics*, 8:55–69, 1992.

[97] M. Cui, H. von Storch, and E. Zorita. Coastal sea level and the large-scale climate state: A downscaling exercise for the Japanese Islands. *Tellus*, 47A:132–144, 1995.

[98] R. Daley. *Atmospheric Data Analysis*. Cambridge University Press, 1991. Cambridge Atmospheric and Space Science Series 2.

[99] P.J. Daniell. Discussion following "On the theoretical justification and sampling properties of autocorrelated time series" by M.S. Bartlett. *J. Royal Stat. Soc. Suppl.*, 8:27–41, 1946.

[100] F.N. David. *Tables of Ordinates and Probability Integral of the Distribution of the Correlation Coefficient in Small Samples*. The Biometrika Office, London, 1938.

[101] R.E. Davis. Predictability of sea-surface temperature and sea-level pressure anomalies over the North Pacific Ocean. *J. Phys. Oceano.*, 6:249–266, 1976.

[102] R.J. Deland. Travelling planetary waves. *Tellus*, 16:271–273, 1964.

[103] M. Dorn and H. von Storch. Identification of regional persistent patterns through Principal Prediction Patterns. *Beitr. Phys. Atmos.*, 72:105–111, 1999.

[104] N.D. Draper and H. Smith. *Applied Regression Analysis*. Wiley & Sons, New York, Chichester, Brisbane, Toronto, Singapore, second edition, 1981. 709pp.

[105] C. Duchêne and C. Frankignoul. Sensitivity and realism of wind-driven tropical ocean models. *J. Marine Systems*, 1:97–117, 1990.

[106] T.J. Dunkerton. A two-dimensional model of the quasi-biennial oscillation. *J. Atmos. Sci.*, 42:1151–1160, 1975.

[107] J. Durbin and J.S. Watson. Testing for serial correlation in least-squares regression. I. *Biometrika*, 37:409–428, 1950.

[108] J. Durbin and J.S. Watson. Testing for serial correlation in least-squares regression. II. *Biometrika*, 38:159–178, 1951.

[109] J. Durbin and J.S. Watson. Testing for serial correlation in least-squares regression. III. *Biometrika*, 58:1–19, 1971.

[110] W. Ebisuzaki. A Method to Estimate the Statistical Significance of a Correlation when the Data is Serially Correlated. *J. Climate*, 10:2147–2153, 1997.

[111] B. Efron. *The Jackknife, the Bootstrap and other resampling plans*. J.W. Arrowsmith, Ltd., Bristol, England, 1982. 92pp.

[112] Sir W.P. Elderton and N.L. Johnson. *Systems of Frequency Curves*. Cambridge University Press, 1969. 216pp.

[113] K.O. Emery and D.G. Aubrey. *Sea Levels, Land Levels, and Tide Gauges*. Springer Verlag, 1991.

[114] E.S. Epstein. *Statistical Inference and Prediction in Climatology: Bayesian Approach*, volume 42. American Meteorological Society, Boston, 1985. Meteorological Monographs 20.

[115] B. Everitt. *Cluster Analysis*. Halsted, New York, second edition, 1980.

[116] J.P. Finley. Tornado predictions. *Am. Meteor. J.*, 1:85–88, 1884.

[117] G. Fischer and H. von Storch. Klima = langjähriges Mittel? *Meteor. Rdsch.*, 35:152–158, 1982.

[118] D.A. Fisher. Stratigraphic noise in time series derived from ice cores. *Ann. Glaciology*, 7:76–83, 1985.

[119] R.A. Fisher. On the mathematical foundations of theoretical statistics. *Phil. Trans. R. Soc. Lond.*, A222:309–368, 1922.

[120] R.A. Fisher. Theory of statistical estimation. *Proc. Camb. Phil. Soc.*, 22:700–725, 1925.

[121] C.K. Folland, T.R. Karl, and K.YA. Vinnikov. Observed climate variations and change. In J.T. Houghton, G.J. Jenkins, and J.J. Ephraums, editors, *Climate Change: The IPCC Scientific Assessment*, pages 195–238. Cambridge University Press, 1990.

[122] C.K. Folland and D.E. Parker. Observed variations of observed sea surface temperature. In M.E. Schlesinger, editor, *Climate-Ocean Interactions*, pages 23–52. Kluwer Academic Press, North Atlantic Treaty Organization, Commission of the European Communities, 1990.

[123] C.K. Folland and D.P. Rowell. Workshop on simulations of the Climate of the Twentieth Century using GISST, 28–30 November 1994. Climate research technical report 56, Hadley Centre, Bracknell (UK), 1995. 111pp.

[124] K. Fraedrich and H. Böttger. A wavenumber-frequency analysis of the 500 mb geopotential at 50^oN. *J. Atmos. Sci.*, 35:745–750, 1978.

[125] K. Fraedrich and Dümmel. On single station forecasting: The geopotential height, its vertical and time structure and 500 mb ARMA prediction. *Beitr. Phys. Atmos.*, 56:221–239, 1983.

[126] K. Fraedrich and E. Kietzig. Statistical analysis and wavenumber-frequency spectra of the 500 mb geopotential along 50°S. *J. Atmos. Sci.*, 40:1037–1045, 1983.

[127] K. Fraedrich, M. Lutz, and A. Spekat. Statistical analysis of 500 mb geopotential along 50°N: Zonal teleconnections in winter and summer. *Beitr. Phys. Atmos.*, 58:345–360, 1985.

[128] C. Frankignoul. Sea surface temperature anomalies. *Rev. Geophys.*, 23:357–390, 1985.

[129] C. Frankignoul. Stochastic climate models: Where do we stand? In M. Latif, editor, *Strategies for Future Climate Research*, pages 137–160. Max-Planck-Institut für Meteorologie, Bundesstrasse 55, 20146 Hamburg, Germany, 1991.

[130] C. Frankignoul. Climate spectra and stochastic climate models. In H. von Storch and A. Navarra, editors, *Analysis of Climate Variability: Applications of Statistical Techniques*, pages 29–52. Springer Verlag, 1995.

[131] C. Frankignoul. Statistical analysis of GCM output. In H. von Storch and A. Navarra, editors, *Analysis of Climate Variability: Applications of Statistical Techniques*, pages 139–157. Springer Verlag, 1995.

[132] C. Frankignoul, C. Duchêne, and M. A. Cane. A statistical approach to testing equatorial ocean models with observed data. *J. Phys. Oceano.*, 19:1191–1207, 1989.

[133] C. Frankignoul and K. Hasselmann. Stochastic climate models. Part II: Application to sea-surface temperature anomalies and thermocline variability. *Tellus*, 29:289–305, 1977.

[134] C. Frankignoul and A. Molin. Analysis of the GISS GCM response to a subtropical sea surface temperature anomaly using a linear model. *J. Atmos. Sci.*, 45:3833–3845, 1988.

[135] C. Frankignoul and A. Molin. Response of the GISS general circulation model to a midlatitude sea surface temperature anomaly in the North Pacific. *J. Atmos. Sci.*, 45:95–108, 1988.

[136] F. Gallagher, H. von Storch, R. Schnur, and G. Hannoschöck. The POP Manual. Technical Report 1, Deutsches Klima Rechen Zentrum, Bundesstrasse 55 20146 Hamburg 13, Germany, 1991. 66pp.

[137] W.L. Gates. AMIP: The Atmospheric Model Intercomparison Project. Technical Report 7, Program for Climate Model Diagnosis and Intercomparison, Lawrence Livermore National Laboratory, L-264, PO Box 808, Livermore, CA, 94550, USA, 1992.

[138] W.L. Gates, A. Henderson-Sellers, G.J. Boer, et al. Climate models—evaluation. In J.T. Houghton et al., editors, *Climate Change 1995*, pages 229–284. Cambridge University Press, 1996.

[139] A. Gelman, J. Carlin, H. Stern, and Rubin. *Bayesian Data Analysis*. Chapman & Hall, 1994. 526pp.

[140] H.R. Glahn and D.A. Lowry. The use of model output statistics (MOS) in objective weather forecasting. *J. Appl. Meteor.*, 11:1203–1211, 1972.

[141] R. Glowienka-Hense and A. Hense. The effect of an arctic polynya on the Northern Hemisphere mean circulation and eddy regime: A numerical experiment. *Climate Dynamics*, 7:155–163, 1992.

[142] W.J. Glynn and R.J. Muirhead. Inference in canonical correlation analysis. *J. Multivar. Anal.*, 8:468–478, 1978.

[143] H.H. Golub and C.F. van Loan. *Matrix Computations*. Hopkins University Press, 1989.

[144] N.R. Goodman. On the joint estimation of spectra, co-spectrum and quadrature spectrum of a two-dimensional stationary Gaussian process. Technical report, Engineering Stat. Lab., New York University, New York, 1957. Sci. Paper No. 10.

[145] W.A. Gough and W.J. Welch. Parameter Space Exploration for an Ocean General Circulation Model Using Isopycnal Mixing Parameterization. *J. Marine Research*, 52:773–796, 1994.

[146] N.E. Graham, T.P. Barnett, R. Wilde, M. Ponater, and S. Schubert. On the roles of tropical and mid-latitude SSTs in forcing interannual to interdecadal variability in the winter Northern Hemisphere circulation. *J. Climate*, 7:148–169, 1994.

[147] F.A. Graybill. *Theory and Application of the Linear Model*. Duxbury Press, Belmont, California, 1976. 704pp.

[148] F.A. Graybill. *Matrices with Applications in Statistics*. Wadsworth, second edition, 1983. 461pp.

[149] E.J. Gumbel. *Statistics of Extremes*. Columbia University Press, 1958.

[150] D.B. Guralnik (Ed.). *Webster's New World Dictionary*. Warner Books, New York, second college edition, 1982.

[151] N.B. Guttmann. The use of L-moments in the determination of regional precipitation climates. *J. Climate*, 6:2309–2325, 1993.

[152] D. Gyalistras, H. von Storch, A. Fischlin, and M. Beniston. How to link GCM simulated climates to ecosystem models? Case studies in statistical downscaling of global climate changes in the Alps. *Climate Research*, 4:167–189, 1994.

[153] H. Günther, W. Rosenthal, M. Stawarz, J.C. Carretero, M. Gomez, I. Lozano, G. Serano and M. Reistad. The wave climate of the Northeast Atlantic over the period 1955–1994: the wasa wave hindcast. *Global Atmosphere Ocean Systems*, 6:121–163, 1999.

[154] F.R. Hampel, E.M. Ronchetti, E.M. Rousseeuw, and W.A. Stahel. *Robust Statistics. The Approach Based on Influence Functions*. Wiley, 1986.

[155] J. Hann. *Handbuch der Klimatologie*. J. Engelhorn, Stuttgart, 1883.

[156] J. Hann and K. Knoch. *Handbuch der Klimatologie. Band I: Allgemeine Klimalehre*. J. Engelhorn, Stuttgart, fourth edition, 1932.

[157] E.J. Hannan. *Multiple Time Series*. Wiley, 1970.

[158] E.J. Hannan. The estimation of the order of an ARMA process. *Ann. Statist.*, 8:1071–1081, 1980.

[159] E.J. Hannan and P.J. Thompson. The estimation of coherence and group-delay. *Biometrika*, 56:469–481, 1971.

[160] E.J. Hannan and P.J. Thompson. Estimating group-delay. *Biometrika*, 60:241–253, 1973.

[161] G. Hannoschöck and C. Frankignoul. Multivariate statistical analysis of sea surface temperature anomaly experiments with the GISS general circulation model. *J. Atmos. Sci.*, 42:1430–1450, 1985.

[162] A.R. Hansen and A. Sutera. On the probability density function of planetary-scale atmospheric wave amplitude. *J. Atmos. Sci.*, 43:3250–3265, 1986.

[163] J.A. Hartigan. *Clustering Algorithms*. Wiley, New York, 1975.

[164] L. Hasse. Wind is not wind! In J.R. Scoggings, editor, *Atmospheric Forcing of Ocean Circulation*, pages 255–260. Inst. for Naval Oceanography, 1989.

[165] K.F. Hasselmann. Stochastic climate Models. Part I. Theory. *Tellus*, 28:473–484, 1976.

[166] K.F. Hasselmann. On the signal-to-noise problem in atmospheric response studies. In D.B. Shaw, editor, *Meteorology of the Tropical Ocean*, pages 251–259. Roy. Met. Soc., London, 1979.

[167] K.F. Hasselmann. PIPs and POPs: The reduction of complex dynamical systems using Principal Interaction and Oscillation Patterns. *J. Geophys. Res.*, 93:11,015–11,021, 1988.

[168] K.F. Hasselmann. Optimal fingerprints for the detection of time dependent climate change. *J. Climate*, 6:1957–1971, 1993.

[169] Y. Hayashi. A generalized method of resolving disturbances into progressive and retrogressive waves by space Fourier and time cross-spectral analysis. *J. Meteor. Soc. Japan*, 49:125–128, 1971.

[170] Y. Hayashi. Space-time spectral analysis and its applications to atmospheric waves. *J. Meteor. Soc. Japan*, 60:156–171, 1982.

[171] Y. Hayashi and D.G. Golder. Tropical 40–50 and 25–30 day oscillations appearing in realistic and idealized GFDL climate models and ECMWF dataset. *J. Atmos. Sci.*, 50:464–494, 1993.

[172] G.C. Hegerl, H. von Storch, K. Hasselmann, B.D. Santer, U. Cubasch, and P.D. Jones. Detecting anthropogenic climate change with an optimal fingerprint method. *J. Climate*, 9:2281–2306, 1996.

[173] P. Heidke. Berechnung des Erfolges und der Güte der Windstärkevorhersagen im Sturmwarnungsdienst. *Geografiska Annaler*, 8:301–349, 1926.

[174] A. Hense. Multivariate statistical investigations of the Northern Hemisphere circulation during the El Niño event 1982/83. *Tellus*, 38A:189–204, 1986.

[175] A. Hense, R. Glowienka-Hense, H. von Storch, and U. Stähler. Northern Hemisphere atmospheric response to changes of Atlantic Ocean SST on decadal time scales: a GCM experiment. *Climate Dynamics*, 4:157–174, 1990.

[176] D.M. Hershfield and W.T. Wilson. Generalizing of rainfall intensity-frequency data. *General Assembly, IASH*, 1:499–506, 1957. Toronto (Ontario).

[177] H.H. Hildebrandson. Quelches recherches sur les centres d'action de l'atmosphère. *Svenska Vet. Akad. Handligar*, 1897.

[178] W. Hoeffding. A class of statistics with asymptotically normal distribution. *Ann. Math. Statist.*, 19:293–325, 1948.

[179] K.A. Hollingsworth, M.S. Tracton, A.C. Lorenc, S. Uppala, and P. Kallberg. The response of numerical weather prediction systems to FGGE level IIb data. Part II: Forecast verifications and implications for predictability. *Quart. J. R. Met. Soc.*, 111:67–101, 1985.

[180] M. Holzer. Asymmetric geopotential height fluctuations from symmetric winds. *J. Atmos. Sci.*, 53:1361–1379, 1996.

[181] J.D. Horel. Complex Principal Component analysis: Theory and examples. *J. Clim. Appl. Meteor.*, 23:1660–1673, 1984.

[182] J.D. Horel and J.M. Wallace. Planetary scale phenomena associated with the Southern Oscillation. *Mon. Wea. Rev.*, 109:813–829, 1981.

[183] J.R.M. Hosking. L-moments: Analysis and estimation of distributions using linear combinations of order statistics. *J. Roy. Stat. Soc.*, B52:105–124, 1990.

[184] J.R.M. Hosking, J.R. Wallis, and E.F. Wood. Estimation of the generalized extreme value distribution by the method of probability weighted moments. *Technometrics*, 27:251–261, 1985.

[185] B.J. Hoskins and D.J. Karoly. The steady linear response of a spherical atmosphere to thermal and orographic forcing. *J. Atmos. Sci.*, 38:1179–1196, 1981.

[186] H. Hotelling. The most predictable criterion. *J. Ed. Psych.*, 26:139–142, 1935.

[187] H. Hotelling. Relations between two sets of variants. *Biometrika*, 28:321–377, 1936.

[188] J.-P. Huang, H.-R. Cho, and G.R. North. Applications of the cyclic spectral analysis to the surface temperature fluctuations in a stochastic climate model and a GCM simulation. *Atmosphere-Ocean*, 34:627–646, 1996.

[189] J.-P. Huang and G.R. North. Cyclic spectral analysis of fluctuations in a GCM simulation. *J. Atmos. Sci.*, 53:370–379, 1996.

[190] P.J. Huber. *Robust Statistics*. Wiley, 1981.

[191] C.M. Hurvich and C.L. Tsai. Regression and time series model selection in small samples. *Biometrika*, 76:297–307, 1989.

[192] R.L. Iman and W.J. Conover. A distribution-free approach to inducing rank correlation among input variables. *Commun. Stat. Part B: Simulation and Computation*, 11:311–334, 1982.

[193] IMSL Inc., 2500 Permian Towers, 2500 CityWest Blvd, Houston, Texas, 77042–3020. *IMSL User's manual: Fortran subroutines for statistical analysis*, 1991.

[194] D. Inaudil, X. Collona de Lega, A. Di Tullio, C. Forno, P. Jacquot, M. Lehmann, M. Monti, and S. Vurpillot. Experimental evidence for the Butterfly Effect. *Ann. Improb. Res.*, 1:2–3, 1995.

[195] G.M. Jenkins and D.G.Watts. *Spectral Analysis and its Application*. Holden-Day, 1968. 525pp.

[196] E.S. Johnson and M.J. McPhaden. On the structure of intraseasonal Kelvin waves in the Equatorial Pacific Ocean. *J. Phys. Oceano.*, 23:608–625, 1993.

[197] R.A. Johnson and D.W. Wichern. *Applied Multivariate Statistical Analysis*. Prentice-Hall International, Inc., London, 1982.

[198] I.T. Jolliffe. Rotation of ill-defined principal components. *Appl. Stat.*, 38:139–147, 1989.

[199] I.T. Jolliffe. Rotation of principal components: Choice of normalisation constraint. *J. Appl. Stat.*, 22:29–35, 1995.

[200] M.C. Jones, J.S. Marron, and S.J. Sheather. A brief survey of bandwidth selection for density estimation. *J. Amer. Stat. Assoc.*, 91:401–407, 1996.

[201] P. Jones. The instrumental data record: Its accuracy and use in attempts to identify the "CO_2"-signal. In H. von Storch and A. Navarra, editors, *Analysis of Climate Variability: Applications of Statistical Techniques*, pages 53–75. Springer Verlag, 1995.

[202] P.D. Jones, S.C.B. Raper, R.S. Bradley, H.F. Diaz, P.M. Kelley, and T.M.L. Wigley. Northern hemisphere surface air temperature variations 1851–1984. *J. Clim. Appl. Meteor.*, 25:161–179, 1986.

[203] P.D. Jones, S.C.B. Raper, R.S. Bradley, H.F. Diaz, P.M. Kelley, and T.M.L. Wigley. Northern hemisphere surface air temperature variations 1851–1984. *J. Clim. Appl. Meteor.*, 25:1213–1230, 1986.

[204] R.H. Jones. Fitting Autoregressions. *J. Amer. Stat. Assoc.*, 70:590–592, 1975.

[205] R.H. Jones. Maximum likelihood fitting of ARMA models to time series with missing observations. *Technometrics*, 22:389–395, 1980.

[206] A.G. Journel. *Fundamentals of geostatistics in five lessons*. Short Course in Geology 8, American Geophysical Union, Washington, 1989.

[207] A.G. Journel and Ch.J. Huijbregts. *Mining Geostatistics*. Academic Press, London, 1978. 610pp.

[208] J.G. Kalbfleisch. *Probability and Statistical Inference, I*. Springer Verlag, 1980. 342pp.

[209] E. Kalnay, M. Kanamitsu, and W.E. Baker. Global numerical weather prediction at the National Meteorological Center. *Bull. Am. Meteor. Soc.*, 71:1410–1428, 1990.

[210] E. Kalnay, M. Kanamitsu, R. Kistler, W. Collins, D. Deaven, L. Gandin, M. Iredell, S. Saha, G. White, J. Woollen, Y. Zhu, M. Chelliah, W. Ebisuzaki, W. Higgins, J. Janowiak, K. C. Mo, C. Ropelewski, J. Wang, A. Leetmaa, R. Reynolds, R. Jenne, and D. Joseph. The NCEP/NCAR 40-year re-analysis project. *Bull. Amer. Meteor. Soc.*, 77:437–471, 1996.

[211] S.K. Kao. Governing equations and spectra for atmospheric motion and transports in frequency-wavenumber space. *J. Atmos. Sci.*, 25:32–38, 1968.

[212] S.K. Kao. Wavenumber-frequency spectra of temperature in the free atmosphere. *J. Atmos. Sci.*, 27:1000–1007, 1970.

[213] T.R. Karl, R.G. Quayle, and P.Y. Groisman. Detecting climate variations and change: New challenges for observing and data management systems. *J. Climate*, 8:1481–1494, 1993.

[214] D.J. Karoly, J.A. Cohen, G.A. Meehl, J.F.B. Mitchell, A.H. Oort, R.J. Stouffer, and R.T. Wetherald. An example of fingerprint detection of greenhouse climate change. *Climate Dynamics*, 10:97–105, 1994.

[215] A. Kattenberg, F. Giorgi, H. Grassl, et al. Climate models—projections of future climates. In J.T. Houghton et al., editors, *Climate Change 1995*, pages 285–358. Cambridge University Press, 1996.

[216] R.W. Katz. On some criteria for estimating the order of a Markov chain. *Technometrics*, 23:243–249, 1981.

[217] R.W. Katz. Statistical evaluation of climate experiments with general circulation models: A parametric time series approach. *J. Atmos. Sci.*, 39:1446–1455, 1982.

[218] K.A. Kelly. The influence of winds and topography on the sea surface temperature over the Northern California slope. *J. Geophys. Res.*, 90C6:11,783–11,798, 1985.

[219] K.A. Kelly. Comment on "Empirical Orthogonal Function analysis of Advanced Very High Resolution Radiometer surface temperature patterns in Sante Barbara Channel" by G.S. Lagerloef and R.L. Bernstein. *J. Geophys. Res.*, 93C12:15,753–15,754, 1988.

[220] Sir M. Kendall. *Time-Series*. Charles Griffin and Co., Ltd., second edition, 1976. 197pp.

[221] V. Kharin. The relationship between sea-surface temperature and atmospheric circulation in General Circulation Model experiments. *Climate Dynamics*, 11:395–375, 1995.

[222] G.N. Kiladis and H.F. Diaz. Global climatic anomalies associated with extremes in the Southern Oscillation. *J. Climate*, 2:1069–1090, 1989.

[223] G.N. Kiladis, H. von Storch, and H. van Loon. Origin of the South Pacific Convergence Zone. *J. Climate*, 2:1185–1195, 1989.

[224] W.H. Klein. Objective specification of monthly mean surface temperature from mean 700 mb heights in winter. *Mon. Wea. Rev.*, 111:674–691, 1983.

[225] W.H. Klein and H.J. Bloom. An operational system for specifying monthly precipitation amounts over the United States from the field of concurrent mean 700 mb heights. *Weather and Forecasting*, 4:51–60, 1989.

[226] W.H. Klein and H.R. Glahn. Forecasting local weather by means of model output statistics. *Bull. Am. Meteor. Soc.*, 55:1217–1227, 1974.

[227] W.H. Klein, B.M. Lewis, and I. Enger. Objective prediction of five-day mean temperature during winter. *J. Meteor.*, 16:672–682, 1959.

[228] R. Kohn and C.F.Ansley. Estimation, prediction and interpolation for ARIMA models with missing data. *J. Amer. Stat. Assoc.*, 81:751–761, 1986.

[229] L.H. Koopmans. *The Spectral Analysis of Time Series*. Academic Press, 1974. 366pp.

[230] K.G. Korevaar. *North Sea Climate Based on Observations from Ships and Lightvessels*. Kluwer Academic Publishers, 1990.

[231] G. Koslowski and P. Loewe. The Western Baltic sea ice season in terms of a mass related severity index: 1879–1992. Part I: Temporal variability and association with the North Atlantic Oscillation. *Tellus*, 46A:66–74, 1994.

[232] A. Kumar, M. Hoerling, M. Ji, A. Leetmaa, and P. Sardeshmukh. Assessing a GCM's suitability for making seasonal predictions. *J. Climate*, 9:115–129, 1996.

[233] P.K. Kundu. Ekman veering observed near the ocean bottom. *J. Phys. Oceano.*, 6:238–242, 1976.

[234] P.K. Kundu, J.S. Allen, and R.L. Smith. Modal decomposition of the velocity field near the Oregon Coast. *J. Phys. Oceano.*, 5:683–704, 1975.

[235] H.R. Künsch. The jackknife and the bootstrap for general stationary observations. *Ann. Statist.*, 17:1217–1241, 1989.

[236] F. Kwasniok. The reduction of complex dynamical systems using principal interaction patterns. *Physica*, D 92:28–60, 1996.

[237] F. Kwasniok. Optimal Galerkin approximations of partial differential equations using Principal Interaction Patterns. *Phys. Rev. E*, 55:5365–5375, 1997.

[238] K. Labitzke and H. van Loon. Associations between the 11-year solar cycle, the QBO and the atmosphere. Part I: The troposphere and stratosphere in the Northern Hemisphere in winter. *J. Atmos. Terr. Phys.*, 50:197–206, 1988.

[239] R.T. Lacoss. Data adaptive spectral analysis methods. *Geophysics*, 36:661–675, 1971.

[240] S.J. Lambert. A comparison of operational global analyses from the European Centre for Medium Range Weather Forecasts (ECMWF) and the National Meteorological Center (NMC). *Tellus*, 40A:272–284, 1988.

[241] S.J. Lambert. A comparison of divergent winds from the National Meteorological Center and the European Centre for Medium Range Weather Forecasts Global Analyses for 1980–1986. *Mon. Wea. Rev.*, 117:995–1005, 1989.

[242] M. Latif and M. Flügel. An investigation of short range climate predictability in the tropical Pacific. *J. Geophys. Res.*, 96:2661–2673, 1990.

[243] M. Latif, A. Sterl, E. Maier-Reimer, and M.M. Junge. Climate variability in a coupled GCM. Part I: The tropical Pacific. *J. Climate*, 6:5–21, 1993.

[244] N.-C. Lau and M.J. Nath. A modeling study of the relative roles of the tropical and extratropical SST anomalies in the variability of the global atmosphere-ocean system. *J. Climate*, 7:1184–1207, 1994.

[245] N.D. Lawley. Tests of significance for the latent roots of covariance and correlation matrices. *Biometrika*, 43:128–136, 1956.

[246] M.R. Leadbetter, G. Lindgren, and H. Rootzen. *Extremes and related properties of random sequences and processes.* Springer Verlag, 1983. 336pp.

[247] D.R. Legates and C.J. Willmott. Mean seasonal and spatial variability in gauge-corrected, global precipitation. *Int. J. Clim.*, 10:111–127, 1990.

[248] C. Leger, D.N. Politis, and J.P. Romano. Bootstrap technology and applications. *Technometrics*, 34:378–398, 1992.

[249] E.L. Lehmann and H.J.M. D'Abrera. *Nonparametrics: Statistical Methods Based on Ranks.* Holden-Day, Inc., 1975. 457pp.

[250] J. Lehmhaus, E. Roeckner, I. Bernhardt, and J. Pankrath. Monte Carlo-model for the simulation of long-range transports of air pollutants. In De Wispelaere, editor, *Air Pollution Modeling and its Application III.* Plenum Publishing Company, 1984.

[251] P. Lemke, E.W. Trinkl, and K. Hasselmann. Stochastic dynamic analysis of polar sea ice variability. *J. Phys. Oceano.*, 10:2100–2120, 1980.

[252] D.P. Lettenmaier. Stochastic modeling of precipitation with applications to climate model downscaling. In H. von Storch and A. Navarra, editors, *Analysis of Climate Variability: Applications of Statistical Techniques*, pages 197–212. Springer Verlag, 1995.

[253] H.W. Lilliefors. On the Kolmogoroff-Smirnov test for normality with mean and variance unknown. *J. Amer. Stat. Ass.*, 62:399–402, 1967.

[254] R.Y. Liu and K. Singh. Moving blocks bootstrap captures weak dependence. In *Exploring the Limits of the Bootstrap*, pages 225–248. Wiley, 1992.

[255] R.E. Livezey. Caveat Emptor!—The evaluation of skill in climate predictions. In U. Radok, editor, *Towards Understanding Climate Change*, pages 149–176. Westview Press, Boulder, London, 1985.

[256] R.E. Livezey. The evaluation of forecasts. In H. von Storch and A. Navarra, editors, *Analysis of Climate Variability: Applications of Statistical Techniques*, pages 177–196. Springer Verlag, 1995.

[257] R.E. Livezey and W.Y. Chen. Statistical field significance and its determination by Monte Carlo techniques. *Mon. Wea. Rev.*, 111:46–59, 1983.

[258] A. Lorenc. Analysis methods for numerical weather prediction. *Quart. J. R. Met. Soc.*, 112:1177–1194, 1986.

[259] E.N. Lorenz. Empirical orthogonal functions and statistical weather prediction. Technical report, Statistical Forecast Project Report 1, Dept. of Meteor., MIT, 1956. 49pp.

[260] E.N. Lorenz. Deterministic nonperiodic flow. *J. Atmos. Sci.*, 20:130–141, 1963.

[261] U. Luksch. Simulation of North Atlantic low frequency SST variability. *J. Climate*, 9:2083–2092, 1996.

[262] U. Luksch, H. von Storch, and Y. Hayashi. Monte Carlo experiments with frequency wavenumber spectra. Max-Planck-Institut für Meteorologie, Bundesstrasse 55, 20146 Hamburg, Germany, 1987. MPI Report 10.

[263] R.A. Madden. Estimates of the natural variability of time averaged sea level pressure. *Mon. Wea. Rev.*, 104:942–952, 1976.

[264] R.A. Madden and P.R. Julian. Description of global-scale circulation cells in the tropics with 40–50 day period. *J. Atmos. Sci.*, 24:1109–1123, 1972.

[265] S. Manabe and K. Bryan. Climate calculations with a combined ocean-atmosphere model. *J. Atmos. Sci.*, 26:786–789, 1969.

[266] S. Manabe and R.J. Stouffer. Low frequency variability of surface air temperature in a 1000 year integration of a coupled ocean-atmosphere model. *J. Climate*, 9:376–393, 1996.

[267] K.V. Mardia, J.T. Kent, and J.M. Bibby. *Multivariate Analysis.* Academic Press, London, 1979.

[268] M.P. McCormick, L.W. Thomason, and C.R. Trepte. Atmospheric effects of the Mt. Pinatubo eruption. *Nature*, 373:399–404, February 1995.

[269] N.A. McFarlane. The effect of orographically excited gravity wave drag on the general circulation of the lower stratosphere and troposphere. *J. Atmos. Sci.*, 44:1775–1800, 1987.

[270] N.A. McFarlane, G.J. Boer, J.-P. Blanchet, and M. Lazare. The Canadian Climate Centre second generation general circulation model and its equilibrium climate. *J. Climate*, 5:1013–1044, 1992.

[271] M.D. McKay, W.J. Conover, and R.J. Beckman. A comparison of three methods for selecting values of input variables in the analysis of output from a computer code. *Technometrics*, 21:239–245, 1979.

[272] W. Mendenhall, R.L. Scheaffer, and D.L. Wackerly. *Mathematical Statistics with Applications.* Duxbury Press, 1981. 686pp.

[273] J. Michaelson. Cross-validation in statistical climate forecast models. *J. Clim. Appl. Meteor.*, 26:1589–1600, 1987.

[274] P.W. Mielke, K.J. Berry, and G.W. Brier. Application of multi-response permutation procedures for examining seasonal changes in monthly mean sea-level pressure patterns. *Mon. Wea. Rev.*, 109:120–126, 1981.

[275] U. Mikolajewicz. Interne Variabilität in einem stochastisch angetriebenen ozeanischen Zirkulationsmodell. Max-Planck-Institut für Meteorologie Examensarbeiten 10, 1991.

[276] U. Mikolajewicz and E. Maier-Reimer. Internal secular variability in an OGCM. *Climate Dynamics*, 4:145–156, 1990.

[277] U. Mikolajewicz, E. Maier-Reimer, and T. Barnett. Acoustic detection of greenhouse-induced climate changes in the presence of slow fluctuations of the thermohaline circulation. *J. Phys. Oceano.*, 23:1099–1109, 1993.

[278] A.J. Miller. Selection of subsets of regression variables (with discussion). *J. R. Stat. Soc. Ann.*, 147:389–425, 1984.

[279] J.F.B. Mitchell, T.J. Johns, J.M. Gregory, and S.F.B. Tett. Transient climate response to increasing sulphate aerosols and greenhouse gases. *Nature*, 376:501–504, 1995.

[280] A.F. Mood and F.A. Graybill. *Introduction to the Theory of Statistics*. McGraw-Hill, second edition, 1963. 443pp.

[281] D.F. Morrison. *Multivariate Statistical Methods*. McGraw-Hill, 1976. 415pp.

[282] R.J. Muirhead and C.M. Waternaux. Asymptotic distributions in canonical correlation analysis and other multivariate procedures for non-normal populations. *Biometrika*, 67:31–43, 1980.

[283] A. Murphy, B.G. Brown, and Y.-S. Chen. Diagnostic verification and temperature forecasts. *Weather and Forecasting*, 4:485–501, 1989.

[284] A. Murphy and H. Daan. Forecast evaluation. In A. Murphy and R.W. Katz, editors, *Probability, Statistics, and Decision Making in the Atmospheric Sciences*, pages 379–437. Westview Press, Boulder, London, 1985.

[285] A. Murphy and E. S. Epstein. Skill scores and correlation coefficients in model verification. *Mon. Wea. Rev.*, 117:572–581, 1989.

[286] A. Murphy and R.L. Winkler. A general framework for forecast verification. *Mon. Wea. Rev.*, 115:1329–1338, 1987.

[287] H. Nakamura and J.M. Wallace. Skewness of low-frequency fluctuations in the tropospheric circulation during the Northern Hemisphere winter. *J. Atmos. Sci.*, 48:1441–1448, 1991.

[288] A. Navarra. Steady linear response to thermal forcing of an anomaly model with an asymmetric climatology. *J. Atmos. Sci.*, 47:148–169, 1990.

[289] A. Navarra. The development of Climate Research. In H. von Storch and A. Navarra, editors, *Analysis of Climate Variability: Applications of Statistical Techniques*, pages 3–9. Springer Verlag, 1995.

[290] A. Navarra. Teleconnection patterns. In H. von Storch and A. Navarra, editors, *Analysis of Climate Variability: Applications of Statistical Techniques*, pages 215–225. Springer Verlag, 1995.

[291] J.M. Nese. Systematic biases in manual observations of maximum and minimum temperature. *J. Climate*, 7:834–842, 1994.

[292] N. Nicholls. A method for predicting Murray Valley Encephalitis in Southeast Australia using the Southern Oscillation. *Aust. J. Exp. Biol. Med. Sci.*, 64:587–594, 1986.

[293] N. Nicholls. The use of canonical correlation analysis to study teleconnections. *Mon. Wea. Rev.*, 115:393–399, 1987.

[294] N. Nicholls, G.V. Gruza, J. Jouzel, et al. Observed climate variability and change. In J.T. Houghton et al., editors, *Climate Change 1995*, pages 133–192. Cambridge University Press, 1996.

[295] G. Nitsche, J.M. Wallace, and C. Kooperberg. Is there evidence of multiple equilibria in the planetary-wave amplitude? *J. Atmos. Sci.*, 51:314–322, 1994.

[296] G.R. North, T.L. Bell, R.F. Cahalan, and F.J. Moeng. Sampling errors in the estimation of empirical orthogonal functions. *Mon. Wea. Rev.*, 110:699–706, 1982.

[297] G.R. North, K.Y. Kim, and S.S.P. Shen. Detection of forced climate signals. Part I: Filter theory. *J. Climate*, 8:401–408, 1995.

[298] Numerical Algorithms Group, 256 Banbury Road, Oxford OX2 7DE, UK. *NAG Fortran Library*.

[299] M. Okamoto. An asymptotic expansion for the distribution of the linear discriminant function. *Ann. Math. Stat.*, 34:1286–1301, 1963.

[300] M. Okamoto. Correction to "An asymptotic expansion for the distribution of the linear discriminant function". *Ann. Math. Stat.*, 39:1358–1359, 1968.

[301] N. Oreskes, K. Shrader-Frechette, and K. Beltz. Verification, validation, and confirmation of numerical models in earth sciences. *Science*, 263:641–646, 1994.

[302] D.E. Parker, C.K. Folland, A. Bevan, M.N. Ward, M. Jackson, and K. Maskell. Marine surface data for analysis of climate fluctuations on interannual to century timescales. In D.G. Martinson, K. Bryan, M. Ghil, M.M. Hall, T.R. Karl, E.S. Sarachick, S. Sorooshian, and L.D. Talley, editors, *Natural Climate Variability on Decade-to-Century Time Scales*, pages 241–250 and figs. 222–228. National Academy Press, Washington (DC), 1995. National Research Council.

[303] D.E. Parker, P.D. Jones, C.K. Folland, and A. Bevan. Interdecadal changes of surface temperature since the late nineteenth century. *J. Geophys. Res.*, 99:14,373–14,399, 1994.

[304] C.L. Parkinson, J.C. Comiso, H.J. Zwally, D.J. Cavalieri, P. Gloersen, and W.J. Campbell. *Arctic Sea Ice, 1973–1976: Satellite passive-microwave observations*. NASA Scientific and Technical Information Branch, 1987.

[305] E. Parzen. Mathematical considerations in the estimation of spectra. *Technometrics*, 3:167–190, 1961.

[306] E. Parzen. Multiple time series modelling. In P.R. Krishnaiah, editor, *Multivariate Analysis*, volume 1, pages 389–409. Academic Press, 1969.

[307] E.S. Pearson and H.O. Hartley, editors. *Biometrika Tables for Statisticians*, volume 2. Biometrika Trust (University College), London, 1976.

[308] E.S. Pearson and H.O. Hartley, editors. *Biometrika Tables for Statisticians*, volume 1. Biometrika Trust (University College), London, 1976.

[309] K. Pearson. On lines and planes of closest fit to systems of points in space. *Phil. Mag.*, 2:559–572, 1902.

[310] J. Pedlosky. *Geophysical Fluid Dynamics*. Springer Verlag, 1987. 710pp.

[311] J.P. Peixoto and A. H. Oort. *Physics of Climate*. American Institute of Physics, New York, 1992. 520pp.

[312] S. Peng and J. Fyfe. The coupled patterns between Sea Level Pressure and Sea Surface Temperature in the Midlatitude North Atlantic. *J. Climate*, 9:1824–1839, 1996.

[313] E.W. Peterson and L. Hasse. Did the Beaufort scale or the wind climate change? *J. Phys. Oceano.*, 17:1071–1074, 1987.

[314] E.J.G. Pitman. Significance tests which may be applied to samples from any populations. *J.R. Statist. Soc.*, Suppl. 4:119–130, 1937.

[315] E.J.G. Pitman. Significance tests which may be applied to samples from any populations: II. The correlation coefficient test. *J.R. Statist. Soc.*, Suppl. 4:225–232, 1937.

[316] E.J.G. Pitman. Significance tests which may be applied to samples from any populations: III. The analysis of variance test. *Biometrika*, 29:322–335, 1937.

[317] G. Plaut and R. Vautard. Spells of low-frequency oscillations and weather regimes over the Northern Hemisphere. *J. Atmos. Sci.*, 51:210–236, 1994.

[318] I. Polyak. *Computational Statistics in Climatology*. Oxford University Press, 1996. 358pp.

[319] R.W. Pratt. The interpretation of space-time spectral quantities. *J. Atmos. Sci.*, 32:1283–1300, 1976.

[320] R.W. Preisendorfer and T.P. Barnett. Numerical model reality intercomparison tests using small sample statistics. *J. Atmos. Sci.*, 40:1884–1896, 1983.

[321] R.W. Preisendorfer, F.W. Zwiers, and T.P. Barnett. Foundations of principal component selection rules. In *Scripps Institution of Oceanography Reference Series 81–4*. Scripps Institution of Oceanography, La Jolla, California, 1982. 192pp.

[322] W.H. Press, S.A. Teukolsky, W.T. Vetterling, and B.P. Flannery. *Numerical Recipes*. Cambridge University Press, second edition, 1992. 964pp.

[323] M.B. Priestley. *Spectral Analysis and Times Series. Vol. 1: Univariate Series. Vol. 2: Multivariate Series, Prediction and Control*. Academic Press, London, 1981.

[324] J.A. Pudykiewicz and A.P. Dastoor. On numerical simulation of the global distribution of sulfate aerosol produced by a large volcanic eruption. *J. Climate*, 8:464–473, 1995.

[325] V.S. Pugachev. *Probability Theory and Mathematical Statistics for Engineers*. Pergamon Press, first edition, 1984. 450pp.

[326] M.H. Quenouille. Approximate tests of correlation in time series. *J. R. Stat. Soc.*, B11:18–84, 1949.

[327] U. Radok. Chance behaviour of skill score. *Mon. Wea. Rev.*, 116:489–494, 1988.

[328] A.E. Raftery. Bayesian model selection in social research (with discussion by A. Gelman, D.B. Rubin, and R.M. Hauser, and a rejoinder). In P.V. Marsden, editor, *Social Methodology 1995*, pages 111–196. Blackwells, 1996.

[329] E.M. Rasmusson, P.A. Arkin, W.-Y. Chen, and J.B. Jalickee. Biennial variations in surface temperature over the United States as revealed by Singular Decomposition. *Mon. Wea. Rev.*, 109:587–598, 1981.

[330] E.M. Rasmusson and T. Carpenter. Variations in tropical SST and surface wind fields associated with the Southern Oscillation / El Niño. *Mon. Wea. Rev.*, 110:354–384, 1982.

[331] M.B. Richman. Rotation of principal components. *Int. J. Climatology*, 6:293–335, 1986.

[332] G.H. Ross. Model output statistics using an updatable scheme. In *11th Conference on Probability and Statistics in Atmospheric Sciences: Oct. 1–5 (Monterey, California)*, pages 93–97. American Meteorological Society, Boston, Mass., 1989.

[333] W.H. Ross. A peaks-over-threshold analysis of extreme wind speeds. *Can. J. Statist.*, 15:328–337, 1987.

[334] D.A. Rothrock. Ice thickness distribution — measurement and theory. In N. Untersteiner, editor, *The Geophysics of Sea Ice*, pages 551–575. Plenum Press, New York and London, 1986.

[335] G. Roussas. *A First Course in Mathematical Statistics*. Addison-Wesley, 1973. 506pp.

[336] D.P. Rowell. Assessing Potential Seasonal Predictabilitu with an Ensemble of Multidecadal GCM Simulations. *J. Climate*, 11:109–120, 1998.

[337] D.P. Rowell and F.W. Zwiers. Sources of atmospheric decadal variability over the globe, tropical Pacific, and southern North America. *submitted*, 1999.

[338] J. Sansom and P.J. Thomson. Rainfall classification using breakpoint pluviograph data. *J. Climate*, 5:765–771, 1992.

[339] B.D. Santer, J.E. Penner, T.M.L. Wigley, U. Cubasch, and P.D. Jones. Towards the detection and attribution of an anthropogenic effect on climate. *Climate Dynamics*, 12:77–100, 1995.

[340] B.D. Santer, T.M.L. Wigley, T.P. Barnett, and E. Anyamba. Detection of climate change and attribution of causes. In J.T. Houghton et al., editors, *Climate Change 1995*, pages 407–444. Cambridge University Press, 1996.

[341] R. Schnur, G. Schmitz, N. Grieger, and H. von Storch. Normal modes of the atmosphere as estimated by principal oscillation patterns and derived from quasi-geostrophic theory. *J. Atmos. Sci.*, 50:2386–2400, 1993.

[342] G. A. F. Seber. *Multivariate Observations*. Wiley, New York, 1984. 686pp.

[343] F.M. Selten. An efficient description of the dynamics of barotropic flow. *J. Atmos. Sci.*, 52:915–936, 1995.

[344] F.M. Selten. Baroclinic Empirical Orthogonal Functions as basis functions in an atmospheric model. *J. Atmos. Sci.*, 54:2099–2114, 1997.

[345] F.M. Selten. A statistical closure of a low-order barotropic model. *J. Atmos. Sci.*, 54:1085–1093, 1997.

[346] A. Shabbar and A.G. Barnston. Skill of seasonal climate forecasts in canada using canonical correlation analysis. *Mon. Wea. Rev.*, 124:2370–2385, 1996.

[347] Sir N. Shaw. *Manual of Meteorology. Volume II: Comparative Meteorology*. Cambridge University Press, second edition, 1936. 472pp.

[348] D.J. Shea, S.J. Worley, I.R. Stern, and T.J. Hoar. An introduction to atmospheric and oceanographic data. Technical report, NCAR/TN-404+IA, 1994. 132pp.

[349] S.S. Shen, G.R. North, and K.-Y. Kim. Spectral approach to optimal estimation of the global average temperature. *J. Climate*, 7:1999–2007, 1994.

[350] B.W. Silverman. *Density Estimation for Statistics and Data Analysis*. Chapman and Hall, London, 1986.

[351] E. Slutsky. The summation of random causes as the source of cyclic processes. *Econometrica*, 5:105–146, 1937. Originally in Russian: Problems in Economic Conditions, **3**, 1927.

[352] B.T. Smith et al. Matrix eigensystem routines—eispack guide. In *Lecture Notes in Computer Science*, volume 6. Springer Verlag, New York, second edition, 1976.

[353] P. Speth and E. Kirk. A one-year study of power spectra in wavenumber-frequency domain. *Beitr. Phys. Atmos.*, 54:186–206, 1981.

[354] P. Speth and R.A. Madden. Space-time spectral analyses of Northern Hemisphere geopotential heights. *J. Atmos. Sci.*, 40:1086–1100, 1983.

[355] H.R. Stanski, L.J. Wilson, and W.R. Burrows. Survey of common verification methods in meteorology. Technical Report 8, World Weather Watch, World Meteorological Organization WMO/TD 358, 1990.

[356] M.A. Stephens. Use of the Kolmogorov-Smirnov, Cramer-von Mises and related statistics without extensive tables. *J. R. Stat. Soc.*, B32:115–122, 1970.

[357] M.A. Stephens. Tests based on EDF statistics. In R.B. D'Agostino and M.A. Stephens, editors, *Goodness-of-fit techniques*, pages 97–193. Marcel Dekker, 1986.

[358] W. Stern and K. Miyakoda. Feasibility of seasonal forecasts inferred from multiple GCM simulations. *J. Climate*, 8:1071–1085, 1995.

[359] M.J. Stevens and G.R. North. Detection of the climate response to the solar cycle. *J. Atmos. Sci.*, 53:2594–2608, 1996.

[360] G. Swartz. Estimating the dimension of a model. *Ann. Statist.*, 6:461–464, 1978.

[361] H.J. Thiébaux. Anisotropic correlation functions for objective analysis. *Mon. Wea. Rev.*, 104:994–1002, 1976.

[362] H.J. Thiébaux and M.A. Pedder. *Spatial Objective Analysis: with Applications in Atmospheric Science*. Academic Press, London, 1987.

[363] H.J. Thiébaux and F.W. Zwiers. The interpretation and estimation of effective sample sizes. *J. Clim. Appl. Meteor.*, 23:800–811, 1984.

[364] R.B. Thomas. *The Old Farmer's Almanac*. Yankee Publishing Inc., 1995. 224pp.

[365] R.M. Thorndike. *Correlational Procedures for Research*. Garnder Press, New York, 1978.

[366] J.E. Tillman, N.C. Johnson, P. Guttorp, and D.B. Percival. The Martian annual atmospheric pressure cycle: years without great dust storms. *J. Geophys. Res.*, E98:10,963–10,971, 1993.

[367] H. Tong. *Threshold models in non-linear time series analysis*. Springer-Verlag, 1983. 323pp.

[368] K.E. Trenberth. Some effects of finite sample size and persistence on meteorological statistics. Part I: Autocorrelations. *Mon. Wea. Rev.*, 112:2359–2368, 1984.

[369] K.E. Trenberth. Persistence of daily geopotential heights over the Southern Hemisphere. *Mon. Wea. Rev.*, 113:38–53, 1985.

[370] K.E. Trenberth and J.G. Olson. ECMWF global analyses 1979-86: Circulation indices and data evaluation. Technical report, NCAR/TN-300+STR, 1988. 94pp.

[371] K.E. Trenberth and J.G. Olson. Evaluation of NMC global analysis 1979–1987. Technical report, NCAR/TN-299+STR, 1988. 82pp.

[372] K.E. Trenberth and D.J. Shea. On the evolution of the Southern Oscillation. *Mon. Wea. Rev.*, 115:3078–3096, 1987.

[373] K.E. Trenberth and W.-T. K. Shin. Quasi-biennial fluctuations in sea level pressures over the Northern Hemisphere. *Mon. Wea. Rev.*, 112:761–777, 1984.

[374] J. Tukey. Bias and confidence in not quite large samples. *Ann. Math. Stat.*, 29:614, 1958.

[375] J.W. Tukey. *Exploratory Data Analysis*. Addison-Wesley, Reading (Mass.), 1977. 688pp.

[376] D.E. Tyler. On the optimality of the simultaneous redundancy transformations. *Psychometrika*, 47:77–86, 1982.

[377] D.A. Unger. Skill assessment strategies for screening regression predictions based on a small sample size. In *13th Conference on Probability and Statistics in the Atmospheric Sciences: Feb. 21–23, 1996, San. Fran., Cal.*, pages 260–267. Amer. Meteor. Soc., Boston, Mass., 1996.

[378] T. van Andel. *New views on an old planet. A history of global change*. Cambridge University Press, second edition, 1994. 439pp.

[379] H.M. van den Dool. A bias in skill in forcasts based on analogues and antilogues. *J. Clim. Appl. Meteor.*, 26:1278–1281, 1987.

[380] R. Vautard. Patterns in time: SSA and MSSA. In H. von Storch and A. Navarra, editors, *Analysis of Climate Variability: Applications of Statistical Techniques*, pages 259–279. Springer Verlag, 1995.

[381] R. Vautard, P. Yiou, and M. Ghil. Singular spectrum analysis: A toolkit for short, noisy chaotic signal. *Physica D.*, 58:95–126, 1992.

[382] R.G. Veryard and R.A. Ebdon. Fluctuations in tropical stratospheric winds. *Meteor. Mag.*, 90:125–143, 1961.

[383] L. Vincent. Technique for the identification of inhomogeneities in annual temperature series. *J. Climate*, 11:1094–1104, 1998.

[384] H. von Storch. A remark on Chervin/Schneider's algorithm to test significance of climate experiments with GCMs. *J. Atmos. Sci.*, 39:187–189, 1982.

[385] H. von Storch. An accidental result: The mean 1983 January 500 *mb* height field significantly different from its 1967–81 predecessors. *Beitr. Phys. Atmos.*, 57:440–444, 1984.

[386] H. von Storch. A statistical comparison with observations of control and El Niño simulations using the NCAR CCM. *Beitr. Phys. Atmos.*, 60:464–477, 1987.

[387] H. von Storch. Spatial patterns: EOFs and CCA. In H. von Storch and A. Navarra, editors, *Analysis of Climate Variability: Applications of Statistical Techniques*, pages 227–257. Springer Verlag, 1995.

[388] H. von Storch and D. Baumhefner. Principal Oscillation Pattern Analysis of the tropical 30- to 60-days oscillation. Part II: The prediction of equatorial velocity potential and its skill. *Climate Dynamics*, 5:1–12, 1991.

[389] H. von Storch, T. Bruns, I. Fischer-Bruns, and K.F. Hasselmann. Principal Oscillation Pattern analysis of the 30- to 60-day oscillation in a General Circulation Model equatorial troposphere. *J. Geophys. Res.*, 93:11,022–11,036, 1988.

[390] H. von Storch, S. Güss, and M. Heimann. *Das Klimasystem und seine Modellierung. Eine Einfürung.* Springer Verlag, 1999.

[391] H. von Storch and G. Hannoschöck. Comment on "Empirical Orthogonal Function analysis of wind vectors over the tropical Pacific region". *Bull. Am. Meteor. Soc.*, 65:162, 1984.

[392] H. von Storch and G. Hannoschöck. Statistical aspects of estimated principal vectors (EOFs) based on small sample sizes. *J. Clim. Appl. Meteor.*, 24:716–724, 1986.

[393] H. von Storch and H.A Kruse. The extratropical atmospheric response to El Niño events - a multivariate significance analysis. *Tellus*, 37A:361–377, 1985.

[394] H. von Storch, M. Latif, and J. Biercamp. Simulation of the Southern Oscillation in a general circulation model. *Phil. Trans. R. Soc. Lond.*, A329:179–188, 1989.

[395] H. von Storch and A. Navarra. *Analysis of Climate Variability: Applications of Statistical Techniques.* Springer Verlag, 1995.

[396] H. von Storch and E. Roeckner. Methods for the verification of general circulation models applied to the Hamburg University GCM. Part I: Test of individual climate states. *Mon. Wea. Rev.*, 111:1965–1976, 1983.

[397] H. von Storch, E. Roeckner, and U. Cubasch. Intercomparison of extended-range January simulation with General Circulation Models: Statistical assessment of ensemble properties. *Beitr. Phys. Atmos.*, 58:477–497, 1985.

[398] H. von Storch, D. Schriever, K. Arpe, G.W. Branstator, R. Legnani, and U. Ulbrich. Numerical experiments on the atmospheric response to cold equatorial conditions ("La Niña") during northern summer. *Global Atmos. Oc. System*, 2:99–120, 1994.

[399] H. von Storch and A. Smallegange. The phase of the 30- to 60-day oscillation and the genesis of tropical cyclones in the Western Pacific. Max-Planck-Institut für Meteorologie, Bundesstrasse 55, 20146 Hamburg, Germany, 1991. MPI Report 64.

[400] H. von Storch, U. Weese, and J. Xu. Simultaneous analysis of space-time variability: Principal Oscillation Patterns and Principal Interaction Patterns with applications to the Southern Oscillation. *Z. Meteor.*, 40:99–104, 1990.

[401] H. von Storch and J. Xu. Principal Oscillation Pattern Analysis of the tropical 30- to 60-day oscillation. Part I: Definition on an index and its prediction. *Climate Dynamics*, 4:175–190, 1990.

[402] H. von Storch and E. Zorita. Aspects of the origin of Iberian drought. In *Proceedings of the 15th Annual Diagnostics Workshop, Ashville N. Carolina*. US Department of Commerce, 1992.

[403] H. von Storch, E. Zorita, and U. Cubasch. Downscaling of global climate change estimates to regional scales: An application to Iberian rainfall in wintertime. *J. Climate*, 6:1161–1171, 1993.

[404] H. von Storch and F.W. Zwiers. Recurrence analysis of climate sensitivity experiments. *J. Climate*, 1:157–171, 1988.

[405] J. von Storch. Multivariate statistical modelling: POP model as a first order approximation. In H. von Storch and A. Navarra, editors, *Analysis of Climate Variability: Applications of Statistical Techniques*, pages 281–297. Springer Verlag, 1995.

[406] H. Wackernagel. *Multivariate Geostatistics*. Springer Verlag, 1995. 256pp.

[407] P. Wadhams, M.A. Lange, and S.F. Ackley. The ice thickness distribution across the Atlantic sector of the Antarctic Ocean in mid-winter. *J. Geophys. Res.*, 92C:14,535–14,552, 1987.

[408] J.M. Wallace and R.E. Dickinson. Empirical orthogonal representation of time series in the frequency domain. Part I: Theoretical considerations. *J. Appl. Meteor.*, 11:887–892, 1972.

[409] J.M. Wallace and D. Gutzler. Teleconnections in the geopotential height field during the Northern Hemisphere winter. *Mon. Wea. Rev.*, 109:784–812, 1981.

[410] J.M. Wallace, G.-H. Lim, and M.L. Blackmon. Relationship between cyclone tracks, anticyclone tracks and baroclinic waveguides. *J. Atmos. Sci.*, 45:439–462, 1988.

[411] J.M. Wallace, C. Smith, and C.S. Bretherton. Singular value decomposition of sea-surface temperature and 500 *mb* height anomalies. *J. Climate*, 5:561–576, 1992.

[412] J.E. Walsh and D. Allen. Testing the Old Farmer's Almanac. *Weatherwise*, 34:212–215, 1982.

[413] D.-P. Wang and C.N.K. Mooers. Long coastal-trapped waves off the West Coast of the United States, Summer 1973. *J. Phys. Oceano.*, 7:856–864, 1977.

[414] X.L. Wang and F.W. Zwiers. Interannual variability of precipitation in an ensemble of AMIP climate simulations conducted with the CCC GCMII. *J. Climate*, in press, 1999.

[415] M.N. Ward and C. Folland. Prediction of seasonal rainfall in the north Nordeste of Brazil using eigenvectors of sea-surface temperatures. *Int. J. Climatology*, 11:711–745, 1991.

[416] W.E. Watt, editor. *Hydrology of Floods in Canada: A Guide to Planning and Design*, page 245. National Research Council Canada: Associate Committee on Hydrology, 1989.

[417] B.C. Weare and J.N. Nasstrom. Examples of extended empirical orthogonal function analyses. *Mon. Wea. Rev.*, 110:481–485, 1982.

[418] A.W. Weaver and T.M.C. Hughes. Stability and variability of the thermohaline circulation and its link to climate. *Trends in Phys. Ocean.*, 1:15–70, 1992.

[419] R.O. Weber and R.A. Madden. Evidence of traveling external Rossby waves in the ECMWF analyses. *J. Atmos. Sci.*, 50:2994–3007, 1993.

[420] S. Weisberg. *Applied Linear Regression*. Wiley & Sons, New York, Chichester, Brisbane, Toronto, Singapore, second edition, 1985.

[421] R. Weisse, U. Mikolajewicz, and E. Maier-Reimer. Decadal variability of the North Atlantic in an ocean general circulation model. *J. Geophys. Res.*, 99C6:12,411–12,421, 1994.

[422] N. Wiener. *Time Series*. The M.I.T. Press, Cambridge, Massachusetts, 1949. 163pp.

[423] D.S. Wilks. Resampling Hypothesis Tests for Autocorrelated Fields. *J. Climate*, 10:65–82, 1997.

[424] E.F. Wood, D.P. Lettenmaier, and V.G. Zartarian. A land-surface hydrology parameterization with subgrid variability for General Circulation Models. *J. Geophys. Res.*, 97 D3:2717–2728, 1992.

[425] S.D. Woodruff, R.H. Slutz, R.L. Jenne, and P.M. Steuer. A comprehensive ocean-atmosphere data set. *Bull. Am. Meteor. Soc.*, 68:1239–1250, 1987.

[426] P.B. Wright. On the relationship between indices of the Southern Oscillation. *Mon. Wea. Rev.*, 112:1913–1919, 1984.

[427] P.B. Wright. The Southern Oscillation — An ocean-atmosphere feedback system. *Bull. Am. Meteor. Soc.*, 66:398–412, 1985.

[428] K. Wyrtki. Sea level rise: The facts and the future. *Pacific Science*, 44:1–16, 1990.

[429] J. Xu. Analysis and prediction of the El Niño Southern Oscillation phenomenon using Principal Oscillation Pattern Analysis. Max-Planck-Institut für Meteorologie Examensarbeiten 4, 1990.

[430] J. Xu. On the relationship between the stratospheric QBO and the tropospheric SO. *J. Atmos. Sci.*, 49:725–734, 1992.

[431] J. Xu. The joint normal modes of the atmosphere-ocean system observed from 1967 to 1986. *J. Climate*, 6:816–838, 1993.

[432] J. Xu and H. von Storch. "Principal Oscillation Patterns"—prediction of the state of ENSO. *J. Climate*, 3:1316–1329, 1990.

[433] K.-M. Xu and S.K. Krueger. Evaluation of cloudiness parameterizations using a cumulus ensemble model. *Mon. Wea. Rev.*, 119:342–367, 1991.

[434] K.-M. Xu and D.A. Randall. Development and evaluation of cloudiness parameterizations for use in climate models. *J. Atmos. Sci.*, 53:3084–3102, 1996.

[435] N. Yacowar. Weather element forecasts using rule based systems. In *11th Conference on Probability and Statistics in Atmospheric Sciences: Oct. 1–5 (Monterey, California)*, pages 75–83. American Meteorological Society, Boston (Mass.), 1989.

[436] J. Zhang and T.J. Crowley. Historical climate records in China and reconstruction of past climates. *J. Climate*, 2:833–849, 1989.

[437] E. Zheng, R.E. Basher, and C.S. Thompson. Trend detection in regional-mean temperature series: maximum, minimum, mean, diurnal range and SST. *J. Climate*, 10:317–326, 1997.

[438] E. Zorita, V. Kharin, and H. von Storch. The atmospheric circulation and sea surface temperature in the North Atlantic area in winter: Their interaction and relevance for Iberian precipitation. *J. Climate*, 5:1097–1108, 1992.

[439] F.W. Zwiers. Case studies in data analysis, no. 3. *Can. J. Statist.*, 13:83–108, 1985.

[440] F.W. Zwiers. A potential predictability study conducted with an atmospheric general circulation model. *Mon. Wea. Rev.*, 115:2957–2974, 1987.

[441] F.W. Zwiers. Aspects of the statistical analysis of climate experiments with multiple integrations. Technical Report 18, Max-Planck-Institut für Meteorologie Report, Max-Planck-Institut für Meteorologie, Bundesstrasse 55, 20146 Hamburg, Germany, 1988.

[442] F.W. Zwiers. The effect of serial correlation on statistical inferences made with resampling procedures. *J. Climate*, 3:1452–1461, 1991.

[443] F.W. Zwiers. Simulation of the Asian Summer Monsoon with the CCC GCM-1. *J. Climate*, 6:470–486, 1993.

[444] F.W. Zwiers. Interannual variability and predictability in an ensemble of AMIP climate simulations conducted with the CCC GCMII. *Climate Dynamics*, 12:825–847, 1996.

[445] F.W. Zwiers. Climate Change Detection: A review of Techniques and Applications. In H. von Storch and E. Raschke and G. Flöser, editors, *Anthropogenic Climate Change. Proceedings of the First GKSS Spring School on Environmental Research*. Springer Verlag, 1999. 44pp.

[446] F.W. Zwiers and G.J. Boer. A comparison of climates simulated by a general circulation model when run in the annual cycle and perpetual modes. *Mon. Wea. Rev.*, 115:2626–2644, 1987.

[447] F.W. Zwiers and K. Hamilton. Simulation of Solar Tides in the Canadian Climate Centre General Circulation Model. *J. Geophys. Res.*, 91D:11,877–11,896, 1986.

[448] F.W. Zwiers and V. Kharin. Changes in the extremes of the climate simulated by the CCC GCMII under CO_2 doubling. *J. Climate*, 11:2200–2222, 1998.

[449] F.W. Zwiers and V. Kharin. Intercomparison of annual variability and potential predictability: An AMIP diagnostic subproject. *Climate Dynamics*, 14:517–528, 1998.

[450] F.W. Zwiers and W.H. Ross. An alternative approach to the extreme value analysis of rainfall data. *Atmosphere-Ocean*, 29:437–461, 1991.

[451] F.W. Zwiers and S.S. Shen. Errors in Estimating Spherical Harmonic Coefficients from Partially Sampled GCM Output. *Climate Dynamics*, 13:703–716, 1997.

[452] F.W. Zwiers and H. von Storch. Multivariate recurrence analysis. *J. Climate*, 2:1538–1553, 1989.

[453] F.W. Zwiers and H. von Storch. Regime dependent auto-regressive time series modelling of the Southern Oscillation. *J. Climate*, 3:1347–1363, 1990.

[454] F.W. Zwiers and H. von Storch. Taking serial correlation into account in tests of the mean. *J. Climate*, 8:336–351, 1995.

Index

Printed in the United Kingdom
by Lightning Source UK Ltd.
109605UKS00001B/5-8